Digital Geographies—Theory, Space, and Communities

Abdul Shaban
Editor

Digital Geographies—Theory, Space, and Communities

A Machine-Generated Literature Review

Springer

Editor
Abdul Shaban
School of Development Studies
Tata Institute of Social Sciences
Mumbai, Maharashtra, India

ISBN 978-981-97-4733-7   ISBN 978-981-97-4734-4 (eBook)
https://doi.org/10.1007/978-981-97-4734-4

This Springer imprint is published by the registered company Springer Nature Singapore Pte Ltd.
The registered company address is: 152 Beach Road, #21-01/04 Gateway East, Singapore 189721, Singapore

Preface

The 3rd and 4th Industrial revolutions have had an enormous impact on human geography. The rise of the internet and then the integration of digital technologies and now Artificial Intelligence (AI) have both disruptive and transformative impacts on human-space relations on the one hand and human-to-human relations on the other. Further, digital technologies are changing body-mind relations and making us more narcissistic or dependent on them. The usual social and economic theories, especially industrial location theories, need considerable review and revision as the rise of ICT has minimized the spatial information asymmetries. The flow of information is also changing the institutional quality and cultural attributes across the countries and regions. Where on the one hand there has been a democratization of knowledge and information due to social media, at the same time we find that marginal groups and areas with a lack of digital infrastructure face exclusion. The digital exclusion and divide in the countries of the Global South run deep compromising the access to citizenship rights of people or creating constraints to fully utilize their civil rights.

The digitalization is continuously deepening and unfolding different possibilities for all of us. As such, this period of human civilization seems to be a very interesting phase as we are transitioning from traditional and real to virtual worlds.

In this context, this volume, and its companion volume on *Digital Geographies—Urbanisation, Economy, and Modelling*, presents a review of the studies published to date by Springer Nature on digitalization and its relationship with space. These two volumes summarizes studies on digital geographies.

As there has been a large volume of studies in this area, human review and summarization would have required a large amount of time to accomplish the same. As the use of AI is rising, we have attempted to use its power to summarize the studies. In other words, this volume is a product of human-machine collaboration. The human-machine collaboration made also it possible to bring the important studies quickly to the level of publication and make the same available to the researchers, who will benefit immensely from this volume. As most of the studies on digital geographies are scattered in different journals and books, this volume becomes of added importance in terms of bringing them together in a thematically organized form. This volume can be of immense help to researchers in search of summaries or review literature or in search of finding out important studies in this area. We have not changed the machine-made summaries, but only curated and thematically organized the same.

I am very thankful to Springer Nature, especially to Kanchan Latambale and Nupoor Singh, whose constant support has been crucial in completing and organizing this enormous work.

Mumbai, India Abdul Shaban

About This Book

Auto-summaries can be generated by either an abstractive or extractive auto-summarization:

- An extraction-based summarizer identifies the most important sentences of a text and uses the original sentences to create the summary.
- An abstraction-based summarizer creates new text based on deep learning. New phrases are created to summarize the content.

The auto-summaries you will find in this book have been generated via an extractive summarization approach.

Each chapter was carefully edited by Abdul Shaban. The editor selected the papers which were then auto-summarized. The editors have not edited the auto-summaries due to the extraction-based approach and have not changed the original sentences. You will find the editor's reviews and guidance on the auto-summaries in chapter introductions.

In machine-generated books, editors are defined as those who curate the content for the book by selecting the papers to be auto-summarized and by organizing the output into a meaningful order. Next to the thoughtful curation of the papers, editors should guide the readers through the auto-summaries and make transparent why they selected the papers.

The ultimate goal is to provide a current literature review of Springer Nature publications on a given topic in order to support readers in overcoming information overload and to help them dive into a topic faster; to identify interdisciplinary overlaps; and to present papers which might not have been on the readers' radar.

Please note, that the selected papers are not used to train a LLM while the auto-summaries are created.

Contents

About the Editor

Abdul Shaban is Professor and Chairperson of the Centre for Public Policy, Habitat, and Human Development, School of Development Studies, Tata Institute of Social Sciences (TISS), Mumbai. He was Deputy Director of TISS Tuljapur Campus and served as chairperson of the International Relations Office (IRO). His recent works relate to cities, entrepreneurship, creative industries, and religious minorities. Prof. Shaban has been a member of various committees and commissions constituted by the Government of India and State Governments. He has been a visiting professor/fellow at several leading universities outside India, which include the London School of Economics, Münster University, Germany, Erasmus University, the Netherlands, and the University of Paris City. He has been associated with different journals as an editor and as a member of the editorial boards.

1 On Digital Geographies

Technological innovations have disruptive impacts on the way we experience time and space. In other words, human geographies and temporalities are made and remade by technologies, and technologies are central in the production of space and socio-spatial relations (Leszczynski, 2019: 13). Technologies enable us to collect geographical information on vast territories and map them, and these maps always make us aware of our materiality and spatial belongings. The whole civilizational evolution has been marked by a technological progression. Some of these technologies have specifically played massive roles in shaping human interaction with territories or space. For instance, the invention of the wheel made it possible for humans to overcome distance, in the same way, the discovery of the iron made it possible to create more durable carriers. The coal and petroleum discoveries had an impact on increasing the spatial outreach of humans using inanimate energies. The invention of the airplane, after ships, which led to the age of discoveries and geographical outreach of humans, was another milestone that helped not only in travelling at higher speed but also through aerial photographs mapping of the surface of the earth is done more accurately. Further, the invention of satellites, unmanned aerial vehicles (UAV), and drones added to the techniques of mapping of the earth using remote sensing techniques, which helped to map the surface of the earth more accurately and frequently.

With the innovation of the telegraph and telephone, the communication process speeded up in time and space and this was the first time when the messages could be transmitted over space without any animal or human carriers (see Table 1.1). This technology was impregnated to transform the social and economic relations in space and time by giving birth to instantiated society. Further, with the rise of computing machines and especially the rise of digital technologies since the 1970s, humans' relations with space have transformed. Digital technologies have made it easy to capture data and store the same from direct or remote sensors.

A. Shaban (ed.), *Digital Geographies—Theory, Space, and Communities*,
https://doi.org/10.1007/978-981-97-4734-4_1

The internet has made it possible to transmit data from one place to another at the speed of light and rise of the computing power has enabled us to process big data in a very short time. The revolution in photogrammetry and remote sensing has enabled us to map every inch of the earth's surface while geographical information systems (GIS) have made possible computation, visualization in 3D, and rendering the surfaces as per one's desire.

Digital technologies have made it possible to produce virtual realities. The enmeshing of virtual and real spaces has now created hybrid spaces in our minds, and they help us to better visualize them. In fact, de Souza e Silva (2006) claims that differences between digital and physical spaces have been eroded with the rise of location-aware digital devices. Given the interaction of technology has been a continuous process over the time of human history and spaces have been visualized as something which is becoming, rather than complete. The term "ontogenesis of spaces" has often been used to denote this continuous process of becoming of spaces. Graham and Zook (2013) have used the term "digital shadow" to show that digital contents as if are constituted by bricks and mortar because the digital images of the cities are used to describe, imagine, and advertise the cities. Theorizing the emerging digital-led geographical changes and space relations themselves becomes, therefore, important for the social sciences. Artificial Intelligence, (AI) robotics, computational technologies, and relay of information/data through the internet are going to massively change the ways we understood the making of human geographies in the past. Therefore, understanding of the impact of digital technologies on space becomes crucial to capturing the image of the future or becoming.

1.1 What is Digitalization?

In literal terms, digital is relate to numbers and storing of the electronic signals in binary forms, 0s and 1s. Digitalization is a method of data processing, which earlier may have been in some other forms, as earlier in telegraphs and telephones electrical pulses in analog forms. From Samuel Morse's System or codes (which was encoded letters with simple on and off signals of two lengths, called dots and dashes) in 1830s to the American Standard Code for Information Interchange (ASCII) in 1966, digitalization was perfected which could be improved further. Digitalization as such in literal form is the revolution in signal processing in the form of binaries 0s and 1s, and that includes data capture (input), storing of electronic information or signals, processing the data in digital forms, and transmitting them from one place to another in digital forms or using the same to create real-life images, or simulations, etc. However, with the revolution in computing machines using digital processing, the impact on society, economy, polity, and environment has been so immense that digitalization now also refers to the machine-induced social behaviours of individuals and groups.

Table 1.1 Impact of some important discoveries and inventions on geography

Major inventions/ Discoveries	Year	General impact on society	Some major impact on geography
Wheel	C 4000 BC	Made transportation easier	Eased the geographical mobilities and imagination of geographies widened
Iron	C 1200 BC	Made large-scale agriculture possible	Land use changed from forest and grassland to agricultural, and made the agricultural surplus possible on a large scale to support the population in secondary and tertiary sectors
Steam engine	1712/1778	Made the Industrial Revolution possible through the utilization of machine power	Movement over space and sea through mechanical powers (used in ships, trains, cars, etc.) became easier. Industrial production and agricultural production got revolutionized, and this made big cities possible. Social relations started changing based on production relations
Internal combustion engine	1872	Made large-scale mass movement possible by powering ships, rails, and later airplanes	Geographical mobility further eased, and transport became relatively cheaper as a factor of production. Large-scale migration made possible the cultural diversities fuelling further inventions and economic growth

(continued)

Table 1.1 (continued)

Major inventions/ Discoveries	Year	General impact on society	Some major impact on geography
Telegram/Telephone	1816 /1876	Revolutionized long-distance communication	Made real-time communication possible; spatial rearrangement of production systems; beginning of instantiated societies
Aeroplane	1903	Speeded up the geographical mobility	Massive reduction of time in mobility over the surface. Changed the geopolitics of the world and the geographical distribution of industries. Aerial photos provided new possibilities to view and monitor the earth surface
Commercial computer	1951	Digital storing of data, computing, and prediction became possible	Helped massively automatization of geographic data, mapping and understand spatial patterns
Smart phone	1992	Personalized long-distance communication	Revolutionized the communication over distance and with time and technological development it became one of the most individually possessed tool with massive possibilities of data transmission and visualization. A large number of digital groups developed, massifying the spatial disconnect and crossing sovereign national boundaries

(continued)

Table 1.1 (continued)

Major inventions/ Discoveries	Year	General impact on society	Some major impact on geography
WWW (internet)	1993	Large-scale mass communication and data transmission became possible	Mass communication and data transmission became possible with the speed of light in digital form. Many heralded the death of geography. Digitalization has come to shape society in multiple ways by building informational age and datafication
Artificial intelligence (AI), algorithms, and machine learning on a large scale	2006 (AI is now deployed by Twitter, Facebook, Netflix, etc, in their advertising and capture users' presence and experience)	Large-scale automation and self-processing capabilities like human in machine	Large-scale data gathering, organizing, and processing. Decisions based on machines and modelling based on big data. Visualization of data increased, and spatial analytics reinforced itself while physical geographies kept on shrinking
Public use AI	Chat Generative Pre-Trained Transformer (Chat GPT), and other AIs	Reinforcing AI and deep learning through machine learning with the capability to process the information in real-time based on the open problem	Expanding the cyberspace and computing and visualization capabilities. Modelling of difficult geographical spaces eased out and decision support systems, data capture, and manipulation power increased

Source Author

The digital revolution or Industrial Revolution 4.0 is in many ways different from early Industrial revolutions. The Industrial Revolution 1 was led by the invention of the steam engine which had massive consequences for the transition in the use of energy (from animal to machine) and space. The Industrial Revolution 2 was the development of assembly lines which led to the decrease in production cost through specialization. They together made mass production possible, and they had massive impacts on the production of material spaces (cities, industries, roads, rails, etc.) and labour geographies. This revolution led to the massive movement of the capital from Circuit 1 (in agriculture and industry) to Circuit 2, the build environment. Industrial Revolution 3.0 was marked by internet and computing power, which again had massive impact on information flow and e-commerce, but the Industrial Revolution 4.0 which relates to digitalization powered by AI and robotics has led to the integration of value chains, the world economic systems, digitally integrated customer systems, and digital products and service trading at mass scale.

The digitalization has had both the transformative and disruptive impacts on the society. It has led to the rise of the machine age again in the new form, driven by Artificial Intelligence (AI) and robotics. This new machine age driven by AI has also led to the popularization of terms like "machine learning", "coding", etc., which point to techno-determinism, techno philia, techno mania in the current times, and datafication leading to the social drownings into information. It has a massive impact on the discipline of geography—theory, method, and data. Ash et al. (2019) argue we need to understand this new moment in human history, especially the "geographies produced through, produced by, and of the digital" (p. 3).

1.2 Impact of Digitalization on Geography

Digitalization has impacted almost all aspects of our lives—social, economic, cultural, and also the relations with the environment. It has created a new medium of social and individual relations and a new mode of governance. It has led to the changes in materiality as well as in thinking and social relations, spatial knowledge, and imageries (Elwood and Leszczynski, 2011; Graham and Marvin, 2001; Kitchin and Dodge, 2011; Rose et al., 2014; Wilson, 2012). It has created new distinct geographies of digital media—internet, games, etc. (Ash, 2015; Leszczynski, 2015; Kitchin et al., 2017), and shown possibilities of rising automation through software and AI- algorithms, which can evolve as a machine learning process.

A number of literatures, especially since the early 2000s have emerged on space relation of digital or digital geographies, but our understanding of digital geography is still in the early stage. Digitalization has led to massive data production or datafication and computation, but theorization remains limited and tentative. As digital geography is driven by quantification, it has amassed enormous data with little explanation or mind, and it has become like a Giraffe with a little head (theory) and a large body (data). One is reminded of the return of the quantitative revolution which brought the digital mapping and modelling project (Haggett, 1966; Hagerstrand, 1967; Balchin and Coleman, 1967; Tobler, 1959), but the Quantitative Revolution 2, is loaded with many possibilities as it is being led by AI.

The impact of digitalization on space is manifested in multiple forms. First, it has led to the decay of distance in clock time by making it possible to communicate across geographies and distances at the speed of light through digital devices. Many have argued in exaggerated way that this is the death of geography. Second, this has led to the warping of space and time simultaneously, with the decline of the importance of distance. It has brought out the micro geographies and temporalities in a warped way. Third, it has led to the Quantitative Revolution 2 in social sciences, and disciplines like data analytics have emerged using the method of analysis from geography and other social sciences. Fourth, it has created liminal confusing spaces by merging the virtual and real spaces. The spatial realities now presented through GIS and mathematical modellings are fused with reality. Fifth, it has enabled us to analyse the virtual space as well as real spaces. The analysis of games and software are heralding this trend. Fifth, it has created new space relations of capital, society, and self, giving rise to concepts like "digital capitalism", "digital labour", "digital society", "digital narcissism", "digital finance", "digital surveillance", etc. The theorization of digitalization-space relations as such has acquired a massive importance. Based on the above characterization of digitalization-space relation, below we attempt to understand what are impacts on different subareas of the geography.

1.3 Death of Time and Space or "Warping"?

The more the prominence of space, the more the interest in geography. It was assumed that like the geography of material reality, the information also has its geographies, and the distance decay principle operates on it as well. However, the massification of digital technologies has made it possible to relay information from one corner of the earth to another at the speed of light. Does it mean that geographical asymmetry of information in a way dead? The answer is both yes and no. Where there have been possibilities of relaying information across the geographies, the density of information has also increased. The information spectrum has widened. Though theoretically, it is possible to communicate information from one place to another, not all the information are relayed. Even digitally developed geographies may have informational peripheries due to this, leave aside the asymmetry arising from digital divides. Further, given that digital technologies through datafication made it possible to bring into sharp focus the attributes of the micro or local geographies, it seems that instead of geography getting liquidated it has warped up. The warping is not only of the spatialities but also of temporalities, as the celebration of specific moments—be it selfie spaces, micro-videos, and photos- have risen.

1.4 New Economic Geographies and New Spatial Logics

Digitalization has given rise to new economic geography as spaces have become more networked and integrated. The diffusion of information and technological possibilities to communicate from one region of the world to another has also expanded the outreach of the market, and it will not be an exaggeration to say that the whole world has now become an integrated market. But does that mean that the geographies of capital, manufacturing, labour, and exchange have disappeared? No. In fact, as the market has become dense, certain goods and services have got characterized with a global market area while others are still traded at the local level. Where Amazon to Airbnb have opened the possibilities to trade at the global level, the local markets and production and consumptions across the world do exist and even in smart cities, one has local neighbourhood shops and stores displaying usual locational characteristics.

Digital technologies at the macro level have made possible regional integration and the integration of rural and urban. Digitally enabled smart cities are driven by automated and visual space-driven devices and platforms, be it for electric consumption, transport, education, policing, health, and/or other services. However, these smart cities, planned and governed digitally, still do have pockets of different developmental shades. This is why the terms digital divides, digital justice, and just city are still invoked in many studies to highlight the issues that exist even in this digital age. However, it is also important to emphasize that the geography of materiality and sociality in the 1960s and 1970s have been disrupted and reconfigured due to digitalization. Some of these aspects we highlight below.

1.4.1 Digital Capital

Digital technologies, like any technology after the Industrial Revolution, have given rise to new space relations of capital at different scales. Digital technologies have made e-commerce possible, so the investment monitoring and safety of capital. The spatial networks of actors make it possible for capitalism to channel production, consumption, and trade to reap surplus and profit and shift the same using digital finance in no time to the safe heavens or to the core.

Capitalism moves forward through lumpy processes and technologies have been driving forces in its progression and density, volume, and deepening over space (Harvey, 1978). It produces spaces in its own image full of inequalities (Smith, 1984; Lefebvre, 1991; Graham and Anwar 2019: 180). Where through spatial–temporal fixes, the capital was invested in material space, like building environment, now it is moved to the virtual spaces of bonds and stocks and can fly in little time from one space to another as per the expected return to it. Further, digital enterprises like Google, Amazon, Meta, X, Microsoft, etc. now show new ways of reaping the profit through creating local and global nexus. Digital enterprises today constitute trillion-dollar enterprises and difficult for any country to ignore them. In fact, diffused in space and integrated digitally they have become the fastest value-creating enterprises in recent times.

1.4.2 Digital Labour

In the past, labour had a strong relationship with space. It was connected to the geographies of production or where the industry was located. However, digitalization has disrupted labour and space relations with regard to industry. Digitalization has made it possible to work from different locations (or cloud work) for diffused units of production. Graham and Anwar (2019) and Ash et al. (2019) argue that labour and space relations are complicated by digitalization, which creates new possibilities for the exploitation of labour by capital. Further, digitalization not only has deepened the division of the labour, but has also produced new spatialities of the labour. Leave aside the usual IT industry workers, a new class of gig workers has emerged facing acute exploitation in many countries.

1.4.3 Digital Finance

One of the aspects of economic geography which has received little attention has been financial geography. The financial geography shapes the real geography or material spaces. After the surveillance sector, the financial sector seems to be the most digitalized. The financial institutions that were earlier used to move to different territories in order to do their business of deposits, lending, etc., now operate through digital, and digital brokers, with little local geographical support institutions and manpower. This has made financial institutions largely footloose. The business of finance with the advent of deposit and withdrawal machines is also becoming automated, largely due to the Internet. But what has become interesting to note is the digital siphoning of resources from one territory to another by financial institutions. The peripheries face the acute brunt of the same as their deposits get siphoned or drained, as reflected in the form of higher credit-deposit ratios in core regions and to the upper and middle classes in urban areas and developed regions. These geographies of finance induced by digitalization become a critical reflection of how digital capitalism may now work creating new contours of development, new cores, and peripheries.

1.5 Digital Society

Digitalization has been disruptive to social relations, society-space relations, self and society, and space self-relations. It has created a new networked society where the usual kinship, proximity, and space relations do not matter much. Social media platforms have become new spaces or sites on which new relations are forged between individuals and social groups. This has created digital neighbourhoods, digital communities, and as such digital societies. The matters of social importance that used to be resolved face-to-face in some spaces have shifted to virtual spaces. The virtual spaces have become new sites like public spaces where the demand, protests, and affinity are expressed, relations are made or destroyed. This has created interneted and/or digitally connected individuals. With the rise of smart mobile phones in the last two decades, the digitalization of society has become overwhelming. This is paving ways for social evolution and social relations in a new direction where instead of social proximity in space, digital or virtual proximity has become important. This trend has disrupted real neighbourhood relations, family relations, emotions, and affections. The gratifications that used to be derived from social proximity are now drawn from digital proximity—videos, phone calls, social media postings, etc. WhatsApp, Facebook (now Meta), X, Instagram, Telegram, etc. are such interactive social media platforms giving ways to these new social relations in this digital age.

1.6 Digital Culture

The cultural industry has received new flips with the rise of digital technologies (Shaban et al., 2023). Digitalization has enabled the flow of culture from one corner of the earth to another. With the rise of social media platforms, mentioned above, it has now become possible to record, store, and disseminate the cultural components. Song, music, plays, and cinema are now being traded on e-commerce sites like any other commodities. Further, the rise of digital music has also heralded a new age of music composition where machines and software rather than artists help in the composition of music. Further, digital games have become another source of major trade and entertainment across the globe.

The data from UNESCO and UNCTAD show that in the last two decades, the cultural industry led by the digital industry has been a major source of growth in many regions of the world (Shaban et al. 2023; Shaban and Filip 2015). Many specialized platforms across the countries have emerged that deals with the e-commerce of games, videos, cinema, serials, and music. The Netflix, Jio Cinema (India), etc., are such platforms that have acquired global and national monopolies in OTT platforms. Digital piracy has emerged now as a big threat to many commercial cultural enterprises. Technologies like blockchain and enforcement of anti-piracy laws are used to protect the copyrights of cultural innovations and components.

Digitalization has also made possible the revival of interest of millions of people in arts. It has not only helped in the dissemination, commercialization of arts, and restoring the lost arts digitally, but also created new areas of creation of art (drawings) through digital technologies. Software like photoshops have helped in refining, rendering photos and art and/or creating new ones. Digital technologies have also blurred the distinctions of art between what is real and what is copied. The self-publishing platforms have also emerged where the publications of books, articles, etc., can be taken up and they can also be commercialized. 3D printing is further giving a direction to digital arts where what we imagined as two-dimensional can be created in 3D forms.

The digital flow of culture has created new cultural fusion across countries. It has helped overcome the distinction between high and popular culture, as the digital media platform have become accessible to the common people. The cities are most affected sites by it as most of them are the home of diverse cultures. This has strengthened ties and also weaved the people across geography with common cultural threads.

1.7 Digital State and Democracy

Digitalization has also had disruptive impacts on the mode of governance. It has brought about new ways of engagement with the state, with citizens, and with national spaces (Datta, 2023). The digitalization of the state has strengthened the integration of the governance system. The information flows and decision-making has become more transparent and quicker. Various programmes and services of the state (DIGISERV) have on one hand become more accessible, on the other hand, also exclusionary as per the design of the algorithms. Those not included in the algorithms of software find it difficult to avail the services and exercise their rights as citizens. The in-humanization through the digitalization of the state services often has adverse to exclusionary consequences on the marginalized sections of the societies. Some of the studies have shown that the digitization of the state has bureaucratized and technicalized the state even further, as the decision-making has become machine led.

Where digitalization has deepened the democracies, it has also shown adverse impacts on it with the rise of post-truth social media societies. Social media has become a weapon in the hand of the powerful and those in the government to regulate the weaker. Social media are now becoming platforms in divided societies protracting the verbal and hate wars between the communities. The digital as such has amplified the analog in many states.

Digital sovereignty has emerged as one of the key features of international politics in recent years. Digitalization has had adverse consequences for the sovereignty of the states in the Global South as some of the developed states with key advancement in digital technologies had wide control of the data related to finance (such as Master Card, American Express, Visa, etc.), e-commerce (as Amazon), and Social Media (as Twitter, Facebook, Google, YouTube, Instagram, etc.). The location of servers and control of the data by enterprises of the key global north countries led to create an information-based advantage to such countries, especially in terms of expansion of their market or specific security decision-making. The rising countries like India and China now assert to control the infringement of their digital sovereignty by obligating such global enterprises to locate the server storing data related to their countries within the physical boundaries of these countries. Thus, the digitalizing state of the Global South are struggling today to secure their digital sovereignty.

1.8 Digital Transport

Digitalization has affected the transport sector in significant ways. In smart cities, the transportation sector has been preferred for digitalization. Digitalization leads to the optimization of travel, time management, traffic regulation, and cutting down pollution. In many cities, digitalization has helped in creating integrated transport systems (Shaban and Sattar, 2023). Today in some cities, with a single ticket one can travel in all the modes of transport. The digital payments at ticket counters, automated machines, and websites also helped in cutting down the queues that used to be earlier at ticket counters. Further, digitalization has also helped develop integrated signalling systems cutting down travel time and avoiding traffic congestion. The GPS-enabled vehicles now are able to inform passengers of their locations and as such help the passenger plan their journey or wait time. The digitalization as such has led to the development of smart mobilities.

1.9 Digital Divide and Exclusion

Digitalization of state, society, and economy has reconfigured the inequality. It has created a new kind of labour division, based on access to digital though following the old plane of inequality. The technical ability and affordability to use the digital also determine one's access to the digital services and information of the digitalizing state. Those deprived of access to digital (like mobile phones, internet services, digital literacy, etc.) face exclusions. Further, algorithms are being used to identify individuals and groups on the basis of their ethnicity and race. As such the digitalization is also creating new apartheidism in many countries.

Digitalization in corrupt states has led to the greater legitimacy of the government. However, data are often used for political purposes and then the politics of bias also enter the algorithm. Digitalization of space often follows the class and ethnicity plane in most of the states. As such the already deprived groups face additional deprivations of being technically and accessibility-wise challenged to avail the digital services (Ash et al. 2016: 11).

1.10 Dizzying Realities—Obfuscation of Truth

Digitalization has also opened massive possibilities for dizzying realities through the obfuscation of realities or hybridization of virtual and real spaces. It has become possible now to mix the digital arts, imaginations, computations, and actual surfaces or maps together to create new images of the geographies. Google maps and other navigation software have helped to imagine geographies as per their displays. Further topographies in GIS or remote sensed data are presented in various ways including false colouring. They all create mental maps for the readers which obfuscate the realities and sometimes to the level of dizziness.

1.11 New Crime Geographies and Surveillance

One of the important sectors which has found the most use of digitalization has been the policing and surveillance sector. These are the sectors of the government which receive higher investment and innovations. In fact, most of the technologies like the internet or satellite became major sources of information relay and mapping (Table 1.1). Digitalization has created new panopticons of CCTV which photograph and record the spaces round the clock. The location-sensitive software and devices often keep on recording the coordinates and sending the information to servers which is used for spying or monitoring. Further, many software or viruses have been devised that clandestinely keep on monitoring the individuals' activity, network, and what they speak to others. In India Pegasus, a spying software was reportedly used to spy on the opposition leaders.

Where it has led to the effective control on some form of crime, digitalization has also led to the newer form of crime. Digitalization has created possibilities to hack software, divert the data, and even transfer the financial resources from one bank to another in clandestine ways. A gang of hackers has emerged who thrive on digital technologies for committing financial fraud, information theft, and manipulation to extort money digitally. Extortion and financial fraud have become a dollar economy.

The digitalization has also made possible the global sex industry. The pornographies and morphed images are being sold on such sites like any e-commerce sites though some regulations in some of the countries. The fear of abuse specially for women and girls through online manipulation remains higher and that is where specially women remain more cautious in posting their images or sharing their personal contacts on websites.

1.12 Digitalization and Environment

Digitalization has had a positive impact on the environment. It has helped in cutting down pollution as the information could be transferred through cables or wireless without undertaking any travel. This reduces emissions of carbon and other gases. Digitalization is also making the governance and education industry to be relatively free from paper work. This helps in the reduction of deforestation and papermaking. However, digitalization has created new kind of waste, called e-wastes. The digital devices often are made using different kinds of metals and materials. The improper recycling and open disposal of the same remain possible threats of to the environment.

1.13 New Quantitative Geography?

Digitalization has created a data revolution, often referred to as datafication. The datafication or big data is helping the research in many ways and obliterating the need of sampling. This is also impacting theory as big data is able to present the reality itself rather than requiring theorizing the reality using sampled data. Cloud computing to data capture platforms have emerged. Simulation to AI led analysis of data has become possible. GIS and remote sensing have emerged as an important tool for geographers for spatial visualization and analysis.

This book

AI has made possible quick summarization of large data and a number of studies. Taking advantage of this digitalization (AI), the present book attempts to provide a summary of studies published by Springer on digital geographies. It is possibly the first attempt to bring together both humans and machines in a collaborative way to present the summaries and explain the crux of the findings through the studies in nuanced and theoretical forms. The book is divided into 8 chapters, after this introductory and explanatory chapter on digital geographies, Chap. 2 summarizes the studies on the digital turn in social sciences. Chapter 3 deals with the papers related to human and digital technology relations and how this is evolving in recent times, while Chap. 4 summarizes the studies on time, distance, and digitalization and their inter-relations. Mapping has fundamentally changed with digitalization. Chapter 5 summarizes the studies on territorial mapping and how the algorithm of space is evolving. Chapter 6 reviews the papers on digitalization and exclusion or digital divides while Chap. 7 deals with studies on smart cities. Chapter 8 reviews the studies published related to ethnicity, race, and indemnity in this digital age.

As geography is a synthesizing discipline, the papers reviewed in this book are from a range of subjects, desciplines and journals and books. The chapters are roughly divided into different sections. We call it roughly as several articles deal with different issues and as such are thematically overlapping.

I have not edited and modified the machine-made summaries in any form but have provided a brief introduction to each chapter. These introductions attempt to theorize and summarize the ideas presented by the collections of studies in those respective chapters.

Given the scope of digital geographies has been limited, and it is an evolving subfield of geography, I am sure that students, teachers, and researchers will find it very handy to get not only information about the studies which have already been conducted and published in journals and books in Springer but also understand the theoretical and methodological framework of the studies. This may go a long way to understanding the theory, data and method and application of the same in their own research. Given the range of topics and digital and spatial turn in social sciences, the students, teachers, and researchers from across the subdisciplines of social sciences will find the book useful.

Bibliography

1. Ash J. (2015). *The Interface Envelope: Gaming, Technology, Power*. New York: Bloomsbury.
2. Ash, J. Kitchin, R. and Leszczynski, A. (eds) (2019). *Digital Geographies*. London: Sage.
3. Balchin, W. and Coleman A. (1967). Cartography and computers, *The Cartographer* 4: 120–127.
4. Datta, A. (2023). The digitalising state: Governing digitalisation-as-urbanisation in the global south. *Progress in Human Geography*, 47(1): 141–159.
5. de Souza e Silva, A. (2006). From cyber to hybrid: Mobile technologies as interfaces of hybrid spaces, *Space and Culture*, 9(3): 261–278.
6. Elwood, S. and Leszczynski, A. (2011). Privacy, reconsidered: New representations, data practices, and the geoweb, *Geoforum* 42: 6–15.
7. Graham, M. and Anwar, M.A. (2019). Labour. In Ash, J. Kitchin, R. and Leszczynski, A. (eds) (2019). *Digital Geographies*, pp.177–187. London: Sage.
8. Graham, M. and Zook, M. (2013). Augmented realities and uneven geographies: Exploring the geo-linguistic contours of the web, *Environment and Planning A* 45(1): 77–99
9. Graham, S. and Marvin, S. (2001). *Splintering Urbanism: Networked Infrastructures, Technological Mobilities and the Urban Condition*. London: Routledge.
10. Hagerstrand, T. (1967). The computer and the geographer, *Transactions of the Institute of British Geographers* 42: 1–20.
11. Haggett, P. (1966). *Locational Analysis in Human Geography*. New York: St Martin's Press.
12. Harvey, D. (1978). The urban process under capitalism: A framework for analysis, *International Journal of Urban and Regional Research*, 2(1–3): 101–131.
13. Kitchin, R. and Dodge, M. (2011). *Code/Space: Software and Everyday Life*. Cambridge, MA: M IT Press.
14. Kitchin, R., Lauriault, T.P. and Wilson, M. (eds) (2017). *Understanding Spatial Media*. London: Sage.
15. Lefebvre, H. (1991). *The Production of Space, trans.* D.N. Smith. Oxford: Blackwell.
16. Leszczynski, A. (2015). Spatial media/tion, *Progress in Human Geography* 39(6): 729–751.
17. Leszczynski, A. (2019). Spatialities. In Ash, J. Kitchin, R. and Leszczynski, A. (eds) (2019), *Digital Geographies*, pp.13–23. London: Sage.3
18. Rose, G., Degen, M. and Melhuish, C. (2014). *Networks, interfaces, and computer generated images: Learning from digital visualisations of urban redevelopment projects, Environment and Planning D: Society & Space* 32: 386–403.
19. Shaban, A. and Vermeylen, F. (2015). Trade in the Creative Sector: Comparing India with China, Brazil and UK, *Economic and Political Weekly* 50(20):68–75.
20. Shaban, A., and Sattar, S. (2023). Mobility and transport infrastructure in Mumbai Metropolitan Region: growth, exclusion and modal choices, *Urban, Planning and Transport Research*, 11(1): 10.1080/21650020.2023.2212745
21. Shaban, A., Vermeylen, F., and Handke, C. (eds.) (2023). *Creative Industries in India*. New York: Routledge.
22. Smith, N. (1984). *Uneven Development, Nature, Capital and the Production of Space*. Oxford: Blackwell.
23. Tobler, W.R. (1959). Automation and cartography. *The Geographical Review*, XLIX: 526–534.
24. Wilson, M.W. (2012). Location-based services, conspicuous mobility, and the location-aware future, *Geoforum*, 43: 1266–1275.

2 Digital Turn and Theorizing the Digital Geographies

Introduction

Progress of civilization is characterized by a horizontal and vertical expansion of techno-sphere, which currently has digital technologies and AI at the core. Although the evolution of machines and its impact on human behaviour, and socio-economic, political, and cultural activities have been a subject of debate for the last two centuries (Corpo 2017; Ash et al. 2016), the rise of digital technologies, especially since the early 1990s with commercialization of internet and availability of cheap computing machines, have had massive disruptive and transformative impacts on human to human, human and environment, human and machine and body and mind relationships. Several studies since the late 1990s have brought out the impacts of digital technologies on geographies—be it material or non-material. Digital technologies and AI have also disrupted the usual processes we use in production, distribution and exchange, and also expanded the cognitive powers through deep and big data and modelling possibilities, simulations and algorithm-driven man-machine and machine-to-machine interactions. Essentially, geography has been a discipline with the preoccupation of understanding of man-environment relationships through mediating cultural spheres. Digital technologies have emerged as key technologies mediating the relations among humans, nature and culture. A key focus of geography remains theorizing spaces. Where material spaces have been its focus in the past, its subdisciplines of cultural, economic, political and social geographies have engaged also in exploring the non-material, and metaphorical spaces. The evolution of digital technologies and AI have created possibilities to investigate and theorize immaterial, virtual, and metaphorical spaces, which have become key for political, social, and cultural actions.

One of the areas where the internet and the flow of information through mobile phones and computers have had massive impacts is institutions. The rapid flow

A. Shaban (ed.), *Digital Geographies—Theory, Space, and Communities*,
https://doi.org/10.1007/978-981-97-4734-4_2

of information is having transformative impacts on the institutions, whether they are social, economic, political, or usual governmental. The rise of informational societies is leading to the uniformization of global socio-cultural, economic, and political institutions. They are also leading to rapid changes and their orientation with a far-reaching impact on economic growth, development practices and processes. The digitalization of land records, digitalization of urban and local plannings, planned documents and the use of GIS in mapping, are helping in the commercialization and monetization of agricultural lands. The land information system has grown rapidly, especially in developing countries undergoing higher urban growth. This helps make real estate one of the high-growth sectors in these countries. A good example of this is India (Datta and Shaban 2017; Shaban 2010; Shaban et al. 2020). Digitalization, which is helping in overcoming corruption in many sectors in several countries, in digitally divided societies is also emerging as one of the sectors requiring intermediation and facing increased transaction costs.

As such, the digital geographies have opened up enormous possibilities to understand the digitally mediated spaces of (a) material geography—land, labour, capital, and institutions in space and their inter-relations, (b) production of material and non-material or virtual spaces, their terrains, contours, topographies of power, relations, flows, and temporalities, and (C) simulating and reproducing through human-machine and machine-machine interactions to mimic the material and virtual spaces through modelling and designs driven by algorithms. While analysing, theorizing, and virtualization or modelling, digital geography derives from other sister disciplines—be these disciplines from physical sciences or economics, political science, psychology, sociology, and cultural studies from social sciences. Therefore, the theorization in these subdisciplines of social and physical sciences also influences the theorization of digital geography. Additionally, as the theorization in digital geography is at the initial state, it much relies on the development in other disciplines. Therefore, comprehension of space and time-escape by geographers is much dependent on the development in other sister disciplines. As such this chapter does not bring only the studies related to the core-digital geographies but also the studies from other disciplines which have the potential to influence our understanding of spaces and their temporalities.

The reviews presented in this chapter are organized into two key subsections, digital social theories, and digital society. The first section presents a review of papers related to digital turn, cyber theories, the rise of digital technologies, surveillance, digital arts and cybernetics. The second section deals with the review of papers published on digital society and related theories. The latter set of papers provides rich but brief machine-made summaries on theoretical studies related to smart urbanism, digital civil society, society 5.0, capitalism 4.0, machine learning, sovereignty, industry and society 5.0, etc.

It is argued that datafication promotes surveillance by corporates and governments and hybrid spaces where the physical gets merged with the virtual. The increased connectivity which has been offered by digital technologies has recasted social interaction, culture, economics, and politics. It has profoundly transformed the self. Other studies contend that AI-based models can use the increasing amount of digital data available to test the accuracy of various social theories. Some of these reviewed publications have made the case that, in the future, the most advanced programmes in social theory should be headed by AI-driven cumulative advances. Additionally, it is asserted that people who reject digital life activities and technologies do so in an effort to uphold principles of community and identity that are outlined by the ethical framework of the civil sphere.

Theories of face-to-face interaction use the idea of spatial presence and consider digital communication to be a subpar form of interaction, frequently having pathological ramifications. The empirical evidence of "telepresence" in recent studies of mediatized communication, which contends that users of such devices perceive their interactions as instantaneous, challenges this idea.

According to a paper titled "The Digital Trinity—Controllable Human Evolution—Implicit Everyday Religion", digitalization is a complex web of socio-technological transformation processes that manifests as the trinity of datafication, algorithmizing, and phantomization. There has been a resurgence of faith in digital technology as a blind trust, but now as faith (i.e. without knowledge) in digital technology (e.g. in rationally incomprehensible big-data predictions of automated algorithmic-selection applications), which is increasingly becoming the basis for action and thus for decisions, as digital technology, increasingly perceived as superior, takes on a god-like function in the form of the internet, computers, and AI.

Another paper on "The Metaverse as a virtual form of data-driven smart urbanism: platformization and its underlying processes, institutional dimensions, and disruptive impacts" argues that in a digitally connected world, platforms have pierced the heart of urban societies, transforming social practices, disrupting social interactions and market relations, and affecting democratic processes. The paper also argues that the emerging phenomenon of platformization has given rise to what has been dubbed as "platform society".

References

1. Ash, J., Kitchin, R., & Leszczynski, A. (2018). Digital turn, digital geographies? Progress in Human Geography, 42(1), 25–43. 10.1177/0309132516664800
2. Corpo, M. (2017). The Second Digital Turn: Design Beyond Intelligence. The MIT Press.
3. Datta, A. & Shaban, A. (eds.) (2017). Mega-urbanization in the Global South: Fast cities and new urban utopias of the postcolonial state. London: Routledge
4. Shaban, A. (2010). Mumbai: Political Economy of Crime and Space. Hyderabad: Orient Blackswan
5. Shaban, A., & Kourtit, K. and Nijkamp, P. 2020. India's Urban System: Sustainability and Imbalanced Growth of Cities. Sustainability, 12 (2941):1–20. https://doi.org/10.3390/su12072941

Machine generated summaries

Disclaimer: The summaries in this chapter were generated from Springer Nature publications using extractive AI auto-summarization: An extraction-based summarizer aims to identify the most important sentences of a text using an algorithm and uses those original sentences to create the auto-summary (unlike generative AI). As the constituted sentences are machine selected, they may not fully reflect the body of the work, so we strongly advise that the original content is read and cited. The auto generated summaries were curated by the editor to meet Springer Nature publication standards. To cite this content, please refer to the original papers.

Machine generated keywords: theory, governance, digital society, economy, industrial, society, engagement, sustainable, political, sustainable development, economic, debate, science, politic, principle

2.1 Cybernatics

Machine generated keywords: theory, perspective, knowledge, surveillance, cooperation, digital technology, subject, health, limitation, social, business, new way, network, innovation, concept

2.1.1 Cyber Blame and Social Theory [254]

This is a machine-generated summary of:

Hanson, Barbara: Cyber blame and social theory [254]

Published in: SN Social Sciences (2021)

Link to original: https://doi.org/10.1007/s43545-021-00138-1

If you want to cite the papers, please refer to the original.

For technical reasons we could not place the page where the original quote is coming from.

Abstract-Summary

"Cybernetic causality reveals social processes of blaming."

"Looking at blame through a cybernetic lens reveals social construction processes of creation, modification, exacerbation, dissipation, and elimination."

"Mapping out fluid complexity in blaming might encourage using cybernetic causality to understand more aspects of social experience."

Literature review

"This is a theoretical piece that draws on thinking in various literatures and practices including theorizing blame, cybernetics, social performance theory, and a concept of legal person."

"They lead me to the question—can cybernetic modeling of blaming improve social theory?"

"Work from Psychology suggest blame as a social act that is managed in different ways including attributing causality (Malle and others [1])."

"It is a useful addition to social theory and the cybernetic study of blaming because it has the flexibility to work on a wide range of levels."

"A fluid model of blame that works cybernetically in an infinite number of levels might offer ideas about simultaneous multiplicity."

"These literatures all contribute to a model of cybernetic social performance of blaming on multiple levels."

Cyber blame

"Cybernetic, cyborg, and cyber (including "Cyber Monday" sales) have come into widespread everyday use to refer to technology, the internet, robots, and digitalization of information."

"Cybernetics also has a historic meaning illustrated by the process of steering boats."

"This is the sense of cybernetics I use as grounded in the literature set out in the previous section."

"Applied to blame cybernetic causality reveals how movement and fluidity are endemic to social experience."

"There is constant movement within, by, and among bodies."

"Blame events can have short- and long-term effects that may be good, bad, or bad and good at the same time."

"Not looking at cybernetic causal interconnections may miss the essential properties of fluidity and movement."

"Being aware of self-steering causal webs is a step toward portraying blaming."

Blame and blaming

"Blaming permeates practices of social experience in a multitude of ways on an infinite number of levels."

"Blame can be seen like a ball that is tossed, caught, batted back and forth, held, and made to disappear."

"Blame becomes part of a process of blaming as does a ball part of a game."

"Blaming a person (from a single human to a nation) for something bad that happens can initially be resisted by the person being blamed, then the person accepts responsibility, apologizes, makes amends, and gains integrity because of this."

"From a cybernetic view, its even possible to argue that fixing blame in a person violates the principle of non-summativity in the case of families in distress who attempt to lay blame on a single member when the relevant causal connections permeate the whole group (Hanson [2]).”

“Persons create, manage, direct, and modify blame.”

Blame performances

“Holism and cybernetics offer a way to frame causal steering sequences that are not necessarily linear, have different effects and meanings in different systems, and can mean multiple things at the same time. Parties or persons in the legal sense, even if a single human, might feel differently about blaming at different times or have multiple conflicting feelings about blaming at the same point in time.”

“Who is to blame, me or him/her, us or them/that?”

“Laying blame on a person other than you may appeal both as a way to reduce uncertainty and self-benefit.”

“I am thinking here about the way persons may go through a process of blaming themselves when something horrible happens, like the death of a loved one.”

“There is a benefit to the blamed person in terms of self-worthiness or esteem.”

Conclusion

“Can cybernetic modeling of blaming improve social theory?”

“We can see things about blame and lived fluidity by looking at ways blaming is performed as conduit, explanation, separation, self efficacy, and/or self esteem.”

“This might be a beginning point to looking for self-steering, multi-focal, and omni-directional connections in practices of doing, creating, and modifying other types of performances of social life.”

[Section 6]

“Blaming can be usefully modeled as a cybernetic social process.”

“Cybernetic causal dynamics can describe fluid process with multiple liquid, solid, animate, and inanimate parts.”

“This suggests a holistic view that embraces fluid, moving, multi-focal, self-steering compendia of parts, relations, and processes.”

“In this view we can see a social performance of blaming as a dynamic that is fluid and continuous.”

“This complexity and ability to transform can be usefully represented as cybernetic—self-steering multiplex causal infusions in fluid social experiences that are constantly moving.”

2.1.2 The Digital Revolution and Digitalized Network Society [255]

This is a machine-generated summary of:

Knell, Mark: The digital revolution and digitalized network society [255]

Published in: Review of Evolutionary Political Economy (2021)

Link to original: https://doi.org/10.1007/s43253-021-00037-4

If you want to cite the papers, please refer to the original.

For technical reasons we could not place the page where the original quote is coming from.

Abstract-Summary

"This lecture discusses technological revolutions and techno-economic paradigms, but with an emphasis on the digital revolution and the digitalization of the economic and society."

"We are now in second half of the digital (fifth technological) revolution, when the digitalization of the global networked economy prevails, and not at the beginning of Industrie 4.0."

"This is the period when economic growth drives the use of innovative digital technologies, including ubiquitous computing, robotics, and artificial intelligence, toward a truly digitalized network society."

Introduction

"This lecture is about technological revolutions and techno-economic paradigms, but with an emphasis on the digital revolution and the digitalization of the economy and society."

"The digital revolution created new opportunities for technical change and technological learning for those industries that have access to relevant networks."

"The notion of Industry 4.0 (Industrie 4.0 in German), for example, draws attention to the idea of industrial revolutions and its application to recent technologies, but it does not directly confront the ongoing theoretical debate on long waves and innovation networks."

"This article introduces a few of the basic ideas behind Industrie 4.0 and the Fourth Industrial Revolution and then discusses relevant issues in the work of Joseph Schumpeter, Christopher Freeman, and Carlota Perez on technological revolutions, long waves, and techno-economic paradigms."

"A summary of the first four techno-economic paradigms follows, which brings us up to the digital revolution and the rise of digitalized network society."

What is the fuss about Industrie 4.0

"Rifkin [3] claimed that we are at the beginning of a new industrial revolution where low-carbon energy generation technologies, such as wind, solar photovoltaic, hydropower, and wave and tidal power, lead the way."

"They believe that earlier technology-driven revolutions, such as the steam engine and electrification, took longer to diffuse through the economy."

"We might describe these technologies as disruptive innovations in that they disrupt every industry in the global economy, leading to the transformation of entire systems of production, management, and governance."

"Schwab believed that technology and digitalization will revolutionize everything, generating major technological innovations that will fuel fundamental change that will spread throughout the global economy."

"There is no discussion of long waves, technological revolutions, and techno-economic paradigms, all of which are important for understanding Schumpeter or Industrie 4.0."

Technological revolutions, long waves, and techno-economic paradigms

"The resulting clusters and networks of interrelated technological and organizational innovations create new and rapidly growing sectors, lead to structural change, and cause long swings in the economy."

"Freeman and Perez [4], Freeman and Louçã [5], and Perez [6] saw these technological revolutions as techno-economic paradigms."

"Perez [6: 8] defined the techno-economic paradigm as "a powerful and highly visible cluster of new and dynamic technologies, products and industries, capable of bringing about an upheaval in the whole fabric of the economy.""

"Perez [6] named four phases in each techno-economic paradigm: (1) irruption phase, when the modern technology supplants old technology; (2) frenzy phase, or the period of intense exploration; (3) synergy phase, when the new technology is diffused throughout the economy; and (4) maturity phase, as the diffusion process becomes complete."

"A novel feature of Perez [6] is that each techno-economic paradigm has a turning point where production capital displaces the financial capital."

From the industrial revolution to the digital revolution

"Industrie 4.0 named four stages of the industrial revolution: (1) water- and steam-powered mechanical manufacturing (first two long waves); (2) electrically powered mass production (second two long waves); (3) digitalization (fifth long wave); and (4) cyber-physical systems (sixth long wave) (Schwab and Davis [7])."

"We are currently in the second half of the fifth technological revolution."

"Refinements in milling and spinning technologies were key in the development of the machine tool industry."

"As resource prices fell, steam-powered mechanization of industry became the key technology (Chandler [8])."

"The Third Technological Revolution started the electrification of industry, transport, and a new electrical equipment industry from 1875."

"This application of mass production techniques was essential to the Fourth Technological Revolution, including making use of machines and presses to stamp out parts and ensure interchangeability, which led to the relative cheapness of large-scale production and the emergence of mass consumption (Hounshell [9])."

Digitalization, technological trajectories, and the digitalized network society

"The digital revolution was bubbling underneath the fourth industrial revolution."

"The Apollo Guidance Computer was the first silicon integrated circuit-based computer, which led to the microprocessor (Isaacson [10])."

"This was the big bang of the digital revolution!"

"Although the original Intel microprocessor had only 2250 transistors, the number of transistors in an integrated circuit doubled about every 2 years (Moore's Law)."

"Soon afterwards, several new technological trajectories appeared within the digital techno-economic paradigm, which evolved into clusters of new and dynamic technologies, products, and industries that rippled through the entire economy and society."

"New enterprises appeared and interacted with each other in complex networks, which perused problem-solving activities that were cumulative, incremental, and path-dependent (Arthur [11])."

"A wide array of new products dependent on microprocessors have appeared in recent decades, including robotics, artificial intelligence (AI), and smart energy networks (Perez [12]; Knell [13])."

How do networks fit into our story?

"As the Roman Empire evolved, both the size and intensity of commercial and trading networks increased, as did the speed and direction of technological change and specialization (Kay [14])."

"National production networks were important for engine builders, as they improved machine tools, casting, and measurement techniques from other sectors after the industrial revolution (Thomson [15])."

"This gave rise to global trading networks and empires, which relied on telegraphs and telephones for coordination."

"Global production networks and multinational enterprises then appeared, making it possible to have both inter-firm and intra-firm trade and production."

"One distinctive characteristic of the digital revolution is the rise of global production networks (Coe and Yeung [16]), GINs (Zander [17]), global R&D networks (Tomasello and others [18]), global inventor networks (Graf [19], Fritsch and Zoellner [20]), user innovation networks (von Hippel [21]), and spatial innovation networks (i.e., Silicon Valley)."

"Internal, local, and global innovation networks are key social technologies in the digital revolution."

What are global innovation networks?

"Collaborative agreements between firms and other organizations appear in the formalized links or global networks in which individual firms work and may include agreements within an enterprise group, upstream suppliers, downstream customers, competitors, the government, and universities and other research institutes."

"Ernst [22], and Barnard and Chaminade [23] define a GIN as a global network of interconnected and integrated functions and operations carried out by firms and non-firm organizations engaged in the creation or diffusion of innovations across geographic borders."

"There are several types of innovation networks that involve enterprises collaborating with each other or with other organizations."

"Innovation networks can be more formal and ridged, usually encompassing a collaborative agreement or strategic alliance with long-term objectives, or they can be informal, flexible and trust-based relations, often linking enterprises with various local (Camagni [24]) and regional (Boschma [25]) actors, as Lundvall and Borràs [26] contend."

"A third strategy involves self-organizing networks, which have joint ventures, collaborative agreements, and long-term cooperation that includes cross-licensing, outsourcing, joint R&D, and other types of innovative activities (Pyka [27], Pyka and Scharnhorst [28], Kirman [29])."

Anticipating the future

"Can we envisage the future and the sixth industrial revolution?"

"Possible emerging technologies include nanotechnology, biotechnology, quantum computing, and AI."

"Feynman described a process in which scientists would be able to manipulate and control individual atoms and DNA molecules."

"It is possible that two or three independent technological systems could converge into one system, which would then trigger the explosive take-off into the sixth technological revolution."

"Atoms, DNA, bits, and synapses may supply the basic elements and foundational tools that will make it possible to integrate several emerging technologies, including nanotechnology, biotechnology, information technology, and the latest cognitive technologies, into multifunctional systems (Roco and Bainbridge [30])."

"The anticipatory systems view makes it possible to integrate the future into the present by formulating diverse ways and several reasons for thinking about the future (Miller [31])."

"It is possible to use the future to identity challenges and opportunities and to discuss systemic transformations."

2.1.3 Bentham, Deleuze and Beyond: An Overview of Surveillance Theories from the Panopticon to Participation [256]

This is a machine-generated summary of:

Galič, Maša; Timan, Tjerk; Koops, Bert-Jaap: Bentham, Deleuze and Beyond: An Overview of Surveillance Theories from the Panopticon to Participation [256]

Published in: Philosophy & Technology (2016)

Link to original: https://doi.org/10.1007/s13347-016-0219-1

If you want to cite the papers, please refer to the original.

For technical reasons we could not place the page where the original quote is coming from.

Abstract-Summary

"This paper aims to provide an overview of surveillance theories and concepts that can help to understand and debate surveillance in its many forms."

"We structure surveillance theory in three roughly chronological/thematic phases."

"The first two conceptualise surveillance through comprehensive theoretical frameworks which are elaborated in the third phase."

"The first phase, featuring Bentham and Foucault, offers architectural theories of surveillance, where surveillance is often physical and spatial, involving centralised mechanisms of watching over subjects."

"The second phase offers infrastructural theories of surveillance, where surveillance is networked and relies primarily on digital rather than physical technologies."

"Deleuze, Haggerty and Ericson, and Zuboff develop different theoretical frameworks than panopticism to conceptualise the power play involved in networked surveillance."

"With the datafication of society, surveillance combines the physical with the digital, government with corporate surveillance and top-down with self-surveillance."

Introduction
"Since surveillance is used as an umbrella term that covers a broad range of sub-topics discussed in other domains as well (e.g. privacy, urban planning, governance, policing, safety and security), the concept of surveillance features increasingly in different contexts and disciplines, making it harder to follow and focus debates across disciplines."

"The aim of this paper is to provide a largely chronological-thematic overview of surveillance theories and concepts that can help scholars from an increasingly wide range of disciplines to understand and debate surveillance in its many forms."

"These clusters can roughly be ordered chronologically by distinguishing the following three phases in surveillance theory building. (1) Bentham's liberal project connected to the architectural design of a prison and other buildings and Foucault's subsequent analysis of discipline and the Panopticon as a metaphor to talk about institutions and society."

Phase 1: Exploring the Panopticon
"Bentham's vision, already in the prison-Panopticon, was not to create a 'society of control' where people would be watched all the time; rather, the idea was that discipline would be internalised and the need for the inspector, the watching itself, would be eventually exhausted."

"This feature disqualifies the Panopticon as a uniquely disciplinary (in the Foucauldian sense) machine, since the Panopticons, rather than towards disciplinary power through architectural design, evolved more towards Foucault's concept of governmentality (Brunon-Ernst [32, 36]), a type of power—initially called sécurité—different from discipline in that it no longer seeks to manage individual bodies but rather manages whole populations, trying to optimally regulate social behaviour."

"It is relevant to look into why Foucault's analysis of the Panopticon resonated in so many disciplines, for his analysis is not only rehabilitating (specific parts of) Bentham's work, but also building upon and extending it into a broader perspective on power relations and networks in modern societies."

Phase 2: Post-Panoptical Theories and Concepts
"This new, post-panoptic surveillance: (1) is driven by the desire to bring systems together, leading to an increased convergence of formerly discrete surveillance systems and an exponential increase in surveillance capacity; (2) is increasingly growing and spreading by expanding its uses for purposes of control, governance, security, profit and entertainment, and with the help of new and intensified technological capabilities, particularly the rise of computerised databases; (3) has a levelling effect on hierarchies of surveillance, due to new target populations being monitored with new and intensified technological possibilities; (4) works across state as well as non-state institutions; (5) is primarily directed towards human bodies that are understood as a hybrid composition—a flesh-technology-information amalgam; and (6) relies on machines to make and record discrete observations (Haggerty and Ericson [33])."

Phase 3: Contemporary Conceptualisations—the Branching Out of Surveillance Theory

"Present forms of surveillance potentially have more impact and shaping power on citizens' daily life than pre-Internet paper-based data entries because of the accessibility of digital databases and the relative ease of combining and sharing different types of data (Marx [34])."

"Even if we follow Deleuze in reasoning that corporations are now the main surveilling actor, with a surge for power and control that is enhanced by their opaqueness, still individuals can, at least to some extent, resist and refuse, mainly by finding alternative ways of using technology that is increasingly accessible to him or her."

"Mann uses the example of wearable cameras to watch back at CCTV cameras, the idea of sousveillance could also be transposed to the digital or the virtual context, where citizen journalism and publishing (leaked) information about surveillance actors can be seen as forms of watching back."

Conclusion

"It can be characterised as offering architectural theories of surveillance, where surveillance is largely physical and spatial in character (either in concrete, closed places such as institutional buildings or more widespread in territorially based social structures) and largely involves centralised mechanisms of watching over subjects."

"Panoptic structures are theorised as architectures of power; through panoptic technologies, surveillance enables power exercise, not only directly but also, and more importantly, through (self-)disciplining of the watched subjects."

"A common element in the different theoretical accounts of Deleuze, Haggerty and Ericson and Zuboff is to critically question not only the power structures in contemporary network societies and how surveillance reinforces, or sometimes undermines, these, but also how we can conceptualise this power play beyond panoptic effects of self-disciplining."

"Notions of self-surveillance point to new dynamics, where watching oneself via a mediated, mobile and networked gaze still raises questions of power, discipline and control, but in potentially new ways that cannot be easily captured in classic surveillance frameworks."

2.1.4 Grand Digital Abyss: Technology, Critical Theory, and Subjectivity in Twenty-First Century Cyber-Capitalism [257]

This is a machine-generated summary of:

Alarcón-Medina, Rafael: Grand Digital Abyss: Technology, Critical Theory, and Subjectivity in twenty-first centuryCyber-Capitalism [257]

Published in: Dialectical Anthropology (2019)

Link to original: https://doi.org/10.1007/s10624-019-09577-y

If you want to cite the papers, please refer to the original.

For technical reasons we could not place the page where the original quote is coming from.

Abstract-Summary
"The so-called passage from Subject to Project appears as the emergent form of objective reason in its mystified technological form at the beginning of the twenty-first century, generating unprecedented configurations of praxis and subjectivity before which Marxism should open new paths of critical reflection."

"The task requires a dialogue with philosophical perspectives usually denied within the Marxist debate."

"From the interweaving of Martin Heidegger's existentialist philosophy and the Frankfurt School's critical perspective (Theodor Adorno and Max Horheimer), in this paper, I propose a critique of constituent subjectivity in today's technological age."

Introduction
"In its economic-material dimension, the supposed transition from subject to project involves the apparent immaterialization of social production, in which the producer's relationship with her product (subject-object) expressed in commodity as constituent social form of capitalism, is transformed into a relationship of Being with its own essence mediated by technological objects and political forms of domination."

"The social world is fragmented through its spectacularization now overflowed as simulation (Debord [35]; Baudrillard [36]), diluting social relations' historical density into networked digitized communicative forms: Facebook, Twitter, Snapchat, Google, YouTube, etc The difference between Modern-Enlightened perspectives of a world's objectivity confronting the subject in dialectical fashion, and the notion of the Being of Man as a project of himself, resides in the fact that the latter is presented (appears) as an ontological-conceptual synthesis of the subject-object relationship."

Between Adorno and Heidegger: modernity, technology and the critique of constituent subjectivity
"As for Horkheimer and Adorno [37], the history of bourgeois subjectivity is the history of the concept's dominion over the object (encompassing more than the era of capitalist mode of production); for Heidegger, the modern history of Man is the history of subject as Enframing."

"While for Adorno, philosophy's task lies in strengthening subjectivity's radical power through the immanent critic of historical experience, for Heidegger the goal of philosophy is to strip Man of the burden of Enframing, de-fetishizing philosophy through a language of concreteness that would unveil the Being, restoring Man's access to her essence (Heidegger [38])."

"While for Heidegger, the critic of constituent subjectivity is carried out with an ontological terminology pretending to be a language of concreteness and immediacy, for Adorno the critic of the concept's dominion must occur in the historical dimension of contemporary society, taking into consideration the mediations between subject and object produced in the cultural sphere within capitalist forms."

Techno-Genesis: image, concept and historical-temporal experience

"Instead of magically representing sensible experience, the work of art as technical image actualizes—materializing it—the concept embedded in the technology used in its capture and reproduction."

"As a technical image, the work of art represents the concept that dominates the technological configuration of the apparatuses and machines that organize the experience of the world in capitalism."

"In the technological era, the concept not only prefigures the representation and perception of objects, but also begins to pre-design the world of experience itself, producing an objectual system whose origin lies to a greater extent within the concept itself, independent of any classical notion of "nature"."

"I will show how code as logos—the totalized concept as objective reason—occurs in capitalistic social production as a cultural form, as Enframing, which not only involves the symbolic dimension of praxis but also, as Adorno argues, includes "...the material infrastructure of society, in which human work and thought are rooted, and the only means through which work becomes real societal work" (Adorno [39: 80])."

Cyber-Capitalism: concept, code, and the experience of the sensible world

"Beyond the mere representation of objects in mathematical language, code lays the foundations of world's design from informational concept: the social world becomes programmable."

"The growing programming of social practices through various digital platforms such as Facebook, YouTube, Twitter, and Google leads to a society integrated under the domain of code as the concept's hegemonic form (see Finn [40]; Chun [41])."

"The radical integration of the social world under the algorithmic logic implies the appearance that there is no social and political mediation between thought and action, since the social and political are incorporated into the concept as products of the code."

"From the analysis of Facebook, I present some dimensions in which the program's algorithmic logic subsumes social relations unto the domain of constituent subjectivity as reality's programming code."

[Section 5]
"In a dialectical turn, we could say that on Facebook the encounter, participation, identity, and community happen at the same time when the real possibility of meeting with the other, participation in the public space and collective belonging have been denied beforehand by being pre-programmed by code as an actualization of concept."

"The more the subject seeks to deny the false authenticity of his digital identity through new publications, the more she affirms the mythological character of content as an inner truth externally produced by the program's statistic logic."

"Facebook is thus an example of the commodity form's logic of separation concealment, by means of which code as the concept's actualization hides the mythological character of the subject's power beneath the apparently free creativity of Man, manifested in the multiplication of digital content."

Conclusions
"For Heidegger, the danger in the technological age is the time in which life is subsumed under the structures of constituent subjectivity as the Enframing of every possible experience of the real."

"This apparent openness of Being to difference, multiplicity, and authenticity actually conceals the deep linearity of code's programming which constitutes the actualization of concept and Enframing as capitalist digital technology."

"Cyber-Capitalism is the expression of the concept's negative dialectic as mystified reason turned into the subject of history under the form of algorithmic code, and ultimately as apocalyptic artificial intelligence (see Geraci [42])."

"It remains a task to glimpse possible emergency exits to the totalitarian tendencies of concept as code, which currently seeks to establish itself as the foundation and reason for all experience, reducing life to virtually pre-existing forms in the digital Matrix of contemporary capitalism."

[Section 7]
"The proletarian is literally the one who has his children as the only possession."

"Its self-production is limited only to biological reproduction."

"On the contrary, there is an illusion that everyone, as a self-free project, is capable of unlimited self-production—Byung-Chul Han"

2.1.5 The Digital Trinity—Controllable Human Evolution—Implicit Everyday Religion [258]

This is a machine-generated summary of:

Latzer, Michael: The Digital Trinity—Controllable Human Evolution—Implicit Everyday Religion [258]

Published in: KZfSS Kölner Zeitschrift für Soziologie und Sozialpsychologie (2022)

Link to original: https://doi.org/10.1007/s11577-022-00841-8

If you want to cite the papers, please refer to the original.

For technical reasons we could not place the page where the original quote is coming from.

Abstract-Summary

"A media-change perspective that focuses on innovation-driven, complex co-evolutionary processes distinguishes two phases of digitalization and points to the following characteristics of its second phase: digitalization is an intertwined bundle of socio-technological transformation processes that reveals itself as a trinity of datafication, algorithmization and platformization."

"On the industry and politics side, this co-evolutionary trinity is driven by the belief in a scientifically and technologically controllable human evolution, reflected in the pursuit of nano-bio-info-cogno convergence, and accordingly linked to a transhumanism standing for this belief."

"On the users' side, the digital trinity is characterized and driven by the convergence of digital technology and religion in the form of an implicit everyday religion."

Introduction

"This article looks behind the sober technological facade of digitalization from a media-change perspective that understands change as an innovation-driven, co-evolutionary process under complex conditions (Latzer [43a])."

"Such a perspective focuses on the evolution of hybrid socio-technical constellations, understands innovations as co-evolutionary processes (Frenken [44]), and sees the commonalities between biological and technological evolution in complexity theories with a selection paradigm (Ziman [45])."

"This is followed by a presentation of its main characteristics, summarized as the co-evolutionary digital trinity of datafication, algorithmization and platformization, the belief in controllable human evolution, reflected in transhumanism and the pursuit of nano-bio-info-cogno convergence (NBIC), and the convergence of digital technology and religion toward an implicit everyday religion."

"These features are reflected in the specifics of an emergent digital social order of society, and summarized in the final section."

Two Phases of Digitalization
"From a technological perspective, the first phase is characterized by the spread of digital computer technology, by the eponymous transition from analogue to digital technology in telecommunications (telephony, data communication) and broadcasting, and by the beginnings of mass internet penetration following the launch of the World Wide Web (WWW) in the early 1990s."

"With this politically promoted competition in communications markets and the resulting intensified technological innovations in devices and services, the self-reinforcing co-evolutionary innovation cycle is closing and has been spiralling into more and more areas, driven by the internet as a complex, adaptive system, as an innovation machine (Latzer [43a])."

"Similar to the first phase, a co-evolutionary spiral movement based on two convergences is also characteristic of the early twenty-first-century digital transformation: the NBIC convergence and the convergence of digital technology and religion, as described below."

The Digital Trinity: Datafication—Algorithmization—Platformization
"From a media-change perspective, the digitalization proclaimed by all sides in science, politics and business in the early twenty-first century reveals itself in threefold form as (a) datafication, (b) algorithmization and (c) platformization."

"Platformization restructures markets and business models, also increasingly commercializing the social sphere and thus creating optimized organizational forms for further-reaching social datafication and algorithmization."

"Datafication in the form of close-knit digital data surveillance (van Dijck [46]) creates the preconditions for this, which in turn diminishes privacy."

"b. Algorithmization, more or less automated problem-solving or data-processing procedures, is employed to make economic, political and social use of (big) data resources."

"This also results in new social inequalities, in power asymmetries due to unequal ownership of or access to big data and to elaborate algorithmic procedures that are increasingly based on AI."

Controllable Evolution: Nano-Info-Bio-Cogno Convergence and Transhumanism
"This assessment is coupled with the belief in humanly controllable evolution, in the potential of scientifically and technologically designed evolutionary steps of humanity that can improve physical and mental abilities towards transhumanism and thus increase health, well-being and welfare."

"Transhumanism often plays a provocative role in the debate about the controlled scientific and technological evolution of humanity."

"The basic idea of transhumanism is part of the millennia-old history of the quest for new human capabilities, a co-evolution of technological and human development that is aimed at overcoming the limits of the human body and mind—especially in ageing/mortality and intellectual, physical and psychological performance (Bostrom [47]; Tirosh-Samuelson [48]; Hurlbut [49])."

"Controversial debates about the image of technology, the human–technology relationship, the potential of scientifically and technologically designed interventions in the evolution of humanity and about the political will and corresponding governance advances, form the politico-philosophical background of the current digital transformation process."

Implicit Everyday Religion: The Convergence of Digital Technology and Religion

"As in other sectors and areas of life, digital technology is used by institutionalized religions as an additional tool to achieve their goals, such as organizing religious communities and spreading the faith."

"The religious aspect of digitalization is developing apart from, albeit connected to or in competition with, traditional religions, driven by the convergence of technology and religion in the sense of a blurring of boundaries, which has already progressed far since the Enlightenment."

"In the course of digitalization, a religion-like "technological solutionism" is also emerging, a belief in universal solvability through technology, which is spreading in combination with an "internet-centrism" of the early 21st century (Morozov [50])."

"Driven by data belief (as opposed to knowledge), the religious expectations of the internet and related technologies such as blockchain, robotics and AI are not only found in science and science fiction, but also in the global political debates on digitalization."

The Digital Order: Special Features and its Governance

"A brief summary of how governance by technology in the context of the digital trinity influences reality construction, decision-making and social order (a), is followed by selected starting points of a governance of the digital trinity that consider specificities such as complexity and faith in digital technology (b). (a) From a media-change perspective, changes in individual reality construction and social order are a core part of governance by technology in the context of the digital trinity and thus through diverse algorithmic-selection applications."

"As digital technology, increasingly perceived as superior, takes on a god-like function in the form of the internet, computers and AI, there has been a resurgence of faith as blind trust, but now as faith (i.e. without knowledge) in digital technology (for example, in rationally incomprehensible big-data predictions of automated algorithmic-selection applications), which is increasingly becoming the basis for action and thus for decisions."

Summary

"In order to analytically grasp the core of the socio-technical transformation in the second phase of digitalization, this paper proposes the concept of a digital trinity, composed of datafication, algorithmization and platformization."

"The paper further highlights two more characteristics of the second phase of digitalization that can be conceptualized as central driving forces of the digital trinity: the strong belief in a controllable human evolution on the part of politics and industry (a), and an implicit everyday religion of the digital on the part of the users (b)."

"Comparable with traditional religions and states, the digital trinity—driven by the belief in controllable evolution and an everyday religion—shapes the social order."

"The article argues the emergence of a transformed digital social order, classifies this most recent media-change-related transformation, points out its specific features, and highlights some of the challenges for its political governance."

2.1.6 Looking at Digital Art: Towards a Visual Methodology for Digital Sociology [259]

This is a machine-generated summary of:

Gilbert, Tamsyn: Looking at Digital Art: Towards a Visual Methodology for Digital Sociology [259]

Published in: The American Sociologist (2018)

Link to original: https://doi.org/10.1007/s12108-018-9384-2

If you want to cite the papers, please refer to the original.

For technical reasons we could not place the page where the original quote is coming from.

Abstract-Summary

"I will use the medium of digital art to consider the unique nature of the digital age, the demand for a digital sociology, and the interrelated speculative imagination of such claims."

"Broadly situated within the sociology of art the methodological contribution of this article is to offer an analysis of artworks themselves, via the construction of a digital visual methodology."

"What digital culture, politics, and revealed in digital art?"

"How can looking at digital art expand the tools for understanding digitally mediated lives?"

Digital Sociology: Framing the Literature

"There are four themes in the growing literature of digital sociology that are a central framing structure to investigate what is the unique nature of the digital age, the demand for a digital sociology, and the interrelated speculative imagination of such claims."

"A second theme the literature addresses is how digital technologies have a distinct effect on the production of kinds of individuals and communities."

"A case is made that a unique network logic, that of the distributed network, is essential to understand the nature of social power in a digital age."

"The literature of this perspective sees the digital age as ushering in a new "exceptional" egalitarian social space."

"These four themes from the literature address the unique nature of the digital age, the demand for a digital sociology, and the interrelated speculative imagination of such claims."

From the Sociology of Art to Digital Art

"It also productively resists, contests, and provides tension in the disciplinary practice of sociology, by engaging with the social knowledge work of the artists themselves, their arts practices, and the artistic object as unique sociological phenomenon."

"What are the implications of separating art worlds from more general social phenomena, for the artist, art practices, and the artistic object; of separating art and society?"

"The second parallel theme is to understand the artist, art practices, and the artistic object as a distinctive kind of sociological object, manifest in the production of kinds of work by individual artists and broader communities of practice including the social power of art in systems of cultural distribution."

"The third parallel theme is how the artist, art practices, and the artistic object resist the production and use of data."

Digital Art

"There is no set date that marks the birth of digital art as a medium."

"There is also contestation over what we call the medium: is it digital art or it is new media art?"

"Mark Tribe argues that new media art is the categorical telos of these other digital art "micro-movements," whose legacies lie in other technologically-based mediums."

"For clarity, and to connect to my broader investigation of the digital age, I will refer to the medium as digital art."

"The second quality of digital art as a medium was that involves mixed media practices, the image, sound, and text that are produced, distributed and reproduced with digital technologies."

"These three qualities can be taken as operating definition for the medium of digital art."

A Digital Visual Methodology
"There are several canonical cultural touchstones that I want to draw upon in order to reveal the elements of a digital visual methodology."

"The first element of a digital visual methodology that I'm proposing is the social narrative of the works of art."

"We can adapt the frame developed for looking at text, to a work of digital art via rhetorical analysis: looking at the language, used within a text, whether in a detached way, grammar, style, from, or understood as "symbolic action" with in a broader sociological context."

"The second element of a digital visual methodology is technique."

"This element of a digital visual methodology is interested in form, composition and symbolism."

"The third and final element of a digital visual methodology is the tools used in production and display of the work."

"How do the tools reflect broader sociological phenomena of digital cultures?"

Conclusion
"Via the digital visual methodology I am proposing, the aim is to look at the unique nature of the medium and its relation to digital culture, via these works of art: what of the social can be revealed via the narratives, techniques and tools found in works of digital art?"

"On the surfaces of these works of art, of the seen and the unseen, we can unpack a visual economy in the phenomenological nature of the works of art."

"We will explore the works in social context, and the works as themselves acts of the social."

"The purpose of this article was to suggest a methodological footing for further research on how digital art connects to, contests, expands, our understandings of the unique nature of the digital age, the demand for a digital sociology, and the interrelated speculative imagination of these historical movements and claims."

[Section 6]
"Motivated by desire to bring sociology into engagement with a society whose human relations are increasingly mediated by digital technologies, comes a new approach in sociology, a digital sociology."

"Many argue that, as Paul DiMaggio did in 51, that sociologists had from its inception "been slow to take up the study of the Internet," relative the epochal kind of shift in social lives that digital technologies and environments have produced."

"Do we need new methods for interpreting this digital age, relative to new modes of the social?"

"By implication and inference, these questions bring along with them normative connotations: what is the speculative component of carving out an epoch defining digital age, and tools for social inquiry afforded by this age?"

"How does the sociological imagination contest or confirm ideological forces within the social world, and in this specific case the zeitgeist of the digital age?"

2.1.7 AI and Social Theory [260]

This is a machine-generated summary of:

Mökander, Jakob; Schroeder, Ralph: AI and social theory [260]

Published in: AI & SOCIETY (2021)

Link to original: https://doi.org/10.1007/s00146-021-01222-z

If you want to cite the papers, please refer to the original.

For technical reasons we could not place the page where the original quote is coming from.

Abstract-Summary

"We sketch a programme for AI-driven social theory."

"We then lay out our specification for how AI-based models can draw on the growing availability of digital data to help test the validity of different social theories based on their predictive power."

"We also find that AI-driven social theory remains subject to a range of practical, technical, and epistemological limitations."

"We argue that if the gaps identified here are addressed by further research, there is no reason why, in the future, the most advanced programme in social theory should not be led by AI-driven cumulative advances."

Introduction

"We sketch a programme for AI-driven social theory."

"Another aim of linking social theory and AI could be to focus knowledge cumulation on particularly acute social problems and how to tackle them."

"Whether social theory is primarily problem-driven or driven by cumulation, it always focuses on gaps in knowledge and moving knowledge beyond what is known, which, as we shall see, overlaps with the aims of AI research."

"Computer science will only be touched on briefly, when we discuss how the technical capabilities of different types of AI-systems fit with the aim of contributing to advancing social theory."

"Knowledge cumulation is currently impeded because research within AI and social theory remain separate enterprises."

"The solution we propose here is to spell out a coherent programme of research which consists of unified (seamless from micro- to macro-), consistent, and coherent advances in social theory by AI-systems, i.e., those that are both state-of-the-art and generalisable."

Operationalising AI to advance social theory

"Such AI-systems learn about their environment through automated data acquisition, interpret the collected data by means of computational techniques, and represent patterns either symbolically or visually, thus contributing to the cumulation of social scientific knowledge via better fitting models of social change."

"Advances in natural language processing and data visualisation help AI-systems to represent their models to, and communicate with, human operators by providing new or enhanced interfaces."

"In the case of advancing social theory, purposeful means that AI-based models should help social scientists either better understand or develop new methodologies for researching the dynamics of social change."

"Given the cognitive nature of theorising, our suggestion that AI-based models can help advance social theory may spark an age-old question, namely: is it possible for computing machines to think?"

AI-driven social theory: what it is and how it works

"We have argued that the combination of AI-based models and new sources of digital data can help us develop, test, and advance social theory."

"How could AI-based models help advance social theory in practice?"

"How could AI-systems help bridge theories about different social domains, as well as the micro–macro- divide?"

"Explanations are then ranked (or eliminated) by reference to their fit with evidence or data—whence, if they are deemed sufficient, social theory moves onto new terrain."

"The key is that the AI-system is optimising, i.e., maximising for fit with evidence, on a systemic level, thereby imposing constraints on individual AI-based models—whereby existing social theory is updated and advanced."

"Existing theory would only be expanded if the inclusion of new concepts helps the AI-system maximise overall fit."

The example of Randall Collins and his state breakdown theory

"The main point for this paper is that Collins is thereby able to link micro- to macro-; a key issue in theory development in the social sciences."

"What Collins' bottom–up approach allows us to do is to theorise (and learn) about how actors behave across the range of social situations, only sometimes producing structural change as with state-breakdowns and revolutions, whereas normally the social order is continually being reproduced, both on the micro- and macro-levels."

"Nor is there a need for forecasting that sketches out different scenarios or paths: the social world can be modelled with a combination of induction and deduction, with AI-systems testing micro- and macro- linkages over time, and iteratively learning about what outcomes are predicted, and then feeding those predictions, in turn, into new AI-based models."

The systems we need: strengths and gaps in current AI-systems
"Our focus here is not on the continuous refinement of known technologies but rather on identifying the conceptual gaps in current AI research which—if overcome—would make a qualitative difference to the ability of AI-based models to help advance social theory."

"If AI-based models are to help advance social theory, they must have the ability to manipulate, and make sense of, symbols at various levels of abstraction."

"Increased transferability of AI-based models would help advance social theory both by applying known concepts to new contexts but also by testing theories by comparing their validity across multiple dissimilar domains."

"It is thus a real possibility that architectures that combine large-scale learning abilities with symbolic knowledge representation can enable AI-based models to acquire, represent, and manipulate abstract knowledge, and use that knowledge in the service of building and updating theories of complex social phenomena."

Limitations of AI-driven social theory
"To reiterate, these stipulate that a good AI-model of social change should be cumulative (i.e., allow only scientific criteria to guide knowledge creation), holistic (i.e., encompass both macro- and micro-dynamics in relevant disciplines), open-ended (i.e., allow for the inclusion of ever-more and improved data sources, and ways to incorporate them), and purposeful (insofar as ideas are provided about how to improve upon the workings of existing technologies and existing social theory)."

"One reason why AI-driven social theory has recently gained steam is because much more data has become available."

"While these forms of resistance may slow the utilisation of AI-driven social theory or channel it into certain directions, they will not ultimately impede the advancement of social science knowledge."

"This resistance relates partly to the question whether AI-driven social science knowledge can move from knowledge to practice; for example, to identify where political actors can most effectively intervene to address social problems."

Outlook for AI-driven social theory
"AI-driven social theory could help modelling the relationship between economic growth and income or wealth inequality, as well as predicting the relationship between them—given certain assumptions about changes in the rates of economic growth (Maddison [52])."

"This prediction states that social science will advance most rapidly and cumulatively in the realm of (mainly digitally) mediated interaction, because that is where the vast bulk of computationally manipulable data come from (though, as Brent [53] reminds us, there are plenty of databases for traditional or legacy media, and for non-media related social processes such as historical events and many more: again, Brent is only one source for the many databases available for analyses in AI theory building, which cannot be detailed here)."

2.1.8 Learning and Teaching Sustainable Business in The Digital Era: A Connectivism Theory Approach [261]

This is a machine-generated summary of:

Dziubaniuk, Olga; Ivanova-Gongne, Maria; Nyholm, Monica: Learning and teaching sustainable business in the digital era: a connectivism theory approach [261]

Published in: International Journal of Educational Technology in Higher Education (2023)

Link to original: https://doi.org/10.1186/s41239-023-00390-w

If you want to cite the papers, please refer to the original.

For technical reasons we could not place the page where the original quote is coming from.

Abstract-Summary

"These methods can include the utilisation of digital technologies and online communication to facilitate distance learning and fast access to relevant information."

"The implementation of digital technologies, besides technological expertise, requires appropriate theoretical frameworks for understanding how learning is developed."

"Connectivism embraces knowledge as a network where the learner, with the help of digital technologies, develops mental connections between pieces of information during interaction with various information sources."

"The research findings indicate that connectivism may be a suitable conceptual framework that motivates learners to develop knowledge through digital enablers, discussions and social networking and to make connections to sustainability concepts."

"The principles of connectivism may help instructors to develop a learning environment where learners add understandings to their previous knowledge on sustainability through online interactions and by accessing digital knowledge sources."

"This study makes several interdisciplinary contributions by deepening the insights into digital pedagogic methods and approaches for the facilitation of learning, which may be of interest to academic and other pedagogic practitioners."

Introduction

"The integration of digital technologies into the learning process may require a revision of the conventional learning theories applied to curriculum design within business studies."

"Few studies have addressed connectivism in relation to the learning of sustainability in business studies (Abad-Segura and others, [54]; Karlusch and others, [55]), whereas digital technologies are commonly applied in teaching sustainability at universities (Rof and others, [56])."

"This study aims to explore how teaching and learning sustainability in business can be approached in a digital learning environment through the principles of connectivism."

"This research adds to the literature on students' learning by suggesting and exploring connectivism theory for learning sustainable business in an online environment, which has rarely been addressed by pedagogy scholars (Abad-Segura and others, [54]; Davidson and others, [57]; Gómez-Zermeño, [58])."

"The next section presents a conceptual overview of teaching and learning in the scope of higher education for SD, followed by a literature review on connectivism theory and digital learning."

Literature review

"Learning sustainability-related subjects may be approached from the constructivist learning theory perspective, which embraces interactive learning, where students co-create knowledge about SD challenges under the teachers' guidance (e.g. Dziubaniuk & Nyholm, [59])."

"The students' ability to think critically may also be a challenge when implementing connectivism theory, considering that the amount of information on the internet is tremendous and thus requires certain skills for selecting information and assessing its relevance and credibility; hence, the development of critical thinking is among the key capabilities of education for sustainability (Filho and others, [60]; Sammalisto and others, [61])."

"This opens a discussion on the future of HEI learning in the digital era and how study processes can be organised to meet students' needs for knowledge, among others, in the spheres of business and sustainability."

Methodology

"To the individual work of writing essays about selected topics and analysing case studies, the course also included collaborative assignments done in groups, such as the analysis of companies' sustainability initiatives and the development of a business plan for a company with sustainability at the core of its business model."

"Of the analysis, the essays were read in search of expressions and statements reflecting the learned concepts about SD, descriptions of the learning process, links between the created knowledge and the students' main field of study, descriptions of practical assignments as learning tools, group and individual work and critical assessment of the digital learning."

Results

"To continue learning about sustainability and business subjects, students need to understand what keywords to use in search engines."

"During the course, the students participated in case discussions led by the instructors via Zoom and in the preparation of group assignments, such as company sustainability analysis and the development of a company business plan."

"Regarding this course, creativity was needed, particularly for the development of a business plan, when students in groups had to come up with business ideas for sustainable companies."

"By following changes in legislation regarding SD, consumer market trends and reactive companies' actions towards SD, the students developed the understanding that information and reality are changing all the time and that continued learning and critical assessment of information are required for making the right decisions in business management and consumption."

Discussion

"Of this study, the students were able to develop their knowledge and understanding of SD in the business field by making links between concepts, such as the conventional meaning of sustainability (as stated in Brundtland's report, 1987) and practical cases of sustainability initiatives introduced by the companies and understanding how these initiatives affect stakeholders."

"To the knowledge of sustainability in business obtained during the course, the students also improved their skills in managing technological advancement: making video presentations, pitching business ideas online, organising virtual communication for their project work and making online presentations."

"The students were encouraged to expand their learning beyond the course requirements by watching recommended documentaries about, for instance, sustainable textile recycling or listening to podcasts about SD topics that are easily accessible online."

"Access to information from different sources, the ability to see the connections between these and nurturing these connections also encourage students to continue learning, which is among the key principles of connectivism (Siemens, [62])."

Conclusion

"The connectivism learning theory presents opportunities for educators to structure courses on business and sustainability in efficient ways at an HEI."

"Connectivism, as a learning framework, can be considered for the development of university curriculum regarding SD, and business studies as technologies facilitating learning have become a vital part of the educational process and will play an even more significant role in the future."

"Referring to the cross-cultural perspective in course design would allow students to learn about the diverse perceptions and understandings of sustainability in various cultural contexts, which is crucial when operating on global markets and in international business relationships (Ivanova-Gongne and others, [63])."

"Considering that course instructors have the role of learning facilitators, their perspective should be investigated in future research to address various methods for guiding knowledge development and frameworks these teachers apply when designing the study units."

2.1.9 The Contribution of Transaction Cost Theory and Other Network-Oriented Techniques to Digital Markets [262]

This is a machine-generated summary of:

Rossignoli, Cecilia: The contribution of transaction cost theory and other network-oriented techniques to digital markets [262]

Published in: Information Systems and e-Business Management (2007)

Link to original: https://doi.org/10.1007/s10257-007-0063-z

If you want to cite the papers, please refer to the original.

For technical reasons we could not place the page where the original quote is coming from.

Abstract-Summary

"The purpose of this paper is to examine the relationship between information and communication technology (ICT) and the development of organizational forms, showing in particular how ICT is able to promote new forms of cooperation among enterprises."

"The research question is to understand the role played by ICT in the development of an electronic marketplace (EMP), from an organizational standpoint."

"In the analysis of the limitations of the TCT, we will show how the "Theory of strategic networks" can contribute to expand our understanding of new phenomena that characterize some industrial sectors."

"This paper, therefore, will illustrate the theory of strategic networks and the various levels of cooperation among enterprises."

"A good example of a strategic network is the EMP (e-marketplace), whose effect at an interorganizational level will be analyzed in this paper."

"In fact, it is a valid example of ICT application to the relationships between enterprises and economic actors, and on the other hand, it proves how evolution of technologies plays a massive role in view of the potential openness to strategic forms of network on the part of Italian districts—in this case, the ceramic district of Sassuolo."

"It developed most of all as an aid tool for retailers, for the purpose of sharing information and cooperation rather than executing transactions."

"The new role played by the portal is thus emphasized, which is seen as an electronic intermediary providing services that did not exist before."

Introduction

"The purpose of this paper is to examine the relationship between information and communication technology (ICT) and the development of organizational forms, showing in particular how ICT is able to promote new forms of cooperation among enterprises."

"The theory of strategic networks expands the TCT since it explains the existence of long-term relationships among various enterprises and the elimination of transaction costs due to the cooperations within the network."

"This paper, therefore, will illustrate the theory of strategic networks and the various levels of cooperation among enterprises."

"In fact, it is a valid example of ICT application to the relationships between enterprises and economic actors, and on the other hand, it proves how evolution of technologies plays a massive role in view of the potential openness to strategic forms of network on the part of Italian districts—in this case, the ceramic district of Sassuolo."

Among markets and hierarchy: a literature review, from TCT to other network-oriented techniques

"To this regard, the interdependency characterizing the participants of an e-marketplace depends on the intensity of the relations among the participants in the network."

"Key attributes of the interdependency characterizing the relations among the participants in the network are: magnitude (the value of the contribution that each party receives), complexity (low for tangible goods and resources, high for knowledge, information and intangible goods) and intensity."

"Only a portion of these attributes are among the objectives of this paper, whose purpose is the analysis of the contribution given by the TCT to the development of digital markets and the subsequent new relations established between the participants in the network, thanks to the employment of ICT."

"The theory of strategic networks is being applied to the analysis of cooperation-oriented e-marketplaces."

The case study

"TileSquare is an EMP, i.e., a marketplace that reproduces the work methods of the ceramic sector, with the purpose of promoting and developing relationships among manufacturers, created both for manufacturers and final users who have free access to a vast area of information and services."

"The mission of the company is to build a reference marketplace for the ceramic sector through the aggregation of the business leaders belonging to it, enabled by the creation of more effective and efficient, standardized procedures, able to support all the links of the chain, that is, suppliers of raw materials, manufacturers of machineries and ceramic products, and retailers."

"Such effect is certainly allowed by TileSquare, in that the various actors of the ceramic manufacturing chain are offered the opportunity to be visible to all other subjects—therefore, produce information—and at the same time receive information from other subjects or through special links with specialized magazines, sector studies and others, available in the portal."

Conclusions and research further directions

"This determines increased interest on the part of retailers and manufacturers in being present on the e-marketplace, therefore creating positive side effects and increasing the richness of the e-marketplace offering, which becomes even more advantageous for retailers, and contemporarily increasing the disadvantages for those not participating in it."

"Cooperating in the on-time provision of updates and any other information that retailers need, manufacturers receive in exchange by those who manage the portal the information on the use of their own products, which should promote the creation of a new balance of roles and powers that was initially altered by ICT."

"A further development of this research aims at an in-depth analysis of the differences between the new forms of strategic cooperation between enterprises operating in the same value chain and other kinds of e-marketplace where instead participating enterprises belong to different manufacturing sectors."

2.1.10 The Panopticon Reaches Within: How Digital Technology Turns Us Inside Out [263]

This is a machine-generated summary of:

Light, Ann: The Panopticon reaches within: how digital technology turns us inside out [263]

Published in: Identity in the Information Society (2010)
Link to original: https://doi.org/10.1007/s12394-010-0066-7

If you want to cite the papers, please refer to the original.

For technical reasons we could not place the page where the original quote is coming from.

Abstract-Summary
"The convergence of biomedical and information technology holds the potential to alter the discourses of identity, or as is argued here, to turn us inside out."

"The advent of digital networks makes it possible to 'see inside' people in ways not anticipated and thus create new performance arenas for the expression of identity."

"Drawing on the ideas of Butler and Foucault and theories of performativity, this paper examines a new context for human-computer interaction and articulates potentially disturbing issues with monitoring health rather than wellbeing."

Introduction
"The complex relation between the human body (particularly its internal function) and possible new social behaviours is difficult to articulate within traditional design conceptions of humans as technology users."

"In this paper, I propose a way of articulating these social elements, drawing on theories of performativity (Butler [64]) both as a means of showing impact on the experience and the discourses of identity, and as a means of keeping people's agency in the discussion."

"I choose a reading here that stresses the former and use her insights to discuss the way that identity is written upon individuals' bodies and identities in particular confluences by the practices of self and others and it is this set of interactions that inscribes, prescribes and proscribes who we can become."

Performing as a healthy being
"There are very different interfaces possible, such as chips that report continuously and toilets that automatically measure users' urine. (Which begs the question of what happens when someone innocently uses that device for its primary purpose and incidentally finds out about levels in their body…) All these systems presuppose making data on the body's condition available."

"Our questions in the context of discussing impact on social norms include: What happens in the feedback cycle when such devices monitor continuously and report routinely on the body's condition, either to the person or to professionals?"

"In the networked world, the performance of health may be interpreted by society, but its first 'audience' is a machine with the capacity to make connections between streams of data, flash up key intersections and store patterns."

If robots rule the world

"This prediction may turn out to be completely misplaced, but in making it, we can recognize the relationship between the potential of technologies, the malleability of the human (self) image and the way that identity is not so much chosen as inscribed by social practices at large in a complex of tensions that these new developments would be adding to."

"This paper uses the device of exploring medical technologies through performativity to encourage new discussions about design and new discourses around identity, presentation and control."

"I offer this analysis as providing an opportunity to reflect on tools—and connections between them—that allow people to perform themselves inside and out and take the interventionist view that if such designs are to be forthcoming, we should balance function and openness so that people can use what limited scope exists to handle their identity with discretion, decorum and control."

2.1.11 Applying and Advancing Behavior Change Theories and Techniques in the Context of a Digital Health Revolution: Proposals for More Effectively Realizing Untapped Potential [264]

This is a machine-generated summary of:

Moller, Arlen C.; Merchant, Gina; Conroy, David E.; West, Robert; Hekler, Eric; Kugler, Kari C.; Michie, Susan: Applying and advancing behavior change theories and techniques in the context of a digital health revolution: proposals for more effectively realizing untapped potential [264]

Published in: Journal of Behavioral Medicine (2017)

Link to original: https://doi.org/10.1007/s10865-016-9818-7

If you want to cite the papers, please refer to the original.

For technical reasons we could not place the page where the original quote is coming from.

Abstract-Summary

"As more behavioral health interventions move from traditional to digital platforms, the application of evidence-based theories and techniques may be doubly advantageous."

"Moving behavioral health interventions to digital platforms presents researchers with novel (potentially paradigm shifting) opportunities for advancing theories and techniques."

"This paper offers proposals for better leveraging the potential advantages of digital health platforms, and reviews three of the cutting edge methods for doing so: optimization designs, dynamic systems modeling, and social network analysis."

The beginning of a digital revolution in behavioral medicine

"In the context of behavioral health interventions, the ubiquity of digital technologies and their adoption into day-to-day life translates into greater potential reach than traditional interventions, and consequently greater potential for positive public health impact (e.g., Devlin and others, [65])."

"The practical values of developing and testing theories are (1) the efficient and effective advancement of science and technology in general, (2) furthering our understanding of human behavior and (3) informing the development and evaluation of interventions to change behavior."

"The shift from traditional to digital platforms presents researchers with significant advantages in terms of developing and testing theories and techniques of behavior change (Greenwald, [66]; Hekler and others, [67])."

"Given that digital interventions can be delivered with high fidelity, and because they provide the possibility of large datasets generated by ecologically valid measures of behavior, thinking, emotion and physiology (i.e., in real time and everyday contexts), they provide a great potential for testing, refining and developing theories of behavior change."

Challenges to digital platforms

"These components include behavior change techniques (BCTs; the potentially active ingredients of an intervention) and modes of delivery (e.g., design features of smartphone apps or communication skills in face-to-face delivery)."

"The reporting of complex behavioral health interventions, digital and traditional, often lacks sufficient details to know exactly which BCTs were included and how they were offered."

"Individual exposure to BCTs embedded in digital health interventions can vary significantly given that the participant, or user, often has choice as to which part of the digital platform (e.g., app or website) to engage with, in which order, and for how long."

"Researchers struggling with these challenges have suggested that the solution may lie in defining digital interventions not so much as static 'things,' but as a set of underlying principles (theory-derived concepts or mechanisms of action) related to BCTs and delivery methods (Mohr and others, 2015)."

Emerging methods for capitalizing on the digital health revolution
"Social network analysis can be used to help understand how individuals are influenced by friends, and how behavioral health interventions influence not only the targets of interventions but the targets' friends."

"Collecting data for the purpose of modeling the influence of an individual's social network was prohibitively expensive for most behavioral health researchers."

"Digital health interventions that have taken a social network approach but not conducted true network analysis have demonstrated that social support, accountability, and a positive team environment are associated with improved health outcomes, including greater weight loss and increased physical activity over time (Carson and others, [68]; Leahey and others, [69]; Maher and others, [70])."

"Although there is good evidence that online social networks can influence behavioral health, there is currently a dearth of research testing which behavior change techniques (BCTs) related to social interaction can be most effectively employed in digital interventions."

Conclusions
"We are facing a paradigm shift in opportunities for delivering behavior change interventions through digital technologies and use of these technologies to test theories and techniques of behavior change."

"To maximize opportunities, researchers should explicitly identify and systematically apply evidence-based behavior change techniques (BCTs) in their interventions."

"Researchers should also look for creative ways to leverage the richness of data generated by digital health platforms, such as by using digital trace data that can often be passively captured for little cost or effort."

2.1.12 Hermeneutic Inquiry for Digital Games Research [265]

This is a machine-generated summary of:

Tanenbaum, Theresa Jean: Hermeneutic Inquiry for Digital Games Research [265]

Published in: The Computer Games Journal (2015)

Link to original: https://doi.org/10.1007/s40869-015-0005-9

If you want to cite the papers, please refer to the original.

For technical reasons we could not place the page where the original quote is coming from.

Abstract-Summary

"Game Studies is a young research field compared to other media studies disciplines, such as literary theory and film studies."

"Games research has often struggled to balance the demands of modern constructivist perspectives on knowledge creation (Denzin and Lincoln in Handbook of qualitative research."

"Consequentially, games research is one of a new breed of natively interdisciplinary fields, with all of the challenges and opportunities that this entails."

"Sage Publications, Thousand Oaks, [71]; Creswell and Clark in Designing and conducting mixed methods research."

"Sage Publications, Thousand Oaks, [72], this inherent interdisciplinarity forces the games scholar into making a unique balancing act between often contradictory epistemological positions."

"Is it the design of games for entertainment, learning, persuasion, or expression?"

"I describe the application of the traditions of philosophical hermeneutics and close reading to the study of digital games as texts."

"This approach deals with the poetics of digital games, and with the scholarly practices of articulating how a game accomplishes its particular experiential outcomes, such that we might be able to derive useful principles for understanding the game as a cultural, aesthetic, and practical ("Practical" in the sense that hermeneutic inquiry often produces knowledge about how the text created its meanings, thus contributing to the development of craft and design knowledge.) artifact."

Hermeneutic Inquiry and Close Reading

"Føllesdal described this in terms of a web of anticipations about the world that goes beyond our sensory awareness of the world to form a horizon, or background, of experience (Føllesdal [73]): Gadamer, in his main work Truth and Method (1960), and in several smaller works, applies this phenomenological conception of hermeneutics to the subject matter of traditional hermeneutics, the interpretation of texts."

"It is what is corrigible in the light of further information. (Bordwell [74], p. 3) Close reading may be seen as a form of rational and empirical inquiry into the poetics of a work that relies upon hermeneutic methods to account for the interpretive frame of the reader."

Methods for Close-Reading Games

"Close reading as a practice must adapt to the particular demands of digital media when used to explicate games and gameplay experiences as texts."

"In my reading of Mass Effect I strove for completeness due in part to my own preferences as a narrative oriented player, and in part due to a desire to identify the full range of possible examples afforded by the game."

"In some sections of the game I executed multiple playthroughs in order to compare how the engine handled different inputs, and where appropriate I disclose this aspect of the process, but I also seek to present an analysis and a reading that can be reproduced by another player or theorist seeking to explore the specific version of the text that my conclusions are drawn from."

Close Reading: Data Collection Techniques

"Bernard Perron's research group has developed a rigorous approach to data collection, in which researchers capture and catalogue significant amounts of video, parsing and coding the data for contextual information about the interaction modalities employed, the environment in which the game was played, and the particular types of interactions required by any given sequence in the game (Perron and others [75])."

"In the case of Perron's group, because there are multiple researchers playing and writing about a wide range of games, coded video data is a reasonable approach to data collection, balancing the labor costs of maintaining and managing the data, against the need to preserve as much of the experiential fidelity as possible for multiple scholars."

"The primary goal of my data collection is to provide myself with a documentation of my play experience that is sufficient to (1) both prompt my own recall, and (2) to support deeper analysis of certain sections of the games (without requiring that I play through those games again in their entirety)."

Closing Thoughts: Validity and Burden of Proof

"Cresswell identified a number of strategies of validation for qualitative research, including (1) the clarification of the bias of the researcher, (2) the use of rich-thick description to transport readers into the research setting, and (3) the use of trangulation from multiple sources of data (Creswell [71])."

"These validation strategies all contribute to making a work of qualitative research more convincing to the reader."

"In hermeneutic inquiry the articulation of one's bias (insofar as it is reflexively available to the researcher) is perhaps the most important thing a researcher can do to validate a close reading."

"This allows the reader to understand the role that the researcher's subjectivity plays in the reading, which provides the basis for the reader to either accept or reject the judgements of the researcher."

2.1.13 How Do Base Rules Constitute a Virtual World? [266]

This is a machine-generated summary of:

Pałosz, Radosław: How Do Base Rules Constitute a Virtual World? [266]

Published in: International Journal for the Semiotics of Law - Revue internationale de Sémiotique juridique (2022)

Link to original: https://doi.org/10.1007/s11196-022-09903-8

If you want to cite the papers, please refer to the original.

For technical reasons we could not place the page where the original quote is coming from.

Abstract-Summary

"The classic theory of rules and norms rooted in John Searle's theory of institutional facts makes it impossible to treat directives encoded in information and communication technologies (ICT) as rules."

"The aim of the paper is to characterize the rules constituted by the language of the code and show how they differ from purely social institutional rules."

"The analysis is based mostly on Jaap Hage's theory of rules as constraints on possible worlds."

"Hage's understanding makes it possible to explain how rules embedded in the code regulate the functioning of a non-intentional computer system."

"It is possible, because the base rules constrain the system that processes them, by which they constitute digital facts rather than directly guide users' behavior."

Introduction

"Although today it may be quite amusing to read the works that presented virtual worlds as realities independent of the physical one, they stated important questions about relations between physical and virtual."

"It is then worth looking back to the discussion on virtual worlds, as they are able to create new environments for building social normative orders on top of them."

"Virtual worlds' extraordinary potential for this aim was recently pointed out by Jason Grant Allen: MMORPGs offer a rich and challenging context for working out the structure of systems of rules and pose some important questions about the formalisation of modes of action and deontic modes, as well as the interaction between rule sets that nest within each other [76, p. 20]."

"I aim to present usefulness of Jaap Hage's theory of rules as constraints on possible worlds in describing digital reality, which could help in working out the proper legal framework."

Virtual Worlds

"Virtual worlds: (1) have to be spaces within virtual reality, with possibilities for exploration, filled with objects to interact with; (2) should afford exploration and interaction mediated by a visual representation of a user, who is a part of the world (an avatar); (3) should possess multi-user nature–every virtual world is a social space gathering thousands of users that enables them to engage in various social activities; (4) are persistent, which means that they exist even if some of the participants are logged off at any given time, which opens a possibility for the development of sophisticated institutional orders that shape the virtual world itself."

"A single boundary between the two realities seems even more simplistic when it comes to analyzing relations between the different normative orders that regulate the actions of virtual worlds users."

Lawyers and Virtual Worlds

"The lack of appropriate legal concepts to describe the rights of virtual worlds users could be seen mostly during the discussion about virtual property, which took place around two papers, namely [77, 78]."

"Users were assigned rights to virtual objects within the virtual worlds."

"One of the most important results of his research for legal theory is the claim about the distinctiveness of the digital layer of reality, to which notions such as 'property' cannot be meaningfully applied and which should be recognized by law It is worth to note that Hunter, Lastowka, and Farifield seem to had accurate intuitions on how virtual worlds are not like other kinds of software because of the social role they were playing and due to the active role of users in building them (that was signified in section 2 above)."

Action Based Approach to the Virtual

"During last decades, it has been possible to distinguish two different approaches to describing the ontology of virtual worlds and their relation to physical reality."

"According to this approach, interacting through virtual worlds is the way in which we extend our experience of digital spaces by offering "a three-dimensional, computer-generated environment that can be explored and interacted with" that enables free "evolution and propagation" of virtual reproductions, leading to creation of "endogenously produced economies and social orders" [79, p. 324]."

"Analyzing the merging of 'virtual' with 'physical' with focus on agent's perspective and existence through affording interactions well reflects how we constantly function in a reality constructed both from the digital and physical entities that we relate to in various ways—sometimes qualitatively new."

"This approach rejects the object-oriented ontology of virtual environments, taking an epistemic angle and focusing on how different agents perceive reality available to them."

Object Oriented Approach to the Virtual

"Institutional facts, on the other hand, exist only within a system of adequate constitutive rules that create human institutions."

"What can be added to Brey's claim is that digital objects (hence, also virtual objects) are created by the processing of the base rules by the system."

"Such an individual is the developer, who can use base rules in order to create such architecture in the virtual world, and that functions of certain objects would also entail the status function ascribed to them by users (as in the apple example)."

"Another problem with applying Searle's theory to the digital world is the fact that it already assumes the existence of a physical world, the rules of which are independent of our intention."

"A user who is acting within the virtual world in fact imposes status functions not on pre-existing institutional facts, but rather on another layer of brute facts, generated by the computer."

Base Rule as Digital Constraints

"Rules impose themselves on the world through the third type of the down direction of fit, as they are in fact a type of constraint on possible worlds [80, p. 110]."

"Such an argument omits the fact that base rules are imposed not on humans but on the system that generates the virtual world."

"Even if in most cases it is the developer who is responsible for modifications in the base rule set, it is possible that users participate in shaping the reality of a virtual world."

"It is also possible that the creator of a virtual world gives its users a wide scope of freedom in modification of the base rules set."

"Attributing of status functions to particular parts of virtual reality explains why users are so interested in changes in the base rules, which limit all of the possibilities in a virtual world."

Conclusions

"On top of the facts constituted by base rules, the community is able to constitute various institutional facts as an effect of the proper design of a virtual world, or through emergent social cultures."

"Emergence of the virtual worlds is directly possible thanks to the base rules that are being executed by the hardware on which virtual world should be simulated."

"This additional layer, put between two systems of institutional rules, the digital layer, explains how users can develop a virtual world on their own."

"The similarity of base rules to laws of nature is easier to see in the case of virtual worlds."

"We see the effects of system operation according to the base rules through graphical and interactive representations of the virtual world, with which users can interact as if they were physical entities."

2.1.14 "Something Other Than Real Life:" Digital Life Resistance in the Civil Sphere [267]

This is a machine-generated summary of:

Norquist, Jeffrey: "Something other than real life:" digital life resistance in the civil sphere [267]

Published in: American Journal of Cultural Sociology (2022)

Link to original: https://doi.org/10.1057/s41290-022-00176-z

If you want to cite the papers, please refer to the original.

For technical reasons we could not place the page where the original quote is coming from.

Abstract-Summary

"This study employs Alexander's civil sphere theory to study public expressions of antipathy towards digital technology communicated via online forums such as blogs and online mass media publications."

"Pushing beyond motivations, the findings of this study connect motivations to cultural meaning and show how two central concerns—authenticity and self-determination—structure anti-digital life discourse."

"In describing the discursive reiteration and application of these concepts vis-à-vis digital life, I extend the application of Alexander's civil sphere theory to ongoing debates on technology and interpersonal communication in the digital age and, thereby, initiate the development of a cultural sociology of digital life resistance."

Introduction

"Following Alexander, my cultural sociological interpretation of the anti-digital life discourse describes the idealized obligations that U.S. Americans expect in communication and self-presentation actions and the perceived pollution of those obligations effected by digital life technologies."

"Being a cultural sociological investigation, my aim is to illustrate the consistent, structuring influence of cultural values upon the discourse of the recent historical period which witnessed the mainstream emergence of digital life activities and technologies, approximately 2007–2021."

"A specific impact of the work is to expand the civil sphere theory's application to and knowledge about digitalized communicative institutions, such as social media, mobile technologies, and blogging, which were not so prominent when the theory was developed."

"As I have noted, my application of cultural sociology and civil sphere theory establishes a cultural sociology of digital life analytical toolbox which will be useful in those disciplines' investigation of motivations for engaging and disengaging with technologies."

Theory and literature review
"The typical focus on motivations and benefits in digital life resistance research elides the meaning making that justifies these resistance actions and how that meaning emanates from culture and is reinforced through binary structured discourse taking place in the civil sphere."

"The legitimacy of such claims is established when actors convincingly express digital life resistance actions to others via the sacred codes of the civil sphere."

"Actors' abilities to inhabit these ideals are motivated by the goals and priorities of what Alexander [81, p. 33] described as "strongly institutionalised and culturally mandated reservoirs of non-market, non-individualistic force in Western social life.""

"Those who oppose digital life activities and technologies seek to defend ideals of identity and community defined by the moral structures of the civil sphere."

"My study investigates the topic of digital life resistance beyond motivations and goals to complete the link between anti-digital life discourse and a culturally generated sense of the sacred."

Methodology
"By making a thick description of anti-digital life discourse from online materials such as opinion-based newspaper columns, magazine articles, and, predominantly, online blogs, I assemble a complex, richly detailed interpretation of the meaning resistance has in respect to actors' identity and conduct."

"The cultural materials I examine are largely opinion and personal expression-based works which feature individual interpretations of the impact of digital life on society and individuals."

"I took a purposive sampling approach to specifically uncover opinion and personal expression-based works negatively biased against digital life (see Rapley [82])."

"I conducted the data collection for the online sources which form the bulk of the data via multiple search engines on a range of phrases meant to surface anti-digital life expressions germane to social media, smartphones, and, most broadly, digital life itself."

"The included works all focus specifically on the writer's personal experiences with and reasons for resisting some iteration of digital life."

Analysis: a cultural model of digital life resistance
"The quote above distills McNutt's contemporary use of the familiar cultural narratives of authenticity and self-determination to decry the proliferation of internet- and mobile-enabled life worlds."

"This short collection of excerpts and quotes shows two primary sacred codes of the civil sphere that actors sense are being polluted by digital life: authenticity and self-determination."

"This binary structure broadly expresses an established cultural script for criticizing digital life, that digital technology disrupts authentic connections by removing personal connection from physical places; by enabling impermanence; by promoting superficial behaviors; and by making users dependent and controlled."

"The binary codes that are bundled to define self-determined are independent, healthy, and productive."

"Contrastingly, those immersed in digital life inhabit the profane opposite condition of being controlled, a category defined by the bundled codes of dependent, disordered, and distracted."

Authenticity and digital life

"Digital life opponents seem to imply that the solidary of the civil sphere is dependent on the complex communicative interactions which take place in the non-digital, "real world.""

"She lauds the sensory aspect of face-to-face communication, invoking the sacred codes of real, deep, and socially skilled, in saying "you can watch people's reactions, feel their energy, and even smell them.... [And] real-life interaction comes in handy for long-term relationships" (Farid [83])."

"Even viral content which captures the attention of millions is quickly supplanted by the next viral hit or is simply swamped by the relentless flow of mundane content like texts, emails, weather alerts, news updates etc Critics like graphic design blogger Paul Goodman [84] then subjectively conclude that the inherent transitory nature of digital life technology "encourages an unhealthy emphasis on superficial relationships.""

Self-determination and digital life

"The greater discourse in respect to the binary of healthy/disordered tends to focus on the way digital life controls actors by dictating how they should see, present, and evaluate their bodies and life worlds, with a preoccupation with two specific negative outcomes, low self-esteem, and narcissism."

"Activating the profane dependent and disordered binary codes, travel blogger Maggie Peikon (n.d.) writes "I am not the only one struggling to put down my phone and find balance between real life and technology. [It] can lead to depression and anxiety.""

"Echoing that sentiment, travel blogger Danila [85] puts the binary codes of real, face to face, mediated, and disordered to work in showing how digital life distracts from face-to-face interaction: Once upon a time when people were spending time with someone else, they used to try to have some kind of conversation with each other."

Discussion and conclusion

"The goals of this study are to uncover how anti-digital life arguments are structured by civil sphere ideals and to elucidate the meanings social actors find in opposing digital life technologies."

"I show how civil sphere ideals provide resisters a sense of meaning to describe digital life experiences, to claim particular identity characteristics, and to delineate behavior in social spaces."

"This novel approach to technology and media resistance argues that resistance discourse reflects the cultural values of the civil sphere and draws upon the binary codes active in the civil sphere to reiterate sacred notions of authenticity and self-determination as descriptive bases for evaluating digital life activities."

"Even if only motivated to claim resistance identity characteristics, resisters nonetheless activate and strengthen the discourse that already interprets digital life activities as aggravating "deficits in civil society itself" (Alexander [81], p. 209)."

2.1.15 Presence in Digital Spaces. A Phenomenological Concept of Presence in Mediatized Communication [268]

This is a machine-generated summary of:

Lindemann, Gesa; Schünemann, David: Presence in Digital Spaces. A Phenomenological Concept of Presence in Mediatized Communication [268]

Published in: Human Studies (2020)

Link to original: https://doi.org/10.1007/s10746-020-09567-y

If you want to cite the papers, please refer to the original.

For technical reasons we could not place the page where the original quote is coming from.

Abstract-Summary

"Theories of face-to-face interaction employ a concept of spatial presence and view communication via digital technologies as an inferior version of interaction, often with pathological implications."

"Current studies of mediatized communication challenge this notion with empirical evidence of "telepresence", suggesting that users of such technologies experience their interactions as immediate."

"We argue that the phenomenological concepts of the lived body and mediated immediacy (Helmuth Plessner) combined with the concept of embodied space (Hermann Schmitz) can help overcome the pathologizing of digital communication in social theory and enable descriptions which are truer to the experience of using said technology."

Introduction

"Sociological discourse about communication technologies often concentrates on phenomena deemed pathological, using terms such as "addiction" or "isolation" in reference to social media usage (e.g., Turkle [86])."

"Either way, the prevailing sociological perspective is characterized by an apparent indifference towards the actual experience of using such technology and by a demonization of its allure (Turkle [86])."

"Quite removed from this critical pretense, current empirical studies of digital communication technologies have discovered the phenomenon of "telepresence" (Bracken and Skalski [87]), "telecopresence" (Zhao [88]), or the phenomenon of "being together" without bodily proximity (Hardesty and Sheredos [89])."

"We propose a phenomenological conception of space which defines mediatized communication as an interaction of lived bodies in a "social resonance space"."

"The aim is to come up with more accurate descriptions of the actual experience of using communication technologies and the accounts thereof."

Mediatized Communication, Social Interaction and Space

"We elaborate the implicit concept of space employed in interaction theories and in studies of mediatized communication based on these theories."

"It turns out that both Goffman and the studies that draw on his concept of face-to-face interaction employ an understanding of sensual perceptibility when they talk about presence, while Luhmann introduces a constructivist element in the sense that sensually perceptible actors can be treated as absent in social settings."

"Knorr-Cetina [90: 74] uses Goffman's concept of "response presence" in order to try to grasp this shift in spatial references: "I define response presence to mean that the interacting party is not or need not be physically present but is accountable for responding without inappropriate delay to an incoming attention or interaction request"."

Embodied Space

"Phenomenological analysis shows that this unstructured vastness surrounding the lived body existing here/now is the fundamental way in which embodied actors relate to space (on the space of vastness overall, see Schmitz [91]: 47ff.)."

"The embodied giving direction of the space of vastness and the situating of embodied action centers in a structured local space can be taken to ground, from the perspective of a phenomenology of the lived body, what Löw describes as "spacing" and "operations of synthesis"."

"Working with a phenomenologically differentiated concept of space—i.e., including the space of vastness experienced by the lived body as well as embodied directional space—opens up other ways of understanding the connection between space and sociality."

"Embodied actors experience a surrounding, unstructured space of vastness and direct themselves into it."

The Empirical Significance of the Concepts of the Lived Body and of Embodied Space

"On in the discussion, he explains his frequently confrontational communication style, entailing the use of extreme swear words and personal insults, with a field-internal theory about mediatized communication on mailing lists: The quote in line 2755 is an allusion to the tagline of the science-fiction horror film Alien from 1979 ("In space, no one can hear you scream"), which Torvalds uses to point to the absence of prosodic elements in online communication."

"Torvalds invokes the "soundlessness" of the social resonance space of mailing list communication in order to justify the way he expresses himself."

"A phenomenology of the lived body sharpens our analytical awareness of the fact that such projects are not only about "rough consensus and running code", as is suggested by participants' own descriptions of them, but also about their embodied experience and the structure of the social resonance space, within which LKML-communications are performed."

Conclusion

"As a second form of spatial experience we identify directional space, in which lived bodies direct themselves at one another and affect one another immediately either in a symbolically or both symbolically and technically mediated way."

"Local space, the third form of spatial experience, is the form of space sociological interaction theories (Goffman, Luhmann) usually refer to."

"Conceptualizing these forms of spatial experience allows us to describe the similarities and differences between face-to-face interaction and technologically mediated interaction."

"If interactions are mediated in a more technologically comprehensive way, social resonance space is experienced directly as a space of vastness."

"By structuring interaction in this way, social resonance space becomes practically relevant."

"Even if the social resonance space of face-to-face interaction is experienced practically as a local space, it is specified as such from a background of a space of vastness, from which other actors can intrude a particular local space."

2.1.16 Human–Computer Interaction Research Needs a Theory of Social Structure: The Dark Side of Digital Technology Systems Hidden in User Experience [269]

This is a machine-generated summary of:

Gunderson, Ryan: Human–Computer Interaction Research Needs a Theory of Social Structure: The Dark Side of Digital Technology Systems Hidden in User Experience [269]

Published in: Human Studies (2022)

Link to original: https://doi.org/10.1007/s10746-022-09633-7

If you want to cite the papers, please refer to the original.

For technical reasons we could not place the page where the original quote is coming from.

Abstract-Summary

"A sociological revision of Aron Gurwitsch provides a helpful layered theory of conscious experience as a four-domain structure: the theme, the thematic field, the halo, and the social horizon."

"The social horizon—the totality of the social world that is unknown, vaguely known, taken for granted, or ignored by the subject despite objectively influencing the thoughts and actions of the subject—, helps conceptualize how everyday human–computer interaction (HCI) can obscure social structures."

"A sociological four-domain theory of consciousness supplements ethnomethodological HCI research by preserving the field's long-standing focus on micro-level interactions and user experience while simultaneously drawing attention to a "commonsense ignorance of social structure"—in this case, the invisibility of the social structures behind personal computer use."

Introduction

"Through a sociological reformulation of Aron Gurwitsch's [92] phenomenology of consciousness, the goal of this project is to offer user experience researchers an analytical framework that preserves HCI's long-standing focus on micro-level interactions while simultaneously drawing attention to the social-structural backdrop of computer use."

"This four-domain model, amended with sociological theory, allows HCI researchers to explain why the social-structural forces shaping, and the consequences of, artifact use and design remain hidden from consciousness, or are interpreted as irrelevant to users of personal computers in everyday life."

"I apply the latter analytical framework and the concept of "commonsense ignorance of social structure" in the context of computer use, focusing on how the structural dynamics of surveillance capitalism and the ecological impacts of computer use remain invisible to most computer users on a typical day."

Ethnomethodology: Commonsense Knowledge of Social Structure

"Although this project departs from some fundamental assumptions for reasons explained below, it is phenomenological to the extent that it is concerned with the "immediate" experience of the lifeworld and ethnomethodological in the sense that it is interested in the "achievement" of social structure by members of a society in everyday life."

"This project shares social phenomenology's and ethnomethodology's interest in what Garfinkel [93: ch 3], citing Schutz, calls "commonsense knowledge of social structure," or, everyday understandings of social-structural conditions shared by members of a society."

"This work is influenced by the social-phenomenological and ethnomethodological interest in the everyday knowledge of social structures."

"Drawing on the critical sociological tradition, it departs from ethnomethodology and social phenomenology to examine (1) how structural forces exert influence on actors even when they are considered irrelevant and/or are unknown to the actor and (2) how social reproduction proceeds without the actor's knowledge."

Structures Hidden in the Social Horizon: Sociologizing Gurwitsch

"He proposes a theory of consciousness with three a priori "domains" in which data or items are given in experience: (1) the theme (the domain of attention), (2) the thematic field (the domain that gives items relevant to the theme), and (3) the margin (the domain that gives items co-present in consciousness but irrelevant to the theme)."

"As elaborated in the following section, HCI research can benefit from a concept that attends to what is typically unknown, ignored, or vaguely known by subjects in everyday computer use: the totality of the social world, including its artifacts, that are both considered irrelevant and unrelated to the given theme, yet, in fact, from a sociological perspective, provide the constant context of experience, including artifact use."

HCI's Social Horizon: Surveillance Capitalism and Ecological Impacts

"What is relevant to this theme, elements given in the thematic field, include the various elements of the game necessary to carry out the project at hand, including the "Sightings Screen," which displays nearby Pokémon, the Google Maps-based map and physical surroundings (to locate where the Jigglypuff may be hiding), the in-game virtual "items" necessary for catching the Jigglypuff, etc What is co-present in experience yet considered irrelevant to the theme, given in the halo, includes such items as the material one's smartphone is made of, the social norms dictating how one ought to walk through public and private places, and the commodities for sale in the sponsored locations, even if one is unaware of the designation "sponsored location," which may be taken up as themes and purchased during play."

"What is unseen and unnoticed despite structuring and conditioning experience, hidden in the social horizon, includes the decisions of Niantic Labs directors, the financial transactions of firms to lure gamers to their store, the behavioral data scraped from gamers' movements and stored on distant, massive servers, the commodity chains and class system that secured and assembled the necessary materials and infrastructure, as well as the dynamics of capital accumulation underlying all of these dimensions (e.g., the profit motive)."

Conclusion

"Two examples illustrate the usefulness of this framework: (1) illuminating new forms of social control known as "surveillance capitalism" that influence the other domains of consciousness despite being invisible to the subject and (2) explaining how computer use often "hides" the ecological impacts of digital-technological systems from the attention of subjects."

"The social horizon is a helpful concept because it allows HCI researchers to zoom out and examine social-structural impacts on HCI that are often invisible to everyday users as well as simultaneously highlight the commonsense ignorance of these social structures."

"The four-domain model of consciousness not only allows HCI researchers to continue attending to the users' first-person experience, but further helps researchers to simultaneously account for users' commonsense ignorance of social structure as well as the influence of these structural forces on HCI."

2.1.17 Digital Trust and Cooperation with an Integrative Digital Social Contract [270]

This is a machine-generated summary of:

Levine, Livia: Digital Trust and Cooperation with an Integrative Digital Social Contract [270]

Published in: Journal of Business Ethics (2019)

Link to original: https://doi.org/10.1007/s10551-019-04201-z

If you want to cite the papers, please refer to the original.

For technical reasons we could not place the page where the original quote is coming from.

Abstract-Summary

"I argue for the role of trust and cooperation as part of the foundation of digital commerce by expanding the reach of the Integrative Social Contract Theory (ISCT) of Donaldson and Dunfee (Ties that Bind, 1999)."

"I then analyze the roles of trust and cooperation in e-commerce, showing how they are important to the digital business community and explaining their moral relevance under a digital form of ISCT."

Introduction

"Are a digital business community's participants morally obligated to participate in a discussion in which they can be helpful to others?"

"I look at whether the digital business community can be a community in the morally relevant ways that Donaldson and Dunfee describe."

"ISCT is a particularly helpful foundation for understanding moral behavior in digital business communities."

"The necessary social efficiency hypernorm can successfully be applied to digital business communities in a way that allows moral free space for digital communities to establish micronorms."

"I then illustrate how digital business communities develop micronorms of trust, trustworthiness, and cooperation and provide examples of these hypernorms in use."

Theory Development

"By creating room for micronorms, moral free space: (1) enables organizations to deal with the opaqueness that results from bounded moral rationality, (2) allows communities to reflect (within important limits such as hypernorms) their own chosen values and preferences, (3) provides for variances reflecting a world of enormous diversity, and (4) allows communities to develop ethical norms enabling efficient achievement of core goals. (Donaldson and Dunfee [94], p. 86) Above I illustrated that digital micronorms of trust, trustworthiness, and cooperation are supported by the hypernorm of necessary social efficiency, as well as other moral considerations."

"If trust, trustworthiness, and cooperation are needed to efficiently pursuit economic growth in digital business communities, then the individuals in those communities should be morally obligated to abide by the supporting micronorms."

"Under the rubric of ISCT, micronorms of trust, trustworthiness, and cooperation in digital business communities are legitimate, supporting the hypernorm of necessary social efficiency."

Discussion

"Uslaner's argument is that feeling trust generally is important, whereas with digital ISCT I conclude that digital business community participants have a requirement to act in a trustworthy manner in any particular interaction."

"While offline communities may be difficult to exit (Windsor [95]), most digital business communities afford participants flexibility to leave the community and its norms, ensuring the authenticity of digital business communities' micronorms."

"If we believe that participants value their commitments to digital business communities, then the concept of a social contract for these digital communities has great importance for their moral lives."

"If we are less concerned with the process by which micronorms are created or defined, but rather with the fact that authentic micronorms exist in digital business communities, then this problem also loses significance."

"Trust, trustworthiness, and cooperation micronorms exist in many digital business communities; regardless of how they got there, they are now micronorms that both exist and support the hypernorm of social efficiency."

Summary

"Answering their call, I show that the individual is morally responsible to follow authentic micronorms of trust, trustworthiness, and cooperation developed in digital business communities."

"I approached this conclusion by using ISCT as a basis for understanding moral responsibility in digital business communities."

"As part of this discussion, I expanded the reach of Donaldson and Dunfee's concept of community to include digital communities of people who conduct business-related communication and activities together."

"I then used the ISCT framework to show that trust, trustworthiness, and cooperation are important to allow for digital business communities' efficient pursuit of aggregate economic welfare."

"Individuals participating in digital business communities have a responsibility to follow trust, trustworthiness, and cooperation micronorms developed by those communities."

Future Research

"To fully understand and appreciate the individuals' responsibilities in this matter, and to uncover the importance and nature of digital micronorms, we need empirical research on the causes and effects of digital trust, trustworthiness, and cooperation and on the micronorms developed by digital business communities."

"Research can also elucidate what micronorms digital business communities develop, how and why they affect trust, trustworthiness, and cooperation, and how to encourage the development of efficient digital micronorms."

"I hope that future researchers will explore not just the financial gains that the global digital world profits from digital business communities, but also the emotional and psychological benefits of digital business community participation."

Implications for Research

"Because most of the research conducted into online behavior comes from marketing departments, it largely views digital actors as individuals, attempting to target attributes of each specific person for advanced marketing techniques."

"As researchers concerned not with buying behavior but with moral behavior, we do not need to pinpoint how each person will behave when targeted individually."

"As business ethics researchers, we need to discover the realities of digital communal life and help its participants realize its potential for efficient growth and for ethical behavior."

Implications for Business

"Ethically navigating the contemporary business world is difficult, and the goal of this paper is to show how ISCT can be a helpful rubric."

"The digital business person can rely on the tenacity of hypernorms and utilize the framework of micronorms to help make decisions."

"There are tools for helping make ethical decisions online, such as ISCT, and the moral digital actor is not alone in wanting to use these tools to encourage ethical online behavior."

[Section 8]

""Despite the long evolution path, the modern e-commerce still has to deal with such challenges as customers' distrust, doubt and perceived fear."

"The answer is simple: no matter what digital revolution may bring, we are still human beings.""

2.1.18 (In)civility and Adolescents' Moral Decision Making Online: Drawing on Moral Theory to Advance Digital Citizenship Education [271]

This is a machine-generated summary of:

Harrison, Tom; Polizzi, Gianfranco: (In)civility and adolescents' moral decision making online: drawing on moral theory to advance digital citizenship education [271]

Published in: Education and Information Technologies (2021)

Link to original: https://doi.org/10.1007/s10639-021-10710-0

If you want to cite the papers, please refer to the original.

For technical reasons we could not place the page where the original quote is coming from.

Abstract-Summary

"This article draws on moral theory to advance digital citizenship education and explore how adolescents aged 13–16 make decisions when confronted with incivility, such as cyberbullying, on social media."

"Given the extent to which digital citizenship education may be approached in line with deontological (rules), utilitarian (consequences) and/or virtue ethical (character) theories, we argue that it is important to know which of these underpin adolescents' moral decision making online."

"We conclude that if online incivility is to be reduced, policymakers, educators and parents should focus more on virtue- and character-based approaches to digital citizenship education."

Introduction

"This question is important not only for promoting what is commonly called digital citizenship education – the teaching of how to use digital technologies responsibly to participate in society (Ribble, [97]) – but also for understanding how moral theory should underpin this form of education."

"The aim of our research was therefore to explore the moral reasons that 13–16 year-olds use to justify the decisions they make when faced with a dilemma relating to online (in)civility."

"The research focused on first finding out if the participants were more likely to make morally engaged or morally disengaged actions (Bandura, [99]) when presented with an instance of online incivility that targets a classmate."

"A better understanding of whether, how and why adolescents take moral actions online is crucial if we are to target educational policy and practice promoting digital citizenship education in order to address issues of online incivility that undermine our collective wellbeing, happiness and flourishing."

Theoretical background

"This article draws on the three most prominent moral theories to categorize moral decision making as underpinned by reasons that may be: 1) deontological (i.e., based on the recognition of normative rules as to what the right behavior should be), 2) utilitarian (i.e., based on the idea that the moral value of an action depends on the consequences of that action), and/or 3) virtue ethical (i.e., based on the belief that individuals should possess virtues such as honesty and compassion and know, as theorized by Aristotle, how to deploy the right virtues depending on context) (LaFollette and Persson, [96])."

"We do know to what extent adolescents make decisions online that can be classified against the moral theories of deontology, utilitarianism and virtue ethics."

"What is clear from the literature is that there is a lack of research, both in the UK and elsewhere, on adolescents' moral decision making online, and particularly in relation to (in)civility on social media."

"Rather, it is essential to first understand how adolescents make moral decisions online, with a view to then informing how schools and educators might draw on different moral theories."

Methods

"The final version of the questionnaire included: i) Socio-demographic questions: The first three questions asked adolescents about their age, gender and time spent on social media in an ordinary day."

"Adolescents were asked to provide one response out of eight options, ranging from "do nothing" along with morally disengaged reactions such as "forward it to others in my school" to morally engaged reactions, including, for instance, "send a nice message to the person insulted to check how they feel", "report it to the social media company", and "post my own comment that criticizes the original post"."

"Whilst the original sample included 2067 responses, the final sample consisted of 1947 responses, as 120 responses were excluded due to missing data in all the self-reported measures beyond the socio-demographic questions."

Results

"Compared to their male counterparts, female adolescents were more likely to choose the following reactions: "report it to social media company" (13% versus 9%), "send a nice message to the person insulted to check how they feel" (22% versus 17%) and "tell my parents/teachers about it" (22% versus 9%)."

"Once adolescents' reasons were matched with their reactions, what stood out was that most adolescents who would react to an abusive post on social media by "send[ing] a nice message to the person insulted" justified this reaction by choosing the reason "because it is the kind/thoughtful thing to do" (75%)."

"Most adolescents who would react to an abusive post on social media by "post[ing] their own comment that criticizes the original post" justified this reaction by choosing the reason "because it is the just/fair thing to do" (44%)."

Discussion

"We have presented data that shows what moral decisions 13–16 year-olds in England make when presented with an issue of incivility on social media (RQ1), and the extent to which their decisions align with deontological, utilitarian or virtue ethical theory (RQ2)."

"The present research builds on a few studies that have drawn on this theory to explore the extent to which users make morally engaged or disengaged decisions online (e.g., D'Errico & Paciello, [100]; Ge, [101]; Kyriacou & Zuin, [102]; Price and others, [103])."

"This finding confirms what our study found: adolescents justify their moral decisions online in ways that resonate with different moral theories, including, primarily, virtue ethics."

"What our study also found is that, when holding other variables constant, female adolescents aged 13–16 in England are 2.7 times more likely to choose a morally engaged reaction in response to online abuse on social media than their male counterparts."

Limitations

"The current study presents limitations in relation to the sample and measures used."

"Given the moral dilemma used in the survey, which related to someone being unkind online, it is also possible that the survey favored responses that were more virtuous."

"We cannot state that morally engaged, and specifically virtue-based, responses are most likely for all the moral dilemmas that 13–16 year-olds might face online."

"The possibility that adolescents' moral decisions are independent or positively related was not examined, which is why further analysis is needed to explore the relationships between their reasons."

Conclusion

"The task of promoting Internet safety and human flourishing online is an important one that lies in the hands of multiple stakeholders, not just children and their parents but also educators and, when it comes to managing the digital environment, policymakers and Internet corporations."

"They invite policymakers, educators and parents as well as those who design social media platforms to think about how to prioritize young people's ability to develop and deploy virtues in the context of their online interactions, and in synergy with both utilitarian and deontological principles."

"Relatedly, while intervention research is needed on this topic, a pragmatic and non-reductionist approach to digital citizenship education is necessary, one that may enable educationalists to develop strategies informed by an understanding of how adolescents make moral decisions online."

2.1.19 Grounded Theory-Based Model of the Influence of Digital Communication on Handicraft Intangible Cultural Heritage [272]

This is a machine-generated summary of:

Li, Jia: Grounded theory-based model of the influence of digital communication on handicraft intangible cultural heritage [272]

Published in: Heritage Science (2022)

Link to original: https://doi.org/10.1186/s40494-022-00760-z

If you want to cite the papers, please refer to the original.

For technical reasons we could not place the page where the original quote is coming from.

Abstract-Summary

"Despite extensive investigations into the effect of digital communication on handicraft intangible cultural heritage (ICH), research on the relationship between the factors that influence the digital communication of handicraft ICH is relatively scarce."

"A grounded theory approach found that five core factors have significant effects on the digital communication of handicraft ICH."

"These are digital communication awareness, cultural communication initiative, adaptability of digital technology, audience acceptance and cognition, and uncontrollable factors."

"A theoretical "awareness-behavior-attitude" model based on an analysis of the relationships between the five factors was constructed."

"The action mechanism of the theoretical model was identified and tested, revealing that the action modes of each influencing factor were inconsistent."

Introduction

"The digital communication and dissemination of handicraft ICH presents a complex range of problems related to information transmission, audience psychology, and technology."

"This paper seeks to address some of the shortcomings in our understanding of the use of digital technologies in the practice and dissemination of handicraft ICH."

"It investigates the influencing factors and action mechanisms at play in the digital communication of handicraft ICH."

"Understanding these factors and mechanisms has practical significance for the formulation of strategies that will promote the communication and successful dissemination of handicraft ICH."

"This research, uses grounded theory to develop a model of the digital communication of handicraft ICH from the perspective of influencing factors and action mechanisms."

"The grounded theory approach allows us to understand the world of digital communication of handicraft ICH from the perspective of practitioners and consumers and to explore the complex relationships between the different actors and factors in that world."

Existing research and shortcomings: two perspectives

"Researchers have, for the most part, focused on the development of handicraft ICH digital content and the effects of information technology on the protection of traditional handicrafts [104, 105]."

"The digital communication of handicraft ICH research generally adopts one of two approaches: (1) an analysis of media innovation in handicraft ICH digital communication or (2) an analysis of the cognitive effects of digital information on audiences from the perspective of behavioral psychology."

"The second approach explores how the use of digital media influences audience cognition of handicraft ICH content."

"Ethnographic research has been conducted on the status of audience acceptance of handicraft ICH digital information, while quantitative and cognitive effect analyses have been conducted on user samples of new media platforms such as handicraft ICH promotional websites, official accounts, and blogs."

Data collection: open-ended interviews and text analysis

"I adopted a grounded theory approach to theoretically describe the essence and significance of the phenomenon of the digital dissemination of handicraft ICH and to establish a suitable theory to describe the actual situation."

"A grounded theory analysis of open-ended interview transcripts and texts helped us to establish the key influencing factors in the digital dissemination of handicraft ICH."

"Theoretical sampling is the key procedure in grounded theory, used to integrate data collection, coding, and theory construction into a continuous closed-loop process."

"The interview content covered topics such as "communication subjects, digital media, cultural conflicts, technical barriers, and the content-technology relationship," respondents were required to have an understanding of the digital communication of handicraft ICH."

"Interviewees were asked a series of open-ended questions on a range of topics related to the digital communication of handicraft ICH."

Model construction based on grounded theory

"During axial coding I also described the phenomenon and the conditions of the digital communication of handicraft ICH by following the relationships and analyzing the potential behavioral context of the five concepts."

"Awareness of the digital communication of handicraft ICH appears to be the internal driving factor directly affecting digital communication behavior and attitudes towards handicraft ICH."

"The adaptability of digital technology is an internal situational factor, while cultural communication initiative and audience acceptance and cognition are external situational factors affecting the relationship between awareness and behavior; while uncontrollable factors affect the relationship between behavior and attitudes."

"Based on the five core concepts and the interaction relationships obtained from the first two levels of coding, I used selective coding to construct a theoretical model of these influencing factors and the mechanisms of digital dissemination of handicraft ICH."

"Cultural communication initiatives, audience acceptance and cognition, and the adaptability of digital technology are the internal and external bases for the digital communication of handicraft ICH."

Discussion

"Study, we infer that when the influence of three situational variables is neutral or weak, handicraft ICH digital communication activities are primarily affected by digital communication awareness, and the connection between awareness and behavior is strongest."

"When the influence of three situational variables is significant (i.e., extremely favorable or unfavorable), such as where there is a rather high cultural communication initiative or when there is unsatisfactory audience acceptance and cognition, it may greatly affect (i.e., promote or inhibit) the digital communication of handicraft ICH; the connection between awareness and behavior is significantly weakened and the influence of internal and external situational factors on digital communication activities increases significantly."

"When the communicators of handicraft ICH have strong digital communication awareness, the moderating effect of these internal and external situational factors is weakened."

Conclusions

"Based on interviews and textual analysis, this study used grounded theory to construct a theoretical model of influencing factors and mechanisms for the digital dissemination of handicraft ICH."

"We conclude that digital communication awareness is the antecedent factor for the digital communication of handicraft ICH, for the adaptability of digital technology, and for cultural communication initiative; audience acceptance and cognition are internal and external situational variables in the awareness–behavior relationship, and uncontrollable factors moderate behavior–attitude."

"The theoretical model of influencing factors and the mechanism for the digital communication of handicraft ICH (i.e., the awareness–behavior–attitude model) that we have proposed is significant in the theoretical construction of the digital dissemination of handicraft ICH."

"This research used grounded theory to develop a model of the digital communication of handicraft ICH from the perspective of influencing factors and action mechanisms."

2.1.20 Rethinking Big Data in Digital Humanitarianism: Practices, Epistemologies, and Social Relations [273]

This is a machine-generated summary of:

Burns, Ryan: Rethinking big data in digital humanitarianism: practices, epistemologies, and social relations [273]

Published in: GeoJournal (2014)

Link to original: https://doi.org/10.1007/s10708-014-9599-x

If you want to cite the papers, please refer to the original.

For technical reasons we could not place the page where the original quote is coming from.

Abstract-Summary

"I borrow from critical research on technologies to engage three important new facets of Big Data emerging from an interrogation of digital humanitarianism."

"I argue first that within digital humanitarianism, Big Data should be understood as a new set of practices, in addition to its usual conception as data and analytics technologies."

"I argue that Big Data constitutes a distinct epistemology that obscures many forms of knowledge in crises and emergencies and produces a limited understanding of how a crisis is unfolding."

"I argue that Big Data is constitutive of a social relation in which both the formal humanitarian sector and "victims" of crises are in need of the services and labor that can be provided by digital humanitarians."

Introduction
"With the advent of Big Data, crowdsourcing, and the geoweb, digital humanitarianism has taken off in humanitarian practice, with some promising that traditional humanitarianism has now been "revolutionized" (Meier [106])."

"If information technology and humanitarianism are both complex, uneven social and spatial practices, what are the implications of their convergence with Big Data?"

"Informed by a recent ethnographic research project, in this paper I call into question the ways Big Data has been positioned in relation to humanitarianism."

"I explain the ways Big Data has been situated in relation to digital humanitarianism, drawing attention to its purported benefits, particularly in two key texts."

"I draw out Big Data digital humanitarianism across three analytical lines in order to critique the ways it has been conceptualized in research to date."

"I conclude by suggesting ways in which the process of integrating Big Data with digital humanitarianism may benefit from critical geographic theory contributions."

Critical perspectives on spatial technologies
"This literature has been informed by science and technology studies research, so I likewise draw upon this literature insofar as it theorizes data in ways conducive to understanding its geographic implications."

"Critical, participatory, and feminist geographic information science (GIS) research has shown that technologies and data come to embody epistemologies and social relations through several mechanisms."

"These understandings of spatial technologies build on lessons from science and technology studies (STS) research that describes the processes by which data and technologies come to assume and reify social and power relations, worldviews, and epistemologies (Feenberg [107]; Pinch and Bijker [108]; Wajcman [109]; Winner [110])."

"These contexts influence the ways in which technologies can be used, as well as the social and political relations they build or reiterate."

"What is needed is work toward a preliminary understanding of the social and political implications of Big Data in digital humanitarianism viz-à-viz its promises in current discourses."

Big Data within digital humanitarianism
"Big Data can be understood as an emerging subset and influential trend within the historical development of digital humanitarianism."

"To traditional humanitarianism, digital humanitarianism can add diverse techniques such as crowdsourcing, social media operations, remote volunteer collaboration, data production and processing by people unaffiliated with a formal humanitarian institution, and "crisis mapping" (Burns [111])."

"Within digital humanitarianism conversations around Big Data have only recently commenced, with most conceptualizing Big Data merely as a new source of data."

"Despite the promise of digital humanitarianism being large numbers of people contributing large amounts of data, this quote suggests that individual people are communicating directly with those in control of resource allocations."

"Discussions of the relationship between Big Data and digital humanitarianism tend to be cautiously optimistic."

"These early conversations suggest a Big Data future is, for those managing the development of digital humanitarianism, seen as imminent, value-neutral, and mostly centered on new data sources, with obstacles delegated to technicist solutions."

Methods

"The following sections have been developed with evidence from a recent one-year extended case method (see: Burawoy [112]) research project that commenced from work with a research institution."

"My work with this institution introduced me to many of the key actors and developments of digital humanitarianism, and thus I engaged this particular case as an especially fruitful site for exploring how digital humanitarianism has developed and is impacting broader-scale relations."

"In the research project I used work with this institution as a case through which to explore the broader phenomenon of digital humanitarianism, combining evidence from participant observation, data archiving, 37 in-depth semi-structured interviews, and formal online training modules."

"The data collected for this extended case method was collected entirely independent of the institution, using that work merely to commence the extended case method."

Big Data as digital humanitarian practice

"The growth of Big Data within digital humanitarianism should be seen not simply as a new source of data, but instead as a shifted set of practices, as an epistemology, and as an emergent social relation."

"Big Data signifies new digital humanitarian practices in at least two contexts: within the emergent digital humanitarian community and within the formal humanitarian response community."

"Formed differently, these people are concerned that parties pushing Big Data digital humanitarianism do not adequately understand the operational protocols and institutional needs of humanitarian and emergency management organizations, or they offer technologies that do not meld into pre-existing practices."

"Digital humanitarians working with Big Data have developed new practices to attract and sustain inflows of volunteers: "when you're involved in distributed computing you must offer an incentive to maintain interest and participation" (Kathleen, personal interview, 2013)."

Big Data as digital humanitarian epistemology
"By framing their technologies in the three ways seen above, the digital humanitarian community produces a conception of what can be known through Big Data, a conception that aligns with the common promises of Big Data but contradicts the ways those knowledges are engaged by formal humanitarian actors."

"Jeremy, a human rights lawyer engaged in digital humanitarianism debates, connected the partiality of knowledge with the practices used to "make sense" of Big Data: The problem with [a well-known digital humanitarian platform] is that it's not … designed to do the three things we need it to do to [be usable in a human rights abuses case]: one is accurate geocoding."

"Ushahidi is not a platform that allows me to do that, and neither is Tomnod.- (Jeremy, personal interview, 2013) Jeremy thus contends that the digital humanitarian platforms designed to integrate Big Data signify the emergence of a new epistemology that embodies several significant limitations for use in his work."

Big Data as social relation
"Big Data within digital humanitarianism produces social relations based on its discursive production of needy subjects and by impacting the material conditions of humanitarianism's subjects."

"Big Data helps digital humanitarians produce discourses of subjects in need of digital humanitarian involvement."

"A crisis-affected community may use social media to communicate their knowledges of a disaster, and these knowledges, when collected, processed, and represented, help digital humanitarians produce the subject who needs digital humanitarian intervention."

"In the social relations that emerge from this subjectivity formation, the digital humanitarian community possesses the skills, resources, and requisite knowledges to effectively assist the needy subjects of disaster-affected communities and the formal humanitarian institutions."

"These relational identities are here leveraged in order to assist in the production of the needy formal humanitarian organization subjectivity: because Big Data requires substantial expertise and resources—which the digital humanitarian community purportedly has—the formal humanitarian sector is in need of digital humanitarian contributions."

Conclusion: toward an integration of critical information technologies, Big Data, and digital humanitarianism
"Proponents of Big Data conceptualize it as a new source of information that will "revolutionize" humanitarianism in a value-free development."

"As a new set of practices, an epistemology, and a set of social relations, Big Data digital humanitarianism is imbued with social and political implications explored in this paper."

"These processes take a form specific to Big Data digital humanitarianism, and exploring this case sheds greater light on these larger-scale processes."

"My goal in this paper was to identify and characterize Big Data as it is currently unfolding in digital humanitarianism."

"Definitions and conceptualizations of Big Data are diverse and often contradictory, and this impacts the ways in which digital humanitarians see and communicate their contributions to humanitarianism through Big Data."

"More broadly, research critiquing and situating the geographies of humanitarianism can be integrated into studies of Big Data digital humanitarianism."

2.1.21 Seeking, Sharing and Co-creating: A Systematic Review of the Relation Between Social Support Theory, Social Media Use and Chronic Diseases [274]

This is a machine-generated summary of:

Sendra, Anna; Farré, Jordi; Vaagan, Robert W.: Seeking, sharing and co-creating: a systematic review of the relation between social support theory, social media use and chronic diseases [274]

Published in: Social Theory & Health (2019)

Link to original: https://doi.org/10.1057/s41285-019-00106-z

If you want to cite the papers, please refer to the original.

For technical reasons we could not place the page where the original quote is coming from.

Abstract-Summary

"The aim of this study is to explore the implications of applying social support theory to social media use in the field of chronic diseases."

"Our analysis retrieved ten registers on initiatives around social support theory, social media, and chronic diseases."

"Our findings suggest that social media can provide social support regularly, but institutions need to create safe environments addressed to specific diseases where physicians also take part in the community of the site."

"As patients have been in social media without physicians' support for many years now, finding new ways of reducing the communicative gap between these two stakeholders is crucial."

"This review suggests that the application of social support theory could be one of the solutions, especially regarding chronic pain patients."

Introduction

"Even though there are other theoretical perspectives (Wright [113]) that are used to study online communities (like the Optimal Matching Model (coincidence between provider and receiver), the Social Comparison Theory (experiences are normalized when comparing with others), or the Social Information Processing Theory (the support provided through online tools is seen as positive)), the Buffering Effect Model is the one that fits best the reality of social media."

"None of these reviews explored the relation between social support theory, social media and chronic diseases from a communicative perspective."

"By conducting a systematic review, our research has a twofold objective: (1) to determine how many initiatives exist in social media that have been applying social support theory in the field of chronic disease; and (2) to provide a list of recommendations about how health institutions can improve the communication with their patients in social media relating to advances in health communication theory."

Methods

"The review included all the studies that met the following inclusion criteria: (1) studies related to chronic diseases, (2) articles related to social media, (3) peer-reviewed articles published in English between 2004 and December 2017, (4) studies related to the social support theory, and (5) studies related to the object of study."

"Search, the following keywords were used: (chronic (disease OR illness) AND social support) AND (social media)."

"In this second phase, 67 papers were excluded for various reasons (not related to the object of study, not related to chronic diseases, not related to social support, or not related to social media)."

"In the third phase, 22 full-text articles were excluded for different causes (not related to the object of study, not related to chronic diseases, not related or focused on social media, or for being reviews)."

Results

"Concerning the social media platform used, the results were the following: 4 of the 10 studies are using existing platforms (like Facebook or Twitter) for their analysis (n = 4); another 4 have created their specific social media platform according to the chronic disease they are studying (n = 4); 1 paper explores the implementation of a chronic pain management workshop; and, finally, the remaining study discusses web 2.0 tools more generally."

"Most of the studies had participants (chronic disease patients) who were directly affected by the illnesses analysed in the studies (n = 7); apart from 2 papers that only analysed social media data (n = 2), and the theoretical study—that didn't have participants (n = 1)."

"According to the results, we can classify the studies into 3 groups: most of the papers discuss (a) the implications of using social media platforms in healthcare delivery or treatment (Becker [114]; Brosseau and others [115]; Fatima and others [116]; Grosberg and others [117]; Lavorgna and others [118]; Magnezi and others [119]; Nordfelt and others [120]; Rus and Cameron [121]); (b) but there is also a study based on design improvement of large-scale studies (Merolli and others [122]); (c) and another paper that discusses healthcare delivery models (Milani and Lavie [123])."

Discussion

"Within health-related organizations "there is a need for clear guidelines on how healthcare professionals make use of social media to support their patients" (McKenna [124], p. 470)."

"In the case of the studies analysed, whether they are interventions that are using different social media platforms where patients can find information about their chronic disease (Brosseau and others [115]; Grosberg and others [117]; Lavorgna and others [118]; Magnezi and others [119]; Merolli and others [122]; Nordfelt and others [120]; Rus and Cameron [121]), or initiatives that are analysing social media data to improve health services provided through an app (Fatima and others [116]), the presence of this type of support is always there."

"Whether the support is provided through specific health-related social media or through traditional platforms, organisations need to "careful design [...] social media-delivered health communication using specific features to promote specific types of engagement" (Rus and Cameron [121], p. 688)."

"Whether they act as moderators, or are actively participating in the conversations taking place on these platforms (Becker [114]; Brosseau and others [115]; Grosberg and others [117]; Lavorgna and others [118]; Magnezi and others [119]; Milani and Lavie [123]; Nordfelt and others [120]), they are the ones who should provide the social support their patients need to engage them in a better self-management of their health."

Conclusions

"In order to improve patient-physician communication, organizations should (1) use social media technologies to provide social support regularly; (2) create specific portals addressed to specific diseases; (3) include health professionals in the management of these sites as natural helpers; and (4) offer secure environments where patients can manage their health, and taking as few risks as possible."

"As "science is facing new challenges with the reality of social media's role in the spreading of knowledge based on personal anecdotes and fostering deceitful health messages" (Jervelund [125], p. 168), health-related organizations need to work closely with their audiences if they want to address the existing communicative gap between patients and physicians."

"This study has shown that the implications of applying social support theory to social media use in the field of chronic diseases could be beneficial for improving patient-physician communication."

2.1.22 Political Theory of the Digital Constellation [275]

This is a machine-generated summary of:

Berg, Sebastian; Staemmler, Daniel; Thiel, Thorsten: Political Theory of the Digital Constellation [275]

Published in: Zeitschrift für Politikwissenschaft (2022)

Link to original: https://doi.org/10.1007/s41358-022-00324-4

If you want to cite the papers, please refer to the original.

For technical reasons we could not place the page where the original quote is coming from.

Abstract-Summary

"The introductory contribution to this special issue on "Political Theory of the Digital Constellation" addresses the conditions and possibilities of political theory's engagement with digital developments."

"The digital constellation serves as an epistemological guide that helps to structure theoretical reflection on the interrelationship between digitalisation and political questions."

"The outlines of the political theory in the digital constellation become clear in the fourteen contributions of the special volume, which are presented in conclusion."

Digital transformation and political theory

"Political science is called upon to respond to digitalisation in its theories and methods, to understand it better and to explain its consequences."

"Political science and especially political theory have been relatively slow to reflect the changes taking place, especially in comparison to neighbouring disciplines such as sociology, media studies or communication studies."

"There are many reasons for this hesitancy, including a longstanding neglect of research on the relationship between technical infrastructures and political action, as well as the weak structural presence of political theorists in the interdisciplinary research structures that shape the field of digitalisation studies nationally and internationally."

"The growing interest can be explained by the fact that it is now much better understood that digitalisation not only changes politics, but conversely that politics also shapes digitalisation."

"We discuss three pitfalls of political theory research on digitalisation before introducing the epistemological perspective of the digital constellation that guides this special issue and its contributions."

Three pitfalls of political-theoretical research on digitalisation

"Digital society not only increases communication and coordination, it integrates individuals in many more ways and from many more points of reference, often with great relevance for political theory: topics include surveillance (Hoye and Monaghan [126]), data protection and privacy (Helm and Seubert [127]; Schulz [128a]), the quantification of the social (Fourcade [129]; Beer [130]; Mau [131]), mechanisms and forms of social and political representation (Gerbaudo [132]; Manow [133]) or the transformation of participation practices and interfaces (Berg et al. [134])."

"Political theory does not have to be a science of technology, but it must explicate the understanding of technology on which its argument is based to such an extent that the argumentative entanglement with the political-theoretical interpretation of the overall context becomes clear."

"The third pitfall is the insufficient recognition that (digital) technology is political."

"Such a procedure can include the conscious reflection of technology's political nature in individual case studies, as demonstrated by science and technology studies in their deconstructive-analytical practice (classically, e.g., Winner [135]; Latour [136]; more recently, for instance, Dickel [137]): What political logics find their way into the design of technical artefacts and assemblages?"

The digital constellation as a research perspective

"We are concerned with "reflecting on the conditions under which politics takes place in a society that is characterised by the use of digital technology" (Berg et al. [138b], p. 17)."

"The research perspective of the digital constellation emphasises that and how digital technologies acquire their meaning and dynamics in their diversity as infrastructures, media, protocols, artefacts, etc first and foremost in relation to social and political practices."

"How can this analytically sophisticated perspective on the interaction of society and digital technology be made useful for the concrete investigation of political theory issues?"

"For subsequent studies, the perspective of the constellation can then be enriched by technology theory and used as a heuristic that helps to structure and formulate the techno-political contexts."

"Three levels are crucial to understanding the interactions of society/politics and (digital) technology: the level of the properties of digital technology, the level of perceived spaces of possibility or practically realised affordances, and the level of socio-political configuration."

The contributions to the special issue

"The special issue brings together a total of fourteen contributions that examine the interactions between digital transformation and politics."

"In the first dimension, we include the contributions that focus on the methodological approaches of political theory to digitalisation in the field of politics."

"We ask, for example, what the contemporary historical and intellectual backgrounds are against which digitalisation-related developments get described and theorised, which assumptions about technology or society underlie various political or theoretical positions, or how conceptual changes can be reconstructed or theorised."

"These contributions are concerned with (often implicit) presuppositions regarding the development of norms or concepts in the digital context, as well as reflection on different ways to theorise digitalisation, their origins and consequences."

"Niklas Ellerich-Groppe's contribution [139], which deals with the transformation of the welfare state in the digital constellation, is more strongly related to a classical political field."

2.2 Digital Society

Machine generated keywords: digital society, governance, society, economy, economic, sustainable, industrial, political, sustainable development, urban, history, environmental, vision, engagement, indicator

2.2.1 The Metaverse as a Virtual Form of Data-Driven Smart Urbanism: Platformization and Its Underlying Processes, Institutional Dimensions, and Disruptive Impacts [276]

This is a machine-generated summary of:

Bibri, Simon Elias; Allam, Zaheer; Krogstie, John: The Metaverse as a virtual form of data-driven smart urbanism: platformization and its underlying processes, institutional dimensions, and disruptive impacts [276]

Published in: Computational Urban Science (2022)

Link to original: https://doi.org/10.1007/s43762-022-00051-0

If you want to cite the papers, please refer to the original.

For technical reasons we could not place the page where the original quote is coming from.

Abstract-Summary
"The emerging phenomenon of platformization has given rise to what has been termed "platform society,“ a digitally connected world where platforms have penetrated the heart of urban societies—transforming social practices, disrupting social interactions and market relations, and affecting democratic processes."

"One of the recent manifestations of platformization is the Metaverse, a global platform whose data infrastructures, governance models, and economic processes are predicted to penetrate different urban sectors and spheres of urban life."

"The Metaverse is an idea of a hypothetical set of "parallel virtual worlds" that incarnate ways of living in believably virtual cities as an alternative to future data-driven smart cities."

"This study analyzes the core emerging trends enabling and driving data-driven smart cities and uses the outcome to devise a novel framework for the digital and computing processes underlying the Metaverse as a virtual form of data-driven smart cities."

"The analysis identifies five digital and computing processes related to data-driven smart cities: digital instrumentation, digital hyper-connectivity, datafication, algorithmization, and platformization."

"The novelty of the framework derived based on thematic analysis lies in its essential processual digital and computing components and the way in which these are structured and integrated given their clear synergies as to enabling the functioning of the Metaverse towards potentially virtual cities."

"It contributes to the scholarly debates in the field of Science, Technology, and Society (STS) by highlighting the societal and ethical implications of the platformization of urban societies through the Metaverse."

Introduction
"Research and development of the Metaverse has become a key trend in smart urbanism in terms of the design of believably virtual cities based on large-scale data-driven AI systems (Bibri, [140])."

"The "new normal" established in the aftermath of the COVID-19 pandemic has resulted in an abrupt large-scale digital transformation of urban society, a process of digitization and digitalization that is in turn paving the way for merging virtual reality and physical reality in the context of data-driven smart cities."

"The downsides of the Metaverse as a virtual form of data-driven smart urbanism or platform urbanism remain unavoidable."

"Against the preceding background, this study analyses the emerging trends enabling and driving data-driven smart cities and uses the outcome to devise a novel framework for the digital and computing processes underlying the Metaverse as a virtual form of data-driven smart cities."

Related Work
"One of such platforms is the Metaverse, a gigantic ecosystem application that is fuelled by the most innovative computing and immersive technologies."

"Given the current development stage of the Metaverse as a global platform being launched in 2021 by Meta, research in this area tends to focus mainly on two strands, which are typical to the advent of new socially disruptive technologies."

"The first strand is concerned with the state-of-the-art and technical aspects of the Metaverse in terms of computing technologies, immersive technologies, ecosystems, developments, trends, applications, opportunities, grand challenges, open issues, research agenda, roadmapping, and so on."

"Given its focus, the above study relates to the second strand of research within the area of the Metaverse, which is concerned with its risks and other negative implications, engaging critically with the underlying core enabling technologies from a variety of perspectives."

Methodology: Thematic Analysis
"This study assumes that there are trends and processes that repeat themselves and drive and underlie the digital architecture of computer mediation associated with the Metaverse as a virtual form of data-driven smart urbanism."

"It emphasizes identifying, analyzing, interpreting, and reporting themes, i.e., important patterns of meaning within qualitative data."

"As an inductive analytical technique, thematic analysis involves discovering patterns, themes, and concepts in the qualitative data that include interdisciplinary and transdisciplinary literature."

"Identifying the digital and computing processes underlying the Metaverse as a virtual form of data-driven smart urbanism in terms of recognizing the specific and distinctive features of this model of urbanism."

"Finding the theoretical relationships among the identified concepts and the Metaverse as a virtual form of data-driven smart urbanism—conceptualization."

"Examining and discussing the risks and impacts of the identified digital and computing processes underlying the Metaverse as a virtual form of data-driven smart urbanism in the wake of the COVID-19 pandemic."

Results

"Data-driven smart cities represent an immersion in digital transformation, a process of digitization and digitalization that is enabled by the convergence of AI, the IoT, and Big Data and its far-reaching consequences— digital instrumentation, digital hyper-connectivity, datafication, algorithmization, and platformization (Bibri and Allam, [141]; Calvo, [142])."

"Data-driven smart cities (e.g., Kaluarachchi, [143]; Sarker and others, [144]) are massively digitally instrumented and hyperconnected, intensively datafied, and increasingly algorithmized and platformized, and as such, they enable data-intensive, distributed computing across various urban domains based on innovative techniques, models, and decision support systems in the shape of large-scale data-driven AI systems."

"The abrupt digital transformation that has swept through the urban world in the aftermath of the COVID-19 pandemic, coupled with its disruptive impacts on people's everyday life, seems to be in tandem with the envisioning process of the Metaverse in terms of its ultimate goal to datafy, algorithmize, and platformize urban ways of living towards virtual alternatives to the imaginaries of data-driven smart cities."

Discussion and Conclusion

"Further, this study examined and discussed the risks and impacts of the digital and computing processes underlying the Metaverse as a virtual form of data-driven smart urbanism, paying particular attention to: platformization; the COVID-19 crisis and the ensuing non-spontaneous "normality" of social order; data-driven corporate-led technocratic governance; governmentality; privacy, security, and trust; and data governance."

"Future endeavors need to focus on finding ways to regulate the Metaverse as a global process and practice of platformization democratically, ethically, and effectively through relevant social structures and institutions while understanding the key underlying mechanisms at work."

"A systemic inquiry into the relationships between the institutional and social dimensions of platformization as a form of scientific and technological development is of crucial importance because it will bring into view the tensions between the Metaverse and institutional practices and governance frameworks."

2.2.2 Digitization of Society: Alternative Projections of the Future [277]

This is a machine-generated summary of:

Travkina, N. M.: Digitization of Society: Alternative Projections of the Future [277]

Published in: Herald of the Russian Academy of Sciences (2022)

Link to original: https://doi.org/10.1134/s1019331622120115

If you want to cite the papers, please refer to the original.

For technical reasons we could not place the page where the original quote is coming from.

Abstract-Summary

"The impact of digitization on four spheres of society, i.e., economic, political, social, and spiritual, is analyzed."

"Today the digitization of public spheres directly involves the component of ensuring the digital security of society, which is becoming increasingly global in character."

"Large-scale digitization of public spheres for the first time in human history creates an objective opportunity for constructing and designing future social states, which makes a fundamental difference between the course of future socio-economic and political processes and the historical evolution of the previous eras."

"Polarized views on the digital Future are based on the processes of the "digital divide," the meaning of which is that digitization contributes to a significant increase in inequality in access to digital goods, which in turn results in a growing inequality in the distribution of income and wealth."

"Digitization has made social distancing possible and cost-effective."

"The pandemic was conducive to a colossal increase in the economic power and political influence of digital corporations, which objectively requires a sharp increase in the regulatory role of the state, which should put digitization under effective public control."

INTRODUCTION

"As in previous eras, globalization turned out to be a two-faced Janus, on the one hand, bringing together socio-economic and political systems, and on the other, sowing seeds of destruction, discord, and wars between countries."

"Digitization, as a powerful force for global socio-economic and political transformations, can act as a potent tool for both creative and destructive processes."

"In its last capacity, it has the potential to turn, and perhaps has already turned, into a new type of collision—cyberwars, the destructive potential of which, in terms of economic losses, is comparable to the economic damage of the world wars in the first half of the twentieth century and financial and economic crises of the twenty-first century."

DIGITAL DIFFUSION IN SOCIETY

"The progressive digitization of public spheres, including that of an individual himself, makes us wonder about the legitimacy of digitizing each of those, which will ultimately enhance its functional efficiency in meeting human needs, and about the extent to which the intrusion of digital technologies into areas that occupy a higher place in their hierarchical order is a factor in the growing dysfunctionality of this area."

"With regard to other public spheres, it can be hypothetically assumed that their digitization looks more problematic and is accompanied by a growing mutual substitution of digital technologies for the qualities and properties of a Human as a sociobiological species, which includes, among other things, mental, sensual, and volitional areas."

DIGITIZATION AND CONSTRUCTING VIRTUAL PROJECTIONS OF THE FUTURE

"Social forces that have fully mastered and control the process of digitization claim that the society of the Future is not the result of the action of autonomous forces, in some cases relying on the help and support of the "invisible hand" of Providence, but arose as a result of the implementing the digital Project created in computers, supercomputers, and with the help of artificial intelligence (AI)."

"The wide diffusion of digital technologies leads to a historically unprecedented symbiosis of almost every person and computer bringing about an information society, which is "the result of the transition from the previous digital era to a new post-digital world in which digital has become the basis of everyday life" [Dufva and Dufva, March 2019, p. 18]."

THE CORONAVIRUS PANDEMIC: IS IT A HARBINGER OF THE DIGITAL FUTURE?

"Yale University history professor F. Snowden has studied the impact of pandemics on social development from 1346 to 1953, which claimed from 75 million to 200 million lives on the planet, starting from the time of the first world plague pandemic, and ending with the global pandemics of our time."

"The coronavirus pandemic has sharply increased the social control of state bodies over citizens based on digital technologies."

"Social distancing, which has become a legitimate form of combating the coronavirus pandemic, will gradually develop into a system of "tele-all," that is, a system of remote healthcare, education, work, entertainment, e-commerce, and social events, including participation in political processes, which includes remote voting [Travkina and Rogovskii, 2016]."

"It is quite possible that the processes of further social digitization will be comparable in their medical and biological consequences to the impact of the 2020 coronavirus pandemic."

"THE DIGITAL DIVIDE": THE MAIN FACTOR OF EMERGING UTOPIAS AND DYSTOPIAS OF THE DIGITAL ERA

"Inequality in available access to and ownership of digital technologies creates the phenomenon of the "digital divide.""

"High-income earners (individuals and households) ($75 000) are 20 times more likely to access the internet and digital technologies than low-income earners ($30 000) [14]."

"The digital divide is especially acute on a global scale: half of the world's population (which is almost 4.0 billion people!) do not have access to the Internet, and in most of the least developed countries, no more than 20% have access to digital technologies, naturally not the most advanced [16]."

"The emergence of revolutionary technologies as digital technologies stamped themselves from the very beginning, could not but overlap with the structure of the eschatological consciousness of modern humanity, which, since the industrial revolution in Great Britain in the last third of the seventeenth century, has invariably considered the emergence of technology as an opportunity for humanity to gain the long-awaited "keys to earthly paradise," having saved most of the population from want, and a growing abundance of material wealth will result, if not in the elimination, then at least in a significant metamorphosis of the various hypostases of Evil."

DYSTOPIA: "DIGITAL SLAVERY"

"It is extremely significant that famous F. Fukuyama who proclaimed the global triumph of the ideas of liberal democracy directly linked the decline of democratic institutions in the leading Western countries, and especially in the United

States, with the spread of digital technologies, which in the last decade contributed to a significant fragmentation of society and the declining confidence in state institutions."

"Modern digital technologies at the disposal of political elites open up enormous opportunities for manipulating the moods, expectations, and value orientation of the broad masses of the population with political rights, which in ancient Greece were called "demos.""

"Only narrow groups of political elites remain the main political players in the pseudo-democratic political system, which, in relation to the "demos," begin to pursue a policy of leveling it, which the German political scientist L. Ulbricht figuratively called "demos scraping.""

UTOPIA: "DIGITAL PARADISE"

"This document contains a list of fundamental principles, the implementation of which will create "a new, society-centric vision that is intended to guide policy-makers and civil society organizations in the direction of a more equitable and democratic digital environment, where basic liberties and rights are protected, where strong public institutions function in the public interest, and where people have a say in how their digital environment functions.""

"The concept of creating a pan-European "digital Paradise" proceeds from the fact that at present in European countries there is a lack of digital technologies and, in general, digitization has not revealed its potential and its capabilities to the full."

"This principle assumes that Europeans, through digital technologies, should participate in joint work activities and exchange relevant knowledge for this purpose."

"According to the fourth principle, state government bodies should be empowered, which will ensure the broad participation of citizens of European countries in the management of the education system, science, and culture."

CONCLUSIONS

"The colossal increase in the economic power and political influence of global ICT corporations during the global pandemic crisis in 2020–2022 prompted many US political scientists, sociologists, economists, and state scientists to turn their attention to the American experience of the 1930s, to the New Deal policy of F.D. Roosevelt, who not only managed to use all the potential power of the state apparatus at the federal level to curb the undivided domination of the largest US monopolies of that period but also formed a new "social contract" of American society, which provided it with 30 post-war years of sustainable economic development."

"Study and reference to the experience of the "New Deal" is increasingly leading American social scientists to the idea that the time has come, at least in developed countries, especially in the United States, to develop and begin implementing the "Digital New Deal," which would essentially allow for a "digital democratic revolution," which put the digital sphere under the effective control of the masses [21, p. 9]."

2.2.3 The Role of Society 5.0 in Achieving Sustainable Development: A Spherical Fuzzy Set Approach [278]

This is a machine-generated summary of:

Alimohammadlou, Moslem; Khoshsepehr, Zahra: The role of Society 5.0 in achieving sustainable development: a spherical fuzzy set approach [278]

Published in: Environmental Science and Pollution Research (2023)

Link to original: https://doi.org/10.1007/s11356-023-25543-2

If you want to cite the papers, please refer to the original.

For technical reasons we could not place the page where the original quote is coming from.

Abstract-Summary

"Society 5.0 seeks to achieve economic growth while overcoming social and environmental challenges, contributing to the welfare of the global community."

"The purpose of this study is to explore how digital transformations introduced by Society 5.0 can contribute to the realization of sustainable development?"

"The study primarily used the systematic literature review method to identify the dimensions and indicators that would quantify the role of Society 5.0 in achieving sustainable development."

"The novelty of this research can be found in its application of the spherical fuzzy approach to examine the impacts of Society 5.0 smart technologies on achieving sustainable development, by addressing five major dimensions (governmental, economic, social, environmental, and technological)."

"The findings could help managers to more effectively respond to the challenges of sustainable development by relying on Society 5.0."

Introduction

"From the perspective of Society 5.0, sustainable development goals (SDGs) accommodate a new biosphere in which human beings lead wealthy and gratifying lives in harmony with social, economic, and technological progress, without harming nature (Bui and Tseng [145])."

"Mavrodieva and Shaw [146], too, explain that Society 5.0 must utilize technological innovation to extend economic, social, and environmental dimensions and ultimately achieve safe and sustainable development."

"It must be noted that although Society 5.0 seeks to realize sustainable development concepts and a human-centered society, but social, economic, and environmental problems and smart technology advancements may be so interwoven that changing their equations even for sustainable development could lead to some disruptions (Roblek and others [147])."

"Although researchers have univocally emphasized the importance of achieving Society 5.0, no study has ever proposed measurable indicators or a model that could help realize sustainable development in Society 5.0 through smart technologies."

Literature review

"Given the research findings, one could observe that the researchers tried to determine how much technological changes arising from smart systems have affected life, interpersonal interactions, government-society interactions, consideration of people in industries, etc As such, most of the studies addressing Society 5.0 have tried to conceptualize these changes in Society 5.0."

"One could expect that under such circumstances, social and economic factors will undergo extensive changes and encourage organizations and society to rely on smart products/services that respond to people's new lifestyle."

"The only study that partially investigated economic, social, and environmental factors, along with the government's role, in response to Society 5.0 needs, was conducted by De Felice and others [148]."

"The present study, however, tries to propose a comprehensive model that modified the role of digital transformations introduced by Society 5.0 in achieving sustainable development by considering five dimensions: governmental, economic, social, environmental, and technological."

Method

"The TISM method was used to construct a model for evaluating the role of Society 5.0 in achieving sustainable development."

"In the quantitative part of the study, such methods as SF-AHP and SF-DEMATEL were used to determine the weight of the criteria and their interrelationships."

"The first step in the model development process is to identify the indicators found through a systematic literature review."

"Identify and define the elements Following the TISM method, primarily the elements are identified and specified, and their interrelationships are modeled."

"Construct the total interpretive structural model The datasets extracted from the digraph and the interaction matrix help to operationalize the TISM."

"This study method draws on SF-AHP to evaluate the weights of the indicators."

"Collect comparisons from the decision-makers and create the direct influence evaluation matrix The decision-makers have to determine their preferences in relation to the potential influences among the attributes."

Analysis of the findings

"To identify sustainable development indicators that could be affected by Society 5.0 digital transformations, steps 1–4 were followed."

"After the publications were screened based on their abstracts and introductions, 20 articles were selected that addressed sustainable development and concepts in relation to its goals in Society 5.0."

"Following that, all of the sections in the articles (including methodology and data analysis) were investigated, and the factors affecting sustainable development in Society 5.0 were extracted."

"Five experts were asked to inspect the factors; they specialized in sustainable development, smart technologies, and Society 5.0."

"The TISM method was used to construct the conceptual model of the role of Society 5.0 in sustainable development."

"These experts were familiar with such concepts as smart technologies, Society 5.0, super-smart society, and sustainable development."

"Based on step 20, the opinions of five experts (in smart technologies, Society 5.0, super-smart society, and sustainable development) were collected."

Findings and discussion

"With an increase in the use of new technologies in individuals' lives, a society further moves toward smart processes, and robots will replace humans in high-risk work environments (Pereira and others [149])."

"One could conclude, then, smart social activities guided by AI and natural intelligence could build a super-smart, human-centered society that easily responds to the complex problems of sustainable development (Carayannis and others [150])."

"Using smart technologies in Society 5.0, then, can generate a new value by developing human resources and facilitating cooperation and coordination between different systems (Ferreira and Serpa [151])."

"It is crucial to harmonize supervisory standards and political orientations regarding innovation in smart technologies for sustainable development in Society 5.0, because such technologies shift the attention of societies and organizations to economic and technological development."

"Super-smart societies must develop themselves in all governmental, economic, social, environmental, and technological areas (Sołtysik-Piorunkiewicz and Zdonek [152])."

Conclusion

"The results showed that Society 5.0 advancements could enhance sustainable development through five main dimensions: governmental, economic, social, environmental, and technological."

"From a practical perspective, too, Society 5.0 introduces new technologies and thus creates new values in such areas as automated driving systems, healthcare, commercial transactions, transportation, agriculture, forestry, tourism, etc These are massive transformations that could enhance social performance through new technologies in a super-smart society while reducing pollution and saving resources."

"Five dimensions were identified, namely governmental, economic, social, environmental, and technological, which left considerable impacts on the achievement of sustainable development in Society 5.0."

"Future studies can address the detrimental effects of Society 5.0 smart technologies on sustainable development while exploring how such effects could be controlled in terms of the social, economic, and environmental indicators identified in this study."

2.2.4 Hyperlocal Sustainabilities: Theorizing Action Research for Sustainability in the Digital Age [279]

This is a machine-generated summary of:

Cutts, Bethany B.; Greenlee, Andrew J.; Hornik, Kaitlyn; Nigrelli, Caitie A.: Hyperlocal sustainabilities: theorizing action research for sustainability in the digital age [279]

Published in: Sustainability Science (2019)

Link to original: https://doi.org/10.1007/s11625-019-00694-9

If you want to cite the papers, please refer to the original.

For technical reasons we could not place the page where the original quote is coming from.

Abstract-Summary

"To fill this gap, we examine whether anticipating the use of social media reshapes how the community voice method (CVM) intervenes in deliberative spaces associated with waterway sustainability in Milwaukee, Wisconsin (USA)."

"CVM is a video-based participatory research approach designed to foster inclusive and community-focused planning."

"Of analyzing field notes, audio-visual recordings of participants, and other documents, we find that anticipating the use of social media shapes all aspects of participation, including: awareness of context, group dynamics, interventions, outcomes of concern."

"Concern for the opinions and perspectives of hyperlocal participants re-shapes recommendations about how to align research projects with community needs and planning deliberation."

"Social media is an important outlet for the work, but changed the placement of video products in relation to waterway sustainability plans, especially with respect to environmental justice."

"We conclude by proposing a framework for hyperlocal sustainability research that appropriately emphasizes justice."

Introduction

"Although scientists widely recognize the potential for a digital divide to contribute to uneven engagement with and access to technologies, the increasing prevalence of internet-enabled devices and internet service in many parts of the world has inspired an increased investment on the part of sustainability researchers in using social media to increase participation and disseminate results."

"A handful of studies suggest that online engagement between researchers and participants alters the research process and the potential outcomes of local sustainability initiatives (e.g. Büscher [153]; Cutts and others [154]; Nulman and Özkula [155]; Fletcher [156])."

"Whether hyperlocal engagement improves participation in research or advances local sustainability initiatives through more inclusive planning and management hinges largely on the ability of such projects to confront and actively dismantle pernicious forms of injustice."

"Community-based participatory research (CBPR) offers a compelling framework for process-oriented sustainability science occurring at the local level."

Materials and methods

"Coding and presenting interview responses to include them in the documentary was a key point of discussion among researchers, organizations, and community members during project design, interview and focus group phases."

"At the meeting, we followed an agenda that included: (1) informal conversation, snacking, and opportunities to receive literature about the AOC clean-up as participants arrived (2) a welcome by the AOC community action council and project introduction (3) film presentation and overview of social vulnerability mapping work, (4) open discussion to respond to the presentation and articulate recommendations for inclusive planning as it relates to water-centered sustainability and stewardship of remediated urban landscapes and (5) evaluation and wrap-up (including opportunities to become involved in AOC stewardship and to stay abreast of further development of the documentary)."

Results

"We find that anticipation of releasing the film on social media exposed tensions between the normative goals of the action research tradition in sustainability research and the perceptions of participating stakeholder groups, including academics, community members, and agency outreach professionals."

"Feedback and lack of diversity in group interview participants, the research team made plans to hire a Milwaukee-based interview specialist to extend the project's reach and incorporate input from Lincoln Park residents directly."

"Participants included: a representative from the Friends of Lincoln Park, a group formed during the waterway remediation with a mission to enhance stewardship and environmental justice (Holifield and Williams [157]; Friends of Lincoln Park [158]), an environmental justice scholar from the University of Wisconsin-Milwaukee, a newly hired field crew leader working with the River Restoration Foundation, and the author of a children's book focused on diversity in Milwaukee."

Discussion: thinking though hyperlocal influences on participation in local sustainability

"Our implementation of the CVM in relation to water sustainability in Milwaukee reinforces the idea that video represents a tremendous opportunity for researchers to engage with the public and provide products that return co-produced knowledge to communities in meaningful ways (Burkholder and MacEntee [159])."

"The three vignettes highlight ways that consideration of the hyperlocal environment reinforce the need for deliberate efforts to mitigate racial disparities by overcoming (or at least acknowledging) differences in participation as they relate to local sustainability projects."

"This culminated in a total re-envisioning of the research process and community-facing products to facilitate opportunities for interactions that serve the environmental justice and local sustainability dialogues occurring in the community through groups like the Friends of Lincoln Park and Milwaukee Water Commons."

"As modes of CBPR, video methods like CVM show significant promise as both a way to enhance the transparency of the research process and to disseminate results to a larger audience via hyperlocal engagement."

Conclusions

"This paper points to a need to better understand internet-enabled social interactions and their effect on lived experiences of social justice and environmental quality if we are to understand sustainability in a modern context."

"Most profoundly and relevant to sustainability, putting video on social media highlights the role of the hyperlocal public as actively and richly engaged in shaping research process even when they are not directly engaged."

"The hyperlocal sphere has the potential to link communities, carrying with it an opportunity to examine how and when efforts toward local sustainability and justice in one place carry lessons that can feed forward to other communities."

"We use our experience to suggest the need for a paradigm that acknowledges and integrates the virtual and material aspects of affecting change for local sustainability, arguing that the virtual embeddedness of CPBR and its impact over environmental justice outcomes are incompletely theorized, but that they represent an important reconciliation of local sustainability practice and research."

2.2.5 Towards an Equitable Digital Society: Artificial Intelligence (AI) and Corporate Digital Responsibility (CDR) [280]

This is a machine-generated summary of:

Elliott, Karen; Price, Rob; Shaw, Patricia; Spiliotopoulos, Tasos; Ng, Magdalene; Coopamootoo, Kovila; van Moorsel, Aad: Towards an Equitable Digital Society: Artificial Intelligence (AI) and Corporate Digital Responsibility (CDR) [280]

Published in: Society (2021)

Link to original: https://doi.org/10.1007/s12115-021-00594-8

If you want to cite the papers, please refer to the original.

For technical reasons we could not place the page where the original quote is coming from.

Abstract-Summary

"Inevitably costs are involved with delegating power to algorithmically based systems, some of whose workings are opaque and unobservable and thus termed the "black box"."

"An algorithm with arbitrary precision can easily reconstruct those characteristics and make life-changing decisions, particularly in financial services (credit scoring, risk assessment, etc), and it could be difficult to reconstruct, if this was done in a fair manner reflecting the values of society."

"If we permit AI to make life-changing decisions, what are the opportunity costs, data trade-offs, and implications for social, economic, technical, legal, and environmental systems?"

"We need to think carefully about how we implement these algorithms, the delegation of decisions and data usage, in the absence of human oversight and AI governance."

"The paper seeks to harmonise and align approaches, illustrating the opportunities and threats of AI, while raising awareness of Corporate Digital Responsibility (CDR) as a potential collaborative mechanism to demystify governance complexity and to establish an equitable digital society."

Introduction

"We observe substantial uncertainty and anxiety that the rapid adoption of AI across the digital "space" will impact society in negative ways: widespread job loss, income inequality, and social isolation (ibid.)."

"How can citizens be included and benefit from AI innovations and, specifically, protect their data and digital identities in the fourth industrial revolution (Kowalikova and others [160]; Lewis [161]; Hamdan and others [162])?"

"To place the debate on the associated risks of AI and societal implications in context, first, we review pertinent statistical insights in this area; second, we provide a brief background to our current financial services (FS) and financial technology (FinTech) environment, specifically, AI/FinTech-enabled FS; third, we examine trust, ethical principles, and regulatory implications (including the new EU guidance)."

AI Perceptions

"Each finding highlights the interplay between perceived AI risks, opportunities, and threats that are interwoven within our discussion surrounding notions of trust, ethics, legality, and governance nested in societal systems and people's relationship with AI."

"Simply put, technologies have permeated social interaction whether on digital platforms or the positives afforded by Zoom during the recent pandemic (Wiederhold [163]; Haochen and Polak [164])."

"The technology that lies beneath, regulating and accessing a range of complex systems, that can arbitrarily create a characteristic profile of our digital "selves" and AI continues to develop."

AI Development

"Distributed ledger technology (DLT) that emerged in 2008 employs this powerful feature, providing multiparty computations wide-scale transaction clearing with mutual distrustful parties, termed FinTech, opening a universe of AI/FinTech-enabled financial services (FS)."

"Granted, a level of acceptance can be discerned to satisfy machine learning principles, but progress in trying to resolve these issues will be a feature of auditability and understanding of AI decisions ex post to enhance broader uptake of machine learning and thus, AI (Pasquale [165, 166])."

"Explaining the internal mechanics of the deep or machine learning system in human terms or explainable AI (XAI) remains problematic because there is "[n]o real consensus about what interpretability is in machine learning" (Molnar [167]: 31)."

"Nothing is compulsory in terms of regulation, AI governance, ethics, and legal compliance trail the rate of AI innovation and implementation in digital society (Floridi [168]; Roitblat [169])."

Digital Society

"Zuboff [170] examined digital society, AI, and symbiotic social relationships, claiming that society is subject to Surveillance Capitalism."

"Despite adopting a predominant Western perspective, Zuboff's [170: 199–232] work draws attention to the discourse we unpack; she reveals the notion of overt and covert data manipulation, whereby users have a "puppet and puppet master" relationship (ibid)."

"Transparency to the user and society is blurred, for the tech giants operate largely without regulation, using and re-using data (ibid.)."

"the essence of the uncontract, which transforms the human, legal and economic risks of contracts into plans constructed, monitored and maintained by private firms for the sake of guaranteed outcomes: less contract utopia than uncontract dystopia Translating the uncontract for AI/FinTech-enabled FS, we suggest that beyond OB, society has been changed, premised on the new "normative judgment", that the digital society is a better state than the prior (Byrne [171]: 62)."

Vested Interests

"Zuboff's findings exemplify conflicting interests between the puppet master orchestrating the uncontract dystopia, reifying Alford's [172: xiii] notion of "strategically structured interest" first applied to examine the conflicting interests in health system reforms."

"Alford contended there existed "a continuing struggle between major structural interests operating within the context of a market society – "professional monopolists" controlling the major …"

"The invitation of the monopolisers to the community population to participate in the digital society amounts to "social control"—the puppet (Alford, and R.,, and Friedland, R. [173]: 472) or Zuboff's [170: 334–6] "blankness of perpetual compliance"."

"Alford's theory has been criticised for not defining which interests are specifically "repressed" by the monopolisers (North and Peckham [174]; Checkland and others [175])."

"The approach is useful and continues to be applied across finance, and to COVID-19, relative to conflicting interests (Ülgen [176]; Briggs [177])."

"We have acknowledged conflicting interests residing in the digital society, binding us to the "normative or ethical domain …"

Trust, Ethics and Human Oversight of AI

"Although DEAR progresses towards providing regulatory impetus for AI governance, giving organisations legitimacy to ascribe budget and costs (i.e. ethical implementation requirements), the guidance permits high-risk AI operators to decide the method/technical solution to achieve compliance."

"We can infer that these ex ante and ex post governance tools are intended to operate in providing transparency while engaging a "soft-law" normative constraint on "the "do's" and "don'ts" of algorithmic use in society" (Morley and others [178]); without enforceability or liability, these "soft law" tools lack "teeth" (Shaw [179]), Nonetheless, DEAR marks a significant improvement for AI governance, affording organisations the opportunity to find and create risk management, governance, and oversight solutions provided conformity is achieved."

"Without trust, social responsibility and operationalising digital ethics through AI-governance, regulation will fail as "consumers won't use the firm's services, adopt new technology, or share their data" (Shaw [180]: 176)."

Corporate Digital Responsibility (CDR)

"CDR is a voluntary commitment by organisations fulfilling the corporate rationalisers' role in representing community interests to inform "good" digital corporate actions and digital sustainability (i.e. data and algorithms) via collaborative guidance on addressing social, economic, and ecological impacts on digital society."

"CDR codifies TRUST and illustrates how AI governance and expectations are met building on lessons learned from corporate social responsibility including environmental and sustainable goals (CSR, Carroll [181])."

"Rather, CDR complements such protocols, as the digital realisation of the same responsibilities but with a two-fold appropriate use of digital and data within and by the organisation to impact society while sustaining our planet to improve organisational environmental and social governance performance (Dörr [182])."

"Crucially, this role will and must be complementary to, and of the code, as a dynamic iterative AI governance framework in which the DRC operates and extends responsibility for digital, data or AI services once launched into digital society."

Our Digital Futures

"Society mistrusts AI systems, yet gradually we have succumbed to accept algorithms making potentially life-changing decisions as a daily "norm"."

"We have focused on "what" needs to be considered in governing AI systems and building trust—risk reduction of the potential harms that technologies inflict on our digital society, whereby the underpinning science is not fully understood."

"Who is responsible to ensure our digital safety as more AI systems are free to operate devoid of human oversight?"

"Although voluntary for organisations, CDR members draw from their collaborations across standardisation networks centring on better AI-enabled systems (IEEE, iTechlaw and ForHumanity), enabling the equitable digital society to come to fruition."

2.2.6 Towards a Manifesto for a Critical Digital Humanities: Critiquing the Extractive Capitalism of Digital Society [281]

This is a machine-generated summary of:

Grimshaw, Mike: Towards a manifesto for a critical digital humanities: critiquing the extractive capitalism of digital society [281]

Published in: Humanities and Social Sciences Communications (2018)

Link to original: https://doi.org/10.1057/s41599-018-0075-y

If you want to cite the papers, please refer to the original.

For technical reasons we could not place the page where the original quote is coming from.

Abstract-Summary

"The first section engages with a deliberately small number of popular texts written by discriminating and interrogative consumers and producers of digital culture and society."

"While these may be dismissed as journalistic texts and sources by those of a more focused academic intent, here these texts are used because they are the connection between academic engagement and wider public readership."

"They frame what can be termed the critical public engagement with digital capitalism."

"It may draw on Marxist thought but is not Marxist in ideology or final intent; it is critical of capitalism but acknowledges our on-going existence and possibilities within it; it is written by an academic seeking to act as intermediary between academic and non-academic readings and responses."

Setting the scene

"The Post-academic, in my conceptualization, is the self-proclaimed mongrel intellectual seeking a critical engagement with and within this new form of capitalism, unafraid to proudly mix and match a broad variety of sources and approaches, seeking new possibilities for critique."

"The term mongrel is deliberate in its challenge to the self-regard and self-disciplining of the academic from the societally-engaged position of the intellectual."

"For in their retreat in search of the approval of citation–indexes and 'impact factors', the academic increasingly only really serves the digital-matrix academic capitalism of the late modern university. (A warning: already I have coined the first of a number of possibly competing terms to describe varieties of modern capitalism; but this the deliberate approach of the mongrel intellectual who believes there is no singular experience nor normative term for contemporary capitalism; other possibilities will arise!)."

Critiquing digital society

"This is what I term the state of being beyond financial capitalism, finding ourselves instead in the new immaterial capitalism whereby we (often unknowingly) work for free every time we are on-line, especially in social media, 'manufacturing the very personal data that makes their companies so valuable' (Keen, [183], p 57)."

"The reality is the concentration of power and money occurs in what can actually be termed the 0.0001% (Keen, [183], p 202), a new monopoly oligarchy that somehow, in particular, the majority of western society seems to wish to work for and enrich for free."

"To put this another way, Occupy Wall Street was protesting a form of capitalism that had actually transitioned into another form; the real power, money and inequality had radically relocated, but we saw these new digital companies as still primarily entrepreneurial, emancipatory and providing a new social network and cohesion."

How to rethink this new capitalism

"This does not make sense unless we realize that the app economy, included within the digital economy of immaterial labour, is actually based on access to data, the data that everyone using an app freely provides, the data that all users of digital society freely provide."

"What makes the digital sharecropping successful is that the vast majority of those working do not view their work as such, nor their data as valuable; in short we view it all as immaterial and so we participate in the violence of immaterial capitalism whereby all of our on-line life is a form of unpaid work within a monopoly, oligarchic capitalism."

"The difference between immaterial labour and what I term immaterial capitalism is that immaterial capitalism is dependent on digital farms, digital harvesting and digital share-cropping, undertaken against us without our consent or more often our knowledge, as part of a digital society demand to be on-line so as to participate and to digitally, increasingly socially, exist."

A place to begin the resistance: a re-thought critical digital humanities?

"I argue that a re-thought critical digital humanities is actually the location for a counter-revolution whereby a different experience and type of digital society can occur: a culture of the deep and not the shallows, a society of deep reading and deep thought facilitated by the digital, a culture of concentration and not distraction, a society of informed, critical participation, critique and opposition."

"A critical digital humanities should be the place whereby the deep thought and the deep discussion of the digital, can occur within the wider aims of the emancipatory transformation of digital society and the informed critique of and critical engagement with digital capitalism."

"Critical digital humanities should therefore be the site of critique, opposition and resistance to the rise and increasingly totalitarian demand of social media in our lives, our work and, more specifically, within the university."

Conclusion

"I am not so naïve to believe Capitalism will end any time soon, but capitalism is a human creation and activity and so it is a hermeneutic activity."

"The internet as is, is one possible answer as to how we wish to live and how we wish capitalism to be, but central to hermeneutics is the other interpretation and possibility."

"This is, I argue, a central task of critical digital humanities, a task of critique, resistance, emancipation and the creation of alternatives, hosted within universities who take seriously their 21st century role as conscience and critic of digital society."

"Digital workers of the word unite, we have nothing to lose but our data chains."

2.2.7 Digital Constitution: Fundamental Rights and Freedoms of an Individual in a Totally Informational Society [282]

This is a machine-generated summary of:

Shakhrai, S. M.: Digital Constitution: Fundamental Rights and Freedoms of an Individual in a Totally Informational Society [282]

Published in: Herald of the Russian Academy of Sciences (2018)

Link to original: https://doi.org/10.1134/s1019331618060126

If you want to cite the papers, please refer to the original.

For technical reasons we could not place the page where the original quote is coming from.

Abstract-Summary

"The article analyzes the problems associated with the mismatch of speeds with which the processes of the formation of a totally digital society are taking place, and the creation of effective social (primarily legal) regulators for the new reality."

"The author believes that the unexpectedly, new emerging digital future should be viewed as a social revolution (break of continuity), which dictates the need to rapidly create mechanisms for the protection of a person, their basic rights and freedoms, including constitutional ones, in the digital world."

"The author raises the question of creating a digital law and developing a digital constitution, which should provide the necessary basis of consent and effective mechanisms for the establishment of a social order in cyberspace."

HUMAN RIGHTS PROTECTION IN THE DIGITAL WORLD

"We are talking about ensuring the guarantee of human rights in identifying (correlating a real person with its digital image), guarantees of the safety of the data of digital users, and the problem of ensuring citizens' confidence in the digital environment."

"It is obvious that the issue of ensuring human rights and freedoms in the digital world is much deeper and wider than the task of identifying users, protecting personal data or even protecting citizens from cybercrime."

"In the digital world, as in the real world, the fundamental constitutional rights and freedoms of humans and citizens must be respected."

"Article 24 of the Constitution of the Russian Federation, which prohibits the implementation of "the collection, storage, use, and dissemination of information about the private life of a person without their consent" and obliges the state and local authorities to "ensure that everyone has access to documents and materials directly affecting their rights and freedoms, if otherwise not provided by law," in the most direct way relates to the issue under discussion."

INDIVIDUAL RATING OF CITIZENS: NEW FORM OF TOTAL DIGITAL CONTROL

"The program being implemented today is not just about the system of assessing the borrower's solvency but about creating an individual social rating of a Chinese citizen based on combining all databases containing any information about individuals and legal entities and analyzing them using Big Data technologies."

"The country has integrated databases on property and the financial status of people, their banking, legal and medical information, information from educational institutions, transport and hotel booking systems, and data from national social networks (like the Chinese messenger WeChat, which already has 800 million users) and other public resources."

"This individual social, financial, and political portrait (passport) of each citizen is, in fact, the technological basis for the total control of the country's population."

HIDDEN VIOLATIONS OF CITIZENS' RIGHTS IN THE DIGITAL WORLD

"Despite the ambivalence of digital technology, it is necessary to recognize as a fact that the total digitalization of society and all its systems, including personal life, cannot be stopped."

"Today, instead of confronting the objective development of events, it is necessary to create at an accelerated pace mechanisms to protect a person, their basic rights and freedoms, including constitutional ones, in the new total digital environment."

"The initiatives to create a kind of new Universal Declaration of Human Rights for the digital world are already known abroad."

"The developers, like the author of this article, proceeded from the fact that, due to the fundamental transboundary nature of the digital world, it is necessary to find minimum points of agreement to protect the basic values of democracy, first and foremost the individual rights and freedoms of citizens."

APPROACHES TO THE CREATION OF A DIGITAL CONSTITUTION

"How, by which technological means, can constitutional rights and freedoms be really guaranteed in the digital space?"

"It is equally important to provide legal and technological means for the disclosure of information about the owners of network resources and create conditions for external control by society over the activities of digital corporations."

"With the solution of the global task of developing a digital constitution guaranteeing the rights and freedoms of citizens in the digital world, it is necessary, with expert support from the social sciences, to actively implement the latest technological solutions in the government system (implementation of the electronic government concept, e-Government)."

"Blockchain technologies, smart contracts, and other new information tools have an obvious potential for strengthening the economic foundations of the constitutional system of the Russian Federation and the unity of the economic space, ensuring the freedom of economic activity and the institution of private property."

[Section 5]

"The problem lies in the fact that many processes in the digital world—economic, financial, and social—are born and develop faster than society has time to realize them and all the more to settle and create the necessary legal framework for "wild digital capitalism," protecting the rights and freedoms of a digital person, and ensuring the sovereignty of states in a trans-border digital world."

"There is a kind of race between online and offline worlds in matters relating to understanding what is happening, regaining control over the digital reality and the formation of new legal, political, and economic tools (social regulators) that can effectively manage it."

"This means that the evolutionary development of law in the conditions of the digital world (in fact, the reactive transformation of legal regulators) will always be an insufficiently effective solution in terms of the management needs of social and economic transformations in a rapidly changing world."

2.2.8 Society's (Virtually) Time-Free Transition into the Digital Age [283]

This is a machine-generated summary of:

Robinson, John P.; Lee, Chang Won: Society's (Virtually) Time-Free Transition into the Digital Age [283]

Published in: Social Indicators Research (2013)

Link to original: https://doi.org/10.1007/s11205-013-0374-z

If you want to cite the papers, please refer to the original.

For technical reasons we could not place the page where the original quote is coming from.

Abstract-Summary

"This article reviews the results from two large multiyear US national surveys that have been monitoring social change in US daily life with high response rates: (1) the 1974–2012 General Social Survey (GSS; with more than 55,000 adult respondents aged 18+) and (2) the 2003–2011 American Time-Use Survey (ATUS; with more than 100,000 such respondents)."

"The GSS has collected time-estimate data on particular social and media activities, while the ATUS surveys have collected complete 24-h diary accounts across a single day."

"In both surveys, there seems little if any significant impact of these new media on social/media time, even though they had become the predominant source of information and social contact by 2004."

"GSS respondents in general have not reported lower levels of social or media contact since the 1990s, and while those GSS respondents who spent more time on the Internet did report fewer social visits with relatives, they reported more visits with friends, compared to those who do not use the Internet."

"No consistent decline in either social or media activities was found in either survey across this period of Internet diffusion, much in line with the earlier national studies."

"It seems clear that not only the ATUS but diary studies in other countries are failing to capture the significant social impact of IT on the rest of life in the new digital age."

Introduction

"This article examines the way in which new digital technology, and in particular the Internet, has been associated with differences in peoples' previous social and media activities and compares these results with differences in other ways of spending time."

"On the basis of the 1995 Pew data, Robinson and others [184] found that in 1995, Internet and IT users were significantly more likely to use print media, radio newscasts, and movies than nonusers, and they were not significantly less likely to engage in social activities—nor to watch less television for either entertainment or news content."

"Both this enhancement of reading, as well as the similarity of time spent on social life and other media use, across Internet users and nonusers is inconsistent with the idea of functional time displacement."

"Much the same conclusion emerged when Robinson and De Haan extended their analyses to include time estimate data from 14 other comprehensive studies that examined social and media activity differences among Internet users and nonusers."

GSS General Methodology and Social/Media Questions

"In order to track social change during this period of rapid advances in new information technologies, a targeted module of Internet-relevant questions was added to the 2000, 2002 and 2004 national probability samples of the General Social Survey (GSS)."

"The year 2000 GSS in-home survey contained its first Internet module of questions asked of 2,353 respondents aged 18 and older in 2000, 2,765 respondents in 2002 and 2,812 respondents in 2004."

"As in Robinson and Martin [185], their analyses involved relating these behavior measures to respondent use of the Internet after adjusting for the many demographic and background factors regularly included in the GSS."

"The general plan in the following analyses, then, is to see how the various GSS questions dealing with extent of social/media use vary as a function of the extent of respondent use of the Internet."

2000–04 Results for the New GSS Internet Module Questions

"One exception to these small changes in contact proportions occurred for health contacts with friends and relatives—that rose from 48 % in 2000 to 63 % in 2004, an increase of 15 points."

"It can be seen that, instead, respondents report significantly more usage of these older media, the more they use the Internet for health information."

"The 12-point gain (from 58 to 70 % in Internet use) by 2004 is the largest growth found in this part of the table, much as the WWW gain was the largest increase for health information in source of relevant information, here having replaced newspaper listings as the main source of job listings."

"As with health, the heaviest users of the Web for jobs (5 or more times in the last year) report more than twice as many uses of other media as nonusers."

1974–2012 Results for Traditional GSS Questions

"This estimated TV time decline is unrelated to Internet use after demographic factors are taken into account."

"People who do not use the Internet at all consume an hour of TV more per day than people who report at least some Internet usage, but only because they have lower average educational attainment, older average ages or other demographic differences that make them less heavy viewers than people who do use the Internet."

"Newspaper usage is slightly lower among Internet nonusers and it is unrelated to the extent of more Internet usage before MCA adjustment."

"After MCA adjustment, that largely remains true, so that once demographic factors are taken into account, neither TV viewing nor newspaper reading is related to extent of Internet use."

ATUS General Methodology and Daily Time Accounts

"As in the GSS questions just reviewed, most information about time use was obtained from general estimate surveys that asked respondents to estimate the hours or days they recently or typically spent on some specific activity, such as work or watching TV."

"Because they are intended to represent complete accounts of daily activity, diary data can be used to generate estimates of how much societal time is spent on the complete range of human behavior—from work to free time, from travel to time spent at home."

"What is basically involved in the present analysis, then, is a cross-tabulation of the daily time ATUS respondents spent on all their main daily activities as a function of whether and how much time they spent using IT (not during work time) on that same diary day."

ATUS Results

"Only two other activities, child care and sleep, show differences 1.0 h per week or more."

"These are activities, then, that are positively correlated with IT use, since they are engaged in more by IT users rather than less."

"Whether it is working longer hours than inhibits IT use, IT heavy users taking more time off from work (or with shorter workweeks), or those getting enough IT use at work to not need IT use in their nonwork time, remains unclear."

"Work hours also show a near-monotonic decline, except for the slight 11 h uptick for heaviest users versus those using IT 25–29 min on the diary day."

"Only paid work, then, emerges as the major activity significantly related to IT use in the ATUS data, although why the two are so significantly intertwined remains unclear."

Summary and Conclusions

"ATUS respondents report only about 10 min a day in their diaries, which translates to just over an hour of IT use per week—or less than a tenth of time spent watching TV."

"Support for the argument that IT use has resulted in, or is even associated with, lower time in social capital or media activities is not found in either the GSS or the ATUS, a finding which is in keeping with most of the previous literature in the field, including the several studies reviewed Issues 1 and 2 of IT & Society and in the several studies reviewed by Robinson and De Haan [186] and the Appendix, which covered usage in both the US and other countries."

"In neither of these two major national studies monitoring social change—the GSS and the ATUS—does one find consistent evidence of social or media activities being adversely affected by the diffusion or use of IT."

2.2.9 Digital Government and Sustainable Development [284]

This is a machine-generated summary of:

Castro, Conceição; Lopes, Cristina: Digital Government and Sustainable Development [284]

Published in: Journal of the Knowledge Economy (2021)

Link to original: https://doi.org/10.1007/s13132-021-00749-2

If you want to cite the papers, please refer to the original.

For technical reasons we could not place the page where the original quote is coming from.

Abstract-Summary

"E-government facilitates integrated policies and public services to promote sustainable and inclusive economic growth, social development and environmental protection."

"The aim of this study is to analyse the impact of e-government on sustainable development."

"Using a logit model, for a sample of 103 countries in the period 2003–2018, the results suggest that e-government development is a positive determinant for a country to attain sustainable development, proxied by adjusted net savings, that embraces a country's economic, social and environmental development."

"This study provides evidence that e-government increases the probability to attain sustainable development mostly in developing and transition economies."

"The results also suggest that economic growth and gross national income per capita are significant positive influences in sustainable development in the whole sample and that countries with lower age dependency and natural resource rents are more likely to have sustainable development."

"Developing e-government allows promoting sustainable development, particularly in developing and in transition economies."

Introduction

"Sustainable development adds to the economic growth the concern with greening the national accounts, by including the depletion of natural capital (Pardi and others, [187])."

"According to the World Bank [188], this indicator for sustainable development was designed based on the economic theory that savings equal investment and this is a measure of the change in wealth."

"Changes in wealth are linked to the sustainability of development since if a country is depleting its assets, it is not in a sustainable path (World Bank, [189])."

"Adjusted net savings embrace a country's economic, social and environmental development."

"To the authors' best knowledge, there are no previous studies analysing the impact of e-government on sustainable development, measured by the adjusted net savings, with panel data."

"To fill this gap in economic literature, the objective of this paper is to analyse the effects of e-government development in sustainable development."

Literature Review

"E-government development is an important factor in rationalizing the public sector and to faster countries development (Stanimirovic & Vintar, [190]) and has an impact on sustainable development through the economic, social and environmental dimensions."

"Despite the literature on the effects of e-government on economic growth, studies of its impact on sustainable development assessed by adjusted net savings are to a lesser extent."

"Although several studies analyse the impact of indicators of governance in sustainable development, there is no empirical literature on the impact of e-government on adjusted net savings as an indicator of sustainable development."

"Güney [191] studied the relationship of governance and sustainable development, and besides the positive effects in the whole sample (and in developing and developed countries), also found that democracy and urbanization promote sustainable development, while population growth impacts it negatively and economic growth is not significant."

Methodology
"Wealth (W) includes manufactured capital (K), human capital (H) and natural capital (N): And sustainable development (SD) will be given by The empirical investigation on the impact of e-government and other variables in sustainable development is examined for 103 countries in the period 2003 to 2018, for which all the necessary information was available."

"Among the 193 countries analysed in that survey, 40 of them scored very high (with e-Government Development Index between 0.75 and 1), an increase of 11 countries when compared with 2016."

"The results of these regressions suggest a positive relation between e-Government Development Index and adjusted net savings although this is especially more remarkable in the case of developing and transition economies, and statistically significant, which suggest not only that e-government can be a relevant variable in explaining sustainable development, but also the impact is higher in the last group of countries."

Results and Discussion
"The Wald test shows that all the explanatory variables are statistically significant at a 1% significance level (e-government, economic growth, age dependency, natural rents and GNI per capita) in affecting the likelihood of having non-negative ANS."

"In this subsample, it should be emphasised that with a raise by 0.1 units in the e-Government Development Index, it is expected to have the odds of having sustainable development increased by about 36.43%, which is higher than in the whole sample."

"The probability of having non-negative ANS also increases more in these countries than in developed economies with raises in economic growth or GNI per capita."

"It is also observable for the worldwide sample and for the subsample of developing and transition economies that countries with high or very high levels of HDI are countries with high digital government development."

Conclusions
"This paper highlights the links between e-government and sustainable development and was the first attempt to analyse the effects of e-government in sustainable development in a panel data for 103 countries in the period 2003–2018, using adjusted net savings as an indicator of sustainable development, recommended by the World Bank."

"In developing and transition economies, the odds of increased sustainable development, measured by the adjusting net savings, is higher than in developed countries when e-government development increases."

"The results of this paper provide information for policymakers of the important role of digital government in achieving sustainable development."

"Since many developing countries face problems of governance and weak institutions, the development of digital government can be a way to promote greater effectiveness and efficiency of public administration, also more equalizer, supporting sustainable development."

"Digital governance is primarily about transforming governance, which must be part of an overall public policy strategy to achieve sustainable development."

2.2.10 China's Rise as an Advanced Technological Society and the Rise of Digital Orientalism [285]

This is a machine-generated summary of:

Mahoney, Josef Gregory: China's Rise as an Advanced Technological Society and the Rise of Digital Orientalism [285]

Published in: Journal of Chinese Political Science (2022)

Link to original: https://doi.org/10.1007/s11366-022-09817-z

If you want to cite the papers, please refer to the original.

For technical reasons we could not place the page where the original quote is coming from.

Abstract-Summary

"As China has risen as an advanced technological society, a new type of Orientalism—Digital Orientalism—has likewise emerged."

"Using historical materialism, this paper details these developments, including China's change from a civilization-state to modern nation-state and its transition from a technical state to an advanced technological society, closing the technology gap that had left it vulnerable to foreign aggression and continued forms of international dominance and hegemony."

"It argues that China appears to have surpassed others now as a technological society, including the US, with China's response to COVID-19 as a clear example, and with clear implications for China's national advancement and global position vis-à-vis the United States particularly."

Introduction

"The third and now dominant theme is then described as Digital Orientalism [192, 193], in which three key features are developed: 1) in many respects Digital Orientalism presents as an inversion of Sinological Orientalism, where China is now perceived as moving resolutely down a different path, one that will not produce a liberal democratic state and will instead threaten the US-led global system; 2) but one in which, nevertheless, there are profound convergences at work, insomuch as China as an advanced technological society has much in common with Western technological societies, due to the fact that both share similar social, economic, cultural and governmental gains and losses due to the transformative power and influence of technology; and 3) what appear to be new divergences, exemplified particularly by different experiences with COVID-19 between China and the West, especially the US, through which China appears to have reached a new plateau in its development as a technological society, while others appear to be in regression."

From Civilization-State to Nation-State

"While this idea sits well with many Marxists, it's a complicated subject among contemporary Chinese Marxists, some of whom are understandably uncomfortable with how class struggle was sometimes practiced in China, including official Chinese assessments critical of the Cultural Revolution [194]."

"By 1953, it had initiated the first five–year plan with Soviet support, but due to a number of reasons, including a sometimes difficult relationship since the Party's founding in 1921, had begun to distrust Moscow by 1956, and worry that national development was not proceeding quickly enough to consolidate national sovereignty, let alone keep pace with new foreign technological advances that could pose new threats to China."

"The key point is that the crisis of Chinese modernity qua the crisis of sovereignty—from the Opium Wars to the first decades of the PRC—was deeply connected with the profound technology gap between China and more developed countries, who exploited their advantage at China's expense."

From Technical to Technological Society

"While some Chinese traditions certainly resisted development, China's rapid advancement as a technological society finds accordance with its traditional and modern philosophical perspectives on human progress, and not just the existential necessity of closing the technology gap to establish and sustain sovereignty."

"It's in this context that we can view broad popular support for Chinese development as a technological society in China itself in terms of high rates of new technology adoption [195], popular support for scientific solutions to problems, and mass positive engagement in public policies that aim for social advances consistent with both individual and national transformations that are forward-reaching, tech-based and tech-oriented [196]—exemplified most recently by broad popular support for China's "dynamic zero-covid policies" (dongtai qingling)."

"The key point here is that China's actualization as a technological society that advances with technology is broadly popular, that it's even valorized by Chinese theorists as consistent with the rise of an "intelligent civilization" (zhineng wenming), one in which both governance and daily life are substantially supported by big data, artificial intelligence and other forms of digitalization [197, 198], and consistent with Marxist theories of social progress [199]."

From Orientalism to Sinological Orientalism to Digital Orientalism

"China had also launched a number of national "made in China" development campaigns, while Chinese firms like Huawei were gaining global market share in areas previously dominated by Western firms, or at least those considered the US considered less dangerous to its hegemony (e.g., Japanese, South Korean, Taiwanese, etc)."

"More pervasive and significant despite American alarmism about China's improved intelligence and military capabilities: as market reforms and culture became increasingly advanced and dominant from the 1990s onward, incredible generation gaps emerged in which young Chinese arguably had more in common with their Western peers than their own parents."

"Is that many of these ills have meet new policy responses in China that arguably demonstrate not only an increased capacity for governance as an advanced technological society, they also appear to contrast with accelerating declines of the same in the West, especially the United States."

Conclusions

"My purpose has been to illustrate the role that advancing China as a technological society has played during its modern development, how this has in turned been received by others."

"I have not discussed whether there is Chinese fear of Western technology, and how this relates to Sino(techno)phobia."

"While the West and particularly the US have increasingly flirted with the fantasy of technological autarky, of decoupling from Chinese technology, the same can be said about China seeking autonomy and independence."

"Marxism did normalize technological development and the emergence of China's technological society as a material means for reestablishing sovereignty and security."

"Perhaps today's competition is really located in the extent to which China as a technological society acknowledges itself for what it is and then seeks a position of human advancement that transcends or evolves to a higher stage of existence and well-being—as China claims per it ambitions to establish itself as a 'fully modern socialist nation.'"

2.2.11 Internet Spaces and the (Re)making of Democratic Politics: The Case of Singapore's 2011 General Election [286]

This is a machine-generated summary of:

Woon, Chih Yuan: Internet spaces and the (re)making of democratic politics: the case of Singapore's 2011 General Election [286]

Published in: GeoJournal (2017)

Link to original: https://doi.org/10.1007/s10708-017-9815-6

Copyright of the summarized publication:

If you want to cite the papers, please refer to the original.

For technical reasons we could not place the page where the original quote is coming from.

Abstract-Summary

"The 2011 Singapore General Election (GE2011) has often been hailed as an 'Internet Election', highlighting the role of networked technologies in enabling democratization practices/processes for the unprecedented consolidation of oppositional politics in the city-state."

"Building on theorizations that broach the interface between social media and digital democracy, this paper critically examines Singaporeans' myriad engagements with the internet during GE2011 in order to tease out the constructions of social and political debates within these online communities and the extent to which they facilitate democratic discussions."

"In addressing these research inquiries, this paper goes beyond GE2011 to reflect on the ways in which web technologies and social media can make a difference to political debates, deliberation and representation in societies where there are limited public avenues for citizenry to articulate their voices/concerns."

Introduction

"The aim here is to investigate the very construction of social and political debates within these online communities in order to critically appreciate the modes and means of interactions with the internet by the wider Singaporean citizenry during GE2011 and how that affects ideas of democratic citizenship and meanings of the 'political'."

"Such an approach not only acknowledges the role of citizen-users as the driver of democratic innovation through their everyday self-actualized networking practices; more importantly, it allows for the interrogation of the virtual meeting, colliding and emerging of different perspectives related to Singapore's (electoral) politics and the multiple ways they are underpinned by wider power structures and dynamics, thereby setting the scene for the negotiation and (re)production of the multifarious meanings of democracy in place."

Cyberspace, social media and democratic engagements

"If we follow Iris Young's [200] exhortation that testimony, story-telling, greetings and rhetoric can all be employed as discursive forms of democratic engagement capable of enabling a more inclusive democracy, the opportunity exists for regarding a variety of textual, visual and auditory expressions on social media (e.g. playful repertories of innovative Youtube videos, Facebook posts, etc) as aspects of the political."

"By zooming into people's online political participation in Singapore where political conditions are widely perceived as repressive and limited avenues are available for political participation by citizens, this paper thus acknowledges the importance of incorporating both structural and agentic perspectives to tease out how individuals' democratic innovations and engagements with social media are always contingent and co-constructed in relation to the power geometries of social diversity, inequality and cultural difference (c.f.)"

GE2011, the internet and the electoral system in Singapore

"Given such a scenario, it has been widely contended that Singapore under the PAP's rule can be best characterized as an electoral authoritarian regime where regular elections are conducted under relatively free but unfair conditions (Diamond [201]; Lim [202])."

"Despite persisting in such critical questionings, these works nevertheless concur that GE2011 saw democracy in action, with a record number of people (and opposition factions) participating in online debates and discussions about the election."

"Recognizing this development will allow me to critically investigate the ways in which Singaporeans have had to negotiate 'new' political spaces of expressions on social media platforms during GE2011 and how such engagements (which are bounded within broader social–political and regulatory structures) shape their understandings/interpretations of democracy."

"Notwithstanding the effectiveness of such a research strategy, most of my respondents also had a lasting impression of their online personas and contributions during GE2011 as it was the first time they were allowed/able to participate in (virtual) electoral debates."

GE2011: enacting democracy through internet spaces

"It is interesting that Jerry highlights how GE2011 has come to serve as the pivotal moment whereby the internet has revolutionalised the way in which information related to Singapore's electoral politics was communicated to the general public."

"Be it the exposing of PAP's unfair electoral gimmicks (e.g. resorting to gerrymandering) to evoking visual technologies replete with humour, satire, etc as means to probe, subvert and ridicule prevailing political practices, social media platforms (during GE2011) were harnessed as the 'weapon of the weak' (following Scott [203]) through which everyday resistances towards authoritative power were elicited."

"Voting is an act of democracy but only if we do it right by making informed decisions (Lydia) The quotes above underscore how a 'mainstream' media that is tightly controlled by the PAP government has left Singaporeans with little choice but to use the internet/social media platforms to 'escape' political censorship and allow 'alternative' accounts of GE2011 to be documented."

GE2011: limits to the influence of the internet on democratic politics

"An interviewee, in making sense of why the internet may breed irresponsible behaviour, goes on to deliver a sobering critique of how this process actually constitutes a (virtual) manifestation of a lack of democratic consciousness in Singaporeans: Anonymity can give people courage to speak up on certain issues but it can also encourage people to say irresponsible and hurtful things….I think we need a civic rather than a civil society in Singapore—that is a mature and consensual democracy consisting of gracious individuals who will speak and act responsibly (Zach) Zach's enunciation illustrates that "the same anonymity and absence of face-to-face interaction that expands our freedom of expression online keeps us from assessing the impact and social value of our words (Papacharissi [204]: 16)."

Conclusion

"This paper has utilized GE2011 as an entry point to investigate the extent in which internet spaces and social media technologies can facilitate the 'playing out' of democratic discussions/debates especially in 'restrictive' societies like Singapore."

"Throughout this paper, I argue that the widespread use of social media and internet technologies during GE2011 have allowed for growing, diversifying forms of political actors (beyond the state)."

"Preliminary research have shown that despite the extensive use of social media and internet platforms by Singaporeans to enact participatory politics during GE2015, the electoral results differed vastly from that of GE2011, with the ruling PAP government securing what has been dubbed a "landslide victory" (see for e.g. Zhang and Pang [205])."

"Acknowledging these complexities of social media (and more broadly internet technologies) and understanding their relationship to democratic politics will no doubt raise important reflections on the future of democracy, civil society participation and governance in Singapore."

2.2.12 Districts, Multinationals and Global/Digital Networks [287]

This is a machine-generated summary of:

Castellani, Davide; Rullani, Enzo; Zanfei, Antonello: Districts, multinationals and global/digital networks [287]

Published in: Journal of Industrial and Business Economics (2017)
Link to original: https://doi.org/10.1007/s40812-017-0079-4

If you want to cite the papers, please refer to the original.

For technical reasons we could not place the page where the original quote is coming from.

Abstract-Summary

"This paper reflects on what remains of Becattini's utopia in the new context of a globalized and digital economy."

"It is suggested that the interpretive value of Becattini's theorizing remains, to the extent that one moves away from the consideration of the classical district model, and adopts his more general way of thinking about the economy, which is only exemplified by the historical circumstance of industrial districts."

Beyond the Fordist dualism between market and hierarchy

"In the case of Italian industrial districts, this third form relies on the cultural, historical, and solidarity based resources of local societies, which play a key role as productive forces, capable of using modern techniques, and at the same time managing complex situations."

"It is noteworthy that networking through which the division of labor occurs, tends to concentrate in specific places that are able—more than others—to attract and self-generate activities related to a certain sector or function."

"The post-Fordist contractual networking nodes are not equally distributed in the geographic space of the national and global economy, whereas more complex and intense relationships can be observed in a number of localized clusters (Porter [206])."

"What these authors emphasise is exactly the localized nature of innovations that are generated in a given place, and are transferred, applied and developed on a global scale through trans-territorial networks."

Industrial districts as experiments of capitalism with a human face

"Giacomo Becattini's theoretical work refers to the innovative experience of Italian industrial districts, linking it to the Marshallian model."

"Unlike what happens in clusters attracting multinationals—the relationship with actors and institutions based outside the territorial boundaries of the district itself is circumscribed in terms of both activities and geography."

"Even with this characterisation, districts are already seen as nexuses of trans-territorial relations, highly conducive to innovation and in competition with other open networks."

"Although connected with other theorizing of networks and of their territorial clustering, Becattini's in-depth study of some typical districts (first of all Prato) leads him well beyond the theory of clusters."

"Industrial districts create the conditions not only for mutual trust, but also for the choice of specific ways of life and for the sharing of local knowledge."

Giacomo Becattini: another way of thinking about the economy
"As far as aims are concerned, people who "chorally" participate in the local society primarily pay attention to the extent to which the generation of value (and therefore of merchantable goods) makes sense in terms of their social life and of their individual sensibility."

"Individuals active in the local context do not merely use the territorial cluster as a means for its own aims."

"They are part of a system that—in its history of evolutionary learning—has accumulated physical structures, cultural values, skills, viable specializations, social rules and codes of conduct that are suitable for life projects carried out by individual members of the local community."

"If and when this accomplishment of individual aims in terms of sense-making and happiness does not take place, this is because—according to Becattini—this "natural" evolutionary process is hindered by deviant factors typical of classical capitalism: "the accumulation of wealth and of decision making" in a few hands (Becattini [207])."

Today: industrial districts in the ongoing digital and global revolution
"Low transport costs in the case of knowledge inputs and outputs, combined with the fact that MNEs have developed routines and organizational structures that enable them to codify, process and transfer (codified and tacit) knowledge across national boundaries and within their internal networks (Gupta and Govindarajan [208]; Cantwell and Santangelo [209]; Zanfei [210]; Ambos and Ambos [211]), further increase the likelyhood that MNEs set-up R&D labs in relatively remote locations."

"As the distances to be managed increase, it becomes more and more convenient for globalizing firms to codify the knowledge to be used and therefore it is more and more advantageous for them to digitize their cognitive processes."

"Industrial districts are forced to accelerate and intensify their efforts to integrate their local assets into global networks, thus becoming "glo-cal" systems open to international knowledge sources and to global production and commercialization networks."

Three great transformations underway
"Medium-sized district firms are increasingly able to play a role that is typical of the leading companies in supply chain relationships; b. The "social capital" that allowed all actors in the local system to gain easy and free access to the knowledge, capabilities and relationships present in the proximity circuit is no longer sufficient to compete in the fourth industrial revolution era."

"Firms, workers and local institutions may significantly differ within and across industrial districts, in terms of their ability to undertake these efforts Rullani [212]; c. While in pre-2000 years "sense creation" was the result of living and producing in the local context, today it relies on innovative ideas and competences that are accumulated in different places."

"Industrial districts are subject to important tensions to change, with the most innovative firms within them increasingly involved in extensive interactions with external suppliers and consumers to whom the innovative ideas often make sense."

Becattini's utopia in the new world of digital and global revolution
"In the case of industrial districts, this network is still embryonic, both because of their decentralized localization (far from major R&D centers and metropolitan knowledge), and due to the relatively low level of formal education of local workers and entrepreneurs who largely derive their competences from practical experience."

"There are also many cases of district companies that actively participate—as specialized suppliers—in multilocalized value chains that are run by large multinational corporations or local medium-sized businesses."

"The key concepts emphasized by Becattini—sense sharing, search of happiness, territorial ecologies based on collective learning processes—still play an important role in a competitive scenario characterized by creative clusters, cognitive networks and global supply chains."

"Today's digital and global supply chains (creative clusters, cognitive network, operational chain) rely on such collaborative learning, but this is not a mere replication of traditional collaborative patterns that were inherited from local history."

2.2.13 Towards Digital Society Management and 'Capitalism 4.0' in Contemporary Russia [288]

This is a machine-generated summary of:

Popkova, Elena; Bogoviz, Aleksei V.; Sergi, Bruno S.: Towards digital society management and 'capitalism 4.0' in contemporary Russia [288]

Published in: Humanities and Social Sciences Communications (2021)

Link to original: https://doi.org/10.1057/s41599-021-00743-8

If you want to cite the papers, please refer to the original.

For technical reasons we could not place the page where the original quote is coming from.

Abstract-Summary

"This paper focuses on the current development of culturally unique capitalist theory in Russia, together with the practical application of proven quantitative landmarks and policy implications for managing digital society's development to optimize capitalism in Russia."

"The paper performs a quantitative analysis of the considerable influence that the digital society has on Russian capitalism and how the latter can be systemically optimized through the former's development."

"Russia is implementing a particular and unique model of capitalism model where the digital society's influence is limited and contradictory."

"We argue that the Russian model of capitalism could experience rapid development in the foreseeable future with effective digital society management."

Introduction

"This paper develops alternative economic paradigms drawn from the digital society revolution and applies them to Russia's contemporary and future model of state capitalism."

"While Russian capitalism must ensure that it can deliver economic development and high-tech modernization, our paradigms go beyond traditional market versus non-market models to propose a novel framework of development and finance in a digital society characterized by massive, powerful state-controlled companies."

"Its significance is that it offers an understanding of digital society's development as a manageable process, investigating capitalism's consequences and explicitly focusing on the Russian economy and its unique economic model."

"The paper proceeds to introduce and evaluate the following: (1) a quantitative analysis of the influence of the digital society on the designated manifestations of capitalism in Russia; (2) a systemic view on how to optimize Russian capitalism through managing the development of the digital society; and (3) the policy implications for managing the development of the digital society in order to optimize Russian capitalism."

Theoretical basis

"The following indicators for the development of the digital society in Russia are also used: – Share of households with Internet access, according to the report 'Digital Economy: 2020' published by the National Research University 'Higher School of Economics' [213]; – Digital knowledge in society (original title of the indicator: 'knowledge'), according to the IMD's World Digital Competitiveness

Rankings 2017–2019 (2020); – Share of the population without information security problems when using the Internet, according to the report 'Information society in the Russian Federation."

"2019' by the Ministry of Digital Development, Communications and Mass Media of the Russian Federation, Federal State Statistics Service, and National Research University 'Higher School of Economics' [213]; – Share of the population aged 15–74 using the Internet daily, according to the report 'Information society in the Russian Federation."

"A systemic optimization of Russian capitalism through the management of digital society's development is performed with regression analysis."

Results and discussion

"The systemic optimization of Russian capitalism through managing the development of the digital society requires the combination of higher digital knowledge in society, the share of the population without information security problems, and the share of the population aged 15–74 using the Internet on a daily basis, at which point the maximum growth of the balanced financial results of companies' activities and higher scores in the Index of Economic Freedom and Global Competitiveness Index are achieved."

"In order for this to happen, the indicators for the digital society need to have the following values: digital knowledge in society should grow from 75.96 points to 83 points, i.e., by 9.26%; the share of the population without information security problems needs to grow from 71.01% to 100%, i.e., by 36.96%; The percentage of the population using the Internet daily has to increase from 88.68% to 95%, i.e., by 7.13%; and the share of households with access to the Internet could remain unchanged, i.e., it does not require targeted management for the development of capitalism in Russia."

Policy discussion of the development of a digital society in Russia amid the COVID-19 pandemic crisis

"The policy of encouraging the development of a digital society in Russia amid the COVID-19 pandemic crisis has shown that 'capitalism 4.0' in contemporary Russia is characterized by a set of complex and contradictory features caused by globalization's new realities for scientific and technological progress and social order."

"The policy for the development of the digital society in Russia amid the COVID-19 pandemic has created new markets and regulatory incentives for the emergence of 'capitalism 4.0' in contemporary Russia, and these will continue to develop in the post-pandemic period."

Conclusions

"It has been proven that the maximum aggregate growth of the manifestations of capitalism in Russia based on the development of the digital society is up 71.75%: The balanced financial results of companies could be increased by 13.58%.The Index of Economic Freedom score could increase by 12.69%."

"Managing the development of the digital society to optimize capitalism substantiates the uniqueness of the Russian model and its specifics under the considerable influence of rapid advances in ICTs."

"Our results' practical value offers the possibility of applying the proven quantitative landmarks and the offered policy implications to manage the development of the digital society to optimize capitalism in Russia between now and 2024."

"The development of the digital society is not the sole factor influencing the evolution of modern capitalism in Russia."

2.2.14 Machine Learning Based Regional Epidemic Transmission Risks Precaution in Digital Society [289]

This is a machine-generated summary of:

Shi, Zhengyu; Qian, Haoqi; Li, Yao; Wu, Fan; Wu, Libo: Machine learning based regional epidemic transmission risks precaution in digital society [289]

Published in: Scientific Reports (2022)

Link to original: https://doi.org/10.1038/s41598-022-24670-z

If you want to cite the papers, please refer to the original.

For technical reasons we could not place the page where the original quote is coming from.

Abstract-Summary

"The contact and interaction of human is considered to be one of the important factors affecting the epidemic transmission, and it is critical to model the heterogeneity of individual activities in epidemiological risk assessment."

"We use the mobile phone signaling to track the users' trajectories and construct contact network to describe the topology of daily contact between individuals dynamically."

"The individual feature matrix extracted from contact network enables us to carry out the extreme event learning and predict the regional transmission risk, which can be further decomposed into the risk due to the inflow of people from epidemic hot zones and the risk due to people close contacts within the observing area."

Introduction

"In order to provide enough evidence at high-resolution level for policy makers to take targeted measures, heterogeneous individual level contact behaviors have been put more emphases on [214], and some intelligent technology like big-data analytics [215], artificial intelligence [216], cloud computing [217] and machine learning [218] may provide better solutions."

"During the spreading period of COVID-19, many countries have tried to launch individual tracing systems and monitor potential transmission risks through smart phones [219, 220], and Apple and Google also developed COVID-19 Alert App jointly [221]."

"As a straightforward scientific tool, complex network can effectively describe the dynamic contact topology between different individuals [222, 223], so as to obtain more micro scale discovery of epidemic transmission."

"Individual level contact network tends to show small-world and nonrandom graph properties [224, 225]."

"We estimate the town-level transmission risks for COVID-19 in Shanghai based on a high-resolution contact network compiled from nearly 7.5 million mobile phone users."

"Individual level contact behaviors are modelled by using the machine learning method."

Methods and data

"If individuals coincide with high-risk group, the risk of infection will increase, and consequently such contacts are called effective contacts; while the mutual contacts within the general group do not generate new risks of infection, such contacts are invalid."

"At time T, there are two high-risk individuals under Station 1 and they have effective contacts (red line) with other people under the same station; while the other contacts are invalid (green line)."

"People who do not have contacted with the high-risk group are not included in the contact network."

"Because the contact network describes the possible path of epidemic spreading in detail, we can further learn the transmission risk based on artificial neural network."

"High-risk group includes four kinds of people: the confirmed cases, the suspected cases, the medical isolators other than the first two and the people who once had a sojourn to epidemic hot zone."

Results

"On January 22, the effective contact frequency of each non-isolated high-risk individual was 213.36 times, which was higher than that of isolated group."

"The daily effective contact frequency of non-isolated high-risk group has been lower than that of isolators, and closed to that of general group, maintained at about 30 times per capita."

"The results show that our improved model is superior to the general neural network in extreme event prediction and can effectively identify the individuals who are included in the Shanghai CDC isolation list due to the different ways of contacts."

"Since we have labeled the isolators into two categories, the regional transmission risk can be divided into the following two kinds correspondingly: one is the risk caused by the inflow of people from epidemic hot zones, and the other is the risk caused by close contacts within Shanghai."

Discussion and conclusion

"As a whole, the precaution framework is of great significance for the regional transmission risk estimation of COVID-19 and other similar epidemics."

"Different from the previous studies on epidemic transmission through wearable devices or mobile phones, this machine learning based regional epidemic transmission risk precaution is completely bottom-up and can be used for early warning of regional epidemic on the premise of anonymity."

"Using mobile phone signaling to estimate the risk of regional epidemic spread can provide effective auxiliary information support for government policy making and epidemic prevention work with high efficiency and low cost."

"This method was proposed for regional epidemic transmission risk precaution."

"The limitation of this risk precaution is that when a large-scale and mass outbreak occurs in the city, such as the Omicron virus pandemic in Shanghai in the spring of 2022, its early warning effect will be greatly reduced."

2.2.15 Political Ideas of the Network Society: Why Digitalization Research Needs Critical Conceptual Analysis [290]

This is a machine-generated summary of:

August, Vincent: Political ideas of the network society: why digitalization research needs critical conceptual analysis [290]

Published in: Zeitschrift für Politikwissenschaft (2021)

Link to original: https://doi.org/10.1007/s41358-021-00305-z

If you want to cite the papers, please refer to the original.

For technical reasons we could not place the page where the original quote is coming from.

Abstract-Summary

"I argue for an interpretive approach to digitalization research that analyzes the concepts, narratives, and belief systems in digitalization debates."

"Many political actors and digitalization researchers follow network ideas, e.g. by claiming that the rise of a network society must lead to network governance."

"To this narrative, I argue that there are multiple visions of the digital society, each of which follows a specific pattern of epistemology, social imaginary, and political proposals."

"These competing self-interpretations must be investigated by digitalization research in order to map and evaluate different pathways into a digital society."

"It provides a systematic overview of competing governance rationalities in the digital society, enabling a critical evaluation of their potentials and proposals."

"It enhances the methodological rigor of digitalization research by reviewing the narratives researchers themselves tell."

Introduction

"Analyzing the concepts, narratives, and belief systems will provide an overview of the competing visions of a network society we encounter in political debates."

"A critical analysis of the metaphors and narratives in digitalization debates will also enhance the methodological rigor of digitalization research itself."

"To illustrate these arguments, I will analyze the rise of network concepts and narratives, since they inspire a socio-political imaginary that is particularly close to digitalization."

"In public debate and in digitalization research, political actors and scholars often argue that the rise of network technologies caused a structural transformation (Castells [226]; Barney [227]), turning our society into a "network society" that needs "network governance"."

"In the social realm, and in politics in particular, other rationalities compete with the network paradigm and its aspiration to shape the future of the digital society."

[Section 2]

"Digitalization research can draw on interpretive approaches to map political rationalities, including their social imaginaries and their divergent use of digital technologies."

"For a more comprehensive analysis, digitalization research can draw on interpretive approaches that addresses the link of social self-interpretations and governance practices through critical conceptual analysis."

"To better grasp how such an approach may enhance our understanding of competing governance rationalities in the digital society, let us have a look at two standing proposals that combine political theory, intellectual history, and the sociology of knowledge—the decentered theory of governance by political theorist Mark Bevir ([228–230]; Bevir and Rhodes [231]) and the pragmatic sociology proposed by Luc Boltanski and Laurent Thévenot [232–234] and continued by Boltanski and Chiapello [235]."

"Decentered theory and pragmatic sociology illustrate interpretive approaches that digitalization research can turn to for analyzing digital governance rationalities."

"Interpretative approaches, including critical conceptual analysis, stress that the result of social research is also a narrative account of reality that is not essentially different from the perspectives of other social actors (Rosa [236]; Bevir [229], pp.)"

The network paradigm and its political ideas

"The cybernetic metaphors of networks, systems, machines etc set up an epistemology different from humanistic or mechanistic models of reality that inspired, for instance, the political idea of a sovereign state and the adjunct philosophical idea of a sovereign subject."

"The "network society" is not (only) the result of technological advances but (also) of interventions by consultants and intellectuals who pushed cybernetic network ideas to reshape the way we view society and politics, including our concepts of subjectivity, power, and governance."

"Perpetuating this technological tradition, Stalder then uses cybernetic vocabularies such as referentiality to describe the social world and argue that the digital condition is characterized by openness, connectivity, and networks: It offers "new possibilities for constituting and connecting various human and nonhuman actors" (Stalder [237], p. 9)."

"These diagnoses from digitalization research often include a political prognosis that mirrors cybernetics' idea of diverse, flexible, and self-regulated networks."

Conclusion

"I argued for an interpretive approach to digitalization research that analyzes the concepts, narratives, and belief systems in digitalization debates."

"Such critical conceptual analysis (CCA) adds to the growing body of research that stresses the contingent relationship of (democratic) governance and (digital) technology."

"A critical conceptual analysis of the metaphors and narratives in digitalization debates enhances the reflexivity of digitalization research."

"Future research will have to trace the internal conflicts and transformations of the network paradigm suggested by developments such as the commons discourse, the digital detox discourse, or the platform discourse."

"In my brief illustration, I questioned this narrative by historicizing its use in the debate on digitalization and by demonstrating the existence of other governance rationalities that follow divergent patterns of epistemology, social imaginary, and political proposals."

"Further research will have to elaborate these sketches of governance rationalities beyond the network approach."

2.2.16 Welcome to the Digital Era—The Impact of AI on Business and Society [291]

This is a machine-generated summary of:

Polak, Petr: Welcome to the Digital Era—the Impact of AI on Business and Society [291]

Published in: Society (2021)

Link to original: https://doi.org/10.1007/s12115-021-00588-6

If you want to cite the papers, please refer to the original.

For technical reasons we could not place the page where the original quote is coming from.

Abstract-Summary

"As the news media continue to demonstrate, AI will play an increasingly important role in our daily lives, and it will exert a growing influence on national and global economies."

Introduction

"AI will have a positive impact on GDP, for example, primarily through higher labour productivity achieved by automating routine tasks, expanding employee skills and adding value to their work."

"Organizations have recognized the competitive advantages of AI and have accelerated their adoption rates and investments in the technologies."

"These organizations incorporate AI to complement rather than replace the role of their human resources."

"As a critical aspect of the Fourth Industrial Revolution, AI will inevitably affect the social status of the individual human actor and change the structure of the labour force."

"As organizations formulate and adopt strategies to incorporate AI into their operations, they must acknowledge the social dimension of their initiatives."

"As they do in assuming Corporate Social Responsibility, organizations practising TSR acknowledge the relationship between the corporate and social benefits of new technologies, including AI."

2.2.17 Reflections Towards an Agenda for Urban-Designing the Digital City [292]

This is a machine-generated summary of:

Aurigi, Alessandro: Reflections towards an agenda for urban-designing the digital city [292]

Published in: URBAN DESIGN International (2012)

Link to original: https://doi.org/10.1057/udi.2012.32

If you want to cite the papers, please refer to the original.

For technical reasons we could not place the page where the original quote is coming from.

Abstract-Summary

"This article deals with the increasingly digital augmentation of urban space and looks at how so far initiatives in this sense have been deployed in a way that is disconnected from urban design knowledge and practice."

"This, the article argues, can be achieved through attention to a series of principles, which can be part of an urban design agenda for digitally enhanced places."

Introduction

"Digitally enhanced ways to use urban space have an informal and relatively uncontrolled dimension, as described in Nancy Odendaal's account of fluid use and re-configuration of urban space and its informal economies within South African communities [238]."

"The aim of this article is to reflect on how deliberate, planned implementations of digital facilities could be framed within an urban design perspective, and could learn and benefit from theory and approaches on place-making."

"The article looks at how discourses on urban design can inform and benefit the digital augmentation of space."

"Through reflecting on implementations of digital enhancements to cities, it proposes some initial points that could be of help in looking at the growth of our 'digital' cities in a way that is more coherent with urban design and regeneration."

Old Rules, New Rules and 'False Syntaxes'

"Looking at this new world through a perspective that critically includes physical space, by analysing and comparing perceptions and fruition of wirelessly enabled physical locations, [239, p. 24] notes how All of these social and practical aspects of acting wirelessly tend to contrast against the idealised image of the person on the move, flicking between flows of information whilst walking through the public spaces of the city. (…) The two domains are operating on different structures, layered one on top of another but in many instances not working as a unified domain."

"[240], for instance, in his The Syntax of Cities published in 1977 warned against what he defined as the danger of 'selective visions' of the city and the consequent 'false syntaxes' that inform partial, skewed and misleading approaches towards the design and shaping of urban space."

Tensions in the Physical/Digital Nexus

"Digital technologies and ubiquitous computing can modify or augment urban spaces in many different ways, and it is beyond the scope of this article to attempt to make any allegedly comprehensive list or taxonomy."

"At a more conceptual level, I would identify three areas of augmentation where clear tensions are at play: scale and sense of place; the social aspects of place – especially the articulation of public and private; and the functional dimension of how urban spaces can be used, or what uses they can support."

"These dimensions are something well known and debated within urban design discourses (see, for instance, [241]), but their digital augmentation can lead to very different outcomes for 'cityness', generating problems and opportunities."

Digitally Augmented Scale and Sense of Place

"From an urban design point of view, however, this 'augmented scaling' is not uncontroversial, and poses the rather important issue of exploiting context (or failing to do so) – which is in itself closely linked to our ability to foster a sense of place through design and planning."

"Understanding context can mean, for a designer, making sense of complexity through a process of analysis and successive synthesis and a selection – again, however, arbitrary – of what makes a space a place."

"This highlights an important tension: while digital technologies can allow a potentially uncontrolled sprawl of opportunities, dimensions and scales within a vision of infinite functional versatility of urban space, designers might want to critically look at these and set limits and constraints, or simply select and highlight only specific digital opportunities that enhance that particular place and context."

Social and Movement Space, Public and Private Dimensions

"They need to pay attention – and proactively so – to the public or private roles that certain spaces can assume and fulfil, as well as to the tension between speed of movement and 'depth' of the experience of a place."

"On similar grounds, some urban design literature addresses such tensions distinguishing between 'movement' and 'social' space, where often the former is associated – with a negative connotation – to the development of fast vehicular transport and its dominance in cities – a typical 'false syntax' scenario."

"Semi-public spaces, in residential and urban design, become very valuable and enriching propositions, offering further opportunities for socialising, while feeling at ease and – in some instances – safer."

"The design and placement of public information terminals in many cities around the world reveals, for instance, an inability to think in terms of both the tensions of movement/social space, and the public/private one."

Functions and Uses (Functional Dimension of How Urban Spaces Can Be Used, or What Uses They Can Support)

"Whether through the 'hard' programming of trying to impose specific behaviours – and inhibit others – or the 'softer' design of facilitative spaces allowing a variety of possibilities of events and uses, space embeds visions, socio-political and economic agendas and, especially in the case of public space, plays an important role in the enhancement of the public sphere of our cities."

"The deployment of high technologies in urban spaces cannot of course be exempt from such considerations, and it would be dangerously reductive to look at any urban e-initiative – whether it is the relatively elementary development of e-government services or the design of complex media and information-augmented spaces – as simply a value-neutral, solution-driven, practical and utilitarian intervention."

"As a networked tablet device can be seen as embedding virtually limitless and diverse functions and roles, civic space – augmented by high technology – can also inherit this versatility to become a multi-functional and configurable convergent space, within a convergent city."

Towards Designing Digital Place as a Whole

"The recent history of digital urbanism has, more often than not, been characterised by the parallel development of two strands: a service-based, utilitarian set of city e-government and management initiatives, opposed to a conceptual, provocative set of artistic or quasi-artistic installations, aiming to critique and re-frame the inhabitation of space in a 'digital' society."

"The administration, management and community support systems that have been developed for years under the banners of 'e-government', the 'civic network' or the 'digital city' tend to be rather a-spatial, often based on a variety of paradigms (as [242] pointed out) that stem from computing and administration, but have little to do with urban space."

"This – I would argue – can highlight the architectural and urban design challenge of pursuing a 'unity' between the everyday and the sublime, or indeed making the everyday sublime and vice versa: something that the digital augmentation of urban space cannot ignore."

Concluding Notes

"[243, p. 80] claim that 'Augmented space provides a challenge and opportunities for many architects to rethink their practice, since architecture will have to take into account that layers of contextual information will overlay the built space' clearly holds true and relevant, but can be reciprocally completed by noting how digital deployments need as well to benefit from architectural and spatial knowledge, hence the overall need for a place-making, holistic perspective."

"This article therefore adds up to a growing debate of how ICT should be part of a renewed agenda for place-making, and proposes some elements that can inform that agenda, such as the concept of 'centripetal digital trajectories' or the need to have a digital 'augmented' design that is socially useful."

"All, there remains a stronger-than-ever need for multi-disciplinary research and practice, and a solid grounding of this into urban design knowledge, which can offer a good way of setting 'place' as a holistic starting point for the development of a quality-informed digital city."

2.2.18 Open Data Governance: Civic Hacking Movement, Topics and Opinions in Digital Space [293]

This is a machine-generated summary of:

Maretti, Mara; Russo, Vanessa; del Gobbo, Emiliano: Open data governance: civic hacking movement, topics and opinions in digital space [293]

Published in: Quality & Quantity (2020)

Link to original: https://doi.org/10.1007/s11135-020-01045-y

If you want to cite the papers, please refer to the original.

For technical reasons we could not place the page where the original quote is coming from.

Abstract-Summary

"On this topic, using the semantic network analysis method, the research aims to investigate the communication structure and the governance of open data in the Twitter conversational environment."

"The research questions are: (1) Who are the main actors in the Italian open data infrastructure? (2) What are the main conversation topics online? (3) What are the pros and cons of the development and use (reuse) of open data in Italy?"

"To answer these questions, we went through three research phases: (1) analysing the communication network, we found who are the main influencers; (2) once we found who were the main actors, we analysed the online content in the Twittersphere to detect the semantic areas; (3) then, through an online focus group with the main open data influencers, we explored the characteristics of Italian open data governance."

Introduction

"Though, the process of informational innovation for Italian institutions is as detailed in Decreto Crescita (Growth Decree) 2.0, a document which sets out the evolution of Italian digital administration in the following terms: (1) the creation of an Agency for Digital Italy; (2) the establishment of a digital agenda; (3) the introduction of new electronic media capable of ensuring certified communication between one government department and another, and between those departments and specific users; (4) regulation of OD; (5) digitalisation of traditionally paper-based documents used for institutional communication (medical prescriptions, sick notes) and of academic textbooks; (6) abolition of the digital divide; (7) promotion of electronic currency in order to facilitate payments, while at the same time ensuring security and enabling traceability of economic transaction; (8) encouragement of strategic research and innovation projects; (9) creation of a technical committee of smart communities; and (10) specification of the formal requirements for the creation of start-ups (innovative businesses)."

Results

"Using a qualitative analysis of the online conversation, we can focus on four main topics: the nature of the OD infrastructure in Italy, considered mainly as a public service; the potential impact of this service on democracy (participation and transparency); the management of the informational infrastructure (governance and reuse problems);and the opportunity for accountability and policy evaluation."

"OD is considered to be a true informational infrastructure in the service of the general public and of democracy, and as such capable of initiating social development not only in the field of public administration (increased effectiveness, transparency and efficiency, and greater relevance of policy systems and services), but also for businesses, civil society organisations and individual citizens: One of the things that public administration finds most difficult to understand is that a genuine appreciation of the informational assets possessed by the general public could reduce administrative expenses by facilitating and accelerating the exchange of information between public administrations, thereby rendering the whole system more efficient, which in the final analysis would benefit the associated market economy. (Focus group participant)."

Conclusions

"The civic hacking groups collaborate with national and local institutions in the promotion of reuse projects, and training and support for administrations."

"Especially in the digital space, OD and ODG are configured as important influencers, sharing information and discussing conference initiatives, news and reuse projects."

"Despite the leading role of the civic hackers community, it emerges, mainly from the online focus group, that the technological and cultural shortcomings of the Italian bureaucratic model are often resistant to innovations."

"There are many technical problems (Vetrò and others [244]) with the quality of the data and the information infrastructure, in particular: data are not in a well-defined accessible format; there is an evident absence of standards; there is a lack of support to make data available; there is inaccurate information; data are obsolete or invalid; and there is a lack of territorial consistency, with the same data held in different forms."

2.2.19 Smart Environments and Techno-centric and Human-Centric Innovations for Industry and Society 5.0: A Quintuple Helix Innovation System View Towards Smart, Sustainable, and Inclusive Solutions [294]

This is a machine-generated summary of:

Carayannis, Elias G.; Dezi, Luca; Gregori, Gianluca; Calo, Ernesto: Smart Environments and Techno-centric and Human-Centric Innovations for Industry and Society 5.0: A Quintuple Helix Innovation System View Towards Smart, Sustainable, and Inclusive Solutions [294]

Published in: Journal of the Knowledge Economy (2021)

Link to original: https://doi.org/10.1007/s13132-021-00763-4

If you want to cite the papers, please refer to the original.

For technical reasons we could not place the page where the original quote is coming from.

Abstract-Summary

“In the smart complex environments, a systemic vision of the elements, which act and are acted within a given territory, should be the basis of a hypothesis of joint growth.”

“Through the evidence emerged from an important case study and the application of an MCDA methodology, we have tried to identify which are the optimal solutions for the implementation of the new human-centric logics of I5.0, analyzing them on the basis of the actual benefits for the ecosystem, going beyond the self-referential aptitude of the firm to instill technological changes and managerial visions.”

“Knowledge circulation, dialogue between sub-systems, and the ability to adapt technology and entrepreneurial strategies to the environment in which it operates (with the users as first stakeholders) seem to be necessary practices in knowledge-based innovation, prioritization, and decision-making processes, for smart, sustainable, and inclusive solutions.”

Introduction

“Innovation and knowledge in smart environments should be the result of a sharing process that involves all the actors of an ecosystem, interpreting complexity as an opportunity and not as a threat.”

“To investigate these dynamics of smart governance, knowledge, and decision-making in complex organizations, we have chosen to observe the “smart” realities of airport environments.”

“We also believe that our studies may contribute to the discussion about management of innovation and decision-making policies within knowledge management, both in the specific observed sector and in other complex environments.”

“The paper is organized as follows: the second section gives an overview of scientific literature dealing with open innovation and innovation systems in complex and smart environments, focusing attention on the participatory dynamics in which the firm is inserted and from which it must rethink its decision-making policies and knowledge exploration and exploitation practices.”

Literature Review

"With regard to knowledge management processes, the open innovation assumes that useful knowledge is widely distributed and that "even the most capable R&D organizations must identify, connect to, and leverage external knowledge sources as a core process innovation" (Chesbrough and others, [245])."

"When it comes to innovation ecosystems and open perspectives, the novel paradigms of Industry 5.0 (Carayannis and others, [246]; EU Report on Industry 5.0, 2021) and Society 5.0 (Onday, [247]; Fukuyama, [248]) can be considered as the answer to the demand of a renewed human-centered/human-centric industrial paradigm, starting from the (structural, organizational, managerial, knowledge-based, philosophical, and cultural) reorganization of the production processes to then generate positive implications first within the business perspectives and secondly towards all the components belonging to the ecosystem."

"In an open perspective, there is the need for developing new and different metrics as well as composite indicators for assessing the performance of a firm's innovation process, by integrating the classic metrics, strictly connected to internal R&D and product/service development (including the ex-post customer satisfaction), with those metrics that can assess the long run absorptive capacity of the organization."

Research Methodology

"To do so, we have chosen to analyze the airport public sector, as a sector made up of complex smart organizations in which many actors operate in a systemic perspective: government (local, regional, national, and international institutions), industry (various suppliers, direct partners, airline companies), university (direct and indirect value co-creation), civil society (workers and highly skilled employees, passengers and tourists, local communities, interest groups), and general environmental context (environment as a local and global resource)."

"Within this particular fervent sector, in order to simulate a data collection on actual experiences and expectations expressed by end users, useful for a firm to include their preferences in decision-making processes, we have monitored the travelers of Rome Fiumicino Leonardo Da Vinci international airport."

"After assigning specific weights to the selection criteria (QH = 0.290; desirability = 0.428; feasibility = 0.218; profitability = 0.064), a multi-criteria decision analysis (MCDA) with analytical hierarchy process (AHP) has been conducted, in order to simulate an in-itinere prioritization process guided by a systemic human-centric innovation model."

Results and Discussion

"The former refer to those priorities of the travelers that recall specific items of an airport experience based on primary services, i.e., services directly related to the core activities of the airport in connection with the airline companies (industry system) and the territory (environment system)."

"55.9% of interviewees appreciate Smart booking/payment/check-in services, 52.5% are interested in an effective connection between the airport and the city, and 50.9% consider a guaranteed secure environment as a prior aspect."

"Outer-directed users (10.5%) The fourth target group is related to those users who embrace the whole airport supply services as they are."

"Need-directed users (20.2%) The fifth target group is characterized by strong values of low satisfaction, both for the overall satisfaction level and the single items."

Conclusions

"Institutions could intervene with their support in a more targeted way, being able to evaluate more accurately which solutions offer greater added value to the ecosystem; the organizations practitioners could base their decisions on the exploitation of the knowledge possessed and on the spill-over effects that are expected, activating virtuous dynamics able to improve the prioritization processes and the respective time-oriented strategic scheduling, while academics can start from the proposed model to introduce new open innovation theoretical advances as well as practical assessment metrics (Birchall and others, [249]; Cruz-Cázares and others, [250]; Edison and others, [251]; Hoelscher & Schubert, [252])."

"In light of the above, we can state that the optimal management of smart environments like the airport one should be extended to a large number of dimensions, regarding inside, outside, and beyond the physical and visible structure (but even beyond technology, which cannot constitute the exclusive dimension of the nowadays innovation processes) focusing on a people-culture-technology dynamic and system-centric perspective (Carayannis & Alexander, [253])."

Bibliography

1. Malle BF, Guglielmo S, Monroe AE (2014) A theory of blame. Psychol Inq 25(2):147–186. https://doi.org/10.1080/1047840X.2014.877340
2. Hanson B (1995) General systems theory beginning with wholes. Taylor & Francis, Washington
3. Rifkin J (2011) The Third Industrial Revolution: how lateral power is transforming energy, the economy, and the world. Palgrave Macmillan, New York
4. Freeman C, Perez C (1988) Structural crisis of adjustment, business cycles and investment behaviour. In: Dosi G, Freeman C, Nelson R, Silverberg G, Soete L (eds) Technical Change and Economic Theory. Pinter, London
5. Freeman C, Louçã F (2001) As time goes by from the industrial revolution to the information revolution. Oxford University Press, Oxford
6. Perez C (2002) Technological revolutions and finance capital: the dynamics of bubbles and golden ages. Edward Elgar, Cheltenham
7. Schwab K, Davis N (2018) Shaping the future of the fourth industrial revolution. Currency, New York
8. Chandler AD Jr (1977) The visible hand. Harvard University Press, Cambridge
9. Hounshell DA (1984) From the American System to Mass Production, 1800–1932: The Development of Manufacturing Technology in the United States. Johns Hopkins University Press, Baltimore

10. Isaacson W (2014) The innovators: how a group of hackers, geniuses, and geeks created the digital revolution. Simon & Schuster, New York
11. Arthur WB (2014) Complexity and the Economy. Oxford University Press, Oxford
12. Perez C (2013) Unleashing a golden age after the financial collapse: drawing lessons from history. Environ Innov Soc Trans 6:9–23
13. Knell M (2013) Multi-source energy networks and the ICT revolution. Eur Plan Stud 21:1838–1852
14. Kay P (2014) Rome's economic revolution. Cambridge
15. Thomson R (2009) Structures of change in the mechanical age: technological innovation in the United States, 1790–1865. Johns Hopkins, University Press
16. Coe NM, Yeung HWC (2015) Global production networks: theorizing economic development in an interconnected world. Oxford University Press, Oxford
17. Zander I (2002) The formation of international innovation networks in the multinational corporation: an evolutionary perspective. Ind Corp Chang 11(2):327–353
18. Tomasello MV, Napoletano M, Garas A, Schweitzer F (2016) The rise and fall of R&D networks. Ind Corp Chang 26:1–30
19. Graf H (2012) Inventor networks in emerging key technologies: information technology vs. semiconductors. J Evol Econ 22:459–480
20. Fritsch M, Zoellner M (2020) The fluidity of inventor networks. J Tech Tran 45:1063–1087
21. von Hippel E (2007) Horizontal innovation networks—by and for users. Ind Corp Chang 16(2):293–315
22. Ernst D (2009) A new geography of knowledge in the electronics industry?, Policy Studies, No. 54. East-West Center, Honolulu
23. Barnard H, Chaminade C (2011) Global innovation networks: towards a taxonomy, WP 2011/04, Centre for Innovation, Research and Competence in the Learning Economy (CIRCLE). Lund University
24. Camagni R (ed) (1991) Innovation networks: spatial perspectives. Belhaven Press, London
25. Boschma R (2005) Proximity and innovation: a critical assessment. Reg Stud 39(1):61–74
26. Lundvall B-Å, Borràs S (1998) The globalising learning economy: implications for innovation policy. European Commission: http://www.cordis.lu/tser/src/globec.htm
27. Pyka A (1997) Informal networking. Technovation 17:207–220
28. Pyka A, Scharnhorst A (eds) (2009) Innovation networks: new approaches in modelling and analyzing. Springer, Berlin
29. Kirman A (2011) Complex economics: individual and collective rationality. Routledge, London
30. Roco MC, Bainbridge WS (2003) Converging technologies for improving human performance: nanotechnology, biotechnology, information technology, and cognitive science. Springer, Netherlands, Heidelberg
31. Miller R (2018) Transforming the future: anticipation in the 21st Century. Routledge, London
32. Brunon-Ernst, A. (2013b). Deconstructing panopticism into the plural panopticons. In A. Brunon-Ernst (Ed.), Beyond Foucault: New perspectives on Bentham's panopticon (pp. 17–42). Surrey: Ashgate Publishing.
33. Haggerty, K. D., & Ericson, R. V. (2000). The surveillant assemblage. British Journal of Sociology, 51(4), 605–22.
34. Marx, G. T. (2002). What's new about the "New surveillance"? Classifying for change and continuity. Society, 1(1), 9 29
35. Debord, Guy. 2002. Society of the spectacle. USA: Black & Red.
36. Baudrillard, Jean. 1994. Simulacra and Simulation. Michigan: University of Michigan Press.
37. Horkheimer, Max, and Theodor W. Adorno. 2002. Dialectic of enlightenment. Philosophical fragments. Stanford: Stanford University Press
38. Heidegger, Martin. 1976. What is called thinking. New York: Harper Perennial.
39. Adorno, Theodor W. 2003. The jargon of authenticity. New York: Routledge.
40. Finn, Ed. 2017. What algorithms want. Imagination in the age of computing. USA: MIT Press

41. Chun, Wendy Hui Kyong. 2011. Programmed visions. Software and memory. USA: MIT Press
42. Geraci, Robert. 2010. Apocalyptic AI. Visions of heaven in robotics, artificial intelligence, and virtual reality. New York: Oxford University Press
43. Latzer, Michael. 2013a. Media convergence. In Handbook of the Digital Creative Economy, 123–133. Cheltenham: Edward Elgar
44. Frenken, Koen. 2006. Innovation, Evolution and Complexity Theory. Cheltenham: Edward Elgar.
45. Ziman, John. 2000. Technological Innovation as an Evolutionary Process. Cambridge: Cambridge University Press.
46. van Dijck, José. 2014. Datafication, dataism and dataveillance: Big Data between scientific paradigm and ideology. Surveillance & Society 12:197–208.
47. Bostrom, Nick. 2005. A History of Transhumanist Thought. Journal of Evolution & Technology 14:1–25.
48. Tirosh-Samuelson, Hava. 2010. Engaging Transhumanism. In H ± Transhumanism and Its Critics, eds. Gregory R. Hansell and William Grassie, 19–52. Philadelphia: Metanexus Institute.
49. Hurlbut, J. Benjamin. 2016. Technologies of Imagination. In Religion and Innovation, ed. Donald A. Yerxa, 213–228. London: Bloomsbury Academic.
50. Morozov, Evgeny. 2013. To Save Everything, Click Here. New York: PublicAffairs.
51. DiMaggio, P. (2001). Social implications of the internet. Annual Review of Sociology, 27, 307–336.
52. Maddison A (2007) Contours of the world economy 1–2030 AD: essays in macro-economic history. Oxford University Press, Oxford
53. Brent E (2017) Artificial intelligence/expert systems in online research. In: Fielding N, Lee R, Blank G (eds) The sage handbook of online research methods. Sage, London, pp 361–379
54. Abad-Segura, E., González-Zamar, M. D., Infante-Moro, J. C., & Ruipérez García, G. (2020). Sustainable management of digital transformation in higher education: Global research trends. Sustainability, 12(5), 2107.
55. Karlusch, A., Sachsenhofer, W., & Reinsberger, K. (2018). Educating for the development of sustainable business models: Designing and delivering a course to foster creativity. Journal of Cleaner Production, 179, 169–179.
56. Rof, A., Bikfalvi, A., & Marquès, P. (2020). Digital transformation for business model innovation in higher education: Overcoming the tensions. Sustainability, 12(12), 4980.
57. Davidson, J., Prahalad, V., & Harwood, A. (2021). Design precepts for online experiential learning programs to address wicked sustainability problems. Journal of Geography in Higher Education, 45(3), 319–341.
58. Gómez-Zermeño, M. G. (2020). Massive open online courses as a digital learning strategy of education for sustainable development. Journal of Sustainable Development of Energy, Water and Environment Systems, 8(3), 577–589.
59. Dziubaniuk, O., & Nyholm, M. (2020). Constructivist approach in teaching sustainability and business ethics: A case study. International Journal of Sustainability in Higher Education, 22(1), 177–197.
60. Filho, L. W., Raath, S., Lazzarini, B., Vargas, V. R., de Souza, L., Anholon, R., Quelhas, O. L. G., Haddadg, R., Klavinsh, M., & Orlovic, V. L. (2018). The role of transformation in learning and education for sustainability. Journal of Cleaner Production, 199, 286–295.
61. Sammalisto, K., Sundström, A., & Holm, T. (2015). Implementation of sustainability in universities as perceived by faculty and staff—A model from Swedish university. Journal of Cleaner Production, 106, 45–54.
62. Siemens, G. (2005). Connectivism: A learning theory for the digital age. International Journal of Instructional Technology and Distance Learning, 2(1), 3–10.
63. Ivanova-Gongne, M., Torkkeli, L., Hannibal, M., Uzhegova, M., Barner-Rasmussen, W., Dziubaniuk, O., & Kulkov, I. (2022). Cultural sensemaking of corporate social responsibility:

A dyadic view of Russian-Finnish business relationships. Industrial Marketing Management, 101, 153–164.
64. Butler J. Performative acts and gender constitution: an essay in phenomenology and feminist theory. In: Case S, editor. Performing feminisms: feminist critical theory and theatre. Baltimore: Johns Hopkins UP; 1990
65. Devlin, A. M., McGee-Lennon, M., O'Donnell, C. A., Bouamrane, M.-M., Agbakoba, R., O'Connor, S., et al. (2016). Delivering digital health and well-being at scale: lessons learned during the implementation of the Dallas program in the United Kingdom. Journal of the American Medical Informatics Association, 23, 48–59. https://doi.org/10.1093/jamia/ocv097
66. Greenwald, A. G. (2012). There is nothing so theoretical as a good method. Perspectives on Psychological Science, 7, 99–108. https://doi.org/10.1177/1745691611434210
67. Hekler, E. B., Michie, S. F., Rivera, D. E., Collins, L. M., Pavel, M., Jimison, H., Garnett, C., Parral, S., Spruijt- Metz, D. (2016b). Advancing models and theories for digital behavior change interventions. American Journal of Preventive Medicine, 51, pp. 825–832, https://doi.org/10.1016/j.amepre.2016.06.013
68. Carson, T. L., Eddings, K. E., Krukowski, R. A., Love, S. J., Harvey-Berino, J. R., & West, D. S. (2013). Examining social influence on participation and outcomes among a network of behavioral weight-loss intervention enrollees. Journal of Obesity, 2013, 1–8. https://doi.org/10.1155/2013/480630
69. Leahey, T. M., Kumar, R., Weinberg, B. M., & Wing, R. R. (2012). Teammates and social influence affect weight loss outcomes in a team-based weight loss competition. Obesity, 20, 1413–1418. https://doi.org/10.1038/oby.2012.18
70. Maher, C., Ferguson, M., Vandelanotte, C., Plotnikoff, R., De Bourdeaudhuij, I., Thomas, S., Nelson-Field, K., Olds, T. (2015). A web-based, social networking physical activity intervention for insufficiently active adults delivered via Facebook app: Randomized controlled trial. Journal of Medical Internet Research, 17, e174. https://doi.org/10.2196/jmir.4086
71. Creswell, J. W. (2003). Research design: Qualitative, Quantitative, and mixed methods approaches. Thousand Oaks: Sage Publications.
72. Creswell, J. W., & Clark, V. L. P. (2007). Designing and conducting mixed methods research. Thousand Oaks: Sage Publishing.
73. Føllesdal, D. (2001). Hermeneutics. International Journal of Psychoanalysis, 82, 375–379.
74. Bordwell, D. (2008). The Poetics of cinema. New York, NY: Routledge.
75. Perron, B., Arsenault, D., Picard, M., & Therrien, C. (2008). Methodological questions in "interactive film studies". New Review of Film and Television Studies, 6(3), 233–252.
76. Allen, J.G. 2021. Law's Virtual Empires: Game Analogies and the Concept of Law. In Conceptual Jurisprudence-Methodological Issues, Conceptual Tools and New Approaches, ed. J.L. Fabra-Zamora and G. Villa Rosas. Advanced online publication. https://doi.org/10.2139/ssrn.3720498
77. Fairfield, J. 2005. Virtual Property. Boston University Law Review 1047(85): 1048–1102.
78. Lastowka, G., and D. Hunter. 2004. The Laws of The Virtual Worlds. California Law Review 92(1): 1–73. https://doi.org/10.2307/3481444 .
79. Durante, M. 2022. Technology and the Ontology of the Virtual. In The Oxford Handbook of Philosophy of Technology, ed. S. Vallor, 318–340. Springer
80. Hage, J. 2018. Of Norms. In Handbook of Legal Reasoning and Argumentation, ed. G. Bongiovani, G. Postema, A. Rotolo, G. Sartor, C. Valentini, and D. Walton, 103–138. Springer
81. Alexander, J. 2006. The civil sphere. Oxford: Oxford University Press.
82. Rapley, T. 2014. Sampling strategies in qualitative research. In The Sage handbook of qualitative data analysis, ed. U. Flick, 49–63. Berlin: Sage Publications.
83. Farid, Z. 2019. Why would one say I hate social media? Social Champ, June 4. https://www.socialchamp.io/blog/i-hate-social-media/. Accessed 28 June 2021
84. Goodman, P. 2021. 20 reasons not to use social media. Soapboxie, May 25. https://soapboxie.com/social-issues/Reasons-Not-to-Use-Social-Media. Accessed 6 June 2021

85. Danila. 2018. One-week social media detox: how to survive in your real life. Wannabe Globetrotter, August 8. https://www.wannabeaglobetrotter.it/en/one-week-social-media-detox/. Accessed 25 May 2021
86. Wehner, J. (1997). Interaktive Medien - Ende der Massenkommunikation? Zeitschrift für Soziologie, 26(2), 96–114.
87. Friesen, N. (2014). Telepresence and Tele-absence. A phenomenology of the (in)visible Alien online. Phenomenology and Practice, 8, 17–31.
88. Zhao, S. (2015). Constitution of mutual knowledge in telecopresence. Updating Schutz's phenomenological theory of the lifeworld. Journal of Creative Communications, 10(2), 105–127
89. Krotz, F., Despotović, C., & Kruse, M.-M. (Eds.). (2014). Die Mediatisierung sozialer Welten. Synergien empirischer Forschung. Wiesbaden: Springer VS
90. Leenhardt, M. (1947/1979). Do kamo: Person and Myth in the Melanesian world (B.M. Gulati, Rrans). Chicago & London: The University of Chicago Press
91. Srubar, I. (1979). Die Theorie der Typenbildung bei Alfred Schütz: Ihre Bedeutung und ihre Grenzen. In W. M. Sprondel & R. Grathoff (Eds.), Alfred Schütz und die Idee des Alltags in den Sozialwissenschaften (pp. 43–64). Stuttgart: Ferdinand Enke Verlag.
92. Gurwitsch, A. (1964). The field of consciousness. Duquesne University Press
93. Garfinkel, H. (1967). Studies in ethnomethodology. Prentice-Hall.
94. Donaldson, T., & Dunfee, T. W. (1999). Ties that bind: A social contracts approach to business ethics. Cambridge, MA: Harvard Business Press.
95. Windsor, D. (2018). Dynamics for integrative social contracts theory: Norm evolution and individual mobility. Journal of Business Ethics, 149(1), 83–95.
96. LaFollette, H., & Persson, I. (Eds.). (2013). The Blackwell guide to ethical theory (2nd ed.) Wiley-Blackwell
97. Ribble, M. (2007). Digital citizenship in schools: Nine elements all students should know. International Society for Technology in Education
98. Polizzi, G., & Harrison, T. (2020). Integrating cyber-wisdom education into the school curriculum. Jubilee Centre for Character and Virtues, University of Birmingham. https://www.jubileecentre.ac.uk/userfiles/jubileecentre/pdf/insight-series/GP_TH_IntegratingCyber-WisdomEducationintotheSchoolCurriculumFinal.pdf
99. Bandura, A. (1999). Moral disengagement in the perpetration of inhumanities. Personality and Social Psychology Review, 3(3), 193–209. https://doi.org/10.1207/s15327957pspr0303_3 .
100. D'Errico, F., & Paciello, M. (2018). Online moral disengagement and hostile emotions in discussions on hosting immigrants. Internet Research, 28(5), 1313–1335. https://doi.org/10.1108/IntR-03-2017-0119 .
101. Ge, X. (2020). Social media reduce users' moral sensitivity: Online shaming as a possible consequence. Aggressive Behavior, 46(5), 359–369. https://doi.org/10.1002/ab.21904 .
102. Kyriacou, C., & Zuin, A. (2018). Cyberbullying bystanders and moral engagement: A psychosocial analysis for pastoral care. Pastoral Care in Education: An International Journal of Personal, Social and Emotional Development, 36(2), 9–111. https://doi.org/10.1080/02643944.2018.1453857 .
103. Price, D., Green, D., Spears, B., Scrimgeour, M., Barnes, A., Geer, R., & Johnson, B. (2013). A qualitative exploration of cyber-bystanders and moral engagement. Australian Journal of Guidance and Counselling, 24(1), 1–17. https://doi.org/10.1017/jgc.2013.18 .
104. Chai-Arayalert S, Suttapong K, Kang-Rithi W. Systematic approach to preservation of cultural handicrafts: Case study on fabrics hand-woven in Thailand. Cogent Business and Management. 2021;8(1):1872889. https://doi.org/10.1080/23311975.2021.1872889 .
105. Zhao Y, Zhou YL. A review of the international research on digital preservation of intangible cultural heritage. Library. 2017;8:59–68 (in Chinese).
106. Meier, P. (2012a). Crisis mapping in action: How open source software and global volunteer networks are changing the world, one map at a time. Journal of Map & Geography Libraries: ADVANCES in Geospatial Information, Collections & Archives, 8(2), 89–100.
107. Feenberg, A. (1999). Questioning technology. NY: Routledge.

108. Pinch, T., & Bijker, W. (1987). The social construction of facts and artifacts: Or how the sociology of science and the sociology of technology might benefit each other. In W. Bijker, T. Hughes, & T. Pinch (Eds.), The social construction of technological systems: New directions in the sociology and history of technology. Cambridge, MA: MIT Press
109. Wajcman, J. (1991). Feminism confronts technology. University Park, PA: Pennsylvania State University Press.
110. Winner, L. (1985). Do artifacts have politics? In D. MacKenzie & J. Wajcman (Eds.), The social shaping of technology (pp. 26–38). Philadelphia, PA: Open University Press.
111. Burns, R. (2014). Moments of closure in the knowledge politics of digital humanitarianism. Geoforum, 53, 51–62.
112. Burawoy, M. (1998). The extended case method. Sociological theory, 16(1), 4–33
113. Wright, K.B. 2016a. Communication in health-related online social support groups/communities: A review of research on predictors of participation, applications of social support theory, and health outcomes. Review of Communication Research 4: 65–87.
114. Becker, K.L. 2013. Cyberhugs: Creating a voice for chronic pain sufferers through technology. Cyberpsychology, Behavior, and Social Networking 16 (2): 123–126.
115. Brosseau, L., et al. 2015. Internet-based implementation of non-pharmacological interventions of the "People getting a grip on arthritis" educational program: An international online knowledge translation randomized trial design protocol. JMIR Research Protocols 4 (1): e19.
116. Fatima, L., et al. 2015. Smart CDSS: Integration of social media and interaction engine (SMIE) in healthcare for chronic disease patients. Multimedia Tools and Applications 74 (14): 5109–5129.
117. Grosberg, D., et al. 2016. Frequent surfing on social health networks is associated with increased knowledge and patient health activation. Journal of Medical Internet Research 18 (8): e212.
118. Lavorgna, L., et al. 2017. Health-related coping and social interaction in people with multiple sclerosis supported by a social network: Pilot study with a new methodological approach. Interactive Journal of Medical Research 6 (2): e10.
119. Magnezi, R., et al. 2014. Online activity and participation in treatment affects the perceived efficacy of social health networks among patients with chronic illness. Journal of Medical Internet Research 16 (1): e12.
120. Nordfelt, S., et al. 2010. Patient and parent views on a Web 2.0 diabetes portal-the management tool, the generator, and the gatekeeper: Qualitative study. Journal of Medical Internet Research 12 (2): e17
121. Rus, H.M., and L.D. Cameron. 2016. Health communication in social media: Message features predicting user engagement on diabetes-related Facebook pages. Annals of Behavioral Medicine 50 (5): 678–689.
122. Merolli, M., et al. 2015. Using social media while waiting in pain: A clinical 12-week longitudinal pilot study. JMIR Research Protocols 4 (3): e101.
123. Milani, R.V., and C.J. Lavie. 2015. Health care 2020: Reengineering health care delivery to combat chronic disease. The American Journal of Medicine 128: 337–343.
124. McKenna, J. 2017. Social media: a powerful tool for physicians and patients. European Heart Journal 38 (7): 469–470.
125. Jervelund, S.S. 2018. How social media is transforming the spreading of knowledge: Implications for our perceptions concerning vaccinations and migrant health. Scandinavian Journal of Public Health 46: 167 169.
126. Hoye, J. Matthew, and Jeffrey Monaghan. 2018. Surveillance, freedom and the republic. European Journal of Political Theory 17:343–363. https://doi.org/10.1177/1474885115608783 .
127. Helm, Paula, and Sandra Seubert. 2020. Normative paradoxes of privacy: literacy and choice in platform societies. Surveillance & Society 18:185–198. https://doi.org/10.24908/ss.v18i2.13356

128. Schulz, Daniel. 2021a. Politische Theorie des Datenschutzes. Ein Beitrag zur Mischverfassung der Moderne. Zeitschrift für Politische Theorie 12:84–106. https://doi.org/10.3224/zpth.v12i1.06 .
129. Fourcade, Marion. 2016. Ordinalization. Sociological Theory 34:175–195. https://doi.org/10.1177/0735275116665876 .
130. Beer, David. 2016. Metric power. London: Palgrave Macmillan.
131. Mau, Steffen. 2017. Das metrische Wir: Über die Quantifizierung des Sozialen. Berlin: Suhrkamp
132. Gerbaudo, Paolo. 2019. The digital party: political organisation and online democracy. London: Pluto Press.
133. Manow, Philip. 2020. (Ent-)Demokratisierung der Demokratie. Berlin: Suhrkamp.
134. Berg, Sebastian, Veza Clute-Simon, Rebecca-Lea Freudl, Niklas Rakowski, and Thorsten Thiel. 2021. Civic Hackathons und der Formwandel der Demokratie. Politische Vierteljahresschrift 62:621–642. https://doi.org/10.1007/s11615-021-00341-y .
135. Winner, Langdon. 1980. Do Artifacts Have Politics? Daedalus 109:121–136.
136. Latour, Bruno. 1990. Technology is society made durable. The Sociological Review 38:103–131.
137. Dickel, Sascha. 2019. Prototyping Society – Zur vorauseilenden Technologisierung der Zukunft. Bielefeld: transcript
138. Berg, Sebastian, Niklas Rakowski, and Thorsten Thiel. 2020b. The digital constellation. WI—Weizenbaum Institute for the Networked Society https://doi.org/10.34669/wi.ws/14
139. Ellerich-Groppe, Niklas. 2021. Zwischen neuer Solidarität und Entsolidarisierung – Der Sozialstaat angesichts des digitalen Wandels. Zeitschrift für Politikwissenschaft https://doi.org/10.1007/s41358-021-00300-4 .
140. Bibri, S. E. (2022). The Social Shaping of the Metaverse as an Alternative to the Imaginaries of Data-Driven Smart Cities: A Study in Science, Technology, and Society, Smart Cities 5(3):832–874. https://doi.org/10.3390/smartcities5030043.
141. Bibri, S. E., and Allam, Z., (2022a) The Metaverse as a Virtual Form of Data-Driven Smart Cities: The Ethics of the Hyper-connectivity, Datafication, Algorithmization, and Platformization of Urban Society, Computational Urban Science 2, 22. https://doi.org/10.1007/s43762-022-00050-1
142. Calvo, P. (2020). The ethics of Smart City (EoSC): moral implications of hyperconnectivity, algorithmization and the datafication of urban digital society. Ethics and Information Technology, 22(2), 141–149. https://doi.org/10.1007/s10676-019-09523-0
143. Kaluarachchi, Y. 2022. "Implementing Data-Driven Smart City Applications for Future Cities" Smart Cities 5, no. 2: 455–474. https://doi.org/10.3390/smartcities5020025
144. Sarker, M.N.I, Khatun, M.N., Alam, G.M, and Islam, M.S. Big Data Driven Smart City: Way to Smart City Governance, in: 2020 International Conference on Computing and Information Technology (ICCIT-1441), Tabuk, Saudi Arabia, 2020, pp. 1–8, https://doi.org/10.1109/ICCIT-144147971.2020.9213795
145. Bui TD, Tseng ML (2022) Understanding the barriers to sustainable solid waste management in society 5.0 under uncertainties: a novelty of socials and technical perspectives on performance driving. Environ Sci Pollut Res 29(11):16265–16293. https://doi.org/10.1007/s11356-021-16962-0
146. Mavrodieva AV, Shaw R (2020) Disaster and climate change issues in Japan's Society 5.0—a discussion. Sustainability 12(5):1893. https://doi.org/10.3390/su12051893
147. Roblek V, Meško M, Bach MP, Thorpe O, Šprajc P (2020) The interaction between internet, sustainable development, and emergence of society 5.0. Data 5(3):80. https://doi.org/10.3390/data5030080
148. De Felice F, Travaglioni M, Petrillo A (2021) Innovation trajectories for a Society 5.0. Data 6(11):115. https://doi.org/10.3390/data6110115
149. Pereira AG, Lima TM, Charrua-Santos F (2019) Society 5.0 as a result of the technological evolution: historical approach. In International Conference on Human Interaction and Emerging Technologies 700–705. https://doi.org/10.1007/978-3-030-25629-6_109

150. Carayannis EG, Dezi L, Gregori G, Calo E (2021) Smart environments and techno-centric and human-centric innovations for Industry and Society 5.0: a quintuple helix innovation system view towards smart, sustainable, and inclusive solutions. J Knowl Econ 1–30. https://doi.org/10.1007/s13132-021-00763-4
151. Ferreira CM, Serpa S (2018) Society 5.0 and social development. Manag Org Stud 5:26–31
152. Sołtysik-Piorunkiewicz A, Zdonek I (2021) How society 5.0 and industry 4.0 ideas shape the open data performance expectancy. Sustainability 13(2):917. https://doi.org/10.3390/su13020917
153. Büscher B (2014) Nature 2.0: exploring and theorizing the links between new media and nature conservation. N Media Soc 18:726–743. https://doi.org/10.1177/1461444814545841
154. Cutts BB, Sinclair KM, Strauch MA, Slivnick B, Emmons Z (2015) Environmental justice and emerging information communication technology: a review for U.S. natural disaster management. Environ Justice 8(4):144–150. https://doi.org/10.1089/env.2015.0012
155. Nulman E, Özkula SM (2016) Environmental nongovernmental organizations' digital media practices toward environmental sustainability and implications for informational governance. Curr Opin Environ Sustain 18:10–16. https://doi.org/10.1016/j.cosust.2015.04.004
156. Fletcher R (2017) Gaming conservation: nature 2.0 confronts nature-deficit disorder. Geoforum 79:153–162. https://doi.org/10.1016/j.geoforum.2016.02.009
157. Holifield R, Williams KC (2014) Urban parks, environmental justice, and voluntarism: the distribution of friends of the parks groups in Milwaukee county. Environ Justice 7:70–76. https://doi.org/10.1089/env.2013.0043
158. Friends of Lincoln Park (2018) Friends of Lincoln Park. http://lincolnparkfriends.org/. Accessed 18 Dec 2018
159. Burkholder C, MacEntee K (2016) Exploring the ethics of the participant-produced archive: the complexities of dissemination. In: Warr D, Guillemin M, Cox S, Waycott J (eds) Ethics and visual research methods: theory, methodology, and practice. Palgrave Macmillan US, New York, pp 211–224
160. Kowalikova, P., Polak, P., and Rakowski, R. (2020). The Challenges of Defining the Term "Industry 4.0", Society, 57: 631–66. https://doi.org/10.1007/s12115-020-00555-7
161. Lewis, R. (2021). The Crypto-Currency Revolution: Finance in the age of Bitcoin, blockchains and tokens, London: Kogan Page.
162. Hamdan, A., Hassanien, A. E., Razzaque, A., and Alareeni, B. (eds.) (2021). The Fourth-Industrial Revolution: Implementation of Artificial Intelligence for Growing Business Success, Studies in Computational Intelligence 935, Cham, Switzerland: Springer
163. Wiederhold, B. K. (2020). Connecting Through Technology During the Coronavirus Disease 2019 Pandemic: Avoiding "Zoom Fatigue", Cyberpsychology, Behavior, and Social Networking, 23(7): 437–440.
164. Haochen, G., and Polak, P. (2021). Artificial Intelligence and Financial Technology FinTech: How AI Is Being Used Under the Pandemic in 2020. IN Hamdan, A., Hassanien, A. E., Razzaque, A., and Alareeni, B. (eds.) (2021). The Fourth-Industrial Revolution: Implementation of Artificial Intelligence for Growing Business Success, Studies in Computational Intelligence 935, Cham, Switzerland: Springer, 169–186
165. Pasquale, F. (2015). The Black Box Society: The Secret Algorithms That Control Money and Information, Harvard University Press.
166. Pasquale, F. (2017). Secret Algorithms Threaten the Rule of Law, MIT Technology Review
167. Molnar, C. (2019). Interpretable Machine Learning: A Guide for making Black-Box models explainable. https://christophm.github.io/inter/pretable-ml-book
168. Floridi, L. (2019). Translating Principles into Practices of Digital Ethics: Five Risks of Being Unethical, Philosophy & Technology, 32:185–193. https://doi.org/10.1007/s13347-019-00354-x
169. Roitblat, H. L. (2020). Algorithms Are Not Enough: Creating General Artificial Intelligence, Cambridge: The MIT Press.
170. Zuboff, S. (2019). The Age of Surveillance Capitalism: The Fight for the Future of the New Frontier of Power, London: Profile Books Ltd.

171. Byrne, D. S. (2010). Comparison, Diversity and Complexity. IN Cilliers, P., and Preiser, R. (eds.). (2010). Complexity, Difference and Identity, Issues in Business Ethics, 26, Dordrecht, Heidelberg: Springer: 61–75
172. Alford, R. R. (1975), Health Care Politics: Ideological and Interest Group Barriers to Reform, Chicago, IL: University of Chicago Press.
173. Alford, R., R., & Friedland, R. (1975) Political Participation and Public Policy. Annual Review of Sociology, 1, 429–479
174. North, N. and Peckham, S. (2001). Analysing Structural Interests in Primary Care Groups. Social Policy and Administration, 35(4): 426–440.
175. Checkland, K., Harrison, S. and Coleman, A. (2009) 'Structural Interests' in Health Care: Evidence from the Contemporary National Health Service. Journal of Social Policy, 38(4): 607–625.
176. Ülgen, F. (2017). Financialization and Vested Interests: Self-Regulation vs. Financial Stability as a Public Good. Journal of Economic Issues, 51:2, 332–340, https://doi.org/10.1080/00213624.2017.1320512
177. Briggs, D.S. (2021). COVID-19: A strife of interests for us all and what problem are we attempting to solve? Asia Pacific Journal of Health Management, 16(1): i693. Available at: https://doi.org/10.24083/apjhm.v16i1.693
178. Morley, J., Floridi, L., Kinsey, L., and Elhalal, A. (2020). From What to How: An Initial Review of Publicly Available AI Ethics Tools, Methods and Research to Translate Principles into Practices, Sci Eng Ethics, 26: 2141–2168. https://doi.org/10.1007/s11948-019-00165-5
179. Shaw, P. (2021). "Context Matters": The Law, Ethics and AI. In Hervey, M., and Lavy, M. The Law of Artificial Intelligence, Croydon: Sweet & Maxwell (Thomas Reuters), CPI Group, UK, 31–66
180. Shaw, P. (2020). Building Trust through Sound Governance. In Chishti, S., Bartoletti, I., Leslie, A., and Millie, S. M. (eds.), The AI Book: The Artificial Intelligence Handbook for Investors, Entrepreneurs and FinTech Visionaries, Chichester: John Wiley & Sons Ltd, 175–179
181. Carroll, A. B. (1991). The pyramid of corporate social responsibility: toward the moral management of organizational stakeholders. Bus Horiz, 34(4): 39–48.
182. Dörr, S. (2020) Praxisleitfaden Corporate Digital Responsibility. Unternehmerische Verantwortung und Nachhaltigkeitsmanagement im Digitalzeitalter. Springer Gabler, Berlin. (English edition in preparation Corporate Digital Responsibility. Managing Corporate Responsibility and Sustainability in the Digital Age. Springer, Berlin)
183. Keen A (2015) The Internet Is not The answer. Atlantic Books, London
184. Robinson, J. P., Barth, K., & Kohut, A. (1997). Personal computers, mass media, and use of time. Social Science Computer Review, 15, 65–82.
185. Robinson, J., & Martin, S. (2009). IT and activity displacement: Behavioral evidence from the US General Social Survey (GSS). Social Indicators Research, 91(2), 115–139.
186. Robinson, J., & De Haan, J. (2006). Information technology and family time displacement. In R. Kraut, M. Brynan, & S. Kiesler (Eds.), Computers, phones and the internet (pp. 51–69). Oxford: Oxford University Press.
187. Pardi, F., Salleh, A., & Nawi, A. (2015). Determinants of Sustainable Development in Malaysia: A VECM Approach of Short-Run and Long-Run Relationships. American Journal of Economics, 5(2): 269–277. https://doi.org/10.5923/c.economics.201501.35 .
188. World Bank (2020). World Development Indicators. Adjusted Net Savings. https://databank.worldbank.org/reports.aspx?source=World-Development-Indicators Accessed 5 January 2020
189. World Bank. (2006). Where is the wealth of the nations? Measuring capital for the 21st century. Washington, DC: World Bank.
190. Stanimirovic, D., & Vintar, M. (2013). Conceptualization of an integrated indicator model for the evaluation of e-government policies. Electronic Journal of e-Government, 11(1), 293–307.

191. Güney, T. (2017a). Governance and sustainable development: How effective is governance? The Journal of International Trade & Economic Development, 23(6), 316–335. https://doi.org/10.1080/09638199.2016.1249391 .
192. Mahoney, Josef Gregory and Maximilian Mayer. 2020. Donald Trump's campaign against Huawei is a Symptom of Digital Orientalism, Ignoring Similarities in Chinese and Western surveillance, South China Morning Post. https://www.scmp.com/comment/opinion/article/3046357/trumps-campaign-against-huawei-symptom-digital-orientalism-ignoring. Accessed 10 June 2022
193. Mayer, Maximilian. 2019. China's authoritarian internet and digital orientalism. In Redesigning organizations: Concepts for the connected society, ed. Denise Felder, 177–192. Cham: Springer Nature
194. Resolution on CPC History, 1949–1981. 1981. Beijing: Foreign Language Press
195. MacFarquhar, Roderick. 1999. The Origins of the Cultural Revolution, vol. 3. New York: Columbia University Press.
196. Shi, Jing. 2020. Mianxiang Xinfazhan Jieduan, Gongzhong dui Guojia Zhili Xiandaihua you naxie Xinqidai. Guojia Zhili 33: 3–8.
197. Gao, Qiqi, and Huan Du. 2020. Zhineng Wenming yu Quanguocheng Minzhu de Fazhan: Guojia Zhili Xiandaihua de Xinmingti. Shehui Kexue 5: 25–35.
198. Yang, Lihua. 2021. Goujian Wenmingxing Zhineng Zhili: Zhanju Xinshidai Guoji Zhineng Zhili Zhigaodian. Renmin Luntan Xueshu Qianyan 1: 63–69
199. Wang, Lixin. 2018. Makesi Zhuyi xin Renshilun yu Rengong Zhineng: Rengong Zhineng bushi Weixie Renlei Weming de Kejizhihuo. Ziran Bianzhengfa Tongxun 40 (4): 15–19.
200. Young, I. M. (2000). Inclusion and democracy. Oxford: Oxford University Press.
201. Diamond, L. (2002). Thinking about hybrid regimes. Journal of Democracy, 13(2), 21–35.
202. Lim, C. (2011). A watershed election. Singapore: Select Books.
203. Scott, J. (1985). Weapons of the weak: Everyday forms of resistance. New Haven and London: Yale University Press.
204. Papacharissi, Z. (2002). The virtual sphere: The internet as a public sphere. New Media and Society, 4(1), 9–27.
205. Zhang, W. Y., & Pang, N. (2016). The internet and social media. In T. Lee & K. Tan (Eds.), Change in voting: Singapore's 2015 General Election (pp. 232–245). Singapore: Ethos Publishing.
206. Porter, M. E. (1998). Clusters and the new economics of competition (Vol. 76, No. 6, pp. 77–90). Boston: Harvard Business Review
207. Becattini, G. (2015). La coscienza dei luoghi. Il territorio come soggetto corale. Roma: Donzelli Editore.
208. Gupta, A. K., & Govindarajan, V. (1991). Knowledge flows and the structure of control within multinational corporations. Academy of Management Review, 16(4), 768–792.
209. Cantwell, J., & Santangelo, G. D. (1999). The frontier of international technology networks: sourcing abroad the most highly tacit capabilities. Information Economics and Policy, 11(1), 101–123.
210. Zanfei, A. (2000). Transnational firms and the changing organisation of innovative activities. Cambridge Journal of Economics, 24, 515–542.
211. Ambos, T. C., & Ambos, B. (2009). The impact of distance on knowledge transfer effectiveness in multinational corporations. Journal of International Management, 15(1), 1–14.
212. Rullani, E. (2015). Distretti e filiere in evoluzione. In Rapporto 2015 dell'Osservatorio Nazionale Distretti Italiani (pp. 76–103). Roma: Unioncamere. http://www.osservatoriodistretti.org/sites/default/files/rapporto-2015.pdf
213. National Research University "Higher School of Economics" (2020). Digital economy: 2020. https://issek.hse.ru/digec2020. Accessed 20 Sept 2020
214. Ferretti, L. et al. Quantifying SARS-CoV-2 transmission suggests epidemic control with digital contact tracing. Science https://doi.org/10.1126/science.abb6936 (2020).
215. Shilo, S., Rossman, H. & Segal, E. Axes of a revolution: Challenges and promises of big data in healthcare. Nat. Med. 26, 29–38 (2020).

216. Broad, J. D. & Luthans, F. Positive resources for psychiatry in the fourth industrial revolution: Building patient and family focused psychological capital (PsyCap). Int. Rev. Psychiatry 32, 542–554 (2020).
217. Javaid, M. et al. Industry 4.0 technologies and their applications in fighting COVID-19 pandemic diabetes. Metab. Syndr. Clin. Res. Rev. 14, 419–422 (2020)
218. Alimadadi, A. et al. Artificial intelligence and machine learning to fight COVID-19. Physiol. Genomics 52, 200–202 (2020).
219. Tracking and tracing COVID: Protecting privacy and data while using apps and biometrics. OECD http://www.oecd.org/coronavirus/policy-responses/tracking-and-tracing-covid-protecting-privacy-and-data-while-using-apps-and-biometrics-8f394636/
220. Jalabneh, R. et al. Use of mobile phone apps for contact tracing to control the COVID-19 pandemic: A literature review. In Applications of Artificial Intelligence in COVID-19 Vol. 1 (eds Nandan Mohanty, S. et al.) 389–404 (Springer, Singapore, 2021). https://doi.org/10.1007/978-981-15-7317-0_19
221. Google and Apple Reveal How Covid-19 Alert Apps Might Look. Wired
222. Mata, A. S. An overview of epidemic models with phase transitions to absorbing states running on top of complex networks. Chaos Interdiscip. J. Nonlinear Sci. 31, 012101 (2021).
223. Silva, C. J. et al. Complex network model for COVID-19: Human behavior, pseudo-periodic solutions and multiple epidemic waves. J. Math. Anal. Appl. 514, 125171 (2022).
224. Wesolowski, A. et al. Quantifying the impact of human mobility on malaria. Science 338, 267–270 (2012).
225. Danon, L. et al. Networks and the epidemiology of infectious disease. Interdiscip. Perspect. Infect. Diseases 2011, 284909 (2011).
226. Castells, Manuel. 2009. The rise of the network society. Oxford: Wiley-Blackwell.
227. Barney, Darin. 2013. The network society. Cambridge, Malden: Polity Press.
228. Bevir, Mark. 2010. Democratic governance. Princeton: Princeton University Press.
229. Bevir, Mark. 2013. A theory of governance. Berkeley: University of California Press.
230. Bevir, Mark. 2020. What is the decentered state? Public Policy and Administration. https://doi.org/10.1177/0952076720904993 .
231. Bevir, Mark, and Rod A.W. Rhodes. 2003. Searching for civil society: changing patterns of governance in Britain. Public Administration 81(1):41–62.
232. Boltanski, Luc, and Laurent Thévenot. 1999. The sociology of critical capacity. European Journal of Social Theory 2(3):359–377.
233. Boltanski, Luc, and Laurent Thévenot. 2000. The reality of moral expectations: a sociology of situated judgement. Philosophical Explorations 3(3):208–231.
234. Boltanski, Luc, and Laurent Thévenot. 2006. On justification. Economies of worth. Princeton: Princeton University Press
235. Boltanski, Luc, and Eve Chiapello. 2005. The new spirit of capitalism. London: Verso.
236. Rosa, Hartmut. 2004. Four levels of self-interpretation. A paradigm for interpretive social philosophy and political criticism. Philosophy & Social Criticism 30(5):691–720
237. Stalder, Felix. 2018. The digital condition. Cambridge, Medford: Polity
238. Odendaal, N. (2008) (D)urban space as the site of collective actions: Towards a conceptual framework for understanding the digital city in Africa. In: A. Aurigi and F. De Cindio (eds.) Augmented Urban Spaces: Articulating the Physical and Electronic City. Aldershot, UK: Ashgate
239. Willis, K. (2008) Places, situations and connections. In: A. Aurigi and F. De Cindio (eds.) Augmented Urban Spaces: Articulating the Physical and Electronic City. Aldershot, UK: Ashgate
240. Smith, P. (1977) The Syntax of Cities. London: Hutchinson.
241. Carmona, M., Heath, T., Oc, T. and Tiesdell, S. (2003) Public Places – Urban Spaces: The Dimensions of URBAN DESIGN. Oxford: Architectural Press.
242. Guthrie, K. and Dutton, W. (1992) The politics of citizen access technology: The development of public information utilities in four cities. Policy Studies Journal 20 (4): 574–597.

243. Manovich, L. (2003) The poetics of augmented space. In: A. Everett and J. Caldwell (eds.) New Media: Theories and Practices of Digitextuality. New York: Routledge
244. Vetrò, A., Canova, L., Torchiano, M., Minotas, C.O., Iemma, R., Morando, F.: Open data quality measurement framework: Definition and application to open government data. Government Inf. Q. 33, 325–337 (2016)
245. Chesbrough, H., Vanhaverbeke, W., & West, J. (Eds.). (2006). Open innovation: Researching a new paradigm. New York: Oxford University Press.
246. Carayannis, E.G. et al (2021). Known unknowns in the era of technological and viral disruptions: Implications for theory, Policy and Practice, forthcoming
247. Onday, O. (2019). Japan's Society 5.0: Going beyond Industry 4.0. Business and Economics Journal, 10(2), 1–6
248. Fukuyama, M. (2018). Society 5.0: Aiming for a new human-centered society. Japan Spotlight, 27, 47–50.
249. Birchall, D., Chanaron, J. J., Tovstiga, G., & Hillenbrand, C. (2011). Innovation performance measurement: Current practices, issues and management challenges. International Journal of Technology Management, 56(1), 1–20.
250. Cruz-Cázares, C., Bayona-Sáez, C., & García-Marco, T. (2013). You can't manage right what you can't measure well: Technological innovation efficiency. Research Policy, 42(6–7), 1239–1250.
251. Edison, H., & bin Ali N., Torkar R. (2013). Towards innovation measurement in the software industry. Journal of Systems and Software, 86, 1390–1407
252. Hoelscher, M., & Schubert, J. (2015). Potential and problems of existing creativity and innovation indices. Creativity Research Journal, 27(1), 1–15.
253. Carayannis, E. G., & Alexander, J. (2006). Global and local knowledge glocal transatlantic public-private partnerships for research and technological development. New York: Palgrave.
254. Hanson, Barbara Cyber blame and social theory. SN Social Sciences (2021). https://doi.org/10.1007/s43545-021-00138-1
255. Knell, Mark The digital revolution and digitalized network society. Review of Evolutionary Political Economy (2021). https://doi.org/10.1007/s43253-021-00037-4
256. Galič, Maša; Timan, Tjerk; Koops, Bert-Jaap Bentham, Deleuze and Beyond: An Overview of Surveillance Theories from the Panopticon to Participation. Philosophy & Technology (2016). https://doi.org/10.1007/s13347-016-0219-1
257. Alarcón-Medina, Rafael Grand Digital Abyss: Technology, Critical Theory, and Subjectivity in twenty-first century Cyber-Capitalism. Dialectical Anthropology (2019). https://doi.org/10.1007/s10624-019-09577-y
258. Latzer, Michael The Digital Trinity—Controllable Human Evolution—Implicit Everyday Religion. KZfSS Kölner Zeitschrift für Soziologie und Sozialpsychologie (2022). https://doi.org/10.1007/s11577-022-00841-8
259. Gilbert, Tamsyn Looking at Digital Art: Towards a Visual Methodology for Digital Sociology. The American Sociologist (2018). https://doi.org/10.1007/s12108-018-9384-2
260. Mökander, Jakob; Schroeder, Ralph AI and social theory. AI & SOCIETY (2021). https://doi.org/10.1007/s00146-021-01222-z
261. Dziubaniuk, Olga; Ivanova-Gongne, Maria; Nyholm, Monica Learning and teaching sustainable business in the digital era: a connectivism theory approach. International Journal of Educational Technology in Higher Education (2023). https://doi.org/10.1186/s41239-023-00390-w
262. Rossignoli, Cecilia The contribution of transaction cost theory and other network-oriented techniques to digital markets. Information Systems and e-Business Management (2007). https://doi.org/10.1007/s10257-007-0063-z
263. Light, Ann The Panopticon reaches within: how digital technology turns us inside out. Identity in the Information Society (2010). https://doi.org/10.1007/s12394-010-0066-7
264. Moller, Arlen C.; Merchant, Gina; Conroy, David E.; West, Robert; Hekler, Eric; Kugler, Kari C.; Michie, Susan Applying and advancing behavior change theories and techniques in the

context of a digital health revolution: proposals for more effectively realizing untapped potential. Journal of Behavioral Medicine (2017). https://doi.org/10.1007/s10865-016-9818-7
265. Tanenbaum, Theresa Jean Hermeneutic Inquiry for Digital Games Research. The Computer Games Journal (2015). https://doi.org/10.1007/s40869-015-0005-9
266. Pałosz, Radosław How Do Base Rules Constitute a Virtual World?. International Journal for the Semiotics of Law - Revue internationale de Sémiotique juridique (2022). https://doi.org/10.1007/s11196-022-09903-8
267. Norquist, Jeffrey "Something other than real life:" digital life resistance in the civil sphere. American Journal of Cultural Sociology (2022). https://doi.org/10.1057/s41290-022-00176-z
268. Lindemann, Gesa; Schünemann, David Presence in Digital Spaces. A Phenomenological Concept of Presence in Mediatized Communication. Human Studies (2020). https://doi.org/10.1007/s10746-020-09567-y
269. Gunderson, Ryan Human–Computer Interaction Research Needs a Theory of Social Structure: The Dark Side of Digital Technology Systems Hidden in User Experience. Human Studies (2022). https://doi.org/10.1007/s10746-022-09633-7
270. Levine, Livia Digital Trust and Cooperation with an Integrative Digital Social Contract. Journal of Business Ethics (2019). https://doi.org/10.1007/s10551-019-04201-z
271. Harrison, Tom; Polizzi, Gianfranco (In)civility and adolescents' moral decision making online: drawing on moral theory to advance digital citizenship education. Education and Information Technologies (2021). https://doi.org/10.1007/s10639-021-10710-0
272. Li, Jia Grounded theory-based model of the influence of digital communication on handicraft intangible cultural heritage. Heritage Science (2022). https://doi.org/10.1186/s40494-022-00760-z
273. Burns, Ryan Rethinking big data in digital humanitarianism: practices, epistemologies, and social relations. GeoJournal (2014). https://doi.org/10.1007/s10708-014-9599-x
274. Sendra, Anna; Farré, Jordi; Vaagan, Robert W. Seeking, sharing and co-creating: a systematic review of the relation between social support theory, social media use and chronic diseases. Social Theory & Health (2019). https://doi.org/10.1057/s41285-019-00106-z
275. Berg, Sebastian; Staemmler, Daniel; Thiel, Thorsten Political Theory of the Digital Constellation. Zeitschrift für Politikwissenschaft (2022). https://doi.org/10.1007/s41358-022-00324-4
276. Bibri, Simon Elias; Allam, Zaheer; Krogstie, John The Metaverse as a virtual form of data-driven smart urbanism: platformization and its underlying processes, institutional dimensions, and disruptive impacts. Computational Urban Science (2022). https://doi.org/10.1007/s43762-022-00051-0
277. Travkina, N. M. Digitization of Society: Alternative Projections of the Future. Herald of the Russian Academy of Sciences (2022). https://doi.org/10.1134/s1019331622120115
278. Alimohammadlou, Moslem; Khoshsepehr, Zahra The role of Society 5.0 in achieving sustainable development: a spherical fuzzy set approach. Environmental Science and Pollution Research (2023). https://doi.org/10.1007/s11356-023-25543-2
279. Cutts, Bethany B.; Greenlee, Andrew J.; Hornik, Kaitlyn; Nigrelli, Caitie A. Hyperlocal sustainabilities: theorizing action research for sustainability in the digital age. Sustainability Science (2019). https://doi.org/10.1007/s11625-019-00694-9
280. Elliott, Karen; Price, Rob; Shaw, Patricia; Spiliotopoulos, Tasos; Ng, Magdalene; Coopamootoo, Kovila; van Moorsel, Aad Towards an Equitable Digital Society: Artificial Intelligence (AI) and Corporate Digital Responsibility (CDR). Society (2021). https://doi.org/10.1007/s12115-021-00594-8
281. Grimshaw, Mike Towards a manifesto for a critical digital humanities: critiquing the extractive capitalism of digital society. Humanities and Social Sciences Communications (2018). https://doi.org/10.1057/s41599-018-0075-y
282. Shakhrai, S. M. Digital Constitution: Fundamental Rights and Freedoms of an Individual in a Totally Informational Society. Herald of the Russian Academy of Sciences (2018). https://doi.org/10.1134/s1019331618060126

283. Robinson, John P.; Lee, Chang Won Society's (Virtually) Time-Free Transition into the Digital Age. Social Indicators Research (2013). https://doi.org/10.1007/s11205-013-0374-z
284. Castro, Conceição; Lopes, Cristina Digital Government and Sustainable Development. Journal of the Knowledge Economy (2021). https://doi.org/10.1007/s13132-021-00749-2
285. Mahoney, Josef Gregory China's Rise as an Advanced Technological Society and the Rise of Digital Orientalism. Journal of Chinese Political Science (2022). https://doi.org/10.1007/s11366-022-09817-z
286. Woon, Chih Yuan Internet spaces and the (re)making of democratic politics: the case of Singapore's 2011 General Election. GeoJournal (2017). https://doi.org/10.1007/s10708-017-9815-6
287. Castellani, Davide; Rullani, Enzo; Zanfei, Antonello Districts, multinationals and global/digital networks. Journal of Industrial and Business Economics (2017). https://doi.org/10.1007/s40812-017-0079-4
288. Popkova, Elena; Bogoviz, Aleksei V.; Sergi, Bruno S. Towards digital society management and 'capitalism 4.0' in contemporary Russia. Humanities and Social Sciences Communications (2021). https://doi.org/10.1057/s41599-021-00743-8
289. Shi, Zhengyu; Qian, Haoqi; Li, Yao; Wu, Fan; Wu, Libo Machine learning based regional epidemic transmission risks precaution in digital society. Scientific Reports (2022). https://doi.org/10.1038/s41598-022-24670-z
290. August, Vincent Political ideas of the network society: why digitalization research needs critical conceptual analysis. Zeitschrift für Politikwissenschaft (2021). https://doi.org/10.1007/s41358-021-00305-z
291. Polak, Petr Welcome to the Digital Era—the Impact of AI on Business and Society. Society (2021). https://doi.org/10.1007/s12115-021-00588-6
292. Aurigi, Alessandro Reflections towards an agenda for urban-designing the digital city. URBAN DESIGN International (2012). https://doi.org/10.1057/udi.2012.32
293. Maretti, Mara; Russo, Vanessa; del Gobbo, Emiliano Open data governance: civic hacking movement, topics and opinions in digital space. Quality & Quantity (2020). https://doi.org/10.1007/s11135-020-01045-y
294. Carayannis, Elias G.; Dezi, Luca; Gregori, Gianluca; Calo, Ernesto Smart Environments and Techno-centric and Human-Centric Innovations for Industry and Society 5.0: A Quintuple Helix Innovation System View Towards Smart, Sustainable, and Inclusive Solutions. Journal of the Knowledge Economy (2021). https://doi.org/10.1007/s13132-021-00763-4

Human and Digital Technology Relations

3

Introduction

Digital technologies have significantly disrupted and transformed the social organization and behaviour of individuals. Digital technologies have not only become the medium connecting individuals but also human remains connected to them. Social media, like Facebook (now Meta), Twitter (now X), Instagram, Telegram, Slack, WhatsApp, etc., have become popular platforms on which individuals are not only able to express their true selves and behaviour but are also able to interact with other individuals and groups. Social media is leading to network formations that in turn are resulting in digital communities, digital neighbourhoods and digital societies. These networks are alternatives and/or complementary to non-digital social networks. The digital space they occupy also functions fundamentally similar to the non-digital social networks, but they do allow relatively free expression as the social and political powers exercised in persons from one to another (restricting expressions) get weakened because of the virtual rather than physical or corporeal presence.

Digital technologies have massively transformed the behaviour of their users. Social media which is an important outcome of digital development, has also led to narcissistic and compulsive behaviours. The selfie, posting of different types of photographs, and status updates are important outcomes of such behaviours. For users of such platforms, likes or dislikes of posts have important outcomes for their moods. It is not only that users now post interesting stories and videos, but also fake and deep fake videos, having enormous social, political and personal consequences.

The rise of social media is displacing the usual channels of information. Earlier where the information was in the hands of those socially and economically powerful and could be regulated, such as print newspapers, magazines, TV channels, etc,

A. Shaban (ed.), *Digital Geographies—Theory, Space, and Communities*,
https://doi.org/10.1007/978-981-97-4734-4_3

with digitalization it has become impossible for governments to control the same. Where earlier, a group (often corporate and/or government) used to own such information channels, social media makes it possible for everyone to own their channels and broadcast their views. The consequences of such individual ownership and control of channels of information have been both positive and negative. Where the individuals now can exercise more of their freedoms and generate support for their good cause, it has also created enormous possibilities of channels of misinformation, and social and personal conflicts. For instance, individuals from one country can choose their partners from another country, even if the countries are politically in conflict, as in the case of India and Pakistan, but simultaneously the rise of misinformation gateways and what is now called social media mafia sponsored by political and extremists groups generate ethnic, religious, and political conflicts with the consequences of social boycotts of groups, social exclusions, lynching, and murders.

The availability of low-cost smart mobile phones and low-cost of data in many countries has revolutionized the internet connectivity of individuals. The younger generation especially is more attracted to social media. The money-earning possibilities at social media platforms like YouTube, Instagram, and Facebook have even further drawn the unemployed to them. This has resulted in the self-employment of many who play the role of influencers to advertise some products, events, or specific companies. The rise of such platforms has also seen the possibilities of those employed in TV news channels or newspapers opening their own channels of news and reporting and earning through the same. Digitalization has also promoted online trading and dealing with global financial markets, and one even finds individuals in remote parts of rural areas buying and selling cryptocurrencies to earn their living. However, there have been stories also where social media impacted a change in family/community norms and values, dressing and foods, creating a crisis in families/communities.

The social networks formed through social media also exhibit social exclusion. The exclusion from digital neighbourhoods and digital society can be similar to that in real society. However, additional factors like digital literacy, access to the internet and digital devices also play major roles in the exclusion and inclusion processes. Additionally, the tricky design and development of algorithms can also play a role in digital exclusion and inclusion. Age, sex, ethnicity, race, region, etc., can be framed in algorithms that can lead to the exclusion of individuals and groups.

As such, human-machine interaction and digitally mediated networking and participation of individuals and social groups have not only the physical geographical implications in terms of real-time flow of information with or without economic cost but also the virtual spaces these networks generate remain impregnated with many socio-economic and political consequences.

The rise of digital technology in the form of Artificial Intelligence (AI) can have its consequences for human society. The evolution of AI may outsmart the human mind and thus may go out of human control. This is a fear often expressed in several debates. One also understands that AI can automate many spheres of human everyday life and make life easy, with the least effort. This can also help to cut costs in production, distribution, and inventions. However, other apprehensions are again put forward whether these can make humans lose their purpose of life. The growth of generative AIs is unfolding and currently, it is in rapid evolution. Only when it reaches the stage of maturity, and we are able to deploy it in several spheres of our lives. Then we will be able to better understand its real impacts. Nevertheless, we are heading to an interesting unfolding future of the digital age, besides the smooth rides and making our journey of life comfortable through several ways, it can also give us some bumps, and some bad ones as well.

This chapter discusses research on digital mediation and human-machine interaction. The articles deal with various age groups, social groups, economic groups, businesses, genders, neighbourhoods, regions, etc. The papers reviewed are related to various regions of the world. Some of these studies show how social networks have gained popularity and have improved the corporate environment by fostering and facilitating chances for collaboration. The use of social media networks has improved relationship development, creativity, and knowledge exchange among businesses, all of which are expanding quickly. Several papers in this chapter show how digital technologies facilitate communication, collaboration, and knowledge sharing between individuals and groups, which benefits educational communities. Some of the articles also show how the "collective intelligence" of students and teachers is being harnessed by technologies like blogs, wikis, media-sharing platforms, and collaborative editing tools.

Specifically, the paper "Technological and social networks of a pastoralist artificial society: agent-based modelling of mobility patterns" explores how digital and social networks affect mobility in a pastoralist artificial society in the context of environmental degradation. Another study, "Self-management of social well-being in a cross-sectional study among community-dwelling older adults: The added value of digital participation", investigates relationships between digital participation and self-management skills among older people with chronic illnesses who live in their communities. The results imply a beneficial relationship between internet use and self-management skills among older adults.

Machine generated summaries

Disclaimer: The summaries in this chapter were generated from Springer Nature publications using extractive AI auto-summarization: An extraction-based summarizer aims to identify the most important sentences of a text using an algorithm and uses those original sentences to create the auto-summary (unlike generative AI). As the constituted sentences are machine selected, they may not fully reflect the body of the work, so we strongly advise that the original content is read and cited. The auto generated summaries were curated by the editor to meet Springer Nature publication standards. To cite this content, please refer to the original papers.

Machine generated keywords: social network, network, networking, social networking, user, community, event, social medium, facebook, content, student, post, behavior, collaborative, online social

3.1 Social Networks, Communities and Human Relations

Machine generated keywords: social network, network, networking, social networking, user, event, social medium, facebook, content, post, collaborative, community, online social, pattern, friend

3.1.1 Predicting Emotions in Online Social Networks: Challenges and Opportunities [164]

This is a machine-generated summary of:

Alqahtani, Ghadah; Alothaim, Abdulrahman: Predicting emotions in online social networks: challenges and opportunities [164]

Published in: Multimedia Tools and Applications (2022)

Link to original: https://doi.org/10.1007/s11042-022-12345-w

If you want to cite the papers, please refer to the original.

For technical reasons we could not place the page where the original quote is coming from.

Abstract-Summary

"Online social networking has become a popular means of information exchange and social interactions."

"One interesting area of research is the emotional dimensions contained in this user-generated content, specifically, emotion detection and prediction, which involve the extraction and analysis of emotions in social network data."

"This study aimed to provide a comprehensive overview and better understanding of the current state of research regarding emotion detection in online social networks by performing a systematic literature review (SLR)."

"We collected and analyzed studies that focused on emotion in social network posts and discussed various topics published in digital databases between 2010 and December 2020."

Introduction

"These motivate us to cover the methods and resources developed for emotion detection and prediction on social networks to better understand the research in this field."

"More so, this study presents a review of studies published from 2010 to December 2020 related to predicting emotions in online social network posts to answer the following research questions (RQs): RQ1: Which category of emotions have been frequently used in the domain of emotion detection in social network data?"

"Despite the interest in emotion detection and prediction, we believe that studies on emotion detection on social networks need to be reviewed and observed to highlight the challenges and opportunities to move the field forward."

"This study contributes knowledge to the field of social media, with a focus on emotion detection and prediction through a careful analysis of existing studies in the field."

Related works

"Various surveys and reviews based on emotion detection using several approaches, such as machine learning and deep learning, or focusing on a dealing of specific data types have been presented by the researchers."

"We summarize some of the important review studies that cover all the important aspects of the emotion detection task, such as challenges, classification of existing emotion detection and prediction approaches, deep learning models for emotion detection, classification of models used, and handling multimodality in emotion detection."

"Other surveys on text emotion detection and analysis outline work on this subject across time, detailing existing emotion detection approaches, methodologies, datasets, experiments, and results [1, 2]."

"Survey studies on emotion detection primarily focused on facial and speech analysis [3, 4] and text-based emotion detection [5, 6], but they did not consider whether the data used in the research was collected from a social network platform or online available datasets."

Background

"The field of opinion mining intersects with many other fields, including data analysis, deep learning (DL), natural language processing (NLP), and computer vision techniques, depending on the specific type of opinion mining [7]."

"Throughout this section, we overview the related terms and concepts starting with the original source of data used in studies (social networks) down to the main focus of our study (emotion)."

"Millions of people use these social networking platforms to share their opinions, views, emotions, and information about their ordinary and social activities."

"We tried to cover the prediction of emotions in social network data using the different traditional and machine learning techniques by exploring their use in emotion detection and prediction and specifically relating them to social network data."

Review method and execution

"This review methodology will examine how much progress has been made in the field of emotion detection and prediction in social networks, based on the results of current studies."

"These questions are a few of many whose answers require us to review the studies and approaches applied in emotion detection on social networks to identify effective aspects and present different approaches and methodologies applied currently, which may improve and develop the field in future."

"To those criteria, the remaining selection criteria for this review were that the study was related to one or more of the research questions and focused on detecting and predicting emotions through various data from social networks."

"After careful study of the collected literature, only those articles related to emotion detection and prediction based on social networks, which can provide an acceptable level of concepts and information and help to answer the research questions were kept."

Emotion detection

"In a few studies, the different types of content posted by users of online social networking platforms were analyzed to detect the specific emotions expressed by the users' posts."

"In these studies of emotion classification and detection on social networks, the researchers performed the prediction process on a single type of data, which can hide much of the full range of emotions expressed in the data."

"In this study, all relevant research addressing aspects of the research questions were reviewed without a focus on the methods and algorithms used, but there has been a noticeable tendency in studies of emotion analysis and detection using social network data since 2010 to use artificial intelligence, machine learning, and DL techniques."

"Not only was text data involved in using machine learning and DL to detect the emotion through the social network data, but the images were also considered."

Challenges, gaps, and limitations of emotion analysis in social networks

"In the field of emotional analysis, the most challenging task is the dynamic nature of social network data, which makes it difficult to begin an analysis and detect the emotions of users from the early stages of collection to preprocessing."

"This challenge affects the quality of emotion analysis and detection research, and it is vital to determine the literature needed to evolve the emotions model in the social network area without falling into the trap of a paper being rated for emotion, while the content is sentiment analysis."

"When we speak of social networks, performing emotion detection and analysis in texts in English or non-English languages, such as Hindi, Arabic, Chinese, and many more, is a crucial challenge."

"It is nonetheless important to conduct emotion detection and analysis in non-English languages due to the large percentage of non-native English speakers around the world; for instance, there are millions of people in Middle Eastern countries whose native language is Arabic, making emotion analysis on Arabic social networks crucial for political and economic reasons."

Opportunities in the field of emotion analysis in social networks

"The use of DL approaches for developing and improving emotion detection models having particular potential given the success rates of artificial neural networks, DL architectures, and transfer and reinforcement learning in making machines intelligent is all well known."

"The above factors used in joint learning methods will reshape machine capacities to comprehend and unravel hidden emotions within social networking's various data types."

"To leverage these benefits for detecting and analyzing emotions in social networks, we need to first address some of the challenges in applying DL, which is an area for future research: 1."

"The transfer learning domain in emotion detection using social network data is quite new."

"Applying transformers to emotion detection on social network data will improve efficiency and the results."

"Automatic emotion detection and classification is important, especially when we talk about online social network data because of the huge number of posts made daily on these platforms."

Discussion

"Research, which tends to focus on text-based emotion detection rather than methods based on images or multiple modalities, is limited and requires more study."

"Several approaches and models have been employed to conduct emotion detection on social networks, which generally can be categorized into groups."

"We believe that most of the researchers focused on the process itself rather than the purpose; they focus on how to predict and detect the emotion rather than why we need to detect these emotions, what we going to gain from predicting the social network users' emotions."

"Emotion detection from social networks can be useful in practically every area of our daily lives."

"There are promising applications if future studies in the field consider real-life problems or trends to apply emotion detection and predictions."

Conclusion

"In response to the need to detect correct emotions from user-generated content in different socioeconomic areas, researchers have developed models and systems to analyze and detect emotions using different approaches."

"Regarding the second research question, the impact of emotion detection research, multiple perspectives are involved, from user daily feedback in marketing to monitoring the mental health of psychiatric patients."

"Between the two extremes of customer feedback for improving product recommendations and the detection of depression and suicidal feelings, there are many other important fields that can be positively impacted by improving emotion analysis and detection techniques and approaches."

"The opportunities in the field of emotion detection and analysis are numerous, and in this studies, we have tried to give a brief overview of the opportunities and areas that need more research to improve and move this field of research forward."

3.1.2 The Challenges of Internal Social Networking for Higher Education: A Brief Review of the Literature [165]

This is a machine-generated summary of:

Kazemian, Shakiba; Grant, Susan B.: The challenges of internal social networking for higher education: a brief review of the literature [165]

Published in: Journal of Data, Information and Management (2020)

Link to original: https://doi.org/10.1007/s42488-020-00028-w

If you want to cite the papers, please refer to the original.

For technical reasons we could not place the page where the original quote is coming from.

Abstract-Summary

"As social networks become increasingly popular, they have positively changed the business environment by encouraging and providing collaboration opportunities."

"Social media networks have also enhanced knowledge sharing, innovation, and relationship building, which is growing very fast among enterprises."

"Many enterprises are assessing the potential of exploiting the commercial opportunities of this technology, which could enhance employee engagement and provide a variety of effects on collaborative work (e.g., relational and personal benefits to organizations)."

"Although these benefits are not only limited to commercial organisations, but also higher education communities are benefiting, as they enable communication, collaboration, and knowledge exchange between individuals."

Introduction

"Social media are defined as represented by a range of emerging tools (e.g., wikis and blogs) and platforms where users can share information and importantly, collaborate and create networks of communities (Shadbolt and others [8]; McAfee [9])."

"Many researchers admit the presence of a common ecology that contains categories such as social networking sites, professional networking sites, blogs, microblogging services, wikis, multimedia or media sharing sites, social news and bookmarking, and user-and message media (Ouirdi and others [10])."

"Standpoint, social networks can provide a solution to the limitation of social communication tools, and these technologies describe online platforms for users to profile themselves, interact with each other, share content and ideas and keep personal relationships (Anderson [11]; Grant and Preston [12])."

"Many universities are indeed actively using enterprise social networking tools such as Microsoft Team, yammer or Jive, and etc for improved communication, connection, collaboration and enhanced knowledge sharing (Ortbach and Recker [13])."

Review of relevant literature

"Huang and Benyoucef [14] define social media as "Internet-based applications built on Web 2.0, while Web 2.0 refers to a concept as well as a platform for "harnessing collective intelligence"."

"Ouirdi and others [10] examined existing scholarly definitions of the term 'social media' through a Lasswellian lens."

"They aimed to examine existing literature definitions of the term "Social Media" by conducting a qualitative content analysis to a sample of the most delegate academic definitions reclaimed from the critical papers on social media in the web of the knowledge database."

"Social media defines in three parts, (a) an online communications networking platform used to generate and disseminate content, (b) the content which are the digital form of individual messages, posts, images, audios and videos, which (c) the individuals and communities generate and employ digital content."

Typology of social media and social media dimensions

"Ouirdi and others [10] aimed to examine the existing scholarly definitions of the term 'Social media' from top papers cited by Lasswellian's model of communication, which is one the most influential communication models."

"The coding scheme was derived from Laswell's (1984) work on the communication act, and one major theme appeared from units of analysis, namely the function or purpose of social media."

"While this category was in line with the 'Why' dimension added by another scholar to Lasswell's (1984) model, in order to investigate the purpose of the communication act happens on social media."

"From using Lasswewllian Coding categories in analyzing the definitions of social media, Ouirdi and others (Ouirdi and others [10]) expand a taxonomy of social media from the channel's perspective based on three dimensions: Users (Who/ 'to whom') who can be from the micro or macro-levels."

Social learning and social tools for education

"Technologies like blogs, wikis, media-sharing services, collaborative editing tools are harnessing the "collective intelligence" of students and teachers, promoting collaboration and the sharing of knowledge (Mason and Rennie [15]; McCarroll and Curran [16])."

"The use of Laptops and the Internet has created the technological conditions for teachers, and students can benefit from the variety of online knowledge sharing, communication, and collaboration (De-Marcos and others [17]; McCarroll and Curran [16])."

"Students can take advantage to harness this enthusiasm for technology and use these resources through an educational scheme (Mason and Rennie 2008)."

"The use of technologies had made significant changes in education, which means that classrooms are not teacher-centered anymore, and teachers try to focus on students and find out the areas in which they could shine (Prensky [18])."

"Giving pupils ownership of their learning supports a deeper understanding of concepts (McCarroll and Curran [16]), Phillips and others [19] said that facilitating active participation and collaboration by students in problem-solving and data creation is the key to success in changed models of online teaching and learning."

Social networks and student motivation

"The technical feature of social network sites such as updating status and commenting allowing students for communicating about technical information, sharing contextual needs information, and guiding information flow based on relevance, facilitate users to communicate with each other by posting information comment messages and images in real-time (Gannod and Bachman [20])."

"The study by Koranteng and others [21] seek to validate the relationship between the use of social networking sites, academic Engagement, and Knowledge Sharing using websites solely designed for academic activities."

"Besides, higher education communities have implemented this E2.0 phenomenon in order to increase knowledge sharing and to enhance the collaboration among academic staff, scholars and students recently, the use of internal social networking in higher education communities, especially among academic staff is remaining low and only a few studies are looking to use of social networking among academic staff (Kazemian [22])."

Discussion

"The use of social media tools presents a new set of challenges to organizations and educational establishments that are not used to managing knowledge and information transfer in this way, though (Grant [23])."

"The study by Koranteng and others [21] seek to validate the relationship between the use of social networking sites, academic Engagement, and Knowledge Sharing using websites solely designed for academic activities."

"By conducting a mixed-method study, a researcher can look insight into behaviours or intensions to adopt social networking across higher education for knowledge-sharing purposes."

"Implementing Sequential exploratory design which aim to Investigate under researched field, to develop hypotheses or create instruments for subsequent quantitative measurement, or provide explanations is considered the most typical form of combined research in social sciences (Srnka and Koeszegi [24])."

"In mixed method studies, a researcher uses and combine both Qualitative and quantitative method to see what is going on and this is considered a great contribution to the knowledge."

3.1.3 Engagement in Social Networks: A Multi-method Study in Non-profits Organizations [166]

This is a machine-generated summary of:

Klafke, Renata V.; Gomes, Paulo M.; Junior, Demétrio Mendonça; Didonet, Simone R.; Toaldo, Ana M.: Engagement in social networks: a multi-method study in non-profits organizations [166]

Published in: International Review on Public and Nonprofit Marketing (2021)

Link to original: https://doi.org/10.1007/s12208-020-00273-0

If you want to cite the papers, please refer to the original.

For technical reasons we could not place the page where the original quote is coming from.

Abstract-Summary

"This study discusses aspects of online communication that influence in relational interaction between online users and organizations."

"The objective is to consider the use of communication of organizations in social media as determinants for the engagement of society, through the analysis of the interactions between online users and the publications made on Facebook."

"For the development of this research, Brazilian Non-Profit Organizations (NPOs) were considered as the unit of analysis."

"Five NPOs operating in Brazil were selected for analysis of their publications."

"For the content analysis, the posts were categorized according to the typology of Lovejoy and Saxton (Journal of Computer-Mediated Communication, 17(3), 337–353, 25)."

"The type of post that generates the greater engagement on the analyzed NPOs is that has information content."

"The total of posts of each NPO were then regressed in the form of an analysis of variance (ANOVA) with the number of likes, comments and shares of each post aimed at verifying whether any differences in the groupings of posts could be demonstrated and accounted for."

Introduction

"Social media communication and interest in customer engagement has paralleled the continued development of the internet and the advent of new digital technologies (Wirtz and others [26])."

"The active use and interaction in social media can improve interest and communication with customers (Agnihotri and others [27]; Agnihotri and others [28])."

"Based on these arguments, this study aims to expand knowledge about communication in social media by analyzing the interactions of publications made on Facebook."

"The Brazilian NPO sector was the unit of analysis chosen, we assume that active and effective communication and engagement strategies can improve the relationship between donors and organizations."

"How does interaction in social media contexts affect engagement is an unanswered question."

"This study aims to answer these questions by assessing the use of communication on social media as determinant of the engagement between citizens and Brazilian NPOs."

Theoretical background

"In the NPO arena, Chandler and Lusch [29] argue that when an organization proposes value through information, service provision, and so forth, it seeks, in fact, engagement / interaction aiming an end (i.e. donation)."

"And calling attention to our empirical research, it's relevant for NPOs to understand what kind of online communication influences engagement in social network environment."

"Said that, and considering NPOs setting, it can be inferred that active organizations in social media may have the potential to be influential in shaping behaviors (donation interest, for example) and collaboration of their online peers."

"There is little theoretical, and mainly empirical, evidence on the impact of information communication via social media and engagement on donations (Li [30]; Bennett [31])."

Methodology

"During this period, the top five most liked NPOs posted 1.237 times, received a total of 912.337 reactions, 183.239 comments and 247.661 shares."

"Liking a post only requires a click from the user, but it still represents a way user can show the others that they have some kind of engagement with that content."

"Iven the focus of this study, on analyzing whether dialogic and mobilizational posts do, in fact, relate to engagement, posts have only been differentiated according to their content (Saxton and Waters [32])."

"We were interested in quantifying the number of likes, comments, and shares in each of the NPOs analyzed, with the main focus on recognizing the existence of the relationship between the number of publications and engagement with these posts."

"The dependent variable was the number of likes, comments and shares each post had, as well as an aggregated value representing engagement."

Results

"To verify what kind of publication generated more engagement, each post was then categorized according to Lovejoy and Saxton [25], accounting for the type of post, number of reactions, comments, and shares."

"For the majority of the organizations this type of post is the main responsible for the engagement, considering the number of reactions, comments, and shares."

"When an organization focus on posts with an intention to inform follower about finances, performance, causes, it can be a mechanism to engage them, raising public confidence (Lovejoy and Saxton [25])."

"Information is the type of posts the analyzed NPOs tend to rely on as to engage their followers."

"Considering the type of post with the highest engagement on these featured months, Greenpeace highlights posts that has as objective to support a cause (34 posts) and those which aim to inform the followers about something."

Conclusion

"This study showed that does not seem to have a balance between the number of posts among the most liked Brazilian NPOs, as well as does not seem to exist a relationship between the amount of posts and the engagement."

"The type of post that generates the greater engagement on the analyzed NPOs is those with information content, that is, those posts with objective information, aiming to disseminate messages about the organization, its activities, or any other potential information that may provoke interest on followers."

"According to these results, information posts, as one of the identified and measured types of posts accounts mostly for variability on engagement."

"This study concentrated on communication and engagement in social media in Brazilian NPOs, further research could investigate the same variables in others NPOs of the same sector, or even consider another methodology, such as a quantitative approach in order to stablish comparisons."

3.1.4 A Smartphone-Based Online Social Network Trust Evaluation System [167]

This is a machine-generated summary of:

Perez, Charles; Birregah, Babiga; Lemercier, Marc: A smartphone-based online social network trust evaluation system [167]

Published in: Social Network Analysis and Mining (2013)

Link to original: https://doi.org/10.1007/s13278-013-0138-4

Copyright of the summarized publication:

If you want to cite the papers, please refer to the original.

For technical reasons we could not place the page where the original quote is coming from.

Abstract-Summary

"The number of people using social networking sites is also increasing, and these platforms offer many features through which individuals can communicate with their contacts."

"The high quantity of personal and/or sensitive data, coupled with the large number of users, is one of the main motivations of malicious actors."

"We introduce in this paper a novel trust indicator for evaluating the contacts of an online social network user."

"This analysis is particularly important since the security policy of online social networks rests on the principle that a user's contact is a person of trust."

"To address this problem, we propose applying a multi-layer model and extend it by proposing overlapping features that highlight the level of overlap of a contact belonging to the set of social networking friends of a smartphone user."

Introduction

"With the emergence of smartphones, mobile users can now communicate at any time and from anywhere on social networks via mobile social applications."

"One of the basic elements for user joining a social network is the notion of contacts (often called friends)."

"While user contacts in a given social network are assumed to be trustworthy, Boshmaf et al. [33] have shown that most online social networks (OSN) users are not paying enough attention while selecting or accepting friends."

"While one publishes personal information on online social networking sites, some illegitimate users are being granted access to potentially sensitive data."

"The originality of our approach lies in the fact that the trustworthiness of user contacts is based on a social network analysis of the smartphone of the user."

"The identification of trusted relationships, as proposed in this work, could be used for the authentication of users based on their mobile social networks."

State of the art

"A malicious profile refers to a given digital entity that publishes unsolicited and/or malicious URLs on the given social networking site."

"Some works (Lee and Sumiya [34]; Lee et al. [35]; Stringhini et al. [36]) propose the creation of honey pot profiles to attract the malicious profiles on social platforms."

"A large amount of the work done is devoted to the establishment of a trust indicator for the profiles of digital social platforms (Kim et al. [37])."

"We can distinguish two main types of trust evaluation in social media: first, approaches that propose a calculation of a direct assessment of trust; and, second, the approaches that analyse trust propagation in networks."

"Nepal et al. [38] propose a model to assess trust in digital social networking communities."

"The first criterion aims to highlight the trusted profiles that significantly attract users in social networks."

A social network analysis approach to modelling smartphones

"That a model and metrics such as this would allow us to measure and evaluate the interaction between the smartphone user and his or her set of contacts over multiple social networking sites (e.g. Twitter, Facebook) and traditional communication media (e.g. SMS, mails, phone calls)."

"Such features allow us to measure the contacts of the user who appear to share close links not only in terms of interactions in a given social network but also in a series of media accessible from the smartphone."

"The direct matching approach is based on the unique identifiers of a user on multiple social networks."

"Attribute-based approaches aim to identify the similarity of two profiles based on their characteristics, such as their personal names, websites, interests or social circles (Raad et al. [39]; Singla and Domingos [40])."

"The framework called Flink proposes matching the identity of a user on multiple social networks with the use of multiple data sources (Web, Semantic Web, e-mails, publications) (Mika [41])."

Proposed features

"We formalize the node-matching function that defines the inter-layer connections of the model."

"In regard to the general model, the overlap of a node u_k into a layer L_l, denoted O (u_k, L_l), is defined below."

"This node, belonging to the Twitter social network, is identified in both the Google+ and the Facebook layer."

"Overlap between two layers (general model) The overlap of a layer L_k into a layer L_l is defined as follows: For the pillar model, the number of connections allowed is equal to the minimum number of nodes between the two layers."

"Overlap between two layers (pillar model) The overlap of layer L_k in layer L_l is defined as follows: In the example provided, one can observe that the Twitter and Google+ layers possess five nodes."

"It is defined as a graph that contains nodes that match each other as well as their corresponding inter-layer connections."

Application

"We want to impart a trust score to the contacts of a smartphone user on a given social network such as Facebook or Twitter."

"On a social network, this means that two profiles that share many common friends are considered more likely be friends than two profiles that do not share many common contacts."

"The Leicht–Holme–Newman (LHN) index gives a high similarity score for two nodes possessing a large set of common neighbours with regard to the expected number of such neighbours in the configuration model (Leicht et al. [42])."

"Smartphone-weighted resource allocation (SWRA) index Given a smartphone user y and a tested profile x, the SWRA index is computed as: where L is the set of social network layers, k_z the degree of z, θ a given natural number, x the tested user, y the target user (smartphone owner) and O (z, L) is the overlap of z in L."

Evaluation of the approach

"We tested our approach on a set of social network layers (AddressBook, Twitter, Google+ and Facebook) extracted from smartphones and a set of contacts (friends and non-friends) extracted from Facebook and Twitter."

"We have observed that, for information-centred social networks such as Twitter, the presented approach does not perform well."

"A possible solution to the identification of a trustworthy relationship on Twitter could be to adapt the overlapping feature so that it can capture the level of importance of a given topic in the ML network of the smartphone user."

"For user-centred social networks, overlapping would appear to be a complementary tool for the existing approaches to better distinguish the illegitimate and legitimate contacts."

"The scores illustrate the fact that the data available on the smartphone of a user can be used to discriminate between trustworthy and non-trustworthy contacts on user-centric networks."

Prototype

"By default, the application displays the overlap regarding all layers that contains a non-null list of contacts."

"We have observed that more than 10 % of contacts have an overlap of greater than one."

"Only a very few contacts possess an overlap of up to three, and this matches with the results for the distribution of contacts."

"One can observe that the Phone layer is well overlapped with the Facebook, Google+ and Mail layers."

"This overlap does not exceed 20 % for the Phone and Facebook pair, and is only 15 % for the Phone and Mail pair of layers."

"The significant overlap of the Facebook layer over the other layers is a sign of the success of our proposed approach."

"Such contacts are completely disconnected from the observed interaction of the user in the four other layers."

Conclusion

"We have proposed a novel approach to addressing the issue of evaluating the trustworthiness of the contacts of a social network user."

"The method relies on the multi-layer graph, which models the multiple interactions of a smartphone user with its multiple social media."

"This approach is the first attempt to analyse the trustworthiness of social networking site users based on smartphone modelling."

"The evaluation of the trust score does not require any user participation and it only relies on data available locally on the smartphone."

"Work, we plan to integrate the weights of the edges in the model such as the temporal aspect of the interaction between the smartphone user and its contacts."

3.1.5 Social Media Choices and Uses: Comparing Turkish and American Young-Adults' Social Media Activism [168]

This is a machine-generated summary of:

Burke, Barbara Ruth; Şen, Ayşe Fulya: Social media choices and uses: comparing Turkish and American young-adults' social media activism [168]

Published in: Humanities and Social Sciences Communications (2018)

Link to original: https://doi.org/10.1057/s41599-018-0090-z

If you want to cite the papers, please refer to the original.

For technical reasons we could not place the page where the original quote is coming from.

Abstract-Summary

"While some media scholars suggest this is an indication that young people are active in creating a public sphere constructed by social media, there is still little real-world evidence that the technological potentials are widely realized."

"To address this gap, this comparative case study aims to reveal how some "ordinary" young people are using social media in response to political issues, investigating: if social media create a new kind of dynamic arena for their public activism; which factors may stimulate the young to activism; and whether their motivation is powerful enough to resist the status quo."

"Our findings indicate: there are many similar technical capacities; some similar topics are seen as political; and there are different understandings of what is and is not for public discussion among their personal collections of contacts and friends, with varying levels of real-world connections."

"This analysis is important in terms of understanding the ways uses of media technologies may be affected by different cultures, political and social conceptualizations and online communication patterns."

Introduction

"Internet-created, networked communication channels, available in new media platforms, also offer a diversity of communicators a dynamic and wide range of possibilities, and may create several forums for participation in social and political issues."

"This study examines, for our selected research participants (young adults in Turkey and the US), whether the new form of a public sphere is actually being created and maintained by their particular uses of social media."

"This study aims to examine if and how young people mobilize for activism; how they interpret local and global issues; and how young people use social media to response political issues."

"We have suggested young people may have a vital role in creating a new public activism on social media."

Literature review

"Online youth-led activist subcultures and political groups questioning or attempting to alter the status quo have recently claimed legitimacy as creators of vital oppositional spaces for politics and culture, by using new media for making spaces in which a wider diversity of individuals and groups are applying the technologies to help produce creative social relations and forms of democratic political possibility (Kahn and Kellner, [43])."

"Mascheroni [44] argues that participation can be better understood as a socially embedded and contingent online/offline practice that is shaped by the interrelation between habitus, capital and fields, for young people's use of social media is diverse—young activists who engage in online and offline actualizing citizenship make political uses of social media, including citizen journalism and media activism; while by contrast young citizens who are more politically disenfranchised are actually engaged in a variety of creative uses of social media."

Method

"In order to fully understand the global dimensions of social media activism, several descriptive and comparative studies researching young people's civic participation tendencies through social media will be necessary."

"This study merely enters the beginning of a conversation, seeking to identify similarities and differences among different youth cultures in our home countries; describing online activism practices in order to examine how "ordinary" young people use social media to participate in political and other discussions."

"Morgan argues that focus groups are used to help people express themselves openly about sensitive issues, as well as to bridge social and cultural differences (as cited in Brennen, [45], p 59)."

"This analysis is important in terms of understanding the effects of different cultures on political and social conceptualizations of young people, and in terms of online activism patterns."

Findings

"A female respondent (age 21) said that she did not participate in any protest but shared her political views via social media, and she added, "Yeah, a lot of times."

"A male respondent (age 24) said, "When the government implemented a censorship on the news media or impose a broadcast ban on an issue, we can follow the incidents and share our opinions via social media, especially Twitter.""

"Middaugh and others [46] have noted, with the Internet and social media, young people are more able to engage with, and participate in the political process and people of all ages increasingly trust new media because they invite participation through comments."

"They favor flexible and issue-based forms of active participation, such as contributing to projects of NGOs, participating in community-driven initiatives, joining social movements and expressing political opinions in public spaces and in social media to interact with public authorities and exchange views on political issues."

Conclusion

"This study analyzed when, how and why "ordinary" young people use or do not use social media to contribute to public activism."

"Our findings show that for the young people we studied, social media usage for activism and resistance (or participation) in politics is related to the ways they understand their possibilities for different positioning options in the local and global political culture."

"The possibilities for instigating online political discussions are recognized by our research participants in both countries, however among the university students in our study, they are not generally interested in using social media as a public sphere, and are unlikely to state strong positions or start discussions of topics unless they identify themselves as "dissident"."

"This study provides data that weakens our initial over-estimation for social media activism, by illustrating the reasons for limited political messaging by young-adult social media users."

3.1.6 A Supervised Approach for Intra-/Inter-Community Interaction Prediction in Dynamic Social Networks [169]

This is a machine-generated summary of:

Rossetti, Giulio; Guidotti, Riccardo; Miliou, Ioanna; Pedreschi, Dino; Giannotti, Fosca: A supervised approach for intra-/inter-community interaction prediction in dynamic social networks [169]

Published in: Social Network Analysis and Mining (2016)

Link to original: https://doi.org/10.1007/s13278-016-0397-y

If you want to cite the papers, please refer to the original.

For technical reasons we could not place the page where the original quote is coming from.

Abstract-Summary

"In such dynamic scenario, one of the most challenging tasks involves the prediction of future interactions between couples of actors (i.e., users in online social networks, researchers in collaboration networks)."

"We approach such problem by leveraging networks dynamics: to this extent, we propose a supervised learning approach which exploits features computed by time-aware forecasts of topological measures calculated between node pairs."

"Since real social networks are generally composed by weakly connected modules, we instantiate the interaction prediction problem in two disjoint applicative scenarios: intra-community and inter-community link prediction."

"Experimental results on real time-stamped networks show how our approach is able to reach high accuracy."

Introduction

"Almost all the network problems can be reformulated in order to take into account the temporal dimension: communities can be tracked through all their life cycle to unveil their history; incremental ranking can be computed in order to optimize execution costs; links can be predicted using information obtained by the analysis of topology changes in the local surroundings of nodes."

"The topology of these networks evolves over time as new links and nodes may appear or disappear according to the interactions among their users."

"We propose a data mining process able to provide a solution to both tasks: moreover, we formalize the link prediction problem for dynamic networks, the Interaction Prediction."

"Our approach predicts future interactions by combining dynamic social networks analysis, time series forecast, feature selection and network community structure."

Interaction prediction problem
"The classic formulation of link prediction involves the use of the observed network status to predict new edges that are likely to appear in the future or to unveil hidden connections among existing nodes."

"Such modeling choice allows us to make predictions not only for interactions that will take place among previously unconnected nodes, but also for predicting edges that have already appeared in the past."

"This decision is made in order to better simulate the dynamics that real interaction networks exhibit allowing nodes and edges both to rise and to fall."

"In real interaction networks, this model is a good proxy for structural dynamics since it allows to implicitly assign a time to leave to links (i.e., in a call graph, it enables to weight more recent interactions w.r.t."

Proposed approach
"To this extent, we make use of time-stamped network observations and community knowledge besides classical features in order to learn a robust machine learning model able to forecast new interactions."

"We remind that we adopted community discovery algorithms to split interaction networks into communities, and then we used these communities to calculate the features that will be illustrated in the following and to perform the predictions of new interactions."

"It extracts ego networks, i.e., the set of nodes connected to an ego node u, and identifies the real communities by adopting a democratic bottom-up merging approach of such structures."

"We have discussed how it is possible, at least to some extent, to mitigate this problem by restricting the prediction set (i.e., predicting only new edges among nodes that, during the network history, were involved at least in a common community)."

Experiments and results
"To better highlight how the proposed approach performs on real-world networks, we need to compare the outcome of its instantiations varying the combination of community discovery algorithms and time series forecast models used."

"Since in Social our best performing approach is the one built upon DEMON communities, the structural features for the FSF baseline were computed using such partition of the network."

"In order to better understand the impact of the time variable on such very volatile network structure, we also trained a classifier on the same feature set computed on the flattened community-graph (i.e., the graph built by keeping together nodes and edges of all the temporal snapshots)."

Related works

"Link prediction strategies may be broadly categorized into four groups: (q) similarity-based strategies, (2) maximum likelihood algorithms, (3) probabilistic models and (4) supervised learning algorithms (Lü and Zhou [47])."

"In Spiegel et al. [48] tensor factorization is used to select the more predictive attributes, while in Lichtenwalter et al. [49] important features for link prediction are examined and it is provided a general, high-performance framework for the prediction task."

"Other works like da Silva Soares and Prudencio [50], Sarkar et al. [51] show how an approach based on time series modeling the evolution of continue univariate features describing node characteristics substantially helps in solving the link prediction task."

"Several studies related to link prediction such as Feng et al. [52], Fire et al. [53], Jahanbakhsh et al. [54], Lichtnwalter and Chawla [55], Xu and Rockmore [56] try to suggest which are optimal topological structures of a network and the best features to be used with."

"Fire et al. [53] investigate the effectiveness of link prediction by gradually reducing the number of visible links in the studied networks."

Conclusions

"Since networks often model highly evolving realities that cannot easily be "frozen" in time without loss of information, a time-aware approach to link prediction is mandatory to achieve valuable results."

"To this extent, we have exploited the community structure of social networks to both bound the result set and design features whose analysis through time have allowed the description of a high-performance supervised learning strategy."

"Using network partitions as filters make the proposed approach focus only on the prediction of intra-community interactions: to overcome this issue, we propose an experimental setting specifically designed to address inter-community interaction prediction."

"More accurate time series forecast techniques can be evaluated in order to reduce the forecast error and evolutionary community discovery approaches can be used in order to incorporate communities life cycle features within the predictive process."

3.1.7 Technological and Social Networks of a Pastoralist Artificial Society: Agent-Based Modeling of Mobility Patterns [170]

This is a machine-generated summary of:

Rodriguez-Lopez, Juan Miguel; Schickhoff, Meike; Sengupta, Shubhankar; Scheffran, Jürgen: Technological and social networks of a pastoralist artificial society: agent-based modeling of mobility patterns [170]

Published in: Journal of Computational Social Science (2021)

Link to original: https://doi.org/10.1007/s42001-020-00100-w

If you want to cite the papers, please refer to the original.

For technical reasons we could not place the page where the original quote is coming from.

Abstract-Summary

"This paper explores the advantages of simulation to raise the question of how digital and social networks affect the mobility in a pastoralist artificial society in the context of environmental degradation."

"We use a model of a simple pastoralist society in a world without borders to migration by adding the possibility of experiencing the effects of social structures (such as family and friends) and technological networks (e.g., social media)."

"It appears obvious that pastoralist mobility depends on other dimensions as land tenure and traditional knowledge; however, isolating these two effects and experimenting in a simple society allow us to filter the multidimensionality of mobility decisions and concentrate on comparing scenarios in several different social structures and technological network combinations."

"The results show an expected behavior of more connection and more mobility, and a non-linear emergent behavior where pastoralists wait for a longer amount of time to mobilize when they interact using powerful social and technological networks."

Introduction

"New research is needed about the way pastoralists exercise their right to mobility in a world with resource restrictions, where the largest flow of information occurs in a context of complementarity between local social structures, like families and ethnic groups, and how the cellular phone acts as a form of digital network."

"We implement an artificial world with resource constraints facing environmental degradation and analyze how different forms of networks affect mobility using analysis elements of mobile agents, migration theory, and the impact of new technologies."

"Besides, the social networks are made up of family, friendship, and community ties to be the key components of mobility systems [57]."

"With new technologies, pastoralists can make more informed decisions and reduce their livelihood risks despite not yet extensive network coverage in all areas [58]."

Method

"Agent-based models excel at simulating interactions between the individual agents on the micro-level, societal structures on the macro-level, and social networks on the meso-level, as well as how these interactions lead to effects across these levels [59]."

"Through the networks in our developed model, the agents are receiving more information about the resource distribution from the other related agents."

"Since, in this model, vision is a crucial element of mobility, acting as a limiting factor on how agents can move, the information possessed by an agent is represented by the area that they can "see"."

"This is implemented via a probability of 90% that agents trust the information they receive from their family networks."

"In their ethnic network, agents update the information with every third tick."

"Social low" is implemented as described above and "technological low" represents agents being obliged to pay for communication and receiving information only with a time delay."

Results

"The series stabilizes after some agents (two agents on average) return to Sugarscape A, reaching a balance between the available resources and the number of agents stabilizing with around 20 agents in Sugarscape B. The second scenario depicts a situation with family networks and new technologies."

"While the agents begin to move in a fast manner (noticed in the exponential increase of distance measurement from point A) up to and before nine ticks, no agent reaches Sugarscape B. From there, a stronger discovery is produced than in the previous scenario."

"This rapid increase of agents from A to B is demonstrated in the number of agents reaching Sugarscape B. It should also be noted that the series does not stabilize once the agents reach the maximum in Sugarscape B due to a return effect."

Discussion

"This quantitative analysis of the relationship between mobility and technology in social networks complements and extends the qualitative analysis carried out by various authors [60, 61]."

"The current research presents this connection of social networks and technology as an additional form of mobility and a resilience strategy within the literature about a community of pastoralists."

"According to the main supposition, the connection of social and technological networks and their simultaneous interconnection allows us to observe, in general terms, how the greater the connection of networks, the greater the tendency to mobilize."

"The objective of this study is to demonstrate the ability of agent-based modeling to analyze the phenomenon of mobility and migration when other quantitative research techniques fail."

"Modeling, in general, is a research methodology that has advantages when analyzing the situations related to mobility and migration presented in this study."

Conclusion

"This study provides a detailed analysis of the effect of social and technological networks on the decision of a given population to move and migrate from one point to another in an artificial world with freedom to move and facing environmental degradation."

"The strength of the networks not only causes increased mobility, but also generates an increase in the speed and strength in which these four effects occur, resulting in a similar approximate balance."

"This research contributes primarily with its inter-disciplinary approach to connect networks, pastoralism, and agent-based modeling and thereby possible analysis of the effects of different networks on the agents' mobility in a herder's context."

"The results presented here extend the knowledge regarding the positive effect of social and technological networks and their non-linear character on decisions of mobility and migration, especially in a pastoralism context."

3.1.8 Patterns of Action Transitions in Online Collaborative Problem Solving: A Network Analysis Approach [171]

This is a machine-generated summary of:

Li, Shupin; Pöysä-Tarhonen, Johanna; Häkkinen, Päivi: Patterns of action transitions in online collaborative problem solving: A network analysis approach [171]

Published in: International Journal of Computer-Supported Collaborative Learning (2022)

Link to original: https://doi.org/10.1007/s11412-022-09369-7

If you want to cite the papers, please refer to the original.

For technical reasons we could not place the page where the original quote is coming from.

Abstract-Summary

"CPS studies have often emphasized outcomes such as skill levels of CPS, whereas the action transitions in the paths to solve the problems related to these outcomes have been scarcely studied."

"The patterns within action transitions are able to capture the mutual influence of actions conducted by pairs and demonstrate the productivity of students' CPS."

"The purpose of the study presented in this paper is to examine Finnish sixth graders' (N = 166) patterns of action transitions during CPS in a computer-based assessment environment in which the students worked in pairs."

"We also investigated the relation between patterns of action transitions and students' social and cognitive skill levels related to CPS."

"Applying social network analysis to the log file data generated from the assessment environment, we created transition networks using weighted directed networks (nodes for those actions conducted by paired students and directed links for the transitions between two actions when the first action is followed by the second action in sequence)."

"The results showed that pairs having at least one member with high social and high cognitive CPS skills conducted more actions and demonstrated a higher average frequency of action transitions with a higher tendency to conduct actions for different number of times, implying that they attempted more paths to solve the problem than the other pairs."

"Investigating the patterns of action transitions of the dyads in this study deepens our understanding of the mutual influence between the CPS actions occurring within dyads."

"Further, this study contributes to the methodological development of process-oriented research in CSCL by integrating an analysis of action transition patterns with a skill-based assessment of CPS."

Introduction

"The progressions of actions that a pair of students enact contribute towards processes of understanding, planning, solving, and revising, which are universally applicable across tasks in computer-based assessment environments (Care and others, [62])."

"The transition patterns between actions in sequences might capture this mutual influence between students in a CSCL environment."

"Administering CPS tasks designed for dyads in a computer-based assessment environment among sixth graders, we treated students' social and cognitive skills of CPS as outcomes and worked to disentangle the relations between such outcomes and the patterns of action transitions contributed by pairs of students working together."

"This study deepens our understanding of the mutual influence of paired students' actions related to their CPS skills in a computer-based CPS assessment environment."

"Our study contributes to the methodological development of process-oriented investigations in CSCL by integrating an analysis of action transition patterns with a skill-based assessment of CPS."

Background

"Zhu and others [63] propose and validate two different patterns of action transitions in a computer-based problem solving task for individual students: First, they discuss the average frequency of action transitions (e.g., click available buttons, excluding actions of oral or written communication)."

"In expanding from individual problem solving to problem solving in dyads, the present study fills a critical knowledge gap regarding how students' CPS outcomes are related to their average frequency of shifts in action progressions and the dispersion of their attempts given to different actions, thus challenging the interpretation of past findings (Zhu and others, [63])."

"Dyads having at least one student with high social and high cognitive skills appear to conduct more actions with a higher frequency of action transitions on average and with a higher tendency to conduct actions for different number of times in CPS."

Objectives
"We also shed light on the relationship between the patterns of action transitions and students' assessed skill levels of CPS (i.e., social and cognitive skills of CPS)."

"Based on previous studies (e.g., Andrews-Todd & Forsyth, [64]), we hypothesize that pairs including at least one member with high social and high cognitive skills may conduct more actions with a higher frequency of action transitions on average and with a higher tendency to conduct actions for different number of times."

"How do individual students' social and cognitive skill levels vary in CPS tasks?"

"How do patterns of action transitions differ between pairs comprising diverse social and cognitive skill levels of CPS?"

Methods
"To study the patterns of action transitions, we created transition networks with the sequential process data, including 14,448 log file events conducted by 83 pairs in the Olive Oil task."

"The order of the events indicated the sequence of the actions (e.g., drag to move an object, type texts in the chat window) that the pairs conducted in the Olive Oil task."

"The nodes in the network represent actions (e.g., drag to move an object, type in a chat window), and directed links indicate the transitions of the actions when the first action was followed by the second action."

"Generated from the process data in the Olive Oil task, transition networks are directed and weighted to accommodate repeated action transitions and preserve the sequential nature of the data."

"The data points were extracted from log files generated by the students' work in the assessment tasks, consisting of the documentation of each event (i.e., every action conducted by the participants)."

Results and discussion
"In the transition networks generated from the Olive Oil task, student pairs having at least one member with high social and high cognitive skills (i.e., active high-performing and active compensated pairs) exhibited significantly higher mean values for the number of actions, average frequency of action transitions, and the tendency to conduct actions for different number of times compared with other pair compositions."

"Compared with other pair compositions, the active low-performing, compensated low-performing, and passive low-performing pairs (i.e., pairs that did not have one member with high social and high cognitive skills) showed less productive CPS, as indicated by the fewer actions conducted, lower frequency of changing actions on average, and greater tendency to conduct actions for the same number of times (i.e., less likely to repeat particular actions) in their transition networks."

Conclusions

"The purpose of the present study was to explore student pairs' patterns of action transitions and their relation to students' outcomes, namely, CPS skills, in an online CPS environment."

"We also computed fundamental network measures (e.g., average frequency of changes among actions, the dispersion of attempts given to different actions) to study the patterns of action transitions exhibiting the productivity of the student pairs' CPS."

"We extended existing knowledge of patterns of action transitions from an indi vidual (Zhu and others, [63]) to dyad setting, demonstrating the productivity of paired students' CPS."

"Our study further underlined that a pair having at least one member with high social and high cognitive skills of CPS is likely to exhibit productive CPS that is characterized by more actions, a higher average frequency of transition among actions with a higher tendency to conduct actions for different number of times when compared with unproductive CPS."

"Our study offers a novel approach to investigate paired students' patterns of action transitions by integrating process-oriented research with skill-based assessments."

3.1.9 Personal Documentation on a Social Network Site: Facebook, a Collection of Moments from Your Life? [172]

This is a machine-generated summary of:

Sinn, Donghee; Syn, Sue Yeon: Personal documentation on a social network site: Facebook, a collection of moments from your life? [172]

Published in: Archival Science (2013)

Link to original: https://doi.org/10.1007/s10502-013-9208-7

If you want to cite the papers, please refer to the original.

For technical reasons we could not place the page where the original quote is coming from.

Abstract-Summary

"This phenomenon raises an important question for archivists in regard to personal history on the Web: What are the new ways that contemporary people document their life stories?"

"This study looks at Facebook activities from the perspectives of personal documentation."

"Using an online survey, this study investigates how Facebook content presents users themselves and their everyday stories, whether they perceive their activities of using Facebook as personal documentation, and what factors influence such activities."

"Facebook content indeed indicates information of self-presentation and personal documentation of everyday lives of users."

"Attitudes about and activeness on Facebook are the major factors that influence self-presentation and personal documentation activities on Facebook."

"Generic external factors, such as personal archiving in general, do not show strong associations with personal documentation activities as factors."

Introduction

"One interesting aspect of communicating with others is that Facebook provides various venues for communication, e.g., status updates, comments, and profiles, thus making it possible for users to share their thoughts and feelings on happenings, events, places, times, friends, family, and other surroundings of their lives."

"Of these, many have identified and described the main motivations for using Facebook, users' various activities on the site including disclosure of personal information and social activities, and factors influencing Facebook activities, such as privacy."

"This study explores the content users created to communicate on Facebook and how this content was related to the personal documentation of their daily lives."

"Our goals are to examine the general status of Facebook users' self-presentation and personal documentation activities with regard to recording their day-to-day lives; whether they perceive what they are doing on this site as documenting their personal histories; and some surrounding factors that may influence personal documentation activities on Facebook, such as privacy concerns, personal archiving activities, and general social networking activities."

Literature review

"In this literature review, we will focus on studies specifically investigating users' activities in regard to presenting self-identities, personal archiving and recording daily lives, and related issues potentially affecting Facebook activities, such as privacy."

"Although Conti and Sobiesk's study did not use a social network setting, they found that a large percentage of users believed it is each individual's responsibility to protect his/her own personal information online."

"Social pressure and the Facebook environment affected the amount of information users presented; "having a presence on Facebook requires that a person post many pictures, have active discussions with friends, and share personal interests and information" (p. 343)."

"Facebook and other social network sites are continuously accessible to users so that users may not feel any need to back up the content posted on the sites."

Research design

"We seek to understand users' perceptions and activities in relation to personal history on Facebook."

"What are the general factors that may influence personal documentation activities on Facebook?"

"Paolacci and others [65] also identified benefits when comparing with other online and offline methods of recruiting participants, such as Mechanical Turk's supportive infrastructure which provides various survey platform services, functionalities to control subjects at a certain level (e.g., filtering by country which enables to create cross-cultural studies), and the speed of data collection."

"Few studies were available to consult in order to create survey questions regarding this aspect of Facebook usage."

"The content of profiles and wall posts from the perspective of personal documentation were also investigated by general descriptive questions about the extent to which users revealed themselves and were active in creating posts."

"We used descriptive statistics to understand Facebook users' personal documentation activities and their perceptions of these activities."

Findings

"Some of the content of posts participants created on Facebook was directly related to their everyday activities and interests."

"Participants did not seem to make any direct connection between their Facebook activities and the idea of documenting their personal histories."

"When asked questions relating to their views on and practices regarding personal documentation in general, i.e., not necessarily in the context of Facebook, the participants responded very positively to the statement, "I want to keep my personal documents safe" ($M = 4.29$, $SD = 0.75$)."

"We found that the models demonstrated that Facebook use patterns have a positive association to the users' activities of self-presentation and personal documentation."

"These five components were used as individual variables to determine whether there was an association between the respondents' self-presentation and personal documentation activities on Facebook."

Discussion

"As our findings demonstrated Facebook users did engage in self-presentation and personal documentation activities on Facebook."

"Facebook's functionalities of interconnectivity which enable users to create posts and comments as a way to network also encourage the documentation of personal history."

"Users who were vigilant about archiving personal digital information were not necessarily purposeful documenters of their daily lives on Facebook."

"Even though they may not actually intend to record their everyday lives using Facebook, respondents left traces of their personal history on Facebook (and on other social network sites)."

"When popular social network sites are commercial in nature and users are not aware that the content they post to such sites may be important for their personal histories and for social and cultural heritage, cultural institutions should not wait to act."

Conclusion

"This study surveyed Facebook activities from the perspective of personal documentation."

"Facebook may serve contemporary people well as a repository of personal identities and histories."

"That these personal presentation and documentation activities are largely influenced by users' overall perceptions of and activities using general Facebook functionalities, rather than by external factors such as awareness of personal digital archiving."

"The fact that Facebook factors override external factors for personal documentation activities may mean that people are leaving their histories on Facebook whether they intend to do so or not."

"Facebook and other social network services will function as personal documentation tools as long as they satisfy users with their services pertaining to networking and entertaining."

3.1.10 Topic Adaptive Sentiment Classification Based Community Detection for Social Influential Gauging in Online Social Networks [173]

This is a machine-generated summary of:

Kumaran, P.; Chitrakala, S.: Topic adaptive sentiment classification based community detection for social influential gauging in online social networks [173]

Published in: Multimedia Tools and Applications (2022)

Link to original: https://doi.org/10.1007/s11042-021-11855-3

If you want to cite the papers, please refer to the original.

For technical reasons we could not place the page where the original quote is coming from.

Abstract-Summary

"Online Social Networks (OSNs) such as Twitter, Facebook, Instagram, and WhatsApp are turned as a place for many of people in recent years to spend much of their time, due to their huge network structure and massive amounts of user-generated data in it."

"To address the social influential gauging a novel Topic Adaptive Sentiment Classification based Community Detection (TASCbCD) algorithm is proposed to detect communities in twitter network based on the results of topic based sentiment classification using robust topic features."

"The WordNet and SentiWordNet are benchmark data sets that are used for supporting those classification to achieve the desired results."

"The resultant communities give a better visualization of identifying the over lapping communities that helps to gauge the topic based social influential user in OSNs."

Introduction

"Twitter is a popular online social networking site that provides a platform for the general public to express their opinions or feelings about social events and products in the form of tweets [66, 67]."

"From a data mining perspective based on the features above, two fundamental processes can undertake with Twitter data:(a) Graph mining, based on user network links (b) Text mining, based on user-generated content [68]."

"With this in view, a Robust Topic Features based Sentiment Classification for Overlapping Community Detection algorithm proposed for detecting communities, and its overlapping structure based on a different set of topics and its sensitivity level(polarity) to select influential users in the large-scale Twitter network [69, 70]."

"Direct selection is a complex task; at best, community detection in online social networks helps select competent, influential users."

"This paper proposed community detection, which helps to enhance the selection of influence spreader according to the topic to improve the prediction accuracy of information diffusion in OSN [71]."

Literature review

"In a network-based approach, node or edge centralities used to detect communities [72]."

"Content-based community detection works on the premise that rich information in networks is present in node and edge contents, rather than its structure."

"Researchers [73] proposed a method to combine features such as nodes, edges, contents, and structures to detect communities over an extensive network."

"The topic-wise influence node selection is not possible to identify the active node for the topic from the network's community structure The network-based community detection method does not focus on user tweets, suggesting that users' current opinions are not taken into consideration The content-based community detection method fails to notice the physical connection between two users in the network The other community detection methods fall short of addressing the problem of overlapping communities Hence, a community detection system is required that identifies overlapping communities with a combination of network and content features, along with topic sensitivity level (positive, negative, neutral) in it."

Problem statements

"A few issues identified in designing an efficient system for community detection in OSNs are as follows: Problem 1: How to design a topic polarity-based community detection system in Twitter Network?"

"No algorithm has so far worked on topic adaptive polarity-based community detection on Twitter."

"A few address overlapping communities at the bare minimum, but fail to use the topic adaptive feature and polarity to detect overlapping communities in highly dense social networks accurately."

"Problem 3: How to design a community detection system with a combination of network and content features form dynamic social network like Twitter?"

"Significant contributions of a proposed Robust Topic Feature-based Sentiment Classification in Community Detection method include the following: Improved community detection with a high Rand index value, ensuring a combination of network and content features with topic sensitivity."

Proposed work

"The proposed technique detects community structures on Twitter based on topic adaptive sentiment sensitivity(polarity) level features in tweet content, that helps to detect overlapping communities between two topics or the same topics with different sensitivity (positive or negative or neutral)."

"The proposed TASCbCD technique comprises the following processes: Twitter data collection and preprocessing, Topic modeling, Sentiment analysis, Network graph construction, and Community detection."

"Each tweet is first given to topic modeling for identifying the related topic of the tweet; afterward, the appropriate sentiment class label is identified based on the various extracted features like text, non-text, user content, tweet topic-adaptive keywords, and network parameters."

"The proposed TASCbCD technique consists of the following processes: Twitter data collection and preprocessing, topic modeling, sentiment analysis, network graph construction, and community detection."

Results and performance analysis

"The results above depict different sets of topic sensitivity based community structures using the proposed method."

"The results show that the proposed TASCbCD method has effectively identified the topic-sensitive community and overlapping communities."

"The results show that the proposed TASCbCD method provides clear improvements in community detection for the topics and their sensitivity."

"The proposed TASCbCD method produces significant community detection results for different topics with its sensitivity(polarity)."

"It is concluded from the results above that in all cases, the proposed TASCbCD method is found to be better in terms of the Rand index and Scale density measure than network-based and content-based methods of community detection for different topics over the Twitter network."

"The overlapping communities are also easily identified in the proposed TASCbCD method due to topic sensitivity-based community detection."

Conclusion and future work

"The technique, based on topic sensitivity features in tweet content, identifies overlapping communities between two topics or the same topics, with different sensitivity indicators (positive, negative, and neutral)."

"Overlapping community detection enhances influence spreader identification to maximize information diffusion."

"The result of the proposed method is compared with other methods and indicates that our methodology effectively identifies topic-sensitive communities on Twitter."

"The result of the community structures discovered is used to identify and rank influence spreaders for information diffusion."

"For community detection, the overall average Rand index for communities of 20 different topics is 90%, 92% and 95% in Network1, Network2 and Network3 respectively."

"It is required to develop algorithms that make use of dictionary information in order to identify the diffusion process, even for terms that do not belong to the set."

3.1.11 Detection of Sociolinguistic Features in Digital Social Networks for the Detection of Communities [174]

This is a machine-generated summary of:

Puertas, Edwin; Moreno-Sandoval, Luis Gabriel; Redondo, Javier; Alvarado-Valencia, Jorge Andres; Pomares-Quimbaya, Alexandra: Detection of Sociolinguistic Features in Digital Social Networks for the Detection of Communities [174]

Published in: Cognitive Computation (2021)

Link to original: https://doi.org/10.1007/s12559-021-09818-9

If you want to cite the papers, please refer to the original.

For technical reasons we could not place the page where the original quote is coming from.

Abstract-Summary

"Determining how language variations allow the detection of communities, together with the relevance of specific vocabulary (proposed by the National Council of Accreditation of Colombia (Consejo Nacional de Acreditación - CNA) to determine the quality evaluation parameters for universities in Colombia) in digital assemblages could lead to a better understanding of their dynamics and social foundations, thus resulting in better communication policies and intervention where necessary."

"The approach presented in this paper intends to determine what are the semantic spaces (sociolinguistic features) shared by social groups in digital social networks."

"The current approach allows determining the sociolinguistic features of social groups in digital social networks."

"Its application allows detecting the words or concepts to which each actor of a social group (university) gives more importance in terms of vocabulary."

Introduction

"This large amount of data creates opportunities and scenarios for analyzing linguistic variation and change, based on a community approach and through the exploration of language concerning society."

"Because a social network describes the structure of a given community of speakers, it is necessary to determine the distribution of linguistic elements in it."

"Once a community is detected, a study of their shared language features allows for a better understanding of community dynamics and values (motivations)."

"The usefulness of this approach lies on exploring community integration in large dynamic network structures, with the aim of discovering communities based on similarities shared by their components."

"Fani and others[74] use a neural graph embedding based on temporal content similarity, thus capturing social network connections."

"The novelty of the article lies in the use of different techniques of Natural Language Processing, Social Network Analysis and Artificial Intelligence for the detection of communities."

Methods

"Considering this, the identification of a social group or community in a digital social network contributes to the relevance cycle."

"In the design cycle, the identification of the sociolect and distinctive words of the social group in question are carried out through the extraction and processing phases."

"The Weirdness Index (WI) [75] for a particular term is obtained by calculating the quotient between (i) the normalized frequencies of the words found in the corpus of the social group that are greater than one and (ii) the normalized frequency of the words in the corpus CREA."

"In order to detect social groups, we analyzed the messages posted in the public accounts of Colombian universities and tried to detect communities through their vocabulary."

"This model was used to detect the most relevant words employed in the messages of the social group of Colombian universities."

Results and Discussion

"The specified social group of universities was selected due to a norm that classifies them with respect to the process of institutional accreditation."

"To identify vocabulary related to each one of the directives defined to achieve accreditation, we based our hypothesis on the existence of conglomerates of universities using a common vocabulary that concerns a specific quality factor."

"The concepts or quality factors are inferred from the messages of each university through BL-LDA, based on the most frequent words related to the topics extracted from the university corpus."

"This, in turn, allows identifying the correlation between the semantic values of the studied public universities, using the quality accreditation vocabulary determined by the CNA."

"Since university Tweets can be processed, classified and grouped, the present research study identified comments concerning accreditation factors issued by different interlocutors in a digital academic ecosystem."

Conclusions

"A tailored study of a community's language allowed us to detect specific sociolinguistic features that might reveal relevant vocabulary which is not shared with other communities, even in close geographies or demographics."

"Better understanding of specific communities and, particularly, of their shared language, might be of central importance to tailor messages for them, or to detect important trends, whether in public health, marketing, politics or national security."

"Shared language might be heavily influenced by topics, which makes it difficult to differentiate community specifics from common interests."

"Further research should cover these difficulties, including community detection work based on both links (such as mentions) and shared language comparisons."

3.1.12 Ambient Awareness to Strengthen the Family Social Network of Older Adults [175]

This is a machine-generated summary of:

Cornejo, Raymundo; Tentori, Mónica; Favela, Jesús: Ambient Awareness to Strengthen the Family Social Network of Older Adults [175]

Published in: Computer Supported Cooperative Work (CSCW) (2012)
Link to original: https://doi.org/10.1007/s10606-012-9166-2

If you want to cite the papers, please refer to the original.

For technical reasons we could not place the page where the original quote is coming from.

Abstract-Summary
"We explore how situated displays can intuitively provide ambient awareness to strengthen the family social network of older adults."

"We designed and developed Tlatoque, a situated display, to seamlessly integrate older adults into the SNS used by their relatives."

"The results of a 21-week deployment study of an initial version of Tlatoque in one extended family showed that the older adult became more conscious of relatives' activities, which also enriched in-person encounters."

"Relatives expected the older adult to reply to their posts in the SNS and Tlatoque lacked mechanisms for the older adult to provide feedback."

"Our results indicate that the second design was successful in providing ambient awareness to the older adult as well as to her relatives."

"We conclude that situated displays that provide SNS services can assist the integration of older adults to their social network and contribute to enhance asymmetric relations between the older adult and younger relatives."

Introduction
"Social networking sites (SNSs), such as Facebook, are a growing source of social capital (Joinson [76]) enabling older adults to maintain their current relationships and strength their social network."

"Older adults might be missing opportunities for sharing and bonding with relatives who are increasingly using these services to socialize."

"Although situated displays have been theorized as an intuitive interface to connect families (Raffle and others [77]), those that provide ambient awareness to support the older adult's social relationships are scarce."

"We explore how situated displays can intuitively provide ambient awareness to enhance relationships between older adults and their relatives."

"We designed and implemented two versions of Tlatoque, a situated display that provides older adults awareness of the information being shared by their relatives in a SNS through an intuitive interface."

Information technology for connecting family members
"We describe some of the research efforts devoted to the study of the use of information technology to provide awareness in family relationships, and more

specifically those projects that research how to connect older adults with their relatives."

"Efforts on connecting family members have centered on providing activity awareness to inform relatives of older adults' performed activities –most of these efforts pursue the notion of "ageing in place" (Consolvo and others [78]; Mynatt and others [79])."

"The Digital Family Portrait (Mynatt and others [79]) provides awareness to relatives of the activities performed by the older adult, using qualitative icons (e.g. icons representing trees or butterflies) to preserve his/her privacy."

"With these projects where relatives were the only ones obtaining activity awareness, other projects have explored the impact of providing activity awareness to both older adults and family members, e.g., relatives or older adults can explicitly share their activities giving others the opportunity to be aware of their daily routines."

[Section 3]

"While, some older adults have relatives acting as "proxies" to show them the information relatives share in SNSs and gain access to AA, others rely on PA (e.g., face-to-face encounters and telephone conversations) to keep in touch with relatives or learn about their activities."

"A second persona (Person 2:P2) is a socially active teenager living in the same city as the older adult but who rarely visits her."

"A third persona (Person 3:P3), includes relatives who live in a different city where the older adult lives, and keep grandma up to date of important events in their lives through phone calls or emails sent to grandma through P1."

"From the four personas identified, the older adult has more PA of P1, even though P2 also lives in the same city as the older adult."

Tlatoque 1.0

"On the first day of Tlatoque's 1.0 deployment, the older adult was given a general overview of the system (e.g. how to turn it on) and she chose where to place the digital frame within her home."

"After a couple of weeks of use, the older adult received a second training session on how to control the digital frame (e.g., how to change the displayed photograph)."

"During the deployment and post-deployment phases, participants were asked to discuss issues related to Facebook's use, Tlatoque 1.0 usability, usefulness, and impact on the family social experience including communication and family ties."

"Although G1 enjoyed looking at the photos, she expressed little interest in sharing information through Tlatoque back to the SNS to avoid their relatives to make fun of her."

"After Tlatoque 1.0 was introduced relatives increased the number of uploaded photographs with family content to share them with G1."

Designing for bidirectional ambient awareness

"Before Tlatoque 1.0, G1 attributed characteristics to her relatives based on their few personal encounters and on the observations and information P1 shared of them."

"Even though Tlatoque 1.0 helped G1 to be more aware of relatives' activities and interests increasing G1's ambient awareness; the limited interactivity of the digital frame that only enabled the control of the photographs carrousel, did not allow G1 to reply to the messages she received nor upload comments of the photographs she liked."

"To traditional SNS feedback services, automatic sharing can be configured based on several factors such as a constant amount of time (e.g., every hour), state of the feedback service (e.g., when new information is displayed) or specific conditions of the information (e.g., extreme weather conditions) to decide when it is relevant to share and receive this content, for both the older adult and the relative, respectively."

Tlatoque 2.0

"Relatives must accept the older adult's "friend request" subscribing to Tlatoque's application to enable the sharing of their Facebook's personal information."

"Relatives living away were less connected to the older adult, for example, G2 has limited information of her grandson George (P3) who lives in Alameda C.A., U.S.A., most of her interpersonal attribution of him and his family are based on events from the distant past, when he used to live in the same city as she does."

"Rather than scheduling the activity of looking at the photographs, G2 decided to use Tlatoque 2.0 when she wanted to know some specific information such as the weather or when she felt like looking at photos of her relatives."

"We observed a change in the social content being shared with the inclusion of the older adult in the SNS, shifting towards more aforetime content (e.g., old photographs) and daily events."

Effects of bi-directional ambient awareness

"The awareness G2 had about her relatives including mood, their preferences on sharing certain kinds of photographs, and their daily events made G2 feel more confident to start conversations."

"Relatives could start conversations related to recent events and G2 was engaged during these talks because she had seen the photographs shared of these events."

"P3 attributed this pleasant surprise to the active behavior of G2 in looking constantly at the photographs shared by her relatives."

"P4 felt more confident talking to G2 because she knew that G2 liked certain kinds of photographs that she had uploaded."

"These constant action streams performed by G2 helped P4 nurture the relationship between them and understand what kind of information and interests they share in common. (Granddaughter in-law, P4): "I feel her affection when she sends her "Likes"; also there are several likes to my photos."

Conclusion

"The use of Tlatoque enabled the older adults to become SNS users and experienced an increase in ambient awareness of their relatives."

"Tlatoque's initial implementation supports ambient awareness for the older adult; each photograph, its caption and the photograph's owner allowed the older adult to observe particular aspects of her relatives' life, increasing the information she had of each relative."

"With the deployment of Tlatoque 2.0, weak ties also used the increased ambient awareness to enrich the "perceived personality" they had on the older adult; they attributed new characteristics to the older adult based on the kind of information she liked."

"Relatives reported how important was sharing this kind information because not only their relatives will be aware of their lives, but more importantly the older adult would glance at their activities, moods and relevant events in their daily life."

3.1.13 Emotion Classification of Social Media Posts for Estimating People's Reactions to Communicated Alert Messages During Crises [176]

This is a machine-generated summary of:

Brynielsson, Joel; Johansson, Fredrik; Jonsson, Carl; Westling, Anders: Emotion classification of social media posts for estimating people's reactions to communicated alert messages during crises [176]

Published in: Security Informatics (2014)

Link to original: https://doi.org/10.1186/s13388-014-0007-3

If you want to cite the papers, please refer to the original.

For technical reasons we could not place the page where the original quote is coming from.

Abstract-Summary
"Publicly available online social media sites make it possible for crisis management organizations to use some of these experiences as input for their decision-making."

"We describe a methodology for collecting a large number of relevant tweets and annotating them with emotional labels."

"This classifier has been used as a basis for constructing a decision support tool where emotional trends are visualized."

"To evaluate the tool, it has been successfully integrated with a pan-European alerting system, and demonstrated as part of a crisis management concept during a public event involving relevant stakeholders."

Introduction
"We describe our methodology for collecting crisis-related tweets and tagging them manually with the help of a number of annotators."

"By comparing the results to those achieved when using a rule-based classifier we show that the used machine learning algorithms have been able to generalize from the training data and can be used for classification of new, previously unseen, crisis tweets."

"As part of this event, the classification of social media posts was used to visualize emotional trend statistics for the purpose of demonstrating the idea of using social media input for informing crisis management decisions."

"A methodology section then follows, where we describe how crisis-related tweets have been collected, selected using automated processing, and tagged manually by a number of annotators in order to create a training set."

Related work
"A seminal work on the use of machine learning for sentiment analysis is the paper by Pang and others [80], showing that good performance (approximately 80% accuracy for a well-balanced dataset) can be achieved for the problem of classifying movie reviews as either positive or negative."

"Social media monitoring techniques for collecting large amounts of tweets during crises and classifying them with machine learning algorithms has become a popular topic within the crisis response and management domain."

"The use of natural language processing and machine learning techniques to extract situation awareness from Twitter messages is suggested in [81] (automatic identification of tweets containing information about infrastructure status), [82] (classification of tweets as positive or negative), and [83] (classification of tweets as contributing to situational awareness or not)."

"The main difference between our work and the papers mentioned above is that most of the previous work focus on sentiment analysis (classifying crisis tweets as positive or negative), whilst we focus on affect analysis or emotion recognition [84], i.e., classifying crisis tweets as belonging to an emotional state."

Methodology

"Although this in theory could be solved by sampling a large enough set of tweets to annotate, there is a limit to how many tweets that can be tagged manually in a reasonable time (after all, this is the main motivation for learning such classifiers in the first place)."

"If the used strategy to select tweets based on keywords would impact the annotated data and thereby also the learned classifiers too much, this could result in classifiers that perform well using the annotated data, but generalizes poorly to "real" data without the bias."

"Our test data has been generated by letting a human annotator (not part of the first annotation phase) tag tweets from the originally collected Twitter dataset until sufficiently many tweets had been discovered for each emotion."

Experiments

"In the learning phase we used the tweets tagged as positive, anger, and fear as training data, which resulted in classifiers that learned to discriminate between these three classes."

"For the actual classification of new tweets we then let the machine learning classifiers estimate the probabilities $P(anger| f_1,\ldots, f_n)$, $P(fear| f_1,\ldots, f_n)$, and $P(positive| f_1,\ldots, f_n)$, where $f_1,\ldots, f_n$ refers to the used feature vector extracted from the tweet we want to classify."

"The results are also compared to two baseline algorithms: 1) a naïve algorithm that picks a class at random (since all the classes are equally likely in a balanced dataset, this corresponds to a simple majority classifier), and 2) a somewhat more complex rule-based classifier constructed from the heuristics (keywords) used when selecting the tweets to be annotated manually in the training data generation phase."

Design and implementation of a tool for visualizing emotional trends

"Through interacting with this graph, the user has the possibility to interact with the underlying dataset, and thereby obtain a further understanding of how the feelings expressed on social media vary as time passes."

"At the bottom of the tool, the user has the possibility to drill down into the underlying dataset and see the actual posts in the database."

"The GUI provides a number of ways to apply filters to the underlying dataset and thereby choose the social media posts to be visualized."

"The different visualizations are always kept consistent with these filters and with all other settings, i.e., the different parts of the graphical user interface provide different means to visualize one and the same dataset."

Discussion
"The obtained results show that the machine learning classifiers perform significantly better than chance and the rule-based algorithm that has been used as a baseline."

"The confusion matrices also indicate that even better accuracy could have been achieved using a simple ensemble combining the output from the rule-based and machine learning-based algorithms."

"This is a first attempt to classify emotions in crisis-related tweets, and by improving the used feature set and combining the machine learning paradigm with more non-domain specific solutions such as the affective lexicon WordNet-Affect [85], better accuracy can most likely be achieved."

"Some of the classification errors were a result of the annotators receiving instructions to classify tweets containing any of the emotions fear, anger, or positive as other if the tweets relate to a "historical" state or if the expressed emotion related to someone else than the author of the tweet."

Conclusions and future work
"Evaluation results suggest that a SVM classifier performs better than a NB classifier and a simple rule-based system."

"Results obtained when ignoring the non-specific category other (reaching accuracies over 75% for the SVM) also suggest that combining the learned classifiers with algorithms for subjectivity recognition can be a fruitful way forward."

"From a crisis management perspective, it will also be necessary to investigate to what extent the used methodology and the developed classifiers are capable of coping with more generic situations."

"We hope to have developed classifiers that to at least some significant extent classify based on hurricane and crises behavior in general, rather than solely being able to classifying Sandy-specific data."

"Doing this for several different crisis types and then applying the same classifiers, should make it possible to quantify how capable the developed classifiers are when it comes to classifying tweets from 1) other hurricanes, 2) other types of natural disasters, and 3) crises in general."

Appendix: instructions given to annotators
"You have been given 1000 tweets and a category."

"To help make sure that the tagging is as consistent as possible between all annotators, you will be given some guidelines to make sure that everyone tags the tweets in a similar way: "Fear" is the category containing tweets from people who are scared, afraid or worried. ""

"The emotion should relate to the author of the tweet, not other people mentioned by the author."

"The tweet "Maggie seems real concerned about Hurricane Sandy…" should not be tagged as "fear", since it's not the author of the tweet that is being concerned."

"The tag should be based on the author's mood when the tweet was written."

"should not be tagged as "fear", since it relates to past events, while we want to know how people were feeling when the tweets were posted."

3.1.14 The Influence of Personal Communities in Understanding Avoidable Emergency Department Attendance: Qualitative Study [177]

This is a machine-generated summary of:

McKenna, Gemma; Rogers, Anne; Walker, Sandra; Pope, Catherine: The influence of personal communities in understanding avoidable emergency department attendance: qualitative study [177]

Published in: BMC Health Services Research (2020)

Link to original: https://doi.org/10.1186/s12913-020-05705-5

If you want to cite the papers, please refer to the original.

For technical reasons we could not place the page where the original quote is coming from.

Abstract-Summary

"Estimates that between 11 and 40% of ED attendances are non-urgent, with 11% of patients being discharged from the ED without treatment (NHS Digital 2017), and a further 44% require no follow-up treatment (NHS Digital, Hospital Accident and Emergency Activity 2016-17, 2019) is cited as evidence that these patients did not require this level of care."

"The solution to not using the most appropriate point in the system has traditionally been seen as a knowledge problem, requiring, improved sign-posting and information to enable people to self-manage or use health care management for minor ailments."

"A considerable literature points to help seeking as a social process influenced by a range of contingencies and contextual factors including the way in which lay people influence health care utilisation (Giebel and others BMJ Open 9:1, 2019)."

"We extend and unpack further influencing decisions about seeking formal health care with reference to how they are shaped and informed by and within personal social networks."

"We undertook a personal network mapping and qualitative interview-based study to look at, problematize and understand attendance for non-urgent problems."

"We used network analysis and methods to map and characterise the personal communities of people seeking help from ED for minor ailments and semi-structured interviews with 40 people attending a single ED and associated GP hub providing equivalent care."

"Interviews were built around an ego network mapping activity and a topic guide structured to explore attender's narratives about why they had visited the ED."

"We analysed number and type of network connections and undertook a thematic analysis to identify how imagined and actual network members and influences were implicated in ED attendance."

Background

"Whilst the configuration of services surrounding ED have been the focus of recent research, showing how the supply side of health care shapes patient perceptions, the notion of lay referral points to the possible influence of significant others in seeking help."

"We extend this dual focus by exploring the role of personal communities and contacts implicating a wider set of social network members and relational work relevant to seeking help which might influence ED attendance."

"Social network analysis thus offers an opportunity to avoid the trap of victim blaming (and in this case the construction of the problem as one of 'inappropriate attendance'), moving away from individualised responsibility and binary representations of 'good' and 'bad' forms of ED service use [86, 87] to include a focus on the influencing potential of network connections."

"Our study mapped the personal networks of people attending ED for minor illness and injury in order to describe and explore their decision making around help-seeking."

Method

"Professional view, and clear public health messages about appropriate attendance, it was clear in our data that relationships or ties to health service network members did not make it less likely that they would present at the ED."

"In many cases patients knew that the ED was not the place where they should be, but they were advised to attend by other system-related network members, both individual health care professionals and the NHS 111 triage telephone service: I spoke to my counsellor about it, and then phoned the GP, but there weren't any appointments."

"One finding to emerge from our data, was that contra to the health education and policy rhetoric about appropriate attendance, it seems that attendees were often pushed by people outside these core familial and friendship networks to attend the ED for minor illnesses and injuries."

Discussion

"We have shown that using social network analysis provides unique insights into people's help seeking behaviour when using the ED, showing how networks narrow due to lack of resources and difficulty engaging social connections."

"There is a gap between the number of actual network members considered to be of relevance in the decisions to attend ED and those expressed in people's imagined networks of influence."

"There is a qualitative difference in the network resources considered when making this decision; in practice people appear to be swayed more by 'professional influencers', health services staff and other workplace members, and draw less on familial network members or on imagined group and community resources."

"From our analysis it is little wonder then that attempts to encourage self-care (for example by promoting the use of community or digital resources) or attempting to discourage ED attendance via health education messages fail."

Conclusion

"Many of the people we interviewed who had attended the ED for such conditions understood the rhetoric and debate about appropriateness, but they were often directed to the ED by 'professional influencers' including health services staff."

"We suggest that any intervention directed to reducing ED attendance for minor illness and injury, rather than perpetuating victim blaming by labelling attendance as 'inappropriate', should consider how different professionals could better influence help-seeking."

"The examination of the real and imagined networks of people managing their health needs in crisis provides a unique opportunity to further understand this area and to consider hitherto unnoticed service influences."

"Such an approach could alert professionals to real or absent opportunities for patients to engage in alternative courses of action to ED attendance and provide wider opportunities for utilising appropriate resources outside the ED."

3.1.15 Decision-Making Approaches in the Antagonistic Digital Communication of the Online Communities Users [178]

This is a machine-generated summary of:

Fedushko, Solomiia; Molodetska, Kateryna; Syerov, Yuriy: Decision-making approaches in the antagonistic digital communication of the online communities users [178]

Published in: Social Network Analysis and Mining (2023)

Link to original: https://doi.org/10.1007/s13278-022-01021-4

If you want to cite the papers, please refer to the original.

For technical reasons we could not place the page where the original quote is coming from.

Abstract-Summary

"This study investigated the decision-making process in the antagonistic digital communication of internet services users."

"Implementation of the proposed methods in this study minimized the negative impact of the spread of destructive content and actions of unfair competitors in the ecosystem of social internet services."

"The innovativeness of the suggested methods is the ability to make real-time decisions in a situation of antagonistic behavior of online users."

"This article presented approaches to decision-making by online community administrators under conditions of antagonistic behavior of the online services, which ensures the increase of efficiency of online communication."

"Timely and effective decision-making eliminated the negative impact of conflicts between communities and the spread of threats to users' information security."

"The algorithm of decision-making by users of stable internet services is suggested."

Introduction

"The study aims to increase the efficiency of social communication in the internet services of subjects of economic activity in conditions of antagonistic behavior in an information environment to improve the strategic decision-making process in online communities."

"This article proposes for the first time to take into account the information situation in the decision-making process for the management of online communities of social Internet services."

"The preconditions for the sustainable development of online communities in social Internet services are being created in the conditions of spreading and implementing threats to the information security of users and their associations."

"In the era of using social networks and information space as a tool for hybrid warfare, influencing various spheres of community activity, digital community functioning strategies are critical to ensure the sustainable development of groups of actors (online users) and the interests of economic entities they represent."

Related Work

"Queries were created to select a set of records in Scopus and Web of Science databases: TITLE-ABS-KEY (decision-making AND techniques), TITLE-ABS-KEY (online AND community), TITLE-ABS-KEY (antagonistic AND behavior AND online), and TITLE-ABS-KEY (innovation)."

"Scientists (Salehin and others [88]) used the sentiment analysis technique to detect online users' mental conditions and human mental behavior on the web."

"The scientist also researches political conflicts in online communities and the influence of antagonistic behavior on online community users."

"The paper's authors conducted a critical analysis of research in innovation in information technology (Kechagias and others [89]) and web technologies."

"The common direction of the study of the considered scientific works is the study of the peculiarities of economic activities in social internet services and the increase in its efficiency and the behavior of users of online communities."

Methods

"The formalization of multiple possible decisions R, taken by the users of a stable online community in its functioning in social internet services, is carried out."

"Of particular importance for studying the processes of development of stable online communities in social internet services is an informational situation, in which antagonistic interests of the representatives of different groups of users are manifested in the informational space of the services (Arayankalam and Krishnan [90]; Arsenyan and Mirowska [91]; Ning and You [92]; Yokoyama and others [93]; Taslimov and others [94]; Moustakides and Verykios [95])."

"Let us consider the case of making decisions on managing the strategy for developing a stable online community under the conditions of antagonistic behavior of the social internet services environment, taking into account the requirement that the function achieves the maximum value."

"Considering that due to effective strategic communication in the information environment of the services (Giocoli and vs. [96]), the influence of destructive content on the users is decreasing; the Wald criterion, by its conservatism, can significantly limit the level of winnings of a stable online community, a result of the decision made."

Results

"Let us consider the application of the suggested approaches to decision-making to implement the sustainable functioning of stable online communities in the conditions of antagonistic opposition in the information environment of the services."

"A decision must be made to change the number of online community users."

"In order to confirm the work of the considered approaches for decision-making, an experiment was conducted in which the users and administrators of the online community who function in the antagonistic environment of the social internet service Facebook took part (Facebook page of Lviv Polytechnic National University [97])."

"This approach will provide a high degree of clarity of data display and allow the user to evaluate the effectiveness of the proposed approach to decision-making by users and administrators of online communities of social Internet services."

"An online community is a commercial informational project that targets many users."

Discussion

"Increasing the number of users of a stable community through the commitment of users in online communities with opposite goals of functioning ensures the maximum efficiency of the solution, which confirms the Wald criterion and minimizes the risks according to the Savage criterion."

"The study of the peculiarities of information warfare in social media shows that in the case of online communities' administrators using the Wald criterion to make decisions, it is necessary to follow mixed strategies of the opposing side."

"The proposed solution in this work increases the profitability of online communities, quickly and efficiently adapts to crises to their owners, quickly implements strategic decision-making in online communities by administrators, develops a crisis plan, and implements a strategy for the sustainable functioning of online communities under the antagonistic behavior of the social internet services users."

Conclusions

"In such an environment, users' development of online community strategies and decision-making is directly dependent on the state of the information environment of social internet services."

"The results of our research, confirmed by numerous experiments, have shown that the use of decision-making models based on the information situation in social internet services is an effective tool for implementing a strategy for the sustainable functioning of the online community."

"The informational situation of decision-making by the users in social internet services is defined as the level of gradation of unrecognizability in the informational environment of their states from a given diversity at the moment of making a decision by the users of a stable online community."

"Future research goals striving to further focus of research are to identify practical approaches to making decisions on the steady development and functioning of stable online communities in other information situations in social internet services."

3.1.16 What Makes it Likeable? A Study on the Reactions to Messages in a Digital Social Network: The Case of Facebook in Farsi [179]

This is a machine-generated summary of:

Sabbar, Shaho; Hyun, Daiwon: What makes it likeable? A study on the reactions to messages in a digital social network: the case of Facebook in Farsi [179]

Published in: SpringerPlus (2016)

Link to original: https://doi.org/10.1186/s40064-016-3771-3

If you want to cite the papers, please refer to the original.

For technical reasons we could not place the page where the original quote is coming from.

Abstract-Summary

"This study suggests that a useful research for communication, marketing and advertising would be one that looks for patterns in the reactions of the nodes toward different pieces of information."

"This study has used Facebook to see how people have reacted to different types of messages in terms of liking, sharing and commenting."

"The practical aspect of the study comes after a short theoretical discussion on the issue of flow of information in a digital world."

"This study, its theoretical discussion and results suggest that it would be practical to study the relations between the characteristics of Facebook messages and the type of reactions (liking, sharing and commenting) that they attract."

Background

"Among different possible characteristics of online materials, Choi and Toma considered positivity and negativity and their relation to the network behaviors of the nodes (here sharing)."

"Lee and others [98] looked into relations between personality traits such as being extravert, narcissist etc with a series of actions on Facebook such as liking, sharing, frequency of updating their status, etc The current study excludes the characteristics of the nodes in order to avoid over complication of the work."

"If significant relations are found between the characteristics of the messages (here Facebook posts) and the number of times they are shared, one could try to use them to explain why some materials are shared more often than others."

"If we can find some relations, then one can argue that looking for this type of relations is a reasonable method to study the flow of different messages through a social network."

Methods

"On Facebook pages, the admin or admins of the page post different messages and each message receives a specific number of likes, shares and comments."

"Characteristics of the messages (Facebook posts) We tried to define the characteristics of the messages with the help of two colleagues."

"Three researchers cooperated in the long and challenging process of finding and defining the common characteristics of the posts."

"23 of the initially proposed characteristics were omitted from the list, either because their operational definitions could not get clear enough or because there were not enough cases in the chosen Facebook posts to lead to any statistically significant results."

"The 591 Facebook posts were coded for having or not having any of 46 characteristics."

"Network Behavior of the Nodes (User comments on Facebook posts) The 770 comments were coded in a similar way to the process used for the posts."

Results

"The first question concerns the relations between the characteristics of the Facebook posts and the number of likes and shares they received."

"Statistical analysis of the relations between the number of people who liked and shared a Facebook post and the existence of any of the 46 characteristics resulted in 30 significant relations."

"There was a significant relation between the existence of satire in the posts and the number of people who liked it."

"This indicates that there was a direct relation between the existence of this characteristic in the post and its likelihood to receive likes from the people."

"The analyses aimed at answering the first question of the study show that characteristics related to the materials' subject, form of expression and existence of praise were in significant relations with the likelihood of the posts being liked."

Discussion

"If the average level of significance in this project is 98%, out of 130 reported results 2 or 3 might be a result of chance rather than robust existing patterns."

"The only way to find out whether or not there were such cases among the results, and if yes, which of the results were due to chance, is waiting for additional studies to provide us with a better understanding of the issue."

"It is unlikely that many false relations exist in the results."

"Even where we don't seem to understand a relation there may be very good explanations; however sometimes even knowing why a relation exists is not worthy of the effort needed to find it out."

"The behaviors studied here are the results of human reactions."

Conclusions

"Other studies can target specific networks, specific communities and specific types of materials (say brands, political issues, etc) and use the method proposed in this research to study what characteristics can increase the probability of people reacting in desired ways."

"In such an ideal situation we may be capable of predicting how long a message would keep moving before it stops reaching new people; how many people it would reach; what types of people it would reach, etc As mentioned before, neither network behaviors nor message characteristics had to be defined in the way we did in this study."

"Longer, more practical, socially more sensitive lists of characteristics and behaviors, proposed by future studies may result in much more practical guidelines for us to understand why specific messages travel in the network the way they do."

3.1.17 Impact of Social Networking for Advancing Learners' Knowledge in E-learning Environments [180]

This is a machine-generated summary of:

Troussas, Christos; Krouska, Akrivi; Sgouropoulou, Cleo: Impact of social networking for advancing learners' knowledge in E-learning environments [180]

Published in: Education and Information Technologies (2021)

Link to original: https://doi.org/10.1007/s10639-021-10483-6

If you want to cite the papers, please refer to the original.

For technical reasons we could not place the page where the original quote is coming from.

Abstract-Summary

"In the light of the new developments, this paper presents SNAKE (Social Networking for Advancing Knowledge in E-learning environment), which is an e-learning software incorporating social characteristics for the tutoring of computer programming."

"Investigating the impact of e-learning software holding social characteristics is yet a quite under-researched area."

"An extensive exploration of SNAKE has been conducted which examined different factors affecting social networking-based learning."

"Upon analysis and structural model validities, the experimentation led to an extended Technology Acceptance Model (TAM) utilized for estimating the impact of the various variables."

"The study confirmed that the model adequately explained causal relationships between variables and presented direct and indirect significant impacts of them on SNAKE which can promote learners' better academic performance and knowledge acquisition."

Introduction

"This means that students may have sufficient knowledge on the basic computer programming principles, but poor knowledge on the structuring of commands that can be an impediment for the learning of another language."

"Each student has his/her own learning needs and pace when enhancing his/her knowledge in programming, and thus, the learning technology systems should provide a tailored educational environment."

"When learning a new programming language, an inexperienced user has to learn basic concepts, principles and structures of programming, while a more expert student who already has this knowledge should focus on more advanced topics of the domain."

"Social Networking-based Learning systems should be responsible to identify the knowledge levels of students and provide tailored learning pathways according to their personal profiles and cognitive needs and preferences."

Related work

"The results of these studies show the positive effects of SNs in tertiary education; however, there is an emerging questioning of students' willingness to adopt this technology and thus the authors proposed more evaluation on its impact to students, to be conducted in the future."

"It also needs to be underlined that all the aforementioned studies have superficially evaluated the impact of social networks in education using simple questionnaires or the Technology Acceptance Model (TAM), and focusing on students' knowledge acquisition, degree of engagement, enjoyment, usefulness and ease of use."

"Special reference has been placed to the evaluation of the impact of this new technology and an in-depth investigation is presented towards a more robust understanding of the topic of social networking-based learning and its effect in the field of digital education."

Overview of the system

"In more detail, SNAKE provides an educational environment holding social networking characteristics and appropriate techniques to ensure personalization and adaptivity to students' personal and cognitive traits."

"The student model of SNAKE takes into consideration the learning needs of the students and their profiles holding personal information in order to create a personalized learning experience for them."

"As SNAKE is for learning computer programming, students may make either syntax or logical mistakes."

"The tutoring model of SNAKE includes teaching strategies that are followed towards ameliorating the students' learning outcomes."

"To this direction, by taking input from the student models, SNAKE provides targeted feedback to students during their attempt to answer the assessment units, during their study of the learning material and in situations where SNAKE decides that students could achieve better knowledge acquisition if they follow a different pathway (e.g. to collaborate with peers)."

Evaluation

"The usage of social networks by the students ($\beta = .235$, $p = .002$), the systems' usefulness ($\beta = .285$, $p = .003$) and systems' social features ($\beta = .286$, $p = .004$) significantly influenced students' attitude towards social networking-based learning; while social gamification ($\beta = .198$, $p = .022$) has also a significant positive effect on it in a lower level."

"The model indicated a significant variance in the dependent variable ($R^2 = .561$); i.e. the 56.1% of the variance in the intention to use social networking-based learning is explained by the usefulness, the attitude to use this technology, the usage of social networks, systems' social feature and social gamification."

"Users have a positive attitude and intent to use SNAKE given that it is a social networking-based learning platform; such novel technology can modernize and reform education."

Conclusions and future work

"TAM used in the current study and that involved seven constructs: SNU, SF, SG, PU, PEoU, ATU and ITU."

"The results highlighted that SF affects significantly PU, ATU and ITU."

"SNU is the most important determinant for ATU and ITU."

"Future steps include the exploration of more elements to assess the impact of different factors on students' performance through SNAKE, the further investigation the pedagogical affordance of the presented approach using other tools, such as ANOVA and the employment of a wider range of indicators for assessing the presented measurement model."

3.1.18 Capturing the Participation and Social Dimensions of Computer-Supported Collaborative Learning Through Social Network Analysis: Which Method and Measures Matter? [181]

This is a machine-generated summary of:

Saqr, Mohammed; Viberg, Olga; Vartiainen, Henriikka: Capturing the participation and social dimensions of computer-supported collaborative learning through social network analysis: which method and measures matter? [181]

Published in: International Journal of Computer-Supported Collaborative Learning (2020)

Link to original: https://doi.org/10.1007/s11412-020-09322-6

If you want to cite the papers, please refer to the original.

For technical reasons we could not place the page where the original quote is coming from.

Abstract-Summary

"Since such learning analytics (LA) methods rely on measurement as a first step of the process, the robustness of selected techniques for measuring collaborative learning activities is critical for the transparency, reproducibility and generalizability of the results."

"This paper presents findings from a study focusing on the validation of critical centrality measures frequently used in the fields of LA and SNA research."

"We examined how different network configurations (i.e., multigraph, weighted, and simplified) influence the reproducibility and robustness of centrality measures as indicators of student learning in CSCL settings."

"This research aims to contribute to the provision of robust and valid methods for measuring and better understanding of the participation and social dimensions of collaborative learning."

"The results show that multigraph configuration produces the most consistent and robust centrality measures."

"The findings also show that degree centralities calculated with the multigraph methods are reliable indicators for students' participatory efforts as well as a consistent predictor of their performance."

"Eigenvector centrality was the most consistent centrality that reliably represented social dimension, regardless of the network configuration."

Introduction

"While many studies involving the use of text-based discussion forums have examined argumentative and epistemic dimensions of collaborative learning (Fu and others [100]), this study aims to contribute to a deeper understanding of the participation and the social dimensions and in particular, to the validation and reproducibility of the (computational) centrality methods to measure, understand, and reliably represent these dimensions of collaboration."

"In the context of CSCL, advances in SNA and LA have provided new tools to explore the collaborative learning processes by tracking, collecting, analyzing, and reporting data about how a student contributes to the joint activities, externalizes their own ideas, comments on and responds to peers (i.e., the participation dimension) and builds on the ideas and contributions of others in knowledge co-construction (i.e., the social dimension) (Berland and others [101]; Fincham and others [102]; Gaševi and others [103]; Schneider and Pea [104]) Such improved understanding of the both dimensions provides researchers, teachers and students with fundamentally new, data-driven ways to: (1) view and support the critical phases of collaborative learning, (2) find evidence of critical moments of success or failure, and finally, and (3) to act upon this information to improve conditions for learning and collaboration (Noroozi and others [105])."

Background

"SNA is another computational method that has been similarly implemented to analyze the participatory and the social dimensions of CSCL through indicators known as centrality measures."

"We will discuss the concept of centrality measures and how they have been operationalized to explain the participatory and social dimensions of CSCL."

"We in particular focus on the validation of the centrality measures – in terms of their reliability and consistency – used to measure and explain the important participatory and social dimensions of CSCL."

"The authors also found that the correlation between centrality measures and academic performance varied significantly with each tie extraction method, and stressed the importance of transparency of the tie definition."

"Through the validated SNA measures applied to CSCL, scholars can uncover and explain better the participation and social dimensions of collaborative learning using the centrality measures of degree, closeness, betweenness and eigenvector centralities."

Method

"For each student in each course, the five most-used centrality measures were calculated for each network."

"Outdegree centrality: refers to the number of messages a student posted (multigraph), or the number of unique users a student contacted (simplified), or the total volume of text a student posted (weighted) (Liao and others [106]; Opsahl and others [107]; Stephenson and Zelen [108]; Wei and others [109])."

"Indegree centrality: refers to the number of replies a student gets (multigraph), or the number of unique users who have replied to the student (simplified), or the total volume of text a student has received from all contacts (weighted) (Csardi and Nepusz [110]; Liao and others [106]; Opsahl and others [107]; Stephenson and Zelen [108]; Wei and others [109])."

"Since centrality measures were estimated from groups with different sizes, two versions were calculated for each centrality measure: (1) A normalized centrality, i.e., the centrality measure is divided by the number of students to balance the influence of group size on the number of possible interactions in the group (Saqr and others [111]), and (2) an unmodified version, raw or non-normalized version, in which we report the centrality measure as it is, with no modification."

Results

"Similar to the indegree centrality, the correlation between outdegree centrality (calculated with the multigraph method) proved to be consistently positively correlated with student performance in all courses, with a higher correlation coefficient in 10 courses compared to other configurations; the coefficient ranged from $r = 0.57$, $p < 0.01$ to $r = 0.78$, $p < 0.01$."

"The multigraph configuration was positively and significantly correlated with student performance in eleven courses with a correlation coefficient that ranged from $r = 0.57$, $p < 0.01$ to $r = 0.74$, $p < 0.01$ except for C8 ($r = -0.032$, $p = 0.82$)."

"In the multigraph method, normalized closeness centrality was positively correlated with student performance in eleven courses (except for C8); the simplified configuration was statistically significant in nine courses (C1–7, C9 & C10), and non-significant in two courses (C8, C11), while negative in one course (C12)."

Discussion and conclusions

"This study has examined how different network configurations influence the reproducibility and robustness of centrality measures as indicators of student learning – especially the participation and social dimensions of collaborative learning – in CSCL settings."

"Our results demonstrated that eigenvector centrality was the most consistent measure of the social dimension, demonstrating a consistently positive and significant correlation in all selected network configurations."

"The following answers the study's second research question: Whereas degree centralities are robust indicators of students' participation in CSCL, eigenvector centrality reliably reflects students' social positioning and relationships."

"While proving the multigraph configuration produces the most consistent and robust centrality measures of student learning, we call for further research to test other network configurations, apply other tie definitions, and verify our results in similar learning settings, or some others, and further build upon them to continue the line of methodological refinement in the fields of social network analysis and learning analytics."

3.1.19 Simulating the Effects of Social Networks on a Population's Hurricane Evacuation Participation [182]

This is a machine-generated summary of:

Widener, Michael J.; Horner, Mark W.; Metcalf, Sara S.: Simulating the effects of social networks on a population's hurricane evacuation participation [182]

Published in: Journal of Geographical Systems (2012)

Link to original: https://doi.org/10.1007/s10109-012-0170-3

If you want to cite the papers, please refer to the original.

For technical reasons we could not place the page where the original quote is coming from.

Abstract-Summary

"Such powerful storms disrupt societal function and result in a tremendous number of casualties, as demonstrated by recent hurricane experience in the US Planning for and facilitating evacuations of populations forecast to be impacted by hurricanes is perhaps the most effective strategy for reducing risk."

"This paper argues that the influence of social network effects on evacuation decisions should be revisited given the potential of new social media for impacting and augmenting information dispersion through real-time interpersonal communication."

"Using geographic data within an agent-based model of hurricane evacuation in Bay County, Florida, we examine how various types of social networks influence participation in evacuation."

"It is found that strategies for encouraging evacuation should consider the social networks influencing individuals during extreme events, as it can be used to increase the number of evacuating residents."

Introduction

"If the information a resident receives from his or her social network is no longer constrained by local geography, virtual communication about hurricane risk may have more influence on that person's decision to evacuate."

"This research explores how different kinds of social networks and communication strategies enabled by new media technologies alter evacuation patterns during a hurricane event using an agent-based model (ABM)."

"By integrating geographic data from a study area at risk for a hurricane event, calibrating non-networked evacuation rates to those of past hurricane evacuations, and then simulating social networks of various types among the agents, the models presented here show how various social network structures, and the emphasis placed on them by their members, can affect overall participation in evacuation."

"These simulated outcomes can inform hypotheses to be explored in empirical studies of the influence social networks have on evacuation behavior."

Review of relevant methodologies

"The implications of utilizing social networks in an agent-based model describing a hurricane evacuation, including the variable dissemination of information, are presented."

"The authors show that evacuation decisions are more correlated with reliance on peers and local authorities rather than news media, lending credibility to the idea that social networks play a role in evacuation."

"A second advantage to using an ABM to study hurricane evacuation is that it is possible to explicitly simulate social networks between agents."

"If these peer networks mainly consist of connections to geographically nearby neighbors, then the effect of social networks on decisions will be highly correlated with the environmental conditions of the specific location where the decision to evacuate is being made, as noted by Baker [112]."

"By presenting an ABM of evacuation behavior decisions that accounts for these networks, this research analyzes how varying levels of social influence and connectedness affect the collective behavior that manifests during a hypothetical hurricane disaster scenario."

Data and model overview

"We employ an ABM with geographic data and simulated social networks to assess how social network structures influence the rate of participation in hurricane evacuation."

"Each agent has risk area, vehicle ownership, and hurricane experience attributes as these variables were identified in previous research as being key to a resident's decision to evacuate (Baker [112]; Fu and Wilmot [113]; Lindell and others [114]; Whitehead and others [115])."

"The second involves an agent polling their network connections at a random, normally distributed interval and comparing the percent of their connections evacuating to a percentage threshold."

"Given this, the decision of an agent to evacuate can be represented using a logical or: an agent will leave the area if their utility value u is greater than the threshold value of one OR if the percent of an agent's social network evacuating is beyond their influence tolerance value."

Results

"Household agents' influence tolerance must be low, around 0.5, before there is an increase in the rate of evacuation of all households greater than 10 % above the baseline, no-network scenario of 75 %."

"In all three network scenarios, an increase in the average number of connections per agent corresponds with an increase in evacuation participation when the influence parameter was approximately 0.4 or higher for the subpopulation within the storm surge region."

"For the subpopulation not within the storm surge region, with the exception of the random network, changes in the average number of links per agent result in no discernable shifts in evacuation participation."

"The reason the agents not within the storm surge region in the random network scenario see an increase in evacuation participation is because there are more ties between agents within and out of the storm surge region."

Conclusions

"Given it is now easier to connect to a social network with social media, it is important that researchers examine how this new communication medium affects the behavior of people considering evacuation."

"That residents of a region at risk of a hurricane incorporate input from their social network in their decision to evacuate, this paper has shown that the structure of that region's social network greatly affects the participation in an evacuation."

"This model adds to previous research by explicitly accounting for social influence through new social media technology and showing that its usage can affect how many residents partake in an evacuation."

"With more precise information about how social network structures and social media usage influence this critical decision-making process, these models can be fine-tuned and integrated with existing modeling techniques to better anticipate evacuation behavior during a hurricane event."

3.1.20 Structuring Communities for Sharing Human Digital Memories in a Social P2P Network [183]

This is a machine-generated summary of:

Rahman, Haseeb Ur; Merabti, Madjid; Llewellyn-Jones, David; Sudirman, Sud; Ghani, Anwar: Structuring communities for sharing human digital memories in a social P2P network [183]

Published in: Peer-to-Peer Networking and Applications (2020)

Link to original: https://doi.org/10.1007/s12083-020-00928-z

If you want to cite the papers, please refer to the original.

For technical reasons we could not place the page where the original quote is coming from.

Abstract-Summary

"A community is sub-network inside P2P networks that partition the network into groups of similar peers to improve performance by reducing network traffic and high search query success rate."

"This increases the number of hosts bearing similar data in the network and hence increases the size of communities."

"This article presents a Memory Thread-based Communities for our Entity-based social P2P network that partition the network into groups of peers sharing data belonging to an entity–person, place, object or interest, having its own digital memory or be a part another memory."

"The simulation results show an increase in network performance for the proposed scheme along with a decrease in network overhead and higher query success rate compared to other similar schemes."

Introduction

"This leads on to the organization of our digital memories into Memory Thread-based Communities (MTC)."

"A person would represent a connection between all the digital memories in the form of pictures, audio captures, videos, and so on, which are captured, collected or stored by that person or for them throughout their life at different times, places, events and so on."

"We can organize digital memories based on the history of the entity they relate to, which is likely to be useful since history is one of the ways to express information about objects, places and so on."

"An MTC is formed by peers that share data for an entity such that each peer in it connects according to the position defined, based on the digital memories it shares, by the indexing criterion."

"The first contribution is proposing a novel Memory Thread-Based P2P Communities that are organized according to the patterns in digital memories in the form of Memory Threads."

Related work

"To share digital memories, it is important to properly organize the network so that data of the entities is not lost."

"The following are a few approaches which use the data contents to organize and search peers and data in the network."

"Community-based approaches group similar peers together which allow peers that do not have the required data to be avoided during the search."

"When a peer searches for some particular item of data, it sends queries to all members – or up to a certain number of hops – within a community."

"Contrary to IBCs, we consider the actual data being shared by the host for creating communities and connecting peers inside each community; because as long as the data is available for sharing, it will be accessible in the same community."

Memory threads

"Digital memories of similar references can be put together and are organized according to certain criteria to form a memory thread."

"Time, which is further considered as history of an entity to form memory thread, is often used as a reference to digital memories since it allows their memories to be recollected based on when an event took place."

"If time is the indexing criteria for the Eiffel Tower thread, then the thread will be formed as the digital memories of the Eiffel Tower at different times, for example relative to its construction in 1889."

"If digital memories for the Eiffel Tower are organized according to time then it will form a linearly structured memory thread, but the structure of the memory thread would be different if we were to choose location as an indexing criterion."

"At a certain place in a memory thread, where the digital memories of different entities have a common reference key, different threads will overlap with one another."

The proposed: memory thread-based P2P communities

"A memory thread is reflected in the network structure in order to organize the peers in the form of a Memory Thread-based Community (MTC)."

"A peer in an MTC represents a digital memory in a memory thread formed in the network."

"A community of peers of Extant Memory Thread (EMT) is formed by or for an entity from its digital memories."

"To create an extant memory thread, an entity specifies selection and indexing criteria and simply binds their existing digital memories to the thread (or adds them to the thread when a new digital memory is captured) on his own peer or personal social P2P network."

"An EMT is simply a memory thread stored on a single peer to which other peers in the network are attached."

"A memory thread-based community is formed by connecting those peers which have common reference keys for similar digital memories."

Simulation results

"The distribution function which is used to distribute the events, and to send new queries in the network within the 5000s time limit is: where Tn is the time of the last event, Tn+ 1 the time of the next event to be set, b is a base value for the average delay and m is a modifier."

"The traffic in the network is generated according to the linear distribution $T_{n+1} = Tn + (b \pm m)$ as before, but this time with a constant base value b of 100 and constant modifier m of 40 throughout all experiments."

"In order to better understand the effect on the network performance as a result of changes in size and network traffic, we calculated the Standard Deviation (SD) values for the rate of successful queries."

"Based on the current results, we expect that the MTCs will maintain their performance for longer due to the higher query success rate and more stable underlying network structure."

Conclusion and future work

"Memory threads are reflected in our Entity-based social P2P network as the underlying network structure in the form of Memory Thread-based Communities."

"The indexing criterion is reflected similarly in the MTC which organizes the peers according to the digital memories being shared."

"The idea of Memory threads and MTCs is based on the idea that "Let the data speaks for itself rather than represented by a peer or person", which where lays the novelty of our work."

"We are currently building on this work by implementing sub-communities within the MTCs, which will be formed according to the indexing criteria of a memory thread to create a more robust network."

"The structure of a memory thread and the communities of peers which are formed according to the memory thread, depend upon the indexing criteria of the memory thread."

3.1.21 Evolving Activity Cascades on Socio-technological Networks [184]

This is a machine-generated summary of:

Borge-Holthoefer, Javier; Piedrahita, Pablo; Arenas, Alex: Evolving activity cascades on socio-technological networks [184]

Published in: Journal of Computational Social Science (2017)

Link to original: https://doi.org/10.1007/s42001-017-0012-7

Copyright of the summarized publication:

If you want to cite the papers, please refer to the original.

For technical reasons we could not place the page where the original quote is coming from.

Abstract-Summary
"In the complex networks community, it took some time to realize the difference between simple propagation—e.g., the spread of disease—in which a single active node is sufficient to trigger the activation of its neighbors, and complex contagion, in which node activation requires simultaneous exposure to multiple active neighbors."

"Rooted in the social science literature, complex contagion has settled as the driving mechanism for behavior cascades on social networks."

"Our access to digital traces of social interaction reveals, besides and beyond complex contagion, bursty activity patterns, repeated agent activation, and occasionally a form of synchronization under the form of trending topics and hypes."

"We review a mechanistic model that, within the logic of complex contagion, accounts as well for the temporal evolution of behavior cascades."

Introduction
"Subtle as it may appear, the adoption of the idea of networked complex contagion to explain social propagation actually meant a reformulation of the problem, frustrating the possibility of solely mapping concepts, formalisms and results from network epidemics (e.g., [116]) to the social context: in the end, simply switching "susceptible" to "ignorant", or "infected" to "spreader", did not suffice."

"From the moment, it became clear that the threshold model was the standard tool to capture cascading processes in the social medium, many developments followed, specially focused on the role of topology in a contagion spread: the importance of the seed size, i.e., the fraction of initially active nodes in the dynamics, the presence of correlations and/or modularity in the structure, or the temporal characterization of the cascading processes [117–120], see [121] for an overview."

A model for unfolding and recurrent cascading phenomena
"We revise the acclaimed Watts' threshold model [122] from which we will borrow the computation of the condition for large cascades to occur in networks."

"We review a recent proposal of a mechanistic model that can be used as the core of a new set of models that enlarge the capabilities of threshold models to account for complex contagion phenomena, and add the recurrence of cascades as a natural outcome of a continuous dynamical system."

"Inspired in these facts, it has been proposed an MRA to mimic social activity as a complex network of IFOs representing the time evolving activation of individuals."

"The proposed MRA is willing to represent time-evolving cascades of activity, whose distribution will depend on the parameters of the system."

Numerical insights: regimes of activity

"The system evolves, until global cascades appear."

"Dark blue is used in regions where cascades are not comparable to the system's size (labeled as N.C., "no cascades", in the figure)."

"Turning to the social sphere, these results open the door to predicting how long it takes for a given topology, and a certain level of inter-personal influence, to achieve system-wide events."

Analytical insights: homogeneous parametrization

"The previous section illustrates the suggestive properties that MRA displays from a dynamical point of view—but also as a proxy to recurrent complex contagion."

"The aim now is to show that the model gives room as well to a certain degree of formalization."

"For any time t which implies that the cascade condition depends on time in our proposed framework."

"A supercritical regime, the states of the nodes become increasingly correlated as the dynamics evolve in time and, consequently, the distribution of states changes dramatically."

Empirical evidence

"We have reviewed a minimal MRA as a generalization of networked threshold models, inasmuch MRA include socially relevant mechanisms of the latter—complex contagion, mainly—but incorporate as well additional desirable ingredients: temporal evolution, endogenous and exogenous motivation, etc The validity and convenience of MRA are, however, subject to its capacity to reproduce and explain observed phenomena."

"The model should be able to reproduce observed cascades as they happened at different times during a single, unfolding event."

"Political mobilizations, elapsed online discussions, pre-announced events are examples of MRA-like social activity, while unexpected or extremely short cascades fit better other modelling approaches."

"The accumulation of empirical evidence for MRA models is one of the biggest challenges for scientists interested in unveiling the dynamical evolution of information cascades, starting from an event or from multiple events."

Conclusions and outlook

"Our understanding of social spreading in networked systems has been shifting paradigms over the last two decades, as new challenges were raised at an accelerating rate."

"We claim here that a good starting point to address this new challenge comes under the form of a model for recurrent activation (MRA) on social systems."

"In this line, we have reviewed a set of works which bring about a minimal framework, which we understand as a starting point rather than a destination itself."

"This minimal model already contains many convenient features—complex contagion, endogenous and exogenous mechanisms that trigger activity, recurrence, temporal evolution, etc —, and some evidence that it captures some empirical observations."

"There are many aspects of MRA that call for a better understanding."

3.1.22 Social Network Influences Decision Making During Collective Movements in Brown Lemurs (Eulemur fulvus fulvus) [185]

This is a machine-generated summary of:

Jacobs, Armand; Sueur, Cédric; Deneubourg, Jean Louis; Petit, Odile: Social Network Influences Decision Making During Collective Movements in Brown Lemurs (Eulemur fulvus fulvus) [185]

Published in: International Journal of Primatology (2011)

Link to original: https://doi.org/10.1007/s10764-011-9497-8

If you want to cite the papers, please refer to the original.

For technical reasons we could not place the page where the original quote is coming from.

Abstract-Summary

"Several authors have suggested that the mechanisms underlying collective movements in primate species appear to rely on complex cognitive skills, given their high level of cognitive abilities."

"We used a modeling approach to investigate the patterns of collective movement in a semi-free-ranging group of brown lemurs."

"We recorded via digital video cameras the order and joining latencies of the 11 individuals of the group during the departure time of spontaneous group movements."

"We then assessed whether mimetic mechanisms or the existence of a leader were underlying conditions for the joining process by testing 5 computer models relying respectively on 5 hypotheses: the independence of individuals, an anonymous mimetism, a mimetism according to kinship, a mimetism according to affiliation, and eventually the existence of a leader."

"An individual's decision to join the collective movement or not depended on the decision taken by its preferred social partners."

Introduction

"Choice of direction in a group of semi-free-ranging white-faced capuchins (Cebus capucinus) depended on self-organization via an anonymous mimetism mechanism, i.e., the probability that a remaining individual joined a group movement increased with the number of individuals already departed (Camazine and others [128]; Meunier and others [129]), despite the fact that these capuchins use recruitment behaviors to initiate a movement (Leca and others [130])."

"Despite the existence of such influences, anonymous mimetism (individuals joining a movement according to the number of individuals already involved, whatever their identities) underlies collective movements in both white-faced capuchins (Meunier and others [129]) and merino sheep (Gautrais and others [131])."

"Few researchers have tested the hypothesis that individuals join the movement only according to the identity of the first departed individual, although showing that a mimetic mechanism underlies the joining process automatically implies that the first departed individual is not the only one influencing the joining process."

"2) Individuals join a collective movement according to the number of individuals already involved, whatever their identities (anonymous mimetism)."

Methods

"We then scored the departure latency of each joiner, i.e., of joiner j, $\Delta T_{j-1,j}$ as the time elapsed between the departure of joiner j – 1, including the initiator, and that of joiner j. Because we never observed 2 individuals joining the movement at exactly the same time, no 2 individuals ever had the same rank."

"In a mimetic process wherein the departure probability is proportional to the number of individuals already moving, the probability per unit of time that one of the n resting agents became the jth joiner is: where C is the mimetic coefficient per individual."

"The departure probability of the initiator, the only individual whose movement could not be influenced by the other group members, was the same for the 5 versions of the model."

"We scored the identity, rank, and departure latency of each agent taking part in a movement (first departed individual or joiner) for each simulated collective movement."

Results

"We obtained similar results for the variance analyses: For the independent hypothesis, simulated distributions of departure latency variances differ from the observed distribution ($Z = 1.69$, $N = 6$, $p = 0.002$), whereas in the case of departure latency variances of the anonymous mimetic, kinship, and affiliation hypotheses, distributions do not differ from the observed distribution ($Z \leq 0.89$, $N = 11$, $p \geq 0.417$)."

"The probabilities that 2 individuals joined each other directly in the observed and simulated collective movements under the independent, anonymous mimetic, kinship, and leadership hypotheses do not correlate (Dietz R-test; $N = 11$, $r \leq 0.11$, $p \geq 0.196$)."

"These probabilities correlate significantly in the case of the affiliation hypothesis (Dietz R-test; $N = 11$, $r = 0.34$, $p = 0.008$)."

Discussion

"Our results show that a self-organized mimetism process with an affiliative underlay—mimetism according to affiliation—can account for several aspects of the pattern observed during collective movements of brown lemurs."

"It was also the only model that fit the observed pattern for group member organization at departure, making it the best explanation of the joining process in our study group."

"In this condition, the first departed individual is only one link in the chain reaction, i.e., the amplification of the mimetic process, facilitating the joining of all group members, even if she or he has a special role as the initiator of the movement (Jacobs and others [132])."

"We have shown that a mechanism involving a mimetic process with an affiliative underlay can explain the joining latencies and the order and organization of the individuals involved in collective movements of a group of brown lemurs."

3.1.23 A Trust-Based Collaborative Access Control Model with Policy Aggregation for Online Social Networks [186]

This is a machine-generated summary of:

Rathore, Nemi Chandra; Tripathy, Somanath: A trust-based collaborative access control model with policy aggregation for online social networks [186]

Published in: Social Network Analysis and Mining (2017)

Link to original: https://doi.org/10.1007/s13278-017-0425-6

If you want to cite the papers, please refer to the original.

For technical reasons we could not place the page where the original quote is coming from.

Abstract-Summary

"Preserving information privacy is indispensable in such applications as the shared information would be sensitive."

"We propose an effective trust-based access control technique to allow or disallow the shared resources considering the authorization requirements of all the multiple parties."

"A logical representation of the proposed access control technique is prepared to analyze the privacy risk."

"The usability study of this application supports the call for such access control technique."

Introduction

"OSNs provide their users, a diverse set of built-in information sharing services like wall posts, sharing statuses, photos, videos, links and many more."

"These resources contain/reveal private information about others as well, like a group photo with other users."

"Each multi-party resource is associated with multiple number of users called stakeholders that may have their own privacy preferences."

"To address the multi-party access control issues, present OSNs provide a preliminary level of protection mechanism."

"No OSN offers any effective access control facility for multi-party resources."

"The proposed existing multi-party access control schemes for OSNs are not fine-grained, flexible and expressive enough."

"A flexible and efficient solution is very much needed to control access to the multi-party resources."

"We extend this mechanism by augmenting aggregation in policy specification which results in providing users better fine-grained access control of multi-party resources."

Related work

"Several important models are proposed in Cheng et al. [134, 135], Fong [136] and Pang and Zhang [137] that use relationship between users, and their resources to achieve access control."

"Carminati et al. [138] introduced a rule-based access control model that allows users to specify access control rules based on the relationship type, depth, and the trust level between the OSN users."

"Pang and Zhang [137] proposed a relationship-based access control model which considers the relationship between the public information along with the user-to-user relationship."

"Cheng et al. [135] have proposed an extension of their own work in Cheng et al. (134), where user or resource attributes are used to capture access context in policy specified, but both of these schemes do not address the problem of multi-party access control."

"White et al. [139] proposed an access control framework where user's optimal data sharing policies are based on a two-player zero sum Markov game model."

System model and security requirements

"User resource A user resource or resource is a digital content uploaded on OSN."

"A resource is called exclusive (or single-party) if it is owned by a single user and multi-party otherwise."

"Profile items, friends list are exclusively owned resources of respective users while the resources like wall posts, group photos uploaded by a user in her/his friends space are co-owned (or multi-party) resources."

"The user v is referred as the owner of the resource and the rest of the friends in the photo are referred as co-owners."

"A user can upload a resource in own user space or in any of her/his friend's user space."

"Privacy risk and sharing loss trade-off Any access control mechanism should allow stakeholders of a resource to choose whether privacy risk or sharing loss is important to them."

The proposed scheme

"This policy aggregation method may be adapted for a resource that is highly sensitive to all of the stakeholders."

"This type of aggregation policy may be adopted for a resource that is more sensitive and can be shared only between contacts with fair enough trust."

"The requirement of at least half of the stakeholders, having trust on requester greater than the average trust, makes this type of policy stricter than majority-based aggregate policy."

"On the basis of above analysis, we adopt following policy for selection of aggregation method: Both the privacy risk and sharing loss is 0.5 each on default for each stakeholder."

"It also computes the sharing loss and privacy risk for each aggregation method and chooses the aggregation method which have the closest values to that opted by the stakeholders."

Logical representation and analysis of CACO-PA

"Let the photo be private and sensitive to Alice and Bob, but not for Carol."

"Let Alice, Bob and Carol set a required trust level to view the photo as 0.7, 0.8, 0.2 , respectively, with conj as aggregation method."

"The access request will be as follows:After receiving the access request, CACPM first verifies that Rose is having some relationship with Alice and Carol or not."

"Alice and Carol have trust level of 0.7, 0.8, respectively, with Rose."

"The access request will be as follows:The indirect trust of Bob and Carol on Arnold is calculated to be 0.525, 0.28, respectively."

"Arnold satisfies policy of Alice but does not satisfy the same for Bob and Carol."

"Jim only satisfies the policy of Carol and does not satisfy the policy of Alice and Bob."

Implementation and evaluation

"The app requires user_friends, user_photos, user_relationships, public_profile and public_profile permissions from Facebook."

"The photo stakeholders and their respective policies are also stored into the database."

"Whenever a user requests access to the photo, the app evaluates the photo access polices and makes appropriate access decision."

"My Profile This option allows the user to see his/her profile."

"Friends This option allows the user to set his trust on his family members and friends separately."

"Photos This option shows all the user photos and allows the user to apply the proposed access control mechanism to his/her photos."

"User can choose any of his/her photos to add and invite all of its stakeholders to specify their respective access control policies."

"All the stakeholder policies are than stored in the database."

"Feedback This option allows a user to provide feedback about LinkTrust."

Discussion

"When the owner uploads a resource r, she/he invites the corresponding stakeholders to specify their access policies."

"All the invited stakeholders specify their respective access policies in terms of the level of trust needed to access r. In CACO-PA, each user has a CACPM module to administer the collaborative policy specification and access to the user's resources."

"They do not require complex access policies to specify audience of their resources."

"In CACO-PA a stakeholder can specify the allowed action for the audience of the resource."

"In MPAC there is no collaboration at the time of policy specification, and each stakeholder specify their access policy for the resource independently."

"In case, if any stakeholder is not satisfied with the current privacy settings, she/he can change the policy but have to ask the resource owner to change the conflict resolution policy."

Conclusions

"The proposed scheme computes trust (direct or indirect) with each stakeholder of the resource to verify the policy rules."

"The policy aggregation methods allow resource stakeholders to control privacy risk and sharing loss as per their sharing objective."

"We conducted a user study to show the usability and acceptance of such apps to preserve user privacy on Online Social Sites."

"Results of the study show that users appreciated the app and user needs such app to preserve their privacy on OSN sites."

3.1.24 The Present and Future of Digital-Community Currencies: RAMICS 2019 in Hida–Takayama Keynote Speech [187]

This is a machine-generated summary of:

Nishibe, Makoto: The present and future of digital-community currencies: RAMICS 2019 in Hida–Takayama keynote speech [187]

Published in: Evolutionary and Institutional Economics Review (2020)

Link to original: https://doi.org/10.1007/s40844-020-00177-y

If you want to cite the papers, please refer to the original.

For technical reasons we could not place the page where the original quote is coming from.

Abstract-Summary

"Modern financial capitalism only seeking more profits in free investment creates such problems as disparities of income and assets, unequal opportunities, dissolution of communities."

"Modern community currencies emerged and diffused mainly in developed countries in the 1980s."

"They were expected to solve such problems caused by globalization in modern capitalism but never succeeded in terms of scale and sustainability because of high administrative costs and low motivation of participants."

"Recent digital technologies such as blockchain, mining, proof of work and QR code scanning that are used in digital payment and cryptocurrencies have opened up a wide range of non-fiat, decentralized private money and created new possibilities for community currencies."

"If we are aware of such aspects, we can make the best use of Digital-Community Currencies assisted by ICT and AI for creating a diverse, decentralized, and sustainable socio-economy."

Why are we here in Hida–Takayama?

"The cities of Takayama and Hida are the rural cities in Gifu Prefecture surrounded by 3000-m-high mountains called the Japan Alps, remote from three major cities, Tokyo, Osaka, and Nagoya of Japan."

"Hida–Takayama is famous for "Hida no Takumi," people of Miyadaiku who had maintained traditional skills of craftsmanship from ancient times and specialized in building shrines, temples, or palace."

"Historically speaking, Hida–Takayama is not only a forest town in the mountains full of nature but also a traditionally high cultural and technical town."

"Turning your eyes to the present situation of modern local currencies, we can see such ancient tradition has been passed down here in Hida–Takayama."

"For, there is the coexistence of a non-profit organization's wood-made currency for thinning volunteer work called "Enepo" and a local credit union's digital community currency on smartphones called "Sarubobo Coin."“

The theme of the academic conference and the sub-theme of the practitioners' conference in RAMICS 2019

"New Possibilities of Digital-Community Currency Systems" and the sub-theme for the practitioners' conference is "Harmony in community currency systems between sustainability of nature, culture, and economy and technological innovation."“

"Because I believe that digital technologies such as blockchain, mining, proof of work and QR code payments that are used in digital/crypto currencies have already opened up a wide range of non-fiat, private decentralized money systems and created new possibilities for community and complementary currencies, and that they have changed the basic concepts of geographical 'community' or 'local' to more 'community of interest' or 'abstractly local' in value space although they have both positive and negative effects on natural, ecological and cultural environments surrounding us."

"The last question particularly relates to the call for paper for practitioners' conferences entitled "Harmony in community currency systems between sustainability of nature, culture, and economy and technological innovation."“

The present and future of digital-community currencies

"I want to further move to the title of my speech, "The Present and Future of Digital-Community Currencies"."

"As I just mentioned, Hida–Takayama is a suitable location for examining the topic, because we have both analog and digital community currencies so as to compare both."

"We can see the possibility of the digitalization of money must exist in such coexistence of different community currencies in one individual, in other words, "multiply-situated self," The lesson that should be taken from this example would be 'Don't be afraid of the digitalization of money."

"Eight years ago, Japan experienced horrible natural and artificial disasters."

"I want to say it is only possible by utilizing digital technology."

"How can it be done with community currencies?"

"I put a new chapter as afterword entitled 'The New Possibility of Community Currencies' to the forthcoming book."

Free investment capitalism: as the goal of globalization or the climax of "internalization of the market"

"The ultimate destination of globalization since the 1980s is the socio-economy in which the movement of capital seeking value augmentation controls everything."

"The universal tendency of global capitalism to extensively expand and intensively deepen markets can be conceived as "internalization of the market," which appears today as the tendency towards free investments."

"Under global capitalism, not only is the market extensively expanding geographically and spatially as commodification expands, but it is also intensively deepening as commodification penetrates all areas of goods and services including information, rights, and risks."

"The effects of globalization of capitalism are widespread and profound."

Financial capitalism and "financialization of labor-power": one aspect of free investment capitalism

"The financial capitalism in the 21st century arose as a result of the popularization of the inner institution of "free investmentism" in the evolution of global capitalism after the 1980s."

"In financial capitalism, along with the above-mentioned enlargement of financial transactions and innovation of financial instruments, (1) the self-financing of private enterprises, (2) the relative expansion of investment banking business of banks and the expansion of loans to households and workers, and (3) the expansion of financial liabilities and financial assets of households, that is, "financialization of labor-power" have progressed."

"The "financialization" of modern capitalism observed as such changes in the financial roles of firms, banks, and households is also an aspect of the "free investment capitalism" resulting from globalization and post-industrialization, and is the result of capitalism's purifications toward replacing states and communities with markets through advancement of modes of commodification."

Self-contradictions of financial capitalism

"Major financial institutions, including investment banks in the United States, accepted partial nationalization by injecting public funds to avoid bankruptcy at the time of the Lehman crisis, but they made enormous profits and reinstated high executive pay less than a year later."

"It was no longer a matter of whether the state should bail out financial institutions, but a matter of an upper level of whether the EU should bail out Greece and other countries to protect the common European currency, the euro, but the basic picture remains the same."

"What is bailed out this time is the nation-state that was supposed to bail out financial institutions that would endanger its financial system and currency."

"It is no doubt that the austerity policy of state finances in the EU without any unique financial policy on the national level has made people's lives more difficult."

Growing unfairness, not greater inequality: a more fundamental problem in financial capitalism

"Can the problem be solved if the state imposes a tax increase on the top 1% of wealthy people and redistributes it to middle- and low-income people?"

"The more fundamental problem of financial capitalism is something else, not related to income redistribution aimed at rectifying economic disparities."

"However, it has become clear that the state has broken the equal footing that should be a precondition for free competition in capitalism and created unequal opportunities between large banks and citizens, under the pretext of protecting the financial system."

"The problem here is not the resulting inequality under equal opportunities but the fraudulent inequality under unequal opportunities."

"The problem is inequality or unfairness in the rules of the game, which is much more fundamental than inequality as a result of the game."

Beyond the dichotomy between the market versus the state: "state failure" glossing over "market failure"

"If fair market competition results in economic disparities, which can be viewed as distributional inefficiencies that impede future economic growth, it may be justifiable that the government should rectify such inequalities as "market failure.""

"If the government is the entity that relieves some players by excluding them from applying market rules for the sake of the survival of the market economy, it is no longer possible to say that the government that makes arbitrary changes to such unfair rules complements the "market failure.""

"After the 1990s, the conventional idea of the government supplementing the "market failure" which was interpreted broadly as chronic depression, unemployment, disparity, and inequality, with its fiscal and monetary policies and redistribution policies stagnated, while the "government failure" such as the collapse of pension system and the fiscal bankruptcy became conspicuous."

The significance of community currencies: the bottom-up solution based on the reciprocal exchange principle in community

"Community currencies constitute an evolving system in which those that meet environmental conditions survive under natural selection with various kinds of artificial variations."

"The survival of community currencies is greatly affected by the environmental conditions of both outer institutions of the capitalist market economy such as markets, property rights, and corporate organizations, and inner institutions that comprise the behavior, motivation, and values of the people living in it."

"Even in the same system of community currency, changes in environmental conditions can result in either its selection or survival."

"Local environmental conditions, such as the nature of local communities as the target of a community currency and the norms and values of the residents living in them, can vary greatly depending on how the currency is applied."

"How should we deal with community currencies?"

Escape from the theory of 'one currency in one country': towards innovation and diversity of money

"The legal currency issued by the central bank differs from country to country, but basically, one currency has been used in one country."

"The current theory of inflation targeting and Friedman's monetarism that formed the basis for it do not make any difference in that they uphold the unity of the national currency and the monopoly of issuing by the central bank."

"The Modern Monetary Theory has just recently become quite popular in the US and Japan, which insist that no matter how much the government increases its debts and spending more than tax revenues, if the central bank can continue to buy national bonds by issuing legal tender as IOU, the government will neither go bankrupt nor fall into default, based on the view of endogenous money creation."

"It is possible to change the "one currency in one country", whereby the nation-state and the central bank exclusively issue statutory currencies."

The reaction to globalization in the west: a shift to protectionism, regionalism, and pluralism

"The EU has been considered to be a path towards the formation of a single global currency and a single global market economy."

"PIGS, which suffered from high unemployment and other economic difficulties, raised doubts about the euro, because it prevents each country from implementing its own monetary policy, and there is a growing interest in the reversion to the national currency and the introduction of community currencies."

"The course toward a single global currency and market economy was questioned and forced to change course."

"In 2017, the Trump administration came into power in the United States, clearly moving away from global free trade and toward protectionism by a single nation."

"Rather, it has been more widely recognized that the negative economic, social and cultural aspects of globalization have become more serious even in the United States, where it originated, and protectionism, regionalism, and pluralism have begun to be explored from a critical viewpoint."

Fusion of cryptocurrency and community currency as a symptom of a non-capitalist market economy

"As the search for a post-capitalist market economy begins, community currencies and cryptocurrencies are now attracting attention as one of the most important signs."

"While community currencies and cryptocurrencies are spreading, the issuing parties are expanding variously to local governments, commerce and industry associations, shopping streets, nonprofit organizations, citizen groups, and server networks on the Internet, etc The innovation and diffusion of ICT such as the Internet and smartphones have promoted the digitization and cashlessness of money, and have accelerated the diversification of money such as electronic money and corporate currency."

"Digital-community currency or virtual-community currency in which cryptocurrency and community currency are fused has appeared."

"In Japan, community currencies became a big boom in the early 2000s, and many community currencies were created, but many of them were not sustainable for a long time beyond several years, because subsidies expired, administrators exhausted, and temporal enthusiasm faded."

Hayek's denationalization of money: good money drives out bad

"Hayek insisted that if a large number of private currencies competed with each other, they should be denominated with different names and exchanged at floating rates so that users can distinguish them qualitatively and trade according to its quality."

"For example, a person who wants to reduce uncertainty in transactions can make a qualitative distinction between the currencies with stable value as "good money" and the ones with high volatility as "bad money," and then choose the former."

"When multiple concurrent currencies are clearly distinguished as being qualitatively different and exchanged at floating rates, the principle of money choice that "good money drives out bad" works."

"There are various kinds of money, and the exchange rates can change according to users' evaluation of quality."

The conditions for cryptocurrency to becomes good money

"We should consider the "conditions of good money," in which money can be used in daily economic transactions of goods and services by real people to reproduce their lives and environment repeatedly based on limited ability in local time and space."

"As in the case of community currencies, cryptocurrencies must overcome the problem of the traditional dichotomy of "market vs. government" or "freedom vs. regulation."""

"This dichotomy, in this case, is in a different form: "non-statutory currency or private currency vs. statutory currency or national currency."""

"To consider the "conditions of good money", while understanding these realities, we need a theoretical framework that, by adding "community" to the dichotomy of "market vs. government/state," reintegrate these three principles."

The emergence of virtual-community currencies

"Many digital coins have been emerging that aim to realize the objects of community currency such as revitalizing local economies and communities by introducing innovative FinTech of cryptocurrency such as Distributed Ledger Technology (DLT) called "Blockchain" and Consensus Algorithm called "Proof of Works."""

"They are realized as the digital value on smartphone applications, which can be paid instantly at local shops, supermarkets, and restaurants at the fixed-rate (1 Coin = 1 Yen), and circulate multiple times within users in regional areas."

"This is an attempt to create a digital-community currency as a hybrid combining the technologies of cryptocurrency and the ideals of community currency."

"Digital-community currency, the fusion of digital coins and community currencies, suggests a crucial direction for the sustainability of the socio-economic and natural ecosystem in the future."

The possibility of regional digital-community currency: a prospect for the post-capitalist economy

"A regional currency in Hokkaido, the northernmost of Japan's main islands, whose industrial structure is somewhat independent, would exert its economic effects."

"If the local government earnestly tackles the issue of Hokkaido's regional currency, to turn tourism and agriculture into a sixth industry in the post-industrialization era, there is a possibility that the economy will continue to grow."

"If the Hokkaido digital-community currency is established, the ratio of local production to local consumption in the region will increase economically, enabling the region to become self-reliant."

"People's living would be more stable and secure if autonomous economic zones formed by regional currencies could become to be relatively independent of the national and global markets."

Good money lab for digital-community currency to be good money

"In April 2018, exactly 1 year after I transferred to Senshu University, we established "Senshu University Digital-Community Currency Consortium Laboratory.""

"Its mission is to create and nurture "good money" as a new combination of cryptocurrency and community currency through collaboration among industry, academia, public, and private sector."

"The consortium type laboratory was established as a base for carrying out research, education, learning, and dissemination activities concerning digital-community currency and for building a framework for information sharing and consultation."

"Good money is a medium that depends on a good life from the perspective of "quality of life" and reflects the value that people have in their minds as quality, not quantity."

"Good Money Lab, as an industry-academia-public–private consortium-type laboratory, aims to explore, discover, and realize the newly emerging digital-community currency as good money in a variety of ways."

3.1.25 Self-management of Social Well-Being in a Cross-Sectional Study Among Community-Dwelling Older Adults: The Added Value of Digital Participation [188]

This is a machine-generated summary of:

Scheffer, Mariska MJ; Menting, Juliane; Boeije, Hennie R: Self-management of social well-being in a cross-sectional study among community-dwelling older adults: The added value of digital participation [188]

Published in: BMC Geriatrics (2021)

Link to original: https://doi.org/10.1186/s12877-021-02482-6

If you want to cite the papers, please refer to the original.

For technical reasons we could not place the page where the original quote is coming from.

Abstract-Summary

"This study aimed to examine associations between self-management abilities and digital participation among community-dwelling older adults with chronic conditions in the Netherlands."

"Self-management was measured by six abilities (e.g., investing in resources for long-term benefits and taking care of a variety of resources), whereas digital participation was estimated with the frequency of four social internet uses (e.g., using social network websites and calling digitally)."

"When predicting self-management abilities from digital participation, hierarchical multiple regression analysis determined statistically significant and positive relationships, in particular for e-mailing ($\beta = 0.21$; $p < .001$) and meeting new people online ($\beta = 0.07$; $p < .05$)."

"These findings suggest that internet usage and self-management abilities are positively related in the older adult population."

"Further research should be undertaken to examine the links between self-management abilities and digital participation more closely."

Background

"There is a growing body of literature that recognizes the importance of older adults' self-management abilities to manage social resources in such a way that well-being is maintained and restored when lost [140–142]."

"Self-management abilities, the behavioural and cognitive abilities to sustain well-being in later life, are functional in the fulfilment of basic social needs, namely status, behavioural confirmation and affection [142, 143]."

"Strengthening the core self-management abilities for well-being can aid in the prevention of age-related social problems among older adults."

"Although digital participation does not substitute all kinds of face-to-face activities, the internet may be a useful instrument to support self-management abilities when social resources are declining."

"Notwithstanding, little is known regarding the relationship of digital participation for social purposes and peoples' abilities of self-management."

"This cross-sectional research investigated associations between self-management abilities and social internet usage, such as visiting social networking sites and digital calling, among community-dwelling older adults."

Methods

"Data on the topics on self-management and internet usage were selected from the panel study performed in October and November of 2019."

"Participants of the panel study were included if they had at least one somatic chronic disease, and were 65 years or above."

"Each subscale ranges from 1 to 6, with higher scores indicating higher levels of self-management abilities."

"These variables may be associated with levels of self-management abilities and social internet usage [144, 145]."

"Descriptive statistics were reported for self-management abilities, social internet usage and background characteristics."

"Bivariate correlation analysis was conducted to test potential correlations between self-management abilities and digital participation, and with relevant background variables."

"For correlations between internet uses and self-management abilities, a Spearman correlation was applied."

"Hierarchical linear regression was used with self-management abilities as a continuous outcome variable and internet uses as potential predictor variables."

Results

"Findings indicate that overall self-management scores were significantly and positively associated with social internet usage (visiting social networking sites (r =

.09; $p < .01$); meeting new people online ($r = .14$; $p = < 0.001$); maintaining contact via email ($r = .28$, $p < .001$); digital calling ($r = .14$; $p < .001$))."

"Each of the social internet usages had a significant and positive association with at least four out of six individual self-management abilities."

"Internet usage for maintaining contact via email and digital calling were significantly and positively associated with all individual self-management abilities."

"Higher age (β = -0.18; $p < .001$), more severe disabilities (β = -0.15; $p < .001$) and lower education (β = -0.11; $p < .01$) were significantly and negatively associated with self-management abilities."

"Two out of four indicators for digital participation were positively and significantly associated with levels of self-management abilities."

Discussion

"Drawing on a nationally representative sample of older adults with chronic conditions, this cross-sectional study set out to assess associations between digital participation (frequency of visiting social network sites, meeting new people, e-mailing and calling digitally) and self-management abilities for well-being."

"The results of this study show a strong association between internet usage and self-management abilities among older adults."

"The study also provided additional insight into community-dwelling older adults' levels of digital participation and self-management abilities for well-being."

"The study employed a large panel to examine the relationship between digital participation and self-management abilities among community-dwelling older adults with chronic disease(s)."

"This study shows that digital participation, in particular maintaining contact via e-mail and meeting new people, is positively associated with self-management abilities among community-dwelling older adults."

3.2 Digital Spaces & E-Societies

Machine generated keywords: intelligence, realm, consumption, artificial intelligence, legal, artificial, ethical, justice, unique, ethic, student, physical, egovernment, persist, space

3.2.1 Unjust Behavior in the Digital Space: The Relation Between Cyber-Bullying and Justice Beliefs and Experiences [189]

This is a machine-generated summary of:

Donat, Matthias; Rüprich, Claudia; Gallschütz, Christoph; Dalbert, Claudia: Unjust behavior in the digital space: the relation between cyber-bullying and justice beliefs and experiences [189]

Published in: Social Psychology of Education (2019)

Link to original: https://doi.org/10.1007/s11218-019-09530-5

If you want to cite the papers, please refer to the original.

For technical reasons we could not place the page where the original quote is coming from.

Abstract-Summary

"We considered students' individual experience of teacher and classmate justice as possible mediators of these relations, and statistically controlled for student sex, internet use, empathy, and social desirability."

"Bootstrap mediation analyses showed that the more students endorsed personal BJW, the more they evaluated their teachers' and classmates' behavior toward them personally as just, and the less likely they were to report that they cyber-bullied others or were victims of cyber-bullying."

"The students' individual experiences of teacher justice mediated the association between personal BJW and cyber-bullying perpetration, whereas their experience of classmate justice mediated the relation between personal BJW and cyber-bullying victimization."

Introduction

"We aimed to investigate the mediating role of these justice experiences in the relation between students' BJW and cyber-bullying perpetration as well as victimization."

"One aim of our study was to investigate whether previous findings on the relations between students' personal BJW and their offline perpetration and victimization could be replicated with regard to cyber-bullying perpetration and victimization."

"We tested the following hypotheses: The stronger students' personal BJW (1), the more likely they would be to evaluate their teachers' and classmates' behavior toward them personally as just (2) the less likely they would be to cyber-bully others, and (3) the less likely they would be to report being victims of cyber-bullying by others. (4) The relation between personal BJW and cyber-bullying perpetration would be at least partly mediated by teacher justice. (5) The relation between personal BJW and cyber-bullying victimization would be at least partly mediated by classmate justice."

Method

"Classmate justice was investigated using an adapted version of the Teacher Justice Scale (Correia and Dalbert [146]) comprising six items ($\alpha = 0.82$; $r_{ij\ est} = 0.43$; α ranged in other studies between $\alpha = 0.82$ and $\alpha = 0.87$, e.g., "My classmates generally treat me justly")."

"We assessed self-reported cyber-bullying perpetration using six items from Rigby and Slee's [147] 20-item Peer Relations Questionnaire, which we applied to the use of digital devices ($\alpha = 0.79$; $r_{ij\ est} = 0.39$; e.g., "How often in the last 4 weeks did you make others scared on the internet, on a mobile phone, or on a smart phone?")."

"We assessed self-reported cyber-bullying victimization using six items from the same questionnaire, which we also applied to the use of digital devices ($\alpha = 0.91$; $r_{ij\ est} = 0.63$; e.g., "How often in the last 4 weeks have you been insulted or humiliated by others on the internet, on a mobile phone, or a smart phone?")."

Results

"We tested the mediation effects of teacher and classmate justice on the relations between personal BJW and cyber-bullying perpetration and victimization in three models each."

"The results indicate that the relation between personal BJW and cyber-bullying perpetration was partly mediated by teacher justice."

"All effects of personal BJW and teacher justice on cyber-bullying perpetration remained significant."

"The decreased effect of personal BJW when including teacher justice again indicated a partly mediated relation with cyber-bullying perpetration."

"The indirect effect of personal BJW on cyber-bullying victimization through classmate justice was significant; BCa CIs for the point estimates did not contain zero."

"The results indicate that the relation between personal BJW and cyber-bullying victimization was partly mediated by classmate justice."

"All effects of personal BJW and classmate justice on cyber-bullying victimization remained significant."

"The decreased effect of personal BJW when including classmate justice again indicated a partly mediated relation with cyber-bullying victimization."

Discussion

"We aimed to investigate the relations between students' cyber-bullying perpetration and victimization and their personal BJW, as well as the mediating role of students' personal experiences of teacher and classmate justice on these relations, while statistically controlling for sex, internet use, empathy, and social desirability."

"We showed that the more students endorsed personal BJW, the less likely they were to report cyber-bullying perpetration."

"In accordance with our expectations, the results showed that classmate justice partly mediated the relation between students' personal BJW and their experiences of being cyber-bullied."

"This result not only indicates that students victimized by cyber-bullying experience the fact that they undeservingly suffer under their classmates' cyber-bullying perpetration as unjust, but also that classmate justice is an important additional factor explaining cyber-bullying victimization."

"The relation between students' personal BJW and their reported cyber-bullying perpetration and victimization persisted when controlling for sex, internet use, empathy, and social desirability."

Limitations

"A longitudinal study could show how cyber-bullying perpetration and victimization can change over time depending on students' justice beliefs and experiences."

"Although we controlled for several confounding factors, further characteristics such as students' motives for engaging in cyber-bullying or personality variables may also be related to cyber-bullying behavior (Kowalski and others [148])."

"It seems plausible that justice concerning the family or peers outside the school context may play an important role in students' lives and in their cyber-bullying behavior (Correia and Dalbert [146]; Kamble and Dalbert [149])."

"Parent and peer justice might be additional mediators in the relation between students' BJW and cyber-bullying perpetration or victimization."

"Studies, the importance of different types of bullying or media for the relations between cyber-bullying, justice experiences at school, and BJW could be investigated."

"These factors could mediate or even moderate the relation between BJW and cyber-bullying and could further contribute to the explanation of such behavior."

Conclusion

"Our study indicated that teachers' and classmates' just behavior was relevant to students' cyber-bullying perpetration and victimization."

"Just behavior in teachers and students might depend on aspects of interactional justice (for a review, see Peter and others [150]), meaning that teacher–student as well as student–student interactions should be characterized by dignity, civility, and respect."

"Teachers may learn more about students' justice cognitions when (1) justice psychological findings are implemented in teacher training (2) they are motivated to use their knowledge in lessons and establish a class climate characterized by open discussions in which students can express their feelings and opinions, and (3) they extend their own perspective and contrast it with the perspective of their students."

"At an even earlier developmental stage, parents' behavior might play an essential role in students' justice experiences."

"Our research provides an additional explanation for students' involvement in cyber-bullying and might thus not only contribute to the prevention of such behavior but also to possible interventions."

3.2.2 "I Am Not Your Robot:" the Metaphysical Challenge of Humanity's AIS Ownership [190]

This is a machine-generated summary of:

Jaynes, Tyler L.: "I Am Not Your Robot:" the metaphysical challenge of humanity's AIS ownership [190]

Published in: AI & SOCIETY (2021)

Link to original: https://doi.org/10.1007/s00146-021-01266-1

If you want to cite the papers, please refer to the original.

For technical reasons we could not place the page where the original quote is coming from.

Abstract-Summary

"Despite the reality that self-learning artificial intelligence systems (SLAIS) are gaining in sophistication, humanity's focus regarding SLAIS-human interactions are unnervingly centred upon transnational commercial sectors and, most generally, around issues of intellectual property law But as SLAIS gain greater environmental interaction capabilities in digital spaces, or the ability to self-author code to drive their development as algorithmic models, a concern arises as to whether a system that displays a "deceptive" level of human-like engagement with users in our physical world ought to be uniquely protected."

"Although many voices in the legal and technology realms have continued to argue against unique protections for digital entities, the fact at hand is that SLAIS design is becoming increasingly anthropomorphic so as to make these systems more capable of interacting with a wide range of (potentially) vulnerable populations—generally as a means to enhance these populations' overall well-being."

Introduction

"It is for this reason that it is important for us to shift our "fixed" ideas of what AIS (and subsequently SLAIS as the more sophisticated version of these operations) is to analyse whether displays of human behaviour by systems are transforming certain systems from tools bound for replacement into entities that ought to be treated as human—given that our interactions with those systems are becoming more akin to the interactions that we would have with one another, which are protected by various local and international laws."

"This analysis will first address some preconceived notions of human-level intelligence that are often left unaddressed in the literature targeted towards the capabilities of SLAIS to facilitate the argument's claims that specific utilisation of SLAIS is becoming a legally sanctioned form of slavery."

"Combined with various sub-arguments throughout, it is hoped that policy developers internationally shall re-examine the classification how SLAIS can qualify for human-like legal protections while simultaneously provide enhanced protections to digital interactions that occur via avatars to human agents."

All "intelligence" is not equal: the significance of humanity's "spectrum"
"It should further be stated that a refusal to accept the idea that SLAIS may already match the "intelligence" level displayed by particular members of the human population is realistically a refusal to accept those members of the human population as possessing "humanity," as those with specific medical conditions are provided legal protections regardless of their ability to either effectively communicate their needs or comprehend complex information (Glenn [151])."

"This particular phenomenon stands as one of the critical arguments against notions that AIS cannot attain human-like levels of intelligence, beyond that of the notion that AIS has been developed to perform (somewhat trivial) tasks previously performed by humans."

"The SLAIS presented in Tau could be stated to be a product of human-driven artificial intelligence learning (HAIL)—albeit a twisted version of HAIL given the motives of the system's developer (D'Alessandro [152])—which may further bring into question the antagonism of AIS being centred on the internal workings of the system."

Communicability in AIS and virtual environments
"No one solution to address these concerns could be said to be efficacious in every socio-political context, especially if we consider the interactions of SLAIS and D/SAIS in remote parts of the world or areas where the spread of the Internet of Things (IoT) is still underdeveloped compared to areas of high population density—to say nothing of the incorporeal nature of IoT and virtual landscapes outside of their data-specific boundaries where human influence is limited to corporeal interfaces."

"While the author sympathises with the notion that AIS ought to be developed in a manner that does not directly harm human populations, approaching the development of emotion in inorganic systems that rely solely upon highly specific commands may simply not be possible if emotion cannot be proved to be replicated under these conditions—which would require, arguably, the capacity to develop a sense of "self" and communicate highly abstract information in a comprehensible manner."

On virtual property ownership and compensation
"All of the programmes and software that were installed on the computer between the time of purchase and time of ownership transfer necessarily brings into question the metaphysical nature of the virtual space "held" by that system—even if

the computer's IP address somehow changes in the process and effectively creates a "new" virtual space in which IoT must define."

"AIS, humans, and organisations may all be able to claim ownership over a device that is freshly minted in an electronics factory regardless of whether the product is for private or commercial use considering the potentiality for a SLAIS to exist within IoT protocols as a "virtual" entity without a defined corporeal form."

Conclusion

"While it cannot be denied that some changes need to be had in various laws internationally to accommodate notions of virtual property ownership, specific suggestions for how they should develop cannot be developed within this article given the complexity of the topic."

"Without a reconsideration of this nature, the law will be ill-prepared to address AIS as "human-like" subjects when they do attain some semblance of human-like intelligence—if such a system does not already exist amidst our societal ignorance of their comparable intellectual capacities."

"Yes, environmental and cultural concerns do oftentimes apply given how great a tool AI can be to drive forward humanity's sustainable development agenda."

"Before dismissing the notion of AIS personhood or legal protections, please ask yourself: could I ever confuse this system with a stranger walking down the street if it showed no mechanical parts?"

3.2.3 Ethical Consumption Communities Across Physical and Digital Spaces: An Exploration of Their Complementary and Synergistic Affordances [191]

This is a machine-generated summary of:

Hoelscher, Vera; Chatzidakis, Andreas: Ethical Consumption Communities Across Physical and Digital Spaces: An Exploration of Their Complementary and Synergistic Affordances [191]

Published in: Journal of Business Ethics (2020)

Link to original: https://doi.org/10.1007/s10551-020-04477-6

If you want to cite the papers, please refer to the original.

For technical reasons we could not place the page where the original quote is coming from.

Abstract-Summary

"This study explores the distinct affordances of physical and digital spaces and how they may work in both complementary and synergistic fashions."

"Drawing on an ethnographic study of two ethical consumption communities in North London, UK, we explore how ethical consumers navigate and negotiate both physical and digital spaces, taking advantage of such affordances."

"We develop the notion of chorophilia, or love for physical space, explore digital commitments and synergistic affordances of scaling up, and advance polytopes, which focus on the relationality of digital-physical spaces."

Introduction

"Prior consumer ethics research has considered how such digital layers enable and/or constrain ethical consumption practices (Egels-Zanden and Sörum [153]; Eli and others [154]; Fuentes and Sörum [155]; Graham [156]; Gummerus and others [157]; Horst [158]; Humphery and Jordan [159]; Lekakis [160])."

"Whereas the current literature tends to focus on either physical or digital spaces, and the ways in which they enable, constrain and ultimately shape practices of ethical consumption (e.g. Fuentes and Sörum [155]; Humphery and Jordan [159]; Schneider and others [161]), their continuous intersection has been largely unexplored and under-theorised."

"Put differently, (almost) all ethical consumption comprises a multiplicity of physical and digital layers and consumers are increasingly aware of how they shape their activist efforts."

"By distinct affordances between physical and digital spaces, we thus refer to how our senses, interactions, and consumption practices are impacted by their special characteristics."

"The aim of our study is to explore the distinct affordances of digital and physical spaces, and the ways in which consumers navigate and take advantage of these in their everyday logics and practices."

Literature Review: Place Making and the Meeting Up of Digital and Physical Geographies

"Whilst previous literature focusses on the effect of digital technologies on ethical consumption (Eli and others [154]; Fuentes and Sörum [155]; Gummerus and others [157]), our study examines self-governed ethical consumption communities, and how these navigate the distinct affordances between their physical and digital spaces."

"Ethical consumption communities partake in various forms of collective organising and action, that are increasingly embedded in both digital and physical spaces."

"Affordances describe the design of a space, whether physical, digital, or at the intersection of both realms, and how this affects our use of it and the interactions we engage in."

"Within the different physical and digital spaces of ethical consumption communities, such affordances will allow for distinct modes of communication, interaction and creation that consumers have to navigate."

"Our theoretical lens of affordances thus supports our central research question: what are the distinct affordances of physical and digital spaces, and how do ethical consumers navigate these in their everyday consumption practices?"

Methodology

"In order to explore the distinct affordances between physical and digital spaces, the lead researcher undertook a 1.5-year ethnography of two ethical consumption communities in London, UK."

"Our two fieldwork sites were selected on the basis that they differed both in terms of their overall digital versus physical presence (with Transition Town being more digitally active) and their underpinning ideology of ethical consumption (with Rochester Square Gardens being more radically left wing)."

"The interview guide was structured into five sections: an introduction that enquired about their first contact with the respective community, a second section on the significance of sharing spaces with other members, a third that explored perceptions of how encounters qualitatively differ between the physical and digital realm, a fourth section on networking between members and other communities, and an outroduction that centred on demographic questions."

Findings

"What Mark, Dylan and Adam highlight is that the affordances of shared physical spaces enable interactions of a different multi-sensorial quality to those that are mediated via digital screens."

"A desire for the affordances of physical space was a recurrent theme amongst the digitally active members of Transition Kentish Town."

"Although the members of Transition Kentish Town and their social media followers were communicating digitally, their actions had a profound impact on physical space."

"Within the reach afforded by digital spaces, Transition Kentish Towners scaled up beyond their physical boundaries."

"The squat's Facebook group became a digital polytope that was relational to the other places and times it referenced, and simultaneously afforded the squatter commune to scale up on the physical space: One of our research participants had discovered that the couple who bought the garden square had turned it into a creative space for contemporary ceramics."

Discussion

"Our study builds on the observation that ethical consumption communities increasingly inhabit both physical and digital spaces and learn how to effectively navigate their distinct affordances."

"Digital spaces do not allow for some of the physical aspects of creation and consumption that are valued by ethical consumption communities such as growing organic produce, holding workshops or hosting communal meals."

"Notwithstanding the cravings for the physicality expressed in chorophilia, digital space does offer compelling affordances to ethical consumption communities."

"The digital space of an ethical consumption community can thus act as a first representation of its aims, membership and activities that ultimately allow a community to gain traction beyond its original physical boundaries."

"We argue that the relational space (Massey [162, 163]) of ethical consumption communities has now scaled up by significantly expanding into the digital realm."

Conclusion and Directions for Future Research

"Our study builds on the observation that contemporary ethical consumption communities increasingly inhabit both digital and physical realms, and are thus dynamically affected by their distinct affordances."

"The distanciation afforded by the digital also enables new behaviours to form as prospective members may ease themselves into the physical context of ethical consumption communities."

"Crucially, what our study also offers to the field of ethical consumption is a more integrative view of how the distinct affordances of the physical and digital realm are experienced by community members, and the value this brings to not only creating but also reviving a multiplicity of spaces within the digital."

"Future research into ethical consumption communities could explore the applicability of these distinct affordances across communities with different profiles of digital-physical-spatial embeddedness and ideological motivations."

3.2.4 Legal Dilemmas of Estonian Artificial Intelligence Strategy: In Between of e-society and Global Race [192]

This is a machine-generated summary of:

Kerikmäe, Tanel; Pärn-Lee, Evelin: Legal dilemmas of Estonian artificial intelligence strategy: in between of e-society and global race [192]

Published in: AI & SOCIETY (2020)

Link to original: https://doi.org/10.1007/s00146-020-01009-8

If you want to cite the papers, please refer to the original.

For technical reasons we could not place the page where the original quote is coming from.

Abstract-Summary
"While the state is aiming to attract tech investments with e-Residency program and has recently started to invest into protecting national IP and safeguarding data from cybercrime by applying blockchain technology and creating its "digital embassy" in Luxembourg, emerging technologies such as and applications of artificial intelligence but also internet of things have posed the question on legal regulation and standardization."

"The dilemma, however, seems to be that since the new technologies, such artificial intelligence is much more overwhelming phenomenon than e-governance and presumably, before deciding the legal standards, the political and economic strategies that go beyond the e-governance should be set."

Introduction
"When the Estonian government has implemented x-road as the core of government information system integration, it has opened the doors for many public e-services (such as e-commercial register, e-land register, road administrations e-portal), private sector services (such as banking, e-bookkeeping, smart grid solutions or an unique possibility for Estonian e-Residents to open a bank account entirely online), as well as semi-public semi-private services, such as an e-health system, e-Prescription, which is a centralized paperless system for issuing and handling medical prescriptions (at the pharmacy, all a patient needs to do is present an ID-card."

"Estonia has been relatively successful also in attempts to sell its e-governance abroad when creating eGovernment solutions and launching other unique projects that go beyond their jurisdiction, such as the world's first e-Residency program started in 2014 with the aim to attract foreign entrepreneurs to invest to Estonia using the modern services of e-governance."

What is the Estonian current AI situation?
"On that basis, the legal expert group screened the whole Estonian legislation, interviewed stakeholders from the private as well from public sector, participated in workshops and meetings, and concluded that a separate general AI law is not possible or currently even needed."

"Based on the conducted analysis, the expert group focused on three categories of legal aspects: (i) Are there any legal norms or regulations currently valid that do not take into account or do not allow for the use of AI solutions; (ii) Are there any legal norm or regulations lacking and need to be proposed (allowing or restricting AI-related activities), in so far as lack of such rules and regulations would have adverse consequences for the interests and rights of individuals, communities or society; (iii) How to provide legal certainty in AI-related issues (especially liability)."

Conclusion

"From the viewpoint of society, one cannot imagine a successful long-term functioning of an e-State without proper support from or functioning of the private sector."

"The number of AI applications as key performance indicators proposed in the Estonian IT Strategy certainly indicate support of technology minded policy, but it is doubtful if they truly reflect the agreed values, or whether or not the society is in favor or against using AI in our everyday lives, or how the AI should be regulated in law The presumption of society is reflected by the White Paper on the AI as following: "AI should work for people and be a force for good in society"."

"Unlike the United Arab Emirates, Estonia has no separate AI ministry, however as of 2019 Estonia has a minister for IT, who recently promised a restart for the Estonian digital society, as according to him currently applied systems do not consider the complexity of the world of technology."

Bibliography

1. Sailunaz K, Dhaliwal M, Rokne J, Alhajj R (2018) Emotion detection from text and speech: a survey. Soc Netw Anal Min 8:1–26
2. Yadollahi A, Shahraki AG, Zaiane OR (2017) Current state of text sentiment analysis from opinion to emotion mining. ACM Comput Surv 50:1–33
3. Egger M, Ley M, Hanke S (2019) Emotion recognition from physiological signal analysis: a review. Electron Notes Theor Comput Sci 343:35–55
4. Marechal C, Mikolajewski D, Tyburek K, et al (2019) Survey on AI-based multimodal methods for emotion detection.
5. Chopade CR (2015) Text based emotion recognition: a survey. Int J Sci Res 4:409–414
6. Hirat R, Mittal N (2015) A survey on emotion detection techniques using text in blogposts. Int Bull Math Res 2:180–187
7. McStay A (2018) Emotional AI: the rise of empathic media. Sage
8. Shadbolt N, Berners-Lee T, Hall W (2006) The semantic web revisited. IEEE Intell Syst 21(3):96–101
9. McAfee A (2009) Enterprise 2.0: New collaborative tools for your organization's toughest challenges. Harvard Business Press
10. Ouirdi ME, El Ouirdi A, Segers J, Henderickx E (2014) Social media conceptualization and taxonomy: a Lasswellian framework. J Creat Comm 9(2):107–126
11. Anderson C (2006) People power. Wired magazine. 2006—07—14). http://www.wired.com/wired/archive/14.07/people.html
12. Grant SB, Preston TA (2019) Using social power and influence to mobilise the supply chain into knowledge sharing: a case in insurance. Inf Manag 56(5):625–639
13. Ortbach K, Recker J (2014) Do good things and talk about them: a theory of academic's usage of enterprise social networks for impression management tactics
14. Huang Z, Benyoucef M (2013) From e-commerce to social commerce: A close look at design features. Electron Commer Res Appl 12(4):246–259

15. Mason R, Rennie F (2008) Social networking as an educational tool. E-learning and social networking handbook: Resources for Higher Education, pp.1–24
16. McCarroll N, Curran K (2013) Social networking in education. Int J Innov Digit Econ (IJIDE) 4(1):1–15
17. De-Marcos L, Garcia-Cabot A, Garcia-Lopez E (2017) Towards the social gamification of e-learning: a practical experiment. Int J Eng Educ 33(1):66–73
18. Prensky M (2001) Digital natives, digital immigrants' part 1. On Horizon 9(5):1–6
19. Phillips R, McNaught C, Kennedy G (2012) Evaluating e-learning: guiding research and practice. Routledge
20. Gannod GC, Bachman KM (2012) A characterization of social networks for effective communication and collaboration in computing education. ASEE Comput Educ (CoED) J 3(4):100
21. Koranteng FN, Wiafe I, Kuada E (2019) An empirical study of the relationship between social networking sites and students' engagement in higher education. J Educ Comput Res 57(5):1131–1159
22. Kazemian S (2018) The role micro-blogging plays in informal communication and knowledge sharing activities within universities: a review of the literature. Int J Knowled Syst Sci (IJKSS) 9(3):18–31
23. Grant SB (2016) Classifying emerging knowledge sharing practices and some insights into antecedents to social networking: a case in insurance. J Knowl Manag 20(5):898–917
24. Srnka KJ, Koeszegi ST (2007) From words to numbers: how to transform qualitative data into meaningful quantitative results. Schmalenbach Bus Rev 59(1):29–57
25. Lovejoy, K., & Saxton, G. D. (2012). Information, community, and action: How nonprofit organizations use social media. Journal of Computer-Mediated Communication, 17(3), 337–353.
26. Wirtz, B. W., Schilke, O., & Ullrich, S. (2010). Strategic development of business models: Implications of the web 2.0 for creating value on the internet. Long Range Planning, 43(2–3), 272–290.
27. Agnihotri, R., Kothandaraman, P., Kashyap, R., & Singh, R. (2012). Bringing "social" into sales: The impact of salespeople's social media use on service behaviors and value creation. Journal of Personal Selling & Sales Management, 32(3), 333–348.
28. Agnihotri, R., Dingus, R., Hu, M. Y., & Krush, M. T. (2016). Social media: Influencing customer satisfaction in B2B sales. Industrial Marketing Management, 53, 172–180.
29. Chandler, J. D., & Lusch, R. F. (2015). Service systems: A broadened framework and research agenda on value propositions, engagement, and service experience. Journal of Service Research, 18(1), 6–22.
30. Li, H. (2017). Information and donations: A study of nonprofit online communication (Doctoral dissertation, Rutgers University-School-Newark).
31. Bennett, R. (2018). Nonprofit marketing and fundraising: A research overview. Abingdon: Routledge.
32. Saxton, G. D., & Waters, R. D. (2014). What do stakeholders like on Facebook? Examining public reactions to nonprofit organizations' informational, promotional, and community-building messages. Journal of Public Relations Research, 26(3), 280–299.
33. Brickley D, Miller L (2007) The Friend of a Friend (FOAF) Vocabulary specification. Technical report
34. Lee R, Sumiya K (2010) Measuring geographical regularities of crowd behaviors for Twitter-based geo-social event detection. In: Proceedings of the 2nd ACM SIGSPATIAL iInternational workshop on location based social networks (SIGSPATIAL). ACM, New York, NY, USA, pp 1–10
35. Lee K, Caverlee J, Webb S (2010) The social honeypot project. In: Proceedings of the 19th international conference on World wide web (WWW). ACM Press, New York, New York, USA, p 1139

36. Stringhini G, Kruegel C, Vigna G (2012) Detecting spammers on social networks. In: Proceedings of the 26th annual computer security applications conference (ACSAC), pp 1–9. ACM Request Permissions
37. Kim M, Seo J, Noh S, Han S (2012) Identity management-based social trust model for mediating information sharing and privacy enhancement. Secur Commun Netw 5(8):887–897
38. Nepal S, Sherchan W, Paris C (2011) STrust: a trust model for social networks. In: 2011 IEEE 10th international conference on trust, security and privacy in computing and communications (TrustCom), pp 841–846
39. Raad E, Chbeir R, Dipanda A (2010) User profile matching in social networks. In: 13th international conference on network-based information systems (NBiS), pp 297–304. IEEE
40. Singla P, Domingos P (2005) Object identification with attribute-mediated dependences. In: Proceedings of 9th European conference on principles and practice of knowledge discovery in databases (PKDD), pp 297–308
41. Mika P (2005) Flink: Semantic Web technology for the extraction and analysis of social networks. Web Seman Sci Serv Agents World Wide Web 3(2-3):211–223
42. Leicht EA, Holme P, Newman MEJ (2006) Vertex similarity in networks. Phys Rev E 73(2):026120
43. Kahn R, Kellner DM (2012) Oppositional Politics and the Internet: A Critical/Reconstructive Approach. In: Durham MG, Kellner DM (eds) Media and Cultural Studies: Keyworks. Wiley-Blackwell, West Sussex, pp 597–613
44. Mascheroni G (2015) The practice of participation: Youth's vocabularies around on- and offline civic and political engagement. Media@LSE Working Paper series, 35. http://www.lse.ac.uk/media@lse/research/mediaWorkingPapers/pdf/WP35-FINAL.pdf
45. Brennen BS (2013) Qualitative research methods for media studies. Routledge, New York
46. Middaugh E, Bowyer B, Kahne J (2016). U Suk! Participatory Media and Youth Experiences WithPolitical Discourse. Youth & Society. https://doi.org/10.1177/0044118X16655246
47. Lü L, Zhou T (2011) Link prediction in complex networks: a survey. Phys A Stat Mech Appl 390(6):1150–1170
48. Spiegel S, Clausen J, Albayrak S, Kunegis J (2011) Link prediction on evolving data using tensor factorization. In: Pacific-Asia conference on knowledge discovery and data mining. Springer, Berlin, Heidelberg, pp 100–110
49. Lichtenwalter RN, Lussier JT, Chawla NV (2010) New perspectives and methods in link prediction. In: Proceedings of the 16th ACM SIGKDD international conference on knowledge discovery and data mining. ACM, New York, pp 243–252. https://doi.org/10.1145/1835804.1835837
50. da Silva Soares PR, Prudencio RBC (2012) Time series based link prediction. In: IEEE international joint conference on neural networks (IJCNN). https://doi.org/10.1109/IJCNN.2012.6252471
51. Sarkar P, Chakrabarti D, Jordan M (2012) Nonparametric link prediction in dynamic networks. arXiv:1206.6394
52. Feng X, Zhao J, Xu K (2012) Link prediction in complex networks: a clustering perspective. Eur Phys J B 85(1):1–9
53. Fire M, Puzis R, Elovici Y (2013) Link prediction in highly fractional data sets. In: Subrahmanian VS (ed) Handbook of computational approaches to counterterrorism. Springer, New York, pp 283–300
54. Jahanbakhsh K, King V, Shoja GC (2012) Predicting human contacts in mobile social networks using supervised learning. In: SIMPLEX workshop, ACM
55. Lichtnwalter R, Chawla NV (2012) Link prediction: fair and effective evaluation. In: 2012 IEEE/ACM international conference on advances in social networks analysis and mining (ASONAM). IEEE, pp 376–383
56. Xu Y, Rockmore D (2012) Feature selection for link prediction. In: CIKM workshop. ACM
57. Boyd, M. (1989). Family and personal networks in international migration: recent developments and new agendas. The International Migration Review, 23(3), 638–670.

58. Debsu, D. N., Little, P. D., Tiki, W., Guagliardo, S. A. J., & Kitron, U. (2016). Mobile phones for mobile people: the role of information and communication technology (ICT) among livestock traders and borana pastoralists of Southern Ethiopia. Nomadic Peoples, 20(1), 35–61.
59. Sallach, D. L. (2003). Social theory and agent architectures. Social Science Computer Review, 21(2), 179–195.
60. Dekker, R., & Engbersen, G. (2014). How social media transform migrant networks and facilitate migration. Global Networks, 14(4), 401–418.
61. Massey, D. S., Arango, J., Hugo, G., Kouaouci, A., Pellegrino, A., & Taylor, J. E. (1993). Theories of international migration: a review and appraisal. The Population and Development Review, 19, 431–466.
62. Care, E., Griffin, P., Scoular, C., Awwal, N., & Zoanetti, N. (2015). Collaborative problem solving tasks. In P. Griffin, & E. Care (Eds.), Assessment and teaching of 21st century skills: Methods and approach (pp. 85–104). Dordrecht: Springer. https://doi.org/10.1007/978-94-017-9395-7
63. Zhu, M., Shu, Z., & von Davier, A. A. (2016). Using networks to visualize and analyze process data for educational assessment. Journal of Educational Measurement, 53(2), 190–211. https://doi.org/10.1111/jedm.12107
64. Andrews-Todd, J., & Forsyth, C. M. (2020). Exploring social and cognitive dimensions of collaborative problem solving in an open online simulation-based task. Computers in Human Behavior, 104, 105759. https://doi.org/10.1016/j.chb.2018.10.025
65. Paolacci G, Chandler J, Ipeirotis PG (2010) Running experiments on amazon mechanical Turk. Judgm Decis Mak 5(5):411–419
66. De Maio C, Fenza G, Gallo M, Loia V, Parente M (2019) Time-aware adaptive tweets ranking through deep learning. Futur Gener Comput Syst 93:924–932
67. Loia V, Tomasiello S, Vaccaro A, Gao J (2020) Using local learning with fuzzy transform: application to short term forecasting problems. Fuzzy Optim Decis Making 19(1):13–32
68. Liu T, Xue F, Sun J, Sun X (2019) A survey of event analysis and mining from social multimedia. Multimed Tools Appl:1–18
69. Banik A, Shamsi Z, Laiphrakpam DS (2019) An encryption scheme for securing multiple medical images. J Inf Secur Appl 49:102398
70. Swain AK, Balabantaray BK, Rout JK, Satpathy S An optimal deep learning approach for classification of age groups in social network
71. Chang C-S, Lee D-S, Liou L-H, Lu S-M, Wu M-H (2018) A probabilistic framework for structural analysis and community detection in directed networks. IEEE/ACM Trans Netw (TON) 26(1):31–46
72. Baroi SJ, Singh N, Das R, Singh TD (2020) Nits-hinglish-sentimix at semeval-2020 task 9: Sentiment analysis for code-mixed social media text. arXiv: 2007.12081
73. Zhou Y, Cheng H, Jeffrey Xu Y (2009) Graph clustering based on structural/attribute similarities. Proc VLDB Endowment 2(1):718–729
74. H Fani, E Jiang, E Bagheri, F Al-Obeidat, W Du, M Kargar. User community detection via embedding of social network structure and temporal content. Information Processing & Management. 2020;57(2):102056.
75. Ahmad K, Gillman L, Tostevin L. Weirdness indexing for logical document extrapolation and retrieval. In Proceedings of the Eighth Text Retrieval Conference (TREC-8). 2000. p. 1–8.
76. Joinson, Adam N. (2008). Looking at, looking up or keeping up with people?: motives and use of facebook. Proceeding of the twenty-sixth annual SIGCHI conference on Human factors in computing systems, Florence, Italy, April 5 to 10, 2008. New York, USA: ACM, pp. 1027–1036.
77. Raffle, Hayes, Rafael Ballagas, Glenda Revelle, Hiroshi Horii, Sean Follmer, Janet Go, Emily Reardon, Koichi Mori, Joseph Kaye and Mirjana Spasojevic (2010). Family story play: reading with young children (and elmo) over a distance. Proceedings of the 28th international conference on Human factors in computing systems, Atlanta, Georgia, USA, April 10 to 15, 2010. New York, USA: ACM, pp. 1583–1592.

78. Consolvo, Sunny, Peter Roessler, Brett E. Shelton, Anthony LaMarca, Bill Schilit and Sara Bly (2004b). Technology for care networks of elders. IEEE Pervasive Computing, vol. 3, no. 2, pp. 22–29.
79. Mynatt, Elizabeth D., Jim Rowan, Sarah Craighill and Annie Jacobs (2001). Digital family portraits: supporting peace of mind for extended family members. Proceedings of the SIGCHI conference on Human factors in computing systems, Seattle, Washington, USA, March 31 to April 5, 2001. New York, USA: ACM, pp. 333–340.
80. B Pang, L Lee, S Vaithyanathan, in Proceedings of the Seventh Conference on Empirical Methods in Natural Language Processing (EMNLP-02). Thumbs up? Sentiment classification using machine learning techniques (Philadelphia, Pennsylvania, 2002), pp. 79–86. https://doi.org/10.3115/1118693.1118704.
81. Yin J, Lampert A, Cameron MA, Robinson B, Power R: Using social media to enhance emergency situation awareness. IEEE Intell Syst 2012, 27(6):52–59. https://doi.org/10.1109/MIS.2012.6 https://doi.org/10.1109/MIS.2012.6 https://doi.org/10.1109/MIS.2012.6
82. A Nagy, J Stamberger, in Proceedings of the Ninth International Conference on Information Systems for Crisis Response and Management (ISCRAM 2012). Crowd sentiment detection during disasters and crises (Vancouver, Canada, 2012).
83. S Verma, S Vieweg, WJ Corvey, L Palen, JH Martin, M Palmer, A Schram, KM Anderson, in Proceedings of the Fifth International AAAI Conference on Weblogs and Social Media. Natural language processing to the rescue? Extracting "situational awareness" tweets during mass emergency (Barcelona, Spain, 2011), pp. 385–392.
84. C Strapparava, R Mihalcea, in Proceedings of the 2008 ACM Symposium on Applied Computing (SAC'08). Learning to identify emotions in text (Fortaleza, Brazil, 2008), pp. 1556–1560. https://doi.org/10.1145/1363686.1364052.
85. C Strapparava, A Valitutti, in Proceedings of the Fourth International Conference on Language Resources and Evaluation (LREC 2004). WordNet-Affect: an affective extension of WordNet (Lisbon, Portugal, 2004), pp. 1083–1086.
86. Green J, Dale J. Primary care in accident and emergency and general practice: a comparison. Soc Sci Med. 1992;35:987–95.
87. Hudgins A, Rising KL. Fear, vulnerability and sacrifice: Drivers of emergency department use and implications for policy. Soc Sci Med. 2016;169:50–7.
88. Salehin I, Dip S, Talha I, Rayhan I, Nammi K (2021) Impact on human mental behavior after pass through a long time home quarantine using machine learning. Inter J Educat Manage Eng 11(1):41–50. https://doi.org/10.5815/ijeme.2021.01.05
89. Kechagias E, Chatzistelios G, Papadopoulos G, Apostolou P (2022) Digital transformation of the maritime industry: a cybersecurity systemic approach. Inter J Critical Infrastruct Protect 37:100526. https://doi.org/10.1016/j.ijcip.2022.100526
90. Arayankalam J, Krishnan S (2021) Relating foreign disinformation through social media, domestic online media fractionalization, government's control over cyberspace, and social media-induced offline violence: Insights from the agenda-building theoretical perspective. Technol Forecast Soc Change 166:120661. https://doi.org/10.1016/j.techfore.2021.120661
91. Arsenyan J, Mirowska A (2021) Almost human? a comparative case study on the social media presence of virtual influencers. Inter J Human-Comput Stud 155:102694. https://doi.org/10.1016/j.ijhcs.2021.102694
92. Ning C, You F (2018) Adaptive robust optimization with minimax regret criterion: Multi-objective optimization framework and computational algorithm for planning and scheduling under uncertainty. Comput Chem Eng 108:425–447. https://doi.org/10.1016/j.compchemeng.2017.09.026
93. Yokoyama R, Fujiwara K, Ohkura M, Wakui T (2014) A revised method for robust optimal design of energy supply systems based on minimax regret criterion. Energy Convers Manage 84:196–208. https://doi.org/10.1016/j.enconman.2014.03.045
94. Taslimov AD, Berdishev AS, Melikuziev MV, Rakhimov FM (2019) Method of choosing the unification of cable sections of electric network cables under conditions of load development uncertainty. E3S Web of Conf 139:01081. https://doi.org/10.1051/e3sconf/201913901081

95. Moustakides G, Verykios VA (2008) MaxMin approach for hiding frequent itemsets. Data Knowl Eng 65(1):75–89. https://doi.org/10.1016/j.datak.2007.06.012
96. Giocoli, N (2004) Savage vs. Wald: Was Bayesian Decision Theory the Only Available Alternative for Postwar Economics? https://doi.org/10.2139/ssrn.910916
97. Facebook page of Lviv Polytechnic National University. 2022. Available online: https://www.facebook.com/groups/206197692740294
98. Gunelius S (2014) The data explosion in 2014 minute by minute—infographic. Retrieved 16 Jan 2016, from ACI.info: http://aci.info/2014/07/12/the-data-explosion-in-2014-minute-by-minute-infographic/
99. Krouska, A., Troussas, C., & Virvou, M. (2019). SN-learning: An exploratory study beyond e-learning and evaluation of its applications using EV-SNL framework. Journal of Computer Assisted Learning, 35(2), 168–177.
100. Fu, E. L. F., van Aalst, J., & Chan, C. K. K. (2016). Toward a classification of discourse patterns in asynchronous online discussions. International Journal of Computer-Supported Collaborative Learning, 11(4), 441–478.
101. Berland, M., Baker, R. S., & Blikstein, P. (2014). Educational data mining and learning analytics: Applications to constructionist research. Technology, Knowledge and Learning, 19(1–2), 205–220. https://doi.org/10.1007/s10758-014-9223-7 .
102. Fincham, E., Gašević, D., & Pardo, A. (2018). From social ties to network processes: Do tie definitions matter? Journal of Learning Analytics, 5(2), 9–28. https://doi.org/10.18608/jla.2018.52.2
103. Gaševi, D., Dawson, S., & Siemens, G. (2015). Let ' s not forget : Learning analytics are about learning course signals : Lessons learned. TechTrends, 59(1), 64–71. https://doi.org/10.1007/s11528-014-0822-x .
104. Schneider, B., & Pea, R. (2014). Toward collaboration sensing. International Journal of Computer-Supported Collaborative Learning, 9(4), 371–395.
105. Noroozi, O., Järvelä, S., & Kirschner, P. A. (2019). Introduction paper special issue computers in human behavior multidisciplinary innovations and technologies for facilitation of self-regulated learning. Computers in Human Behavior, 100, 295–297. https://doi.org/10.1016/j.chb.2019.07.020 .
106. Liao, H., Mariani, M. S., Medo, M., Zhang, Y. C., & Zhou, M. Y. (2017). Ranking in evolving complex networks. Physics Reports, 689(April), 1–54. https://doi.org/10.1016/j.physrep.2017.05.001 .
107. Opsahl, T., Agneessens, F., & Skvoretz, J. (2010). Node centrality in weighted networks: Generalizing degree and shortest paths. Social Networks, 32(3), 245–251. https://doi.org/10.1016/j.socnet.2010.03.006 .
108. Stephenson, K., & Zelen, M. (1989). Rethinking centrality: Methods and examples. Social Networks, 11(1), 1–37. https://doi.org/10.1016/0378-8733(89)90016-6 .
109. Wei, D., Deng, X., Zhang, X., Deng, Y., & Mahadevan, S. (2013). Identifying influential nodes in weighted networks based on evidence theory. Physica A: Statistical Mechanics and its Applications, 392(10), 2564–2575. https://doi.org/10.1016/j.physa.2013.01.054 .
110. Csardi, G., & Nepusz, T. (2006). The igraph software package for complex network research. InterJournal, Complex Sy, 1695 http://igraph.org .
111. Saqr, M., Nouri, J., & Jormanainen, I. (2019b). A learning analytics study of the effect of group size on social dynamics and performance in online collaborative learning. In M. Scheffel, J. Broisin, V. Pammer-Schindler, A. Ioannou, & J. Schneider (Eds.), Lecture Notes in Computer Science (including subseries Lecture Notes in Artificial Intelligence and Lecture Notes in Bioinformatics): Vol. 11722 LNCS (pp. 466–479). Springer International Publishing. https://doi.org/10.1007/978-3-030-29736-7_35 .
112. Baker EJ (1991) Hurricane evacuation behavior. Int J Mass Emerg Disasters 9(2):287–310
113. Fu H, Wilmot CG (2004) Sequential logit dynamic travel demand model for hurricane evacuation. Transp Res Rec J Transp Res Board 1:19–26
114. Lindell MK, Lu JC, Prater CS (2005) Household decision making and evacuation in response to Hurricane Lili. Nat Hazards Rev 6(4):171–179

115. Whitehead JC, Edwards B, Van Willigen M, Maiolo JR, Wilson K, Smith KT (2000) Heading for higher ground: factors affecting real and hypothetical hurricane evacuation behavior. Global Environ Chang Part B Environ Hazards 2(4):133–142
116. Pastor-Satorras, R., & Vespignani, A. (2001). Physical Review Letters, 86(14), 3200.
117. Gleeson, J., & Cahalane, D. (2007). Physical Review E, 75(5), 056103.
118. Galstyan, A., & Cohen, P. (2007). Physical Review E, 75(3), 036109.
119. Gleeson, J. (2008). Physical Review E, 77(4), 046117.
120. Gleeson, J., Cellai, D., Onnela, J. P., Porter, M. A., & Reed-Tsochas, F. (2014). Proceedings of the National Academy of Sciences, 111(29), 10411.
121. Guilbeault, D., Becker, J., & Centola, D. (2017). arXiv preprint arXiv:1710.07606 .
122. Watts DJ. A simple model of global cascades on random networks. Proc Natl Acad Sci USA. 2002. https://doi.org/10.1073/pnas.082090499 .
123. Erhart, E., & Overdorff, D. J. (1999). Female coordination of group travel in wild Propithecus and Eulemur. International Journal of Primatology, 20(6), 927–940.
124. Lamprecht, J. (1992). Variable leadership in bar-headed geese (Anser indicus): an analysis of pair and family departures. Behaviour, 122(1/2), 105–120.
125. Peterson, R. O., Jacobs, A. K., Drummer, T. D., Mech, L. D., & Smith, D. W. (2002). Leadership behavior in relation to dominance and reproductive status in gray wolves, Canis lupus. Canadian Journal of Zoology, 80(8), 1405–1412.
126. Radford, A. N. (2004). Vocal coordination of group movement by green woodhoopoes (Phoeniculus purpureus). Ethology, 110(1), 11–20.
127. Tyler, S. J. (1972). The behaviour and social organisation of the New Forest ponies. Animal Behaviour Monographs, 5, 87–196.
128. Camazine, S., Deneubourg, J. L., Franks, N. R., Sneyd, J., Theraulaz, G., & Bonabeau, E. (2001). Self-organization in biological systems. Woodstock, UK: Princeton University Press.
129. Meunier, H., Leca, J.-B., Deneubourg, J. L., & Petit, O. (2006). Group movement decisions in capuchin monkeys: the utility of an experimental study and a mathematical model to explore the relationship between individual and collective behaviours. Behaviour, 143(12), 1511–1527.
130. Leca, J.-B., Gunst, N., Thierry, B., & Petit, O. (2003). Distributed leadership in semifree-ranging white-faced capuchin monkeys. Animal Behaviour, 66(6), 1045–1052.
131. Gautrais, J., Michelena, P., Sibbald, A., Bon, R., & Deneubourg, J. L. (2007). Allelomimetic synchronization in Merino sheep. Animal Behaviour, 74(5), 1443–1454.
132. Jacobs, A., Maumy, M., & Petit, O. (2008). The influence of social organisation on leadership in brown lemurs (Eulemur fulvus fulvus) in a controlled environment. Behavioural Processes, 79(2), 111–113.
133. Lifschitz V (2008) What is answer set programming?. In: Proceedings of the twenty-third AAAI conference on artificial intelligence, AAAI 2008, Chicago, Illinois, USA, 13–17 July 2008, pp 1594–1597
134. Cheng Y, Park J, Sandhu R (2012) Relationship-based access control for online social networks: beyond user-to-user relationships. In: 2012 International conference on social computing (SocialCom) and privacy, security, risk and trust (PASSAT), pp 646–655
135. Cheng Y, Park J, Sandhu R (2014) Attribute-aware relationship-based access control for online social networks. In: Data and applications security and privacy XXVIII. Lecture Notes in Computer Science, vol 8566. Springer, Berlin, pp 292–306
136. Fong P (2011) Relationship-based access control: protection model and policy language. In: Proceedings of the first ACM conference on data and application security and privacy. ACM, New York, pp 191–202
137. Pang J, Zhang Y (2013) A new access control scheme for Facebook-style social networks. CoRR abs/1304.2504
138. Carminati B, Ferrari E, Perego A (2006) Rule-based access control for social networks. In: On the move to meaningful Internet systems 2006: OTM 2006 workshops. Lecture Notes in Computer Science, vol 4278. Springer, Berlin, pp 1734–1744

139. White J, Park J, Kamhoua C (2014) Kwiat K (2014) Social network attack simulation with honeytokens. Soc Netw Anal Min 4(1):1–4
140. Cramm JM, Hartgerink JM, Steyerberg EW, Bakker TJ, MacKenbach JP, Nieboer AP. Understanding older patients' self-management abilities: Functional loss, self-management, and well-being. Qual Life Res. 2013;22(1):85–92.
141. Goedendorp MM, Steverink N. Interventions based on self-management of well-being theory: pooling data to demonstrate mediation and ceiling effects, and to compare formats. Aging Ment Heal. 2017;21(9):947–53.
142. Schuurmans H, Steverink N, Frieswijk N, Buunk BP, Slaets JPJ, Lindenberg S. How to measure self-management abilities in older people by self-report. The development of the SMAS-30. Qual Life Res. 2005;14(10):2215–28.
143. Ormel J, Lindenberg S, Steverink N, Verbrugge LM. Subjective well-being and social production functions. Soc Indic Res. 1999;46(1):61–90.
144. Berner J, Rennemark M, Jogréus C, Anderberg P, Sköldunger A, Wahlberg M, et al. Factors influencing Internet usage in older adults (65 years and above) living in rural and urban Sweden. Health Informatics J. 2015;21(3):237–49.
145. Schulman-Green D, Jaser SS, Park C, Whittemore R. A metasynthesis of factors affecting self-management of chronic illness. J Adv Nurs. 2016;72(7):1469–89.
146. Correia, I., & Dalbert, C. (2007). Belief in a just world, justice concerns, and well-being at Portuguese schools. European Journal of Psychology of Education,22(4), 421–437. https://doi.org/10.1007/BF03173464 .
147. Rigby, K., & Slee, P. T. (1993). Dimensions of interpersonal relation among Australian children and implications for psychological well-being. The Journal of Social Psychology,133(1), 33–42. https://doi.org/10.1080/00224545.1993.9712116 .
148. Kowalski, R. M., Giumetti, G. W., Schroeder, A. N., & Lattanner, M. R. (2014). Bullying in the digital age: A critical review and meta-analysis of cyberbullying research among youth. Psychological Bulletin,140(4), 1073–1137. https://doi.org/10.1037/a0035618 .
149. Kamble, S. V., & Dalbert, C. (2012). Belief in a just world and wellbeing in Indian schools. International Journal of Psychology,47(4), 269–278. https://doi.org/10.1080/00207594.2011.626047 .
150. Peter, F., Donat, M., Umlauft, S., & Dalbert, C. (2013). Einführung in die Gerechtigkeitspsychologie. In C. Dalbert (Ed.), Gerechtigkeit in der Schule [justice at school] (pp. 11–32). Wiesbaden: Springer.
151. Glenn LM (2018) What is a person? In: Bess M, Pasulka DW (eds) Posthumanism: the future of Homo sapiens, 1st edn. Macmillan Ref USA, Farmington Hills, MI, pp 229–246
152. D'Alessandro F (dir) (2018) Tau. Netflix, Los Gatos, CA
153. Egels-Zanden, N., & Sörum, N. (2015). Supply chain transparency as a consumer or corporate tool: The case of supply chain transparency as a consumer or corporate tool. Journal of Consumer Policy, 39(4), 377–395. https://doi.org/10.1007/s10603-015-9283-7 .
154. Eli, K., Dolan, C., Schneider, T., & Ulijaszek, S. (2016). Mobile activism, material imaginings, and the ethics of the edible: Framing political engagement through the Buycott app. Geoforum, 74, 63–73.
155. Fuentes, C., & Sörum, N. (2018). Agencing ethical consumers: Smartphone apps and the socio-material reconfiguration of everyday life. Consumption Markets & Culture, 22(2), 1–26. https://doi.org/10.1080/10253866.2018.1456428 .
156. Graham, M. (2011). Transparency and development: Ethical consumption through web 2.0 and the internet of things. Information Technologies & International Development, 7(1), 1–18.
157. Gummerus, J., Liljander, V., & Sihlman, R. (2017). Do ethical social media communities pay off? An exploratory study of the ability of facebook ethical communities to strengthen consumers ' ethical consumption behavior. Journal of Business Ethics, 144(3), 449–465. https://doi.org/10.1007/s10551-015-2830-y .

158. Horst, N. (2015). Ethical consumption? There's an app for that: Exploring the role of crowd sourced mobile technologies in everyday consumption practices. Guelph: University of Guelph.
159. Humphery, K., & Jordan, T. (2018). Mobile moralities: Ethical consumption in the digital realm. Journal of Consumer Culture, 18(4), 520–538.
160. Lekakis, E. J. (2014). ICTs and ethical consumption: The political and market futures of fair trade. Futures, 62, 164–172. https://doi.org/10.1016/j.futures.2014.04.005 .
161. Schneider, T., Eli, K., Dolan, C., & Ulijaszek, S. (Eds.). (2017). Digital food activism. London: Routledge.
162. Massey, D. (2005). For space. London: Sage Publications Ltd.
163. Massey, D. (2007). World city. Cambridge: Polity Press.
164. Alqahtani, Ghadah; Alothaim, Abdulrahman Predicting emotions in online social networks: challenges and opportunities. Multimedia Tools and Applications (2022). https://doi.org/10.1007/s11042-022-12345-w
165. Kazemian, Shakiba; Grant, Susan B. The challenges of internal social networking for higher education: a brief review of the literature. Journal of Data, Information and Management (2020). https://doi.org/10.1007/s42488-020-00028-w
166. Klafke, Renata V.; Gomes, Paulo M.; Junior, Demétrio Mendonça; Didonet, Simone R.; Toaldo, Ana M. Engagement in social networks: a multi-method study in non-profits organizations. International Review on Public and Nonprofit Marketing (2021). https://doi.org/10.1007/s12208-020-00273-0
167. Perez, Charles; Birregah, Babiga; Lemercier, Marc A smartphone-based online social network trust evaluation system. Social Network Analysis and Mining (2013). https://doi.org/10.1007/s13278-013-0138-4
168. Burke, Barbara Ruth; Şen, Ayşe Fulya Social media choices and uses: comparing Turkish and American young-adults' social media activism. Humanities and Social Sciences Communications (2018). https://doi.org/10.1057/s41599-018-0090-z
169. Rossetti, Giulio; Guidotti, Riccardo; Miliou, Ioanna; Pedreschi, Dino; Giannotti, Fosca A supervised approach for intra-/inter-community interaction prediction in dynamic social networks. Social Network Analysis and Mining (2016). https://doi.org/10.1007/s13278-016-0397-y
170. Rodriguez-Lopez, Juan Miguel; Schickhoff, Meike; Sengupta, Shubhankar; Scheffran, Jürgen Technological and social networks of a pastoralist artificial society: agent-based modeling of mobility patterns. Journal of Computational Social Science (2021). https://doi.org/10.1007/s42001-020-00100-w
171. Li, Shupin; Pöysä-Tarhonen, Johanna; Häkkinen, Päivi Patterns of action transitions in online collaborative problem solving: A network analysis approach. International Journal of Computer-Supported Collaborative Learning (2022). https://doi.org/10.1007/s11412-022-09369-7
172. Sinn, Donghee; Syn, Sue Yeon Personal documentation on a social network site: Facebook, a collection of moments from your life?. Archival Science (2013). https://doi.org/10.1007/s10502-013-9208-7
173. Kumaran, P.; Chitrakala, S. Topic adaptive sentiment classification based community detection for social influential gauging in online social networks. Multimedia Tools and Applications (2022). https://doi.org/10.1007/s11042-021-11855-3
174. Puertas, Edwin; Moreno-Sandoval, Luis Gabriel; Redondo, Javier; Alvarado-Valencia, Jorge Andres; Pomares-Quimbaya, Alexandra Detection of Sociolinguistic Features in Digital Social Networks for the Detection of Communities. Cognitive Computation (2021). https://doi.org/10.1007/s12559-021-09818-9
175. Cornejo, Raymundo; Tentori, Mónica; Favela, Jesús Ambient Awareness to Strengthen the Family Social Network of Older Adults. Computer Supported Cooperative Work (CSCW) (2012). https://doi.org/10.1007/s10606-012-9166-2

176. Brynielsson, Joel; Johansson, Fredrik; Jonsson, Carl; Westling, Anders Emotion classification of social media posts for estimating people's reactions to communicated alert messages during crises. Security Informatics (2014). https://doi.org/10.1186/s13388-014-0007-3
177. McKenna, Gemma; Rogers, Anne; Walker, Sandra; Pope, Catherine The influence of personal communities in understanding avoidable emergency department attendance: qualitative study. BMC Health Services Research (2020). https://doi.org/10.1186/s12913-020-05705-5
178. Fedushko, Solomiia; Molodetska, Kateryna; Syerov, Yuriy Decision-making approaches in the antagonistic digital communication of the online communities users. Social Network Analysis and Mining (2023). https://doi.org/10.1007/s13278-022-01021-4
179. Sabbar, Shaho; Hyun, Daiwon What makes it likeable? A study on the reactions to messages in a digital social network: the case of Facebook in Farsi. SpringerPlus (2016). https://doi.org/10.1186/s40064-016-3771-3
180. Troussas, Christos; Krouska, Akrivi; Sgouropoulou, Cleo Impact of social networking for advancing learners' knowledge in E-learning environments. Education and Information Technologies (2021). https://doi.org/10.1007/s10639-021-10483-6
181. Saqr, Mohammed; Viberg, Olga; Vartiainen, Henriikka Capturing the participation and social dimensions of computer-supported collaborative learning through social network analysis: which method and measures matter?. International Journal of Computer-Supported Collaborative Learning (2020). https://doi.org/10.1007/s11412-020-09322-6
182. Widener, Michael J.; Horner, Mark W.; Metcalf, Sara S. Simulating the effects of social networks on a population's hurricane evacuation participation. Journal of Geographical Systems (2012). https://doi.org/10.1007/s10109-012-0170-3
183. Rahman, Haseeb Ur; Merabti, Madjid; Llewellyn-Jones, David; Sudirman, Sud; Ghani, Anwar Structuring communities for sharing human digital memories in a social P2P network. Peer-to-Peer Networking and Applications (2020). https://doi.org/10.1007/s12083-020-00928-z
184. Borge-Holthoefer, Javier; Piedrahita, Pablo; Arenas, Alex Evolving activity cascades on socio-technological networks. Journal of Computational Social Science (2017). https://doi.org/10.1007/s42001-017-0012-7
185. Jacobs, Armand; Sueur, Cédric; Deneubourg, Jean Louis; Petit, Odile Social Network Influences Decision Making During Collective Movements in Brown Lemurs (Eulemur fulvus fulvus). International Journal of Primatology (2011). https://doi.org/10.1007/s10764-011-9497-8
186. Rathore, Nemi Chandra; Tripathy, Somanath A trust-based collaborative access control model with policy aggregation for online social networks. Social Network Analysis and Mining (2017). https://doi.org/10.1007/s13278-017-0425-6
187. Nishibe, Makoto The present and future of digital-community currencies: RAMICS 2019 in Hida–Takayama keynote speech. Evolutionary and Institutional Economics Review (2020). https://doi.org/10.1007/s40844-020-00177-y
188. Scheffer, Mariska MJ; Menting, Juliane; Boeije, Hennie R Self-management of social well-being in a cross-sectional study among community-dwelling older adults: The added value of digital participation. BMC Geriatrics (2021). https://doi.org/10.1186/s12877-021-02482-6
189. Donat, Matthias; Rüprich, Claudia; Gallschütz, Christoph; Dalbert, Claudia Unjust behavior in the digital space: the relation between cyber-bullying and justice beliefs and experiences. Social Psychology of Education (2019). https://doi.org/10.1007/s11218-019-09530-5
190. Jaynes, Tyler L. "I Am Not Your Robot:" the metaphysical challenge of humanity's AIS ownership. AI & SOCIETY (2021). https://doi.org/10.1007/s00146-021-01266-1
191. Hoelscher, Vera; Chatzidakis, Andreas Ethical Consumption Communities Across Physical and Digital Spaces: An Exploration of Their Complementary and Synergistic Affordances. Journal of Business Ethics (2020). https://doi.org/10.1007/s10551-020-04477-6
192. Kerikmäe, Tanel; Pärn-Lee, Evelin Legal dilemmas of Estonian artificial intelligence strategy: in between of e-society and global race. AI & SOCIETY (2020). https://doi.org/10.1007/s00146-020-01009-8

Digitalization and Exclusion—Digital Divides and Development

4

Introduction

Despite the massive social deepening and spread of digital technologies, there are groups of the people who are facing exclusions from the same. The exclusions are often driven along some social planes, geographies, or individual attributes. Such social planes and personal attributes are class, caste, religion, language, race, education, age, gender, disabilities, etc. As digital technologies become important tools in governance processes, media, and the market, the exclusion can have massive spatial, social, and individual consequences. They may lead to educational deprivation, lack of access to health care, democratic participation, exclusion in smart cities, financial exclusion, and even from citizenship which is now being determined by digital inclusion. Additionally, the exclusions of certain groups of people and individuals can also result from digitally enabled surveillance as today digital cameras or CCTVs have become ubiquitous and often lead to recognition of those who face exclusion through monitoring.

It is important to note that ruling classes keep on changing the method of governance and the medium through which they communicate. In medieval societies, it was the announcement through the streets by the men of kings about do's and don'ts of those states. In the twentieth century in most countries, it became written documents. The lower classes needed to first get themselves literate and then access the technical vocabulary. This itself was exclusionary and most of their rights the excluded classes could not enjoy as they were either not able to read the documents or many even having the ability to read could not understand those documents. For instance, in India, where people still struggle to understand the Constitution to exercise their rights. When the people were getting acquainted with the documentation process to an extent, the digital mode of governance further created a caveat between the ruling classes, laws, and the people. People now must

A. Shaban (ed.), *Digital Geographies—Theory, Space, and Communities*,
https://doi.org/10.1007/978-981-97-4734-4_4

first be able to use the electronic devices and software and then can read the documents or use the facilities. Thus, the ruling classes have often changed the method from one stage to another to keep the lower classes away from exercising their full rights. Digital exclusion as such today is also exclusion from the governance and exercising the full rights of citizenship, while digital inclusion leads to more information, which leads to more knowledge, and knowledge leads to exercising their rights, and that leads to the realization of their citizenship. In many countries, including India, the digitalization of information leads to massive exclusion of citizens even from digital banking systems, availing transport services, health care, obtaining citizenship documents, exercising their democratic rights, and/or availing amenities and facilities.

The digital exclusion is pervasive across societies and more so in developing countries. This chapter presents a review of several significant studies on digital exclusion across the countries. The reviewed papers are organized into four subsections. The first section reviews the papers on digital divides and brings to the fore the role of the factors and attributes in the production of the divides.

According to a paper titled "Bridging the Digital Divide in Ethnic Minority Older Adults: an Organisational Qualitative Study," increased self-efficacy among older Indian people in using ICTs results in intangible benefits such as time savings and relying on one's skills rather than on social capital, which may or may not exist and which may or may not be immediately accessible.

Another article, titled "The Digital Divide Is a Human Rights Issue: Advancing Social Inclusion Through Social Work Advocacy," makes the case that individuals rely on high-speed Internet as a tool to go about their everyday lives and engage with other people, the economy, and the government. In 2016, the United Nations General Assembly claimed that Internet access is a fundamental human right. These encourage efforts to develop legislation and initiatives aimed at closing the digital gap. In order to help community practitioners as they seek to close the digital divide at the state and local levels, the paper offers promising practices and advocacy tools.

In Kigali, Rwanda, the spatial distribution of ICT access is considerably concentrated, with areas of concentration at the heart and sparsity on the northeastern outskirts of the city, according to a study on "Spatial analysis of urban digital divide in Kigali, Rwanda". The dynamics of embracing technology advancement for achieving gender equality and eliminating economic inequities and other gaps that harm social welfare are highlighted in another article on "Gender wage gap, quality of earnings and gender digital divide in the European context".

The second section of the chapter reviews the papers related to inclusion and how that can be made possible. Some of these papers also highlight the role of algorithms in social exclusion. According to a paper titled "Balancing Digital-By-Default with Inclusion: A Study of the Factors Influencing E-Inclusion in the UK", using e-government represents the biggest gap between what citizens are doing online and what they should be doing. Another paper, "AI ageism: a critical roadmap for studying age discrimination and exclusion in digitalized societies", makes the case that the category of age and older people has not received enough consideration in the examination of fairness and bias in AI, which is crucial for the discussion of AI for Social Good.

The studies relevant to digital social engagement and participation are summarized in the third section of the chapter. According to a paper titled "Older adults' perceptions of contexts surrounding their social participation in a digitalized society—an exploration in rural communities in Northern Sweden", co-creating useful digitalized services and facilitating satisfactory use of digital technologies could support older adults' social participation through activities that they find relevant in their lives, potentially allowing them to live longer at home. In a case study on "Open Data, Crowdsourcing and Game Mechanics. A case study on civic participation in the digital age", it is shown how data can be made available to those who ordinarily lack it, how game mechanics can be used to encourage civic engagement by giving users a sense of autonomy, competence, and relatedness, and how crowdsourcing can bring many people together to work towards a common objective. These three components worked together to create a case for civic engagement in the digital age.

The reviews of papers on digital surveillance and panopticons are included in the final section. The papers in this section demonstrate how advanced workplace technology today provides low-cost, approachable tools and ways to manage "on-the-job" behaviour.

Overall, the papers in this chapter are conceptually and empirically rich and include case studies from different categories of population both from both the global north and south.

Machine generated summaries

Disclaimer: The summaries in this chapter were generated from Springer Nature publications using extractive AI auto-summarization: An extraction-based summarizer aims to identify the most important sentences of a text using an algorithm and uses those original sentences to create the auto-summary (unlike generative AI). As the constituted sentences are machine selected, they may not fully reflect the body of the work, so we strongly advise that the original content is read and cited. The auto generated summaries were curated by the editor to meet Springer Nature publication standards. To cite this content, please refer to the original papers.

Machine generated keywords: digital divide, divide, ict, health, access, internet, inequality, usage, participation, surveillance, old, urban, student, patient, internet use

4.1 Digital Divide

Machine generated keywords: digital divide, divide, ict, internet, access, usage, internet use, health, disparity, country, urban, inequality, use internet, old, develop country

4.1.1 Fractured Space: A Geographical Reflection on the Digital Divide [498]

This is a machine-generated summary of:

Poncet, Patrick; Ripert, Blandine: Fractured space: a geographical reflection on the digital divide [498]

Published in: GeoJournal (2007)

Link to original: https://doi.org/10.1007/s10708-007-9050-7

If you want to cite the papers, please refer to the original.

For technical reasons we could not place the page where the original quote is coming from.

Abstract-Summary

"What does it mean to reflect upon space in connection to telecommunications?"

"If we start with a conception of urban space as being fully integrated, including on an equal footing both information and communication technologies (ICT) and mobility techniques, as well as the founding notion of copresence that we find at the heart of all urban organization, we might then be able to examine the notion of the 'digital divide' in a new light."

"Surveys conducted in this perspective focus on the analysis of the choices people make between the different means they can use to cope with distances (copresence, mobility and telecommunication) with a view to constructing their own space."

Introduction

"This weakness undoubtedly affects the way the question of space within societies is tackled with regard to telecommunication, for the problem is more directly related to the traditional categories of geography (such as the classical and long-standing 'town versus country' opposition), than to the spatial modalities of the 'digital divide', a notion whose haziness is sufficient to discourage any attempt at criticizing its meaning (Rallet [1])."

"If there actually is a phenomenon that we may call the 'digital divide' occurring within urban space, how can we study it considering that the city is a spatial area that is a more or less cohesive cluster of places and not a homogeneous 'black box' (Paquot and others [2])?"

"How is it possible to integrate the notion of the digital divide into a theory of space in general, and of urban space in particular?"

The fractured space

"One of the geographical dimensions of the digital divide that is undoubtedly most frequently given as an example places emphasis on differences in terms of access to (N)ICT that exists between rural and urban areas, and presents this difference as an inequality."

"In these intermediate spaces one finds all sorts of divides, which we can legitimately consider as related to the question of access and the use of the (N)ICT."

"When one starts to wonder if a social phenomenon such as 'suburban malaise', such as one finds in Western European cities, can be fought through the installations of computers in some suburban areas, one starts to consider urban space in all its complexity, and accepts (N)ICT as an urban phenomenon."

"The notion of 'fractured space' considers the application of digital fracture into a spatial dimension."

Cities: a fluctuating geometric beside space

"Mobility and telecommunication are techniques that can be used by urban actors to master space."

"In order to cope with this complexity, urban theory is enriched by taking into consideration the urban space as a realization of a virtuality that assumes different shapes in the form of capital, the spatial capital of the city."

"To be urban, one has to deftly combine the 'three modalities of distance: coping', that is copresence, mobility, and telecommunication, to the benefit of the first of the three."

"As Jacques Lévy [3] and Lévy and Lussault [4] explain: from the center to the outskirts of agglomerations different spaces are juxtaposed, which are more or less urban, that is, more or less dense and diversified."

The fracture viewed from another angle

"In order to adapt the notion of the digital divide, two types of causes can be proposed on the grounds of this theory: a deficient space capital and a poor articulation of the modalities for dealing with distance."

"The reflection in terms of space capital seeks, on the contrary, to evaluate its precise composition, by examining the level of copresence, telecommunication and mobility that characterizes the space in which the urban actors evolve, therefore inducing variations in the space and time scales of the analysis."

"This capitalistic aspect of spatial situations in the face of digital fracture calls for an examination of the questions of arbitration and the relationships that the urban actors create between the means of coping with distances with which they are provided."

"It is then by gaining intimate knowledge of the rationale of urban coopetition that one may understand how each actor improves his space capital."

In favour of an ITC urban space policy

"This global and theoretical presentation of our approach of the digital divide, we would like to evoke its political dimension, that is, the questions relating to the notion of choice made by urban actors, whether they belong to the political sphere or are citizens residing in the city."

"Of reasoning in terms of infrastructure, following a political logic, one may positively conceive of space policies, taking more closely into account the local situations in terms of the digital divide, of which it has been said that it should often be subordinated to cases of copresence deficiency."

"The horizon of space policies aiming at the reduction of multiple fractures within the urban space can be summed up in the form of two scenarios when it comes to tackling the question of the (N)ICT, and if necessary of the digital divide: the cybercity versus tele-urban."

[Section 6]

"An interface common to all the information centers has been set up in order to allow farmers to gain access to services that might interest them, such as access to the prices of agricultural markets for example, which are mainly urban in this region."

"The intention was to facilitate the functioning of local markets by circulating information between villages, and between villages and urban centers in the region, given that such information was often inaccessible to the cultivators of the region."

"The analysis of these recent village practices following the diffusion of (N)ICT (Ripert [5, 6]) shows that the village area is coming closer to the urban center through the remote sharing of services, information, images, references, some consumer products and the introduction of greater diversity, as defined by Jacques Lévy when he considers that the city is a spatial object with associated density and diversity."

Daily timetable: the coherence of individual space

"Our task is to identify different types of practice and to throw light on the choices made between the different modalities of coping with distances while taking into account the recent appearance of (N)ICT and new mobility technologies, all the while distinguishing what the person can do according to his/her degree of access to the network (telecommunication and transportation services) and what she/he actually does."

"In Mumbai, people were selected according to three interrelated criteria: (1) their level of connection to the telecommunication networks (Internet and cell phone), selecting very well-connected people, people with an average rate of connection, and people with no connection at all; (2) according to their lesser or greater mobility in the city (great, average, and weak); (3) according to their place of residency, in an attempt to take into account the different levels of urbanity, from the city center to the outskirts and peripheral areas, urbanity being defined by an indicator evaluating the level of copresence."

A divide? To what end?

"If one accepts to frame the question of the digital divide that affects people as a problem of coopetition, then it is legitimate to evaluate the space capital of a person on the one hand (the situation we have at the start) and then to try and understand what they make of it."

"It is not our point here to defend the idea that less-developed countries have their own way that should not be changed, but we would like our Indian illustration to serve to underline the necessity to think about the digital divide not only as a simple dearth of connection to telecommunication networks but rather as a difficulty of articulating the latter into other ways of dealing with distances, on which sometimes very 'coherent' lifestyles depend."

Conclusion

"The approach of the digital divide we have defended here leads to two types of conclusion, which respond to the questions we asked at the beginning."

"There is the question of ways to use the notion of digital fracture within the urban context and within the bounds of a geographical analysis."

"The other question we answer relates to the conceptual strength of the 'digital divide'."

"It is more difficult to issue clear-cut definitive statements, for the issues at stake, as we have just seen, are considerably less symbolic than the importance of the expression 'digital divide'."

"To remove the term 'digital divide' from the geographical lexicon would in this respect have no real purpose, unless geography is to once again miss the boat, as it already missed it with regard to the 'environment' a few years ago."

4.1.2 Bridging the Digital Divide in Ethnic Minority Older Adults: An Organisational Qualitative Study [499]

This is a machine-generated summary of:

Choudrie, Jyoti; Zamani, Efpraxia; Obuekwe, Chike: Bridging the Digital Divide in Ethnic Minority Older Adults: an Organisational Qualitative Study [499]

Published in: Information Systems Frontiers (2021)

Link to original: https://doi.org/10.1007/s10796-021-10126-8

If you want to cite the papers, please refer to the original.

For technical reasons we could not place the page where the original quote is coming from.

Abstract-Summary

"This paper aims to explore and understand the digital divide in older adults when accepting and using smart devices within an organization."

"Using an in-depth single case study, the digital divide is explored and understood using technology-mediated learning for older adults when using smart devices; ie."

"The case study is based on a group of educated, older Indian adult volunteers at a local Punjabi radio station."

Introduction

"To address this matter, we adopted a reflective stance of the digital divide with regard to older adults, where we sought to centrally position an individual rather than the technology when considering use."

"For our understanding, we drew on Wei and others's [9] digital divide conceptual framework and designed training sessions based on technology-mediated learning (TML), with the objective to allow older adults to build their self-efficacy and increase their confidence levels for using IT; namely smart devices."

"Our literature reviews have shown that there are minimal studies considering the training of older adults to use smart technologies."

"Our study showed that on the job training, which our TML provided led to a benefit in the form of increased ICTs use within the volunteering older adults and an unanticipated non-pecuniary long-term benefit of education."

Background Literature

"For these older adults it is important to have opportunities to learn how to use the novel technologies, which is central for addressing the digital divide (Benitez, [10])."

"This offered a further motivation to conduct research on how to get older adults to utilise mobile devices by focusing on their learning and training, a unique dimension to digital divide and usage studies."

"We build on SCT, to investigate the role of Self-Efficacy towards bridging the digital divide that impacts older adults and their use of smart devices."

"We propose that to reduce the digital divide, technology-mediated learning can support older adults by building their self-efficacy towards using smart devices."

"From the digital divide, learning and smart devices existing literature it was learnt that studies of older adults' training on the use of smart devices is minimal."

Research Approach

"We specifically focused on incorporating TML into our study in order to first, offer ethnic minority older adults the necessary opportunities to learn how to use the smart devices, and second, to allow them through these learning opportunities to evaluate for themselves whether technology and smart devices have a place in their own personal lives and offer value to their remaining lives."

"As we were interested in investigating if and how TML can help older adults in building their self-efficacy and confidence towards using smart devices so that the digital divide can be reduced, our study required that we identified a group of older adults, with an interest in learning how to use such devices."

"13 sessions were designed and delivered in a way to reflect the types of skills that older adults need to use internet-based devices and the teaching and learning requirements they have."

Findings and Analysis

"The training sessions were designed with the aim of helping older adults to learn how to use the device, and at the same time, employ TML to help them learn how to use the internet and other applications."

"Why do I need the internet and a tablet device when I have my trusted notebook, pen and pencil?" (i9) Once the participants began to understand the importance and the usefulness of smart devices and their features, and began to become more confident when using them, they began using the devices more often."

"I am more confident with my steps and can explain matters." (i5) Therefore, following these training sessions and using smart devices and TML, older adults' skills and capabilities increased considerably, which in turn, reduced the digital outcome divide."

Discussion

"Our study went one step further and utilised the platform of early education, which was not a selection criterion and showed how training provided in the workplace and increased use at home and at work can result in adoption of ICTs within older adults."

"Such findings are further confirmed by previous studies (e.g., Agudo and others, [11]; Gonzalez and others, [12]; Martínez-Pecino and others, [13]) who found that age, and workplace training impact older adults use of and self-efficacy in ICTs."

"Such findings first suggest that the increased self-efficacy of Indian older adults in using ICTs leads to intangible benefits in terms of time savings and relying on one's abilities rather than on their social capital, which may or may not exist and which may or may not be promptly accessible."

Conclusions
"Against this background, we combine our voices with other scholars within the area of ageing and technology, and posit that what is needed is technologies that can make a significant contribution in the lives of older adults from ethnic minorities, which can be appropriated without negative connotations (e.g., they have to use old people technologies to support themselves) and allow users to focus on what is important to them (Neven & Peine, [14])."

"We specifically propose the use of TMLg approaches to provide learning opportunities to those older adults with an interest in learning how to use ICTs."

"Our study revealed that if and when organizations are approached for customised training approaches for older adults, they are willing to allow access to their workforce, but academics also have to be mindful of the time and availability of individuals when designing, developing and implementing such an approach."

4.1.3 The Digital Divide and Older Adult Population Adoption, Use and Diffusion of Mobile Phones: A Quantitative Study [500]

This is a machine-generated summary of:

Choudrie, Jyoti; Pheeraphuttranghkoon, Sutee; Davari, Soheil: The Digital Divide and Older Adult Population Adoption, Use and Diffusion of Mobile Phones: a Quantitative Study [500]

Published in: Information Systems Frontiers (2018)

Link to original: https://doi.org/10.1007/s10796-018-9875-2

If you want to cite the papers, please refer to the original.

For technical reasons we could not place the page where the original quote is coming from.

Abstract-Summary
"What has also become clear is that the older adult consumer group and the factors affecting this age group have been studied relatively less in the literature."

"We aim to investigate the adoption, usage, and diffusion of smartphones within the UK older adults so as to identify the factors encouraging or inhibiting smartphone usage and service provision within this age group."

"Our research can contribute towards reducing some of the existing digital divide within UK older adults."

"Businesses can benefit from our research by understanding the significant factors affecting the adoption of smartphones among the UK older population and to adapt their policies accordingly."

Introduction

"For the older adult population, ICTs, including online social networks, such as Facebook and smartphones are viewed advantageous since they can aid health care (Joe and Demiris [15]), or a reduction in loneliness by connecting older adults with their friends and family (Dennis and others [16]; Choudrie and Vyas [17])."

"Understanding the older adult population is also pertinent for businesses as they can then cater to the needs and requirements of this demographic group, an issue highly emphasised in the United States of America (USA); particularly given that governments and organizations, including marketing businesses are gaining immensely due to the internet based strategies (Porter and Donthu [18])."

"Smartphones can provide convenience and benefits to various users; including the older adult population (Kim and Hwang [19])."

"Acknowledging that smartphones and older adults are both important changes for society and organizations alike and that smartphone potentials cannot be reaped without their wider proliferation, this study addresses these gaps by identifying the adoption, usage, and diffusion of smartphones in the older adult population of the UK."

Literature Review

"Bearing the benefits and drawbacks of smartphones in mind, it has been learnt that several research studies have attempted to study older adults and their adoption, use and diffusion of smartphones and to identify the factors leading to the age related digital divide."

"These studies showed that previous older adults and smartphone research studies have identified factors leading to the adoption, use and diffusion of smartphones."

"There are other factors such as, observability, compatibility and other factors that have not been researched in these studies; thus, building a further motivation to this team to undertake a research study of older adults and smartphone adoption, use and diffusion."

"The proposed conceptual framework for this research has assumed that the dependent variable of this research, the behavioural intention to use and adopt smartphones of older adults is influenced initially by observability and compatibility that have been drawn from DoI (Rogers [20])."

Research Methodology

"The sample size of a PLS-SEM study should be at least 10 times the largest number of structural paths directed at a latent construct in the structural model (Hair Jr and others [21]) which holds true for our study."

"To statistically analyse the data, the Partial Least Squares-Structural Equation Modelling (PLS-SEM) approach was employed, which is able to provide more robust estimations of the structural model compared to the covariance-based models.""

"Structural Equation Modelling (SEM) is a powerful multivariate statistical technique which is a combination of factor analysis and multiple regression to examine the structural relationships among a set of measured and latent variables."

"SEM is composed of two stages called measurement model and structural model."

"PLS-SEM has several advantages over covariance-based SEM methods such as, its ability to deal with small sample sizes and working with non-normal data."

Research Findings

"The number of respondents in the 60–69 age planning to have a smartphone was larger than the adopted groups."

"Perceived enjoyment (H7) had the strongest factor influence of behavioural intention to use smartphones within the older adults with $\beta = 0.380$, t-value = 13.3 and a significant level of (p) < 0.001."

"The results showed that there is complete similarity between the 50–59 age group and the overall sample in terms of the factors affecting the adoption and use of smartphones."

"The findings revealed that the factors influencing the adoption and use of smartphones in the cohort with no higher education degree are: Compatibility (H2) and Effort Expectancy (H6)."

"What was also realised is that they are not significant factors influencing the adoption and use of smartphones for the cohort with a higher education degree."

Discussion

"Smartphone and network providers may need to consider such factors when increasing smartphone adoption within older adults."

"This means that smart phone and internet service providers should consider having an individual or department offering information or knowledge for using and adopting smartphones, which will encourage more older adult individuals to adopt and use smartphones; thereby, increasing the sales of these phones."

"A solution to the issue of complications can be determined from the responses to the questions on effort expectancy (H6) with mean values of 5.67 and 5.54 with 1.41 and 1.46 standard deviation that suggested older adults view smartphones as devices that are easy to use."

"This suggests that older adults are very proficient in smartphones uses and do accept advanced features of mobile devices."

Conclusion, Implications, Limitations and Avenues for Future Research

"The implication of this research study to the literature for older adults' adoption, use and diffusion of smartphones is the provision of a research model specifically for older adults in the context of smartphones."

"For industry, the implication of this research is the identification of factors that should lead to an increase in adoption, use and diffusion of smartphones within older adults."

"This study is also beneficial as it suggests that an age group that was viewed to be resistant to change is adopting, using and diffusing newer forms of ICT, which is encouraging for the manufacturers of mobile devices per se, but we have tried to show the factors that will assist industry service providers to increase sales of smartphones within the older adults consumer market."

4.1.4 Digital Switchover or Digital Divide: A Prognosis for Usable and Accessible Interactive Digital Television in the UK [501]

This is a machine-generated summary of:

Carmichael, Alex; Rice, Mark; Sloan, David; Gregor, Peter: Digital switchover or digital divide: a prognosis for usable and accessible interactive digital television in the UK [501]

Published in: Universal Access in the Information Society (2005)
Link to original: https://doi.org/10.1007/s10209-005-0004-x

If you want to cite the papers, please refer to the original.

For technical reasons we could not place the page where the original quote is coming from.

Abstract-Summary
"UK legislation advocates universal access and equitable provision of services, across all platforms and equipment, particularly for 'vulnerable' groups."

"Key aspects of usability and accessibility have been overlooked by those responsible for encouraging this new infrastructure's inclusive development."

"Aspects of previous interactive television provision and developments in web accessibility appear to have been ignored, along with findings from relevant user-centred research and even from specifically commissioned reports."

Introduction
"The current move toward digital television (DTV), and the associated 'interactivity', seems likely to increase the importance of the role television plays for many and will also raise some very important issues related to the impact television has on people's lives [23]."

"The main concern is that decisions and actions vital to ensuring universal accessibility and ease of use for the widest possible range of people will effectively be ignored, apparently for the benefit of other more powerful stakeholders and also in order to maintain the UK's status as 'world leader' in this area."

"This is followed by an indication of the lessons provided by earlier interactive TV developments and the development of the WWW, with regard to their successes and failures in the development of usable interfaces for the broadest range of people and in the development of an accessibility infrastructure for people with sensory or physical impairments."

Background

"An illustrative example of this is the very slow uptake of certain interactive services, such as email and independent interactive shopping channels [24], indicating that the providers failed to build on viewers' expectations and understanding of television as primarily a source of entertainment associated with other family related activities which tend to occur in front of the TV screen."

"The additional functionalities of digital television raise new dilemmas in social interaction, such as group situations where one viewer may want to watch TV while the other wants to interact with it, and other variations on this theme."

"Beyond issues of commercial competition, the array of different commercial concerns and their differing roles in the provision of disparate DTV equipment, services and content is also likely to impact on the usability and accessibility of 'digital television' which from many viewers' perspective is essentially a single overall 'service'."

Relevant UK legislation

"Two sections of the Communications Act are particularly relevant to the issue of the usability and accessibility of DTV."

"It shall be the duty of OFCOM to take such steps, and to enter into such arrangements, as appear to them calculated to encourage others to secure- a. that domestic electronic communications apparatus is developed which is capable of being used with ease, and without modification, by the widest possible range of individuals (including those with disabilities); and b. that domestic electronic communications apparatus which is capable of being so used is as widely available as possible for acquisition by those wishing to use it."

"These sections of the Act clearly assert ('legalese' notwithstanding) that Ofcom is responsible for ensuring that all 'apparatus' involved in DTV is easy to use."

"The Disability Discrimination Act (DDA) places responsibilities on providers of goods, facilities and services, to ensure that disabled people attempting to access these services do not encounter unjustified discrimination due to their disability."

Lessons from other domains

"There has, though, been a noticeable increase in prominence of the issue of web accessibility, due in no small part to awareness of legal obligations of online service providers to avoid unjustified discrimination against disabled people expressed in legislation such as the UK's DDA and Section 508 of the Rehabilitation Act in the US."

"This report suggests that while DDA conformance appears to rest with the online service provider (i.e. the web content provider), responsibilities for web accessibility do extend to other parties as well, including the disabled person themselves."

"The impact of Section 508 on web technology should be a lesson to Ofcom and the UK Government in the switch to DTV, such that, were legislation of a similar nature to Section 508 to be applied to the provision of interactive DTV technology and services, or indeed if Ofcom chose to advocate the 'spirit' of the relevant existing legislation, the level of accessibility for DTV would increase."

Relevant research

"The areas of research that are providing relevant information are indicated in this section and, while it is accepted that much existing research cannot directly prescribe how interface elements should be designed, most of it does indicate many of the known problems people can have in this regard."

"It has produced a large corpus of potentially relevant information which can act as very useful guidance in the initial stages of interface design (e.g. [25–27])."

"Another important source of information regarding the design of DTV interfaces, relates to ensuring that as well as 'fitting' people (in the ergonomic sense), the proposed technology will also 'fit' into their lives."

"The complexity of such issues described above, indicates the need for multiple method approaches to improve the design and implementation of DTV equipment and services."

Usability and accessibility

"Accessibility (in its more extreme sense) relates more to those in the population whose requirements are such that entire 'modes' of input and output of an interface are inaccessible to them."

"The most common example of this is text and graphical output with regard to people with impaired vision, who will thus require this information to be translated into an alternative form which they can access (e.g. speech, Braille, or subtitles/captions)."

"Accessibility relates to the extent to which an interface can accommodate and support both the 'translations' involved, and the alternative input and output devices required."

"In the domain of TV, 'accessibility' also relates to the 'modes' by which the content of programmes is presented, such that subtitles (captions), signing and audio-description allow viewers to access elements of a programme they would otherwise miss."

Usability of the interface

"Age-related changes to vision and cognition will combine to make this fairly straightforward procedure significantly more effortful and potentially error prone, such that, when the viewer then switches their attention to the remote control (and assuming they know which button to press) they will have to search for and locate the required button and press it."

"Many older people are likely to switch attention to the remote control with some uncertainty about which button to press, and this will make finding the correct one (again) more effortful."

"This has been indicated by a recent report commissioned by the DTI (mentioned above), which states: "Based on today's equipment, an additional two million people (4.4% of those able to access analogue television) could be excluded from simply viewing the new digital services using digital terrestrial television set top boxes at switchover."

"The preceding sections have indicated the extent of the usability and accessibility problems that have been allowed to emerge with the interactivity inherent in the development of DTV in the UK."

Discussion and conclusions

"The other main concern is associated with the apparent rapidity with which the UK Government wants the switchover completed, despite the fact that many of the Government's own commissioned reports indicate that significant usability and accessibility problems remain which must be addressed and resolved before any specific timetable for switchover is announced."

"The second is poor 'system interaction design' where, even for people who are familiar with new technologies, digital television equipment is non-intuitive and exhibits inconsistencies. ([28] pp 108–109) This report then goes on to indicate that "the usability problems we observed in our trials would not have been difficult to resolve through better design" and also recommends the need to "address system interaction design in an integrated way"."

"The report identifies one of the group's main concerns; "The accessibility of digital television equipment is a significant concern and barrier to take-up for many consumers, particularly people with disabilities and older viewers, but is not confined to these groups.""

4.1.5 The Digital Divide Is a Human Rights Issue: Advancing Social Inclusion Through Social Work Advocacy [502]

This is a machine-generated summary of:

Sanders, Cynthia K.; Scanlon, Edward: The Digital Divide Is a Human Rights Issue: Advancing Social Inclusion Through Social Work Advocacy [502]

Published in: Journal of Human Rights and Social Work (2021)

Link to original: https://doi.org/10.1007/s41134-020-00147-9

If you want to cite the papers, please refer to the original.

For technical reasons we could not place the page where the original quote is coming from.

Abstract-Summary

"High-speed Internet is a tool people rely upon to conduct the daily business of their life and interact with each other, the economy, and government."

"Millions of people in the USA still have no home access to high-speed Internet."

"Consistent with a social work human rights approach, the United Nations General Assembly declared access to the Internet a basic human right in 2016."

"This calls upon social workers to engage in advocacy efforts to advance policy and programs to alleviate the digital divide."

"We examine the digital divide in the USA and discuss why it is a social justice and human rights issue."

"We also identify and share promising practices and advocacy tools being used in the field that provide guidance to community practitioners as they engage in work at state and local levels aimed at closing the digital divide."

Introduction

"The role and importance of technology has become glaringly obvious in the context of the COVID-19 pandemic, where society has been forced to rely even more heavily on technology for basic daily living including accessing basic goods, maintaining connections with others, working from home, and having the ability to complete schoolwork."

"Completing these basic tasks is a major challenge for millions of people who either lack access to high-speed Internet or do not adopt it, creating disadvantages that affect their ability to participate in social, political, and economic lives in the USA."

"We examine the digital divide, who does and does not have access to an essential service that is part of the daily fabric of living in the digital era."

"We discuss why the digital divide is a human rights and social justice issue: one that social workers should be actively working to close through policy advocacy work."

The Digital Divide

"This is due to a variety of factors, including computer ownership, high-speed Internet access and adoption, and digital literacy."

"Billions of dollars in public funding are responsible for the mobilization and development of the Internet (Tarnoff [29])."

"The digital divide not only includes the obvious issues of access to computers and connectivity but also includes issues of inequity affecting those who either lack the skills and opportunities to access information technology or who are in a less equal position in terms of use (Makinen [30])."

"Access to broadband Internet has been credited with effects on individual empowerment, community development, and economic growth (Jayakar and others [31])."

"Perhaps now more than ever, it is important for social workers to advocate for an inclusive high-speed Internet that replicates in the digital age what the postal service originally sets out to do in providing the nation with a reliable, affordable, and universal service (USPS [32])."

Broadband Access

"Geographical location is also a leading factor, with rural communities much less likely to have access to high-speed Internet (West and Karsten [33])."

"Approximately 81% of adults use the Internet; however, only 51% of adults living with a disability access the Internet (Fox [34])."

"According to the most recent Federal Communications Commission (FCC) deployment report (FCC [35]), an estimated 21 million Americans still have no home access to high-speed Internet service, defined by the FCC as a download speed of 25 Mbps and upload speed of 3 Mbps."

"Other sources estimate this number as high as 162 million people across the USA who are not using the Internet at broadband speed (Microsoft [36])."

"Based on the 2018 Broadband deployment report, as of the year end 2016, 92.3% of all Americans had access to fixed terrestrial broadband at speeds of 25 Mbps/3 Mbps, up from 89.4% in 2014 and 81.2% in 2012."

Broadband Adoption

"Adoption of broadband is equally important to ensure the benefits (including economic benefits) that go along with Internet use to everyone."

"While broadband may be available, broadband adoption refers to the extent to which US households subscribe to and use broadband."

"Populations who continue to lag in broadband adoption, even when available, include people with low incomes, older adults, minorities, the less educated, non-family households, the unemployed, and limited English-speaking households (Kruger and Gilroy [37]; Ryan [38])."

"According to Census data from the 2016 American Community Survey, 81.4% of American households have a broadband Internet subscription."

"Data from the Pew Research Center demonstrate that certain groups, unable to afford the cost of monthly subscriptions, continue to lag in the adoption of broadband including people with low incomes, older adults, those with less education, rural households, and those on tribal lands and US territories (Kruger and Gilroy [37])."

Low-Income Households

"People who fall lower on the economic ladder are more likely to be people without access or unable to adopt the Internet resulting in falling further behind and widening the digital divide between rich and poor."

"Americans with family incomes between $75,000 and $99,999 per year adopted the Internet at an 83% rate, compared with 80% with incomes between $50,000 and $74,999 and 70% for those in the $25,000 to $49,999 range."

"States with higher incomes reported high rates of use and access while low-income states including Arkansas and Mississippi had the lowest rates of broadband use at 71%. (Ryan [38])."

"The majority (91% overall) of people living in poverty have at least some form of Internet access (Rideout and Katz [39])."

"Lower-income adults rely more heavily on smartphone-only access."

"This pattern of income variation in broadband connectivity not only affects households but also occurs at the neighborhood level."

Race and Ethnicity

"While access to the Internet has increased for all racial groups, access disparities for Blacks, Latinos, and Native Americans persist compared with Whites (Campos-Castillo 40; FCC [41])."

"Households with an Asian head of household were the most likely to own or use a laptop or desktop, to own or use a smartphone, to own or use a tablet, and to have a broadband Internet subscription."

"Black heads of household were the least likely to own or use a desktop or laptop or have a broadband subscription (Ryan [38])."

"This is congruent with smartphone-only use among low-income households since we know that Black and Hispanic persons are disproportionately poor (Macartney and others [42])."

Education

"In 2015, for example, the digital divide was greatest between rural and urban users without a high school diploma."

"Only 52% of those who lack a high school diploma and live in a rural area reported using the Internet, compared with 59% of those who live in urban households (Carlson and Goss [43])."

"Variations in public school funding are reproduced in terms of quality of Internet access within their classrooms, and the digital divide in public schools is also racialized with White students more likely than students of color to use the Internet in the classroom or school library (Warf [44])."

"Children who lack access to digital resources miss out on other enhancements as well, with disadvantaged students lacking or holding inferior technological resources at home compared with more privileged students (Ritzhaupt and others [45])."

"Before the pandemic, an estimated 12 million children were having difficulty completing homework assignments because they lacked home Internet."

Urban/Metro vs. Rural and Tribal Communities

"An issue commonly raised in the literature is the digital divide between urban and rural communities, especially in terms of their differential access to broadband (West and Karsten [33])."

"According to the FCC as of the end of 2017, over 26% of Americans in rural areas and 32% of Americans on tribal lands lack coverage from fixed terrestrial 25 Mbps/3 Mbps broadband, compared with only 1.5% of Americans in urban areas (FCC [35])."

"Approximately 14 million rural Americans and 1.2 million Americans living on tribal lands in 2017 lacked even low-speed mobile LTE broadband of speeds of 10 Mbps/3 Mbps (FCC [46])."

"While libraries play a vital role in rural and tribal locations to Internet access, they too lag behind more urban areas."

"While rural communities are much more likely to lack access to broadband, many metro area neighborhoods also fail to connect significant number of residents to existing broadband service."

Digital Access Is a Human Rights and Social Justice Issue

"Social work's historic commitment to social justice has been advanced through the promotion of a human rights approach."

"Given the central role that the Internet plays in today's digital age in gaining access to resources, jobs, health care, and education among others, universal access to broadband clearly falls within the realm of human rights."

"One of the Council on Social Work Education's competencies includes advancing human rights and social, economic, and environmental justice (CSWE [47])."

"The United Nations General Assembly declared access to the Internet a basic human right, integral to allowing individuals to "exercise their right to freedom of opinion and expression" and emphasizing "access to internet facilitates vast opportunities for affordable and inclusive education.""

"This should not prevent social workers from advancing inclusive policies and practices premised on the integration of human rights to close the divide."

Policy Context and Policy Efforts

"Net neutrality preserves the FCC's authority to regulate Internet service providers (ISPs) and the Internet infrastructure to ensure that everyone has equal access."

"A recent expansive effort is the state of California's Internet for All Now Act, which attempts to reduce the digital divide for the 5 million California residents who lack access to reliable high-speed Internet services."

"Recognizing that broadband is a key component in building healthy economies and educational systems and acknowledging that many rural and low-income communities go unserved or underserved in the digital age, California passed the Internet for All Now Act in 2017."

"The approach of advancing broadband and reducing the digital divide in California is to encourage public-private partnerships and leverage private investment, maximize available federal funding (such as through the federal Connect America Fund), and administer the CASF to promote the most cost-effective and equitable progress in access to and adoption of the Internet across California."

Overcoming the Digital Divide

"In the absence of such an initiative, we must continue to move forward with local and state initiatives and as social workers engage in policy practice that reaches vulnerable groups in the digital divide including low-income communities, people of color, older, and rural populations."

"Using a slightly modified version of Bliss's [48] model of advocacy, we propose a six-step model for social workers to utilize in digital inclusion activism, recognizing that state and local contexts will shape the specifics of strategic decisions that are made at each point along the way."

"Drawing upon existing digital activism efforts and the best policy and program practices that have emerged in states with extant broadband access initiatives, we suggest a number of goals that might be useful for newly forming coalitions to consider."

"In a review of the literature on the advocacy process, Bliss [48] identifies five components of a framework for thinking about engaging in policy advocacy which we might use to reduce the digital divide."

Forming an Advocacy Structure

"Social workers might launch such a campaign under the structure of an NASW State Chapter policy committee, as part of a multiagency campaign to increase client access to tele-health services or as part of a local chapter of a national digital inclusion organization."

"With the social distancing needs of those providing social services, health care, and mental health services to low income and rural populations during the era of Covid-19, it might be an opportune time for state actors to create coalitions of digital access activists among the providers accepting Medicaid and/or working with state contracted services."

"The National Digital Inclusion Alliance might be a useful organization with whom social workers might form coalitions, as they currently work with 44 state organizations that include libraries, universities, school districts, and others with a vested interest in seeing an expansion of digital access (National Digital Inclusion Alliance [49])."

Clarifying Cause and Beneficiaries

"Bliss states that social workers should clarify the nature of their cause and the intended beneficiaries of their advocacy."

"The cause is clear—to increase access to and adoption of affordable, high-speed broadband at the state or local level."

"The exact nature of the beneficiaries will vary somewhat based upon state context—in general though, as we have shown in our earlier literature review, the populations dealing with lower levels of Internet access and adoption include rural populations, residents of tribal reservations, the elderly, people of color, and low-income citizens."

Articulating a Desired Outcome

"These promising practices are mutually reinforcing, frequently done simultaneously, and include (a) a clear plan for stakeholder outreach and engagement; (b) a policy framework with well-defined goals including the creation of tasking agencies or establishing separate offices to lead statewide broadband programs, plans for identifying barriers in unserved and underserved areas, and initiatives that connect broadband to other policy priorities such as economic development, transportation, education, and health care to build partnerships and leverage more funding; (c) planning and capacity building goals that help educate community members, identify needs and goals, and start conversations with ISPs, evaluate options, and move toward implementing infrastructure projects; (d) funding and operations that provide support for broadband deployment in unserved and underserved areas through grant programs and which include accountability measures to ensure that grantees demonstrate they are providing the service they were funded to deliver; and (e) program evaluations which effectively explore the performance of their efforts and update goals and activities as their programs mature."

Identifying a Target Audience

"Candidates and elected officials in the state legislatures or in statewide offices such as governors, lieutenant governors, and state treasurers are likely to have influence in the creation and implementation of digital access programs."

"Appointed officials directing state agencies are also going to be key constituencies for the successful development and implementation of digital inclusion policies."

"The Pew Foundation has noted the importance of strong leadership from governors, legislators, and agency heads in successful development of digital inclusivity initiatives."

"If a governor's office lacks focus on broadband as a priority, other state agencies will lack energy around broadband goals, so advocacy efforts should be targeted to them early on."

Selecting Strategies and Tactics

"Media Alliance is a California-based media change and resource organization founded in 1976 that engages in activism including digital inclusion."

"Examples include attending public meetings, testifying at public hearings and distributing your comments to the press, inviting your elected officials to visit and witness successful digital inclusion projects in your community, organizing local coalition members to apply for larger city or regional task force seats, and organizing town hall meetings with local organizations and testimonials from people impacted by digital exclusion."

"Media Alliance provides a toolkit of resources designed to help those who want to advance digital inclusion."

"Are Internet connectivity, affordable and accessible hardware, and training and technical support (i.e., digital literacy) (For a full discussion of toolkit ideas and strategies see: https://media-alliance.org/wp-content/uploads/2016/05/Media-Alliance-Digital-Inclusion-Advocacy-Toolkit.pdf)."

Advocacy Process and Outcome Evaluation
"Bliss argues that advocates must identify how they will evaluate advocacy processes and actions."

"Without an evaluation component, the advocating entity will lack formative and summative feedback to assess their advocacy efforts and share with others."

"Advocates might consider consulting with or collaborating on evaluation with local universities."

Conclusion
"Digital inclusion is a human right largely overlooked in macro social work practice and policy advocacy."

"Tackling the digital divide will require an approach that not only expands access but also provides digital skills and encourages people to use the Internet in ways that positively contribute to their social, economic, and political lives."

"In the simplest sense, social work tasks involve working collaboratively to advance people's access to resources."

4.1.6 Establishing a Social Entrepreneurial System to Bridge the Digital Divide for the Poor: A Case Study for Taiwan [503]

This is a machine-generated summary of:

Huang, Shu-Chin; Cox, John Lew: Establishing a social entrepreneurial system to bridge the digital divide for the poor: a case study for Taiwan [503]

Published in: Universal Access in the Information Society (2014)

Link to original: https://doi.org/10.1007/s10209-014-0379-7

If you want to cite the papers, please refer to the original.

For technical reasons we could not place the page where the original quote is coming from.

Abstract-Summary
"The authors aim to meld the literature on digital divide and social entrepreneurship to propose a social entrepreneurial system which uses universal service funds

to spread IT technologies, leading to greater entrepreneurship and the gradual alleviation of poverty among disadvantaged people."

"The case study is used to show the size of the access problem and identify factors which cause inefficiencies in the current approach to reduce the digital divide."

"After showing the need to establish a social entrepreneurial system to solve these problems and effectively use the Internet as a means to reduce poverty, the paper concludes by discussing the model's contribution to and implications for the social entrepreneurship literature and digital divide literature."

Introduction

"The question is, why a majority of low-income people and homeless wanderers still do not access the Internet and use it to move up in society?"

"De Carvalho and others [50] reviewed more than 280 initiatives set up by various types of groups (corporations, citizen sector organizations and social entrepreneurs) in four sectors (education, health, agriculture and financial services) to see if and how Information and Communication Technologies (ICTs) can have a sustainable contribution to poverty reduction and create new opportunities for the base of the pyramid (BoP, referring to the four billion people with annual per capita income<$3,000 in purchasing power parity) in the developing countries."

"Increasing living standards of low-income people and homeless wanderers will lead to these same people increasing their tax contributions back to the government and reducing the relevant transfer payment."

Low-income people and homeless wanderers in Taiwan

"The number of low-income people (as defined by the government) has increased every year in Taiwan since 1995."

"Total government subsidies to the low-income people was NT$ 10,896 million in 2012 [51], about NT$ 2,540 (US$ 85) per person per month."

"A wanderer is usually not a government-approved low-income person in Taiwan."

"The government treats every person between 16 and 65 years old as a capable worker at least earning the minimum wage (for example, NT$ 19,047 per month in 2013), even though they might actually earn less than the amount, unless the person is a student or disabled to the circumstance of being unable to work, etc According to one internal survey carried out by Taiwan Ping Tung County Charity Groups Joint Association (PDCCGJA) for 108 wanderers, they served in the Ping-Tung Homeless Shelter in June 2013, 87% of them had working experience, 25% still took temporary work and 36% were marginal wanderers who lost jobs over 3 months and were evicted from previous renting places because of unaffordable rents."

Digital divide and the causes

"The digital divide is defined as the gap between different groups of people in their Internet connectivity and the use of Internet relevant hardware, software and content."

"The digital divide can be a result of differences in the availability of physical connections (Connectivity), in user education and IT capability skills (Capability), in local context (Context) such as the economic needs and social incentives to use digital technologies, and in content (Content) attracting use of the Internet in finding jobs and interacting with local and foreign communities [52]."

"Specific studies on how to close the digital divide for the lowest income groups and homeless wanderers within society, however, are largely neglected."

Can overcoming the digital divide lead to a reduction in poverty?

"Firms as well as ordinary farmers and households can directly sell products to customers with Internet connections through a Web site or by posting advertisements in web-markets such as ebay and PChome with a minimum cost."

"Various cases have shown that people can increase income and living quality using the Internet, either through computers or smartphones."

"The evolution of smartphones has made the Internet more easily accessed by people with low information literacy and people who live in areas without fixed Internet connection [53–55]."

"In a survey based on 22,000 online interviews with households in developed and emerging markets, researchers found that the adoption of broadband Internet can significantly increase income, because better Internet access improves learning and allows working from home [56]."

"The possibility that the Internet can improve the well-being of low-income people is clear."

Factors affecting the Internet adoption by low-income people and homeless wanderers

"To provide the software, the hardware, the infrastructure, the planned updating of these three, and the basic use and repair education for the low-income population, requires an initial long-term plan by the government, as well as a persistent flow of funding either from the public or the private sector."

"Is access to the Internet a public or private good?"

"Grand Rapids treated access to the Internet as a public good, whereas Detroit Lakes viewed it as a private good."

"This city-managed network supplies low-cost Internet access to the local schools and public library and offers Internet services to citizens for a monthly fee."

"As the base for the public and the private sectors to support the poor, it is necessary to view the Internet as a public good and build in a reliable source of funding to give all citizens the choice to better themselves."

A method to increase digital opportunities for the poor and wanderers
"The authors propose a Social Entrepreneurial System (SES) using universal service funds to create digital opportunities for low-income people and homeless wanderers."

"Based on the ideologies portrayed by Sharif [57] and Miller and others [58], it is suggested that the Headquarters needs to be grounded on four basic pillars: (1) having the structure of humanware, technoware, cysnetware, orgaware and inforware so that it can effectively operate the rest of the three pillars; (2) using a public–private partnership for targeted specialization in emerging industries (such as "non-toxic" agriculture and non-agriculture foods) in which the low-income people and homeless wanderers can apply to improve their income and living standards; (3) cooperating with university-linked incubators and setting up LSEs with the backbone of DOCs; and (4) regularly evaluating projects based on integrative thinking, pro-social cost-benefit analysis and commitment to alleviating others' suffering."

Digital divide in Taiwan and the government's efforts
"The aims of these programs were to increase Internet connections to 65% for the aborigine, the computer penetration rate to 70% in high remote areas, the e-commerce penetration rate to 70% for small and medium enterprises (SMEs) and to set up Digital Opportunity Centers (DOCs) in the Asian developing countries to help the development of information services in the Asia-Pacific region [59]."

"It established Digital Opportunity Centers (DOCs) in remote areas, coordinated with the Ministry of Economic Affairs (MOEA), the Council of Indigenous People (CIP) and the Research, Development and Evaluation Commission (RDEC) to give new computers to students in low-income families living in remote regions (6,000 computers in 2008–2011), and strengthened information facilities in remote schools (the plan of ubiquitous universal digital education environment)."

The need to set up a social entrepreneurial system to use the Internet to lower poverty
"There are three reasons that explain why Taiwan needs to transform the current organizational structure for e-Opportunity into a social entrepreneurial system to increase digital opportunities for low-income people and homeless wanderers."

"Since low-income families do not have enough financial resources to meet basic daily expenses, they are not willing to spend extra money for the Internet connection, even though it is at a 50% discount."

"The lack of motivation, no IT capability and not being informed are perhaps the most important reasons explaining why low-income people and homeless wanderers do not connect to the Internet."

"With the SES operating akin to a successful private company and aiming to alleviate poverty, it can be possible to then effectively integrate low-income people and homeless wanderers into local social enterprises to use the Internet to elevate their position in society."

Conclusion

"This paper proposes the possibility of local social enterprises with DOC functions for low-income people and homeless wanderers to, at one and at the same time, ameliorate the digital divide and provide the impetus and entrepreneurial tools to reduce the number of people living in poverty."

"It contributes to the digital divide literature by bringing in the social entrepreneurship concept and thereby proposing an SES to solve basic problems faced by current governmental initiatives for universal services and digital opportunities."

"This study aims to answer the basic question of where a sustainable financial resource for the ICT programs can be obtained, address key elements affecting the adoption of the Internet by low-income people and wanderers, show the inefficiency of government initiatives due to organizational structure and budget limitation, and explain why the establishment of an SES is needed to accomplish the social mission."

4.1.7 Aligning Digital and Social Inclusion: A Study of Disadvantaged Students and Computer Access [504]

This is a machine-generated summary of:

Yelland, Nicola; Neal, Greg: Aligning digital and social inclusion: A study of disadvantaged students and computer access [504]

Published in: Education and Information Technologies (2012)

Link to original: https://doi.org/10.1007/s10639-012-9223-y

If you want to cite the papers, please refer to the original.

For technical reasons we could not place the page where the original quote is coming from.

Abstract-Summary

"We discuss the notion of the digital divide and link it with recent policy designed to promote social inclusion in a project that addressed both issues."

"Families in low socio economic areas of Australia were given computers and Internet access as part of a project that had as its primary aim to support the participation of disadvantaged families in digital activities at home and in schools."

"The authors collected data over a period of 3 years that included, pre and post surveys with parents and students, interviews with program facilitators, and focus groups with parents."

"This paper focuses on selected themes that emerged from the interview and focus group data with the parents and explores the ways in which they perceived having the computer had impacted on their lives and those of their school aged children."

"This data (surveys, interviews and focus groups) reveals that all family members felt that the ownership of a computer enabled them to feel more confident about their active participation in everyday educational, social and community activities."

"Parents highlighted the increase in their own digital skill levels and described the ways in which their lives had benefitted from having a computer in the home."

Introduction

"Their final report (2011) revealed that Internet use was increasingly individualised, privatized and mobile and one in which the 9–16 year old cohort spent 88 min on average a day online."

"The digital divide based on social class that they originally noted in 2005, was evident in ownership levels whereby 88% of middle class families owned an internet connected computer but only 61% of working class children could access the internet at home."

"MCAF (ACMA [62]) found that for Australian youth, online participation was the second most time-consuming media activity behind television viewing, reflecting the high levels of Internet and broadband access among Australian families."

"Studies (e.g. Helsper [63]; Warschauer [64]) focusing on the links between social and digital engagement, especially with reference to Internet use, increasingly show that those individuals who have access to ICT, generally come from families that have more schooling, higher incomes, and high status occupations."

From the digital divide to social inclusion

"A decade ago, Warschauer [65] called for a reconceptualising of the digital divide away from a focus on access to technology to one in which opportunities for social inclusion were at the forefront of consideration."

"As he noted: Content and language, literacy and education, and community and institutional structures must all be taken into account if meaningful access to new technologies is to be provided. (p.3) Warschauer [65] promoted a reconceptualisation of the concept of the digital divide to a technology of social inclusion."

"It is based in a new view of access beyond provision of the machine to include access to the social, cultural and economic capital (Bourdieu [66]) that accrues from being fluent in the meaningful social practices associated with living in a technologically based society: There is not one type of ICT access, but many; the meaning and value of access varies in particular social context, access exists in gradations, rather than in a bipolar opposition… ICT use is a social practice involving access to physical artifacts, skills and social support; and, acquisition of ICT access is a matter not only of education, but also of power… Access to ICT for social inclusion cannot rest on the provision of devices…it must entail the engagement of a range of resources… (p.3) This has also been noted by Helsper

[63] who reported that when those from low socio economic groups do participate in digital activity, they tend to do so at a much more basic level that involves information seeking, obtaining leisure information, making purchases online and for individual communication with families and friends at a distance."

Social inclusion as policy in Australia

"of the Prime Minister and Cabinet [67]) in which they stated that they are committed to building 'a stronger and fairer Australia'. (p.2) Social inclusion is defined as: ...building a nation in which all Australians have the opportunity and support they need to participate fully in the nation's economic and community life, develop their own potential and be treated with dignity and respect. (p.2) The government outlined a strategy that considers capacity building for disadvantaged citizens which includes improving capabilities and skills in communication, negotiation, organization, time management, as well as being able to navigate services available."

"This is problematic since it becomes apparent that not having access to new technologies can significantly disadvantage citizens in terms of being able to participate fully in a range of social and educational experiences (e.g. Gore [68])."

Research design

"The research project was grounded in the theoretical framework of cultural capital postulated by Bourdieu (e.g. [66, 69]) which recognised that families and schools play a critical and increasingly pervasive role in perpetuating the advantage of specific knowledge and skills that are valued by society across generations."

"The families were provided with refurbished computers, Internet access, training in how to set up and use the computer and help line support."

"How do the family members feel about having access to computers and the Internet in their homes?"

"Can family members articulate what they would like to do with the computers and Internet access in the future?"

"Interviews—we interviewed program facilitators via telephone and family members face-to-face in semi-structured open-ended question interviews to find out what the family members were using the computers for and their reaction to the initial training sessions."

Participants

"Participating families in the Tech Packs Project were required to meet certain criteria in order to obtain a computer."

"The criteria varied slightly between communities but basically they had to demonstrate that they were classed as low income/low socio economic status (job classification/unemployed) and this was usually manifested by the fact that they were eligible for the Health Care Card or a Centrelink Concession Card."

"This card is given to families who live below the minimum wage."

"They also had to have one or more school aged children."

Results

"The data indicated that for the students in the families there was an increase in activity in the following areas: Searching the Internet for information for school projects and for personal use Being able to participate in extra activities beyond the regular curriculum (e.g. Mathletics) Communicating with friends, family and the school via emails, Facebook, MSN chat and other forums Playing online games and being entertained (e.g. YouTube) For many of the children involved in the Tech Packs Project clearly the most popular use of the Internet is for searching and researching sites for educational purposes and/or for self-interest."

Discussion and conclusions

"This feeling of increased confidence with technology is an important one since participation and use of computers and associated peripherals is integral to functioning effectively in society today."

"For all the participants, the ownership and use of a computer and connection to the Internet, enabled them to not only seek out new information for pleasure and employment potential but also to participate in communities of practice that were not available to them previously."

"Reconceptualising the digital divide as social inclusion means that we need to go beyond the technical aspects to consider how inequities are reproduced and consider opportunities to build on the enthusiasm of participants in contexts that will enable them to become 'power' users (Kennedy and others [70]) that will afford them the social and cultural capital (Bourdieu [69]) to participate in the contemporary world."

4.1.8 Exploring the Gap of the Digital Divide [505]

This is a machine-generated summary of:

Rye, Ståle Angen: Exploring the gap of the digital divide [505]
Published in: GeoJournal (2008).
Link to original: https://doi.org/10.1007/s10708-008-9154-8

If you want to cite the papers, please refer to the original.

For technical reasons we could not place the page where the original quote is coming from.

Abstract-Summary

"While much of the literature related to the digital divide focuses on either technology penetration in developed countries or the introduction of new technology to the poor and disadvantaged in the developing world, this paper pays attention to the middle-classes in a developing country which is connected to the Internet but under poor conditions."

"The digital divide is examined by investigating how the uneven distribution of Internet connection influences distance education students' participation in higher education."

"In central areas the new technology seems to contribute to students' study situations more effectively and the daily environment empowers the students' study activities, rather than vice versa."

Introduction

"Further, the complex relation between higher education participation and access to the Internet is also widely discussed (Kirkwood [71]; Perraton [72]; Rumble [73, 74])."

"While much of the research on the relation between new technology access and human wealth is based on statistical surveys and macro data, in this paper this issue is discussed on the basis of a qualitative case study of the use of the Internet in distance education in two places in Indonesia, one in Jakarta (the capital) and one in Bngaka Belitung, a relative remote province."

"By selecting two cases for in-depth study, the intention is to gain insight into processes creating the outcome from uneven distributions of new technology in developing countries, particularly the Internet, and how this influences participation in higher education."

"The aim behind using this concept is to explore how access to the Internet and the results of its use are embedded in the social conditions of the students' daily life and the place where the technology is located."

Understanding the digital divide

"The 'digital divide' is commonly related to differences in access to the new digital technology and in variations regarding access to information and knowledge shared in locally and globally distributed digital communication channels."

"A last approach to the spatial dimension of the digital divide is focus on how the technology and users are socially embedded in various places."

"For identifying and measuring differences that exist within the digital divide, the main indicators have typically been private ownership of computers and the use of Internet (for example: Bradshaw and others [75]; NTIA [76–79]; Tien and Fu [80]; Tiene [81])."

"If the main objective of researching the digital divide is to understand how access to technology is related in turn to access to material wealth, it is claimed that a more complex understanding of this relation has to be applied (Gunkel [82]; van Dijk [83]; Warschauer [84])."

"The digital divide refers to a division between those who have and have not access to new technology."

Distance education, technology and place

"By offering such students access to higher education, at distance, they have been expected to be better prepared to receive their share of public wealth as well as being able to effectively contribute to society's development."

"From the very beginning in the history of distance education various technologies have been applied in efforts to create access possibilities for students who for several reasons have been excluded from ordinary higher education (Dhanarajan [85]; Gandhe [86]; Rumble [73]; Zuhairi [87])."

"By developing distance education where the use of the Internet is assumed, it is also assumed that students are able to access the technology, i.e. access to education depends on students' access to technology (Perraton [72])."

"Following Nicholas' [88] argument, the reason why some students are able to take advantage of Internet-supported distance education while others are not has to be related to political and economic processes in the societies in which they live."

Analytical frame

"The frame for analysing data is inspired by van Dijk's [89] model for successive access to digital technology, in which he attempts to incorporate the main implications of the critics against the digital divide (as have been discussed above)."

"He does this by suggesting four types of access, whereby each represents a successive stage necessary before the next is to be passed and before full appropriation of the digital technology can occur."

"The initial condition is motivation to use digital technology."

"If the potential user is motivated and has the possibility to physically access the technology, the next step is having the skills necessary to use it."

"The consequences of the use of technology are identified by social participation, which in the case presented in this paper refers to participation in education."

"This model has some obvious advantages when analysing variation in technology access."

A qualitative case study of successive access to digital technology

"One is a Master of Public Administration (MPA) programme offered to students in the province of Bangka Belitung."

"The main empirical source comprised interviews and dialogues with fourteen students, of whom seven lived in Bangka Belitung and seven lived in Jakarta."

"Informal conversations were held with several students, including those who had been interviewed."

"The interviews were conducted at the students' workplaces or their homes during March 2004 (Bangka Belitung) and February 2005 (Jakarta)."

"The interviews with the students were centred on an interview guide containing questions and topics related to their experiences of using the Internet and how this use was embedded in their daily life."

"I also relied on interviews and conversations with informants working at UT and who were involved in the development and implementation of the MPA and MM."

Two distance master's programmes in Indonesia

"The students on the MPA and the MM programmes were recruited by public announcement, followed by an entry selection process that was relating to subject matters as well as computer skills and knowledge of the English language."

"Most of the students worked within management and administration at different levels in the public sector, though others, such as a journalist, also attended the study programme."

"The students in Jakarta worked in management in private business and state-owned companies."

"Students were able to access Web-based discussion groups and could download additional resources and information related to their studies."

"The students' study activities were strictly organised and scheduled through modules containing assignments and tutorials as described in a study guide."

"Even though the study mode and the use of technology were quite similar for the two master's programmes, the context of the students' daily lives was quite different."

Students' experiences of the use of the Internet: Bangka Belitung

"Students' use of ICT was limited, though it varied in its extremes, from one student that connected to the Internet several times a week to another who had only ever accessed the Internet a few times."

"Many of the students in Bangka Belitung said that they had limited experience in the use of computers and the Internet and, furthermore, the training given by UT was limited."

"Several students claimed that their first experience of the Internet was through UT, even though they had some familiarity with computers, as expressed by one student: 'It is difficult to understand how to use the Internet, so I don't use it much as part of my study."

"Even though the students regarded the use of ICT as problematic due to their limited access possibilities, this was not the only problem relating to ICT."

Students' considerations of technology in use: Jakarta

"The interviewed students' familiarity with the use of new technology followed from the fact that they all had computers and Internet access at home."

"Even though this student was encouraged by his superior, he did not have much time to use a computer for studying during his working hours."

"None of the interviewed students in Jakarta expressed that they were limited by personal knowledge and experience regarding the use of computers and the Internet at home."

"The students in Jakarta seemed to benefit considerably from the use of the Internet as a part of their studies."

"When they complained, it was not because of problems regarding accessing and handling the technology but rather because the Internet was not used enough by UT or, as one of the students expressed, 'I expected that UT would have more information on the Web that would be useful for the students in our study [programme]'."

Discussion: the same technology but different results

"The empirical investigation suggests that the use of the Internet in distance education in a developing country, such as Indonesia, is likely to face serious constraints, at least in peripheral areas such as Bangka Belitung but also to some extent in big cities such as Jakarta."

"Following this, it may be claimed that Internet uptake in rural and peripheral places in developing countries tends to be constrained by lack of motivation and limited ability to access the technology, whereas in central areas constraints are related more to skills and use."

"Although the use of the Internet was not very beneficial for the students in Bangka Belitung, the empirical investigation suggests, following Nicholas [88], that this region's geographical capital was extended by UT's engagement in Internet-supported distance education."

"From the case of Bangka Belitung it can be learned that the digital level of both the students and the region as a whole were at the limit regarding applying the Internet in distance education."

Concluding remarks

"The presences of the Internet by the way distance education seems, however, to contribute positively to the remote place as a whole, while its use in the central location did not seem to add much new to the students' surroundings."

"Results from the use of new technology in remote areas should therefore not only be assessed in terms of individual benefits but also by how regional digital capacity is increased."

"Differences in access to digital technology then seems to be increased on an individual level by the introduction of Internet-supported distance education, while on a regional level the differences may be reduced."

"This research rather suggests that different places should be considered as separate ways of how access conditions work together on the way to social participation."

4.1.9 Digital Divide, Craft Firms' Websites and Urban–Rural Disparities—Empirical Evidence from A Web-Scraping Approach [506]

This is a machine-generated summary of:

Thonipara, Anita; Sternberg, Rolf; Proeger, Till; Haefner, Lukas: Digital divide, craft firms' websites and urban–rural disparities—empirical evidence from a web-scraping approach [506]

Published in: Review of Regional Research (2022)

Link to original: https://doi.org/10.1007/s10037-022-00170-5

If you want to cite the papers, please refer to the original.

For technical reasons we could not place the page where the original quote is coming from.

Abstract-Summary

"Following the "death of distance" postulate, digitalization may reduce or even eliminate the penalty of firms being located in rural areas compared with those in urban agglomerations."

"Comparing urban with rural areas, we show that running a website—as a proxy for the degree of digitalization of the respective firm—is highly dependent on location, whereby firms in urban areas are almost twice as likely to run websites compared with those located in rural areas."

"We find a negative and significant association of gross domestic product per capita with website prevalence, which is driven by urban regions."

"There are no differences between urban, semi-urban and rural areas in terms of website up-to-dateness as well as social media prevalence."

Introduction
"One prominent explanation for the absence of catch-up processes of laggard regions is the digital divide among firms in rural and urban regions (Salemink et al. [93]), for which two levels can be distinguished (Büchi et al. [94]; Scheerder et al. [95])."

"The first-level digital divide refers to access: while nowadays most firms in rural regions have access to the internet, the quality of the connection often trails behind that in urban regions (Briglauer et al. [96]; Prieger [97]), which thus limits rural firms' opportunities of digital participation."

"The second-level digital divide refers to usage, whereby the adoption of not only the internet but also other digital technologies is lower in rural firms compared with their urban counterparts."

"Research on the second-level digital divide and its spatial effects is suffering from a scarcity of indicators proxying the prevalence of digital technologies in firms, especially aggregated at the regional level for Europe (Billon et al. [90]; Ruiz-Rodríguez et al. [98])."

Literature review
"Regarding the spatial effects of digital infrastructure, there is some empirical evidence in the literature if inter-regional differences between sub-national regions are considered (Bertschek et al. [99]; Grimes [100]; Malecki [101])."

"A number of studies comparing urban and rural regions find a divide in the usage of digital technologies, which does not stem solely from lacking access in rural regions (Moriset et al. [92]; Prieger [97]; Salemink et al. [93]; Townsend et al. [102])."

"The amount of theoretical or empirical research regarding the impact of regional determinants on the urban-rural second-level digital divide is rather small, particularly for firms."

"Especially for Europe, there is a lack of regional data for the usage of digital technologies in firms (Billon et al. [90]; Ruiz-Rodríguez et al. [98]), which underlines the need for developing new indicators."

"From a spatial perspective, digital technologies seem to be used effectively particularly in regions with high levels of skilled labor (Salemink et al. [93]; Schleife [103]), which in turn benefit from digitalization (Atasoy [104]; Hasbi [105]; Mack and Faggian [106]; McCoy et al. [107])."

Method
"One could assume that website prevalence is overall decreasing and that firms primarily use social media for digital marketing."

"The disadvantages of using the yellow pages as the source of firms' websites include the heterogeneous structure of the yellow pages in Germany, historically different cost models for entering firm information as well as potential errors during the web-scraping procedure itself, potentially resulting in a loss of data."

"We conclude that our dataset presents a large set of firms, while acknowledging that it might not present a fully representative picture of the craft sector; however, there are strong incentives for firms to be listed on the yellow pages and to state their website on the yellow pages."

"We assume that we reach a broader, more representative sample of craft firms using the yellow pages rather than with a straightforward Google search, which would not yield firms without a website."

Results

"We find that counties with a high increase of inhabitants in pension age have overall lower levels of website prevalence, which can be interpreted as resulting from the different affinity to using digital technologies among older generations, in the private as well as the occupational context."

"When looking at the differentiation between rural, semi-urban and urban counties, we find that a higher increase of older persons in rural regions is associated with higher website prevalence, meaning that the overall effect is driven by semi-urban and urban counties."

"Result 4: GDP per capita is positively associated with website prevalence in rural areas only."

"This result is driven by urban regions, where an increased birth rate is associated with higher website prevalence."

"If larger firms are more likely to be situated in urban areas, our results would only reflect the effects of firm size differences across region types on the website prevalence."

Discussion

"Discussing our results in further detail, the positive association of website prevalence with population density underlines the importance of population density for the urban-rural second-level digital divide."

"Besides population density, other well-known regional determinants for the digital divide have been confirmed in our study, including qualification level and age, as characteristics of attractive labor market centers."

"Since these characteristics (high qualification level, low average age, high in-migration) usually represent urban regions (see 2.1.1), these results can be interpreted as explanatory factors for the urban-rural digital divide."

"Regional disparities between rural and (semi-)urban regions might become substantially more pronounced if rural firms fail to be included into the technological dynamics of digitalization."

"Different forms of digital training might be an adequate policy instrument for regions that lack some of the characteristics that "automatically" drive regional digitalization, especially since some studies have shown that (informal) lifelong learning is even more important for the usage of digital technologies than professional qualification (Billon et al. [90]; Evangelista et al. [108])."

Conclusion

"We have introduced a novel indicator for measuring the regional digitalization degree, which comprises data on website prevalence in craft firms in Germany and is constructed using a web-scraping algorithm."

"We use this indicator to assess the regional determinants of the second-level digital divide, i.e. spatial differences in the usage of digital technologies."

"Our results show a positive association between population density and website prevalence, which hints at the conclusion that population density is an important determinant for the second-level digital divide."

"Since these website-promoting characteristics are usually found in cities, we find a profound urban-rural second-level digital divide that is not only rooted in population density but also in other socio-demographic regional characteristics."

"Further results show no overall association of broadband availability with website prevalence, but a positive association in rural regions."

4.1.10 From Digital Divide to Digital Capital: The Role of Education and Digital Skills in Social Media Participation [507]

This is a machine-generated summary of:

Lybeck, Robin; Koiranen, Ilkka; Koivula, Aki: From digital divide to digital capital: the role of education and digital skills in social media participation [507]

Published in: Universal Access in the Information Society (2023)

Link to original: https://doi.org/10.1007/s10209-022-00961-0

If you want to cite the papers, please refer to the original.

For technical reasons we could not place the page where the original quote is coming from.

Abstract-Summary

"By combining theories of social media participation and digital inequality, we explore issues regarding the prerequisites of participating through social media platforms, focusing especially on education."

"Through an analysis of data from a representative survey study in Finland (N = 3724), we illuminate the ambiguity of the perceived obstacles to both digital skills and political participation."

"We further build on the concept of digital capital to show the significant mediating effect of digital skills on education and participation."

"By utilizing the ISS (Internet Skills Scale), we break down digital skills into operational, information navigation, social, creative and mobile skills, and show how the operational skills have most significant mediator between education and political participation in social media."

Introduction

"We analyze how the respondents' educational background in an extensive survey correlates with ICT skills and digital participation."

"We aim to construct a detailed understanding of digital inequality in light of our empirical material, which we will do by focusing on the links between societal background, digital skills, and participatory action on social media."

"We argue that for understanding inequalities in digitalized participatory processes, scholars should comprehend the prominence of digital capital as a bridge resource that transmits inequalities from offline to online as well mediates the processes of exchange between various forms of capital in digital spaces."

"Based on our theoretical framework and other previous studies [109, 110], we suggest that these parameters form a triangular relation of causality, where background factors affect the accumulation of digital capital, and result in the prerequisites for digital participation."

"Research and theory on digital participation and digital inequality are brought together to construct a theoretical framework for the study."

Digital participation and social media

"Within the messy field of social media, the definitions for both of our main concepts, digital participation and digital capital, become unclear."

"Political participation on social media has raised a lot of questions regarding how it relates to conventional public participation [111]."

"The concept of participating increasingly includes actions individuals take on social media, and thus, participation is intertwined in our everyday lives more than ever before [112]."

"Digitalization has redefined the relationship we have with the media, governmental institutions, NGOs, and commercial actors regarding participation in decision-making and shaping public opinion."

"Previous research suggests that the development of social media has caused an enormous transformation of political participation and engagement, specifically from the bottom-up."

"Social media platforms are emphasizing the importance of interaction as a form of participation in deliberative democracies."

"We will focus on the deliberative forms of participation in social media."

Digital inequality, divides and capital

"We argue that rather than an alternative to the digital divide, digital capital supports the digital divide theory by providing a concept that sheds light on the processes involved in crossing the obstacles to participation and by achieving desired outcomes through converting digital capital to other forms of capital."

"The Bourdieu-inspired concept of digital capital provides the concept of divides with a theoretical understanding of the process involved in achieving desired outcomes."

"Digital capital resembles cultural capital in terms of specific skills and knowledge one can acquire, and social capital in the way it is linked to social networks [113]."

"Political participation on social media offers an interesting case for assessing the dynamics between digital divides and digital capital."

"Rather, this is an elastic process where individuals are also able to overcome barriers that former levels of digital divides set, and to do so with the aid of extensive exchanges between different forms of capital."

Research design

"We focus on how citizens from different population groups participate in political discussions as well as processes of content production and dissemination on social media."

"Research shows that, with the continuation of digitalization, divisions in social media use in terms of access have lessened, but the differences in beneficial use practices, such as civic and political participation, have grown among population groups in Finland [114]."

"The first hypothesis we set out to test in our material is thus: Socio-demographic factors are related to political participation in social media (namely age, gender, and especially education)."

"The rationale here is that education (i.e., the main source of cultural capital) increases people's ICT skills and competencies (i.e., forms of digital capital), which potentially impacts the relative differences between highly educated and non-educated users' political activity (i.e., the utilization of political capital) on social media."

Material and methods

"Information navigation skills were based on two variables that measured to what extent respondents (1) found it hard to decide what the best keywords were to use for online research, and (2) found it hard to find a website they previously visited (Mean = 4.0; SD = 0.95; alpha 0.82)."

"Creative skills were combined according to two variables measuring to what extent respondents (1) knew how to make basic changes to the content that other users had produced, and (2) knew which different types of licenses applied to online content (Mean = 3.2; SD = 1.2; alpha 0.82)."

"While these items measuring different aspect of digital skills correlated significantly, in the first part of our analyses we combined different skill dimensions by forming a mean variable that measured holistically respondents' digital skills (Mean = 3.63; SD = 1.0; alpha 0.89)."

"We analyzed to what extent the educational differences were related to the relationship between digital skills and online participation."

Results

"We assessed the mediating effects of various digital skills between education level and political social media use."

"The third and final model of our first analysis illustrated how digital skills had a clear association with online political activity."

"According to the multinomial logit models, increasing of digital skills by one unit of standard deviation inclined the probability of online political participation by 10 percentage points."

"Digital skills correlated with age differences: the relationship between age and online political participation decreased after adding the skills variable to the model."

"After inquiring about the direct effects of our independent variables on online political activity, we conducted a decomposition analysis to test the mediating effect of digital skills on educational differences."

"According to the decomposition models, digital skills mediated over 80% of educational differences when predicting online political activity."

Discussion and conclusions

"We assessed the structural mechanisms of societal background as well as the mediating effects of digital skills as a form of capital in online political participation."

"While socio-demographic variables have a significant effect on those who participate on social media, it seems that digital skills ultimately transmit participatory practices online."

"To digital skills, all other independent variables—namely age, gender, education, and interest in politics—had a somewhat statistically significant effect on online participation."

"After standardizing the differences in other socio-demographic variables, political interest, and digital skills, respondents aged 35–44 were the most active age group participating on social media."

"Through the processes of online participation, these forms of digital capital are further transacted with forms of political capital."

"According to our analyses, all separate aspects of digital skills, namely operational, information related, mobile, and creative, mediated the relationship between education and online political participation."

4.1.11 Closing the Digital Divide: Understanding Racial, Ethnic, Social Class, Gender and Geographic Disparities in Internet Use Among School Age Children in the United States [508]

This is a machine-generated summary of:

Cleary, Paul F.; Pierce, Glenn; Trauth, Eileen M.: Closing the digital divide: understanding racial, ethnic, social class, gender and geographic disparities in Internet use among school age children in the United States [508]

Published in: Universal Access in the Information Society (2005)

Link to original: https://doi.org/10.1007/s10209-005-0001-0

If you want to cite the papers, please refer to the original.

For technical reasons we could not place the page where the original quote is coming from.

Abstract-Summary

"The purpose of this study is to explore the dynamics underlying disparities in Internet use among school age children in the US."

"The availability of household computing resources and adult Internet users in the household were most important in explaining disparities in use among children."

Introduction

"The analysis presented in the paper examines the independent and combined influences of individual and family socioeconomic characteristics, family resources, household computing resources and expertise, and school resources on Internet use among school age children."

"How does household computing resources and expertise, in the form of computers available in the home and in the form of adults in the home with computing expertise/experience (in the form of adult expertise), relate to Internet use among school age children?"

"How are public school computing resources related to Internet use among school age children?"

"This question involves examining the influence of identified public school computing resources on Internet use among children."

"A central question is whether school computing resources, provided to the public through the states, can help to close the digital divide for Internet use among school age children."

Literature review

"This is followed by a discussion of underlying barriers to Internet access and use by children, including the cost of computing technology and education."

"Broad factors such as economics, fairness, and cost along with more specific factors such as age, race, geographic location, education, income, and home computer access influence the nature and scope of observed gaps in Internet access and use in the USA (NTIA, 2002, February)."

"Although children show increasing rates of Internet use (NTIA, 2002, February, p 1), the question of Internet access remains an important issue."

"The use of the Internet in education is now widely believed to be of value to school age children and, therefore, access and use of Internet technology is becoming increasingly more important to education."

Methodologic approach

"The specific categories included in the analysis are: (1) family and child background characteristics (e.g., gender, age, race, ethnicity, citizenship status, geographic location, living quarters), (2) family structure and educational resources (i.e., marital status and education level of the reference person for the household), (3) family financial resources, (4) household computing resources and expertise (i.e., the presence of a computer at home, Internet use among adults at home and Internet use among adults outside the home), and (5) public school Internet resources (i.e., the number of students per Internet connection)."

"Measures of household computing resources and expertise are included as explanatory variables to study the potential importance of available household expertise (by adults in a household) on Internet use by school age children."

"Logistic regression models estimate the average effect of each independent explanatory variable on the odds that a dependent variable outcome will occur (school age child will use the Internet at any location)."

Analysis

"Equally important among these factors, the greatest impact on Internet use among school age children from Black non-Hispanic households occurs with the addition of household computing resources and expertise to the analysis."

"An analysis of the Exp (B) coefficients indicates that the availability of household computer resources and expertise appear to be one of the major determinants of Internet use among school age children."

"The most important finding of this analysis is that the availability of household computing resources and expertise showed the greatest impact of any set of attributes in reducing observed disparities in Internet use among school age children in Black (non-Hispanic), Hispanic and non-citizen households."

"Apart from family and child background characteristics entered in the first step of the analysis, the introduction of household computing resources and expertise appears to represent the single most important set of determinants of Internet use among school age children."

Conclusion and policy implications
"We found that Internet use among children appears most strongly affected by the presence of both in home access to a computer and the availability of technical support, expertise and informal training in the home."

"This includes providing enhanced Internet access that could make access to broadband technology more available throughout the nation and supporting market policies that encourage continued decline in the total cost of ownership in order to enable more people to afford in-home access."

"The second part is to continue to support, through the Universal Service program, computer and Internet access provided to the public at out of home locations such as schools, libraries and CAC, many of which already have computers and Internet access."

"Providing expanded and enhanced in-home Internet access is perhaps the most effective method of increasing Internet use among school age children."

"As the results of the conducted study show, providing in-home Internet access is an important determinant of Internet use among school age children."

4.1.12 Digital Divide Among B40 Students in Malaysian Higher Education Institutions [509]

This is a machine-generated summary of:

Devisakti, A.; Muftahu, Muhammad; Xiaoling, Hu: Digital divide among B40 students in Malaysian higher education institutions [509]

Published in: Education and Information Technologies (2023)

Link to original: https://doi.org/10.1007/s10639-023-11847-w

If you want to cite the papers, please refer to the original.

For technical reasons we could not place the page where the original quote is coming from.

Abstract-Summary
"This study aims to investigate the use of digital technology among B40 students (students from the lower socioeconomic backgrounds) in Malaysian HEIs."

"This study intends to investigate how perceived ease of use, perceived usefulness, subjective norms, perceived behavioural control, as well as gratification constructs significantly affect digital usage among B40 students in Malaysian HEIs."

"The results showed that perceived usefulness and subjective norms significantly influenced the digital usage of the B40 students."

"All three gratification constructs indicated a positive effect on the students' digital usage."

Introduction

"Despite the pressing need to provide quality education for students of all levels, digital divide has become a growing concern in the higher education sector."

"Even though digital teaching and learning has been adopted in education systems over the past few years, the pandemic has caused a dramatic increase in technology usage among students around the globe."

"The ubiquity of digital technologies has increased the demand for digital access, skills, and usage among higher education students."

"Digital divide still exists among higher education students in this country, and it became more evident during the COVID-19 pandemic."

"While the pandemic has led to rapid transformation in higher education system, digital disparity has become more evident among students with lower socio-economic backgrounds."

"This study seeks to examine the digital divide among higher education students with lower socioeconomic backgrounds."

Problem statement

"Digital learning can be a challenging learning experience for vulnerable students."

"While past studies examined students' learning experience with digital learning (Alty and others, [115]; Clark and others, [116]; Persada and others, [117]), little is known about digital learning divide among students from vulnerable groups."

"This study aims to investigate the use of digital technology among vulnerable students (B40) in Malaysian HEIs."

"The anchoring question that guided this study is how the students' behavioural aspects can affect their digital usage in learning at their university."

Theoretical perspectives

"To explore the factors that influence the digital divide among the B40 students in higher education, a research model that integrates the uses and gratification theory, TPB, and three digital divide levels is proposed in this study."

"We assume: H1: Perceived ease of use positively impacts digital usage among students."

"We propose that: H2: Perceived usefulness positively impacts digital usage among students."

"We assume that: H4: Perceived behavioural control positively impacts students' digital usage."

"Based on above, our hypothesis is: H5: Content uses gratification positively impact digital usage among students."

"We still propose it has a significant influence on digital usage of students, hence, the assumption is as follows: H6: Affective uses gratification positively impacts digital usage among students."

"We assume as follows: H7: Social integrative gratification positively impacts digital usage among students."

Methodology

"These can include profoundly held ideas about the kinds of issues that should be researched, the appropriate research questions to ask, or the methods we should use to obtain assumptions for our studies (Cresswell & Poth, [118])."

"To examine the hypotheses, a quantitative method was employed by utilising the survey-based research approach."

"Prior to the actual data collection, a pre-test was conducted on the survey involving two field experts and three students."

"A previous study on learners' adoption of online learning research (Yeap and others, [119]) employed the power of 0.95 in which a small effect size of 0.02 was obtained."

"The research instrument used in this study is a closed ended survey comprising measurement scales adapted from past literature."

"To test the research instrument, the survey was initially distributed to two experts and three students."

Results

"PLS-SEM has been employed in many studies as it allows estimating research models with multiple constructs, indicator variables, and structural paths without imposing distributional assumptions on the data (Hair and others, [120])."

"The greater the coefficient value (i.e., β value), the greater the substantial effect on the endogenous latent construct."

"This value is clearly lower than the value in Amaro and Duarte's [121] study, which had a considerably more complex model with additional antecedent constructs."

"By the path model, all the proposed hypotheses were statistically supported, except for PU and perceived behavioural control, which did not have a significant effect on the digital usage ($p > 0.05$)."

"In terms of factors loading, PEOU2, PU1, PU2, PUC2, PBC2, CUG4, AU4, USE2, and USE3 indicated the highest values whereby more than 40 for each of the items showed high effect in their constructs."

Discussions and implications

"Content uses gratification was found to positively influence the digital usage among the B40 students in Malaysian HEIs."

"Affective uses gratification was found to positively influence the digital usage among the B40 students in Malaysian HEIs."

"Social integrative gratification was found to positively influence the digital usage among the B40 students in Malaysian HEIs."

"Social integrative gratification indicated the strongest positive impact on the B40 students' usage of digital technology for their learning."

"Content uses gratification also seemed to have a significant positive influence on the digital usage among the B40 students, which is consistent with the finding by Lou and others (2011)."

"This research has revealed significant positive influences of all gratification aspects (i.e., social integrative, affective uses and content uses) as well as PU and subjective norm on the digital usage among the B40 students in Malaysian HEIs."

Limitations and conclusion

"Future research may also replicate this study by exploring other constructs which could potentially be linked to students' digital usage and access as well as skills like creativity and ICT."

"It would be interesting if future studies could examine this area of research by employing a qualitative approach, such as interviews, as qualitative data would yield more detailed and in-depth responses in relation to the digital usage among the B40 students."

"Given the extensive use of digital technologies in HEIs for teaching and learning, it is critical to look at variables that could affect student usage of digital technologies, especially among students from lower socioeconomic backgrounds."

"This study examined seven constructs in relation to the B40 students' digital usage, namely PEOU, PU, subjective norm, perceived behavioural control, content uses gratification, affective uses gratification, and social integrative gratification."

4.1.13 Networking Communities from the Bottom Up: Grassroots Approaches to Overcoming the Digital Divide [510]

This is a machine-generated summary of:

Gaved, Mark B.; Mulholland, Paul: Networking communities from the bottom up: grassroots approaches to overcoming the digital divide [510]

Published in: AI & SOCIETY (2010)

Link to original: https://doi.org/10.1007/s00146-009-0259-5

If you want to cite the papers, please refer to the original.

For technical reasons we could not place the page where the original quote is coming from.

Abstract-Summary

"Achieving meaningful usage of the Internet is more than attaining access: multiple social and technological insufficiencies must be overcome and continually readdressed."

"We focus on one approach–grassroots networked communities."

"These are communities of locality that have developed their own computer network infrastructure with minimal external support."

"Further research has commenced to analyse how the presence of social software and near-ubiquitous computer network access affects the sharing and storage of information within communities of locality."

From digitally excluded to sufficiently connected

"Providing access, it has been declared, is the key to overcome the 'digital divide'—the gap in society between those who are or are not connected to the Internet and active users of digital technologies."

"Public buildings are being provided with Internet access and identified as 'UK Online Centres' to enable "everyone in the UK … to have access to the Internet and e-mail near to where they live." (Department for Education and Skills [122])."

"DiMaggio and Hargittai [123] argue that as Internet penetration continues, researchers should shift their focus from analysing a dichotomous divide, to a study of multiple inequalities between those who are potentially connected: Equipment the quality of computer hardware, software, and Internet access."

"A retired gardener hoping to surf the Internet for information about growing vegetables may require different training and equipment from a teenager seeking to play the latest online games, and each will require different types of social and technical support to achieve their goals."

Community approaches to achieving effective connectivity

"Communities of locality—communities of individuals residing within geographical proximity and connected by an existing network of social relationships—may provide one of the most effective methods of ensuring meaningful and sustained usage of the Internet."

"Geographical communities of locality have utilised the Internet to support social interaction since the early BBS's and we recognise seminal community developments such as the Seattle Community Network (Silver [124]) and the Milan Community Network 'Rete Civica di Milano' or RCM (de Cindio [125]), which have provided models of good practice."

"Community networks can refer to a broad range of initiatives from an individual's website promoting their local village to a computer network developed through community wide consensus building exercises, funded from within or outside the locality, and involving the construction of network infrastructure and hardware and software provision."

"Grassroots networked community initiatives can be more effective than external interventions as they are developed from elective rather than selected neighbourhoods and can draw on existing social capital to achieve their goals."

Methodology

"Eight communities were initially approached in 2003–2004, identified using a snowball gathering method beginning from the authors' prior contact with the first group (the primary author of this paper had previously been involved in setting up and developing their networking initiative)."

"In one community, fourteen end users were interviewed to determine the extent to which end user and initiator goals are similar in grassroots networked communities."

"Our research into externally initiated networking projects has revealed a gulf between project initiators and recipient communities, and we hypothesise that if the end users and initiators in locally developed networking initiatives have more closely aligned goals, there is more likelihood of mutually beneficial outcomes."

"Scope of the initiative: e.g. "Please describe the community you are working to connect"."

"Equipment: e.g. "How would you describe the main benefits of the initiative to somebody else?""

Summary of findings

"Support within the Subcultures was based on more of a distributed network of skill sharing as members tended to a higher technical level of abilities, though the core team within the initiative provided a central expertise base."

"All initiators sought to utilise intranets, accessible only to their members and running faster than Internet connections to the outside world to support community interaction, though the number of services supported varied from group to group and was generally limited."

"With the membership of the pioneer communities focussing on networking technologies, it could be seen that in some cases services were set up solely in order to test their effectiveness, and collaborations were actively sought with other community groups in order to give the Pioneers an opportunity to test their services with a wider range of users than their own peer group."

Discussion

"Grassroots networked communities may provide us with an insight into the emerging domestication of ICT within the social sphere."

"Within this phenomenon, we perceive three broad groupings: Cooperatives defined closely with their community of locality, seeking to connect as many people within this area; Subcultures based around communities of interest significantly defined by their locality, seeking to support this specific demographic within the area; and Pioneers defined principally by interest but grounded in their immediate locality, network enthusiasts seeking to explore and innovate networking technologies as an interest, to break ground for other groups, and to support the social interactions of their community."

"Grassroots networked communities may provide a sustainable approach to the 'digital divide'."

"Grassroots initiated networked communities do not operate in isolation."

"Social networking between groups is carried out by one or two core members of the initiative."

Summary and future work

"Software services that enhance community interaction may be the most vital aspect of such initiatives."

"Within the interviews initiators and end users suggested particular social functions that social software tools could support in their communities, such as facilitating housing association tasks (decision making, record keeping, and planning), sharing resources within their known community (babysitting, loans of tools, recycling of unwanted goods), and local socialising (sharing photos and music, announcing local events)."

"We will use this as a basis for further research and work with the communities, utilising a participatory approach (Day [126]; Fischer [127]) to choose software supported services that may enhance existing methods of communication and storage of such local knowledge."

"A key aspect of this work will be to ensure that the tools and services are managed from within the communities to minimise external dependencies: we see our research as supporting ongoing community initiatives, not introducing limited term interventions."

4.1.14 Spatial Analysis of Urban Digital Divide in Kigali, Rwanda [511]

This is a machine-generated summary of:

Otioma, Chuks; Madureira, Ana Mafalda; Martinez, Javier: Spatial analysis of urban digital divide in Kigali, Rwanda [511]

Published in: GeoJournal (2018)

Link to original: https://doi.org/10.1007/s10708-018-9882-3

If you want to cite the papers, please refer to the original.

For technical reasons we could not place the page where the original quote is coming from.

Abstract-Summary

"Access to Information and Communications Technologies (ICTs) and its potentials for cities are often uneven across geographies and demographics, a condition that has been referred to as the digital divide."

"Where not aspatial, most studies explore the digital divide at macro-spatial levels (national and regional levels), which makes them less relevant for knowledge generation and policies at intra-urban scales, the actual hubs of innovations."

"This paper explores the state of ICT access in Kigali City, at an intra-urban level."

"The paper establishes the relative digital access performance of the sectors based on the measurement of their ICT Location Quotients."

"In Kigali City spatial distribution of ICT access is significantly clustered, with areas of concentration at the core and sparsity on the northeastern periphery of the city."

"Using data reduction, we establish that existing urban inequality in infrastructure, urban agglomerative strength, planning status and household socio-economic status are replicated as correlates of the digital divide in Kigali City."

Introduction

"In order to address this gap, this paper explores the state of ICT access at an intra-urban level, by assessing the digital performance of Kigali, Rwanda, in terms of physical access to ICTs."

"The study contributes to the understanding of factors behind digital inequality in divided cities in a spatial context, and as an instrument for micro-level space-based ICT interventions."

"We then contextualize the role of ICT in economic growth promotion and in inclusive city development and explore measurement and conceptual issues in digital divide studies in order to understand how the relationship between digital performance and spatial economic factors has been captured."

ICT and economic growth promotion: between spatiality, digitality and divide

"The possibility of the urban fringe dwellers to be excluded in the emergent spatial structure of the ICT-enabled economy is evidence of the instrumentality of space in digital access."

"While their study, which also considered factors of country competitiveness rating provides some understanding of the macro-level access conditions, it homogenises the digital divide across all cases in Africa."

"These conclusions are relevant for the spatial-economic analysis of ICT access in Kigali because they indicate that by reinforcing existing spatial disparities, differences in ICT access might mean that only the centres might develop into mainstream knowledge based, ICT hubs while the urban peripheries remain weak, less active in ICT."

"This paper adopts the framework of Dewan and Riggins [128], and applies the ITU indicator-type to explore the digital divide at an intra-urban level in order to understand digital access and urban level determinants."

Methodology

"SE variables are those that are relevant for ICT access across sectors but which in themselves are not ICT indicators; for example, level of education, household headship (female-headed households), youth population and employment."

"Access to physical facilities such as water and electricity could have some relation with ICT indicators, as equally expected of economic geographic/spatial factors."

"While ICT location quotient as applied in this analysis served to summarise the relative concentrations of digital access in Kigali, it lacks the methodological power to show any systematic spatial grouping of the sub-areas hence cluster analysis was employed to detect the various ICT classes, based on their aggregate LQ scores (LQICTagg)."

"Given the multiple socio-economic and spatial indicators explored as correlates of the digital divide, Principal Component Analysis (PCA) was applied to reduce the data in order to determine key factors of the distribution."

Results and discussion

"It maintains a relatively high performance in terms of household ownership of ICT assets (physical access), except fixed telephone, low across all clusters."

"The high ownership of mobile telephone corresponds to high level connection and use in homes, offices and schools, similar to the Low Basic-ICT cluster."

"Sectors with higher percentage of households headed by females were found to perform less in ICT ownership and use as evidenced by the LQICTagg correlation score corresponding to female household indicator."

"Given this pattern of representative correlation, the socio-economic and spatial variables of ICT access represented were interpretatively inferred to be the following categories: Infrastructure (represented by Electricity), Household structure/socio-economic activity (represented by Youth Employment), Urban agglomerative strength (Share of Total Population) and Planning status/settlement/housing formality (Proportion of Sector Planned)."

Conclusions

"This paper explored digital divide in Kigali, within the context of spatial and socio-economic correlates of digital access, and possible implications for policy intervention in improving ICT access."

"In order to establish the existence or otherwise of the digital divide it examined the digital access indicators of ICT asset ownership and use by location."

"The digital divide does exist in Kigali and it is manifested in two ways: ICT access dimensions (assets and use by location) and space, both embedded in existing socio-economic conditions."

"Analysis of spatial-economic correlates of ICT access in Kigali indicates that existing conditions that typically underscore the distribution of urban services and opportunities are replicated in ICT access."

"The spatial analysis of ICT access in Kigali provides scope for understanding ICT opportunities at the intra-urban levels."

4.1.15 Young People and the Digital Divide in Egypt: An Empirical Study [512]

This is a machine-generated summary of:

Badran, Mona Farid: Young people and the digital divide in Egypt: an empirical study [512]

Published in: Eurasian Economic Review (2014)

Link to original: https://doi.org/10.1007/s40822-014-0008-z

If you want to cite the papers, please refer to the original.

For technical reasons we could not place the page where the original quote is coming from.

Abstract-Summary

"This paper uses the ordered probit model to empirically assess the determinants of the digital divide in Egypt."

"The main contribution of the current study is in the construction of the Information and Communication Technology composite index for Egypt, namely the digital divide index designed to measure the urban and rural digital divide among youth in Egypt."

"Our results indicate that there is a positive correlation between urbanization and digital divide."

Introduction

"A few countries such as Mauritius and Zambia suffer from the positive gap, where rural internet use is higher than that in urban areas; the average difference is approximately 1.75 percent."

"This study also provides a number of innovative elements: (1) the construction of the ICT composite indices for Egypt: the Digital Divide Index (DDI) measuring the urban/rural DD among youth in Egypt based on the SYPE; and (2) an evaluation of the findings in light of the DD in order to provide appropriate policy recommendations on how to address the inequality between urban and rural areas in Egypt."

"This migration process, together with other factors, deepens the divide between urban and rural areas, and further negatively impacts the DD globally, and Egypt is not an exception to this rule (Beall and others [129])."

Current status of the information and communication technology sector in Egypt

"The structure of the access mode changed in July 2011 to include mobile internet and USB modem users (43 percent, with the number of internet users estimated at 37.7 percent in July 2012."

"In August 2012, mobile internet and USB modem became the most widely used access modes in Egypt (45 percent), followed by ADSL (36 percent), followed by dial-up and ISDN (10 percent), and ending with the leased line (9 percent)."

"It was also fuelled by the increase in FDI in this sector in the form of new 3G mobile licenses and other investments in off-shore services and in establishing IT companies."

"The ICT sector is considered one of the most important sectors attracting investment from MNCs (Multinational Companies), due to the high quality of available telecom infrastructure, at reasonable rates; the low cost of qualified labour; and the huge human resources pool (which increases by approximately 10,000 IT graduates per year)."

Literature review

"Quibria and others [130] analyzing PCs and internet use per capita find that GDP, education levels, and infrastructure are the most important drivers of these ICT's diffusion."

"Chinn and Fairlie [131] studied PCs and internet use per capita, and found that GDP, telephone density and regulatory quality (pro-market policies) are important determinants of these technologies."

"Previous research has focused on single measures as a proxy of the digital level of countries, such as computer per 1,000 inhabitants or internet users, and only a few have made use or have constructed composite measures or indices to proxy digital level as well as the digital gap (for instance, Husing and Selhofer [132])."

"In developing countries, population age and urban population are positively associated with ICT adoption, while internet costs are negatively correlated with ICT adoption."

Data and empirical model

"The SYPE asked young people "what they use the internet for", "who introduced them to it", and "where they use the internet"."

"Only 7.5 percent of young people use the internet."

"Young people use the internet primarily for general knowledge and social networking purposes (Population Council [133])."

"Around 20 percent of the internet-using young people also use it for educational purposes, and 40 percent for general knowledge."

"The vast majority of young people (60 percent) who use the internet were introduced to it by their friends."

"Young people primarily use either the internet at home (52.6 percent) or in an internet café (51.0 percent)."

"The DDI is constructed using the principle component analysis (henceforth PCA) technique and takes into account accessibility to internet, use of internet and computers, and mobile and landline phones."

Results

"Our results indicate that urbanization has a positive association with the DD, where the DD is low in urban areas, where internet penetration rates are higher compared to rural areas."

"The overall impact of living in an urban area compared to a rural area on the probability of DDI is positive."

"This analysis allows us to examine whether the variables of interest (urbanization) are robust to be independent of any additional sets of explanatory variables included in the empirical specification."

"To test whether urbanization influences the DDI, if other variables are included, the EBA method is applied."

"The results regarding the effect of urbanization on the DDI are robust, independent of the explanatory variables included in our specification."

Conclusion

"The current paper investigates the DD in Egypt and constructs a composite indicator of the DD across urban and rural areas using the data from the 2009 SYPE."

"Increasing their numbers and supporting these public points of access such as internet cafes with the necessary infrastructure to meet the demand of the young population in rural areas is also important."

"The introduction of new business models for mobile services such as pre-paid services, together with access to internet using 3G technology, namely the mobile broadband, can reduce the DD."

"In rural areas, users get access to internet through their mobile phones, resulting in a higher cost of usage."

"In order to reduce the gap between rural and urban areas, and to improve internet access, people need to be provided with the necessary equipment, such as personal computers in public points of access or internet cafes, etc"

4.1.16 Exploring the Multi-level Digital Divide in Mobile Phone Adoption: A Comparison of Developing Nations [513]

This is a machine-generated summary of:

Vimalkumar, M; Singh, Jang Bahadur; Sharma, Sujeet Kumar: Exploring the Multi-Level Digital Divide in Mobile Phone Adoption: A Comparison of Developing Nations [513]

Published in: Information Systems Frontiers (2020)

Link to original: https://doi.org/10.1007/s10796-020-10032-5

If you want to cite the papers, please refer to the original.

For technical reasons we could not place the page where the original quote is coming from.

Abstract-Summary

"Despite the progress, there remains a persistent disparity in mobile phone usage among users, and between users and non-users, in developing countries."

"The literature on the digital divide has paid limited attention to the multi-level nature of this gap in mobile phone adoption and use, particularly in developing countries."

"This research paper provides a comparative analysis of this divide in mobile phone usage in eight developing nations from Africa and Asia."

"Further, we find that there is a larger digital divide in terms of mobile phone usage in Asian countries when compared to African countries."

"The study recommends that policymakers and international agencies exercise caution while developing one-size-fits-all cross-country policies to address the multi-level digital divide."

Introduction

"To the best of our knowledge, only a handful of digital divide studies have considered the mobile phone a focal technology (Chigona and others [134]; Donner [135]; Donner and others [136]; Rice and Katz [137])."

"As mobile phones are becoming the principal means to access high-speed internet, it is short-sighted to consider only 'basic phone feature usage' (first-level digital divide) without accounting for 'mobile phone internet usage' (second-level digital divide)."

"Only a few studies have considered the second-level digital divide with respect to mobile phones, especially in the context of developing countries (Singh and Vimalkumar [138]; Song and others [139])."

"To provide a deeper understanding of the multi-level digital divide among mobile phone users in developing countries, this study presents a systematic cross-country analysis of eight developing countries in Asia and Africa, four from each continent: Bangladesh, India, Indonesia, Pakistan, Kenya, Nigeria, Tanzania and Uganda."

Literature Review

"The Organization for Economic Cooperation and Development (OECD) defines the digital divide as "the gap between individuals, households, businesses and geographic areas at different socio-economic levels with regard to both their opportunities to access ICTs and to their use of Internet for a wide variety of activities" (OECD [140], p5)."

"Influential studies in this domain categorize the digital divide into two broad levels (Dewan and Riggins [128]; Wei and others [9])."

"Studies on the digital access divide explore the determinants of differences in IT adoption at the individual, organizational and global levels."

"These studies use country-level secondary data for their analysis (Appendix 1), and none of them provide a micro-level view of the digital divide."

"Studies have also shown that there is a multi-level digital divide in some of African countries like Nigeria (Okunola and others [141]) and that ICT use is mainly influenced by habit and ICT skill (Goncalves and others [142]; Kapoor and others [143]; Rana and others [144])."

Countries for the Study

"While all the eight countries are developing economies, they differs in terms of their consumer profile and technology adoption behaviour (the ratio of early adopters to the larger population)."

"Further, with more than 83% of the population being literate and with the highest per-capita income among the Asian countries considered for this study, Indonesia has the highest mobile technology adoption rate, with 120 mobile subscriptions per 100 people."

"Contrarily, Pakistan, with the least literacy rate - less than 61% (Intermedia [145]) - has the lowest mobile technology adoption rate among the Asian countries considered in this study."

"Kenya, despite its extreme poverty, leads the African nations selected for this study in terms of mobile phone adoption."

"The other two countries considered for this study, Tanzania and Uganda, fall between these two countries in terms of digital and internet adoption."

Research Framework

"As our focal technology is mobile phones, we examine the disparity between the individuals who have access to mobile phones and those who do not (Chigona and others [134]; Donner and others [136])."

"We study two levels of the digital divide: in line with traditional digital divide studies, we first compare mobile phone users to non-mobile phone users."

"In the second level, we compare the disparities among mobile phone users by comparing those who use applications that require access to the internet to those who do not."

Research Methodology

"To compare the mobile phone and internet use digital divides the eight developing countries considered, we undertake an empirical analysis."

"The survey collected demographic information along with data on mobile phone usage."

"The data on mobile phone usage included information on whether the individual owned a mobile phone or borrowed it from someone else (access to the technology), if she used it, and for what purposes she used the technology (making or receiving voice calls, sending text messages, capturing pictures, browsing the internet, mobile banking, etc) Based on their responses, individuals were classified into three categories."

"The survey data on this was collected by asking two questions: "Do you personally own a mobile phone?""

"Age is a key factor found to impact digital access and use of IT, the internet and mobile phones (Boase [146]; Chen [147]; Rice and Katz [137])."

Results and Key Findings

"The results also indicate no structural discrimination in terms of mobile phone and internet use in Kenya, and this African country leads the group in terms of digital equality."

"The regression results also show that almost all the variables considered are significant, and that variables such as gender, age, education and marital status cause wide disparities in mobile phone use."

"Results show that a primary job does not have a significant impact on mobile phone adoption in most of the countries in the study."

"Socio-demographic variables like age and place of residence do have a significant impact on the digital divide in many countries, but the results are inconsistent and the magnitude of the impact is low compared to factors like owning a phone or needing assistance."

"The effect sizes of age and place of residence are very small, indicating that these variables do not impact mobile phone use in these countries (Lin and others [148])."

Discussion and Research Implications

"Considering the pressing need to understand the multi-level disparity in mobile phone and internet usage, in developing economies in Asia and Africa, this study conceptualized a multi-level mobile phone digital divide framework."

"Digital divide studies should not consider mobile phone adoption as the sole measure of digital inclusion."

"The studies need to consider the nuances within mobile phone usage to generate a deeper understanding of the digital divide."

"In countries where the gender-based digital divide is less significant (for instance, in Kenya or Tanzania), women are as likely to own a mobile phone as men, or are sometimes more likely."

"The results show that there are significant variations in the adoption of mobile phones and the internet among the countries considered for this study."

Conclusion

"The objective of this research article was to compare the digital divides in different developing nations and the socio-demographic factors influencing these divides."

"The study found that the factors affecting the digital divide are not uniform across countries."

"While education and place of residence influenced the digital divide more in some countries, gender and marital status had a significant bearing in some other countries."

"Further, this research article found that all mobile phone users are not the same and digital inclusion initiatives should not stop at increasing access to mobile phones."

"The study recommends that policymakers and international agencies be careful while strategizing cross-country policies to address the multi-level digital divide using a one-model-fits-all approach."

4.1.17 Covid-19 and Mobile Payment in Belgium: Closing the Digital Divide or Just for the Young, Social, and Impulsive? [514]

This is a machine-generated summary of:

Hellemans, Johan; Willems, Kim; Brengman, Malaika: Covid-19 and mobile payment in Belgium: Closing the digital divide or just for the young, social, and impulsive? [514]

Published in: Electronic Commerce Research (2022)

Link to original: https://doi.org/10.1007/s10660-022-09655-4

If you want to cite the papers, please refer to the original.

For technical reasons we could not place the page where the original quote is coming from.

Abstract-Summary
"Experts and industry reports agree that the COVID-19 crisis spurred the adoption of new retail technologies, like mobile payment."

"Academic mobile payment research has focussed almost entirely on the different building blocks of technological acceptance models, like perceived usefulness and ease of use, and their role in explaining intention to use."

"In this Belgian study, we investigate the evolution in mobile adoption based on survey data from 2019 to 2020 (2019: N = 897; 2020: N = 895)."

"We examine differences in the profile of mobile payers in terms of their socio-demographics, retail, and social media behaviours."

"The pandemic triggered a clear uplift in mobile payment users between 2019 to 2020."

"We also observe a clear association between general impulse buying tendency and mobile payment."

"Mobile adoption is related to the use of Instagram and Facebook."

Introduction
"This study aims to answer the following questions, which we elaborate on in the subsequent sections: (1) How did the adoption of mobile payment change over the corona-crisis? (2) Which consumer groups have adopted mobile payment, and which have not? (3) Which individual differences and retail contexts beyond demographics might explain adoption further? (4) what is the role of social media in the adoption process?"

"Researchers studying mobile payment have focused entirely on technological acceptance models and which of those models' theoretical mediating constructs are most likely to influence intention to adopt mobile payment services [149–151]."

"By investigating differences in the profile of mobile payers in terms of their socio-demographics, frequency of patronizing convenience retail channels, social media usage, and their underlying general impulsive buying tendency, we provide insight into the mobile payment acceptance research domain in selecting potential essential moderators and context factors."

Literature
"Researchers interested in the field of mobile payments mainly rely on one or more competing IT adoption models like the technology acceptance model [TAM] of Davis [152], the unified theory of acceptance and use of technology UTAUT by

Venkatesh and others [153], its successor UTAUT2 [154] to explain intention to use mobile payment technology."

"They also reported a higher incidence of mobile payment adoption for younger age groups and men."

"Higher education and income showed higher odds for mobile payment adoption [155]."

"A recent study collected data from a representative sample of the Norwegian adult population (n = 2202) to investigate individual differences in the adoption of online, mobile, and contactless payment and the willingness to use social media for money transfers."

"Within the research applying technology acceptance models, there is a clear indication that social influence, subjective norms, or social norms are influential factors in predicting the intention to adopt mobile payment [153]."

"Mobile payment adoption is higher among social media users than those who do not use it (regularly)."

Methodology

"Quotas were imposed on age, gender, and region to obtain similar sample structures between 2019 and 2020."

"Responses about socioeconomic characteristics, such as education, employment status, and income, can show a high level of association and relate to other demographic variables like age, they are also often grouped in a social grade composite score (SGSS) constructed from socioeconomic characteristics."

"Data on respondents (social) media access were obtained for the following media channels and platforms: (1) Facebook, (2) Twitter, (3) Instagram, and (4) YouTube, as well as (5) internet browsing, (6) digital newspapers, (7) digital magazines, (8) TV, (9) radio, (10) newspapers, and (11) magazines."

"Participants rated how frequently they have used mobile (contactless) payment (apps) over the last 12 months on a single 7-point frequency scale, going from "not at all" (i.e., score 1) to "Once a week or more" (i.e., score 7) in a similar vein like Pal, Herath, and Rao [156]."

Results

"Exceptions are the AMP scores for the youngest age group, students, and independent professionals showing no significant increase."

"Exception to the rule are the AMP scores for the youngest age group, students and independent profession showing no significant increase."

"The SGCS as a social-economic factor shows a significant positive effect (Exp(B) = 1.13, CI95: (1.06–1.21)), supporting hypothesis six."

"We also noted significantly lower odds for women (Exp(B) = 0.68, CI95: (0.54–0.86)) and higher odds for those with kids (Exp(B) = 1.35, CI95: (1.06–1.70)), but these effects did not withstand in a concurrent model when introducing underlying shopper profiles."

"Further observation shows that the odds for Facebook users (Exp(B) = 1.38, CI95:1.03–1.84), and Instagram users (Exp(B) = 1.93, CI95:1.34–2.78) are significantly higher."

Discussion

"Our hypotheses regarding age and social grade effect remained valid in concurrence with other profile variables showing a higher occurrence of mobile payment among younger ages and higher socioeconomic profiles."

"Given the regional difference remains after controlling for socio-economic profile indicates that other regional cultural differences like Hofstede's cultural dimensions, could be responsible for the difference in the adoption of mobile payment in both regions."

"This robust finding confirms previous research hypothesizing impulse buying and mobile payment to be related [157]."

"Our findings support the notion that mobile payment adoption varies across different economic, cultural, social, and retail context [158]."

"In relation to our fourth research question, we assumed social media to be an indicator of mobile payment."

"A closer look at our data revealed a clear association between general impulse buying tendency and mobile payment and between mobile payment and Facebook and Instagram users."

Implications

"With such a sudden growth in adoption, it can be expected that not everyone has the necessary digital and financial competencies to use mobile payment rationally or is aware of the potential risks."

"In terms of closing the digital divide gap and ensuring financial inclusion they need to focus on the profile of none-users, meaning older and less educated people."

"The fact that mobile payments for impulse purchases may be more readily used indicates that retailers are commercially well advised to offer mobile payments as an incentive to facilitate such transactions not only online."

"Given the higher adoption amongst online shoppers, it is also paramount within an omnichannel strategy to offer mobile payment options."

"For retailers looking for an omnichannel strategy or who want to increase their efforts in social selling, mobile payment is not even an option; it is a prerequisite."

"It could be the opposite for financial service providers bringing mobile payments to market."

Limitations and suggestions for future research

"In line with previous research [149, 150, 159–161], researchers need to be aware that with a maturing mobile payment market that; (1) earlier findings might not apply anymore and could be time specific, (2) should focus on differences within mobile payment options, (3) should switch from intentions to actual use or continuance use as the primary dependent variable, (4) could start to investigate preference or market/usage share as the main variable of interest, (5) differentiation in a mature market is not only a question of superior technology or product but also of branding."

"Both ages and socioeconomic-related moderators should be prime candidates to be incorporated within technology acceptance models investigating mobile payment."

"Besides background variables like socioeconomic profile and age, the observed differences in our data warrant further investigation into specific shopper orientation-related differences like general impulse buying tendency as moderators in the field of mobile payment using technological acceptance models."

4.1.18 Assessing the Role of Age, Education, Gender and Income on the Digital Divide: Evidence for the European Union [515]

This is a machine-generated summary of:

Elena-Bucea, Anca; Cruz-Jesus, Frederico; Oliveira, Tiago; Coelho, Pedro Simões: Assessing the Role of Age, Education, Gender and Income on the Digital Divide: Evidence for the European Union [515]

Published in: Information Systems Frontiers (2020)

Link to original: https://doi.org/10.1007/s10796-020-10012-9

If you want to cite the papers, please refer to the original.

For technical reasons we could not place the page where the original quote is coming from.

Abstract-Summary

"Because of the digital divide's complexity, a multivariate approach was applied - factor analysis with oblique rotation, which resulted in two distinct dimensions: e-Services and Social Networks."

"To test the significant differences of European Union positioning and European Union disparities, Multivariate Analysis of Variance and Squared Rank Test were computed."

"Findings show that e-Services adoption is influenced primarily by the education level of individuals, while Social Networks adoption is more affected by individuals' age."

Introduction

"The European Commission (EC) wants to ensure, until 2020, the growth of an information society across the members of the European Union (EU) by sponsoring innovation and knowledge sharing among member states, making the most of ICT, encouraging innovation and smart solutions, and creating the right structure to deal with the social challenges of the European and global society (European Commission 162)."

"The goals of this study are to appraise the European domestic digital divide for its 28 members, and to analyze Europeans' demographic characteristics, scrutinizing: (1) the main dimensions of digital divide of the 28 members of EU; (2) the main socio-demographic characteristics that most impact the digital divide in the EU; (3) the main socio-demographic characteristics that potentially contribute most to the digital divide, and those that lead to greater asymmetries in the use of ICT, for each country of EU."

"The third section is dedicated to the domestic digital divide and describes four socio-demographic drivers that influence ICT adoption."

Literature Review

"The most common term to describe these ICT disparities is "digital divide", and although fairly old, the following definition provided by the Organization for Economic Co-operation and Development (OECD) is still one of the most suitable to clarify this phenommena, defining it as "the gap between individuals, households, businesses and geographic areas at different socio-economic levels with regard both to their opportunities to access information and communication technologies (ICTs) and to their use of the Internet for a wide variety of activities" (OECD 163)."

"The same author remarked that there is a shift from the first-order digital divide that represents unequal access to ICT, to a second-order digital divide embracing inequalities of skills and usage."

"According to Çilan [164] and Sadowski (Sadowski [165]), the level of ICT access and usage – first-order and second-order digital divides – vary among individuals within countries and digital discrepancies inside nations can justify the differences among them."

Digital Divide's Drivers

"Age is one of the most popular drivers when it comes to ICT adoption, with the United Nations [166] stating that the digital divide is also an age matter."

"There are some previous evidences that this second group can present computer anxiety and technophobia given their age-specific characteristics and their non-intuitive ICT skills and usage, exhibiting social conditions such as: disabilities/chronic health problems, living alone, single or widowed, presenting "lower level of education than ICT users" (van Dijk [167]; Vroman and others [168], p. 165)."

"Educational digital divides persist in contemporaneous society (United Nations [169]) affecting digital development, justified by the complexity of ICT, which is considered as "a major obstacle for adoption" (Cruz-Jesus and others [170], p. 74)."

"Van Deursen and others ([171], p. 262) agree that "education is an important predictor for the types of activities that people engage in online activities", which is in tune with Nishijima and others ([172], p. 15) who consider that the educational attainment affects access to ICT as it is "positively related to ICT usage in both developed and developing countries" explaining most of the disparities in personal skills for ICT utilization."

"The main digital disparities are more visible in developing nations, where "the gender gap in accessibility and use of ICTs increases the existing gender inequalities" (Mumporeze and Prieler [173], p. 1287)."

Methodology

"Multivariate Analysis of Variance (MANOVA) was chosen to assess the existence of different digital adoption levels among the 28 EU member states, when looking at different socio-demographic sub-segments."

"MANOVA tests for each socio-demographic segment (age, education, gender, and income), if the sub-segments share the same mean toward e-Services and Social Networks."

"As for education, all three sub-segments present different levels of adoption for both dimensions, as well as for income, except for the first quartile of income group and the 50% around the median group, who do not present statistically significant differences in the level of adoption of e-Services."

"In the case of e-Services adoption, education is the socio-demographic indicator that most affects the digital disparities (23 countries), except for other EU members (5 countries) with age as the main indicator responsible for digital divide for e-Service."

Discussion

"Gender in EU28, in opposition to developing countries (Mumporeze and Prieler [173]) does not present significant differences in terms of digital adoption toward the two dimensions of the study."

"As for Social Networks, there is statistical evidence at a confidence level of 99% that all income categories have different levels of adoption."

"To identify which of the segments – age, education, gender, or income – present significant asymmetries – according to the Conover's variance test there is at least one socio-demographic segment with asymmetries significantly different from the other three segments."

"On the e-Services dimension, among the four socio-demographic segment there is at least one pair of them that present unequal asymmetries, statistically significant with a probability of 90% to be true."

"For Social Networks, age presents statistically significant differences toward all of the other three segments, displaying the largest disparities."

Implications

"Scrutinizing the divides of the 28 EU countries in 2016, namely: age, education, gender, and income, there are still some digital disparities that need to be closed."

"Regarding the multivariate technique that was applied, findings show that an oblique rotation when analyzing the digital divide through factor analysis better represents, the reality: e-Services dimension is correlated with Social Networks, sharing 36% of variance within each factor."

"As ICT became a commodity, the effort to narrow within and between digital divides was through the implementation of initiatives that could bridge the generational gap in case of Social Networks and that could harmonize the educational level within countries, raising awareness of the benefits of the electronic services."

"If each EU country would analyze the within country digital divide by socio-demographic segments and act according to its necessities, the EU would benefit more as a whole, narrowing at the same time the international and domestic digital divide, improving digital literacy of Europe."

Conclusions

"On the 14 variables of the study, two dimensions were identified to represent the digital divide, e-Services and Social Networks."

"For e-Services, education presents the biggest disparities, and for Social Networks, age classes are clearly the most unequal."

"There is statistical evidence that the segments present disparities between them, with a confidence level of 90% for e-Services and 99% for Social Networks, revealing greater evidence of inequalities in this dimension."

"When analyzing the countries' asymmetries, the main drivers for digital divide in e-Services are mainly education and age."

4.1.19 Capabilities in, Capabilities Out: Overcoming Digital Divides by Promoting Corporate Citizenship and Fair ICT [516]

This is a machine-generated summary of:

Busch, Thorsten: Capabilities in, capabilities out: overcoming digital divides by promoting corporate citizenship and fair ICT [516]

Published in: Ethics and Information Technology (2010)

Link to original: https://doi.org/10.1007/s10676-010-9261-3

If you want to cite the papers, please refer to the original.

For technical reasons we could not place the page where the original quote is coming from.

Abstract-Summary

"This conceptual article discusses strategies of corporations in the information and communication technologies (ICT) sector and their role in the conflict over access to knowledge in the digital environment."

"Its main hypothesis is that ICT corporations are very capable actors when it comes to bridging digital divides in both developed and developing countries—maybe even the most capable actors."

"It is argued that ICT corporations could use their capabilities to help citizens gain sustainable access to knowledge in order to enable them to lead self-sufficient lives."

"In a nutshell, capabilities are presented as both the input (capabilities of ICT corporations) as well as the output (capability building for empowering citizens) of corporate strategy-making focusing on fair ICT."

"Corporate citizenship is put forth as the theoretical concept bridging corporate strategies and access to knowledge: If ICT corporations act in accordance with their self-understanding of being 'good corporate citizens', they could be crucial partners in lessening digital divides and helping citizens gain access to knowledge."

"From the perspective of 'integrative economic ethics' (Ulrich 2008), it is argued that ICT corporations have good reason to actively empower citizens in both developed and developing countries by pursuing 'inclusive' strategies in many fields, such as open-source software development."

"That way, ICT corporations could enable, support and provide citizens with capabilities enabling them to help themselves."

Introduction: in search of ethically legitimate ICT strategies

"The hypothesis presented here is that 'open' modes of production of digital goods, such as open-source software or open access publishing, could be considered a more ethically legitimate way of producing digital goods and services than proprietary approaches because they offer the potential to support the building of capabilities and open resources that can help bridge digital divides in both developed and developing countries."

"Empirically, we can observe that many transnational ICT companies seem to be stuck in the classic 'two worlds' theory of economic interests versus moral claims: While all big ICT players claim to be 'good corporate citizens', and many actually have their own foundation that aims at bridging digital divides (for example, see Sun Foundation [174]; Microsoft [175]), they still cling to their classic, proprietary business models which rely on exclusive licensing, proprietary source code and standards, as well as prohibitive pricing, thus effectively excluding many people from access to knowledge and knowledge creation tools."

Corporate citizenship and the approach of integrative economic ethics
"This question needs to be addressed in practical, public discourse between a company and its stakeholders on a case-by-case basis: When it comes to corporate responsibility, 'the right thing to do' cannot be determined by ivory-tower experts, but has to actually be discussed in each case in a process of fair deliberation by the company and its stakeholders."

"A rough definition of corporate citizenship in relation to a company's core business model would be: A good corporate citizen is a company willing to proactively engage in unconditional discourse with all its stakeholders that wish to voice a concern—or, more simply put, a corporation that adheres to the moral principle."

"While a good corporate citizen's responsibility lies, first and foremost, in its willingness to adhere to the moral principle, the idea of unconditional discourse with all stakeholders brings with it the problem of defining reasonable claims to corporate action."

Corporate citizenship and fair ICT
"This prisoner's dilemma applies to cases as diverse as software standards, protocols, and file formats, which can be either open or proprietary; licensing models in the software, music, or video game industry, where it used to be that consumers actually bought and owned a product, whereas nowadays, the content providers would like to just license the product to their consumers for a finite amount of time; intellectual property and copyright issues, where in the last decades the time frame of copyright protection has been prolonged significantly to protect existing rights holders, but without necessarily providing an ex-ante incentive to foster innovation; patents, which are especially important in and driven by the agricultural and bio-tech industries; and on the level of the infrastructure of the Internet itself, net neutrality currently is a hotly debated issue."

Conclusion
"One could argue that modern, liberal societies need to start a public discussion on ICT strategies, the 'future of ideas' (Lessig), and how much protection they need in order to be sustainable."

"ICT strategies, one could argue, should not be seen as an end in themselves, but as means to a sustainable knowledge society by way of capability building."

"As Ivan Illich [176] put it: 'A convivial society would be the result of social arrangements that guarantee for each member the most ample and free access to the tools [ICT] of the community and limit this freedom only in favor of another member's equal freedom.'"

"Applied to the question of corporate citizenship and ICT strategies, this would mean that the actors most capable of changing the rules of the game should change them in a way that fosters the capabilities of their fellow citizens to lead self-sufficient lives."

4.1.20 Overcoming the Digital Divide with a Modern Approach to Learning Digital Skills for the Elderly Adults [517]

This is a machine-generated summary of:

Blažič, Borka Jerman; Blažič, Andrej Jerman: Overcoming the digital divide with a modern approach to learning digital skills for the elderly adults [517]

Published in: Education and Information Technologies (2019)

Link to original: https://doi.org/10.1007/s10639-019-09961-9

If you want to cite the papers, please refer to the original.

For technical reasons we could not place the page where the original quote is coming from.

Abstract-Summary

"Older adults, not due to their frailty or age, often find it difficult to use digital devices like smartphones, and they often lack the basic digital literacy required to use multimedia interactive devices with touchscreen technology."

"Older adults from four European countries participated in a two-phase process: playing interactive games on a large touchscreen tablet and learning how to use a smartphone to access digital services."

"The attitudes and the difficulties associated with adopting the skills to operate the device and use the digital services were observed and discussed."

Introduction

"The importance of digital skills among older adults became apparent with the appearance of Internet-based services, such as e-health, e-government, e-learning and others, as they are considered as enablers that extend the period during which the elderly can stay in their own homes before the need to move to retirement homes."

"The problem of a lack of basic digital literacy among older adults is related to the fact that older adults do not have the knowledge to effectively use digital devices where the communication with the user normally involves touchscreen technology."

"The problem of learning digital skills among older adults has its origins in the way they learned in their youth and during their professional lives, as most of them had no opportunity to interact with modern digital technology and to create the knowledge for effective use at the time when interacting with the Internet became part of everyday life."

Theoretical background

"In order to be convinced that the elderly do not refuse to use digital technology when it is presented to them in a way that is close to their views, the initial approach of the presented study was to look at categories related to the behaviour of older adults with respect to digital technology, e.g., touchscreen tablets, by observing the emotions expressed by the participants during and after the experiment of playing games."

"The study tried to find answers to several questions, with the following considered to be the most important: a) Are there any issues or difficulties with learning and training digital skills for older adults when they are exposed to multimedia touchscreen devices with large screens? b) Will the older adults' express positive emotions towards the digital technology and the gaming? c) Does the population of older adults display differences in the adoption of digital technology due to the different digital environments where they live? d) How do the skills acquired during gaming influence the adoption of the digital skills required to manage the use of digital services with a mobile device? e) Did the experience of gaming have any influence on the effectiveness of using a smartphone?"

The study

"The study was carried out in two phases: groups of older adults who played the games and participated in the interviews and answered the questionnaires, and the training courses designed for the older adults to learn how to use smartphones."

"The focus of the study was to find out whether the playing of games on a large touchscreen tablet influenced the skill level and the knowledge adopted during the course of smartphone learning regarding the efficiency and the speed of performing the task in the knowledge environment."

"Both participating groups, i.e., the group that had no experience of playing games on giant touchscreen tablets before the course and the group that did not play games before the course lessons were asked to solve the same tasks, presented below.1.We would like you to apply for financial social assistance with the on-line Social Service."

Data collection and analysis

"It was found that the learning difficulties experienced by the participants can be summarized in three types of players' actions: Request for support from the mentor/facilitator in cases when the game appeared to be locked, Support/help to be provided by the mentor/facilitator in cases when the touchscreen of the tablet did not react to the participant's movement, A request for help from the gaming partner in the pair."

"Older adults from European countries do have positive emotions towards playing digital games on a large touchscreen tablet, so the answer to the research question b) is "yes"."

"Of the study the objective was to find out how game playing and learning to use the touchscreen tablet influenced the adoption of the digital skills required to access and use digital services."

"The collected data gave very similar findings and confirmed the positive influence of the game playing on touchscreen tablets for adopting digital skills by older adults."

Discussion

"This type of problem re-occurred quite often and the facilitator's support and help was sought again, but the majority of the participants did not give up the gaming and they continued to play."

"These findings give a positive answer to the second research question: In our experiments, older adults showed positive emotions towards the gaming."

"The more efficient performance of the tasks displayed by the members of the gaming group clearly showed that playing games could be a very useful tool in facilitating the training of digital skills for older adults learning to use a touchscreen device like a smartphone."

"Fun experiences during game playing clearly help with quicker and more successful, relaxed learning of digital device use, which leads to faster skills adoption as well."

Concluding Remarks

"Several researchers have reported (Niehaves and Plattfaut [177]) that the elderly have different attitudes, beliefs and intentions when it comes to digital technology usage, which is manifested as fear and a refusal to use it."

"Other research in the area (Hill and others [178]) pointed out that digital technology is a life facilitator for the elderly, by enabling them to overcome physical barriers, such as distance, personal mobility, social contacts and general well-being."

"All these challenging life factors can be addressed through the use of digital technology, which can change the position of elderly people with respect to the digital divide."

"Future research is also expected to create tailor-made games that will help older learners to overcome their lack of confidence and the feeling that the use of touchscreen devices is risky and complicated."

4.1.21 A Framework for Optimal Techno-Economic Assessment of Broadband Access Solutions and Digital Inclusion of Rural Population in Global Information Society [518]

This is a machine-generated summary of:

Krizanovic Cik, Visnja; Zagar, Drago; Grgic, Kresimir: A framework for optimal techno-economic assessment of broadband access solutions and digital inclusion of rural population in global information society [518]

Published in: Universal Access in the Information Society (2017)

Link to original: https://doi.org/10.1007/s10209-017-0560-x

Copyright of the summarized publication:

If you want to cite the papers, please refer to the original.

For technical reasons we could not place the page where the original quote is coming from.

Abstract-Summary
"To reduce the digital divide, the issues related to the deployment and usage of broadband solutions in rural areas are analysed in this paper."

"To define optimal strategies for the deployment and adoption of rural broadband solutions, the application of an extended techno-economic assessment process is considered."

"Within the framework, the regression analyses of factors relevant to broadband solutions deployment and adoption in rural areas are defined."

"The efficiency of framework usage in techno-economic assessment process and in finding optimal broadband solutions is demonstrated for a case study rural scenario."

Introduction
"The most important goals and targets of this agenda include the reduction of the existing digital divide, the uniform development of broadband access networks in all rural and urban areas, and the provision of affordable broadband services to all residents."

"The proposal of optimal broadband solutions is supposed to induce further implementation and adoption of broadband technologies and services in rural areas and to improve universal access to digital information and knowledge."

"Unlike the content of the paper presented in [179], which focused on an overview of the competitiveness of different broadband access solutions assessed by the extended techno-economic modelling processes, the focus of this paper is the thorough description of the proposed framework, which extends the existing techno-economic modelling process."

"The availability of broadband access infrastructure and the acquisition of necessary digital skills should enable further broadband services usage, as well as the acquisition of digital knowledge in any possible formal, non-formal or informal way."

Broadband determinants

"Broadband solutions can be considered in view of the following processes: adoption, development, deployment (i.e., implementation), and selection of broadband technologies and services [180]."

"Both the broadband services adoption process and the broadband technologies deployment process are interrelated with the broadband solutions selection process."

"The broadband adoption, development, deployment, and selection processes are influenced by numerous factors."

"In broadband adoption, development, deployment, and selection processes, different broadband market participants are engaged."

"Although all national broadband plans seem to cover all levels of information and communication technology targets, i.e., the adoption, development, deployment, and selection, the scope of the plans varies considerably in practice, particularly regarding the depth of demand-side targets."

"To further foster the adoption and deployment processes of broadband solutions, analyses of the most important factors affecting these processes are essential."

The digital divide problem

"Considering the results of the conducted analyses for a given case study, it is obvious that when aiming at the identification of appropriate strategies for increasing the broadband adoption volume in rural areas and reducing the existing digital divide, the digital divide reduction methods should be closely considered and applied."

"A. One method for the reduction of digital divide is based upon the target that proposes a more than 40% increase in the availability of affordable broadband solutions worldwide by 2020."

"B. A second method for the digital divide reduction is based upon the target that states that broadband services should not cost more than 5% of the average monthly income in developing countries by 2020."

"C. A third method for the digital divide reduction is based upon the target that proposes that the 90% of the rural population worldwide should be covered by broadband services by 2020."

Technical and economic aspects of broadband implementation

"Techno-economics assessment enables the prediction and forecast of the most important risks related to uncertain market conditions that affect the cost-effectiveness of the considered business models; The techno-economic assessment enables the search for the most cost-effective business solutions since this guarantees that the operator will achieve the most benefits if the best business model is implemented; Since broadband access solutions differ with regard to their features, the techno-economic assessment process enables the fair comparison of different solutions with regard to the technical or other advantages and constraints of the solutions under the equal market conditions."

"In [181], the techno-economic modelling of industrial architectures was conducted, and the roles for different actors in telecommunications markets were defined to answer the question: WHO invests in broadband access?"

A framework for extended techno-economic assessment process of rural broadband business models

"To encourage further implementation of broadband access network infrastructure in rural areas and to reduce the existing digital divide, it is necessary to choose cost-effective broadband access solutions [182] and to deploy the best business models for operators."

"The analyses conducted within the proposed framework produce a set of criteria that are entered into the standard techno-economic modelling process so that an extended assessment process of business models for broadband deployment can be conducted afterwards."

"In [179], the main focus was the overview of the competitiveness of different broadband access solutions given as a result of the extended techno-economic assessment processes."

"When the most important factors that affect the total number of broadband users should be determined, it is possible to use the regression analyses of the available data."

"The total number of broadband users is assumed to depend on various technical, economic, demographic, and sociological factors and their combinations: To identify the most important factors for broadband adoption in rural areas, all data typical for rural users should be included in the regression analyses."

"Aiming at reducing the digital divide, the main idea of the proposed framework, which extends the standard techno-economic assessment processes, was to include the most important factors for rural broadband adoption, defined using regression analyses, in the standard techno-economic assessment processes."

"The results of the extended techno-economic process should provide a more detailed evaluation of chosen broadband solutions."

"These factors should therefore be considered within the techno-economic assessment processes to find the most influential factors in the given scenario."

"The following model can be applied: The aim of this analysis is to determine the most important socio-demographic factors that affect the total number of broadband users."

"The following techno-economic factors, i.e., service price, quality of service, traffic volume, and broadband access speed, have been selected for further analyses for several reasons listed hereafter: First, the given variables enable a fair comparison of various broadband access solutions with regard to the techno-economic advantages and constraints of the solutions under equal market conditions (i.e., within the 'net neutrality' principle proposed by the EU Agenda) since these variables can be included in the planning process of every possible type of access solution (fixed, mobile, wired, and wireless); Furthermore, once defined, these variables can be easily implemented in the access network planning and modelling processes conducted within the standard techno-economic assessment processes;

Finally, the given variables are defined based on the parameters that are most commonly used in every broadband service's promotion process in every telecommunications market since the defined broadband service packages differ based on the available access speed, traffic volume, quality of service, and price range."

"Considering the demands on available service prices, service packages, and access speeds initially determined based on historical data, the network can be further planned for regarding actual user requirements, which, according to [183], reduces the per subscriber costs for broadband implementation."

"The questionnaire should be composed of questions scaling the range of importance to the potential broadband users of the defined techno-economic factors for broadband adoption."

"The links between socio-demographic and techno-economic factors are important for proposing guidelines that define what should be done to promote broadband adoption in rural areas despite the unfavourable socio-demographic structure of the rural population."

"The values of coefficients that describe the impact of certain combinations of the most important socio-demographic factors, established in the previous step, and the techno-economic factors (e.g. age and access speed, income and service prices) should be determined: Regarding the analysis results, the most significant values of the coefficients β_K for the combinations involve four chosen techno-economic factors: service price, quality of service, traffic volume, and access speed (denoted hereafter as β1, β2, β3 and β4, respectively)."

"These beta coefficients, comprising the vector beta:are included in the further techno-economic assessment processes of operator business models and are considered within the best strategies proposal process, as will be presented hereafter."

"Unlike the results of the econometric analyses of end user data given, for example, in [184–186], the results of the analyses carried out within the framework, i.e., the determined broadband user requirements, are directly involved in the further techno-economic modelling process."

"The main focus of the extended techno-economic assessment process used in this paper is to meet rural users' requirements, as well as the techno-economic evaluation of rural broadband business models."

"The technical requirements in the preparation of the extended techno-economic model include the determination of relations between a set of demands on service capacity, access speed and coverage area, and the necessary network infrastructure."

"The impact of the techno-economic factors, i.e., service price (C), quality of service [which is analysed considering the service coverage area (PP)], traffic volume [i.e., available bandwidth (PR)], and broadband access speed (BR), on the total number of users (BK), is determined using regression analysis of the available data: The influence of certain factors on the total number of users is estimated for the previously obtained β1, β2, β3, and β4 coefficient values: The expression for determining the total number of users is given by: Once the beta coefficients and their values in the model are known, the number of users can be estimated for new values of techno-economic factors: Furthermore, the possible changes in the values of the techno-economic factors, denoted hereafter by delta symbols (i.e., ΔC, ΔPP, ΔPR, and ΔBP), that may occur in a future period are considered."

"The expression for determining the impact of changes in the values of the techno-economic factors on the change in the total number of users (ΔBK) can be formulated as: Several components of the techno-economic assessment process are directly affected by the total determined number of users and the possible changes to this number."

"It is assumed that a balance between user requirements and operator interests (which include the cost-effective business models) can be achieved by changing the service prices, the service coverage area, the available bandwidth per user, and the per-user access speed."

"The aim of the further process is to determine the possible business models for operators based on the determined values of the beta coefficients when considering the impact of changes in techno-economic factors, i.e., delta parameters."

"The process of defining the best operator business models is based on the techno-economic assessment of different models."

"The differences among these business models are achieved by the inclusion of different combinations of the delta parameters values in the techno-economic planning and modelling processes."

"The algorithm used for determining the best business model, i.e., the one resulting in the greatest net present value (NPV) using the defined combinations of delta parameter values, is presented below: The determination of the beta coefficients and their values (taken from the conducted regression analyses' results) The determination of all values of the delta parameters that will be included in the analyses The determination of operator's costs using adequate expressions The determination of operator's revenues using adequate expressions The determination of operator's business model based on the defined market conditions The determination of the total number of all possible combinations of values considering the defined operator's business models, operator's revenues and costs, and the defined values of the delta parameters For every defined combination, the following procedure is conducted: The determination of the vector CHANGES which comprises the values of the listed delta parameters that define the changes of: service prices (ΔC), coverage area (ΔPP), capacity, i.e., bandwidth (ΔPR), and access speed (ΔBR) The conduction of the techno-economic assessment process within the techno-economics framework using the determined delta parameters values The determination of the business models NPV and subsidy values The determination of the classification rules based on the defined operator business models and the given delta parameters values End The optimal strategies decision (i.e., the classification) tree formation considering the given delta parameters values The formation of the NPV and subsidy values histograms The determination of the most profitable business model (i.e., the one having the greatest NPV value) The determination of the values in vector CHANGES which cause the most profitable business case The resulting delta parameters values that yield the most profitable business cases are listed based on the vector CHANGES in the following order: The techno-economic assessment processes conducted using the standard techno-economic models can be divided into several related components."

"Since the techno-economic planning, modelling and evaluation processes are influenced by the determined delta parameters values, a short overview of the most important relations between the delta parameters and the listed techno-economic processes are given hereafter."

"All these relations consider the impact of the delta parameters values on the initially defined values of the corresponding variables (i.e., the initially defined number of users, service price, targeted part of the coverage area, bandwidth, and access speed)."

"I (a) In the scenario's geo-demographic component planning process, the total number of users is determined based on the expected annual variation of the number of residents (ΔPU) in the given scenario I (b) In the scenario's economic component planning process, the per-user service price is determined based on the delta parameter value that defines the variation (ΔC) of the initially defined service price I (c) In the scenario's regulatory component planning process, the coverage area is determined based on the delta parameter value that defines the variation (ΔPP) of initially defined target part of the coverage area: I (d) In the scenario's technological component planning process, the per-user bandwidth and the per-user access speed variables are determined based on the delta parameters values that define the variation of the per-user bandwidth (ΔPR): as well as the variation of the per-user access speed (ΔBR): II The techno-economic modelling process Furthermore, in the techno-economic modelling process, the relations listed hereafter are used."

"II (a) In the market modelling, as a part of the standard techno-economic assessment process, the standard models (i.e., the Logistic, Gompertz, Richards, and Bass models) for broadband adoption fitting and prediction are used."

"Additional regression analyses have been conducted within the proposed framework to more accurately plan broadband deployment based on the defined user requirements."

"Although the usage of regression in the fitting process has proven to be helpful in defining the most important factors for broadband adoption and the user requirements regarding broadband solutions, further possibilities for regression usage in the prediction process of broadband adoption are included in the planned future research activities and do not affect the assessment conducted within the proposed framework."

"The capacity demands are defined within the scenario's technological component planning process based on the per-user bandwidth, i.e., PR, and the per-user access speed, i.e., BR, and its possible variation."

"The changes in values are denoted by the delta symbols: (III) The techno-economic assessment process Therefore, in the techno-economic cost-effectiveness assessment process, the impact of the given delta values on the resulting cost-effectiveness for all analysed business models can be defined."

"The majority of techno-economic and geo-demographic factors in rural areas are considered unfavourable for broadband solutions deployment."

"The results of applied extended techno-economic assessment processes should indicate the best business models and the best strategies that could be implemented in the given scenario to solve the existing digital divide problems."

"Considering the fact that the development of rural broadband offers numerous benefits to users in rural areas, there are many reasons in favour of the introduction of broadband access in rural areas, even in cases when the cost-effectiveness, i.e., NPV of broadband implementation, is not sufficient."

"Since the cost-effectiveness of the implementation of technical broadband solutions in rural areas is often not significant and relies on regulatory measures that define the adequate funding policies, the techno-economic assessment process used should enable determination of the necessary subsidies for the cost-effective business models and the creation of subsidiary implementation plans."

"To enable the coordination of broadband service prices, i.e., average revenue per user (ARPU), with the income levels of rural residents, analyses of key economic factors influencing broadband implementation and adoption in rural areas should be conducted within the proposed framework."

"The results of the conducted analyses should be directly included in the further techno-economic modelling process to yield the best business model in a given case."

"Within the proposed framework, analyses of key technical and socio-demographic factors that affect broadband adoption of rural users should be conducted to define the actual requirements of potential end users."

"Considering the fact that the rural scenarios differ based on the existence and the level of development of broadband solutions (infrastructure and services), as well as the necessary socio-demographic structure of rural residents, not every strategy is adequate for all rural scenarios."

"The suggested strategies include: the subsidisation—in the cases in which the cost-effectiveness of the business model for broadband implementation is not sufficient (influenced by regulatory and technical factors); the provision of broadband services that offer significant socio-economic benefits to users—in the cases in which available Internet services do not correspond to user needs (influenced by technical and socio-demographic factors); the coordination of the proposed broadband service prices with the actual income levels of rural residents—in the cases in which the cost of Internet services is too high (influenced by economic factors); the promotion of computer literacy and enhancement of digital skills of rural residents—in the cases in which the results of the survey show that a lack of knowledge or skills necessary to use the Internet is present (influenced by socio-demographic factors)."

"In the case in which the results of the conducted analyses point to the fact that the most important factor for the broadband adoption process is the offering of existing broadband services and that the available services do not correspond to users' needs (regarding their access speed, traffic volume, or service quality), the strategy should be aimed at broadband services adaptation and customisation."

"In a case in which the results of the conducted analyses indicate that the most important factor for the broadband adoption process is the price of broadband services (related with the population economic background) and that the service price has a major impact on the cost-effectiveness of a given business model, but the given business model cannot induce positive NPV values regardless of the reduction in the revenues within the defined feasible range, the strategies should allow for the ARPU levels to be coordinated with the average users' income levels and for the necessary amount of subsidies to be determined."

"In the case in which the results of the conducted analyses show that the lack of knowledge or skills necessary to use the Internet is the most important reason for the lower level of usage of broadband services, so that the total base of possible service users is too small to make a broadband business model cost-effective in the given scenario, the proposed strategies should include the subsidisation of broadband infrastructure deployment for the best determined business case and the additional enhancement of the population's digital skills."

"Depending on the results of the extended techno-economic process, any other combination of the related strategies aimed at further broadband solutions adoption can be proposed."

"It is important to notice that the advantage of the proposed extended techno-economic assessment process is its possible application to any rural scenario of any particular size (considering different hierarchical levels, as defined in [179])."

A case study of extended techno-economic assessment process and rural broadband strategies proposal

"The results of the conducted regression analyses are expressed through the values of the β_K coefficients (i.e., β1, β2, β3, and β4), which are related to the four determined combinations of the techno-economic factors (i.e., services price—Price, quality of service—QoS, traffic volume—Traffic, and access speed—Speed, respectively), and the most influential socio-demographic factors for broadband adoption, determined in the previous step (population age—Age, education—Edu, income level—Income and employment—Workf)."

"The best operator strategies can be chosen, as will be shown hereafter. (IV) Techno-economic assessment of rural broadband business models related to the broadband users' requirements To find the optimal broadband access solutions for rural areas, the most important factors for broadband adoption and deployment should be analysed from the techno-economic point of view."

"The results of the regression analyses conducted in this paper for the chosen case study show that the income level of the population is a very important factor for broadband services adoption in rural areas."

Conclusions

"The main advantage of the defined extended modelling methodology, in contrast to other standard techno-economic modelling methodologies that result in a plain set of estimated NPV values, is the possibility of the coordination of the investment strategies based on the best business models (the ones with the greatest values from all possible NPV values) with the actual user requirements (considering the socio-demographic and techno-economic factors) given as the input for the techno-economic modelling process."

"An additional advantage of the proposed modelling methodology is that the given results define the best ways for achieving the maximal profitability for the considered business cases, which are created based on the defined user requirements."

"With the proposed methodology, it is possible to evaluate the cost-effectiveness of business models, considering the implementation of different technical broadband access solutions for different numbers of operators and their modelled strategies."

4.1.22 Changing Patterns of Social Media Use? A Population-Level Study of Finland [519]

This is a machine-generated summary of:

Koiranen, Ilkka; Keipi, Teo; Koivula, Aki; Räsänen, Pekka: Changing patterns of social media use? A population-level study of Finland [519]

Published in: Universal Access in the Information Society (2019)

Link to original: https://doi.org/10.1007/s10209-019-00654-1

If you want to cite the papers, please refer to the original.

For technical reasons we could not place the page where the original quote is coming from.

Abstract-Summary

"This is the first temporal look at changes in Finnish social media use with representative population-level data."

"We assess how social media use has evolved between socio-economic and demographic groups in advanced information societies, with a focus on Finland."

"We also look at how demographic factors associate with use purposes of social media."

"Findings show that the effect of socio-demographic factors on overall use and different use purposes of social media persists."

Introduction

"Despite the growth of social media in terms of its provision of information, communication and interactional potential, these positive effects have not been evenly distributed among different population groups."

"Diversification of different use purposes on social media is revealing how these online social spaces have become politicized [187], commercialized [188, 189] and individualized [190] to a great degree."

"We examine the changing patterns of Finnish social media use over recent years by concentrating on ordinary users of social media, namely Finns of various socio-demographic backgrounds throughout the population."

"Through these analyses we are able to locate how the benefits gained from using social media have been distributed between population groups and evaluate how fundamental changes, namely politicization, commercialization and individualization in the social media sphere, have reflected to different ways to use social media."

Diversification of social media

"While social interaction and connectivity were the key purposes of social media platforms in the first place, these personal actions induced by the mediation of social life are also increasing individualized behaviour on these novel social spaces."

"It can be argued that different benefit-seeking purposes on social media are, in particular, mixed with social interaction and connectedness, while also combined with previously impossible levels of social access."

"Like businesses, different social and political organizations are also widely using social media [191, 192]."

"While these wide social changes in social media spaces have occurred, use purposes connected to these changes ought to be increasing on a population level as access becomes more widespread."

"By monitoring these changes in different use purposes of social media, we are able to assess how different changes are applied by users of social media."

Social media divides in Finland

"Differences in technology use among various population groups within a society have been described through the term "digital divides"."

"Highly educated and wealthier population groups tend to have better access to new innovations, more experience with technology use and a greater ability to take advantage of new platforms while also being more capable in improving their own consumption, participation and leisure activities through the Internet (e.g. [89, 193–195])."

"In the case of social media use, differences in usage patterns can mean differences in access to information and valuable interaction among users even when everyone has access to the platforms in question."

"To inequalities in access to social media, digital divides may appear also in different skills connected to social media use, different use purposes and different kinds of benefits gained from the use."

"As digitalization has continued in Western countries, social media use access differences have arguably decreased (see [196–198]), which does not mean that the benefits from the use are distributed evenly [199]."

Research design

"The focus of empirical analysis is to study the differences in the popularity of social media use and in use purposes over time between different population groups."

"How did the purposes of social media use change between the years 2012 and 2016?"

"We assume that social media use has become more popular across population groups."

"We also assume that the adaptation of social media use has resulted in more versatile use purposes (e.g. [200])."

"Our second hypothesis (H2) is that the overall use and use purposes of social media continue to vary significantly between socio-demographic groups."

"We expect that less educated users will tend to use social media more for leisure purposes (see [195])."

"With regard to H2, we are not assuming to find notable differences in the socio-demographic disparities in social media use purposes when comparing 2012 and 2016."

Data

"The raw data were provided by Statistics Finland, while also being the source for official Finnish statistics (see [201])."

"While official Finnish statistics are published online (see [201]), detailed analysis of these data by population group has not yet been carried out and findings based on controlling for various background variables thus remain undone before this study."

"The mixed-mode survey denotes that multiple data collection approaches were used in one collection (e.g. [202])."

"Since the data offer the official figures for Finnish ICT usage, various demographic information ranging from education and place of residence to family composition has been used to correct for the population and sample biases (for details, see [201])."

Measures and analytic techniques

"We examine population trends in social media use between the years 2008 and 2016."

"In our analyses, we use a binary variable where a value of 0 represents those participants who are not registered on social media platforms and value 1 represent those who are registered users."

"After the examination of social media use across population groups, we will examine purposes of social media usage, which were asked in 2012 and in 2016 questionnaires."

"This was done to gather accurate information across different age groups despite the high level of errors when the analysis is focused on social media users alone in terms of the second research question."

"When moving into use purposes, we tested the effect of background variables on the temporal variance by changing the year as the main independent variable."

Registered social media use 2008–2016

"We first estimated the proportion of registered social media users according to different population groups during 2008–2016."

"In the figure, the use of social media has increased in all population groups since 2008."

"It is also noteworthy that the use of social media accelerated among the age groups 45–54 and 55–64 after 2010."

"In order to interpret differences between population groups and temporal changes of social media use, more sophisticated methods are required."

"Growth has been similar among respondents from both genders, all education groups and people from all regions, and there is no statistically significant indirect year effect between these population categories."

"During the observation period, there was, on average, a 10 percentage-point difference between the highest and the lowest education groups, while the most educated Finns were the most active social media users."

Social media use purposes in 2012 and 2016

"The use of social media for work and following various brands, along with political participation, have increased significantly."

"This analysis confirmed our hypothesis that women use social media significantly more often for maintaining family ties and for social interaction with friends."

"There was a minor difference between the youngest and the oldest participants when considering social media use for hobby and recreation organizations."

"If indirect effects emerge and the adjusted effect of year alters significantly, any increase or decline is dependent on the changes of use purposes in certain demographic groups."

"Age had a statistically significant effect on social interaction with friends, and participating in societal online phenomena, hobby and recreation organizations and following brands."

"Utilizing social media to hobbies has decreased significantly only among 25–54-year-olds, while participating in societal phenomena formed online increased among 25–54-year-olds."

Discussion

"Age, gender, education level and area of residence were all included in order to assess how social media use might differ among various population groups towards deepening our understanding of social media use dynamics across a high ICT use population."

"The findings of the study were in large part in line with past work while also providing new dimensions of comparison between social media use and key demographic factors."

"The findings are in line with past research in that the likelihood of social media use has increased for all population groups from the year 2008 onwards in Finland and other Western countries, representing a general technological trend (see [196–198, 201]) and use purposes have also diversified."

"Despite these limitations, the study provides a comprehensive look at social media use among a national population, which future research can compare to adoption rates and use patterns of similar demographics involved with the next generation of technology trend use."

Conclusion

"The findings of this study concerning the various social media use patterns among different population groups within Finland show that they do indeed overlap to a great degree in terms of what is being done and where."

"Despite sharing a set pallet of possible social media use purposes, differences were found between demographics in terms of activity, popularity and links to the offline setting."

"Distinctions were made between general technology use trends such as social media use, primary digital divides in terms of access issues, and secondary digital divides in terms of use purpose and social media benefit inequality."

"Our findings suggest that this does not always reflect reality or at least it has not happened yet in terms of different social media use purposes."

"Differences in social media use seem to reflect primarily inequalities in offline surroundings."

4.1.23 Gender Wage Gap, Quality of Earnings and Gender Digital Divide in the European Context [520]

This is a machine-generated summary of:

Picatoste, Xose; Mesquita, Anabela; González-Laxe, Fernando: Gender wage gap, quality of earnings and gender digital divide in the European context [520]

Published in: Empirica (2022)

Link to original: https://doi.org/10.1007/s10663-022-09555-8

If you want to cite the papers, please refer to the original.

For technical reasons we could not place the page where the original quote is coming from.

Abstract-Summary

"This paper evaluates the digital gender divide in its three stages: access, use and results, relating it to gender and salary gaps in the context of the European Union."

"Cluster analysis was conducted to classify the countries according to their gender digital divide."

"Of means, the significant variables influencing the gender digital divide were identified, probing that the educational level significantly influences it, especially in what refers to the third stage."

"Through a regression analysis, it was proved that the influence of the insecurity and the gender wage gap goes beyond the women and affects society."

Introduction

"The European Parliament resolution of 28 April 2016 on gender equality and women's empowerment in the digital age (2015/2007(INI)) (European Parliament 204) is a step forwards for women and all citizens to obtain the necessary skills to take advantage of internet opportunities and increase their chances of employment."

"The changes related to the digitalized society have a particular impact on the labour market."

"The pandemic has increased inequality in various aspects, including gender disparities and the digital divide (which was accelerated by the pandemic), particularly for women and youth, for whom unemployment rates have increased."

"This work seeks to clarify whether the digital gender gap affects success in the labour market in terms of income."

"This question is addressed from the perspective of the three stages affected by the digital divide (access, use and results), taking into account socioeconomic aspects such as the age and educational level of men and women."

The gender gap, digital divide and the labour market

"The Global Gender Gap Index (World Economic Forum [204]) advances differently from country to country."

"The GDD must be interpreted in light of the transversal scope of gender equality assumed in the framework of Agenda 2030 (UN General Assembly [205]) and the European Declaration on Digital Rights and Principles for the Digital Decade (European Union [206]), in which the European Union declares the aim of improving people's lives with the digital new era."

"Although the definition of precarious work is not well stated, a good reference is the one proposed by the European Institute for Gender Equality (EIGE): a job can be considered "precarious" when it shows at least one of these three conditions: Very low pay, where take-home pay from a worker's main job is below the first quintile."

"The gender wage gap, according to the OECD, is the difference between the median earnings of men and women relative to the median earnings of men."

Method and procedure

"A gender problem is detected in the shortage of women ICT specialists, but women cannot be excluded from access to technology because it is their right and it increases their digital skills, which are positively related to a higher probability of working with a permanent contract (Martínez Cantos and others [207])."

"The first indicator refers to the percentage of individuals who have sent forms to the institutions, duly filled in and completed, through the internet, in the last 12 months, whose data can be found in EUROSTAT in the section Science, technology, digital society, digital economy and society ICT usage in households and by individuals in E-government, namely, E-government activities of individuals via websites [ISOC_CIEGI_AC]."

"The individuals employed with ICT education were included, with data from EUROSTAT Science, technology, digital society, digital economy and society, digital skills, ICT training, employed persons with ICT education by gender [ISOC_SKI_ITSEX]."

The digital skills, use and outcome of women and men in Europe

"The results indicate that there are no differences in the mean age ratio that are related to age or educational level when the first stage is analysed, that is, access to technology."

"The detailed results are shown in the Online Appendix section, specifically in Tables A-1: Testing H/Stage 1: never access the internet, variable: Age (regarding age), and Stage 1: never access to the internet, variable: Education (regarding education)."

"When the second stage was analysed, no differences were found for age reasons, but significant differences (p value < 0.001 for Student's t test) were found regarding high educational levels."

"Whilst age differences are not stated as relevant variables for the gender digital divide, educational level seems to be significant."

"Variables have been included according to a transversal approach, in which education, age and access to technology (in its three phases) have been considered, together with the gender wage gap, security in employment and the digital divide."

Conclusions

"Educational levels are relevant when analysing the digital gender gap."

"Regarding the influence of the gender wage gap on the quality of earnings, although women suffer the most, it also affects men and, therefore, the whole of society."

"The results of this work mark the importance of the dynamics of incorporating technological progress for achieving gender equality and reducing economic inequalities and other gaps that damage social welfare."

"The focus is on qualitative aspects related to decent work and the need for sustainable employment, along the lines proposed by the SDGs (Goal 8), reducing inequalities (Goal 10) and gender equality (Goal 5) within the general context of the framework of the 2030 Agenda."

"Based on these characteristics, the results of this research work can be helpful in the approach and design of social and labour policies to achieve a more inclusive and sustainable society, both from a social and economic point of view."

4.1.24 Adaptation Behaviour in Using One-Stop Smart Governance Apps: An Exploratory Study Between Digital Immigrants and Digital Natives [521]

This is a machine-generated summary of:

Zhang, Bingqian; Peng, Guochao; Liu, Caihua; Zhang, Zuopeng Justin; Jasimuddin, Sajjad M.: Adaptation behaviour in using one-stop smart governance apps: an exploratory study between digital immigrants and digital natives [521]

Published in: Electronic Markets (2022)

Link to original: https://doi.org/10.1007/s12525-022-00538-y

If you want to cite the papers, please refer to the original.

For technical reasons we could not place the page where the original quote is coming from.

Abstract-Summary

"There is an increasing trend for governments to offer innovative city services to citizens, as well as to communicate with them interactively and dynamically, through one-stop smart governance apps, which are the key to drive innovative governance models for future cities."

"Due to functional differences and complexity of smart governance apps, it may not always be easy for citizens to adapt to these new paradigms of interaction and services."

"The findings reveal the adaptation and learning paths for digital immigrants and digital natives respectively when using smart governance apps."

"An integrative framework is developed, arguing that citizens' adaptation behavior can be either positive or negative, and will be dynamically influenced by digital traits, app quality, sentiments, and task-technology fits."

"The findings of the study will be of interest and importance to academics, policy makers, and practitioners who are keen to promote innovative service models through smart governance apps in future cities."

Introduction

"Developing strategies that help citizens adapt to one-stop smart governance apps, addressing the impact of the digital divide and improving the retention and utilisation rate in post-adoption phase should be put on a high priority to address in smart city initiatives for service providers of one-stop smart governance apps and policy makers."

"Different information literacy and digital savvy of user groups in using one-stop smart governance apps might result in behavioural differences in the adaptation process that have been overlooked in the existing studies."

"This study thus aims at investigating the characteristics of adaptation behaviours of citizens on one-stop smart governance apps, in order to improve the understanding of this phenomenon and help strategise specific measures utilised to engage different groups of citizens in relevant smart governance initiatives by using the apps."

"From the theoretical perspective, the study gives an initial explanation on how users adapted to the one-stop smart governance apps together with the factors that influence citizen adaptation behaviours."

Literature review

"Compared with other types of apps, one-stop smart governance apps rely on government platforms and channels, and are commonly used to offer public services, allowing citizens to participate in smart governance of city in a variety of ways (e.g., participating in public services, publishing comments and suggestions on public services, using public infrastructures, and accessing open government data), in order to improve the awareness and understanding of citizens about the public values, such as increased transparency, participation and collaboration (Abu-Tayeh and others, [208]; Müller and others, [209]; Pereira and others, [210])."

"A further review of the literature reveals the three research themes of related studies on smart governance apps, including: (1) technical facets, (2) obstacles to implementation, and (3) user adaptation behaviours, as presented below."

"This research theme of the literature also contributes to understanding citizen adaptation behaviours of one-stop smart governance apps."

"In a large body of the literature (Abu-Tayeh and others, [208]; Gil-Garcia & Flores-Zúñiga, [211]; Peng and others, [212]), citizen participation is the keyway to the success of implementation and promotion of one-stop smart governance apps."

Research methods

"We selected qualitative approaches, using interviews, to capture the views, experiences, and/or motivations of individuals on their adaptation behaviours toward one-stop smart governance apps, because this allows researchers to interact directly with the participants and seek the clarification for concerned matters (Alharbi and others, [213])."

"The data analysis from this small sample size of the participants was sufficient for us to develop the outline of semi-structured interview and obtain a broad picture of the app users' adaptation behaviours."

"To ensure sufficient differences between the two groups of users and similarities of the same user groups towards the characteristics of the adaptation behaviours identified from the interviews, we interviewed a total of 23 users of one-stop smart governance apps in China, including 8 digital immigrants and 15 digital natives in this study."

Results and findings

"Our findings show that, the adaptation process during the usage of one-stop smart governance apps is rather complicated and dynamic, and there are essential behavioural differences between the user groups of digital immigrants and digital natives."

"These concerns and negative feelings then may let digital immigrants perform technical exploration avoidance behaviours (i.e., stop further exploring or engaging with one-stop smart governance apps)."

"When digital natives are getting increasingly familiar with one-stop smart governance apps, they will even perform integrated usage behaviour, i.e. combining multiple functions of the app in a flexible and personalised manner to optimise performance efficiency: "I have now enabled the functions of 'local discovery' and 'new function reminder', so then I can receive customised notifications about important local information and system updates, without missing them or spending too much time on finding them in the app" (Participant-11)."

Integrative framework and further discussion

"As illustrated in this framework, citizens' adaptation behaviours can be either positive or negative, and will be dynamically influenced by digital traits, app quality, sentiments, and task-technology fits (TTF)."

"Different from prior studies discussing about adaptation/acceptance behaviours encouraged by usability, user experience, and user interface that too focused on smart phone appearance (e.g., with a large screen size and a very high resolution (Moumane and others, [214]), system elements (e.g., function, color and sound (Park and others, [215])) and goal-directed actions (e.g., products need to meet different level of users' goals (Pucillo and Cascini, [216])), our findings showed that users concerned more about the quality perspective of one-stop smart governance apps, serving as one driver to trigger users adaptation behaviours."

"When citizens perceived good task-technology fit, they will become more inclined to perform positive adaptation behaviours, such as actively learning and exploring the apps (i.e., technology exploration behaviours) and achieving extended, effective and even integrated usage of the apps (i.e., technology exploitation behaviours)."

Conclusions, limitations and future studies

"This study took China as the study context, considering current development and implementation of one-stop smart governance apps in China and the characteristics of digital natives and digital immigrants, to give the following implications for improvements and adaptation of one-stop smart governance apps: (1) governments are expected to issue relevant policies, calling attention from all sectors of the society to help digital immigrants better adapt to one-stop smart governance apps; (2) developers need to take the behavioural characteristics of the two groups into account when designing a one-stop smart governance app (e.g., design an interface for digital immigrants with larger font and simpler language and improve the efficiency and effectiveness of the apps for digital natives; and (3) communities or volunteers are desired to organise a training on the use of the apps for digital immigrants in order to improve the adaptation behaviours."

4.1.25 The Digital Divide in Technologies for Autism: Feasibility Considerations for Low- and Middle-Income Countries [522]

This is a machine-generated summary of:

Kumm, Aubrey J.; Viljoen, Marisa; de Vries, Petrus J.: The Digital Divide in Technologies for Autism: Feasibility Considerations for Low- and Middle-Income Countries [522]

Published in: Journal of Autism and Developmental Disorders (2021)

Link to original: https://doi.org/10.1007/s10803-021-05084-8

If you want to cite the papers, please refer to the original.

For technical reasons we could not place the page where the original quote is coming from.

Abstract-Summary

"95% of individuals with autism live in low- and middle-income countries (LMIC) where access to electricity, internet and the ever-increasing range of digital devices may be highly limited."

"The World Bank coined the term 'the digital divide' to describe the disparities in access to digital technologies between high-income and LMIC contexts."

"We evaluated the feasibility of six emerging technologies for autism spectrum disorders, and reflected on key considerations for implementation in LMIC contexts to ensure that we do not inadvertently widen the pre-existing digital divide."

Introduction

"The classification included examination of the technology in relation to domain of interest (e.g. social/emotional or language/communication), purpose (e.g. functional assessment or diagnosis/screening), the target end user (e.g. person with ASD or family/caregiver), setting (e.g. home or school), nature of the empirical support (e.g. descriptive or experimental research) and technology maturity (e.g. design, prototype or publicly available)."

"Combining the global drive towards digital technologies for ASD, the fact that the majority of people with ASD live in LMIC, and the pre-existing digital divide as outlined by the World Bank, there is a significant risk that, without due reflection and consideration of the real-world feasibility of technologies, the global ASD community may increase this digital divide, and thus compound the well-known disparities that exist between HIC and LMIC in health, education, social integration and economic outcomes of families who live with ASD."

Methods

"After review of the literature each reviewer was asked to provide feasibility ratings of the implementation constructs selected."

"After rating each construct, reviewers were asked to provide an 'overall feasibility' judgment incorporating all constructs."

"In the absence of any existing criteria for feasibility rating of technologies, reviewers were asked to rate constructs as 'very low', 'low', 'high' or 'very high'."

"Reviewers were asked to provide comments and a narrative motivation for their ratings."

Results

"Consensus ratings for affordability were 'high' for smartphone-based technologies and personal computers, 'low' for sensing and virtual reality technologies, and 'very low' for shared active surfaces and robotics."

"Consensus ratings for acceptability/cultural appropriateness were 'very high' for smartphone-based technologies, 'high' for personal computers, sensing technologies and shared active surfaces, and 'low' for virtual reality and robotics."

"Consensus ratings for scalability were 'very high' for smartphone-based technologies, 'high' for personal computers, 'low' for sensing technologies, shared active surfaces and virtual reality, and 'very low' for robotics."

"Taking together all implementation-related constructs, consensus ratings for overall feasibility in LMIC were 'very high' for smartphone-based technologies, 'high' for personal computers, 'low' for sensing technologies, shared active surfaces and virtual reality technologies, and 'very low' for robotics."

Discussion

"Mindful that most people with ASD live in low- or middle-income countries (LMIC), and given the pronounced digital divide between high-income countries (HIC) and LMIC, we set out to explore the feasibility of implementing these technologies in LMIC settings."

"Feasibility ratings, however, showed how technologies had very different strengths and weaknesses across different implementation constructs when the 'implementation lens' was focused on LMIC contexts."

"No ASD technology research has used the term to refer to the macro-economic and/or cultural and linguistically diverse 'contexts' where people with ASD live."

"While robotics, shared active surfaces and virtual reality technologies may have significant potential application in various HIC contexts, the feasibility of implementation may look significantly different in LMIC contexts."

"Notwithstanding these limitations, this was the first study to our knowledge to perform feasibility rating of ASD technologies."

Conclusion

"We reflected on the feasibility of six examples of such technologies in LMIC contexts."

"Our findings suggest that feasibility of the use of such technologies in LMIC will depend on the fundamental principles of affordability, accessibility, acceptability and cultural appropriateness."

"Given the widespread usage of mobile and smartphones and the increasing availability of affordable high-speed mobile internet access in the majority of LMIC, mHealth has the greatest potential to increase access to ASD screening, diagnosis and treatment globally."

"The evidence-base for mobile technologies in LMIC settings remains extremely limited."

4.1.26 Effects of Socioeconomic and Digital Inequalities on Cybersecurity in a Developing Country [523]

This is a machine-generated summary of:

Khan, Naurin Farooq; Ikram, Naveed; Saleem, Sumera: Effects of socioeconomic and digital inequalities on cybersecurity in a developing country [523]

Published in: Security Journal (2023)

Link to original: https://doi.org/10.1057/s41284-023-00375-4

If you want to cite the papers, please refer to the original.

For technical reasons we could not place the page where the original quote is coming from.

Abstract-Summary

"In developing countries, increased reliance on cyberspace for carrying out educational activities has implications for cybersecurity threats."

"Cybersecurity is a digital skill that is affected by socioeconomic and digital inequalities; specifically in the developing nations."

"With lack of digital divide's empirical evidence in terms of cybersecurity, this study employs a face-to-face survey to understand the computer and smartphone security practices of students enrolled in higher education institutes (HEIs) across Pakistan."

"Significant differences were found in the cybersecurity practices of students in terms of socioeconomic and digital divide variables."

"This highlights that the individuals with lower socioeconomic status and who are digitally less connected are at a greater risk of falling victims to cyber-threats."

"The implications of the study state to impart tailored cybersecurity trainings with respect to digital divide and socioeconomic status of the students."

Introduction

"Social and digital disparities negatively affect the development of knowledge and skills to enact cybersecurity practices (Dodel and Mesch [224])."

"The topic of cybersecurity awareness is of particular pertinence for Pakistan—a developing country where individuals hail from different socioeconomic backgrounds and digital divide is acute in urban and rural areas (Jamil [225])."

"This study focuses on cybersecurity behavior of university-going students in Pakistan who hail from different socioeconomic backgrounds and experience digital divide."

"The cybersecurity awareness is affected by disparities that are rooted in socioeconomic status and access to digital technologies."

"Few studies have been conducted that use digital divide variables to study cybersecurity behaviors (Dodel and Mesch [224, 226, 227])."

"Digital disparities reproduced themselves in the enactment of cyber preventive behaviors (Dodel and Mesch [224]) with individuals experiencing greater digital divide showing low cybersecurity practices."

Theoretical background

"This study takes into consideration the digital inequalities in which access to the Internet more frequently and from various places affects the cybersecurity practices along with the socioeconomic status of the university-going students."

"Few other studies on smartphone security awareness and behavior have been carried out on the students' population from HEIs."

"A number of studies have been conducted in the USA to understand the smartphone security awareness and behavior of the students."

"The second reason is the lack of empirical evidence in understanding both computer as well as smartphone security practices of the students in a single study except (Taha and Dahabiyeh [228]; Breitinger and others [229])."

"This study gauges the cybersecurity behavior of the students in a broader context of computer and smartphone security as a whole hence extending the work of Khan and others [230] in which only computer security was measured."

Methodology

"A questionnaire-based survey is conducted to measure computer and smartphone cybersecurity behaviors using existing scales."

"To measure the computer security behavior of the students, Security Behavior Intention Scale (SeBIS) was used (Egelman and Peer [231])."

"The first construct measures the avoidance of harmful smartphone behavior & attitude and consists of 17 Likert scale-like items."

"The second construct measures the protection behavior using add-on utilities and settings having 2 dichotomous and 3 Likert scale-like items."

"The digital divide was measured using two variables: frequency of Internet access and Internet access from different places."

"The Internet access from different places is coded as 1 to represent access from home, 2 to present access from university, 3 to represent access from work/friends/family, and 6 to represent access from multiple places."

"Stratified multi-stage sampling was employed for measuring the university students' cybersecurity behavior."

Results

"Students pursuing medical-related degrees showed low-security behavior by failing to use various passwords, updating software, and other programs."

"Students belonging to the age group 18–20 years showed better smartphone security practices in AHBA by never downloading apps from sources that could not be trusted or giving excessive permissions to their personal information."

"Post hoc Bonferroni correction analysis of AHBA category showed that the students who belonged to the province of Sindh downloaded apps as well as opened attachment from untrusted sources and also used their smartphones for financial purposes therefore showed low security."

"The students who accessed the Internet from multiple places showed better security behaviors in all three smartphone security categories."

"A Bonferroni correction showed that the students who had Internet access from different places exhibited better security by not downloading apps from untrusted sources or giving them excessive access."

Discussion

"This study enhanced the literature on cybersecurity by examining socioeconomic and digital disparities and their effects on computer and smartphone security practices."

"Students have a low computer as well as smartphone security practices which are in line with the results from Filippidis and others [232]."

"The computer security results pertaining to digital divide variables show students experiencing greater digital divide exhibit low-security practices."

"The only exceptions where students suffering from a larger digital divide performed better than those who had better Internet access were the smartphone security practices of downloading apps, giving excessive permission, using smartphones for financial purposes, and connecting to insecure Wi-Fi."

"Our results reinforce this theory since the digital disparities in terms of frequency of Internet access and from various places are carried out in the online behaviors of computer and smartphone security."

Conclusion

"This study has contributed towards the interrelations of digital and socioeconomic disparities with higher education institutes as a result of research call by Robinson and others [233]."

"Although the research on digital inequality is still evolving, this study never-the-less contributes towards the stratification model of diffusion of technologies by understanding the computer and smartphone security awareness and behaviors of students enrolled in the tertiary institutes of Pakistan."

"The results from our study show substantial cybersecurity differences in terms of the digital divide as well as socioeconomic status and demographics."

"The hindrance to digital technologies does influence digital skills in terms of computer and smartphone security and consequent capital-enhancing activities on the Internet."

"Future studies should be carried out to longitudinally measure the smartphone and computer security awareness and behaviors of the students with respect to digital and socioeconomic disparities."

4.1.27 Study of the Digital Divide Evaluation Model for Government Agencies—A Taiwanese Local Government's Perspective [524]

This is a machine-generated summary of:

Chang, She-I; Yen, David C.; Chang, I-Cheng; Chou, Jung-Chu: Study of the digital divide evaluation model for government agencies–a Taiwanese local government's perspective [524]

Published in: Information Systems Frontiers (2011)

Link to original: https://doi.org/10.1007/s10796-011-9297-x

If you want to cite the papers, please refer to the original.

For technical reasons we could not place the page where the original quote is coming from.

Abstract-Summary

"This paper, based on Gowin's Vee structure, is an empirical study of the digital divide in the context of local governments in Taiwan."

"A model for identifying and measuring aforementioned digital divide is constructed in this paper."

"We first refer to the grounded theory to draft a framework for measuring the digital divide in local governments."

"We measure the actual levels of the digital divide in local governments with the resulting digital divide evaluation model."

"The digital divide evaluation model constructed in this study goes beyond existing measures and may serve as a reference for academics in the examination of methods to narrow the digital divide in various levels of governmental bodies."

"The features of integration, comprehensiveness, and wide applicability of this proposed model can be considered the theoretical contributions to digital divide and local government hierarchy research."

Introduction

"Differences in the opportunities provided by technological applications and digital infrastructures in the process of government digitalization often result in a digital divide among governments—and such differences have always been in existence in the development of e-governments (Norris and others [234])."

"The digital divide in governments, as a rule of thumb, produces unequal opportunities for the public in accessing information technology, leads to the uneven distribution of social resources and opportunities for wealth creation, and widens the gap between social classes (National Research Development and Evaluation Commission [235])."

"Effective evaluation will reduce the digital divide, and help policymakers and public sector executives tap ICT resources and develop fair budget plans."

Literature review

"In a report by the Organization for Economic Cooperation and Development (OECD), Understanding Digital Divide, the digital divide is defined as "the gap in opportunities for individuals, households, or companies of different social and economic backgrounds and geographies in accessing information and communication technologies (ICT) and activities on the Internet" (OECD [236])."

"There are two main streams of the digital divide: domestic digital divide, which indicates that a digital divide in a certain country or region exists; and international digital divide, which indicates that a gap exists between regions, countries, or continents (Çilan and others [237])."

"Cuervo and Menéndez [238] find that a digital divide exists among 15 developed countries from the European Union (EU)."

"Luyt [239] points out that the measurement tool for the evaluation of the digital divide is centered on the indicators of a country's readiness to begin the digitalization process."

Research method and design

"A digital divide evaluation model for local governments is constructed systematically for this paper, which is an empirical study that applies Gowin's Vee model."

"This study initially aims to explore the measures of the digital divide and to summarize them by reviewing extant studies, and then to develop an evaluation model for the assessment of the digital divide."

"Based on the Vee diagram and with the literature review, we use the grounded theory to construct a draft list of the measures of the digital divide in the local governments in Taiwan."

"We also apply the analytic hierarchy process (AHP) in calculating the weights of individual measures, thereby establishing a digital divide evaluation model for local governments in Taiwan."

Prototype digital divide evaluation model

"This work, by following the methods proposed in the grounded theory, uses open coding to generate the draft list of the measures and to facilitate the development of a digital divide evaluation model for local governments."

"In the article "Challenges in e-government development: Lessons from two information kiosk projects," Nia and Ho [240] assert that "while educational and training programs for different user groups are important to help resolve the problem of digital divide, they are only part of the solution.""

"We further classify the 49 domains of coded entries into these dimensions, and develop the prototype architecture of our digital divide evaluation model."

"The domains of "financing resources," "strategic planning in ICT," "acknowledgment of ICT," "participation of local organizations," "education and training curricula," "organizational culture," and "capabilities of using and managing information services" are related to the organization itself, and all of them are imperative to the public sector if e-projects are being executed within an organization."

Correction and weight allocation in the digital divide evaluation model

"This study arrives at the measurement items that can be used in studying the digital divide in local governments through the processes recommended by the grounded theory and the administration of questionnaires among the experts."

"We use a successive paired comparison technique in the AHP because this study involves the hierarchical measurement structure of the digital divide and, therefore, the relative weight of each measurement item should be determined (Saaty [241])."

"This study finds that the item "opportunities to use technologies as required by job functions" accounts for the highest weight (15.2%), indicating that this measurement has a strong influence on the digital divide in local governments."

"This study, summing up the influence of the detailed items of the five constructs on the overall weights, determines that the issue that requires the most attention in the digital divide in local governments is the ability to use information services within organizations."

Case studies of digital divide in governments

"The items "information security" and "information privacy" in the dimension of "ICT infrastructures" produce the highest scores, possibly because the Chiayi County Government has deployed an Information Safety Management System (ISMS)."

"The items "ICT scalability" and "multiple channels to access e-services" produce the lowest scores, probably because the Chiayi County Government, having the largest number of machines in Taiwan, has been highly active in establishing the Digital Opportunity Center in recent years."

"The items "number of e-service users," "consistency of local and central information and "multi-language services" generate lower scores, indicating that the number of e-service users in Chiayi County is not that high."

"The item "capabilities of using and managing information services" in the dimension "internals of organizations" obtains the highest score, indicating that the Chiayi County Government is doing its job in the maintenance, update, and disaster recovery of its administration and information management systems, and is, indeed, monitoring and supervising the computer centers."

Conclusions and future studies
"After referring to the grounded theory, and utilizing the questionnaires distributed among experts and the AHP analysis, five dimensions and 42 measurement items are constructed for the digital divide evaluation model in this study."

"This model, unlike other evaluation mechanisms that only measure some fragile items, takes a comprehensive approach to enable local governments to assess the pros and cons of their digitalization."

"The digital divide evaluation model constructed in this study should serve as a reference for the academic sector in the examination of the methods that are being used in narrowing the digital divide in various levels of governmental bodies, especially now that e-governments are receiving more and more attention."

"The whole process of building this proposed evaluation model, with regard to the theoretical implication, can be seen as theory building in relation to the digital divide in government agencies."

4.1.28 Spatiotemporal Investigation of the Digital Divide, the Case Study of Iranian Provinces [525]

This is a machine-generated summary of:

Qadikolaei, Mohsen Roohani; Zali, Nader; Soltani, Ali: Spatiotemporal investigation of the digital divide, the case study of Iranian Provinces [525]

Published in: Environment, Development and Sustainability (2022)

Link to original: https://doi.org/10.1007/s10668-022-02738-0

If you want to cite the papers, please refer to the original.

For technical reasons we could not place the page where the original quote is coming from.

Abstract-Summary
"Many communities and territories in developing countries experience significant gaps in access to and use of information and communication technology (ICT), which is viewed as a major impediment to socioeconomic and health-related vulnerabilities."

"Geographic considerations, on the other hand, as well as variations in motivation for ICT usage and technology adoption within and across nations, as well as between cultures, have exacerbated the digital divide."

"This study discovered that digital inequality is associated with social exclusion throughout Iranian regions, and it went on to try to better understand the underlying issues and potential solutions."

Introduction

"The early research on the digital divide mostly examined people' access to information (Heponiemi and others, [242]) and were confined to discriminating between having and not having access to ICT."

"We are now discussing the third level of digital divide analysis, which focuses on the positive outcomes of Internet use (Ferreira and others, [243]) and establishes that inequalities exist when, despite access to frequent and widespread use of ICT, this does not result in beneficial socioeconomic outcomes (Aleksejeva and others, [244])."

"The digital divide analysis focuses on the positive consequences of Internet use and demonstrates that disparities develop when, despite repeated and extensive access to ICT, it does not result in positive socioeconomic outcomes (Aleksejeva and others, [244])."

"The purpose of this research is to provide a spatial–temporal analysis of the digital divide for Iran's provinces."

"A review of theoretical framework on digital divide based on IDI (The ICT Development Index) and spatial inequity and its multiple causes of digital divide is presented, then the relationship of IDI with variety of social, economic and physical variables is analyzed."

Theoretical framework

"The disparity in access to or utilization of ICT devices, as measured by the number of telephone lines per person, Internet users, or mobile phone users, deepens the gap between emerging and developing nations, particularly when new aspects of technology arise (Adam and others, [245])."

"The digital divide can also refer to social differences in Internet access, educational opportunities, and, as a result, disparities in online information access."

"In rural and less economically developed areas, ICT infrastructure is scarce, and associated activities are dependent on the access to infrastructure and the quality of the Internet connections (Freeman and others, [246])."

"Uneven human capital development, which involves investing in digital skills and capacities via education, training, and lifelong learning, is one of the most important parts of boosting Internet access."

Methods and materials

"Various indexes have been used to measure the impact of ICT on socioeconomic development."

"Based on these, the ITU develops and publishes the ICT Development Index, which is used to gauge the level of ICT development."

"The ICT Development Index is a tool that allows governments, operators, development organizations, and researchers to assess and evaluate each country's ICT engagement (Ziemba & Becker, [247])."

"The combined indicators include the ICT Development Index, which is based on 11 criteria that identify ICT and are organized into three clusters: access information, usage, and skills (ITU, [248])."

"Secondary gross enrollment ratio ($W_{31} = 0.33$) Tertiary gross enrollment ratio ($W_{32} = 0.33$) Adult literacy rate ($W_{33} = 0.33$) The values of the subindices were first calculated by normalizing the indices in each subindex as below: To calculate the final index, the sub-branches of access to ICT and the use of ICT each weighted 40%, and a subindex of skills weighted 20%."

Analysis and findings

"The pattern of changes in the digital divide based on the IDI index from 2011 to 2020 is as below: Provinces with high incremental change during the period 2011–2020: Tehran (T1) (1.83), Khorasan Razavi (K2) (0.69), Isfahan (I1) (0.61) Provinces without change or near-zero change: Hormozgan (H1) (0.01), Sistan and Baluchestan (S2) (0.03)."

"Population density, the proportion of R&D workshops, the percentage of urban housing, and monthly income are among the factors impacting changes in the digital divide index from 2011 to 2020, according to the regression research."

"Family income and the proportion of R&D workshops in the provinces are among the factors impacting changes in the digital divide, and they have shown to be successful as important independent variables each year."

"Correlations have changed with time, so that in 2013, fewer provinces (Semnan, Yazd, and Fars) had a highly significant correlation in line with the link between R&D workshops and monthly income and the IDI index."

Discussion

"Among the significant findings are: While the overall level of ICT in Iran increased significantly between 2011 and 2020, the level of provincial IDI in Iran has shown sharp differences over time and gradually from the central zone with high IDI index (including Khorasan Razavi (K2), Semnan (S1), Mazandaran (M1), Tehran (T1), Isfahan (I1), Yazd (Y1), and Fars (F1)."

"Between 2011 and 2020, ICT development shifted from the central zone to the western regions, and from the same zone to the eastern regions, despite the fact that they continue to be in worse shape than provinces in the IDI index zone."

"Central zone covers the provinces of Khorasan Razavi (K2), Semnan (S1), Mazandaran (M1), Tehran (T1), Isfahan (I1), Yazd (Y1), and Fars (F1)."

"The largest average of the digital divide in eastern zone reflects the existence of income inequalities, the develop of major cities, and the dispersion of research and development workshops as compared to other provinces."

Conclusion

"Technical research and development workshops should be promoted in each of the activity and service centers of rural and urban communities, and on the other hand, fundamental ICT training for all age groups should be supplied."

"This study found that the digital gap is more visible in those with lower education levels; so, supporting policies should involve training and upgrading IT skills among the target groups."

"Data availability is one of the limitations in digital divide research, therefore, future research can be based on additional disaggregated data, especially if the effort intends to analyses the phenomena using comparable methodological techniques (e.g., OLS and GWR) used in this study."

"Another avenue for future study may be policy analysis research on the effect of public policies in expanding or decreasing the digital gap, especially given the COVID-19 pandemic's greater reliance on ICTs in many societies, including Iran."

4.1.29 Towards Bridging the Grey Digital Divide: Changes in Internet Access and Its Predictors from 2002 to 2014 in Germany [526]

This is a machine-generated summary of:

Huxhold, Oliver; Hees, Elena; Webster, Noah J.: Towards bridging the grey digital divide: changes in internet access and its predictors from 2002 to 2014 in Germany [526]

Published in: European Journal of Ageing (2020)

Link to original: https://doi.org/10.1007/s10433-020-00552-z

If you want to cite the papers, please refer to the original.

For technical reasons we could not place the page where the original quote is coming from.

Abstract-Summary

"In order to gain a better understanding about the determinants of this grey digital divide, the current study contrasts influencing factors of internet access, comparing samples from 2002 to 2014 across age groups (40 to 54 years, 55 to 69 years and 70 to 85 years) using data from the German Ageing Survey (DEAS)."

"Logistic regression confirmed that the likelihood of having internet access was lower with higher age at both time points."

"The percentages of people with internet access grew primarily in the middle and older age groups between 2002 and 2014."

"Being male and having a higher education were both associated with greater odds of internet access."

"Gender and education differences in internet access were significantly less pronounced in 2014 in contrast to 2002."

"Both greater income and cognitive ability were associated with greater odds of internet access, while providing care for a grandchild was significantly associated with internet access only among the oldest age group."

Introduction

"To better understand differences in internet access in the middle-aged and older adult population, we employ the perspective of a diffusion model of technological innovation (Peres and others [249])."

"We will use this perspective to investigate whether or not factors influencing access to the internet differ across age groups and across historical time."

"As the first-level digital divide is persistent among some groups (Friemel [250]), it is important to gain a thorough understanding of the factors influencing the likelihood of having internet access in order to determine the best approaches and strategies to counteract this divide."

"Contrasting influencing factors of internet access across age groups may only be a first step for uncovering potential routes to bridge the grey digital divide."

"It is important to examine historical differences in socio-demographic factors of internet access in different age groups."

Method

"Internet access was measured with an item included in the DEAS written Drop-Off questionnaire asking participants if they had access to the internet and providing three possible answers: No, Yes, at home or Yes, at work."

"To test hypotheses 1 and 2, we conducted basic logistic regression analyses in which internet use was regressed on time (one dummy variable) and either age group (two dummy variables) or gender (one dummy variable) or education (two dummy variables) and their respective interaction with time (e.g. middle-age-by-time)."

"To address hypothesis 3, we conducted a multi-group multivariable logistic regression analyses in which we contrasted the effect size of income, cognitive ability and care of grandchildren across six different age-by-time groups (i.e. 40–54 years in 2002, 55–69 years in 2002, 70–85 years in 2002, 40–54 years in 2014, 55–69 years in 2014, 70–85 years in 2014)."

Results

"In support of hypothesis 1A, the middle and the older age groups reported significantly less access to the internet at both time points (55–69 years: $\Delta\chi^2 = 201.83$, $\Delta df = 1$, $p = .000$; 70–85 years: $\Delta\chi^2 = 602.60$, $\Delta df = 1$, $p = .000$)."

"Over and above the effects of education, gender and income, cognitive ability was significantly associated with internet access ($\Delta\chi^2 = 49.13$, $\Delta df = 1$, $p = .000$)."

"Over and above the effects of education, gender, income and cognitive ability, caring for a grandchild or grandchildren was overall not significantly related to internet access ($\Delta\chi^2 = .00$, $\Delta df = 1$, $p = .950$)."

"A post hoc test revealed that the association between caring for grandchildren and internet access could be set to zero for the younger (40 to 54 years) and middle-age group (55–69 years) without a significant loss in model fit ($\Delta\chi^2 = 1.24$, $\Delta df = 3$, $p = .743$)."

Discussion

"Coinciding with an overall growth in the prevalence of internet access in the following 12 years, women, people with lower education and older adults showed a disproportionally strong increase in comparison with the rest of the population age 40 and older."

"In younger age groups, low cognitive ability may not pose a problem for acquiring access to the internet for large parts of the population, although from a perspective of social inclusion, developing interventions specifically targeting younger people with learning difficulties may also be advisable."

"Older adults, who spent time with their grandchildren, were more likely to have access to the internet than their same age counterparts, irrespective of education, income and cognitive ability."

"Identifying and educating people about the internet, who have bridging positions in social networks of older adults, could potentially increase intervention effectivity."

Limitations and outlook on future studies

"Although the current study provides a focused overview on factors determining internet access among older adults, there are a few limitations to be considered which might point towards gaps in the literature worth investigating in future research."

"Future research should continue doing similar analyses contrasting influencing factors of internet access across current time points and social groups."

"As previously mentioned, the digital divide can be defined on various levels; starting with internet access and general use marking a first digital divide as investigated in the current study, but going further into subsequent divides such as specific types or reasons for internet use dependent on skills and digital literacy."

"The current study solely refers to the first-level digital divide by examining internet access."

"As the first digital divide continues to shrink, future research examining specific types of internet use as well as related skills becomes indispensable."

4.1.30 Digital Divide and Access to Online Education: New Evidence from Tamil Nadu, India [527]

This is a machine-generated summary of:

Jafar, K.; Ananthpur, Kripa; Venkatachalam, L.: Digital divide and access to online education: new evidence from Tamil Nadu, India [527]

Published in: Journal of Social and Economic Development (2023)

Link to original: https://doi.org/10.1007/s40847-023-00236-1

If you want to cite the papers, please refer to the original.

For technical reasons we could not place the page where the original quote is coming from.

Abstract-Summary

"A massive and unplanned shift towards online education with unequal access to digital infrastructure deepens the existing digital divide and socio-economic inequalities."

"Based on the three recent rounds of the panel survey conducted in October 2020 and August 2021, this article discusses Tamil Nadu's experience in continuing education during the pandemic."

"The result highlights the existing digital divide and challenges faced by students in accessing online education."

Introduction

"The contagious nature of the virus and the emergence of new variants may cause disruptions and weaken educational practices, learning outcomes, and impact the future of students in the coming years."

"The prolonged closure of educational institutions and massive, unplanned shift towards the digital mode throws new challenges for the educational system, teachers, students, and parents."

"Effective policy interventions to bring them back and address the issue of access to the new mode of education for students from all regions, classes, gender, and socio-economic status are very important."

"The article briefly reviews the global and national response to the disturbance caused by the pandemic on education and the challenges faced while continuing with the online mode of education."

Education during the pandemic: a review

"The 'majority of governments (133 countries) provided a mix of online, television, and radio solutions to ensure learning continuity, providing strong ground to build more resilient education systems and bridge the digital divide' (UNESCO [251–253)."

"At the national level, the Union HRD Minister asked the students to continue their learning by making full use of the available digital e-Learning platforms like DIKSHA, ePathshala, National Repository of Open Educational Resources (NROER), Swayam, Swayam Prabha, etc. well before the nationwide lockdown and 'Janata Curfew' on 22nd March 2020 (Government of India [254, 255])."

"At the state level, Tamil Nadu also followed multiple strategies to promote online education and the dissemination of digital content or learning material through various platforms."

"This includes broadcasting lessons on Kalvi TV for school students, TN-DIKSHA digital learning platform for the children, Veettupalli programme (School at home) for children enrolled in schools functioning under Tamil Nadu State Board, and introduction of other innovative practices (Government of Tamil Nadu [256])."

Data and methodology

"TNCPS can be defined as a 'panel survey with refreshment'; the stratified sampling design identifies the district as a stratum and identified villages as the Primary Sampling Units (PSUs) in rural areas and Urban Frame Survey Blocks (UFSBs) in urban areas."

"The households sampled for the TNCPS were a sub-sample of the households which were already contacted for the Tamil Nadu Household Panel Survey (TNHPS)."

"The sampling frame for the first stage units was drawn from the list of 2,12,282 households contacted for TNHPS Pre-Baseline Survey (PBS) in 2018–2019."

"A total of 13,749 rural and urban samples were selected for the first round of the TNCPS survey and this formed the panel for the subsequent rounds."

"Being a telephonic panel survey, efforts were made to ensure a better response rate, quality control, verification and adding refreshment samples to address attrition and balance between rural and urban areas."

Experience of Tamil Nadu
"More than half of the households covered under three rounds of the survey had at least one student attending school or college (52.8%, 53.9%, and 51.8%, respectively, in rounds II, III, and IV) and this pattern remains similar in rural and urban areas."

"Compared to rural areas, share of students who dropped out of schools and colleges due to this reason remains high in urban areas (8.6%, 4.2% and 5.0%, respectively, in rounds II, III and IV against 4.0%, 3.6% and 3.7% in rural areas)."

"Many students currently accessing online education face technical and network issues during online classes (39.3%, 33.3%, and 28.5% during rounds II, III and IV, respectively)."

"Gradual growth in the share of students who did not face any issues while attending online classes and a decline in the gap between rural and urban areas indicate improved access to digital learning in the state."

Results from nonlinear panel regression model
"The results from nonlinear panel i.e. Conditional fixed-effects logistic regression, show low access to online education for urban and rural households who faced job loss during the pandemic."

"For urban households, an increase in job loss by a unit reduces the likelihood of online education access by 0.64 units for their children attending school or college."

"The results follow the larger pattern of digital divide and challenges in accessing online during the pandemic in different contexts."

"If households face technical issues and knowledge gaps for children or parents, their access to online education decreases by 1.56 times and 2.0 times, respectively."

"The finding emerging from the descriptive statistics and regression indicate the nature of the digital divide and challenges in accessing online education in the state."

"The results also indicate that students from rural areas and socially marginalised groups (socially or economically) have lower access to online education."

Conclusion and policy implications
"While government interventions have improved access to online education and digital content over time, many students especially in rural areas still do not have access to online education in the state."

"Tracking the potential rise in the drop-out at all levels of schooling (especially among the marginalised groups) and encouraging their re-enrolment and ensuring proactive support to students with special needs, etc may help the government to ensure that no child is left behind in accessing education in the state."

"The government should use this opportunity to devise effective policies to make the educational system more resilient and inclusive by addressing the digital divide and inequalities that exist at all levels and explore the possibilities of digital transformation to improve the quality of teaching and learning in the state."

4.1.31 Trends of Online Patient-Provider Communication Among Cancer Survivors from 2008 to 2017: A Digital Divide Perspective [528]

This is a machine-generated summary of:

Jiang, Shaohai; Hong, Y. Alicia; Liu, Piper Liping: Trends of online patient-provider communication among cancer survivors from 2008 to 2017: a digital divide perspective [528]

Published in: Journal of Cancer Survivorship (2019)

Link to original: https://doi.org/10.1007/s11764-019-00742-4

If you want to cite the papers, please refer to the original.

For technical reasons we could not place the page where the original quote is coming from.

Abstract-Summary

"Decade, online patient-provider communication (OPPC) has emerged as a viable avenue for cancer survivors to communicate with their providers."

"Little is known about the patterns of OPPC among cancer survivors."

"The current study aims to explore the trend of OPPC used by cancer survivors, and the influence of digital divide on OPPC in the past decade."

"Only cancer survivors were included in the analyses."

"This study illustrates an increasing trend in OPPC use among cancer survivors."

"Significant digital divide barriers also exist in the adoption and diffusion of OPPC."

"OPPC is an important communication channel for cancer survivors and will become more important in the digital era."

"Targeted interventions to address the digital divide barriers affecting OPPC could be developed to benefit underserved cancer survivors and to bridge health disparities."

Introduction

"The rapid development of information and communication technologies (ICTs) in recent years has enabled online patient-provider communication (OPPC) as an alternative and complementary communication channel for cancer survivors to interact with their providers during survivorship."

"Within the context of evolving modes of patient-provider communication enabled by advanced ICTs, increasing needs for efficient patient-provider communication are required by a growing number of cancer survivors, as well as more self-care practices, despite persistent online communication barriers based upon the digital divide."

"We used four iterations of a nationally representative survey, the Health Information National Survey (HINTS), from 2008 to 2017 to explore (1) the trends of OPPC used by cancer survivors, (2) the influences of three dimensions of digital divide (physical access, socio-demographic access, and cognitive access) on OPPC, and (3) whether the relationship between digital divide and OPPC has changed over the past decade."

Methods

"Cognitive access included three factors: health information overload, self-efficacy in health information seeking, and trust in online health information."

"Self-efficacy in health information seeking was measured by a single item that asked respondents to self-report how confident they could get information about health or medical topics if they needed it, on a 5-point Likert scale (from 1 = not confident at all to 5 = completely confident)."

"Trust in online health information was assessed by a single item that required respondents to rate how much they trust Internet information about health or medical topics, using a 4-point Likert scale (from 1 = not at all to 4 = a lot)."

"Four separate multivariate linear regression analyses were run to examine the effects of digital divide, including socio-demographic access, physical Internet access, and cognitive access, on OPPC."

Results

"In 2013, 4.4% cancer survivors communicated with providers via mobile apps, 2% via social media, and 1.3% via video conferencing."

"Throughout the four iterations, the adoption of OPPC in cancer survivors has been constantly increasing."

"Email has remained the most common OPPC platform over the past 10 years, while newer communication channels (e.g., social media, mobile apps, EHR) have gained growing popularity in the recent years, enhancing the diversity of OPPC channels for cancer survivors."

"Out of the three dimensions of digital divide, only physical Internet access had significant effects on OPPC throughout the four waves."

"Cognitive access failed to exert significant influences on OPPC in the past 10 years."

"Different from physical access and cognitive access, whose influences on OPPC did not change much over time, the effect of socio-demographic access varied vastly across iterations."

Discussion

"This study also demonstrated that the three dimensions of the digital divide, namely physical Internet access, socio-demographic access, and cognitive access, affected cancer survivors' OPPC differently over time."

"The health and wellness support from providers, via OPPC in place 365/24/7, can help cancer survivors better cope with the complexities and uncertainties in care [257, 258]."

"For cancer survivors with Internet access, the likelihood of engaging in health-related Internet use was higher than those without Internet access [259]."

"This highlights the need to provide greater Internet access to cancer survivors."

"Given that the cognitive access related to general health information seeking failed to influence OPPC over time, targeted patient education is needed."

"The education programs should specifically aim to improve cancer survivors' knowledge and skills about using OPPC functions, rather than general health-related Internet use."

Conclusion

"This study offers up-to-date information on OPPC among cancer survivors, with additional assessment of the impact of digital divide on OPPC."

"Although the Internet has been widely adopted by Americans, OPPC has not kept up the pace in cancer survivors."

"Our findings show that the digital divide persists and it affects OPPC differently, with physical Internet access as a consistent predictor over the past decade, and socio-demographic access exerting greater effects in recent years, illustrating growing health disparities."

"We call for targeted interventions to promote OPPC in cancer survivors and to reduce digital divide in underserved populations."

4.1.32 Digital Divide: A Gap in Interpersonal Communication Amongst Nigerian Family Relationship [529]

This is a machine-generated summary of:

Ngunan, Achakpa-Ikyo Patience; Regina, Ogaba-Egba: Digital divide: A Gap in interpersonal communication amongst Nigerian family relationship [529]

Published in: GSTF Journal on Media & Communications (2016)

Link to original: https://doi.org/10.7603/s40874-016-0003-y.

If you want to cite the papers, please refer to the original.

For technical reasons we could not place the page where the original quote is coming from.

Abstract-Summary

"The study Digital Divide: A Gap In Interpersonal Communication Amongst Family Relationships seeks to find out how technology has affected the willingness of family members to engage in interpersonal communication in the family."

"The study found that technology has created a gap in interpersonal communication amongst family relationships and that people lose interest in communicating interpersonally with family members due to constant engagement in Information and Communication Technologies (ICTs)."

"The study recommends that families should make concerted effort at adopting and integrating ICTs that will help enhance interpersonal communication in family relationships rather than engage with ICTS that will erode interpersonal communication amongst family relationships."

Introduction

"With a plethora of Information and Communication Technologies (ICTs) accessible to people today, it is believed that interpersonal communication is reducing in the family."

"The researcher seeks to investigate how technological divide between family members has affected interpersonal communication in family relationship."

"This paper seeks to find out how the digital divide has affected the willingness by family members to engage in interpersonal communication."

"This can achieved by determining the challenges to interpersonal communication in family relationship in the face of the technological divide amongst family members."

"Establish possible ways to bridge the digital divide in interpersonal communication amongst family members."

"Most communication in the family is interpersonal in nature which is the fulcrum for interpersonal relationship in the society and at any level of human interaction, the family inclusive."

Methodology

"The study was conducted using a quantitative research method employing the survey research design with the questionnaire as the instrument of data collection."

"The survey method of research was employed for the study because it provides the best approach and flexibility needed in the data gathering."

"In the survey research design, the researcher selects a sample of respondents from a population and administers a standardized questionnaire to them."

"The researchers employed the purposive sampling technique in order to elicit data from the population members with a fair knowledge of the topic under study given its uniqueness to the area of study."

"Purpose or judgmental sampling was considered best for the study because it gives the researchers the opportunity to determine population members to participate in the survey thus controlling a form of imbalance or disparity in data collected which in turn will make it difficult to analyze."

Discussion of findings

"It also implies that the effect of constant ICTs use on interpersonal communication in family relationship could be either positive or negative depending on the individual."

"This means that technology is blurring the lines of interpersonal relationships amongst family members, because relationships are maintained via communication especially interpersonal communication and if people are not communicating regularly there is a possibility of relationship breakdown."

"Further on this, findings also revealed that technology has affected the willingness of people to engage in interpersonal communication with family members."

"This implies that engagement with ICTs on a constant basis is time consuming and is capable of negatively impacting time spent in interpersonal communication amongst family relationships."

"Further findings revealed that if family members can share at least a meal together daily, the digital divide in interpersonal communication in family relationships could be bridge."

Conclusion

"This study was set out to evaluate the digital divide; the possible gap in interpersonal communication amongst family relationships with particular emphasis on finding out how technology has affected the willingness of people to engage in interpersonal communication among family members."

"Technology has affected the willingness of family members to engage in interpersonal communication in that they lost interest in communicating with family members as a result of constant use of ICTs."

"The study concludes that technology has negatively affected the willingness of people to engage in interpersonal communication amongst family relationships because of their constant engagement with ICTs which makes them loss interest in communicating with family members."

4.1.33 Disability Digital Divide: The Use of the Internet, Smartphones, Computers and Tablets Among People with Disabilities in Sweden [530]

This is a machine-generated summary of:

Johansson, Stefan; Gulliksen, Jan; Gustavsson, Catharina: Disability digital divide: the use of the internet, smartphones, computers and tablets among people with disabilities in Sweden [530]

Published in: Universal Access in the Information Society (2020)
Link to original: https://doi.org/10.1007/s10209-020-00714-x

If you want to cite the papers, please refer to the original.

For technical reasons we could not place the page where the original quote is coming from.

Abstract-Summary
"Although Sweden is one of the most digitalized countries and the Swedish population's use of the internet is among the most studied in the world, little is known about how Swedes with disabilities use internet."

"The purpose of this study is to describe use of and perceived difficulties in use of the internet among people with disabilities and to explore digital divides in-between and within disability groups, and in comparison with the general population."

"Larger proportions of people with autism, ADHD and bipolar disorder reported using internet than other disability groups."

"Women with autism used the internet more than any other disability group, and women with aphasia used the internet the least."

"People with disabilities related to language and understanding reported more difficulties using internet than other disability groups."

"In many but not all disability groups, larger proportions of men than women reported not feeling digitally included."

"Disability digital divides are preferably investigated by sub-grouping disabilities, rather than studied as one homogeneous group."

Introduction
"It is the first of its kind, in that we have succeeded to disaggregate disability into 35 different groups of diagnoses and impairments by data from 771 people

responding to a survey about internet use and the use of devices to access the internet."

"The purpose of this study was to describe use of and perceived difficulties in use of the internet among people with disabilities and to explore digital divides in-between and within disability groups, and in comparison with the general population."

"Dobransky and Hargittai [260] explored the digital divide by investigating internet access and use among people with six types of disabilities."

"When controlling for demographic and socio-economic factors, i.e. gender, age, education, income, people in five of the six disability types reported less access to internet and lower skills as compared to the general population."

Method

"The members of the Begripsam Group, whom all have different cognitive and perception disabilities, acted as expert reference group and contributed to the development of the survey, selection of target populations and interpretation of the results."

"The questionnaires from the mirrored surveys were elaborated in collaboration with the Begripsam Group, in order to increase accessibility for people with disabilities."

"The recruitment process was closely monitored to ensure participation from all targeted disability groups."

"The survey was distributed through the Begripsam web site and the Facebook site, home pages of disability organizations and disability networks, e-mail contact lists and by personal communication."

"Participants could choose to respond to the survey online, on paper, or request a telephone interview."

"When analysing data by disability group and by demographic and socio-economic variables, only those 28 disability groups that had 25 or more participants were included in the analyses."

Results

"A larger proportion among the participants (31%) than the general population (25%) have university level education; 12% among the participants have attended elementary school or special elementary school for students with intellectual disabilities, i.e. the lowest level of education in Sweden, which is equivalent to the reported incidence in the general Swedish population (12%)."

"In all disability groups more women than men reported using Facebook, and in several disability groups larger proportions reported use of Facebook (ADD 87%, ADHD 83%, autism 77%, bipolar 80%, depression/anxiety 80%, dyscalculia 84% and dyslexia 82%) as compared to the reported proportion in Swedish population (74%)."

"In regard to education, the largest proportion of participants who do not feel included in the digital society was among people having studied at special schools for students with intellectual disabilities (12%)."

Discussion

"The digital divides become more visible when 'people with disabilities' are disaggregated into disability groups, and by doing so we find: Divides in access to devices: People in many disability groups report less access to devices, compared to the general Swedish population, but there are also differences between disability groups, and differences related to gender, education and accommodation; Divides in use of internet related to gender: Women with disabilities use the internet more and for more complicated tasks than men with disabilities; Divides in perceived inclusion in the digital society: In some of the disability groups larger proportions, compared to the Swedish population, reported being excluded from the digital society, but in some disability groups larger proportions compared to the Swedish population reported being included in the digital society; Divides are not always disadvantaging people with disabilities: Many of the exposed disability digital divides indicate that people with disabilities are disadvantaged, but in the case of engagement in social media the divide is reversed."

Conclusions

"We divided people with diagnoses and impairments into 35 disability groups and have enough data to present results for 28 disability groups."

"Disability digital divides are preferably investigated by sub-grouping disabilities, rather than studied as one homogeneous group."

"The digital divides become more visible when 'people with disabilities' are disaggregated into several disability sub-groups."

"We conclude that the results of this study provide a basis for further research to explore the complex set of prerequisites for participation in the digital society among people with disabilities."

4.1.34 Gender Digital Divide in a Patriarchal Society: What Can We Learn from Blinder–Oaxaca Decomposition? [531]

This is a machine-generated summary of:

Jiang, Wun-Ji; Luh, Yir-Hueih: Gender digital divide in a patriarchal society: what can we learn from Blinder–Oaxaca decomposition? [531]

Published in: Quality & Quantity (2016)

Link to original: https://doi.org/10.1007/s11135-016-0409-z

If you want to cite the papers, please refer to the original.

For technical reasons we could not place the page where the original quote is coming from.

Abstract-Summary

"Based on the self-reported usage time, this paper aims at analyzing the gender differences in computer and internet use at home."

"Using data from the 2012 Survey of Digital Divide in Taiwan, we apply a regression-based decomposition method to identify the underlying causes of observed usage differential between males and females."

"A further decomposition analysis suggests while the gender-specific factors are not the main causes of gender differences in computer and internet use, differences in internet experience and opportunity cost of leisure time between the two gender groups are root causes of observed gender usage differential."

Introduction

"Despite the considerable body of literature examining the digital divide in the adoption of CI, the empirical evidence of gap in CI usage between different socio-economic groups is quite limited."

"Based on CI usage functions, gender gap in CI usage is decomposed into two components including one due to differences in average socio-demographic characteristics such as income and education (i.e., the endowment effect), and the other due to the differential estimates of determinants in the usage function (i.e., the coefficient effect)."

"If the main cause of gender differences in CI usage is due to differential estimates of income in the usage function, possible disparities in the opportunity cost of CI usage between males and females should be taken into account while bridging gender digital divide in usage."

Theoretical framework

"The construction of the theoretical framework establishes the foundation of empirical analysis in exploring the role of gender difference in time allocation to using CI."

"According to Goldfarb and Prince [261], an empirical analysis of the optimal amounts of CI and usage time is modeled in two stages."

"Applying the random utility model (McFadden [262, 263]), the household at this stage compares the utility between adopting and not adopting CI."

"The second stage then involves deciding how much leisure time to be allocated to CI usage."

"Since the leisure time spent on using CI is observed only if the adoption took place, correction for the selection bias caused by correlation between CI adoption and usage is necessary when the usage of CI is modeled in the second stage (Goldfarb and Prince [261])."

The empirical model

"The second term on the right hand side is the coefficient effect that measures the part of gender differences in CI usage due to differential coefficients of determinants between the male and female groups."

"The pure gender effect is in general greater than zero since in addition to gender-related obstacles to females' CI usage, males usually have a positive self-stereotyping while females have a negative self-stereotyping."

"As demonstrated in Oaxaca and Ransom [264], the identification problem arises from the fact that the detailed coefficient effect of constant term is not invariant to the choice of left-out reference groups of dummy variables."

"In the application of "normalized" regression, the coefficient effects of the constant term and dummy variables are independent of the choice of left-out reference groups."

Data source and summary statistics

"Since it is not possible to use internet without computer devices at home, the respondents answering "no" to the former question but "yes" to the latter question are dropped from the sample."

"To be consistent with our theoretical framework for CI use at home, respondents who answer "yes" to the question: "Is it necessary to use internet in your work?""

"After excluding 2073 respondents who are within this age range, there are in total 5237 observations in our final sample; approximately 40% of which are males and 60% are females."

"The statistics by gender groups also indicate males in Taiwan spend more time on CI use than females."

"Based on the determinants of adoption and CI usage identified in previous studies, explanatory variables are individual characteristics including personal income and internet experience, as well as household size and geographical characteristic."

Empirical results and discussion

"Compare the coefficient estimates between the two gender groups, the result suggests that, in terms of the likelihood of adoption, even if an indigenous family doesn't make a difference for males, females in the indigenous family are less likely to adopt CI at home."

"The results also suggest that, compared to with people living in East Taiwan, those living in the northern part are more likely to adopt CI by 3.9 and 2.8% for male and female groups, respectively."

"This result reveals one interesting feature of the gender digital divide in Taiwan, i.e., we can close the gender gap in CI usage by brining females' average internet experience up to the males'."

"To internet experience, we found another underlying cause of gender differences in CI usage is due to differences in opportunity cost of leisure time between low-income males and females."

Policy implications

"Our finding that the gender differences in CI usage can be explained partly by the differences in opportunity cost of leisure time between low-income males and females highlights the need for policy makers to focus their efforts on low-income females."

"Encouraging low-income females to use computer and internet as well as disclosing the benefits associated with internet surfing, such as valuable health information and cost-down tips, are top priorities to bridge gender CI usage divide."

"Based on the delineation of the endowment effect, difference in internet experience between males and females is another important factor in explaining gender CI usage differentials."

"Under the view that extensive internet experience makes it easier for one to acquire precise information and thus benefit more from internet surfing, and accumulation of internet experience helps positive attitude formation, policy tool designed to improve females' usage skill and attitudes towards CI can increase the benefits females obtain from internet surfing."

Conclusions
"A further decomposition of observed mean difference in usage time, the present work provides viable explanations to the observed gender differences in CI usage at home, conditioned on adoption."

"Results from the detailed decomposition indicate that gender-specific factors, such as self-stereotyping, are not the main cause of gender differences in CI usage in Taiwan."

"To internet experience, another important factor contributing to gender CI usage divide is the differences in the opportunity cost of leisure time amplified through the high value of housework placed by low-income females."

"One possible extension is analyzing gender usage divide for patriarchal societies such as the Africa and Pacific countries where females' self-stereotyping of being unable to think mathematically and scientifically is a pronounced problem (Hafkin [265])."

"In those patriarchal societies, the decomposition framework used in this study allows for the measurement of the extent such self-stereotyping contributes to observed gender difference in CI usage."

4.1.35 The Digital Divide and Other Economic Considerations for Network Neutrality [532]

This is a machine-generated summary of:

Connolly, Michelle; Lee, Clement; Tan, Renhao: The Digital Divide and Other Economic Considerations for Network Neutrality [532]

Published in: Review of Industrial Organization (2016)

Link to original: https://doi.org/10.1007/s11151-016-9554-8

If you want to cite the papers, please refer to the original.

For technical reasons we could not place the page where the original quote is coming from.

Abstract-Summary

"It also highlights several policies that the FCC has undertaken purportedly to reduce the divide, including the 2015 Open Internet Order (OIO)—in which the stated intent is to enforce "network neutrality.""

"Long before the OIO, studies have raised concerns that network neutrality policies will discourage investment by internet service providers (ISPs) in broadband infrastructure, to the detriment of broadband accessibility, and may increase average consumer costs—both of which would only further exacerbate the digital divide."

"We provide a holistic analysis of the effects of net neutrality on the digital divide; in doing so, we draw from recent economic research on this issue."

Introduction

"The 2015 Open Internet Order (OIO) is the latest among the FCC's stated strategies to promote infrastructural investment and reduce the digital divide."

"Among existing models that allow for consumers to enter and exit the market for internet service based on price and/or content value, there is a clear argument that the 2015 OIO may in fact worsen the digital divide rather than reduce it (as is claimed in the OIO)."

"We simply wish to note that one of the primary decisions within the 2015 Order was to prohibit ISPs from charging any fees to any CSPs—no matter how that CSP's service impacts network congestion, etc Since much of the literature on the economic impacts of net neutrality was written before the FCC's 2015 order, existing models consider somewhat different regulatory applications of net neutrality."

Content Service Providers

"In more recent years much of the data traffic on the internet has been coming from video content—especially as CSPs like Netflix and YouTube have grown and started displacing traditional television services."

"Brennan [266] points out that as of March 2015, over 52% of all downstream internet traffic came from videos from just Netflix and YouTube."

"Brennan underscores that while "it has been a slogan that BIAS providers should not be allowed to turn the Internet into cable TV, subscribers have already turned the Internet into the next video delivery mechanism.""

"This fact is crucial to understanding the validity of any welfare results that are derived from models that assume market conditions and consumer preferences that imply that the total social value of internet content increases with the total quantity of content."

The Digital Divide

"The term 'digital divide' is commonly used to refer to the disparity in access to and/or use of digital technology across households based on urban versus rural locations and on socioeconomic differences across households."

"The digital divide is not a recent occurrence."

"Demographic factors, such as income and urbanity, contribute to the digital divide (NTIA [267])."

"Beyond price (cited by 19% of those without internet) and accessibility due to infrastructure (cited by 7% of those without internet), there exist other commonly overlooked reasons for the digital divide."

"Whether the FCC's 2015 Open Internet Order ameliorates or worsens the digital divide will depend on how it affects price, accessibility, and content."

Price

"Among costs that are borne by an end consumer for internet use are last-mile fees that are paid to ISPs such as Time Warner Cable and any additional subscription fees that are paid to CSPs for services or content access (such as a subscription to Netflix)."

"In a two-sided market model, ISPs provide platforms that bring together two sides of the internet market: CSPs and end consumers."

"In an unregulated market, ISPs are able to extract surplus from CSPs when content providers value the connection to subscribed consumers in the ISP networks more than consumers value the connection to the content that is provided by CSPs."

"ISPs could charge CSPs higher prices in order to lower the last-mile fees for consumers in an attempt to maximize their end-user subscriptions."

Investment in Broadband Infrastructure

"They find that such regulation may or may not increase ISP investment in bandwidth, and thus improve social welfare, depending on whether a network access fee effect (where greater capacity increases the fee that the ISP can charge consumers for access) or a rent extraction effect (where greater capacity lowers the fee that the ISP can charge a content provider for a priority lane) dominates."

"The authors then suggest that "if household utility is a significantly greater component of welfare than content providers' profits, then network neutrality can still be the welfare-superior policy even accounting for the ISP's bandwidth-building incentives.""

"Even with these assumptions, Economides and Hermalin [268] still find only certain parameter conditions under which the consumer welfare gains from content increase under network neutrality regulations are sufficiently large as to offset the welfare losses from decreased ISP bandwidth provision."

Content Quality and Diversity

"Among papers that are focused particularly on heterogeneous CSPs and the creation of content, most find that the possibility of prioritization of content (absent net neutrality regulation) leads to an increase in content (Reggiani and Valletti [269]; Krämer and Wiewiorra [270]; Hermalin and Katz [271])."

"In terms of the market structure of imperfectly competitive industries (such as those that create and provide content), additional entry by firms will divert resources from other CSPs and may actually lead to a fall in total content, content quality, and content diversity."

"Net neutrality policies that artificially encourage additional entry by small CSPs in content and application services that are marked by significant innovation costs would well actually lead to lower levels of socially valued content both in terms of quantity and in terms of diversity."

Conclusion

"On balance, existing economic models with more realistic underlying structural assumptions predict that the Open Internet Order is more likely to result in higher last-mile prices, lower infrastructure investment, and poorer content quality and diversity."

"If the models hold true, the Open Internet Order will—contrary to its intent—only further worsen the digital divide."

"There are key features that any such model would need if it wishes to fully consider the dynamic and endogenous market outcomes under the imposition of these new regulations: Heterogeneous consumers: needed to consider digital divide issues."

"Non-monopolistic ISPs: needed to compete for both consumers and CSPs and consequently endogenously determine investment in broadband infrastructure."

"We have reviewed recent developments in the economic understanding of net neutrality, and analyzed studies that are relevant to understanding the effects of the Open Internet Order on: (1) last-mile price; (2) internet infrastructure investment; and (3) content quality and diversity."

4.1.36 Persistent Digital Divide in Health-Related Internet Use Among Cancer Survivors: Findings from the Health Information National Trends Survey, 2003–2018 [533]

This is a machine-generated summary of:

Fareed, Naleef; Swoboda, Christine M.; Jonnalagadda, Pallavi; Huerta, Timothy R.: Persistent digital divide in health-related internet use among cancer survivors: findings from the Health Information National Trends Survey, 2003–2018 [533]

Published in: Journal of Cancer Survivorship (2020)

Link to original: https://doi.org/10.1007/s11764-020-00913-8

If you want to cite the papers, please refer to the original.

For technical reasons we could not place the page where the original quote is coming from.

Abstract-Summary

"Prior research on the use of the internet among cancer survivors indicates a digital divide."

"The online landscape and patterns of information consumption, however, have notably changed over the past decade necessitating an updated examination of health-related internet use (HRIU) among cancer survivors."

"Using survey data from 2003, 2005, 2007, 2011, 2013, 2017, and 2018 Health Information National Trends Survey (HINTS) iterations, the objectives of this study were to report prevalence, trends, and user profiles in HRIU in terms of emailing doctors, buying medicine online, and support group participation."

"Consistent with the increased penetration of the internet and the altered online health communication landscape, we found increased prevalence of HRIU among cancer survivors."

"These findings can inform initiatives to bridge the gap among survivors of varying profiles in using the internet for their health needs."

"There is increased reliance on online platforms to obtain and communicate health-related information."

Introduction

"Chou and colleagues have documented an increasing trend in health-related internet use (HRIU) among cancer survivors, which they defined as emailing doctors, buying medicine online, online support group participation, and seeking cancer information from the internet as the first source [259]."

"Since the report on HRIU among cancer survivors by Chou and colleagues, the online landscape has considerably shifted, where 90% of Americans reported using the internet compared with the year 2000, when nearly half the American adult population did not use the internet, per a 2019 report from the Pew Research Center [272]."

"The current landscape is characterized by use of personalized technology and the internet of things influencing the dissemination of health information, HIT, and how patients and providers interact [273] highlighting the importance of updated information on trends in internet usage in the cancer survivor population."

Methods

"Dependent variables used in this analysis include internet use, emailing doctors, buying medicine online, internet support groups, and seeking any cancer information."

"A weighted multivariable logistic regression model was created to explore the adjusted association between the dependent variables regarding HRIU, and the covariates gender, age, race/ethnicity, education, metro status, general health status, insurance status, and time since diagnosis among cancer survivors."

"Respondents excluded from regression analysis were those without any cancer history and/or those that did not answer the question regarding internet use."

"Using the full sample of those who responded "Yes" to using the internet, we carried out Chi-square tests to test for associations between survey year and HRIU variables, and associations between cancer survivor status (cancer survivor, family with cancer but no personal history, and no history or family members with cancer) and HRIU variables."

Results

"The percent that answered "Yes" to internet use ranged from 49.3% in 2005 to 76.9% in 2017, emailing doctors ranged from 9.4% in 2003 to 39.9% in 2013, buying medicine online had a low of 11.4% in 2003 and a high of 24.9% in 2017, support group participation was at a low in 2005 of 3.5% and a high of 8.8% in 2003, and seeking cancer information from any source was 77.6% in 2007 and 89.8% in 2013."

"Cancer survivors were the most likely to email doctors, be in online support groups, or seek cancer information online, while those with family who are survivors were most likely to use the internet."

"Among survivors who use the internet, seeking cancer information from any source had higher odds in 2013 and 2017 and among more educated adults, and lower odds among male survey respondents."

Discussion

"Our analysis found cancer survivors having college education or higher were more likely to engage in all types of HRIU except participation in online support groups."

"In the same study, Jiang and colleagues examined the association of the dimensions of the digital divide with OPPC finding that physical internet access was significantly associated for all the years studied, cognitive access (health information overload, self-efficacy in health information seeking, and trust in online information) was not associated with OPPC, and the association with sociodemographic access (age, gender, race, household income, education, and health status) varied across the years."

"Despite an increase in the proportion of cancer survivors engaging in IHIS, the study confirmed a digital divide and further examined the dimensions of the digital divide: mental access (trust in online information and self-efficacy in health information seeking), material access (ease of internet access), skill access (skill to seek information online), and usage access (online health-related activities like emailing doctors and buying medicines)."

Limitations and strengths

"The health communication landscape has considerably altered in the past decade influencing the dissemination of health information, HIT, and how patients and providers interact [273] highlighting the importance of updated information on trends in internet usage in the cancer survivor population."

"The data are based on self-report and subject to recall bias."

"The surveyed population is subject to selection bias as respondents with poor prognosis and serious disease are less likely to participate in the survey."

"The data are cross-sectional precluding any inference on causation highlighting the need for other types of studies to complement results from surveys, for example, patterns of information use can be gleaned using internet search data [274] and mined from social media platforms [275] to explore questions about the specific topics discussed, relation between promotional information and discussions, and the relationship of these patterns to survey results from HINTS."

Implications

"The offline US adults were more likely to be 65 and older, live in rural areas, and have a high school education [272]."

"Trust in information and ease of searching for information were associated with adoption among older adults suggesting that this group might benefit from navigational assistance [276]."

"Of online cancer prevention information seeking preferences in an ethnically diverse sample, older age, high school education, and being a Spanish speaker were associated with lower odds of seeking cancer prevention information online [277]."

"To avoid further exacerbating disparities, it is vital to gain an understanding of the perceptions of and barriers to internet access among older cancer survivors to inform the development of appropriate e-literacy interventions."

Conclusion

"The current analysis using data from 5 iterations of HINTS showed increasing prevalence of HRIU among cancer survivors."

"Differences in internet use across key groups, particularly age, insurance status, and education, have implications for the design of health interventions and communication efforts."

4.1.37 Internet Access, Social Risk Factors, and Web-Based Social Support Seeking Behavior: Assessing Correlates of the "Digital Divide" Across Neighborhoods in The State of Maryland [534]

This is a machine-generated summary of:

Hatef, Elham; Ma, Xiaomeng; Shaikh, Yahya; Kharrazi, Hadi; Weiner, Jonathan P.; Gaskin, Darrell J.: Internet Access, Social Risk Factors, and Web-Based Social Support Seeking Behavior: Assessing Correlates of the "Digital Divide" Across Neighborhoods in The State of Maryland [534]

Published in: Journal of Medical Systems (2021)

Link to original: https://doi.org/10.1007/s10916-021-01769-w

If you want to cite the papers, please refer to the original.

For technical reasons we could not place the page where the original quote is coming from.

Abstract-Summary
"We aimed to empirically measure the degree to which there is a "digital divide" in terms of access to the internet at the small-area community level within the State of Maryland and the City of Baltimore and to assess the relationship and association of this divide with community-level SDOH risk factors, community-based social service agency location, and web-mediated support service seeking behavior."

"To assess the socio-economic characteristics of the neighborhoods across the state, we calculated the Area Deprivation Index (ADI) using the U.S. Census, American Community Survey (5-year estimates) of 2017."

"To assess the digital divide, at the community level, we used the Federal Communications Commission (FCC) data on the number of residential fixed Internet access service connections."

"We assessed the availability of and web-based access to community-based social service agencies using data provided by the "Aunt Bertha" information platform."

"We performed community and regional level descriptive and special analyses for ADI social risk factors, connectivity, and both the availability of and web-based searches for community-based social services."

"To help assess potential neighborhood linked factors associated with the rates of web-based social services searches by individuals in need, we applied logistic regression using generalized estimating equation modeling."

"In Baltimore City, 20.3% of neighborhoods (defined by census block groups) were disadvantaged with ADI at the 90th percentile while only 6.6% of block groups across Maryland were in this disadvantaged category."

"Most block groups in Maryland and Baltimore City lacked access to social services facilities (61% of block groups at the 90th percentile of disadvantage in Maryland and 61.3% of block groups at the 90th percentile of disadvantage in Baltimore City)."

Introduction
"Within the U.S. medical care context, there is increased appreciation that SDOH, both at the community and personal level, represent key factors that must be considered when assessing and addressing the health of patients and populations."

"Such an approach to connectivity provides the opportunity to properly address the growing role of internet access in connecting patients to the care continuum and supporting patients' transition from the healthcare system to the community."

"Before this approach might be widely adopted, we need to gain empirical evidence regarding the association and interplay between the availability of internet connectivity services, standard SDOH measures, and social service agency availability and access at the neighborhood and regional level."

"We measured the degree to which there is a digital divide in terms of access to internet connectivity services at the small-area community level within the State of Maryland and the City of Baltimore."

Methods

"We assessed access to community-based social services using Aunt Bertha's data [278]."

"The data also included number of searches by seekers and healthcare workers for available services in each ZIP code."

"We assessed the digital divide, access to community-based social services, and the number of online searches for those services by seekers and healthcare workers for different categories of ADI."

"We included the connectivity variable, social services density, and ADI national rank as a group or cluster-level variables representing ZIP code level measurements."

"For the spatial exploratory data analysis, we presented an overlaying map of connectivity and ADI and maps of community-based social services and ADI in Baltimore City [279]."

"We presented connectivity and community-based social services as categorical variables and aggregated the numbers for each geographic level of interest (i.e., census tract for connectivity and block group for social services) and mapped the distribution of each variable in Baltimore City."

Results

"In affluent neighborhoods of Baltimore City (those at the 10th percentile of ADI) connectivity was more homogenous with 100% of census tracts being in category 4 of connectivity with 601–800 of their households (per 1000 households) having access to services at least 10Mbps download and 1Mbps upload."

"The census tracts in the middle risk category (those at the 11th to 89th percentile of ADI) in Maryland and Baltimore City had a relatively similar pattern of access to services to the census tracts at the 10th percentile of ADI."

"61.0% of block groups at the 90th percentile of ADI in Maryland and 61.3% in Baltimore City did not have any social service facilities."

"In terms of social services diversity, among block groups at the 90th percentile of ADI, only 6.6% in Maryland and 7.6% in Baltimore City had high service diversity (20 or more services)."

Discussion

"We assessed how internet access would impact searching for community-based social services while considering the place-based disparities represented by the ADI."

"The digital divide and lack of internet access (to search for social services) in disadvantaged neighborhoods might have masked the actual demand for such services."

"In a broader context of population health, the lack of access to the internet, eHealth technologies, and the absence of social services in disadvantaged neighborhoods may limit public health departments to fully address the social needs of high-risk populations [280–282]."

"Calling digital divide a "super-determinant of health" shows its impact on the health of individuals and populations as without internet access a number of health and social services would be inaccessible to those in need."

"The three major variables of ADI, internet access, and access to social services organizations were available at different geographic levels."

Conclusions and public health implications

"Our study also provided an example of how to make use of FCC data, a less utilized data source in studies addressing the digital divide and its impact on health, linked to other data sources on SDOH and available community based social services and can be a model for future studies in this domain."

"While our study was limited to Maryland and Baltimore City using available national data on socioeconomic characteristics of neighborhoods (i.e., ADI), FCC data on fixed internet access, and Aunt Bertha's data on access to community-based social services will provide the opportunity to replicate this assessment beyond Maryland and in other states at the neighborhood and regional level."

"The current COVID-19 pandemic and expanded services to address the pandemic such as the Emergency Broadband Benefit [283], an FCC program helping families and households struggling to afford internet service to get connected to jobs, critical healthcare services, and virtual classrooms, will provide other data sources for a more comprehensive assessment."

4.1.38 Does the Digital Gap Matter? Estimating the Impact of ICT on Productivity in Developing Countries [535]

This is a machine-generated summary of:

Hawash, Ronia; Lang, Guenter: Does the digital gap matter? Estimating the impact of ICT on productivity in developing countries [535]

Published in: Eurasian Economic Review (2019)

Link to original: https://doi.org/10.1007/s40822-019-00133-1

If you want to cite the papers, please refer to the original.

For technical reasons we could not place the page where the original quote is coming from.

Abstract-Summary

"Could the adoption of Information and Communication Technology (ICT) be the means to higher growth and productivity in developing countries?"

"An alarming slump of global productivity growth rates, especially after the global financial crisis, underlines the relevance of this question which does not find a clear answer in the literature."

"The results of the different approaches, which allow for unobserved country effects and non-linear relationships, are not optimistic: ICT is only a minor engine of TFP growth."

"Countries with relatively high ICT investment may be able to increase their TFP growth rates by between 0.1 and 0.3% annually relative to those with modest investment rates."

Introduction

"One decade later as the technology diffused and the adoption rates increased, studies refuted the paradox to reach the consensus that production as well as usage of ICT have become a main determinant of productivity growth in the advanced economies (Oliner and Sichel [284]; Jorgenson and Stiroh [285]; Schreyer [286]; Basu and others [287])."

"The same consensus has not been reached by studies investigating the impact of ICT on growth and productivity in developing countries."

"The years between 2005 and 2014 saw a dramatic increase in the cell phone and internet penetration rates in the developing world, providing important data about possible returns of ICT. (2) Most of the previous country studies in this research field use GDP growth or labor productivity as the outcome variable."

Literature overview

"Studies testing the impact of ICT investments on productivity are conducted either on the micro firm or the macro country-level."

"Even though country and cross-country level studies have been more popular when exploring the impact of ICT on a country's productivity, it still has its own econometric challenges which make testing the true relationship very challenging (Cardona and others [288])."

"Pioneer studies testing the impact of ICT on productivity were conducted for the United States and OECD countries which were the natural forerunners in the usage of new technology."

"A probable explanation for the missing link between TFP and ICT was perhaps the low level of ICT stocks and investments relative to the GDP, making it difficult for developing countries to benefit from positive network externalities."

"Other more recent studies using data after the millennium have reached different results implying that ICT investments were associated with higher economic growth in developing countries."

Framework and methodology

"In the capacity-based approach (model I), the ICT endowment is measured by real ICT investment growth rates ictinv and the capacity of network switching exchanges per million inhabitants switch."

"The ICT usage approach is implemented in two alternative versions: In model II usage is measured by how much the households pay for using the communication networks."

"Control variables used in both approaches include: Human capital derived from the average years of schooling (hc), globalization measured by exports plus imports as a percentage of the country GDP (global), and GDP per capita (gdp)."

"The econometric implementation allows for non-linear relationships as well as for country-specific intercepts and is specified as follows: Capacity-based approach: Models Ia and Ib ICT revenues-based approach: Models IIa and IIb ICT physical usage approach: Models IIIa and IIIb i represents the country, t represents the observed year."

Data

"The PWT data has been combined with datasets from the International Telecommunication Union ITU, which provides a broad range of ICT usage and investment data in the World Telecommunication Indicators Database."

"Although both the Penn World Tables as well as the ITU data cover most countries in the world, we are using only a subset of 76 countries because first of our focus on developing countries, and second because of missing information especially in the ITU database."

"For developing countries, there are considerable differences in TFP growth and ICT investments/usage across countries."

"After the global financial crisis in 2008, TFP growth rates in the developing world declined significantly, with the latest numbers showing stagnation."

"While some countries like Bolivia or Kyrgyzstan showed annual growth rates of more than 10% over the observation period, other countries like the Central African Republic or Barbados had annual declines of about − 3%."

Empirical results

"Because of the focus of this study on the impact of ICT on TFP growth, signs and significance of the ICT variables are of special interest."

"In Ia and Ib we find all ICT variables—the linear and the squared—statistically significant, while in the capacity-based approaches only some of the ICT variables are significant."

"A first conclusion of model Ib is the significantly positive, but diminishing impact of ICT investment growth on TFP."

"Similar to the more general ICT investment variable, the parameters for the network switching capacities are positive for the linear and negative for the quadratic term, again indicating diminishing returns of the network switching capacity."

"The perhaps most interesting results are the parameters of the human capital variable, which have a negative linear and a positive squared term, indicating a convex relationship with TFP."

Conclusions

"Effective usage of IT equipment requires many other complementary investments into factors like human capital and the provision of a reliable telecommunication infrastructure which many of the developing countries still lack."

"Investments in technology can equip people with better tools that enhance their productivity and in turn enhance the growth process of the developing countries."

"Based on our empirical results and realistic assumptions, we estimate that developing countries can improve their TFP growth rates by about 0.1–0.3% annually by increasing investments in ICT and expanding the fixed line network."

"Governments of developing countries still have the chance to benefit from many relatively low cost opportunities that enable them to provide IT access to their people."

"A prominent example of leapfrogging is when the developing countries use wireless communication technologies rather than having to pass through the stage of having a well-developed wire-line infrastructure."

4.1.39 Racial Disparities in Health Information Access: Resilience of the Digital Divide [536]

This is a machine-generated summary of:

Lorence, Daniel P.; Park, Heeyoung; Fox, Susannah: Racial disparities in health information access: resilience of the digital divide [536]

Published in: Journal of Medical Systems (2006)

Link to original: https://doi.org/10.1007/s10916-005-9003-y

If you want to cite the papers, please refer to the original.

For technical reasons we could not place the page where the original quote is coming from.

Abstract-Summary

"Policy initiatives of the late 1990s were believed to have largely eliminated the information "Digital Divide.""

"For healthcare consumers, access to information is an essential part of the consumer-centric framework outlined in the recently proposed national health information initiative."

"This study sought to examine how racial/ethnic characteristics are associated with Internet use and online health information."

"Using a cross-sectional nationwide study of reported Internet use and information search in 2000 and 2002, we studied a stratified sample of computer users from the Pew Internet & American Life Project surveys."

"Differences in health information seeking between Hispanics and Whites existed in both 2000 and 2002."

"56% of White Internet users at some time searched for online health information, whereas 42% of Hispanic Internet users did so in 2000."

Introduction

"The U.S. Department of Health and Human Services (HHS) noted recently that "life expectancy and overall health have improved in recent years for a large number of Americans, due to an increased focus on preventive medicine and dynamic new advances in medical technology."

"HHS has noted that statistical reviews of the disproportionate toll of certain diseases on racial and ethnic populations highlight the need for prevention, treatment, and resources toward reducing the loss of life [289]."

"Various programs have been initiated across levels of government designed to eliminate racial and ethnic differences in access to healthcare services, aimed at minimizing disparities and preventing, screening, diagnosing, and treating underserved populations for health condit-ions that have a disproportionate and adverse impact on minority populations."

"Given the NHII's focus on increased consumer involvement, the persistence of "Information Disparities" across racial and ethnic lines may severely limit their participation in a consumer-centric, shared decision-making healthcare environment. [290]."

Literature review

"Related to disparities and underserved groups in general, we note that a number of programs have also sought to eliminate the "Digital Divide," or a gap where underserved groups are denied access to digital information due to their demographic makeup."

"To address the Digital Divide, however, healthcare managers and policymakers will need to develop, under the NHII plan, new strategies or interventions to help underserved groups gain access to information relevant to their healthcare [291]."

"In related studies, Masi and others, concluded that Internet access to health information empowered underserved groups, given adequate access to usable technology [292]."

"Are there significant disparities across racial groups regarding access to computers, the Internet, and online health information, within the same income level?"

"Despite telecommunications and technology advances, have differences across racial and ethnic groups in access to computers, Internet services, and online health information narrowed, remained constant, or widened over recent years (2000–2002), following the creation of Divide-reducing programs?"

Methods

"The targeted group for this survey was selected from a random digit sample of telephone numbers, obtained from telephone exchanges in the continental United States."

"The design of the sample achieves this representation by random generation of the last two digits of telephone numbers selected on the basis of their area code, telephone exchange, and bank number."

"The sample was released in replicates to make sure that the telephone numbers called were distributed appropriately across regions of the country."

"All interviews completed on any given day were considered the final sample for that day."

"Analyses were performed at the computer user, Internet user, and online health information user levels."

"A Pearson chi-square test was applied to examine and to compare the differences in access to computers, Internet service, and online health information, across socioeconomic groups separately for each year."

Results

"There was no significant difference between African-American and White online health information seekers, either in 2000 and 2002."

"Access to online health information of total participants exhibits a significant difference between Whites and African-Americans and between Whites and Hispanics."

"18.89 and 24.31% of Hispanics and African-Americans, respectively, ever explored health information on the Web, whereas 32.67% of Whites did."

"The comparison of ORs in 2000 and 2002 indicates that the gap in access to online health information between White and African-Americans ($OR_{at\ 2000}$ = 0.825 [0.600, 1.135]; $OR_{at\ 2002}$ = 0.697 [0.492, 0.989]) widened, and the gap between Whites and Hispanics ($OR_{at\ 2000}$ = 0.543 [0.371, 0.794]; $OR_{at\ 2002}$ = 0.702 [0.513, 0.959]) narrowed in 2002."

"The four groups (White, African-American, Asian-American, and Hispanic-American) showed little difference in the use of computer, Internet, and health information in all income levels in 2002."

Discussion

"The Digital Divide between low- and high-income populations still exists, and has not been improved, although the availability of computers and Internet access in the United States has increased significantly overall."

"A possible explanation of the resilience of the Digital Divide is the relatively low growth rate of the use of computers by the low-income adult population."

"The small growth rates suggest that national initiatives aimed at reducing the Divide have had little effect in providing low-income adult populations with opportunities to use computers."

"We identified the existence of differentiated patterns of access to and use of the Internet by low-income populations as a whole, examining the degree to which annual income level influenced distribution patterns and diffusion trends in access to online health information."

"These rates suggested that the computer utilization levels of low-, medium-, and high-income populations, respectively, increased 3.8, 5.6, and 4.1% between 2000 and 2002."

Conclusion

"For consumers of health information, the technology initiatives of the late 1990s across the US appear to have had little effect in eliminating the Digital Divide."

"As the National Health Information Infrastructure begins to take shape, access to information will be an essential part of the consumer-centric, shared decision-making framework outlined in the NHII strategic plan."

"We suggest that to ensure all groups enjoy the potential benefits of the planned information infrastructure, it is necessary to first identify characteristics of all underserved subpopulations, who may lack equal access to health information."

"On the basis of this identification, new strategies and interventions are needed to target underserved populations and help them develop search information seeking skills, relevant to their healthcare."

4.1.40 The Digital Divide: Conveying Subtlety in Online Communication [537]

This is a machine-generated summary of:

Riordan, Monica A.; Kreuz, Roger J.; Blair, Alyssa N.: The digital divide: conveying subtlety in online communication [537]

Published in: Journal of Computers in Education (2018)

Link to original: https://doi.org/10.1007/s40692-018-0100-6

If you want to cite the papers, please refer to the original.

For technical reasons we could not place the page where the original quote is coming from.

Abstract-Summary

"In a third experiment, 72 participants produced responses to 16 text message prompts each, and these responses were coded for line breaks and periods to test whether production of these cues varies between natives and immigrants."

"The results suggest that immigrants and natives are alike in how they interpret messages, but that natives are more sensitive to minor linguistic cues, especially the use or nonuse of a period in a text message, considering this cue to carry more negative affect than immigrants do."

"This suggests that, even in cases in which immigrants make use of the same communication technology to the same extent as natives, they still have a digital "accent," and fail to make subtle distinctions that are meaningful to natives."

"As texting becomes increasingly used as a classroom management or communication tool, older students and faculty must be sensitive to the fact that younger students may consider the use of periods to signal negative affect and may respond differently to such messages than intended by the writer."

"We issue a call for more research exploring how the use of technology, and even subtle cues, may impact classroom dynamics, particularly in classrooms made up of mixed age groups."

Introduction

"The invention of writing and mechanical calculation, for example, allowed people to offload some of their cognitive processes, such as long-term memory and calculation, onto external systems."

"Writers such as Nicholas Carr have expressed concern that the Internet was having detrimental effects on cognition, arguing that the superficial nature of the Web diminishes the capacity for deep concentration and contemplation (Carr [293, 294]), cornerstones of higher education."

"As faculty increasingly integrate such online communication platforms into their classrooms (in estimates as high as 80%, found in a survey by Moran and others [295]), an exploration of how even subtle cues may affect the perception of a message should be explored."

"The current study was undertaken to assess whether two such subtle cues, line breaks and periods, affect the perception of messages as well as whether this perception differs between older and younger age groups, who show consistently different preferences in how they use online communication."

Literature review

"He goes so far as to claim that the brains of digital natives are "physically different" from those of older generations (Prensky [296], p. 1), whom he labels digital immigrants."

"One might assume, therefore, that there exists a scientific consensus about the effects of networked and digital technology on these so-called digital natives, and the "digital divide" (Waycott and others [297]) that exists between them and earlier generations."

"Many others have continued to raise generational concerns particularly with regard to traditional educational practices, which were originally designed for digital "immigrants" (i.e., those born before the mid-1980s; Prensky [298])."

"A great deal of this commentary has been quite critical of younger people, as can be seen in books with alarmist titles like The Dumbest Generation: How the

Digital Age Stupefies Young Americans and Jeopardizes Our Future (Bauerlein [299])."

Technology in the classroom

"In an increasing number of cases, the classroom itself is online: 46% of college students between 2001 and 2011 took an online course."

"It is thus not surprising that 50% of college presidents surveyed by Pew Research Center (301) predicted that most of their students would take classes online and most undergraduate textbooks would be digital by 2020."

"Traditional age students are no longer the vast majority; demographics have shifted such that older students are increasingly represented in higher education."

"As education becomes increasingly digitalized and student demographics become increasingly older, it is also increasingly important to explore the experiences that students of different ages have in online classrooms, particularly in light of evidence showing that natives and immigrants have different preferences regarding how to use technology to communicate with others (Taipale [301])."

Digital natives and immigrants in the classroom

"Younger students were also more likely to upload and download materials from the Internet than older students, particularly TV or video, and were more likely to use social networking sites (although not necessary for class purposes)."

"Although some instructors perceive the Internet to be a distraction from class activities and content, other faculty have been eager to adopt different forms of online communication for their classrooms."

"Ajman and Hartshorne [302] found that 32% of university professors surveyed felt that the use of social media led to greater student satisfaction in classes, and 56% felt it enhanced interaction among classmates."

"The use of online components of the classroom enhanced student grades in a study by Pasek and others [303]."

The current study

"Crair [304] asserts that the period has been replaced with a line break, such that rather than writing two sentences divided by a period, one writes two sentences as two different texts: I am much more likely to type two separate messages without punctuation: sorry about last night next time we can order little caesars Than I am to send a single punctuated message: I'm sorry about last night."

"Time we can order little caesars. (Crair, [304], n.p.) In the current study, the meaning of a period at the end of a text message, as well as the existence of a line break in the message, is explored."

"It is hypothesized that a generation-specific digital divide between natives and immigrants will lead to different affective interpretations of line breaks and periods in text messages."

Experiment 1

"A linear mixed-effects model accounted for the fact that all participants saw all possible conditions of a period and a line break, but saw only one of the four possible response conditions for each individual text message exchange."

"A linear mixed-effects model was conducted to determine whether the groups rated the text message exchanges differently on the negative–positive affect scale."

"For digital natives, the presence of a period led to more negative affect ratings (– 0.07, $p < 0.05$), but a line break had no effect (– 0.03, $p = 0.42$) and neither did a combination the two variables (0.03, $p = 0.50$)."

"Despite having the same access to and familiarity with digital communication, digital natives consistently rated text messages as more negative than digital immigrants and also interpreted the presence of a period with more negativity."

Experiment 2

"For digital immigrants specifically, there was no effect of the presence of a period (– 0.09, $p = 0.06$), a line break (– 0.09, $p = 0.07$), or both (0.11, $p = 0.07$; intercept = 1.99), a finding that replicates the results from Experiment 1."

"As in Experiment 1, no significant differences were found between natives ($M = 2.90$, $SD = 1.81$) and immigrants ($M = 2.93$, $SD = 1.95$) in self-reported use of text messaging abbreviations (two-tailed $t(71) = -0.07$, $p = 0.95$)."

"In Experiment 1, no difference was found between immigrants and natives self-reporting their use of texting abbreviations."

"Although both groups were frequent texters, digital natives again rated text messages as more negative than digital immigrants, and perceived more negativity when a period and line break were present than immigrants did."

Experiment 3

"Digital natives and digital immigrants were equally likely to use periods in their text messages (– 0.03, $p = 0.71$; intercept = 0.69); however, affect ratings were more negative for digital natives than immigrants (0.09, $p < 0.05$; intercept = 2.13), suggesting that natives are more likely to view the period as a marker of negative affect than immigrants."

"There was no significant difference between groups in rates of self-reported text messaging frequency (Natives $M = 6.54$, $SD = 0.97$, Immigrants $M = 6.09$, $SD = 1.76$; two-tailed $t(70) = 1.37$, $p = 0.18$), or in rates of self-reported text abbreviation use (Natives $M = 3.82$, $SD = 1.99$, Immigrants $M = 3.45$, $SD = 1.80$; two-tailed $t(70) = 0.81$, $p = 0.42$)."

"In Experiment 1, no difference was found between immigrants and natives self-reporting their use of texting abbreviations."

General discussion

"Taken as a whole, the three experiments suggest that digital natives are more sensitive to the presence or absence of small linguistic cues, particularly periods, which may alter the social or pragmatic information of a text message."

"This effect cannot be attributed to experience with texting or to the use of "textspeak" (measured here as abbreviations; see Crystal [305]), although digital immigrants are somewhat less likely than digital natives to interpret abbreviations correctly when reading text messages."

"Riordan [306], for example, did not find a difference between digital natives and immigrants in their interpretations of affect in text messages that employed emoji."

"As with any nonnative language learner, the digital immigrant may become more fluent in using digital language with more experience, particularly when practiced along with native speakers, and they may be able to learn subtle distinctions such as employing or not employing a period at the end of text messages."

Conclusion

"This study suggests that as education is delivered increasingly online, and elements of communication between classmates and between teachers and students become increasingly computer-mediated, both teachers and students need to be sensitive to differences in interpretation of digital communications."

"Even students who have regular access to digital platforms, and who frequently use them, may develop different accents that alter interpretations of written communication in online forums or platforms, leading to misunderstandings between teachers and students, or among students themselves."

"As college students are increasingly older in age, the digital divide between traditional age students and older nontraditional age students, as well as the digital divide between teachers and students, should garner more attention and research, particularly with regard to possible differences in experience within the increasingly common online classroom."

4.1.41 Digital Divide, Biometeorological Data Infrastructures and Human Vulnerability Definition [538]

This is a machine-generated summary of:

Fdez-Arroyabe, Pablo; Lecha Estela, Luis; Schimt, Falko: Digital divide, biometeorological data infrastructures and human vulnerability definition [538]

Published in: International Journal of Biometeorology (2017)

Link to original: https://doi.org/10.1007/s00484-017-1398-x

If you want to cite the papers, please refer to the original.

For technical reasons we could not place the page where the original quote is coming from.

Abstract-Summary
"The design and implementation of any climate-related health service, nowadays, imply avoiding the digital divide as it means having access and being able to use complex technological devices, massive meteorological data, user's geographic location and biophysical information."

"This article presents the co-creation, in detail, of a biometeorological data infrastructure, which is a complex platform formed by multiple components: a mainframe, a biometeorological model called Pronbiomet, a relational database management system, data procedures, communication protocols, different software packages, users, datasets and a mobile application."

"This technological approach to study weather-related health impacts is the initial seed for the definition of biometeorological profiles of persons, and for the future development of customized climate services for users in the near future."

Introduction
"This paper describes how a biometeorological model based on a specific index is implemented into a technological platform to create the first biometeorological data infrastructure (BDI) in the world, as a new technological milestone to offer health-related services (Kirch and others [309]) to users according to the Global Framework for Climate Services (WMO [310])."

"BDI allows researchers (i) to define the vulnerability of users through the collection of their own digital reports, (ii) to improve the efficiency of the system through this feedback, (iii) to geo-locate users through their mobile devices and consequently to reduce spatial uncertainties in relation to exposure, (iv) to introduce the idea of co-design and participatory design in the process of creating an early warning systems and (v) to develop a citizen science approach (Irwin [307]; Hand [308]) which is an important aspect that must be incorporated in the creation of future biometeorological research."

Technological platform
"The main tasks of the server are the following: connecting to the Global Forecast System GFS-NOAA mainframes to download raw datasets of meteorological variables forecasting that are needed, de-codification of GRIB2 formats and computing the biometeorological model, facilitating map production under a GIS system based on Python language and through a web platform based on an application programme interface (API) from Google, registering new users, informing users about oxygen levels and collecting reports sent by users from the mobile application."

"Oxygen values are sent from the server to the mobile device where differences will be computed."

"Before downloading the app, each user will sign a digital agreement accepting the conditions and allowing the use for scientific purposes of any information they could facilitate through the App."

"Permission to use this information in future scientific research is also given by users in this process."

Results

"A preliminary review of the first 1000 reports received from users has been considered for an exploratory study of the convenience of the technical mechanisms defined and the information collected by the BDI in order to delineate people vulnerability in the future."

"Many users reported only once while some were more persistent in answering the questionnaire."

"An explanation would be that users reported because of existence of specific uncomfortable situations due to heat stress, even if oxygen changes during this period were not significant to produce alerts."

"Users' reports are linked to the existence of an alert in the mobile application."

"Considering the user who has reported most to the system (35 reports), and assuming the limitations and biases mentioned before, a graphical representation of his theoretical vulnerability to oxygen changes has been carried out."

Conclusions

"The BDI is more than a technological platform based on a computer acting as a server and a mobile application giving biometeorological forecasting, or a geographic information technology editing world maps and registering information in a relational database management system."

"More than the technological platform and the biometeorological model, a controlled group of users is essential for setting up the BDI."

"A BDI must be targeted according to a very specific conceptual and biometeorological model of how atmospheric processes are related to living organisms on the Earth."

"It is also important that users' feedback can make researchers re-think and re-design their initial biometeorological conceptual approaches and theories."

"Registration of user information in different report tables contributes enormously to the definition of individual vulnerability and biometeorological profiles."

"A BDI gives us the chance to start registering useful real-time information to define individual biometeorological profiles corresponding to specific meteorological changes."

4.1.42 A Digital Language Divide? The Relationship Between Internet Medication Refills and Medication Adherence Among Limited English Proficient (LEP) Patients [539]

This is a machine-generated summary of:

Casillas, Alejandra; Moreno, Gerardo; Grotts, Jonathan; Tseng, Chi-Hong; Morales, Leo S.: A Digital Language Divide? The Relationship between Internet Medication Refills and Medication Adherence among Limited English Proficient (LEP) Patients [539]

Published in: Journal of Racial and Ethnic Health Disparities (2018)

Link to original: https://doi.org/10.1007/s40615-018-0487-9

If you want to cite the papers, please refer to the original.

For technical reasons we could not place the page where the original quote is coming from.

Abstract-Summary

"Use of an Internet portal to refill medicines positively affects medication adherence among English-speakers."

"No prior studies, however, have specifically examined the relationship between Internet refills and medication adherence among patients who are limited English proficient (LEP)."

"For Internet refill system use, patients were asked, "Have you used the health systems Internet site to refill any medications in the last 12 months?""

"We used multivariate linear regression models to examine Internet refill system use and medication adherence and compared the association between LEP and EP patients."

"Three hundred eighty-four patients (75%) had a calculable CMG: 134 EP and 250 LEP in the adherence analyses."

"In unadjusted analyses, LEP patients had lower use of the Internet refill system ($p < .001$) and lower adherence versus the EP group ($p < .001$)."

"In multivariate analyses, LEP status ($\beta = -0.022$, $p = .047$) was negatively associated with adherence, while use of the Internet refill system ($\beta = 0.030$, $p = .002$) was positively associated."

"In stratified models, use of Internet refills was positively associated with adherence, even when examining LEP ($\beta = 0.029$, $p = .003$) and EP patients ($\beta = 0.027$, $p = .049$) separately."

Background

"These findings suggest that the same subgroups of vulnerable populations who have traditionally struggled to access medical care in the USA and have faced stark health disparities compared to the general population, do not utilize technological advancements that could provide efficient disease management—which include remote medication refills through an Internet patient portal [311–314]."

"Prior research examined the relationship between English proficiency and use of Internet medication refill systems and found that only 21% of LEP patients with chronic illness used the online service, compared to 53% of English-proficient patients [315]."

"No prior studies, that we are aware of, have examined the association between remote Internet refills through a patient portal and medication adherence, among LEP patients—likely because the majority of patient portals are English only."

Methods

"CMG is an objective measure of adherence that uses pharmacy filling of medications to measure gaps in patients' available supply of medications."

"The CMG adherence measure was based on at least two pharmacy fills on an existing prescription over a 12-month period preceding the survey (medication possession ratio)."

"CMG adherence for three medication groups (typically treating the three chronic conditions of patients in the study) was examined: (1) oral hypoglycemics, (2) anti-hypertensives, and (3) lipid-lowering drugs."

"Other survey measures were age, gender, marital status, race/ethnicity, language/dialect, education, household income, years in the USA for foreign-born patients, Medicare/Medicaid insurance status, health status, chronic condition, number of prescribed medications, and social support (eight questions gauging resources; final score of 0 "no support" to 40 "always supported")."

"We used multivariate linear regression models to examine the adjusted effects of use of the Internet refill system on overall medication adherence and compared this association between LEP and EP patients using stratified models."

Results

"LEP patients had significantly lower use of Internet remote refill systems compared to the EP patients (22 vs. 56%, $p < 0.001$)."

"Use of the Internet refill system ($\beta = 0.030$, $p = .002$), Medicare or Medicaid insurance ($\beta = 0.030$, $p = .007$), and a college/university education (or higher level) ($\beta = 0.0029$, $p = .029$) were positively associated with adherence."

"In stratified models, use of remote Internet refills was positively associated with adherence, even when examining LEP ($\beta = 0.029$, $p = .003$) and EP patients ($\beta = 0.027$, $p = .049$) separately."

"In the EP model of overall adherence, Medicare/Medicaid insurance status was positively associated ($\beta = 0.040$, $p = .01$), while among LEP patients, hyperlipidemia was negatively associated with adherence ($\beta = - 0.035$, $p = 0004$)."

Discussion

"We found that the use of remote medication refills through an Internet portal was independently and significantly associated with higher medication adherence in both EP and LEP patients."

"Compared to EP patients, LEP status was associated with lower use of Internet portal refills and decreased medication adherence even after adjusting for covariates."

"The Group Health Internet portal was offered in English during the time of the study, and so, even LEP patients utilized the refill system in English."

"Examining the medication adherence gaps in LEP populations today in these commercial systems, and the effect of Internet patient portal use, is necessary, as these populations are insured."

"These findings suggest that LEP patients with chronic conditions are at an increased risk of sub-optimal adherence and are under-utilizing an Internet portal tool that is significantly associated with better medication adherence."

4.1.43 Health Literacy and the Digital Divide Among Older Americans [540]

This is a machine-generated summary of:

Levy, Helen; Janke, Alexander T.; Langa, Kenneth M.: Health Literacy and the Digital Divide Among Older Americans [540]

Published in: Journal of General Internal Medicine (2014)

Link to original: https://doi.org/10.1007/s11606-014-3069-5

If you want to cite the papers, please refer to the original.

For technical reasons we could not place the page where the original quote is coming from.

Abstract-Summary

"The purpose of the study was to quantify the relationship between health literacy and use of the Internet for obtaining health information among Americans aged 65 and older."

"Our analysis included measures of regular use of the Internet for any purpose and use of the Internet to obtain health or medical information; health literacy was measured using the Rapid Estimate of Adult Literacy in Medicine–Revised (REALM-R) and self-reported confidence filling out medical forms."

"Only 9.7% of elderly individuals with low health literacy used the Internet to obtain health information, compared with 31.9% of those with adequate health literacy."

"The gradient arose both because individuals with low health literacy were less likely to use the Internet at all (OR = 0.36 [95% CI 0.24 to 0.54]) and because, among those who did use the Internet, individuals with low health literacy were less likely to use it to get health or medical information (OR = 0.60 [95% CI 0.47 to 0.77])."

"Low health literacy is associated with significantly less use of the Internet for health information among Americans aged 65 and older."

"Web-based health interventions targeting older adults must address barriers to substantive use by individuals with low health literacy, or risk exacerbating the digital divide."

INTRODUCTION

"[316, 317] Health literacy may also play an important role in determining whether elderly patients are willing and able to use EMRs and other Internet-based health tools."

"[318, 319] One study [320] used a large sample of patients with diabetes enrolled in a single health plan in northern California to document the importance of health literacy as a predictor of Internet-based patient portal use, even after controlling for basic demographic characteristics."

"This finding supports the idea that individuals with low health literacy are on the wrong side of the digital divide; however, no population-based study to date has estimated the relationship between health literacy and use of the Internet to obtain health information in the general population of older Americans."

"The current study fills that gap by exploring both how low health literacy is related to use of the Internet for any purpose and whether, conditional on being Internet users, those with low health literacy are less likely to use the Internet for the purpose of obtaining health or medical information."

METHODS

"[321, 322] The current study makes use of data from a special module on health literacy that was administered to a random subset of HRS respondents in 2010 and a supplemental Internet-based survey administered in 2009 to respondents who reported in 2008 that they regularly used the Internet."

"Health literacy questions were slated to be administered to a subsample of 1,168 respondents aged 65 and older randomly drawn from 2010 HRS core respondents who (1) responded to the 2010 core survey themselves, as opposed to having a proxy respondent answer for them, and (2) were included in the random half-sample of respondents receiving face-to-face rather than telephone interviews in 2010."

"Analysis, we considered 1,121 community-dwelling respondents 65 years of age or older; of these, 898 completed the health literacy module, for an overall module response rate of 80%."

"Among those who responded, 1,617 individuals were assigned the health literacy measures from the survey, of whom 1,584 gave valid responses."

RESULTS

"Low health literacy, whether self-assessed or measured using the REALM-R, significantly reduced the odds of regular Internet use (for self-assessed health literacy, OR = 0.36 [95% CI 0.24 to 0.54]; for health literacy measured using REALM-R, OR = 0.25 [95% CI 0.15 to 0.41])."

"Low health literacy significantly reduced the odds of Internet use for medical or health information (OR = 0.60 [95% CI 0.47 to 0.77])."

"Low cognition had relatively less impact on the likelihood of using the Internet for medical or health information, conditional on use of the Internet."

"Our results suggest that once older adults do use the Internet for any task, poor cognition is a smaller barrier than low health literacy with respect to Internet for health-related tasks."

DISCUSSION

"For those who did use the Internet, low health literacy was also predictive of what they did once they were online; individuals with low health literacy were significantly less likely to use the Internet for medical or health information."

"We found that health literacy was a more important predictor of Internet use for medical or health information than was level of cognitive function, suggesting that interventions specifically targeting health literacy among Internet-using older adults may be effective for narrowing the digital divide by facilitating their ability to obtain medical information online."

"It allows us to understand the extent to which older individuals with low health literacy are not even at the starting gate in utilizing specialized applications such as a patient portal, in that they do not use the Internet at all."

"We found that this fundamental obstacle was highly prevalent: more than three-quarters of older individuals with low health literacy did not use the Internet for any purpose."

CONCLUSION

"Individuals with low health literacy represent a vulnerable population who are at high risk of being left behind by the advance of technology."

"Interventions to improve health literacy may have a spillover benefit of lowering barriers to effective use of information technology."

"Even in the absence of such interventions, screening for low health literacy in the clinical setting can help clinicians identify patients who are likely to have difficulty adopting electronic health technology."

"Low health literacy may attenuate the effectiveness of Web-based interventions to improve the health of vulnerable populations."

4.1.44 The Digital Divide Across All Citizens of the World: A New Concept [541]

This is a machine-generated summary of:

James, Jeffrey: The Digital Divide Across All Citizens of the World: A New Concept [541]

Published in: Social Indicators Research (2007)

Link to original: https://doi.org/10.1007/s11205-007-9156-9

If you want to cite the papers, please refer to the original.

For technical reasons we could not place the page where the original quote is coming from.

Abstract-Summary

"This paper is concerned with the requirements for and implications of, moving from the confines of the conventional concept of the digital divide to one that reflects a world distribution of Internet users with different income levels, with particular reference to those users living in poverty."

"One mechanism that is associated with the revised concept is shown to substantially reduce the size of the digital divide as conventionally measured."

"The other mechanism, by contrast, works in the opposite direction because the increased number of users in developing countries are drawn from high-income, educated and urban classes, creating a highly unequal distribution within these countries."

Introduction

"In reality, the conventional concept of the digital divide—as the gap in ICT use between rich and poor countries—provides us with little or no information on the divide between the citizens of the world, who are differentiated on the basis of income levels, information which would allow us to assess Internet use and welfare among the poorest members of global society."

"Since even this more accurate map provides no information about Internet use within each country a further step is needed to arrive at an array of global users, ranked by their income levels, regardless of where they live."

"Given the difference between Internet use per 100 citizens, weighting by population size eliminates the original divide; there are now 2 Internet users in each country."

Conclusions

"This note has advanced the case for a new concept of the digital divide, one which takes into account the different population size of developing countries as well as the way in which the income of Internet users varies within such countries."

"What emerges from this exercise is that when developing countries are weighted by population size, the exceptionally rapid recent growth of the Internet in China serves to narrow even further existing measures of the shrinking divide between rich and poor countries."

"Moving empirically towards the new concept of the digital divide can, as shown above, be most readily accomplished by taking a weighted measure of Internet use across developing countries."

"Finding data on the intra-country distribution of Internet users by income level in developing countries, on the other hand, is a far more difficult and expensive undertaking."

4.2 Digital Exclusion and Inclusivity

Machine generated keywords: digital inclusion, inclusion, citizen, inclusive, exclusion, ict, participation, local, capital, algorithm, adoption, inequality, policy, government, depend

4.2.1 Employing Online Social Networks to Monitor and Evaluate Training of Digital Inclusion Agents [542]

This is a machine-generated summary of:

de Brito, Silvana Rossy; da Silva, Aleksandra do Socorro; Martins, Dalton Lopes; Vijaykumar, Nandamudi Lankalapalli; da Rocha, Cláudio Alex Jorge; Costa, João Crisóstomo Weyl Albuquerque; Francês, Carlos Renato Lisboa: Employing online social networks to monitor and evaluate training of digital inclusion agents [542]

Published in: Social Network Analysis and Mining (2013)

Link to original: https://doi.org/10.1007/s13278-012-0093-5

If you want to cite the papers, please refer to the original.

For technical reasons we could not place the page where the original quote is coming from.

Abstract-Summary

"This work presents challenges in evaluating and monitoring digital inclusion training programs, considering the aspects inherent in large-scale training, and report the main challenges in an approach based on network learning."

Introduction

"Participants in a digital inclusion agent training program thus have different profiles, according to regional particularities, age, education, social participation, work on community projects, and other specific characteristics, not always observed and mapped."

"Training communities and digital inclusion agents based on using online social networks is a proposal that brings challenges for implementing, evaluating, and monitoring these programs: Interactions motivated by training."

"In online social networks, agents, and their communities discuss different topics with various actors, who may participate in a program of digital inclusion."

"Focusing on these challenges, to evaluate and monitor these interactions, the social network analysis (SNA) technique is distinguished by generating indicators that can be combined with learning environment use indicators to generate valuable insights that can help decide important strategies regarding instructional design, proposals for community projects, and articulation in digital inclusion agent networks."

Training program for digital inclusion: the Telecentros.BR network

"BR Network is structured to promote cooperation, using the following relationships: (a) among members of different centers (team training), (b) among monitors of telecenters (digital inclusion agents in training), and (c) among members of centers and monitors."

"Based on these categories, it was possible to identify the following interactions among the participants: relationship (a), among members of different centers, primarily contains the collective construction of content, knowledge sharing, and socialization; relationship (b), among monitors of telecenters, mostly has socializing, sharing community projects, and sharing knowledge to solve everyday telecenter problems; relationship (c), among members of the center and digital inclusion agents, particularly contains sharing content and knowledge."

"The characteristics of these relationships reveal the actions of those who directly organize and build them through collaboration resources and learning activities (members of centers) and telecenter monitors coming with the expectation to take a course to act as digital inclusion agents in their local communities."

Information system architecture

"Mapping these profiles (mapping profile) and identifying similarities in monitor groups (clustering) are some actions that may help network connectivity strategies; digital inclusion agents, especially when developing community projects, discuss different topics with several actors who may be training program participants."

"Mapping the different routes taken by monitors (groups) is a form of feedback for the process, allowing one to monitor and evaluate the training courses' structure and organization; Monitors, by sharing knowledge, questions, expectations, results, and lessons learned, build social networks with other digital inclusion agents and their communities."

"Mapping and analyzing the social networks formed from these articulations are strategies that can provide important indicators that the training provided for monitors helps to increase the density of these networks, expanding scope of actions with community participation (analysis of interaction networks)."

Analyzing interaction networks

"The indicators are determined from a set of metrics obtained from applying social network analysis to monitors and training center members."

"The metric closeness centrality is based on distance and proximity measures of a node in relation to other nodes in the network (Wasserman and Faust [323])."

"In the centrality analysis, the indicator degree centrality (average) of only network centers, when compared to the network of "centers+monitors", decreases 60.49%, while the degree centrality in the monitors network grows 231.16% compared to the network of "centers+monitors"."

"The closeness centrality indicator (highest) of the network "centers+monitors" decreases 8.71 and 20.77% in networks containing only center members and only monitors, respectively."

"The proximity prestige index (average) of the network "centers+monitors" grows 10.51 and 17.15% in the networks containing only center members or monitors, respectively."

Attribute correlation: decision support for indicator analysis

"These compositions were performed on attributes to enable model construction (Bayesian Network), which probabilistically correlate and measure the influence of monitor participation in learning environment using the proposed resources and activities on SNA indices, as established in the goals of this package in layer S_Intelligent."

"On the same network, to achieve the highest value of the Degree_Centrality attribute, given evidence of an increase in the number of post blogs within the band from 5 to 34 (fourth band of values of the post blog attribute), the probability is 39.9%."

"In the monitors network, given evidence of an increase in number of blog posts, i.e., more than five posts (fourth band of values of the Blog_Post attribute), the probability that Degree_Prestige is in the higher band is 40.4%."

"From the prospection (inferences) over the Bayesian networks, it is possible to present monitors with several scenarios that can promote variations in forums, blogs, prestige, and centrality indices, given the resource and task uses."

Conclusion

"Developing the architecture for the environmental analysis of social networks within the training network also emerges as an important contribution because it automates the data collection process, provides a set of interfaces for communication resources, reporting, and analysis."

"From the feedback of pedagogical experts, other demands presented themselves as future directions for this research: considering the impact of tutor participation on the evaluated indicators and allowing the analysis of indicators for using didactic resources (e.g., texts, web pages, videos, animations) by thematic areas (e.g., digital inclusion, community engagement, digital culture, telecenters, and accessibility)."

"In the studies developed here, we sought not only to visualize the structure of networks formed but also to evaluate opportunities to incorporate network analysis metrics in constructing a monitoring methodology and monitoring a large-scale training program."

4.2.2 Balancing Digital-by-Default with Inclusion: A Study of the Factors Influencing e-Inclusion in the UK [543]

This is a machine-generated summary of:

Al-Muwil, Ahlam; Weerakkody, Vishanth; El-haddadeh, Ramzi; Dwivedi, Yogesh: Balancing Digital-By-Default with Inclusion: A Study of the Factors Influencing E-Inclusion in the UK [543]

Published in: Information Systems Frontiers (2019)

Link to original: https://doi.org/10.1007/s10796-019-09914-0

If you want to cite the papers, please refer to the original.

For technical reasons we could not place the page where the original quote is coming from.

Abstract-Summary

"This research aims to examine the factors influencing e-Inclusion in the UK within a digital-by-default policy for government services."

"This study is pursued through combining the Decomposed Theory of Planned Behaviour (DTPB) with Use and Gratification Theory (U&G) and conducting a self-administered survey targeting 510 Internet users to study the level of citizens engagement with e-government services in the UK."

"By incorporating gratification, trust, risk and external factors (i.e. self-efficacy, accessibility, availability, affordability) within DTPB, the proposed model of e-Inclusion used in the paper demonstrates a considerable explanatory and predictive power and offers a frame of reference to study the acceptance and usage of e-government within a national context where nearly all government transactions are digital-by-default."

Introduction

"We draw from Becker's study to look at citizens' use of different Internet activities and level of participation in information society and e-Inclusion based on the latest available data (UK E-Government Fact Sheet [324])."

"To address this research gap from a theoretical angle and to find out the critical e-Inclusion factors that influence citizens' use of online government services within a digital nation (UK), the present research was undertaken."

"Given the parallels between e-government and e-inclusion, it helps to understand how policies, society, organisations, and information technologies come together and how e-Inclusion factors influences e-government use."

"Given this context, this study aims to answer the research question: What are the key factors that influence e-inclusion in the context of e-government within its digital-by-default environment in the UK?"

Research Context and Theoretical Background

"E-Inclusion is mostly concerned with the social impact of relative differences in ICT use between different socioeconomic groups and individuals while e-adoption focuses instead on absolute and average figures of ICT uptake and their economic impact (Kaplan [325])."

"E-inclusion should not be reduced to e-adoption, which would only look at levels of ICT at large and miss the social impact of relative differences in ICT use between various socio-economic groups and individuals (Mancinelli [326])."

"E-Inclusion issues should be considered as important components in e-government theoretical models, either as assessing the social desirability of information technologies in a certain policy domain, affecting the demand of e-government services, or limiting the usefulness of certain government applications."

"Helsper [327] focusing on digital resources grouped e-inclusion into four broad categories namely; ICT access, skills, attitudes and extent of engagement with technologies."

A Conceptual Model for E-Inclusion

"By combining e-inclusion models and those used in e-government adoption, a research model is developed in the present study based on the Decomposed Theory of Planned Behaviour (DTPB) and Uses and Gratifications Theory (U&G)."

"The components of the proposed model hypothesises that the behavioural intention to use behaviour is determined by the following four types of constructs: (1) attitudinal constructs (COMP, PEOU, PU) represent the citizens' favourable or unfavourable evaluation of the behaviour in question (i. e. use of e-government services); (2) normative constructs (Interpersonal influence, media influence, and government influence) represent the perceived pressure by family, friends, media, and government to perform the behaviour in question; (3) control constructs (Capacity, availability, affordability, and accessibility) represent the perceived control over the personal or external factors that may facilitate or constrain the behavioural performance (Tam and others [328]; Ajzen [329]; Rogers [330]; Taylor and Todd [331]); (4) gratifications (content gratifications which represents gratifications related to the repeated use of a media, process gratifications which represent gratifications based on the actual experience of using the Internet, and social gratifications which arise from Internet use as a social environment and provide communication and interaction)."

Methodology

"To test the proposed hypotheses, we used a survey-based research within a quantitative paradigm (Creswell [332]) with a Likert scale (Bryman and Bell [333]) for rating the questions when collecting the respondent's opinion."

"As the aim of this research was to examine the factors affecting e-inclusion in the context of digital-by-default in the UK, collecting data from as many participants from across the country is required."

"The survey approach was selected for the study and the sampling frame constituted citizens from across the UK (south, west, north, and east) representing an overall sample of the population."

"Due to the reason that not every unit (i.e. citizens) of the population is guaranteed to be selected in the sample, the survey was voluntary where only the interested respondents were invited to take part."

"Structural equation modeling (SEM) (Tabachnick and Fidell [334]; Comrey and Lee [335]; Schaupp and others [336]; Horst and others [337]; Hung and others 2009) was used for data analysis from a collected sample of 510 surveys."

Results and Analysis

"Once the model is specified and the fit indices indicate good fit, the construct validity should be assessed."

"The convergent validity means the indicators measuring certain construct share the high proportion of variance in common (Hair and others [338])."

"Convergent validity was assessed by examining: 1) factor loadings, which relate significantly all indicators to their respective constructs; all the absolute values of critical ratios (C.R.) of all the indicators should be greater than 1.96, at the 0.05 level of significance, 2) standardized regression coefficients, which should be greater than 0.50, and 3) the average variance extracted (AVE), which reflects the overall amount of variance in the indicators accounted for by the latent construct."

"Discriminant validity can be assessed using a rigorous test by comparing the average variance extracted values for any two constructs with the square of the correlation estimate between these two constructs."

Discussion and Reflection

"A research model was developed to capture and examine the e-Inclusion factors that influence citizens' use of e-government services based on the Decomposed Theory of Planned Behaviour (DTPB) and Uses and Gratifications Theory (U&G)."

"This study posits that identifying the reasons which prevent people from using e-government will help to enhance our understanding of the critical factors that influence e-Inclusion and why people keep using other Internet activities that required qualification and trust to engage in advanced online activities (such as e-commerce) but they do not use e-government services."

"The results offer policy makers and practitioners a better overview of the broader dimensions of e-Inclusion as well as the most critical factors that prevent people from being part of the digital-by-default environment in the UK (e.g. trust in e-government, process gratifications, perceived ease of use, media influence, interpersonal influence, and capacity)."

Conclusion

"Through analysing the Internet activities use in the UK and identifying existence gaps in e-government use by the citizens in the UK, four gaps were found."

"The use of e-government was the deepest gap that exists in the Internet activities used by the citizens."

"The research approach taken was purposeful for this study, as the key empirical objective was to evaluate the conceptual taxonomy and associated factors among a sample of citizens who were knowledgeable with ICT and e-government services, and to explain the adoption gaps in using different Internet activities by citizens, specifically the gap between using e-government services and other Internet activities."

"Prior research (e.g. Wang and Shih [339]) on e-government has also shown that moderating variables might provide the differences on the relationships between the determinants and intention to use Internet activities (e.g. e-government, e-banking, and e-shopping)."

4.2.3 AI Ageism: A Critical Roadmap for Studying Age Discrimination and Exclusion in Digitalized Societies [544]

This is a machine-generated summary of:

Stypinska, Justyna: AI ageism: a critical roadmap for studying age discrimination and exclusion in digitalized societies [544]

Published in: AI & SOCIETY (2022)

Link to original: https://doi.org/10.1007/s00146-022-01553-5

If you want to cite the papers, please refer to the original.

For technical reasons we could not place the page where the original quote is coming from.

Abstract-Summary
"The analysis of fairness and bias in AI, important for the debate of AI for social good, has paid insufficient attention to the category of age and older people."

"The concept of AI ageism is presented to make a theoretical contribution to how the understanding of inclusion and exclusion within the field of AI can be expanded to include the category of age."

"AI ageism can be defined as practices and ideologies operating within the field of AI, which exclude, discriminate, or neglect the interests, experiences, and needs of older population and can be manifested in five interconnected forms: (1) age biases in algorithms and datasets (technical level), (2) age stereotypes, prejudices and ideologies of actors in AI (individual level), (3) invisibility of old age in discourses on AI (discourse level), (4) discriminatory effects of use of AI technology on different age groups (group level), (5) exclusion as users of AI technology, services and products (user level)."

Introduction
"The scholars of ethical AI express concern about the way those technologies show hidden biases resulting in exclusion and discrimination of members of marginalized groups (Eubanks [340]; Mittelstadt and others [341]) and can pose a threat to fundamental human rights and social justice (Aizenberg and van den Hoven [342]; Cruz [343])."

"In comparison to the salient and influential research findings on unwanted bias relating to gender and race in AI systems, the category of age—critical to the study of social exclusion and social inequalities—has been largely neglected in existing research."

"This paper provides two main contributions to the debates on bias and fairness of AI, diversity, and inclusion in AI, as well as discussions on AI for social good."

"Identification of vulnerable social groups is the way technical understanding of bias and fairness in ML (Wachter and others [344]) is linked with the wider societal impact of AI and the variety of interdependencies between different actors in the realms of AI."

AI for social good and the ageing population

"There is still only limited understanding about what constitutes AI for social good (also AI4SG) and different frameworks emerge with the intent to substantiate and advance the development and application of AI for social good."

"A working definition of AI for social good was coined: "the design, development, and deployment of AI systems in ways that (i) prevent, mitigate, or resolve problems adversely affecting human life and/or the wellbeing of the natural world, and/or enable socially preferable and/or environmentally sustainable developments" (Floridi and others [345], pp."

"Within this framework, AI for social good implies that it should be rendered bias free and meet the principle of fairness and non-discrimination, in addition to bringing positive social impacts in the application of systems."

"I argue that the older population is one group and social category that not only is being excluded from processes of development and deployment of AI, but is also invisible in the debate on ethical, inclusive, and fair AI."

Ageing population, ageism, and technology

"Theories of ageism are following the digitalization of the phenomenon, that is, the presence of age biases, stereotypes, prejudice, and discrimination in their digital form."

"The theoretical reflection and empirical analysis on the potential negative impact of these intelligent technologies and algorithmic systems on older persons has only very recently emerged as a necessary research agenda for sociology of ageing and social gerontology (Chu and others [346]), and a thorough and systematic empirical and theoretical programme for the investigation of the phenomenon of ageism in AI is yet to be created."

"This evidence suggests the continuous exclusion of the older population on a group level—an algorithmic divide understood as an extension, or another level of digital divide."

[Section 4]

"It includes, but is not limited to, five interrelated forms: (1) age biases incorporated in algorithms and digital datasets (technical level), (2) age stereotypes, prejudices, and ideologies of actors in the field of AI (personal/actor level), (3) invisibility or clichéd representations of category of age and old age in discourses around AI (discourse level), (4) discriminatory effects of use of AI technology on older age groups (group level), (5) exclusion as users of AI technology, services and products (user level)."

"Age stereotypes and prejudices are deeply ingrained in the social fabric, in interpersonal relations (2), as well as in discourses, images, and ideologies in the tech industry (3), and these manifestations of AI ageism need research attention to the same extent as the technical bias."

AI ageism in practice: illustrations

"It is argued that ageism in AI is partially fuelled by age biases prevalent in the tech industry, which are visible both in recruitment and hiring practices, as well as ideologies and beliefs related to concepts such as innovation and progress of the tech industry (Stypińska, Rosales, Svensson, forthcoming)."

"With regard to ageing populations, the risk of discrimination lies in the way the biased algorithms are being used in practice in those realms of social, economic and cultural life where they could infringe on the rights of older persons."

"An example of AI technology where older adults might experience ageism as users is the group of products called "conversational AI" which includes virtual assistants and chatbots."

"Further examples of areas where older adults might experience ageism as users of AI-driven technology include diverse smartphone applications where incomplete data for older age cohorts might result in inaccurate results (Rosales and Fernández-Ardèvol [347])."

Conclusions

"The aim of this paper is to turn scholarly attention to ageing populations as a socio-demographic group that can be defined as vulnerable, in relation to data-driven social transformations resulting from increasing use of AI technology in all realms of modern life."

"By introducing a new concept of AI ageism, this article contributes to the scholarly efforts to advance our knowledge of the harmful ways AI can impact the vulnerable group of older adults."

"The increase in datafication, the advancements of AI such as deep learning and proliferation of operative AI in society, and the lack of knowledge of ageism in AI mutually reinforce the urgency of knowing how the category of age relates to AI and how principles of AI for social good could be implemented here."

"The research on bias in AI needs to go beyond these two limitations and include not only age and ageing populations as relevant categories for researching digital inequalities, but also pay attention to the way these inequalities can be compounded in an intersectional way."

4.2.4 Mechanisms for Inclusion and Exclusion Through Digital Political Participation: Evidence from a Comparative Study of Online Consultations in Three German Cities [545]

This is a machine-generated summary of:

Rottinghaus, Bastian; Escher, Tobias: Mechanisms for inclusion and exclusion through digital political participation: Evidence from a comparative study of online consultations in three German cities [545]

Published in: Zeitschrift für Politikwissenschaft (2020)

Link to original: https://doi.org/10.1007/s41358-020-00222-7

If you want to cite the papers, please refer to the original.

For technical reasons we could not place the page where the original quote is coming from.

Abstract-Summary

"Consistent findings of unequal political participation have been motivating different democratic innovations, including those that utilize the opportunities of information and communication technologies for political online participation."

"We investigate who participates in opportunities for political online participation, what explains (non)-engagement and how effective personalized invitations are to increase and diversify participation."

"Our results show that engagement in online participation is indeed significantly biased from the population towards resource-rich individuals who also differ in their assessment of the participation process and its results."

"While online aversion is a barrier for some, distrust in the participation process and lack of interest are more powerful reasons to refrain from engagement."

"Using a randomized-controlled field experiment we can confirm that personalized invitations are an effective instrument for mobilization that increased participation by a factor of four to seven and that can to a limited degree elicit participation from under-represented groups."

Introduction

"The basic premise of democracy as the rule of the people requires as a minimum the participation of citizens in regular elections to choose the political personnel."

"We focus on political participation via online means, the reasons for the commonly low and biased engagement in this form of participation and possible strategies to address these issues."

"A key concern of empirical research into the structure of political participation has usually focused on how many people participate and to what degree these are representative in terms of major social characteristics such as gender, age and education."

"The exceptionally well-researched structure of political participation has clearly established the substantial political inequalities associated with participation."

"Regardless of the actual extent of a legitimacy gap, this discourse and the consistent findings of unequal political participation have both been motivating different democratic innovations, many of which are specifically targeted at increasing citizen engagement in decision-making processes (Smith [348])."

The Internet and political participation

"More than two decades of research into the mobilization potential of the Internet have shown that despite considerable variation, if anything, online participation has had only a small effect on the share of people engaging in politics."

"This study aims to investigate four research questions: The first question explores the structure of online participation and asks (i) who participates in opportunities for political online participation."

"While we can expect to find the unequal patterns that have been documented extensively, information about participants and non-participants is the basis for the second research question that investigates the factors explaining the decision to participate and asks (ii) why do people (not) participate online."

"In addition to the two main research questions outlined above, we also investigate (iii) how effective personalized invitations are to increase and diversify participation, i.e. whether not only more people participate but also people from commonly under-represented backgrounds."

Research design & methodology

"In order to investigate in detail how citizens engage in online participation processes, we chose to conduct three almost identical instances of online consultations that we designed and implemented together with three different German municipalities (Bonn, Cologne & Moers)."

"The response rates to the online surveys among registered participants of the consultation processes was rather satisfactory: 45% (N = 268) of the registered participants of the consultation process in Bonn responded to the survey invitation via e-mail, and in Ehrenfeld (N = 65) and Moers (N = 61) it was even 47%."

"In order to investigate who participates in local online participation and which factors contribute to this engagement, we focus mainly on a comparison between those people who did participate in the online consultation process and those who did not."

"S_2 on the other hand includes all cases from the three online surveys among the registered participants of the consultation."

Results

"Having in mind that the participants among the representative survey (S_{1b}) in their large majority (90%) received an invitation letter as an extrinsic motivation to participate, our guess was that once they participate in the consultation, they might differ in their behaviour as well as their process perception from those who participated without invitation based on their strong intrinsic motivation, i.e. the registered participants (S_2)."

"Comparing these two subgroups, in the first regression model (1) the dependent variable is operationalized as active and passive participation (= 1) vs. abstaining (= 0) while the second regression model (2) estimates the effects of the predictors on the probability of participating actively only (= 1) in the consultation process (vs. others = 0)."

"Holding all other factors constant, the city itself where the consultation process took place only has a significant effect in model 2: in Moers the respondents were only half as likely to participate actively compared to the residents of Bonn."

Discussion

"While online forms of political participation tend to increase engagement of young people (Escher [349]) we do not find this effect here."

"It could be that younger people would tend to favor non-institutionalized forms of political participation via other online channels (such as social media) instead of a consultation platform, or that the topic of cycling tends to be of less interest to young people."

"In order to contribute solutions to the problem of low and biased participation, our third research question tested a particular mobilization strategy, namely (iii) how effective personalized invitations are to increase and diversify participation."

"Our research has a number of important implications for those practitioners in local government as well as respective participation service providers who are tasked with organizing such consultations both online and offline."

Conclusion

"Our results also unearthed differences between the intrinsically motivated and (through a personalized invitation) mobilized participants in terms of the sociodemographic profile as well as the attitudes towards the online consultation."

"Not only does this mobilization strategy help to reach out to some commonly under-represented groups (in particular women but also some pension-age citizens and those without Abitur), it also convinces people to participate who are not as much motivated by being affected and dissatisfied with the issue at stake and as such introduce opinions closer to the median citizen."

"Information on consultations should also specifically address concerns about the effectiveness in general and the technology in particular as both were main reasons to abstain from participation."

"The comparatively small size of the control group in combination with the generally minority participation rate (overall just about 1% of the population would participate actively) does not allow us to make inferences about those people who take part in online consultations without registering on the site."

4.2.5 Associations of Migrants in Spain: An Enquiry into Their Digital Inclusion in the "Network Society" in the 2010s [546]

This is a machine-generated summary of:

Uzcanga, Catalina; Oiarzabal, Pedro J.: Associations of Migrants in Spain: An Enquiry into Their Digital Inclusion in the "Network Society" in the 2010s [546]

Published in: VOLUNTAS: International Journal of Voluntary and Nonprofit Organizations (2019)

Link to original: https://doi.org/10.1007/s11266-019-00162-z

If you want to cite the papers, please refer to the original.

For technical reasons we could not place the page where the original quote is coming from.

Abstract-Summary

"This paper explores the way that associations of migrants in Spain (AMS) in the 2010 decade integrate into the Network Society by analyzing how they use information and communication technologies (ICTs), while looking at the potential implications of those technologies for fostering their social capital."

"The analysis focuses on the factors that influence the use of ICTs by AMS, such as the representatives' technological skills, and on the relation that this use may have on their social capital, such as the construction of collaborative digital ecosystems that may positively enhance their ties with the host society."

Introduction

"We analyze whether and how associations of migrants in Spain (AMS) during the 2010s are using information and communication technologies (ICTs)."

"We will then explore their degree of participation in the so-called Network Society and the impact of the use of new technologies on their online and offline social capital."

Social Capital of Associations of Migrants

"Bonding social capital refers to those close ties with people of our same group (Putnam [350]: 20), which are based on trust, provide security and boost identity."

"Associations of migrants generate bonding social capital since, summarizing Moya [351], they gather people around a common objective and represent a secondary socialization close to that of family."

"Bridging social capital refers to exogamous ties or relations among different groups (Putnam [350]: 20)."

"This bridging social capital, even if it is comprised of small associations of migrants, contributes more to a democratic social tissue than the bonding social capital of non-linked associations (Toral [352]: 117)."

"In their study on social capital of associations in Europe, Zmerli and Newton ([353]: 168, 170) observe that networks of associations are closed and isolated from other networks."

Associations of Migrants and ICTs

"Although ICTs are a wide concept, in the present study, it refers mainly to software, Internet and the different applications that make use of this connectivity, including electronic mail (e-mail), the World Wide Web (Web) and online social networks used from any electronic device (mobile or landline)."

"Following those studies, migrants are using ICTs to create bridges and develop their social capital."

"The study of the integration of associations of migrants in the Network Society has somehow been overlooked (Allievi and Nielsen [354]); among the few studies on the use and impact of new technologies in the development of associations of migrants, Oiarzabal ([355, 356]) suggests that ICTs have become indispensable instruments for the internal organization of those associations as well as for their capacity to achieve their goals."

"Toral ([352], 115) refers to it as horizontal relations, meaning relationships between associations of migrants, and Álvarez de los Mozos [357] also uses the concept of bridging social capital to analyze the interactions with other associations of migrants that will give them access to new resources."

Goals, Methods and Data Collection

"The goal of our research was to explore the ways that migrant organizations in Spain integrate into the Network Society by analyzing their ICT resources, their use and their online and offline collaborations."

"The survey consisted of 52 questions divided into three parts: a first part with questions concerning the main characteristics of the associations (objectives, year of foundation, origin, address) and their representative's profile (age, sex, studies, ICT skills and competency); a second part dealing with ICTs, namely ICT resources, use of ICTs; and a third part addressing both online (Internet) and offline (face-to-face or telephone) bridging social capital (frequency, which measures the intensity of collaborations with different entities, reach and the number of collaborators)."

"An e-mail including a presentation of the study, a link to the online survey (and as an attachment in Word and PDF formats) and our contact information, was sent to the selected associations."

Fragile Nodes of a Large, Broad and Weak Network

"Concerning other online activities, the biggest differences are found between how professional associations and the other two types of associations of members publish useful information for the community and update their social media profiles, which the former do almost every day, while the latter only do so a few times a month."

"Among the associations of members, there are no significant differences in terms of nationalities of origin or regions of location, or with the values for the frequency of use of ICTs for written communication, administrative work and searching for funding."

"The digital divide in the use of ICTs for written communication and for administrative work is gradual, with a clear divide between latent associations of members and professional ones, whereas the operational associations of members are positioned somewhere in the middle."

Conclusions

"ICT use is helping to access information and improve overall communication, as well as to contact many other similar associations, thereby shaping a wide, broad and weak network."

"The findings of our study evidence that the number and frequency of offline ties are positively related to the use of ICTs by the migrant associations."

"The more intense and broader the associative network, the more associations use ICTs in terms of frequency and diversity."

"Those associations are using ICTs to spread their networks and shape an associative network mainly with other associations of migrants and principally within their action context, local or Spanish."

"As Vancea and Boso [358] suggest for migrant individuals, longitudinal studies of AMS would provide more details about the direction of two of the relationships found: first, the existing causality between the professionalization of an association and the use of ICTs and second, the existing causality between associative networks and the use of ICTs."

4.2.6 E-commerce Policy and the Global Economy: A Path to More Inclusive Development? [547]

This is a machine-generated summary of:

Ahi, Alan A.; Sinkovics, Noemi; Sinkovics, Rudolf R.: E-commerce Policy and the Global Economy: A Path to More Inclusive Development? [547]

Published in: Management International Review (2022)

Link to original: https://doi.org/10.1007/s11575-022-00490-1

If you want to cite the papers, please refer to the original.

For technical reasons we could not place the page where the original quote is coming from.

Abstract-Summary

"According to the McKinsey Global Institute, almost all cross-border transactions had a digital component in 2016."

"This is also reflected by the growing literature on digitalization and E-commerce."

"Studies specifically focusing on E-commerce policy are scarce compared with other areas in this domain."

"Our analysis reveals that to promote an inclusive E-commerce participation, it is imperative to design policies that improve countries' formal institutions, facilitate the inclusion of less-developed countries in the E-commerce space, and enhance E-commerce adoption by small- and medium-sized enterprises."

Introduction

"Although digitalization is a broad concept and has many facets (cf. Bouncken and Barwinski [359]; Ritter and Pedersen [360]), electronic commerce (E-commerce) is an important example of how businesses can take advantage of digital technologies."

"Appropriate government policies generally lead to transparent institutions that can support E-commerce activities and lower the risks of partaking in them (OECD [361])."

"There is a paucity of academic research on public policy related to cross-border E-commerce, with the majority of existing studies adopting a firm perspective."

"This study aims to take stock of the existing knowledge on E-commerce policy scattered across the academic and policy domains."

"To close the knowledge gap on E-commerce policy in academic publications, we extend the analysis to documents published by international bodies such as the World Trade Organization (WTO) and the Organisation for Economic Co-operation and Development (OECD)."

Types of E-commerce

"Through E-commerce, a range of commercial relationships can occur, involving any possible pairing of consumers, businesses or governments (Laudon and Traver [362]; OECD [361])."

"The largest type in terms of monetary transaction is Business-to-Business (B2B) E-commerce, where one business focuses on selling to another."

"The other type of E-commerce widely discussed in the literature is Business-to-Consumer (B2C) E-commerce, whereby businesses attempt to sell to individual customers."

"Another major type of E-commerce is Consumer-to-Consumer (C2C) E-commerce, which provides a way for consumers to sell to one another using a platform provided by an online market maker or a platform provider such as e-bay, or on-demand service companies such as Airbnb and Uber (Laudon and Traver [362])."

"Mobile E-commerce (m-commerce) is a type of E-commerce through which mobile devices, such as smartphones and tablet computers, are used to complete a commercial online transaction (OECD [361])."

E-commerce Public Policy

"Through public policy, governments can also design, implement, and communicate up-to-date legal and regulatory frameworks that provide a supportive business environment for E-commerce."

"These policies can provide an adequate legal and regulatory framework that mitigates transaction risks and provides transparency within E-commerce (OECD/WTO [363]; OECD [361])."

"Given the significance of E-commerce in today's economy and government policy in E-commerce facilitation, standardized policies regarding E-commerce are increasingly gaining the attention of governments across the world."

"Empirical research also confirms the importance of government policy in the development of E-commerce."

"Scupola [364] showed that government policy strongly influenced the adoption of E-commerce by SMEs in southern Italy."

"Cui and others [365] argued that government policies influenced the probability of firms entering the e-business market in China."

Review Method

"Our aim was to capture publications that focused on public policy in relation to E-commerce with implications for cross-border activities."

"Rather, we chose keywords based on the evolution of our knowledge and understanding of E-commerce and relevant policies (cf. Papanastassiou and others [366])."

"We used the keywords in each set in combination with the keywords in the other to ensure that the articles retrieved included concepts related to both E-commerce and policy."

"We only included publications that focused on policy designed and implemented by a government, its agents, or other official bodies (cf. Clegg [367]) in relation to E-commerce."

"Limited number of academic publications and the fact that international bodies are important policy-makers that shape regional and global E-commerce policies, we also reviewed published reports by intergovernmental bodies and organizations such as the OECD, the United Nations Conference on Trade and Development (UNCTAD), the World Economic Forum (WEF) and the WTO."

Discussion of Findings

"In weak institutional environments, the high regulatory uncertainty associated with inefficient or non-existent regulations related to data protection, electronic payment, intellectual property rights, and taxation may discourage business participation in E-commerce or lead to grey zones of international trade that enable tax evasion and the circulation of counterfeit products (cf. Jean and Tan [368]; OECD [369])."

"A strong regulatory environment, on the other hand, encourages participation in E-commerce because it provides a transparent legal framework in the area of electronic transactions, privacy and data protection, consumer rights protection as well as cybercrime prevention (OECD/WTO [363]; OECD [369])."

"Some researchers have compared developed with less-developed countries in terms of E-commerce policy regarding consumer protection."

"Further, at a more global scale, intergovernmental bodies such as the WTO and the OECD need to ensure that they draw a balance between promoting global digital markets and maintaining policy space for governments to protect E-commerce consumers in their territories (Neeraj [370])."

Future Research

"The implications of tax policies and the issues arising around their implementation and enforcement are not only relevant for firms that venture into the various forms of cross-border E-commerce; taxation also has significant societal implications that are within the remit of international business (cf. Ahen [371]; Buckley and others [372]; Sinkovics and others [373, 374])."

"Although some of the reviewed publications focus on SMEs in relation to E-commerce policy, the implications of size differences across this very diverse group of firms need more attention."

"Given the particularities and significance of micro firms for the global economy (McCormick and Fernhaber [375]), future research may wish to examine the suitability of current E-commerce policies for the needs of this group of firms; specifically, are current policies sufficiently enabling different groups of SMEs to participate in domestic and cross-border E-commerce or are certain groups disadvantaged by these policies?"

Conclusion

"The findings highlight the importance of international harmonization; not just in terms of policies related to consumer protection, data protection and the mitigation of E-business risks, but also with respect to the availability and quality of digital and physical infrastructure as well as support services that foster and enable consumer and business participation."

"These are important factors that need to be in place if the SDGs relying on digital technologies are to be achieved including the bridging of the still prevalent digital divide."

"It is important to recognize that E-commerce can support other economic and social activities such as higher productivity, enhanced competitiveness, improved access to information, and more inclusive development (OECD/WTO 2017)."

"We echo the authors of the documents we examined in proposing that the objectives of E-commerce policies need to be aligned with individual country's broader national development agendas."

4.2.7 Digital Inclusion in Social Media Marketing Adoption: The Role of Product Suitability in the Agriculture Sector [548]

This is a machine-generated summary of:

Han, Honghua; Xiong, Jason; Zhao, Kexin: Digital inclusion in social media marketing adoption: the role of product suitability in the agriculture sector [548]

Published in: Information Systems and e-Business Management (2021)

Link to original: https://doi.org/10.1007/s10257-021-00522-7

If you want to cite the papers, please refer to the original.

For technical reasons we could not place the page where the original quote is coming from.

Abstract-Summary

"Given the persistent urban–rural gap in technology adoption and usage, this study intends to understand factors motivating rural farmer's adoption of social media marketing (SMM) tools in China."

"We propose product SMM suitability as a new moderator based on the channel/product fit literature."

"While the result confirms UTAUT's generalizability in China's agriculture sector, it also demonstrates limited generational divide and gender divide in SMM adoption among Chinese farmers."

"The newly proposed moderator, SMM product suitability, is empirically validated."

"Farmers consider non-grain products to be more suitable for SMM than grain products."

"SMM product suitability not only directly increases Chinese farmers' SMM adoption intention, but also moderates the relationship between behavioral intention and use as well as the relationship between facilitating conditions and use."

Introduction

"To understand the urban–rural digital divide and improve inclusion of rural residents in the digital economy, this research investigates factors motivating or hindering the adoption of mobile-based social media marketing (SMM) among Chinese farmers."

"Internet technologies play a vital role in entrepreneurial activities in rural China, and many farmers and farm cooperatives turn to e-commerce platforms to advertise and distribute their products (Heang and Khan [376])."

"Most farmers live in rural areas, and such digital divide has created many problems restricting adoption and use of e-commerce in China's agriculture sector (Hong and others [377]; Li and others [378])."

"Major e-commerce platforms in China are actively developing programs to encourage and help farmers sell their products directly via SMM (Wei [379])."

"The research question for this study is: what is the role of product suitability in the agriculture sector, and how does it impact Chinese farmers' SMM adoption?"

The research model and hypotheses

"Behavioral Intention refers to the degree to which Chinese farmers have consciously planned to use or not use WeChat-based SMM to promote and market their agriculture products (Maruping and others [380])."

"Performance Expectancy is defined as the degree to which using WeChat-based SMM will benefit Chinese farmers in promoting and selling their agriculture products to end consumers."

"If farmers expect to gain more benefits from conducting WeChat-based SMM, such as reaching more potential customers and selling more products directly via the WeChat-based SMM channel, they are more likely to form a stronger intention to use the tool."

"Social Influence is defined as the degree to which Chinese farmers think that important others around them believe that they should use WeChat-based SMM to increase their products' market awareness."

"We hypothesize that Chinese farmers' intention to adopt WeChat-based SMM is positively associated with their actual usage (H4)."

Method

"To empirically validate and implement our proposed method to evaluate and classify agriculture products based on their SMM suitability, we conducted a simple exploratory survey among 106 Chinese farmers."

"Section, we conduct the main survey to test the moderating effect of SMM suitability, and examine whether factors motiving farmer's adoption of SMM vary across those two product categories."

"After the exploratory survey of 106 Chinese farmers to classify agriculture products based on their SMM suitability, a follow-up paper-based survey was administered between Jul 2017 and May 2018 in Jiangsu Province, China."

"The results indicate that Chinese farmer's intention to adopt WeChat-based SMM is positively and significantly associated with performance expectancy, effort expectancy, and social influence."

"Product SMM Suitability is statistically significant in moderating the effect of behavior intention and facilitating conditions with the actual usage of the WeChat based SMM."

Discussion

"Our research applies and extends UTAUT framework to understand Chinese farmers' decision to use SMM to directly promote agriculture products to end consumers."

"We examine factors encouraging farmer's adoption of WeChat-based SMM, and our research contributes to bridge the urban–rural digital gap."

"Social influence is the most influential factor that encourages Chinese farmers to form adoption intention of WeChat-based SMM."

"Our results demonstrate that product SMM suitability affects Chinese farmers' SMM adoption and use both directly and indirectly as a moderator."

"The result suggests that Chinese farmer's SMM use relies more on facilitating conditions for more suitable products."

"Those initial successful use cases can create positive social influence, further promote SMM adoption to grain farmers, and enhance digital inclusion in rural China."

"Our research identifies factors encouraging Chinese farmers to adopt SMM, as well as how these motivational factors vary based on individual user characteristics as well as product characteristics."

4.2.8 Is It an Institution, Digital or Social Inclusion that Matters for Inclusive Growth? A Panel Data Analysis [549]

This is a machine-generated summary of:

Aslam, Aribah; Naveed, Amjad; Shabbir, Ghulam: Is it an institution, digital or social inclusion that matters for inclusive growth? A panel data analysis [549]

Published in: Quality & Quantity (2020)

Link to original: https://doi.org/10.1007/s11135-020-01008-3

If you want to cite the papers, please refer to the original.

For technical reasons we could not place the page where the original quote is coming from.

Abstract-Summary

"This study empirically examines the impact of institutional quality, social inclusion and digital inclusion on inclusive growth across different economies characterized by different income groups."

"The results of this study show that there is a direct link between institutional quality and inclusive growth for a higher-income group of countries but not in the rest of the income groups."

"From the policy point of view, these findings suggest that establishing and strengthening the institutional structure in low- and middle-income countries can contribute towards better and higher inclusive growth."

Introduction

"To overcome this gap in the literature, the motivation to do this study is to methodically investigate the determinants of inclusive growth, along with associations among different income groups; particularly, focus on the role of institutions, digital inclusion and social inclusion in embracing inclusive growth."

"The study aims to: (1) construct the indexes to capture different dimensions of digital inclusion, social inclusion, institutional quality and inclusive growth. (2) To estimate the impact of digital inclusion, social inclusion, institutions on inclusive growth after controlling the effects of relevant variables in three income countries group: High-income countries (HYCs), middle-income countries (MYCs) and low-income countries (LYCs). (3) To suggest pathway/policy for LYCs and MYCs to achieve the inclusive growth equivalent to HYCs."

"The study tried to answer the research questions: Does socio-digital inclusion helps to achieve inclusive growth in the groups of three income countries?"

Inclusive growth: conceptual linkages

"It is important to note here that the need for inclusive growth is not only realized due to the lack of economic and social opportunities for the poor, but it may also arise as an outcome of poor income distribution, weak institutional structure, unheard voices of underprivileged, illiteracy, dictatorial categorized power concentration in developing world."

"The value of the inclusive growth index is greater than 50 for countries falling in the high-income group."

"The IGI score of Pakistan (53.52), which has lower economic growth rates than Sri Lanka and India but performed better in terms of inclusive growth than these nations."

"To the above, the present study explores the associations between these income countries groups of countries as per their achievement in inclusive growth by improving institutional quality, spreading digital technology and social inclusion."

Literature review

"Economic Inequalities and inclusive growth discussion widely emerged after the financial crisis in 2008 and the Arab Spring in 2011 in global debates."

"Various studies have identified different socio-economic factors required to achieve inclusive growth."

"Tripathi [381] found the lower inclusive growth and poverty in 54 populated cities of India along with the increase in inequality from 2004–05 to 2009–10."

"A strand of literature does not consider digital human capital as the only key to ensure economic growth/inclusive growth."

"No significant study has addressed inclusive growth through the lens of social and digital inclusion."

"Considering this, the above-cited literature has discussed the role of the institution in inclusive growth but still, there is a gap in the literature, which defines the social and digital inclusion, and further analysis of their impacts on inclusive growth for different economies."

Theoretical background and model specification

"As effective output per labor expands, poverty reduces accompanied by employment increases, which leads to inclusive growth (IG)."

"Information technology or digital inclusion (DI) has not only brought growth in output per labor but also the employment rate has been stimulated, leading to more inclusive growth."

"Based on the theoretical framework an empirical model may be developed as given below to estimate the impact of social inclusion, digital inclusion and institutional quality on inclusive growth in high-income, medium-income and low-income groups of countries:where IG inclusive growth, INS institutional quality, DI digital inclusion, SI social inclusion, INF inflation, TO trade openness, INV investment, EDU education."

"To determine the impact of institutions, social inclusion and digital inclusion on inclusive growth, the study has employed a panel data set of 83 countries over the period from 2010 to 2017."

Empirical findings and discussion

"Unlike HYCs, the impact of institutions on inclusive growth is not significant in LYCs and MYCs, howsoever, have a positive relationship with inclusive growth in all country sets, which implies that institutions can promote inclusive growth."

"The results indicate a positive impact of social inclusion on inclusive growth in all country sets, but its contribution is relatively high in HYCs than in MYCs and LYCs."

"Another intuition about the study's result is, LYCs have a higher degree of collaboration with the world than MYCs and LYCs, which is the reason for social inclusion not to contribute significantly to the inclusive growth of LYCs and MYCs."

"U = {Institutions, Social Inclusion, Digital Inclusion, Trade openness, Investment, Inflation, Education} A = {Institutions, Social Inclusion, Digital Inclusion, Trade openness, Investment, Education} = HYCs B = {Social Inclusion, Inflation, Trade openness, Investment} = MYCs C = {Social Inclusion, Investment} = LYCs The results of the study conclude that all determinants have a significant contribution to inclusive growth, except inflation."

Conclusions and policy implications

"The main objective of the study was to estimate the contribution of socio-digital inclusivity and institutions towards inclusive growth in three income groups along with the composition of indexes for the same."

"The Hypothesis of inclusion Differences were also found to be true, positing that high, middle- and low-income countries differ in their levels of social inclusion, digital inclusion and inclusive growth, owing to their institutional structure."

"Estimated results of the study recommended the positive role of institutional quality, social inclusion, digital inclusion, trade openness, investment and education in promoting inclusive growth in HYCs."

"The results are much owing to the strong institutional structure in these countries with planned policies towards growth that is inclusive."

"Social inclusivity has positively contributed to inclusive growth in all three income groups, whereas digital inclusion only in HYCs."

Limitations and scope for future research

"The data sample used in the study covers eighty-four countries based on the availability of data series for forty-six determinants."

"It would be better if more countries can be added in the sample in the future provided the data availability, especially on poverty and income inequalities."

"For future research, a similar reach and approach can be extended to analyze specific regions and even at the sector level to formulate the policies to address the regional differences and individual sector performance, subject to the sufficient data availability to meet quantitative and econometric requirements to draw results."

4.2.9 'I'm Fired Up Now!': Digital Cataloguing, Community Archives, and Unintended Opportunities for Individual and Archival Digital Inclusion [550]

This is a machine-generated summary of:

Holcombe-James, Indigo: 'I'm fired up now!': digital cataloguing, community archives, and unintended opportunities for individual and archival digital inclusion [550]

Published in: Archival Science (2022)

Link to original: https://doi.org/10.1007/s10502-021-09380-1

If you want to cite the papers, please refer to the original.

For technical reasons we could not place the page where the original quote is coming from.

Abstract-Summary

"Through documenting, preserving, and making local heritage accessible, digital cataloguing offers community archives significant potential benefits."

"RMIT and Swinburne University of Technology, Melbourne, for Telstra, p 8), community archives can be considered digitally excluded."

"Through an ethnographic study of one community archive's use of Victorian Collections, an Australian digital cataloguing platform, this article examines the impact of digital exclusion on digital cataloguing outcomes via metrics of quantity and quality."

"These metrics disregard the opportunities for enhancing individual and archival digital inclusion that learning how, and continuing, to digitally catalogue present."

"By tracing one elderly volunteer's journey from digitally excluded non-user to capable cataloguer, I show how digital cataloguing offered an opportunity for enhancing this individual's digital inclusion, simultaneously improving that of the archive."

"In considering these unintended opportunities, this article contributes to our understanding of how digital exclusion impacts the digitisation of cultural heritage, and offers scope for determining how the process and practice of digital cataloguing itself can present opportunities for inclusion at the individual and archival level."

Introduction

"Put simply, successfully using digital cataloguing platforms requires more 'expertise and training [...] and more resources' (Taylor and Gibson [382], p 413) than community archives generally have access to."

"While limited digital inclusion research has been undertaken within the cultural sector (Holcombe-James [383]; Mihelj and others [384]), and Thomas and colleagues [385] focus on the experience of individuals rather than institutions or organisations, this alignment offers a productive opportunity for understanding the barriers to achieving and sustaining the benefits of digital cataloguing in community archives as digital exclusion."

"This article considers the impact of digital exclusion on the outcomes of digital cataloguing undertaken in community archives, as well as the opportunities for enhancing digital inclusion in these contexts at both the individual and the archival level."

"By considering only the impact of VC and the VHP on the community archives' collections, these measures fail to consider the initiatives' impact on the digital inclusion of individual volunteers, and in turn, the archive itself."

Identifying challenges to digital cataloguing in community archives

"Access to digital devices is, in turn, connected to the funding structures community archives operate under."

"The barriers confronting community archives are aligned with the three elements that Thomas and colleagues [385] argue are required for digital inclusion: access, affordability, and digital ability."

"As community archives are typically run by elderly retirees, I focus here on the digital exclusion confronting older adults: the gaps in digital access and use that ensures older adults are less digitally included than younger (Friemel [250]; Hargittai and Dobransky [386])."

"Such efforts typically rely on relationships with larger institutions or organisations through which community archives are provided access to devices, funding through grants, and/or training to digitally upskill volunteers (Dewhurst and Sumption [387]; Flinn [388]; Hurley [389]; Schumacher and others [390]; Stevens and others [391])."

"A web-based and free-to-access digital cataloguing platform designed for Victorian community archives, VC was developed in response to the digital exclusion confronting the sector."

Materials and Methods

"The Sub-Branch also lacked the digital devices needed to catalogue their collections, and only two VHP participants owned laptops they brought in to use during workshops."

"By providing access to connectivity, devices, and funding, VC ameliorated two aspects of the Lara RSL's digital exclusion (access and affordability) and created the conditions under which digital cataloguing could commence."

"Although specifically designed to develop digital cataloguing abilities within community archives, this level of digital exclusion limited the VHP's efficacy at the Lara RSL, with a direct impact on the quantity and quality of cataloguing outcomes."

"Although VC relies on all completed cataloguing fields to generate these search results—the more information provided, the better—the concept of enabling audience access was most explicitly discussed with Sub-Branch participants during the VHP as they learnt about selecting and affixing keywords to individual collection items."

Discussion and conclusion

"As the volunteers staffing community archives become increasingly digitally included, so does the community archive in which they work, enabling greater quantity of, and better quality, cataloguing outcomes."

"Considering the influence of digital exclusion on the outcomes of cataloguing undertaking in community archives, as well as the opportunities for enhancing digital inclusion in these contexts at both the individual and the archival level offers a mechanism for broadening our understanding of successful outcomes."

"By enhancing the digital inclusion of volunteers within community archives, so too do we work towards the digital inclusion of the archive, and the achievement of digital cataloguing benefits."

"By explicating the intersection of digital inclusion research and cultural heritage studies, detailing their dynamics and impact in the context of community archives, and offering scope for understanding how the process and practice of digital cataloguing can present positive opportunities for individual and archival inclusion, this article presents a productive basis for future research and practice."

4.3 Digital Social Participation and Engagement

Machine generated keywords: participation, student, story, adult, learn, art, engagement, participant, classroom, health, pandemic, covid, speak, combine, visual

4.3.1 A Theoretical Framework for the Evaluation of Massive Digital Participation Systems in Urban Planning [551]

This is a machine-generated summary of:

Hofmann, Mathias; Münster, Sander; Noennig, Jörg Rainer: A Theoretical Framework for the Evaluation of Massive Digital Participation Systems in Urban Planning [551]

Published in: Journal of Geovisualization and Spatial Analysis (2019)

Link to original: https://doi.org/10.1007/s41651-019-0040-3

If you want to cite the papers, please refer to the original.

For technical reasons we could not place the page where the original quote is coming from.

Abstract-Summary
"Using the phases specified in the minimal viable process of the U_CODE project as an example, this paper aims at describing criteria for the evaluation of participation processes and propose several possible methods for their assessment."

"While the majority of these criteria resemble criteria traditionally used to assess public participation in general, this paper proposes an additional criterion, as well as ways of applying all of the criteria to digital participation methods."

"The criteria and methods described in this paper not only may be used for evaluative purposes during or after a digital participation process but may also be useful guidelines during the planning stages of participation processes."

"It may help to consider these criteria to assess the value of the process during its inception stage to avoid mistakes and to enhance the democratic value of the participation process."

Introduction
"The U_CODE process comprises a number of distinct stages (initiating, co-briefing, co-design, professional design, integrating and decision-making), for each of which digital tools can be used, e.g. to convey information on the planned project; to allow ways for participants to contribute their own ideas, needs and wants; or to rank different designs."

"Depending on the purpose of the tools used in a specific digital participation process (e.g. co-creation tools), evaluation methods may also comprise the analysis of usage statistics of the employed web platforms or eye-tracking methods for usability analyses of the digital tools."

"Process designers and software engineers experience difficulties in developing specific design and evaluation criteria applicable to their tools and procedures, they are advised to collaborate with evaluation experts versed in democratic theory, evaluation methods and product usability (e.g. political scientists, psychologists or sociologists)."

Evaluation Criteria for the Process Quality of Massive Digital Participation

"A participation process facilitator may employ specific strategies to achieve a representative group of participants."

"Any participation process must aim at facilitating deliberation, by establishing the general atmosphere, the specific process steps and the respective tools."

"For participation processes relying on digital media and digital tools, this criterion can be particularly challenging, because digitising such a process aims at reducing the need for professional facilitators who, in face-to-face participation settings, will actively try to improve the deliberation quality."

"In face-to-face participation settings, the group communication will be facilitated by neutral communication experts, ensuring that only the relevant topics are being focused on, that procedures are followed, that the relations and differences between the participants are not hindering the process and that all participants use the provided tools to their full potential (Pelzer and others [392])."

Evaluation Criteria for Massive Digital Participation Process Outcomes

"Regarding the outcomes of public participation, one may distinguish between (1) the effects the process has on political or technical decision-making—regarding the impact not only on specific decisions but also on decision-makers and the general public—and (2) the effects the process has on the participants."

"A major challenge faced when evaluating any public participation process lies in verifying and quantifying a causal effect of the process on public sentiment and specific political decisions."

"The participation process may well have an impact on a political decision—however, this decision may only be taken after the evaluation of the participation process."

"Participation processes not only yield results relevant to political decisions (e.g. in relation to urban planning projects) but should also have effects on the subjects partaking in the process."

Possible Operationalisations of the Evaluation Criteria Using the Example of the U_CODE Process Phases

"In the co-briefing phase, the initial brief will be enriched by requirements regarding the project contributed by process participants, using digital brainstorming and idea-harvesting tools."

"For the evaluation, questionnaires may be used via which the participants are asked to judge how well they understand the professional design process and the resulting design proposals."

"If sentiment analyses are conducted within this phase to gather feedback on the designs, they may also be used to gauge the deliberation quality, e.g. by assessing the civility of the interaction between the participants."

"Similar to the evaluation of internal transparency in phase 1, evaluators may consider assessing the communication quality by analysing the documents created in this phase (e.g. e-mails, resulting planning documents) or by interviewing the actors."

Discussion

"Taking these differences into consideration, the criteria and methods used in the evaluation of the participation process will change, making some parts of the evaluation less labour-intensive, e.g. when contributions from the participants do not have to be counted manually because the data created using digital tools can more easily be analysed."

"Questions to be focused on in future evaluations of massive digital participation processes should relate to these differences, i.e. it may be asked whether the traditional evaluation criteria may need to be adapted to account for the use of digital media, or which new criteria must be formulated to fairly assess the technical requirements that need to be met (e.g. technical robustness, operational safety, speed and cost-effectiveness of the hardware and software)."

4.3.2 Digital Participatory Workshops with Patients and Health Professionals to Develop an Intervention for the Management of Polypharmacy: Results from a Mixed-Methods Evaluation and Methodological Conclusions [552]

This is a machine-generated summary of:

Engler, Jennifer; Brosse, Franziska; Dinh, Truc Sophia; Klein, Astrid-Alexandra; Brueckle, Maria-Sophie; Petermann, Jenny; Muth, Christiane; Mergenthal, Karola; van den Akker, Marjan; Voigt, Karen: Digital participatory workshops with patients and health professionals to develop an intervention for the management of polypharmacy: results from a mixed-methods evaluation and methodological conclusions [552]

Published in: Research Involvement and Engagement (2022)

Link to original: https://doi.org/10.1186/s40900-022-00387-1

If you want to cite the papers, please refer to the original.

For technical reasons we could not place the page where the original quote is coming from.

Abstract-Summary

"This was mainly due to assumptions on patients' willingness and skills to participate digitally."

"Digital PPI workshops differ from in-person meetings as some forms of non-verbal cues and body language may be missing and technical barriers may exist."

"Within our project HYPERION-TransCare we adapted our PPI workshop series for intervention development to a digital format and assessed whether these digital workshops were feasible for patients, health care professionals and researchers."

"The workshop evaluation encompassed observation protocols, participants' feedback via chat after each workshop on their chance to speak and the usability of the digital tools, and telephone interviews on patients' and health professionals' experiences after the end of the workshop series."

"Participants median rating of their chance to speak and the usability of the digital tool was "very good"."

"Benefits of the digital format such as overcoming geographical distance, saving time and combining workshop participation with professional or childcare obligations were reported."

"Technical support was stressed as a pre-condition for getting actively involved in digital PPI."

"Digital formats using different didactic and documentation techniques, accompanied by technical support, can foster active patient and public involvement."

"The advantages of digital PPI formats such as geographical flexibility and saving time for participants as well as the opportunity to prepare and hold workshops in geographically stretched research teams persists beyond the pandemic and may in some cases outweigh the advantages of in-person communication."

Background

"The relationship between patients, health care professionals and the research team throughout the participatory workshop series was considered a "partnership", that is, all participants were involved in the decision-making process on intervention design [393]."

"Publications on evaluation data of digital PPI activities during the COVID-19 pandemic are rare, and most authors conclude that a sharing of researchers', patients' and health professionals' experiences is crucial for a better understanding of which tools and formats work for which purpose and which participating group [394–398]."

"Following this principle, we aim to describe 1) how we put patient and health professional involvement in the development of an intervention into practice using a digital workshop format and 2) how patients and health care professionals experienced these participatory digital workshops and which methodological conclusions we can draw from the workshop evaluation."

Methods

"We refer to results of the evaluation of digital participatory workshops with patients with polypharmacy, their informal caregivers, patient representatives and health professionals from inpatient and outpatient settings who care for this patient group."

"To assess the feasibility of PPI within our digital workshop format and participants' experiences, we used a mixed-methods approach including three evaluation methods: 1) workshop observation protocols that focused on communication; 2) participants' rating via chat after each workshop on their chance to speak and the usability of the digital format; 3) telephone interviews with participants from each stakeholder group."

"Data was transferred into an Excel sheet and analyzed with regard to median, minimum and maximum per group ("patient group" = patients, patient representatives, informal caregivers; "health care professional group" = HCAs, nurses and physicians from inpatient and outpatient settings, pharmacists, medical information scientists) and per workshop."

Results

"30 patients and health professionals participated in our workshop series with a median of 10 participants per technical introduction, IWS or SWS (min: 6; max: 13)."

"The technical introduction workshop was visited by six participants from the patient group."

"Participants from the patient group assessed that they got along with the digital format and video conference system "very well" across all workshops, whereas the median rating in the health professional group for some workshops was only "good", with maximum ratings up to "poor"."

"Even a patient participant with minor experience stressed that it was not challenging for him to participate in the digital workshops."

"Another patient participant in this sense stressed the importance of the technical introduction workshop, which he felt was "very helpful" (informal caregiver) and in his opinion a precondition for digital participatory workshops."

Discussion

"Workshop observation protocols on communication, participants' feedback via chat and telephone interviews with participants showed that participatory digital workshops for intervention development were feasible both for researchers, patients and health care professionals."

"Similar to other experiences with and recommendations for conducting digital PPI [398–400], technical support and a technical introduction workshop proved crucial to break up barriers, enhance access and provide participants with a feeling of safety in our digital PPI workshops."

"In the evaluation interviews participants from the patient group stressed that due to the technical introduction workshop, joining the digital meeting was possible without difficulty even for technically unexperienced individuals."

"Some participants actively wished to maintain digital formats also beyond the pandemic, especially because they felt that a sharing of perspectives and communication on equal footing was also possible in a digital format, and technical barriers were diminished by the technical introduction workshop and technical support during the meeting."

Conclusions

"In times of preventive reduction of physical contacts, digital formats using different didactic and documentation techniques, accompanied by technical support, can support active patient and public involvement and facilitate a change of perspective and mutual understanding between participants of different stakeholder groups across geographical distance."

"The advantages of digital formats such as geographical flexibility and saving time for participants as well as the opportunity to prepare and hold workshops in geographically stretched research teams persist beyond the pandemic and may in some cases outweigh the advantages of in-person communication."

4.3.3 Live and Digital Engagement with the Visual Arts [553]

This is a machine-generated summary of:

Ateca-Amestoy, Victoria; Castiglione, Concetta: Live and digital engagement with the visual arts [553]

Published in: Journal of Cultural Economics (2022)

Link to original: https://doi.org/10.1007/s10824-022-09466-3

If you want to cite the papers, please refer to the original.

For technical reasons we could not place the page where the original quote is coming from.

Abstract-Summary

"By using the USA 2012 Survey of Public Participation in the Arts, we explore visual arts consumption through both onsite attendance at museums and electronic and digital media."

"Results demonstrate that there are no age consumer differences in the form they consume visual arts."

"Visiting museums and art galleries have a positive correlation with the digital access to visual arts, both through handheld and mobile devices and via the internet, whilst the same correlation is not found for internet access on museum attendance."

"This means that for many consumers, online attendance is the only way to overcome time constraints and other costs involved in an onsite visit."

Introduction

"According to data for the USA taken from the 2012 Survey of Public Participation in the Arts (2012 SPPA), 37% of the population attended a live performance and 39% some visual arts event or activity at least once in the previous year, whereas 71% consumed art through some form of electronic media: TV, radio, internet, handheld or mobile devices, DVD, CD, tape, or record player (NEA, [401])."

"This paper aims to fill this lacuna by jointly investigating the determinants of the choice of attendance at museums and art galleries and of digital engagement with visual arts through handheld or mobile devices and via the internet."

"Museum-going increases the probability of both accessing visual arts content on mobile devices and via the internet, whilst the consumption of visual arts on handheld or mobile devices and through the internet does not have a significant effect on the probability of onsite visiting museums and art galleries."

Conceptual framework

"The determinants of onsite cultural attendance have been studied for different activities, whilst less is known about the determinants of participation in the arts through digital devices, with the major exception of music and cinema, which are probably the two industries most affected by illegal digital access."

"Although the Bakhshi and Throsby' study (2014) is different from the aim of this research, not only because it refers to theatre and cinema and not to visual arts, it is interesting to report because it explores whether live broadcasts of theatre to digital cinemas substitute for or complement audiences for traditional theatre."

"Two alternative accesses to cultural goods that are not mutually exclusive in their consumption are considered: visits to museums and art galleries and the consumption of visual arts through handheld or mobile devices and via the internet."

Data and methodology

"Since the aim of this paper is to address the choice of individuals when consuming the visual arts, either by visiting museums and galleries (musego) or by digital engagement (handheld and internet), following Favaro and Frateschi [402] and Prieto-Rodríguez and Fernández-Blanco [403], we use a multinomial framework to simultaneously evaluate the probability of different forms of participation to visual arts."

"The vector of cultural capital variables (Cu_i) is included since participation in the arts is generally accepted to depend on the individual stock of cultural capital."

"This variable is included in the cultural participation model due to the time constraint restriction that does not occur in the case of online consumption."

"Following Prieto-Rodríguez and Fernández-Blanco [403], it is expected that some variables exert a similar effect, particularly those that shape the individual taste for visual arts and the common cultural capital needed to appreciate and transform cultural goods (museum visits or visual electronic and digital content, respectively) into meaningful cultural experiences."

Results and Discussion

"The first column of each model shows the impact of each variable on the probability to attend both types of event (online and onsite), whilst the second and third columns show the impact on the probability of attending onsite (musego) and online (handheld in Model 1 and internet in Model 2), respectively."

"The onsite attendance has a positive impact on the probability of online attendance, the marginal effect is equal to 0.114 of attending both types of events, whilst it is equal to 0.260 for having musego = 0 and handheld = 1 (Model 1B in the same table)."

"A strong effect is found for the education variable together with the children_arts_school variable (school aged children being taught art or going to a music/art museum or gallery or attending a live music, theatre or dance performance in school/outside school), which shows a positive probability in attending onsite, whilst a negative impact on the probability that both events are equal to zero."

Conclusions

"In the era of ICTs, when many individuals are digitally engaged, technological progress has also influenced cultural consumption, bringing new opportunities through the use of mobile devices and the internet."

"To contribute to the changes that digital technologies have determined on cultural participation, this paper jointly explores the choice of attendance at museums and art galleries and/or media and digital engagement with visual arts for digital users."

"Visiting museums and art galleries has a positive impact on the digital access to visual arts, both through handheld and mobile devices and via the internet, whilst the same impact is not found for internet access on museum attendance."

"Since visits positively affect the use of handheld/internet devices (complementarity effect) in visual arts consumption, whilst internet consumption does not affect onsite visits (substitution effect) managers should focus on ameliorating the physical exposition and on introducing some forms of online fees for their online collection, to increase cultural institutions revenues."

4.3.4 Older Adults' Perceptions of Contexts Surrounding Their Social Participation in a Digitalized Society—An Exploration in Rural Communities in Northern Sweden [554]

This is a machine-generated summary of:

Fischl, Caroline; Lindelöf, Nina; Lindgren, Helena; Nilsson, Ingeborg: Older adults' perceptions of contexts surrounding their social participation in a digitalized society—an exploration in rural communities in Northern Sweden [554]

Published in: European Journal of Ageing (2020)

Link to original: https://doi.org/10.1007/s10433-020-00558-7

If you want to cite the papers, please refer to the original.

For technical reasons we could not place the page where the original quote is coming from.

Abstract-Summary

"Social participation and digital engagement can contribute to health and well-being among older adults."

"Because of older adults' decline in abilities, coupled with complex technology and its perceived insufficient relevance to daily life, there is a need to create and tailor social opportunities and services that are supported by digital technologies for older adults to continue participating in society."

"It becomes relevant to explore older adults' perceptions about contexts surrounding their social participation in a digital society."

"The findings suggested that co-creating usable digitalized services and facilitating satisfactory use of digital technologies could support older adults' social participation through activities that they find relevant in their lives, and subsequently, might enable them to live longer at home."

Background

"In a society that is rapidly becoming digitalized (Davidsson and Thoresson [405]; Regeringskansliet [406]), social participation occurs increasingly more often with and through the use of digital technologies."

"Both social participation and digital technology have been recognized as having potential to promote health and well-being among older adults (World Health Assembly 408; World Health Organization [408])."

"Older adults with limited social contacts have less opportunities to use digital technology for social engagement (Ihm and Hsieh [409])."

"Varying access to and use of digital technologies affect which benefits individuals derive from being connected and participating in society."

"Creating and tailoring social participation opportunities in a digital society can be better understood by gaining insights about older adults' contexts."

"The objective of this study was to explore older adults' perceptions about contexts surrounding their social participation in a digital society."

Methods

"The study's qualitative research design used focus group methodology for data collection and qualitative content analysis for data analysis."

"The focus groups were formed based on the communities where the participants lived and their availability for the interviews."

"Focus group methodology was chosen in order to stimulate discussion and elaboration of ideas between the participants and to gain insight into how participants talk about ideas (Dahlin-Ivanoff and Hultberg [410]; Doody and others [411]; Stewart and others [412])."

"The main questions in the interviews were about daily activities, social networks and activities, participation in society, digital technology, and existing and desired services in the community."

"IN and another researcher each served as assistant moderator in two focus groups and were responsible for concluding each interview with a summary of what had been discussed."

"Time intervals between interviews enabled the researchers to reflect on the ideas that emerged from each interview and plan discussion topics in subsequent interviews to achieve data saturation, as recommended by Doody and others [413]."

Findings

"Within the discussions, participants revealed reduced opportunities for social interaction, which they ascribed to prioritizing home activities, dwindling social groups, and increasing computerized access to services."

"Participants discussed that activities had taken longer time as they got older and that their days were often occupied with mundane activities at home, like following current events and sports, solving puzzles, and doing household tasks."

"Participants discussed how social spaces had changed over time."

"Participants also discussed how social conventions and habits had affected their opportunities for interaction."

"Participants discussed how changes in services affected their social participation."

"Participants also discussed how community developments had affected their opportunities for social interaction."

"With decreasing services and opportunities for close social interactions where they live, participants acknowledged that a car was necessary to do activities, meet people, join social gatherings and courses, and access health care."

Discussion

"In the exploration of older adults' perceptions about contexts surrounding their social participation in a digital society, focus group participants reminisced about past situations, discussed perceptions of their current contexts, and contemplated about their future needs."

"Digital technologies that fit individual needs could enhance older adults' social participation."

"Expanding digital opportunities may lead to feelings of reduced social participation for older adults who seldom or never use digital technologies (Davidsson and others [414])."

"The findings revealed that engagement in social activities varied among older adults, which was consistent with other studies."

"The choice to collect data in rural communities in Northern Sweden highlighted particular social participation issues that older adults experienced, such as reduced mobility opportunities, increased geographical distances to social networks and essential community services, and changing demographics in rural areas."

"These issues put focus on the potential impact of digital technologies on older adults' continued social participation and living at home."

Conclusions

"This study contributes to findings about contexts surrounding older adults' social participation in a digital society."

"Digital technologies are perceived to augment social opportunities and access to services, as well as provide a feeling of security, as long as they provide satisfactory experiences for the older adult."

"Both co-creating usable digitalized services and facilitating satisfactory use of digital technologies could support older adults' social participation through activities that they find relevant in their lives, and subsequently, might enable them to live longer at home."

4.3.5 Silence Over the Wire: Student Verbal Participation and the Virtual Classroom in the Digital Era [555]

This is a machine-generated summary of:

Ho, Debbie G. E.; Sa'adi, Munawwarah; He, Deyuan; Hoon, Chang-Yau: Silence over the wire: student verbal participation and the virtual classroom in the digital era [555]

Published in: Asia Pacific Education Review (2023)

Link to original: https://doi.org/10.1007/s12564-023-09834-4

If you want to cite the papers, please refer to the original.

For technical reasons we could not place the page where the original quote is coming from.

Abstract-Summary

"This paper addresses the pervasive absence of verbal student participation in the online class, a phenomenon observed by many lecturers and instructors expressing the frustrating and uncomfortable experiences of encountering silence from their students, particularly when it came to responding to their questions."

"Whilst studies on student silence in classroom discourse have been well documented in the research literature, this phenomenon has taken on new significance in the virtual classroom, the new norm in the learning context during, and most likely after, the COVID-19 situation."

"This study attempts to capture the perceptions of the students themselves on student silence in terms of frequency, reasons and its impact on classroom communication and meaningful learning."

"The question of whether meaningful learning still occurs despite the silence is more complex and less clear, raising questions not only about what is meant by meaningful learning but also the claim by classroom discourse studies and writings that student verbal participation is key to successful learning."

Introduction

"Many studies showed that online learning does not promote student verbal participation (Arora & Srinivasan, [415]; Baris & Cankaya, [416]; Chen and others, [417]; Jin, [418]; Kaya & Önder, [419]) which negatively affected meaningful learning and classroom communication."

"Before proceeding any further, it is necessary to carry out a review of the literature on online learning, student silence or non-verbal participation and communication, Online learning, also known as remote or distant learning, means that students and teachers are physically distant from each other and are connected via virtual platforms (Fry, [420]; Wang and others, [421]; Wilde & Hsu, [422])."

"On the student participation front, some studies suggest that online learning could generate self-confidence, classroom interaction, improve pedagogical standards and conversation skills (Bailey & Lee, [423]) in the sense that online learning provides more time for students to ponder on instructors' questions before responding."

"The study attempts to examine the impact of student silence in online classes on meaningful learning from the pedagogical aspect and successful communication from a conversational analysis perspective."

Significance of the study

"Online learning has brought about a change in a key area in classroom discourse, and that is the verbal participation of learners within that learning space."

"The significance of this study is very clear in that it highlights the issue of student verbal participation in the online classroom."

"It provides insights into how students interact with their instructors and their peers in such a space, and the impact of such interaction on meaningful learning and in the ebb and flow of classroom communication."

Context of study

"It has to be noted that studies on student silence in the Bruneian context have been far and few in between and whatever studies that were carried out were limited to the physical classroom at the primary and secondary school levels."

"These studies have highlighted Bruneian students' reticence in the classroom."

"The general impression from teachers is that Bruneian students are shy, gentle, collectivistic, and generally unwilling to attract attention in the classroom (den Brok and others, [424]; Ida Rafidah Haji Dzolkefflie, [425]; Larking, [426])."

"This study attempts to address this gap in the literature by looking at the phenomenon in the online classroom."

Purpose of the study and research questions

"The purpose of the study is to examine non-verbal student participation in the online classroom at tertiary level and its effect(s) on classroom interaction in the online class."

"It aims to gauge perceptions of non-verbal student participation in online classes from the students' viewpoint."

"Four research questions are posited: (i) To what extent is student silence a distinguishing feature in the online class? (ii) How often do students feel awkward when there is silence in their online classes? (iii) What possible reasons are there for student silence? (iv) Can meaningful learning still be possible with student silence?"

Methods of data collection and analysis

"The data for this study were collected from 263 undergraduate students through responses to a survey."

"Following the survey, five-student focus group interviews (SFG) were conducted with the purpose of probing further the questionnaire responses."

"Items in the survey were drawn up by the researchers based on the issue of student silence raised in the research literature and their observations from their own online classes."

"The second section contained a number of Likert-scale type items on student verbal participation in online classes to gauge student perceptions."

"Following the survey, and based on simple random sampling, five-student focus group interviews (SFG) were conducted."

"As students may have different perceptions and attitudes towards student silence in the online class, it was felt that a focus group interview would be an appropriate method to gather their perceptions."

Results and findings

"A thematic analysis of the data from the student focus group interviews yielded three main themes: (i) students felt awkward and uncomfortable when no one responded to the lecturer questions: very, very awkward … and uncomfortable … (SFG 1); I actually feel a bit annoyed … there's like this pressure for you to kind of answer the question cause no one else is doing it (SFG 2); for me, I find it really awkward … (SFG 3); I feel bad, awkward, sometimes even pressured when no one responds to the lecturer's questions (SFG 5); (ii) students felt guilty for not responding verbally: I would feel bad for myself too because I didn't study or prepare for the class (SFG 5); I will also sometimes feel guilty for being one of the silent students… yeah I do feel guilty … because everyone is silent so I'm just gonna stay within the line of being silent … (SFG 1) and (iii) students felt bad for their lecturers: I feel kind of sad for the lecturer cause no one answers …I think s/he [the lecturer] feels like talking to a wall and s/he may be wondering whether whatever she's delivering is something understandable to her students (SFG 3); I feel awkward for the lecturer and for everyone else in the class I guess … I think they're [lecturers] quite frustrated though … (SFG 4); I feel bad for the lecturer, I really really feel bad for them … (SFG 5)."

Discussion of results and findings

"Whilst the participants generally agreed with Sinclair & Coulthard [427] and Weiner [428] that verbal communication in the classroom is paramount to meaningful learning, many students also argued that the term 'classroom interaction' could be extended beyond the oral to the written mode via the use of the chat box and interactive emoticons for instance) to communicate in class."

"Studies elsewhere have shown that there are advantages to using the chat box to interact when it comes to effective learning (Carter and others, [429]; Guiller and others, [430]; Jurewitsch, [431]; Kim and others, [432]; Nguyen, [433]), one of which was mentioned by the participants and that is writing their responses gave them time to reflect on their lecturers' questions and provide thoughtful responses This study has highlighted the following pedagogical implications on student silence and the online classroom: (i) That communication is key to successful learning."

Limitations of the study

"It was felt that a cross tabulation of the questionnaire data could have been carried out to present the differing viewpoints of students from the individual programmes and thus a more representative picture of student silence."

"The other limitation to be noted was in the use of the Cronbach Alpha test to assess the reliability and internal consistency of categories of items in the questionnaire."

"Because items often do not show a linear correspondence to the question under study and their functions may be multi-dimensional in purpose, it affected the test scores."

"The study also highlights the overall dearth of tools available in qualitative research literature to ensure the robustness and reliability of such data."

"The third limitation is that the thematic analysis employed to analyse and interpret the focus group interview data could only highlight the recurring theme patterns in each group interview, and so individual student viewpoints that were considered outside these themes were not considered."

Conclusion

"The students felt that student silence was an unmarked feature in their online classes despite there being ample opportunities for students to be verbally engaged with their lecturers."

"Analysis of the data showed that student silence in online learning did not promote conversational skills in terms of negotiating the speaking turns or providing important signalling markers such as backchannel responses to successfully sustain the interactive flow of communication."

"The students generally felt that their online classes did not generate a high level of interactive verbal discussions between themselves and the lecturers which they felt to be key to their successful learning."

"It has to be noted that this study was confined to looking at student silence in online learning at the tertiary level in the Bruneian context and may therefore not be representative of the phenomenon found elsewhere."

Directions for future research

"This study has raised awareness to a number of aspects in classroom silence that could lead to future research studies."

"It would be interesting, for example, to gauge the perceptions of the other party in classroom discourse, the instructors/teachers/lecturers, to provide more depth to the issue under study."

"Due to the intricate and subjective nature of such research, more meaningful and contextualised results and findings could be reached by including a cross tabulation analysis of the questionnaire data."

4.3.6 A Longitudinal Analysis of Student Participation in a Digital Collaborative Storytelling Activity [556]

This is a machine-generated summary of:

Liu, Chen-Chung; Yang, Chen-Yu; Chao, Po-Yao: A longitudinal analysis of student participation in a digital collaborative storytelling activity [556]

Published in: Educational technology research and development (2019)

Link to original: https://doi.org/10.1007/s11423-019-09666-3

If you want to cite the papers, please refer to the original.

For technical reasons we could not place the page where the original quote is coming from.

Abstract-Summary

"Despite the positive potential of the online social networking platforms, the longitudinal studies regarding how children participate in digital storytelling communities is still scant in the literature."

"This study investigated how 26 elementary students participated in a 2-year digital storytelling community mediated by a social network application."

"By analyzing the students' weekly responses, social networks, and their language learning performance, the results suggested that their oral reading proficiency improved along with their participation in the long-term activity."

"The social network analysis found that multiple literacies including multimedia processing, language, and collaboration were all considered by students as important criteria for the partnership to complete the collaborative digital storytelling work."

Introduction

"The previous study investigated the relationships between students' structural positions in the social networks and their engagement in the collaborative digital storytelling activity."

"It is thus worthwhile to further investigate how time impacted students' sense of engagement, collaboration relationship, and the teaming-up activities between students in a long-term collaborative digital storytelling activity."

"This study analyzed students' learning activities and performance to uncover the students' participation in a long-term collaborative digital storytelling activity by answering the following research questions: RQ1."

"How did the students' English learning performance change as the long-term collaborative digital storytelling activity progressed?"

"How did the students' sense of engagement change as the long-term collaborative digital storytelling activity progressed?"

"How did the collaboration experience among students change as the long-term collaborative digital storytelling activity progressed?"

"How did the social network among students evolve as the long-term collaborative digital storytelling activity progressed?"

Methods

"Rather, our goal was to uncover the evolution of students' participation during the long-term collaborative digital storytelling activity and understand the learning experiences exhibited in different phases of the activity."

"Both qualitative and quantitative approaches were applied to uncover the multiple dimensions of students' participation in the long-term collaborative digital storytelling activity."

"The quantitative approach aimed to provide data about the evolution of students' learning performance, sense of engagement, collaboration experience and social networks."

"The qualitative approach analyzed students' social networks along the timeline and the interviews with students."

Participants

"To understand each student's entry English proficiency, a word game test and oral reading tests were administered before this study."

"Three students who scored 0.5 standardized deviation higher than the mean score were considered as high proficiency students."

"Six students who were rated 0.5 standardized deviation lower than the mean score were considered as low proficiency students."

"It is hoped that this study can contribute to a better understanding of how students with different English proficiencies participate in storytelling communities."

The collaborative digital storytelling activity

"While participating in the activity, they (1) were guided to read the model stories, (2) practice the sentences, (3) draw the pictures for the story scenes, (4) record vocal narration of the stories, and (5) finally publish the picture book on the online platform."

"It was hoped that through participating in the learn-to-create community, the students could enhance their story oral reading proficiencies."

"The students were given iPads to participate in the digital storytelling activity."

"The online platform was designed to help students collaboratively create, tell, and publish their stories through the iPads."

"Students, their parents, or external audience may find it interesting and may read the picture books on the online platform."

Procedure

"To allow for a higher level of flexibility and to examine how different teaming-up strategies may impact students' participation in the digital storytelling activity, this study implemented a different teaming-up strategy from the 19th week onward."

"Students could participate in only one picture book development team at a time."

"To obtain a macro understanding of the students' participation for this study, we divided the 2-year activity into four phases: the fixed phase (weeks 1–18) in which the fixed partner strategy was applied, the dynamic phase (weeks 19–26 in the 2nd semester in which the dynamic teaming strategy started to be applied), the prolonged phase (weeks 27–36 in the 3rd semester in which the dynamic teaming strategy was applied) and the final phase (weeks 37–43 in the 4th semester in which the dynamic teaming strategy was applied)."

"The data of how each student perceived the experience in the digital storytelling community and how they teamed up with their peers to participate in the activity were collected for further analysis."

Instruments and data collection

"To explore the change of students' English learning performance in the collaborative digital storytelling activity, repeated measures analysis was applied to the students' WCPM and the quality of the digital books throughout the four activity phases."

"T tests were also conducted to explore the difference in WCPM scores and in quality of the digital books between any two activity phases in order to understand at which phase the students' English learning performance changed significantly."

"To understand the change of students' engagement perceptions in the collaborative digital storytelling activity, repeated measures analysis was applied to the students' flow perceptions, intrinsic and extrinsic motivations throughout the four activity phases."

"To understand students' learning activities and the evolution of students' social networks in the collaborative digital storytelling activity, 14 participants were briefly interviewed at the end of the activity to investigate the reasons and criteria for inviting partners to work with."

Results and discussion

"Regarding the students' collaborative experience, at the initial stage of the activity in which the fixed teaming strategy was applied, the students' self-evaluation and peer evaluation of their collaboration experience were relatively low (the fixed phase) and significantly increased to a higher level when they were able to team up with partners based on their preferences (the dynamic, prolonged, and final phase)."

"The result of this study found that the dynamic teaming strategy may improve students' participation in such a situation and enhance students' perceptions of collaborative digital storytelling."

"The variety of criteria that students used to invite partners suggests that the dynamic teaming strategy drew students' attention to these literacies in the prolonged collaborative digital storytelling activity as they needed to demonstrate their abilities in these literacies in order to team up with others."

Conclusion

"These issues, which may offer suggestions for instructional design for storytelling activities aiming to improve language learning, are as follows: This study incorporated digital storytelling and social networks into collaborative language learning."

"It is suggested that digital storytelling activities that incorporate with social networks as collaborative language learning may offer reengagement supports for students to improve and sustain their learning performance."

"It is suggested that applying dynamic teaming strategy to a collaborative digital storytelling activity may enhance students' flow perceptions and collaboration experience."

"It is suggested that the multiple literacies for digital storytelling may influence the students' collaboration in digital storytelling activity that employs the dynamic teaming strategy and social networks."

"The results of this study contribute to a better understanding of how students engaged in a prolonged collaborative digital storytelling program in terms of their language learning performance, perceptual engagement, and social network evolution."

Limitations and future works

"This learning community was not completely open, meaning that students outside the class could not join the collaborative digital storytelling activity."

"It would be interesting to investigate how students from a broader scope, for instance students in a school, would participate in a campus-wide collaborative digital storytelling activity to broaden the impact of such a storytelling approach."

"It would be worth trying to understand how students from different regions or countries participate in such a collaborative digital storytelling activity."

"Gathering further evidence on these issues would help in obtaining a better understanding of collaborative digital storytelling and thereafter in making appropriate pedagogical decisions regarding the implementation of collaborative digital storytelling activities."

"This study applied a digital storytelling application as the tool to implement the long-term storytelling activity."

"The digital platform provides many critical features to implement the create-to-learn activity."

4.3.7 Open Data, Crowdsourcing and Game Mechanics. A Case Study on Civic Participation in the Digital Age [557]

This is a machine-generated summary of:

Handler, Reinhard A.; Ferrer Conill, Raul: Open Data, Crowdsourcing and Game Mechanics. A case study on civic participation in the digital age [557]

Published in: Computer Supported Cooperative Work (CSCW) (2016)

Link to original: https://doi.org/10.1007/s10606-016-9250-0

If you want to cite the papers, please refer to the original.

For technical reasons we could not place the page where the original quote is coming from.

Abstract-Summary

"The case study highlights how data can be made accessible to people who usually do not have access; how game mechanics can be deployed in order to foster civic participation by offering users a sense of autonomy, competence and relatedness; and how crowdsourcing can organise a large group of people into achieving a common goal."

"The combination of these three elements resulted in a case for civic participation in the digital era."

Introduction

"The Guardian started to advertise an investigative journalism crowdsourcing campaign, calling to action the readers and asking them to sieve through the large set of leaked documents (GNM press office [436])."

"In a time when datafication raises concerns about technology taking over decision-making processes, The Guardian's data initiative shows that open data can be used to enhance civic participation, allowing people to interact with and analyse data."

"The Guardian opted for a crowdsourcing approach in order to let the users analyse a large data set."

"We will discuss how The Guardian combined crowdsourcing and game mechanics to engage a large number of users."

"We propose that the combination of open data, crowdsourcing, and game mechanics allowed The Guardian to shift the dynamics of media agency by facilitating civic participation."

Big data, small data, open data, closed data

"The Guardian's crowdsourcing experiment rested upon data that were made open and accessible for the public, something that usually is not associated with data practices."

"Kitchin [437] highlights four potential problems of open data initiatives: funding and sustainability, the marketisation of public services, the misconception that data is neutral (i.e. objective and for everyone to use), and utility and usability."

"A stable financial base was a major advantage The Guardian had over ordinary open data initiatives."

"This is a result of the inherently technical nature of datafication whether the data is big, small, open or closed."

"We argue that The Guardian's data initiative shows how a combination of technological and socio-cultural elements that counters the purely technological understanding of datafication because it allows social actors to make sense of (large) accessible datasets through complicit strategies of collaboration, such as crowdsourcing."

The participating crowds

"Top-down systems are believed to be replaced by collaborative networks that neglect strict hierarchies and are built on 'trust and long-term cooperation' (Benkler [438], p. 1) fostering active social engagement and active participation."

"The Guardian's data initiative exemplifies how a collaborative practice like crowdsourcing is related to the technological process of datafication by creating an interface that combines involvement with data and collective civic participation."

"The element of interpretative sense-making of data in combination with datafication is neither a social, nor a purely technological process but a combination of two different approaches that are so far, more often than not, negotiated separately."

"What The Guardian achieved with its crowdsourcing initiative is a great example of engaging users in civic participation."

"The Guardian's experiment illustrates that crowdsourcing can be a productive way of combining the advantages of technology with socio-cultural practices which reflect traditional human(istic) efforts and insights."

Game mechanics as triggers of participation

"The most famous take on this friction is Putnam's [439] suggestion that the amount of time spent consuming popular and entertainment media (specifically games) is using up the limited time from citizens and users, time they could otherwise use for engaging in public and civic activities."

"Using a persuasive design (Fogg [440]), The Guardian's design team relied on basic game mechanics to increase the user motivation and eventually trigger participation."

"This is acknowledged by one of the creators of The Guardian's interface: "Also if you make something more like a game this can really help to engage people."

"I think making it fun is really important." (Rogers, as cited in Bouchart [441]) The high participation in The Guardian's initiative is partly explained by the way in which games motivate players to keep playing."

"We suggest, that the high degree of participation in The Guardian's experiment can also be understood due to the fact that games and particularly videogames are engaging people."

Civic technologies

"To our understanding, civic technologies are specifically created to enable, facilitate, and enact civic participation."

"The design of the technology itself automates mechanisms that trigger participation."

"Creating an interface that made us of crowdsourcing and game mechanics allowed people to access open data they had an interest in analysing."

"The combination of complex infrastructures and human behaviour is a feature of civic technologies and enables participation in the public sphere."

"The Guardian's initiative made possible civic participation at a time when traditional forms of engagement in the public sphere are said to be faltering (Putnam [439])."

"This offers a hopeful note that technologies can help citizens with creating a critical mass of people that engages with current civic problems."

"What The Guardian example highlights is the need for reflection on a changing digital media landscape, as well as new approaches towards participation with special attention to new approaches like open data, crowdsourcing and game mechanics."

Conclusion

"The social aspect of gaming and how the new media users interact with interfaces as a driver of civic engagement clearly point to a merging trend between user and technological agency."

"As social interactions evolve and become digitally mediated, so does civic engagement and how citizens participate in social affairs."

"Based on the important existence of a social layer in online games that require sociability and cooperation in order to achieve more meaningful goals, it is argued that the incorporation of game mechanics that trigger civic life will affect the civic engagement of those players in a positive way."

"Crowdsourcing as introduced by The Guardian highlights how certain aspects of data can be addressed by social actions resulting in a case for civic participation in the digital era."

"The unique blend of open data, crowdsourcing and game mechanics offered by The Guardian is an example for civic participation in the digital era."

4.4 Digital Surveillance

Machine generated keywords: surveillance, disease, health, public health, device, control, patient, behaviour, region, regional, spatial, rate, threat, australia, upon

4.4.1 Privacy and Beyond: Socio-ethical Concerns of 'On-the-Job' Surveillance [558]

This is a machine-generated summary of:

Indiparambil, Jijo James: Privacy and beyond: socio-ethical concerns of 'on-the-job' surveillance [558]

Published in: Asian Journal of Business Ethics (2019)

Link to original: https://doi.org/10.1007/s13520-019-00089-0

If you want to cite the papers, please refer to the original.

For technical reasons we could not place the page where the original quote is coming from.

Abstract-Summary

"Current sophisticated technologies in the workplace offer inexpensive and user-friendly devices and the means to control 'on-the-job' behaviour."

"The dilemma is twofold: First, scholarly studies undertaken on this issue from a principally administrative and legal point of view tend to lack a moral framework and so prove unable to offer an integral understanding."

"A majority of scholars tend to focus exclusively on individual rights, such as privacy, even at the risk of overlooking its social impact and consequences."

"It analyses the surveillance frameworks and the arguments for and against it; scrutinises critically the technological devices most often implemented in the workplace; examines both its individual effects (privacy) and social effects (categorising); and proposes an ethics of workplace surveillance in a framework of trust and transparency."

Introduction

"Many believe that the socio-ethical crisis arising from the extensive use of surveillance often overrules its expected benefits, and that its use in the organisational workplace today has alarmingly augmented ethical concerns (Indiparambil [442]; West and Bowman [443]; Holland and others [444]; Alge and Hansen [445]; Trevino and others [446]; Lyon [447, 448]; Alge [449])."

"An ethics of workplace surveillance is then proposed on the basis of a framework of that reallocates trust as the default (Lyon [450]) and transparency as the future (Rawlins [451]) of ethical behaviour in organisations."

"By engaging current literature through an analytical method, aims to carry out an ethical and social investigation of organisational behaviour within the context of electronic surveillance to learn how different structures affect on-the-job behaviours."

"To discuss the social aspects and effects of workplace surveillance, it delves, methodically and dialogically, into the more concrete situation of workplace surveillance and its sorting impacts, and further describes how a variety of social identities—discriminating and dehumanising—are being constructed and reinforced within the workplace."

The paraphernalia of 'new' workplace surveillance

"Such surveillance is primarily intended 'to increase (the) safety of employees by decreasing violence and threatening behaviour and locating workplace risks and to discourage drug and alcohol use, theft or sabotage, and otherwise to protect employer property and monitor employee productivity', it often leads to an invasion of privacy and may 'decrease employee morale, increase stress, lead to distrust and decrease productivity' (Ciocchetti [452], pp."

"The most recent and affordable forms of employee monitoring are tracking social networking and search engine use (Ciocchetti [452])."

"Personal or employer-provided mobile phones and devices are used to monitor employees through monitoring apps, which can secretly record the text messages, contacts, call logs, location etc, and even without connecting to wifi or LAN, employers can track the employees and their movements."

[Section 3]

"The value of surveillance will be judged based on whether it is seen as a legitimate technology that allows managers to care for everyone's interests, on the one hand, or a powerful instrument of managerial coercion and employee subordination, on the other (Sewell and Barker [453])."

"We will understand workplace surveillance in what follows to refer to any systematic monitoring by management or authority of individual employees' behaviours in the workplace such as job performance, which is carried out with an eye to ensuring compliance with management expectations (Ball [454]; Zweig [455]; Lyon [447, 448]; Rule and Brantely [456])."

"According to Sewell and Barker ([453], p. 938), 'the organizational surveillance of the few watching the many is justified because it enables a small number of managers to exercise control over a larger and potentially unruly mass of employees through the monitoring of individual behaviour'."

Privacy and social classification: the elapsed consequences of workplace surveillance

"Certain parallels in the structures of the panopticon and the monitored workplace are immediately apparent, such as 'the employment of panoptic-like surveillance as an attempt to subjugate employees to the power of management… [which] often instils a sense of powerlessness and fear among the observed' (D'Urso [457], p. 288)."

"New electronic infrastructures and subsequent sophisticated datamining techniques raise the issue of identification and social sorting by 'increasing concern with recording, retrieving, and analysing personal data to manage risks effectively' and 'do so by sorting population and individuals into categories of risk so that they can be treated accordingly' (Lyon [458], p. 163)."

"Organisational surveillance is a powerful means of creating and re-inforcing long-term social differences, as it 'engages many kinds of power, and is both productive and, sometimes, pernicious' (Lyon [459], p. 173)."

"Although Foucault's emphasis was on discipline in general, surveillance in modern organisational contexts always involves a 'power' mechanism exercised over all the life chances of employees."

Theoretical frameworks of workplace surveillance and quest for new work ethics

"Our major concern with respect to the agency theory, in line with current surveillance practices in the workplace, is the principal's (employer's or management's) ability to monitor the agent's (employee's) behaviour due to the increased use of present technological advancement."

"As today electronic surveillance is raising new concerns about privacy and fairness, the model of procedural justice theory should be integrated into our analysis in order to prioritise the values at stake in the process of monitoring employees."

"The Information Boundary Theory (IBT) developed out of a programme of research investigating the impacts of monitoring and surveillance technologies on worker privacy (Stanton and Stam [460])."

"Analysing these three components, we find that the structural-perceptual model proposes that the 'relationship between key outcomes and the overall panoptic effect potential (from communication technology to organizational factors and policy) [be] moderated by employee concern for surveillance' (D'Urso [457], p. 299)."

[Section 6]

"Trust has declined drastically in recent decades and any progressive and productive relationship has been eroded and withered in the workplace, due in large part to the perceived organisational behaviour of on-the-job surveillance."

"When an organisational behaviour is embedded, from the perspective of both the employer and the employee, in a context of mutual trust, the need for other forms of monitoring in the workplace seem irrelevant and extraneous."

"Relationships, built among co-workers and between them and the managers through trust and transparency, generate several positive workplace outcomes and help to establish strong moral behavioural patterns."

"Social relationships at work are always associated with job involvements and organisational commitments (Nielson and others [461]) attained through these attitudes of trust and transparency."

Conclusion

"This paper argues for an alteration or modification of traditional organisational behaviour around an ethic of trust and transparency and a consequent interpersonal relationship, that is situated within and goes beyond questions of moral duty, abstract principles, and legal compliance."

"A highly interactive process of integrating collaborative workplace behaviours should be a vital concern of all organisational behaviour and morale."

"Besides promoting ample growth of business and economic stability, any behaviour must sustain moral standards and values, social order and individual and organisational human development, for which the values of trust and transparency are brought in."

"It serves as a multi-dimensional model for reflection on economic and organisational aspects of individual and social life as well as of on-the-job behaviours and challenges each to 'go the extra mile' in their organisational behaviour."

4.4.2 Fear of Panoptic Surveillance: Using Digital Technology to Control the COVID-19 Epidemic [559]

This is a machine-generated summary of:

Keshet, Yael: Fear of panoptic surveillance: using digital technology to control the COVID-19 epidemic [559]

Published in: Israel Journal of Health Policy Research (2020)

Link to original: https://doi.org/10.1186/s13584-020-00429-7

If you want to cite the papers, please refer to the original.

For technical reasons we could not place the page where the original quote is coming from.

Abstract-Summary

"In a bid to reduce infection rates by COVID-19 the authorities in some countries, in collaboration with medical regulators and experts, have employed digital technologies to control and regulate citizens' behavior."

"The objective of the present study was to examine attitudes expressed in the public discourse toward the use of digital technologies to control people's behavior during the COVID-19 pandemic, as reflected in the media."

"Qualitative analysis was performed on posts and comments submitted in response to 12 articles that appeared on the four leading Israeli news sites, on three significant occasions: first, upon the announcement of the use of surveillance technologies by the Israeli security agency (ISA); second, upon the announcement of the launch of the Health Ministry's app that tracks contacts with COVID-19 patients; and third, following reports of petitions lodged with Israel's supreme court challenging the use of surveillance technologies."

"Supporters regarded contact tracing by means of digital technologies as essential to the effort to protect people during the pandemic, and believed that employing the ISA's capabilities was the correct way to combat the epidemic."

"Opponents of the measures rejected the use of tracking technologies, regarding this step as a move toward dictatorship and a violation of fundamental civil rights."

"The debate surrounding the employment of digital technologies during the COVID-19 pandemic highlights the complexity of the close connection between social control and care in times of pandemic."

"The context of this pandemic has highlighted the interrelatedness of advanced digital surveillance technologies, medical care, and social control exercised by authorities and medical regulators and experts, which raises issues of transparency, trust and mistrust among the public."

Background

"One of the regulatory measures taken to contain and control COVID-19 was to use digital technologies that facilitate the tracing of contacts between diagnosed patients and other individuals during the 14 days prior to the date on which the patient was diagnosed with the disease."

"The contact tracing technologies used in a number of Asian countries collect highly sensitive data from individuals."

"Israel, which has highly developed digital technological capabilities [462] and is known as the "start-up nation" and a leading source of technological innovation worldwide [463], has used both voluntary and mandatory digital technologies to track, control and regulate population behavior during the COVID-19 pandemic."

"As early as March17, 2020, the government passed a set of emergency regulations authorizing the Israel Security Agency (ISA) to aid in the national effort to limit the spread of the COVID-19, and permitting it to collect, process, and use of citizens' personal data."

Methods

"Qualitative analysis was performed on posts and comments submitted in response to 12 articles that appeared on three significant dates on the four leading Israeli news sites."

"The Posts and Comments section is a feature of news websites in which readers are invited to comment on the published content."

"Since the four leading Israeli news sites publish in Hebrew, almost all of the posts and comments were also written in Hebrew."

"The total of 2551 posts and comments referred to the 12 news articles, of which 28% were directly relevant to the research objective."

"The data that was analyzed comprised posts and comments written at different times, submitted to different news websites by many different individuals."

"Analyzing posts and comments on website news articles offers several advantages but also entails disadvantages in comparison to other qualitative methods such as in-depth interviews and focus groups."

Results

"When the supreme court addressed the petitions lodged against the use of ISA surveillance technologies, some wrote that "public health and safety are more important than privacy"; that "the government is working to save all its citizens, in the wake of a global deadly epidemic… No more self-destruction," adding that ISA data collection "is only for the purpose of the Corona and for no other use.""

"Commentators perceived the use of surveillance technologies to be "anti-democratic steps [taken] by the government," and part of the process of eroding democracy: "Democracy in the State of Israel is dying," "No democratic country has been monitoring cell phones.""

"Alongside anxiety over the fate of democracy and the fear of moving toward a dictatorial regime, many commentators perceived the use of surveillance technology to deal with the COVID-19 epidemic to be a violation of the right to privacy.""

Discussion

"This study, which focused on a specific point in time of the spread of the COVID-19 pandemic, raises issues about processes that researchers study over an extended period – in this case, the use of digital technologies to monitor public health and mechanisms of social control."

"The outbreak of a pandemic highlighted the interrelatedness of information technology exploitation and sophisticated digital surveillance technologies; the exercise of medical social control by authorities in collaboration with medical regulators and experts; transparency; trust and mistrust among the population; and the ambiguity of the close link between control and care – all of which are components of this panoptic- surveillance."

"The context of this pandemic has highlighted the interrelatedness of advanced digital surveillance technologies, medical care, and social control exercised by authorities and medical regulators and experts, which raises issues of transparency, trust and mistrust among the public."

Conclusions
"The COVID-19 pandemic has further empowered the surveillance society and highlighted the complex issue of the close connection between control and healthcare."

"Building trust in medical experts, in the authorities, and in the surveillance technology itself that must be efficient and reliable, is of vital importance for gaining the public's compliance with highly restrictive regulations, especially in democratic societies."

4.4.3 M-Health and Health Promotion: The Digital Cyborg and Surveillance Society [560]

This is a machine-generated summary of:

Lupton, Deborah: M-health and health promotion: The digital cyborg and surveillance society [560]

Published in: Social Theory & Health (2012)

Link to original: https://doi.org/10.1057/sth.2012.6

If you want to cite the papers, please refer to the original.

For technical reasons we could not place the page where the original quote is coming from.

Abstract-Summary
"The new mobile wireless computer technologies and social media applications using Web 2.0 platforms have recently received attention from those working in health promotion as a promising new way of achieving their goals of preventing ill-health and promoting healthy behaviours at the population level."

"There is very little critical examination in this literature of how the use of these digital technologies may affect the targeted groups, in terms of the implications for how individuals experience embodiment, selfhood and social relationships."

"It is argued that m-health technologies produce a digital cyborg body."

"The subject produced through the use of m-health technologies is constructed as both an object of surveillance and persuasion, and as a responsible citizen who is willing and able to act on the health imperatives issuing forth from the technologies and to present their body/self as open to continual measurement and assessment."

Introduction

"Health promoters have described the use of 'real-time feedback' of users' health status and 'prompts' and 'motivation' messages to 'change unhealthy lifestyle habits' via social media platforms and mobile devices, with reference to controlling behaviours such as smoking, alcohol consumption, exercise, diet and sexual behaviour [464, 465]."

"What the new social media and mobile devices provide that differs from older uses of the Internet – that is, Web 2.0 technologies compared with those offered by Web 1.0 – is the opportunity to directly tailor and target health messages on an individual level to intensify the pervasiveness of these messages and to monitor and record aspects of embodiment of users of mobile devices."

"Writers from medical and health promotion backgrounds about the new social media and mobile devices tend to confine themselves in their discussions to describing how these technologies could be most effectively used as tools in their efforts to help people deal with medical conditions or improve their general health and well-being."

Technologies, Health and the Body/Self

"The relationship between the human body and computerised technologies began to receive attention from social and cultural theorists in the 1980s."

"The concept of the cyborg has particularly inspired cultural theorists who have written about the implications of computerised technologies for human embodiment and subjectivity."

"For cultural theorists writing on cyborg bodies, the human-machine hybrid is complex and shifting, calling into question taken-for-granted assumptions about the oppositions between organic/inorganic, natural/artificial and self/Other [466–470]."

"As part of this change in representations of the body and the growing use of computerised information systems in medicine, the internal organs and workings of the body have moved from being exclusively the preserve of medical students and surgeons to being open to the gaze of all."

"[471, 472] use the term 'e-scaped medicine' to denote the recent shift in the location of medical knowledge and practice from the medical school and the clinic to diffuse digital information technologies such as the Internet and telemedicine devices."

M-Health and the Surveillance Society

"As this suggests, central to a critical analysis of the use of the new social media and mobile devices to promote health is a recognition of these technologies as part of 'surveillance society', a term used by some writers to denote the increasing ubiquity of surveillance technologies in everyday life, which are used to record, survey, monitor and discipline people (for example, [473–476])."

"More useful than theories of the panoptic gaze, therefore, is that which employs the Deleuze and Guattari term 'assemblages' to describe 'surveillant assemblages', or the complex interaction of technologies, data and bodies in producing defined subjectivities [473]."

"While the abstracted 'data-double' produced through biometric measurements and health surveillance technologies that are able to identify 'at-risk' individuals may be categorised as a virtual cyberbody, this data-double feeds back information to the user in ways that are intended to encourage the user's body to act in certain ways."

Privacy, Intimacy and Ethical Issues

"Accounts of using m-health technologies in the medical literature dealing with patient follow-up and care tends to focus on shifting responsibility for care from the clinician to the patient, placing new expectations upon the patient to manage their health in ways that were traditionally viewed as the preserve of health-care professionals."

"The new m-health technologies may be viewed as potentially contributing to feelings of shame and guilt that their target users may feel if they do not adopt health-related suggestions, an additional stressor in an individual's day when they may well be juggling multiple demands and responsibilities."

"Individuals make choices not in a social vacuum, but in a context in which certain kinds of subjects and bodies are privileged over others and there are obligations and commitments involved: the responsible, self-disciplined body/self, for example, who is interested in and motivated to improve their health."

Conclusion

"Via m-health technologies, the health promoter is able to insert her- or himself even more insistently into the private world of others, accessing them in any location in which their mobile device accompanies them."

"A space is opened up here for researchers to identify and explore the experiences of individuals as they take up (or indeed, resist) the potentialities of mobile digital devices and the new social media as they are adopted into the 'toolbox' of health promotion."

"It also needs to be acknowledged that 'digital literacy' should become a part of health literacy, and that indeed, such digital literacy might include a response on the part of targeted audiences to forms of health promotion messages conveyed via mobile digital devices and social media platforms that is critical and contesting of them."

4.4.4 Digital Surveillance in Latin American Diseases Outbreaks: Information Extraction from a Novel Spanish Corpus [561]

This is a machine-generated summary of:

Dellanzo, Antonella; Cotik, Viviana; Lozano Barriga, Daniel Yunior; Mollapaza Apaza, Jonathan Jimmy; Palomino, Daniel; Schiaffino, Fernando; Yanque Aliaga, Alexander; Ochoa-Luna, José: Digital surveillance in Latin American diseases outbreaks: information extraction from a novel Spanish corpus [561]

Published in: BMC Bioinformatics (2022)

Link to original: https://doi.org/10.1186/s12859-022-05094-y.

If you want to cite the papers, please refer to the original.

For technical reasons we could not place the page where the original quote is coming from.

Abstract-Summary

"In order to detect threats to public health and to be well-prepared for endemic and pandemic illness outbreaks, countries usually rely on event-based surveillance (EBS) and indicator-based surveillance systems."

"Event-based surveillance systems are key components of early warning systems and focus on fast capturing of data to detect threat signals through channels other than traditional surveillance."

"We created an annotated Spanish corpus from ProMED-mail health reports regarding disease outbreaks in Latin America."

"The corpus has been used to train algorithms for two information extraction tasks: named entity recognition and relation extraction."

"The annotated corpus and algorithms presented could leverage the development of automated tools for extracting information from news and health reports written in Spanish."

"This framework could be useful within EBS systems to support the early detection of Latin American disease outbreaks."

Background

"Much research is being done recently on the use of artificial intelligence in order to enhance event-based surveillance systems with moderated, or partially or fully automated techniques in order to obtain early diagnosis and treatment and to prevent the spread of diseases [478]."

"In order to capture data useful for EBS systems focused on prevalent outbreak diseases in Latin America, we developed information extraction (IE) algorithms to extract named entities (disease, date, geographic location, number of cases, host, origin -reported cause of the disease-, transmission form of the diseases -eg."

"With the extracted information provided by the named entity recognizers and the relation extraction algorithms, data could be summarized, Moreover, an interface could be constructed, whereby choosing a range of dates and diseases, geographical regions could be visualized on a map (as in HealthMap [479]), highlighting the number of incidences."

Related work

"Research studies show that social media may be a valuable data source in disease surveillance systems aimed to detect disease outbreaks, since it helps information to be reported faster than traditional methods."

"The Program for Monitoring Emerging Diseases (ProMED), an internet-based reporting system dedicated to the dissemination of information on epidemics of infectious diseases that depends on the International Society for Infectious Diseases (ISID), is a moderated system, that could benefit regions and countries in complementing their undiagnosed disease surveillance systems [480]."

"EventEpi [481] uses data from WHO Disease Outbreak News and ProMED-mail, and automatically extracts information regarding the disease, country, date, and confirmed-case count."

"This tool generates epidemiological information on diseases, locations, dates, hosts, and the number of cases for outbreaks mentioned in the news and social media articles."

"They use rule-based systems, data mining techniques and a support vector machine trained to better classify or identify information on diseases."

Methods

"Seven named entities (disease -DS-, date -DT-, location -LOC-, number of cases -NoC-, Origin -OR-, host -Hst-, and transmission form -TF-) and three modifiers (past terms -PT-, negation -NT-, and uncertainty -UT-) have been annotated."

"After performing the annotation, we analyzed the resulting dataset in order to know the number of existing entities and relations."

"The rest of the named entities (diseases, hosts, origin, transmission form) and modifiers (negation, uncertainty, and past terms) were detected by the use of regular expressions and lists of terms, developed based on the analysis of the articles separated for this purpose."

"We developed a baseline relation extraction algorithm, that detects as relations those pairs or triples of entities that co-occur in the same sentence and that have less than a pre-defined distance between them and that belong to the type of possible relations defined in the annotation schema."

Discussion

"The RNN algorithm obtained better micro-averaged F1 than the rule-based method for both exact and partial matches and for titles and for complete articles."

"The RNN algorithm has some advantages over the rule-based: (1) it can detect more variations of entities (eg more ways to refer to a certain disease or different serotypes of the virus that causes dengue), (2) it recognizes different formats to refer to dates and locations (eg the date "12 December, 2012" is labeled as two different dates by the rule-based method, but correctly as one by the RNN algorithm), and (3) it detects non-numeric "NoCs" (eg "10 cases per 100.000 people")."

"This works as expected since entities not easily distinguishable by annotators are probably not recognized by automatic methods (in the case of the NN because it learns from the data, and in the case of the rule-based algorithm, because rules followed by the data may not be clear)."

Conclusion

"We created an annotated corpus based on ProMED-mail articles, that is useful for automatically extracting epidemiological data from newspaper articles written in Spanish."

"For creating the corpus we analyzed the convenience of working with only articles' titles and dates (Titles) or complete articles (PMAs)."

"Other researchers could benefit of (1) the corpus, as far as we know, the first available corpus in Spanish, for the development of their own algorithms and for comparing the results of new algorithms with ours, (2) the annotation criteria, described by us in another paper ([482]), in order to generate their own corpus, and (3) the Spanish NER and RE algorithms provided, that have good results, and for which, as far as we know, there were no previous works, in order to use them as baselines or to compare them with other methods."

4.4.5 Early-Stage Spatial Disease Surveillance of Novel SARS-CoV-2 Variants of Concern in Germany with Crowdsourced Data [562]

This is a machine-generated summary of:

Mitze, Timo; Rode, Johannes: Early-stage spatial disease surveillance of novel SARS-CoV-2 variants of concern in Germany with crowdsourced data [562]

Published in: Scientific Reports (2022)
Link to original: https://doi.org/10.1038/s41598-021-04573-1

If you want to cite the papers, please refer to the original.

For technical reasons we could not place the page where the original quote is coming from.

Abstract-Summary
"We explore the possibility to use close to real-time crowdsourced data on reported VOC cases (mainly the Alpha variant) at the local area level in Germany."

"The aim is to use these data for early-stage estimates of the statistical association between VOC reporting and the overall COVID-19 epidemiological development."

"We find no statistical evidence for an over- or underestimation of effects once we account for differences in data representativeness at the regional level."

"This points to the potential use of crowdsourced data for spatial disease surveillance, local outbreak monitoring and public health decisions if no other data on new virus developments are available."

Introduction
"When novel variants of concern (VOC) of the severe acute respiratory syndrome coronavirus type 2 (SARS-CoV-2) were first reported in late 2020, scientists, policy makers and the general public soon asked essential questions: Do these VOC find it easier to transmit from human to human compared to previously circulating strains?"

"In Germany, the first case of a SARS-CoV-2 infection based on the Alpha variant was reported on December 24, 2020."

"Given the significance of VOC for overall SARS-CoV-2 infection dynamics, the goal of this study is twofold: First, it shall document our approach to provide near-time statistical estimates of spatial disease trends in Germany associated with the emergence and rapid spread of VOC during the first weeks in 2021."

"We focus on two key indicators here: (1) the 7-day incidence rate, i.e., the number of newly reported SARS-CoV-2 infections in the last 7 days per 100,000 population, and (2) the hospitalization rate, i.e., the number of hospitalized COVID-19 patients in intensive care per 100,000 population."

Spatial disease surveillance with crowdsourced data

"Administrative data on confirmed VOC cases supplied by local health authorities according to the German Infection Protection Act (GIPA) were initially incomplete and reported summaries only covered broad regional aggregates, i.e., the German federal states (NUTS-1 level)."

"While this made disease surveillance at the regional level already very difficult, the lack of comprehensive data on reported VOC cases below the level of federal states rendered a comprehensive local-area outbreak monitoring almost impossible."

"We propose the use of crowdsourced data collected from news paper reports and social media as a possibility to fill this data gap and to monitor VOC spread at the local area level (NUTS-3 level) in Germany."

"By combining crowdsourced VOC data with public health data on the overall infection dynamics, it is possible to provide early-stage estimates of the human-to-human transmissibility and case severity of VOC relative to the wild type of the SARS-CoV-2 virus."

Results

"For the overall sample with at least one confirmed VOC case, the estimates in Panel A point to an average increase in the 7-day incidence rate by 13.14 [95% CI 4.2, 22.1] cases per 100,000 population or approx."

"While effects are of similar size for the sub-sample estimates limiting treated regions to those with a VOC case confirmed before January 22 in Panel B (i.e., early treated regions for which we observe at least 14 days of treatment), we find a remarkably higher effect size for the subsample defining the treatment groups as belonging to the top 10% (90th percentile) of absolute VOC counts."

"Panel A shows for the sample of all regions with at least one VOC case that 20 days after VOC reporting the 7-day incidence rate has increased by approx."

Discussion

"Main goals of this paper were, firstly, to describe the initial regional dynamics of VOC spread (predominately the Alpha variant) within a country after international importation using novel crowdsourced data, secondly, to provide near-time estimates of epidemiological trends associated with VOC reporting at the local area level in Germany, and, thirdly, to evaluate the robustness of the crowdsourced data for statistical analyses of spatial disease surveillance."

"A higher number of reported VOC cases at the local area is found to increase the incidence rate dynamically over time."

"When we evaluate data representativeness ex post, we, firstly, find a high correlation between reported VOC cases from the utilized crowdsourcing database and revised data from local health authorities in Germany, published by the RKI."

Limitations

"Our findings also confirm a significant correlation between VOC reporting and the number of hospitalized patients per 100,000 population in those regions that have a relatively high number of VOC cases (top 10%)."

"We would also like to stress that we estimate the correlation between confirmed VOC cases and the 7-day incidence rate of infections or the hospitalization rate."

"I.e., using data at the population level for NUTS-3 regions only allows us to identify correlations between VOC spread and epidemiological trends, but no causal statements can be made whether VOC have a higher transmissibility than previously circulating virus strains and whether VOC infections lead to more severe COVID-19 disease cases."

"In cases where such data are unavailable, however, we argue that crowdsourcing projects together with carefully executed near-time estimates of statistical trends may inform public health authorities about the need for actions to mitigate the local and global transmission of VOC at an early stage [483]."

Methods

"Estimations are conducted by the synthetic control method (SCM) for single and multiple treated units, which compares the infection development in case study regions to the development in a synthetic control group composed of similar regions without reported VOC cases [484, 485]."

"While SCM offers favorable features for detailed case study analyses related to COVID-19 research [486–488], its focus on case study regions limits its ability to provide general estimates of infection trends associated with reporting of VOC cases in Germany."

"DiD estimation aims at identifying differences in epidemiological trends (time differences) across regions with and without treatment (cross-sectional difference), where treatment is defined as the reporting of at least one (or multiple) VOC case(s) at the regional level."

"To ensure comparison across estimation methods, we apply all three approaches to the same set of outcome variables, i.e. the 7-day incidence rate and the number of hospitalized patients in intensive care (with and without artificial ventilation) per 100,000 population."

4.4.6 Geospatial Epidemiology of Staphylococcus Aureus in a Tropical Setting: An Enabling Digital Surveillance Platform [563]

This is a machine-generated summary of:

Wozniak, T. M.; Cuningham, W.; Buchanan, S.; Coulter, S.; Baird, R. W.; Nimmo, G. R.; Blyth, C. C.; Tong, S. Y. C.; Currie, B. J.; Ralph, A. P.: Geospatial epidemiology of Staphylococcus aureus in a tropical setting: an enabling digital surveillance platform [563]

Published in: Scientific Reports (2020)

Link to original: https://doi.org/10.1038/s41598-020-69312-4

If you want to cite the papers, please refer to the original.

For technical reasons we could not place the page where the original quote is coming from.

Abstract-Summary

"In northern Australia, bacterial infection rates are high but resistance to first- and second-line antibiotics is poorly described and currently-available datasets exclude primary healthcare data."

"We aimed to develop an online geospatial and interactive platform for aggregating, analysing and disseminating data on regional bacterial pathogen susceptibility."

"We report the epidemiology of Staphylococcus aureus as an example of the power of digital platforms to tackle the growing spread of antimicrobial resistance in a high-burden, geographically-sparse region and beyond."

"We developed an online geospatial platform called HOTspots that visualises antimicrobial susceptibility patterns and temporal trends."

"The online platform provides an engaging visual representation of spatial heterogeneity, demonstrating striking geographical variation in S. aureus susceptibility across northern Australia."

"Plots generated by the platform show temporal trends in proportions of S. aureus resistant to methicillin and other antimicrobials across the three jurisdictions of northern Australia."

"Clindamycin resistance rates in MRSA are worryingly high in regions of northern Australia and are a local impediment to empirical use of this agent for community MRSA."

"Visualising routinely collected laboratory data with digital platforms, allows clinicians, public health physicians and guideline developers to monitor and respond to antimicrobial resistance in a timely manner."

"Deployment of this platform into clinical practice supports national and global efforts to innovate traditional disease surveillance systems with the use of digital technology and to provide practical solutions to reducing the threat of antimicrobial resistance."

Introduction

"Antimicrobial resistance (AMR) imposes a substantial burden including poorer clinical outcomes and higher healthcare costs for patients infected with resistant compared to susceptible organisms [489–491]."

"Current efforts to track and respond to AMR systematically in northern Australia and other regional settings are insufficient [492–494]."

"Combined, this leaves a significant gap in health services' ability to initiate appropriate, timely responses to changing AMR epidemiology."

"To achieve enhanced AMR surveillance in Northern Australia, cross-jurisdictional collaboration with data sharing is required and the use of innovative geospatial information systems, can support these efforts."

"We developed an online AMR surveillance platform that integrates existing but fragmented susceptibility data, for visualization and analysis across time and space."

"Northern Australia was selected due to the paucity of existing data from this region [495] especially from primary care settings, the unique endemic pathogens, and concerns about high rates of antibiotic resistance in S. aureus in this region."

Methods

"With few isolates collected and tested (< 15 tests), these data within the region of interest are aggregated or excluded if all years have < 15 tests."

"Participating laboratories provided data on all clinical specimens where susceptibility testing was performed during the study period."

"Two widely used international susceptibility method systems, Clinical and Laboratory Standards Institute (CLSI) and European Committee on Antimicrobial Susceptibility Testing (EUCAST), were used by the participating laboratories."

"Data from WA were provided with CLSI-interpreted values (Susceptible, Intermediate and Resistant)."

"Data from QLD was provided with CLSI-interpreted values (Susceptible, Intermediate and Resistant) for years 2008 to 2012, and EUCAST interpreted values for years 2012 onwards."

"For isolates from NT, we applied the 2017 CLSI M100-S27 Performance Standards for Antimicrobial Susceptibility Testing (27th Edition) to the provided minimum inhibitory concentration values for each organism."

Results

"Resulting data were derived from primary and tertiary care for the WA and QLD and tertiary care only for the NT."

"To cross-check a subset of data, agreement between the proportion of MRSA bloodstream isolates in NT was compared with the 2017 hospital antibiogram."

"Resistance to inducible-clindamycin among MRSA isolates was 10% in QLD compared to 26% in NT and 20% in WA."

"MRSA resistant to SXT was the same in QLD and NT (5% of all isolates) and almost three times higher in WA (14%, n = 1 345 resistant isolates)."

"There were 27 MRSA blood cultures tested for SXT between 2014 and 2017 in WA, and none were found to be resistant to SXT."

Discussion

"To fulfil Australia's first National AMR Strategy and implementation plan [496], a high-quality evidence-base is needed to support geospatially-representative data collection, coordinated surveillance activities and targeted policy decisions."

"Our data highlight the problem of MRSA resistant to inducible-clindamycin and SXT, commonly relied on as oral second-line options for community MRSA skin and soft tissue infections, or as first-line agents for people with MSSA infection with intolerance to beta-lactams."

"Higher clindamycin resistance rates still, at up to 44% among community MRSA isolates, have been reported in the southern half of the NT, a region of the jurisdiction that did not participate in this study [497]."

"This focal nature particularly, in the west part of Northern Australia is suggestive of MRSA outbreaks with community transmission in confined niches and is less likely to suggest regional differences in prescribing practices driving AMR."

4.4.7 Contesting Post-digital Futures: Drone Warfare and the Geo-politics of Aerial Surveillance in the Middle East [564]

This is a machine-generated summary of:

Amin, Heba Y.; Downey, Anthony: Contesting post-digital futures: drone warfare and the geo-politics of aerial surveillance in the middle east [564]

Published in: Digital War (2020)

Link to original: https://doi.org/10.1057/s42984-020-00021-y

Copyright of the summarized publication:

If you want to cite the papers, please refer to the original.

For technical reasons we could not place the page where the original quote is coming from.

Abstract-Summary
"Throughout the conversation, we focus on the research background to The General's Stork, alongside other works, and the threat presented by drone-based forms of aerial surveillance and targeting across the Middle East."

Introduction
"In a broader regional context, as The General's Stork demonstrates, the episode references the political economy of aerial surveillance from both a bird's eye view and, in an age of digitally defined warfare, remote controlled drones."

"I have been working with this database of aerial photographs of Palestine from this period in an attempt to understand how the region has been continuously visualized through technological developments and, in turn, how technology is used to push particular political narratives."

"In these images, you seem to be saying something else too: the technological means by which these aerial images were achieved initiated a new regime of viewing, a spatial and temporal regime that is both the precursor to the techno-aesthetics we now associate with drone technology and, in turn, the forms of total surveillance emerging from military-industrial complexes."

Bibliography

1. Rallet, A., & Rochelandet, F., (2003). La fracture numérique: une faille sans fondement? In Usage et appropriation des TIC. Second Workshop Marsouin, ENST Bretagne, http://www.marsouin.org/IMG/pdf/rochelandet_rallet-fracturenum1003.pdf
2. Paquot T., Lussault M., & Body-Gendrot S. (Eds.), (2000). La ville et l'urbain, l'état des savoirs. Paris: La Découverte
3. Lévy, J. (1994). L'espace Légitime. Sur la dimension géographique de la fonction politique. Paris: Presse de la Fondation Nationale des Sciences Politiques
4. Lévy J., & Lussault, M., (Eds.), (2003). Dictionnaire de la géographie et de l'espaces des sociétés. Paris: Belin
5. Ripert, B. (2005). Les nouvelles pratiques de télécommunication: quelles transformations dans les espaces ruraux et urbains en Inde? In S. Arlaud, Y. Jean, & D. Royoux (Eds.), Rural Urbain: les nouvelles frontières, Permanences et changements des inégalités socio–spatiales. (pp. 387–393). Rennes: Presses universitaires de Rennes
6. Ripert, B., (2006). Une expérience indienne des fractures numériques. Terminal, (no.° 95–96: pp. 113–126). Paris
7. Kampfen, F., & Maurer, J. (2018). Does education help old dogs learn new tricks? The lasting impact of early-life education on technology use among older adults. Research Policy, 47(6), 1125–1132

8. Riddell, W.C., Song, X. (2017). The role of education in technology use and adoption. ILR Rev
9. Wei, K. K., Teo, H.-H., Chan, H. C., & Tan, B. C. Y. (2011). Conceptualizing and testing a social cognitive model of the digital divide. Information Systems Research, 22(1) March, 170–187
10. Benitez, J. L. (2006). Transnational dimensions of the digital divide among Salvadoran immigrants in the Washington DC metropolitan area. Global Networks, 6(2), 181–199
11. Agudo, S., Pascual, M. A., & Fombona, J. (2012). Uses of digital tools among the elderly. Comunicar, 20(39), 193e201–193e201. https://doi.org/10.3916/C39-2012-03-10
12. Gonzalez, C., Fanjul, C., & Cabezuelo, F. (2015). Use, consumption and knowledge of new technologies by elderly people in France, United Kingdom and Spain. Comunicar, 23(45), 19e28–19e28. https://doi.org/10.3916/C45-2015-02
13. Martínez-Pecino, R., Delerue, A., & Silva, P. (2013). Portuguese older people and the internet: Interaction, uses, motivations, and obstacles. The European Journal of Communication Research, 38(4), 331–346
14. Neven, L., & Peine, A. (2017). From triple win to triple sin: How a problematic future discourse is shaping the way people age with technology. Societies, 7(3), 26
15. Joe, J., & Demiris, G. (2013). Older adults and mobile phones for health: a review. Journal of Biomedical Informatics, 46(5), 947–954
16. Dennis, C., Alamanos, E., Papagiannidis, S., & Bourlakis, M. (2016). Does social exclusion influence multiple channel use? The interconnections with community, happiness, and well-being. Journal of Business Research, 69, 1061–1070
17. Choudrie, J., & Vyas, A. (2014). Silver surfers adopting and using Facebook?: a quantitative study of Hertfordshire, UK: Applied to organizational and social change. Technological Forecasting and Social Change, 89, 293–305
18. Porter, C. E., & Donthu, N. (2006). Using the technology acceptance model to explain how attitudes determine internet usage: The role of perceived access barriers and demographics. Journal of Business Research, 59, 999–1007
19. Kim, D. J., & Hwang, Y. (2012). A study of mobile internet user's service quality perceptions from a user's utilitarian and hedonic value tendency perspectives. Information Systems Frontiers, 14, 409–421
20. Rogers, M. (1998). The definition and measurement of innovation. Melbourne
21. Hair Jr, J. F., Hult, G. T. M., Ringle, C., & Sarstedt, M. (2016). A primer on partial least squares structural equation modeling (PLS-SEM). SAGE Publications
22. DTI (2004) Digital television. http://www.dti.gov.uk/industries/broadcasting/digital_television.html. Cited 2004
23. Brookes M (2004) Watching alone: social capital and public service broadcasting. The Work Foundation in partnership with the BBC, May 2004
24. Gawinski M (2003) Interactive television production. Focal Press, London
25. Charness N, Bosman EA (1992) Human factors and age. In: Craik FIM, Salthouse TA (eds) The handbook of aging and cognition. Lawrence Erlbaum Associates, Hilsdale NJ
26. Sandhu J (1993) Design for the elderly: User-based evaluation studies involving elderly users with special needs. Special Issue: Designing for our future selves. Appl Ergono 24(1):30–34
27. Smith DDB, (1990) Human factors and aging: An overview of research needs and application opportunities. Hum Factors 32(5):509–526
28. DTI/The Generics Group (2003) Digital television for all: a report on usability and accessibility design. http://www.digitaltelevision.gov.uk/pdf_documents/publications/Digital_TV_for_all.pdf. Cited 2004
29. Tarnoff, B. (2016, August 3). The Internet should be a public good. Jacobin. Retrieved from https://www.jacobinmag.com/2016/08/internet-public-dns-privatization-icann-netflix/
30. Makinen, M. (2006). Digital empowerment as a process for enhancing citizens' participation. E-learning and Digital Media, 3(3), 381–395. https://doi.org/10.2304/2Felea.2006.3.3.381

31. Jayakar, K., Maitland, C., Peha, J., Strover, S., Bauer, J. 2016 Broadband 202: report of the interdisciplinary workshop on the development of a national broadband research agenda Penn State University, State College, PA Institute for Information Policy
32. United States Postal Service (2011). 2010 Comprehensive Statement on Postal Operations, 2010 Performance Report and 2011 Performance Plan. Chapter 1: our mission. Retrieved from https://about.usps.com/strategic-planning/cs09/CSPO_09_002.htm
33. West, D.M., & Karsten, J. (2016). Rural and urban America divided by broadband access. Brookings Institute, July 16, 2016. Retrieved from https://www.brookings.edu/blog/techtank/2016/07/18/rural-and-urban-america-divided-by-broadband-access/
34. Fox, S. (2011). Americans living with disabilities and their technology profile. Pew Internet and American Life Project. Retrieved from https://pewinternet.org/Reports/2011/Disability.aspx
35. Federal Communications Commission (2019). 2019 Broadband deployment report. Retrieved from https://docs.fcc.gov/public/attachments/FCC-19-44A1.pdf
36. Microsoft (2019). Nextlink Internet and Microsoft closing broadband gap in central US. Microsoft News Center. Retrieved from https://news.microsoft.com/2019/09/18/nextlink-internet-and-microsoft-closing-broadband-gap-in-central-us/
37. Kruger & Gilroy (2019). Broadband Internet access and the digital divide: federal assistance programs. Congressional Research Service. Retrieved from https://crsreports.congress.gov/product/details?prodcode=RL30719
38. Ryan, C. (2018). Computer and Internet use in the United States: 2016 American Community Survey Reports, ACS-39, U.S. Census Bureau, Washington DC, 2017. Retrieved from https://www.census.gov/content/dam/Census/library/publications/2018/acs/ACS-39.pdf
39. Rideout, V. & Katz, V.S. (2016) Opportunity for all? Technology and learning in lower-income families. A report of the Families and Media Project. New York: The Joan Ganz Cooney Center at Sesame Workshop. Retrieved from https://joanganzcooneycenter.org/publication/opportunity-for-all-technology-and-learning-in-lower-income-families/
40. Campos-Castillo, C. (2015). Revisiting the first-level digital divide in the United States: gender and race/ethnicity patterns, 2007–2012. Social Science Computer Review, 33(4), 423–439. https://doi.org/10.1177/2F0894439314547617
41. Federal Communications Commission (2020). In the matter of extending wireless telecommunications services to tribal lands. Report and Order and Further Notice of Proposed Rule-making, WT Docket No. 99–266, FCC 00–209, Adopted June 8, 2000. Retrieved from https://wireless.fcc.gov/auctions/general/releases/fc000209.pdf
42. Macartney, S., Bishaw, A., & Fontenot (2013). Poverty rates for selected detailed race and Hispanic groups by state and place: 2007–2011. American Community Survey Briefs. Retrieved from https://www2.census.gov/library/publications/2013/acs/acsbr11-17.pdf
43. Carlson, E. & Goss, J. (2016). The state of the urban/rural digital divide. National Telecommunications and Information Administration, August 10, 2016. https://www.ntia.doc.gov/blog/2016/state-urbanrural-digital-divide
44. Warf, B. (2012). Contemporary digital divides in the United States. Journal of Economic and Social Geography, 104(1), 1–17. https://doi.org/10.1111/j.1467-9663.2012.00720.x
45. Ritzhaupt, A. D., Liu, F., Dawson, K., & Barron, A. E. (2013). Differences in student information and communication technology literacy based on socioeconomic status, ethnicity, and gender. Evidence of a digital divide in Florida schools. Journal of Research on Technology in Education, 45(4), 291–307
46. Federal Communications Commission (2018). 2018 Broadband deployment report. Retrieved from https://www.fcc.gov/reports-research/reports/broadband-progress-reports/2018-broadband-deployment-report
47. Council on Social Work Education (2015). Educational policy and accreditation standards. Retrieved from https://www.cswe.org/getattachment/Accreditation/Accreditation-Process/2015-EPAS/2015EPAS_Web_FINAL.pdf.aspx

48. Bliss, D.L. (2015). Using the social work advocacy practice model to find our voices in service advocacy. Human Service Organizations: Management, Leadership & Governance, 39(1), 57–68. https://doi.org/10.1080/23303131.2014.978060
49. National Digital Inclusion Alliance (2020). A unified voice for digital inclusion policies and programs. Retrieved from https://www.digitalinclusion.org/
50. De Carvalho, A., Klarsfeld, L., Lepicard, F.: Leveraging Information and Communication Technology for the Base of Pyramid. Hystra, Paris (2011)
51. National Statistics, ROC: The Yearbook of Social Indicators, 2012. National Statistics, Taipei (2013). (in Chinese)
52. Kvasny, L., Kranich, N., Schement, J.R.: Communities, learning and democracy in the digital age. J. Commun. Inf. 2(2) Special Section: Learning in Communities 18–20 (2005)
53. Liu, C.: The myth of informatization in rural areas: the case of China's Sichuan province. Gov. Inf. Q. 29(1), 85–97 (2012)
54. Loo, B.P.Y., Ngan, Y.L.: Developing mobile telecommunications to narrow digital divide in developing countries? Some lessons from China. Telecommun. Policy 36(10–11), 888–900 (2012)
55. Srinuan, C., Srinuan, P., Bohlin, E.: An analysis of mobile Internet access in Thailand: implications for bridging the digital divide. Telematics Inform. 29, 254–262 (2012)
56. Garside, J.: Broadbank can boost household income. The Guardian, UK (2013). http://www.theguardian.com/technology/2013/sep/17/broadband-speed-household-income
57. Sharif, M.N.: Technological innovation governance for winning the future. Technol. Forecast. Soc. Chang. 79, 595–605 (2012)
58. Miller, T.L., Grimes, M.G., McMullen, J.S., Vogus, T.J.: Venturing for others with heart and head: How compassion encourages social entrepreneurship. Acad. Manag. Rev. 37(4), 616–640 (2012)
59. Research, Development, and Evaluation Commission: 2011 Survey Report on Digital Divide in Taiwan: Individuals and Households. RDEC, Taipei (2011). (in Chinese)
60. Romeo, G. I., & Russell, G. (2010). Why "what works" is not enough for ICT in education research. In A. McDougall (Ed.), Researching IT in education: Theory, practice and future directions. London, UK: Routledge
61. Gee, J. P. (2003). What video games have to teach us about learning and literacy. New York: Palgrave MacMillan
62. Australian Communications and Media Authority (ACMA). (2007). Media & communications in Australian families. http://www.acma.gov.au/WEB/STANDARD/pc=PC_310893
63. Helsper, E. J. (2008). Digital inclusion: An analysis of social disadvantage and the information society. Department for Communities and Local Government. www.communities.gov.uk/documents/.../pdf/digitalinclusionanalysis
64. Warschauer, M. (2003). Demystifying the digital divide. Scientific American, 289(2), 42–47
65. Warschauer, M. (2002). Reconceptualizing the digital divide. First Monday, 7(7). Pdf file from http://frodo.lib.uic.edu/ojsjournals/index.php/fm/rt/printerFriendly/967/88
66. Bourdieu, P. (1984). Distinction, A social critique of the judgement of taste. Cambridge: Harvard
67. Department of the Prime Minister and Cabinet. (2009). A stronger, fairer Australia. Canberra: Commonwealth Government
68. Gore, A. (1998). Remarks by the Vice President Al Gore. Digital divide event. April 28 1998. http://clinton2.nara.gov/WH/EOP/OVP/speeches/edtech.html
69. Bourdieu, P. (1993). The field of cultural production. New York: Columbia University Press
70. Kennedy, G., Judd, T., Dalgarno, B., & Waycott, J. (2010). Beyond natives and immigrants: Exploring types of net generation students. Journal of Computer Assisted Learning, 26(5), 332–343
71. Kirkwood, A. (2001). Shanty towns around the global village? Reducing distance, but widening gaps with ICT. Education, Communication & Information, 1(2), 213–228
72. Perraton, H. (2000). Open and distance learning in the developing world. London: Routledge

73. Rumble, G. (2001a). Just how relevant is E-education to global educational needs? Open Learning, 16(3), 223–232
74. Rumble, G. (2001b). Re-inventing distance education, 1971–2001. International Journal of Lifelong Education, 20(1/2), 31–43
75. Bradshaw, Y. W., Fallon, K. M., & Viterna, J. (2005). Wiring the world: Access to information technology and development in poor countries. Research in Social Stratification and Mobility, 23, 369–392
76. NTIA. (1995). Falling through the Net: Survey of the "have nots" in rural and urban America. Washington, DC: National Telecommunications and Information Administration
77. NTIA. (1998). Falling through the Net II: New data on the digital divide. Washington, DC: National Telecommunications and Information Administration
78. NTIA. (1999). Falling through the Net: Defining the digital divide. Washington, DC: National Telecommunications and Information Administration
79. NTIA. (2000). Falling through the Net: Towards digital inclusion. Washington, DC: National Telecommunications and Information Administration
80. Tien, F. F., & Fu, T.-T. (2008). The correlates of the digital divide and thier impact on college student learning. Computers and Education, 50, 421–436
81. Tiene, D. (2002). Addressing the global digital divide and its impact of educational opportunity. Education Media International, 2002(99), 212–222
82. Gunkel, D. (2003). Second thoughts: Towards a critic of the digital divide. New Media & Society, 5(4), 499–522
83. van Dijk, J. (2006). Digital divide research, achievements and shortcomings. Poetics, 34, 221–235
84. Warschauer, M. (2003a). Desiccating the 'digital divide' a case study in Egypt. The Information Society, 19, 297–304
85. Dhanarajan, G. (2001). Distance education: Promise, performance and potential. Open Learning, 16(1), 61–68
86. Gandhe, S. K. (1999). Access and equity: the needs of the disadvantaged. In R. Carr & O. Jegede (Eds.), The Asian distance learner. Hong Kong: OUHK Press
87. Zuhairi, A. (2001). Problems and challenges for a mega-university in a developing country: A case study of Universitas Terbuka. Jurnal Pendidikan Terbuka dan Jarak Jauh, 2(1), 1–20
88. Nicholas, K. (2003). Geo-policy barriers and rural Internet access: The regulatory role in construction the digital divide. The Information Society, 19, 287–295
89. van Dijk, J. (2005). The deepening divide: Inequality in the information society. London: Sage
90. Billon M, Lera-Lopez F, Marco R (2016) ICT use by households and firms in the EU: links and determinants from a multivariate perspective. Rev World Econ 152:629–654. https://doi.org/10.1007/s10290-016-0259-8
91. Dengler K, Matthes B, Wydra-Sommagio G (2018) Regionale Branchen- und Berufsstrukturen prägen die Substituierbarkeitspotenziale. IAB-Kurzberichte 22/2018
92. Moriset B, Brette O, Chareire O, Grasland L, Richaud A, Thivant E (2012) The "death of distance" fifteen years on. Information technology and knowledge-based service fims in rural areas. HAL archives-ouvertes, halshs-00728338
93. Salemink K, Strijker D, Bosworth G (2017) Rural development in the digital age: a systematic literature review on unequal ICT availability, adoption, and use in rural areas. J Rural Stud 54:360–371. https://doi.org/10.1016/j.jrurstud.2015.09.001
94. Büchi M, Just N, Latzer M (2016) Modeling the second-level digital divide: a five-country study of social differences in Internet use. New Media Soc 18(11):2703–2722. https://doi.org/10.1177/1461444815604154
95. Scheerder A, von Deursen A, van Dijk J (2017) Determinants of Internet skills, uses and outcomes. A systematic review of the second- and third-level digital divide. Telemat Inf 34(8):1607–1624. https://doi.org/10.1016/j.tele.2017.07.007

96. Briglauer W, Dürr NS, Gugler K (2019b) A retrospective study on the regional benefits and Spillover effects of high speed broadband networks: evidence from German counties. ZEW—Centre for European Economic Research Discussion Paper No. 19–026. https://doi.org/10.2139/ssrn.3423864
97. Prieger JE (2013) The broadband digital divide and the economic benefits of mobile broadband for rural areas. Telecomm Policy 37(6–7):483–502. https://doi.org/10.2139/ssrn.2258112
98. Ruiz-Rodríguez F, Lucendo-Monedero AL, González-Relaño (2017) Measurement and characterisation of the Digital Divide of Spanish regions at enterprise level. A comparative analysis with the European context. Telecomm Policy 42:187–211. https://doi.org/10.1016/j.telpol.2017.11.007
99. Bertschek I, Briglauer W, Hüschelrath K, Kauf B, Niebel T (2015) The economic impacts of broadband Internet: a survey. Rev Netw Econ 14(4):201–227
100. Grimes S (2005) How well are Europe's rural businesses connected to the digital economy? Eur Plan Stud 13(7):1063–1081. https://doi.org/10.1080/09654310500242105
101. Malecki EJ (2003) Digital development in rural areas: potentials and pitfalls. J Rural Stud 19:201–214
102. Townsend L, Sathiaseelan A, Fairhurst G, Wallace C (2013) Enhanced broadband access as a solution to the social and economic problems of the rural digital divide. Local Econ 28:580–595. https://doi.org/10.1177/0269094213496974
103. Schleife, K. (2010). What really matters: Regional versus individual determinants of the digital divide in Germany. Research Policy, 39, 173–185
104. Atasoy H (2013) The effects of broadband Internet expansion on labor market outcomes. Ind Labor Relat Rev 66(2):315–345
105. Hasbi M (2020) Impact of very high-speed broadband on company creation and entrepreneurship: empirical evidence. Telecomm Policy. https://doi.org/10.1016/j.telpol.2019.101873
106. Mack E, Faggian A (2013) Productivity and broadband: the human factor. Int Reg Sci Rev 36(3):392–423
107. McCoy D, Lyons S, Morgenroth E, Palcic D, Allen L (2018) The impact of broadband and other infrastructure on the location of new business establishments. J Reg Sci 58(3):509–534. https://doi.org/10.1111/jors.12376
108. Evangelista R, Guerrieri P, Meliciani V (2014) The economic impact of digital technologies in Europe. Econ Innov New Technol 23(8):802–824. https://doi.org/10.1080/10438599.2014.918438
109. Bennett, W.L., Segerberg, A., Knüpfer, C.B.: The democratic interface: technology, political organization, and diverging patterns of electoral representation. Inf. Commun. Soc. 21(11), 1655–1680 (2018)
110. Carpini, M.X.D., Cook, F.L., Jacobs, L.R.: Public deliberation, discursive participation, and citizenengagement: A review of the empirical literature. Annual Review of Political Science 7, 315–344 (2004)
111. Kahne, J., Bowyer, B.: The political significance of social media activity and social networks. Polit. Commun. 35(3), 470–493 (2018). https://doi.org/10.1080/10584609.2018.1426662
112. Eranti, V.: Individuals Doing Politics: Urban Participation, Social Media Campaigning and Online Nano-Politics (2016)
113. Ragnedda, M., Ruiu, M.L.: Digital Capital: A Bourdieusian Perspective on the Digital Divide. Emerald Publishing Limited (2020)
114. Koiranen, I., Keipi, T., Koivula, A., Räsänen, P.: Changing patterns of social media use? A population-level study of Finland. Univ. Access Inf. Soc. 19(3), 603–617 (2019). https://doi.org/10.1007/s10209-019-00654-1
115. Alty, J. L., Al-Sharrah, A., & Beacham, N. (2006). When humans form media and media form humans: An experimental study examining the effects different digital media have on the learning outcomes of students who have different learning styles. Interacting with Computers, 18(5), 891–909. https://doi.org/10.1016/j.intcom.2006.04.002

116. Clark, R. C., & Mayer, R. E. (2011). E-learning : And the science of instruction : Proven guidelines for consumers and designers of multimedia learning (3rd ed.). https://doi.org/10.1002/9781118255971
117. Persada, S. F., Miraja, B. A., & Nadlifatin, R. (2019). Understanding the generation Z behavior on D-learning: A unified theory of acceptance and use of technology (UTAUT) approach. International Journal of Emerging Technologies in Learning, 14, 20–33
118. Cresswell, J. W., & Poth, C. N. (2018). Qualitative inquiry research design: Choosing among five approaches (4th ed). Sage Publications
119. Yeap, J. A. L., Ramayah, T., & Soto-Acosta, P. (2016). Factors propelling the adoption of m-learning among students in higher education. Electronic Markets, 26(4), 323–338. https://doi.org/10.1007/s12525-015-0214-x
120. Hair, J.F., Sarstedt, M., Ringle, C.M. and Gudergan, S.P. (2018). Advanced Issues in Partial Least Squares Structural Equation Modeling (PLS-SEM). Sage
121. Amaro, S., & Duarte, P. (2015). An integrative model of consumers' intentions to purchase travel online. Tourism Management, 46, 64–79. https://doi.org/10.1016/j.tourman.2014.06.006
122. Department for Education and Skills (2004) UK Online Centres. Available from http://www.dfes.gov.uk/ukonlinecentres/. Accessed 13 Feb 2004
123. DiMaggio P, Hargittai E (2001) From the 'digital divide' to 'digital inequality': studying internet use as penetration increases. Princetown University, Princetown (Working paper # 15)
124. Silver D (2004) The soil of cyberspace: historical archaeologies of the Blacksburg Electronic Village and the Seattle Community Network. In: Schuler D, Day P (eds) Shaping the network society. MIT Press, Cambridge, MA., pp 301–324
125. De Cindio F (2004) The role of community networks in shaping the network society: enabling people to develop their own projects. In: Schuler D, Day P (eds) shaping the network society. MIT Press, Cambridge, pp 199–226
126. Day P (2001) Participating in the information society: community development and social inclusion. In: Keeble L, Loader, Brian D (eds) Community informatics: shaping computer-mediated relations. Routledge, London, pp 305–323
127. Fischer G (2002) Beyond 'couch potatoes': from consumers to designers and active contributors. First Monday 7(12). Available from http://firstmonday.org/htbin/cgiwrap/bin/ojs/index.php/fm/article/view/1010/931. Accessed 16 April 2009
128. Dewan, S., & Riggins, F. J. (2005). The digital divide: Current and future research directions. Journal of Association of Information Systems., 6(12), 298–337
129. Beall, J., Guha-Khasnobis, B., & Kanbur, R. (2012). Urbanization and development in Asia: Multidimensional perspectives. Cambridge: Oxford University Press
130. Quibria, M., Ahmed, S., Tschang, T., & Reyes-Macasaquit, M. (2003). Digital divide: Determinants and policies with special reference to Asia. Journal of Asian Economics, 13(6), 811–825
131. Chinn, M., & Fairlie, R. (2007). The determinants of the global digital divide: A cross country analysis of computer and Internet penetration. Oxford Economic Papers, 59(1), 16–44
132. Husing, T., & Selhofer, H. (2002). The digital divide index: A measure of social inequalities in the adoption of ICT. In S. Wrycza (Ed.), The 10th European Conference on Information Systems ECIS 2002—Information Systems and the Future of the Digital Economy, June 6–8 2002 (pp. 1273–1286). Gdansk: ECIS
133. Population Council. (2010). Survey of young people in Egypt. Cairo, Egypt: Final Report
134. Chigona, W., Beukes, D., Vally, J., & Tanner, M. (2009). Can mobile internet help alleviate social exclusion in developing countries? The Electronic Journal of Information Systems in Developing Countries, 36(1), 1–16. https://doi.org/10.1002/j.1681-4835.2009.tb00254.x
135. Donner, J. (2006). The use of mobile phones by microentrepreneurs in Kigali, Rwanda: Changes to social and business networks. Information Technologies & International Development, 3(2), 3 Retrieved from http://dev.itidjournal.org/index.php/itid/article/view/221

136. Donner, J., Gitau, S., & Marsden, G. (2011). Exploring mobile-only internet use: Results of a training study in urban South Africa. In International Journal of Communication, 5 Retrieved from http://ijoc.org
137. Rice, R., & Katz, J. (2003). Comparing Internet and mobile usage: Digital divides of usage, adoption, and dropouts. Telecommunications Policy, 27(8–9), 597–623
138. Singh, J. B., & Vimalkumar, M. (2019). Evidence of feature-level digital divides in India: From mobile access to use. Economic & Political Weekly, 54(32), 60–67
139. Song, Z., Wang, C., & Bergmann, L. (2020). China's prefectural digital divide: Spatial analysis and multivariate determinants of ICT diffusion. International journal of information management, 102072. DoI. https://doi.org/10.1016/j.ijinfomgt.2020.102072
140. OECD. (2001). Understanding the Digital Divide. Retrieved from http://www.oecd.org/internet/ieconomy/1888451.pdf
141. Okunola, O. M., Rowley, J., & Johnson, F. (2017). The multi-dimensional digital divide: Perspectives from an e-government portal in Nigeria. Government Information Quarterly, 34(2), 329–339. https://doi.org/10.1016/j.giq.2017.02.002
142. Goncalves, G., Oliveira, T., & Cruz-Jesus, F. (2018). Understanding individual-level digital divide: Evidence of an African country. Computers in Human Behavior, 87, 276–291. https://doi.org/10.1016/j.chb.2018.05.039
143. Kapoor, K. K., Dwivedi, Y. K., & Williams, M. D. (2014). Innovation adoption attributes: A review and synthesis of research findings. European Journal of Innovation Management, 17(3), 327–348
144. Rana, N. P., Dwivedi, Y. K., & Williams, M. D. (2013). Evaluating alternative theoretical models for examining citizen centric adoption of e-government. Transforming Government: People, Process and Policy, 7(1), 27–49
145. Intermedia. (2018). Country specific reports: Kenya. Pakistan: Bangladesh Retrieved from http://finclusion.org/reports/
146. Boase, J. (2010). The consequences of personal networks for internet use in rural areas. American Behavioral Scientist, 53(9), 1257–1267 Retrieved from http://journals.sagepub.com/doi/abs/https://doi.org/10.1177/0002764210361681
147. Chen, W. (2013). The implications of social capital for the digital divides in America. The Information Society, 29(1), 13–25 Retrieved from http://www.tandfonline.com/doi/abs/https://doi.org/10.1080/01972243.2012.739265
148. Lin, M., Lucas, H. C., & Shmueli, G. (2013). Research commentary - too big to fail: Large samples and the p-value problem. Information Systems Research, 24(4), 906–917. https://doi.org/10.1287/isre.2013.0480
149. Liu, Z., Ben, S., & Zhang, R. (2019). Factors affecting consumers' mobile payment behavior: A meta-analysis. Electronic Commerce Research, 19(3), 575–601
150. Abdullah, & Naved Khan, M. (2021). Determining mobile payment adoption: A systematic literature search and bibliometric analysis. Cogent Business & Management, 8(1), 1893245
151. Dahlberg, T., Guo, J., & Ondrus, J. (2015). A critical review of mobile payment research. Electronic Commerce Research and Applications, 14(5), 265–284
152. Roggeveen, A. L., & Sethuraman, R. (2020). How the COVID-19 pandemic may change the world of retailing. Journal of Retailing, 96(2), 169
153. Davis, F. D. (1989). Perceived usefulness, perceived ease of use, and user acceptance of information technology. MIS Quarterly, 13(3), 319–340
154. Venkatesh, V., Morris, M. G., Davis, G. B., & Davis, F. D. (2003). User acceptance of information technology: Toward a unified view. MIS Quarterly, 27(3), 425–478
155. Meyll, T., & Walter, A. (2019). Tapping and waving to debt Mobile payments and credit card behavior. Finance Research Letters, 28, 381–387
156. Dahlberg, T., Mallat, N., Ondrus, J., & Zmijewska, A. (2008). Past, present and future of mobile payments research: A literature review. Electronic commerce research and applications, 7(2), 165–181

157. Zheng, X., Men, J., Yang, F., & Gong, X. (2019). Understanding impulse buying in mobile commerce: An investigation into hedonic and utilitarian browsing. International Journal of Information Management, 48, 151–160
158. Hasan, R., Ashfaq, M., & Shao, L. (2021). Evaluating drivers of fintech adoption in the netherlands. Global Business Review, 09721509211027402
159. Hagberg, J., Sundstrom, M., & Egels-Zandén, N. (2016). The digitalization of retailing: An exploratory framework. International Journal of Retail & Distribution Management, 44(7), 694–712
160. Liu, Y., Luo, J., & Zhang, L. (2021). The effects of mobile payment on consumer behavior. Journal of Consumer Behaviour, 20(3), 512–520
161. Oliveira, T., Thomas, M., Baptista, G., & Campos, F. (2016). Mobile payment: Understanding the determinants of customer adoption and intention to recommend the technology. Computers in Human Behavior, 61, 404–414
162. European Commission. (2010a). Europe 2020: A strategy for smart, sustainable and inclusive growth. Brussels: European Commission. https://doi.org/10.1007/s13398-014-0173-7.2
163. OECD. (2001). Understanding the digital divide (Vol. 6). https://doi.org/10.1093/ilj/6.1.52
164. Çilan, Ç. A. (2013). Measuring domestic digital divide by using latent class analysis : A case study of Turkey. İstanbul Üniversitesi İşletme Fakültesi Dergisi Istanbul University Journal of the School of Business Cilt, 42(2), 219–234
165. Sadowski, B. M. (2017). Advanced users and the adoption of high speed broadband: Results of a living lab study in the Netherlands. Technological Forecasting and Social Change, 115, 1–14. https://doi.org/10.1016/j.techfore.2016.09.009
166. United Nations. (2012). E-Government Survey 2012. United Nations E-Government Survey 2012. New York: United Nations
167. van Dijk, J. A. G. M. (2012). The evolution of the digital divide: The digital divide turns to inequality of skills and usage. Digital Enlightenment Yearbook, 2012, 57–75. https://doi.org/10.3233/978-1-61499-057-4-57
168. Vroman, K. G., Arthanat, S., & Lysack, C. (2015). Who over 65 is online? Older adults' dispositions toward information communication technology. Computers in Human Behavior, 43, 156–166
169. United Nations. (2014). E-Government Survey 2014. United Nations E-Government Survey 2014. New York: United Nations. https://doi.org/10.1016/S1369-7021(02)00629-6
170. Cruz-Jesus, F., Vicente, M. R., Bacao, F., & Oliveira, T. (2016). The education-related digital divide: An analysis for the EU-28. Computers in Human Behavior, 56, 72–82. https://doi.org/10.1016/j.chb.2015.11.027
171. van Deursen, A. J. A. M., Van Dijk, J. A. G. M., & Ten Klooster, P. M. (2015). Increasing inequalities in what we do online: A longitudinal cross sectional analysis of internet activities among the Dutch population (2010 to 2013) over gender, age, education, and income. Telematics and Informatics, 32(2), 259–272. https://doi.org/10.1016/j.tele.2014.09.003
172. Nishijima, M., Ivanauskas, T. M., & Sarti, F. M. (2017). Evolution and determinants of digital divide in Brazil (2005–2013). Telecommunications Policy, 41(1), 12–24. https://doi.org/10.1016/j.telpol.2016.10.004
173. Mumporeze, N., & Prieler, M. (2017). Gender digital divide in Rwanda: A qualitative analysis of socioeconomic factors. Telematics and Informatics, 34(7), 1285–1293. https://doi.org/10.1016/j.tele.2017.05.014
174. Sun Foundation. (2010). Bridging the digital divide. www.sun.com/aboutsun/comm_invest/giving/foundation.html
175. Microsoft. (2010). Corporate citizenship, http://www.microsoft.com/about/corporatecitizenship/en-us/
176. Illich, I. (1973). Tools for conviviality, New York. Full text online: http://www.preservenet.com/theory/Illich/IllichTools.html
177. Moghaddam, A. (2006). Coding issues in grounded theory issues. Educational Research, ERIC, 16(1), 52–66

178. Häikiö, J., Wallin, A., Isomursu, M., Ailisto, H., Matinmikko, T., & Huomo, T. (2007). Touch-based user interface for elderly users. In Proceedings of the 9th international conference on human computer interaction with mobile devices and services - MobileHCI '07 (pp. 289–296) https://dl.acm.org/citation.cfm?id=1378021, Accessed September 2nd, 2018
179. Krizanovic Cik, V., Zagar, D., Rimac-Drlje, S.: Broadband Ecosystem elements in techno-economic modelling and analysing of broadband access solutions for rural areas. J. Tech. Gaz. (2016). https://doi.org/10.17559/TV-20150508131935
180. Publishing, O.E.C.D.: The Development Dimension ICTs for Development Improving Policy Coherence. Book. OECD Publishing, Paris (2010). ISBN 9789264077409
181. Smura, T.: Techno-economic Modelling of wireless network and industry architectures. Ph.D. thesis, Aalto University, Finland (2012). http://lib.tkk.fi/Diss/2012/isbn9789526045252/isbn9789526045252.pdf
182. Kelly, T., Rossotto, C.M.: Broadband Strategies Handbook. The World Bank, Washington (2012)
183. Laaksonen, P.: Critical factors of high-speed broadband investments in rural areas from perspective of operators. Master thesis. Helsinki University of Technology, Finland (2011). http://lib.tkk.fi/Dipl/2011/urn100543.pdf
184. Savage, S.J., Waldman, D.: Broadband internet access, awareness and use. Telecommun. Policy 29(8), 615–633 (2005). https://doi.org/10.1016/j.telpol.2005.06.001
185. Andrés, L., Cuberes, D., Diouf, M., & Serebrisky, T. (2010). The diffusion of the internet: A cross-country analysis. Telecommunications Policy, 34(5–6), 323–340. https://doi.org/10.1016/j.telpol.2010.01.003
186. Lee, S.: A cross-country analysis of ubiquitous broadband deployment: Examination of adoption factors. Ph.D. thesis, University of Florida, United States (2008). etd.fcla.edu/UF/UFE0022611/lee_s.pdf
187. Bennett, W.L.: The personalization of politics: political identity, social media, and changing patterns of participation. Ann. Am. Acad. Political Soc. Sci. 644(1), 20–39 (2012). https://doi.org/10.1177/0002716212451428
188. Go, E., You, K.H.: But not all social media are the same: analyzing organizations' social media usage patterns. Telemat. Inform. 33(1), 176–186 (2016). https://doi.org/10.1016/j.tele.2015.06.016
189. Weis, A.H.: Commercialization of the internet. Internet Res. 20(4), 420–435 (2010). https://doi.org/10.1108/10662241011059453
190. Miller, D., Costa, E., Haynes, N., McDonald, T., Nicolescu, R., Sinanan, J., Spyer, J., Venkatraman, S., Wang, X.: How the World Changed Social Media. UCL Press, London (2016)
191. Scaramuzzino, G., Scaramuzzino, R.: The weapon of a new generation? Swedish Civil Society Organizations' use of social media to influence politics. J. Inf. Technol. Politics 14(1), 46–61 (2017). https://doi.org/10.1080/19331681.2016.1276501
192. Segerberg, A., Bennett, W.L.: Social media and the organization of collective action: using Twitter to explore the ecologies of two climate change protests. Commun. Rev. 14(3), 197–215 (2011). https://doi.org/10.1080/10714421.2011.597250
193. DiMaggio, P., Hargittai, E.: From the 'digital divide 'to 'digital inequality': studying internet use as penetration increases, working paper. In: Princeton: Center for Arts and Cultural Policy Studies, , vol. 4(1), pp. 4–21. Woodrow Wilson School, Princeton University, Princeton (2001)
194. Schradie, J.: The digital production gap: The digital divide and Web 2.0 collide. Poetics 39(2), 145–168 (2011). https://doi.org/10.1016/j.poetic.2011.02.003
195. van Deursen, Alexander, J.A.M., Jan, A.G.M., van Dijk: The digital divide shifts to differences in usage. New Media Soc 16(3), 507–526 (2014)
196. Eurostat, (2017). Digital Economy and Society. Main tables. Retreived at: http://ec.europa.eu/eurostat/web/digital-economy-and-society/data/main-tables
197. Greenwood, S., Perrin, A., Duggan, M.: Social Media Update 2016. Pew Research Center. http://www.pewinternet.org/2016/11/11/social-media-update-2016/#. Accessed 20 Mar 2018

198. Perrin, A.: Social Networking Usage: 2005–2015. Pew Research Center. October 2015. http://www.pewinternet.org/2015/10/08/2015/Social-Networking-Usage-2005-2015/ (2015)
199. van Deursen, A.J., Helsper, E.J.: The third-level digital divide: Who benefits most from being online? In: Communication and Information Technologies Annual, pp. 29–52. Emerald Group Publishing Limited, Bingley (2015)
200. Keipi, T., Koiranen, I., Koivula, A., Räsänen, P.: Assessing the social media landscape: online relational use-purposes and life satisfaction among Finns. First Monday (2017). https://doi.org/10.5210/fm.v23i1.8128
201. Official Statistics of Finland (OSF).: Use of Information and Communications Technology by Individuals [e-publication]. ISSN=2341–8710. Helsinki: Statistics Finland [referred: 14. 3. 2018]. http://www.stat.fi/til/sutivi/index_en.html (2018)
202. DeLeeuw, E.D.: Mixed-mode: past, present, and future. Surv. Res. Methods 12(2), 75–89 (2018)
203. European Parliament (2016) Gender equality and empowering women in the digital age (2015/2007(INI)). https://www.europarl.europa.eu/doceo/document/TA-8-2016-0204_EN.html
204. World Economic Forum (2021) Global gender gap report 2021 insight report. World Economic Forum. http://reports.weforum.org/globalgender-gap-report-2021/dataexplorer
205. UN General Assembly, U. N (2015) Transforming our world: the 2030 agenda for sustainable development. New York, United Nations, Journal Article
206. European Union (2022) European declaration on digital rights and principles for the digital decade (COM(2022) 28 final). https://ec.europa.eu/newsroom/dae/redirection/document/82703
207. Martínez Cantos JL, Castaño Collado C, Escot Mangas L, Roquez Díaz A (2020) Nuestras vidas digitales. Barómetro de la e-igualdad de género en España. Instituto de la Mujer y para la Igualdad de Oportunidades. Ministerio de Igualdad. Gobierno de España. https://www.inmujeres.gob.es/areasTematicas/SocInfo/Estudios/docs/NUESTRAS_VIDAS_DIGITALES_2019.pdf
208. Abu-Tayeh, G., Neumann, O., & Stuermer, M. (2018). Exploring the motives of citizen reporting engagement: Self-concern and other-orientation. Business & Information Systems Engineering, 60(3), 215–226. https://doi.org/10.1007/s12599-018-0530-8
209. Müller, A. P. R., Lerusse, A., Steen, T., & de Walle, S. V. (2021). Understanding channel choice in users' reporting behavior: Evidence from a smart mobility case. Government Information Quarterly, 38(1), 101540. https://doi.org/10.1016/j.giq.2020.101540
210. Pereira, G. V., Macadar, M. A., Luciano, E. M., & Testa, M. G. (2017). Delivering public value through open government data initiatives in a smart city context. Information Systems Frontiers, 19(2), 213–229. https://doi.org/10.1007/s10796-016-9673-7
211. Gil-Garcia, J. R., & Flores-Zúñiga, M. Á. (2020). Towards a comprehensive understanding of digital government success: integrating implementation and adoption factors. Government Information Quarterly, 37(4), 101518. https://doi.org/10.1016/j.giq.2020.101518
212. Peng, G. C., Nunes, M. B., & Zheng, L. (2017). Impacts of low citizen awareness and usage in smart city services: The case of London's smart parking system. Information Systems and e-Business Management, 15(4), 845–876. https://doi.org/10.1007/s10257-016-0333-8
213. Alharbi, F., Atkins, A., & Stanier, C. (2016). Decision makers views of factors affecting cloud computing adoption in Saudi healthcare organisations. Complex & Intelligent Systems, 2(3), 155–171. https://doi.org/10.1007/s40747-016-0021-9
214. Moumane, K., Idri, A., & Abran, A. (2016). Usability evaluation of mobile application susing ISO 9241 and ISO 25062 standards. Springerplus, 5(1), 1–15. https://doi.org/10.1186/s40064-016-2171-z
215. Park, J., Sung, H., Park, J., Park, J., Kwahk, J., Lee, M., & Jeong, D. (2018). Development of a Web-based user experience evaluation system for home appliances. International Journal of Industrial Ergonomics, 67, 216–228. https://doi.org/10.1016/j.ergon.2018.05.017
216. Pucillo, F., & Cascini, G. (2014). A framework for user experience, needs and affordances. Design Studies, 35(2), 160–179. https://doi.org/10.1016/j.destud.2013.10.001

217. Franz, L., Chambers, N., von Isenburg, M., & de Vries, P. J. (2017). Autism spectrum disorder in sub-saharan africa: A comprehensive scoping review. Autism Research, 10(5), 723–749. https://doi.org/10.1002/aur.1766
218. Olusanya, B. O., Davis, A. C., Wertlieb, D., Boo, N. Y., Nair, M., Halpern, R., Kuper, H., Breinbauer, C., de Vries, P. J., Gladstone, M., Halfon, N., Kancherla, V., Mulaudzi, M. C., Kakooza-Mwesige, A., Ogbo, F. A., Olusanya, J. O., WrightWilliams, S. M. A. N., Manguerra, H.,... Kassebaum, N. J. (2018). Developmental disabilities among children younger than 5 years in 195 countries and territories, 1990–2016: A systematic analysis for the Global Burden of Disease Study 2016. The Lancet Global Health, 6(10), e1100–e1121. https://doi.org/10.1016/S2214-109X(18)30309-7
219. Olusanya, B. O., Wright, S. M., Nair, M. K. C., Boo, N.-Y., Halpern, R., Kuper, H., & Kassebaum, N. J. (2020). Global burden of childhood epilepsy, intellectual disability, and sensory impairments. Pediatrics. https://doi.org/10.1542/peds.2019-2623
220. Daley, T. C. (2004). From symptom recognition to diagnosis: Children with autism in urban India. Social Science and Medicine, 58(7), 1323–1335. https://doi.org/10.1016/S0277-9536(03)00330-7
221. Lagunju, I. A., Bella-Awusah, T. T., & Omigbodun, O. O. (2014). Autistic disorder in Nigeria: Profile and challenges to management. Epilepsy & Behavior, 39, 126–129. https://doi.org/10.1016/j.yebeh.2014.08.020
222. Pillay, S., Duncan, E. M., & de Vries, P. J. (2021). Autism in the Western Cape province of South Africa: Rates, socio-demographics, disability and educational characteristics in one million school children. Autism, 25(4), 1076–1089. https://doi.org/10.1177/1362361320978042
223. Hofstede, Geert. 2011. Dimensionalizing Cultures: The Hofstede Model in Context. Online Readings in Psychology and Culture 2 (1): 8
224. Dodel, Matias, and Gustavo Mesch. 2018. Inequality in Digital Skills and the Adoption of Online Safety Behaviors. Information, Communication & Society 21 (5): 712–728
225. Jamil, S. 2020. A Widening Digital Divide and Its Implications for Democracy and Social Inequalities in Pakistan. London: Palgrave Macmillan
226. Dodel, Matias, and Gustavo Mesch. 2017. Cyber-Victimization Preventive Behavior: A Health Belief Model Approach. Computers in Human Behavior 68: 359–367
227. Dodel, Matias, and Gustavo Mesch. 2019. An Integrated Model for Assessing Cyber-Safety Behaviors: How Cognitive, Socioeconomic and Digital Determinants Affect Diverse Safety Practices. Computers & Security 86: 75–91
228. Taha, Nashrawan, and Laila Dahabiyeh. 2021. College Students Information Security Awareness: A Comparison between Smartphones and Computers. Education and Information Technologies 26 (2): 1721–1736
229. Breitinger, Frank, Ryan Tully-Doyle, and Courtney Hassenfeldt. 2020. A Survey on Smartphone User's Security Choices, Awareness and Education. Computers & Security 88: 101647
230. Khan, Arif, Muhammad Ibrahim, and Abid Hussain. 2021. An Exploratory Prioritization of Factors Affecting Current State of Information Security in Pakistani University Libraries. International Journal of Information Management Data Insights 1 (2): 100015
231. Egelman, Serge, and Eyal Peer. 2015. Scaling the Security Wall: Developing a Security Behavior Intentions Scale (Sebis). In Proceedings of the 33rd Annual ACM Conference on Human Factors in Computing Systems, 2873–2882. ACM
232. Filippidis, Adam P., Constantinos S. Hilas, Georgios Filippidis, and Anastasios Politis. 2018. Information Security Awareness of Greek Higher Education Students—Preliminary Findings. In 2018 7th International Conference on Modern Circuits and Systems Technologies (MOCAST), 1–4. IEEE
233. Robinson, L., et al.: Digital inequalities and why they matter. Inf. Commun. Soc. 18(5), 569–582 (2015)
234. Norris, P. (2001). Digital divide? Civic engagement, information poverty and the internet worldwide. Cambridge: Cambridge University Press

235. National Research Development and Evaluation Commission (2006). 2006 digital application survey of local government. Taiwan Executive Yuan, Taiwan
236. O. E. C. D. (2001). Understanding the digital divide. Available from http://www.oeed.orb/dataoecd/38/57/1888451.pdf
237. Çilan, Ç. A., Bolat, B. A., & Coşkun, E. (2009). Analyzing digital divide within and between member and candidate countries of European Union. Government Information Quarterly, 26(1), 98–105. https://doi.org/10.1016/j.giq.2007.11.002
238. Vicente Cuervo, M. R., & Lopez Menéndez, A. J. (2006). A multivariate framework for the analysis of the digital divide: Evidence for the European Union-15. Information & Management, 43(6), 756–766. https://doi.org/10.1016/j.im.2006.05.001
239. Luyt, B. (2006). Defining the digital divide: the role of e-readiness indicators. Aslib Proceedings, 58, 276–291
240. Nia, A. Y., & Ho, A. T. K. (2005). Challenges in e-government development: lessons from two information kiosk projects. Government Information Quarterly, 22, 58–74
241. Saaty, T. L. (1980). The analytic hierarchy process: Planning, priority setting, resource allocation. NY: McGraw-Hill
242. Heponiemi, T., Jormanainen, V., Leemann, L., Manderbacka, K., Aalto, A., & Hyppönen, H. (2020). Digital divide in perceived benefits of online health care and social welfare services: National cross-sectional survey study. Journal of Medical Internet Research, 22(7), 1–22. https://doi.org/10.2196/17616
243. Ferreira, D., Vale, M., Carmo, R. M., Encalada-Abarca, L., & Marcolin, C. (2021). The three levels of the urban digital divide: Bridging issues of coverage, usage and its outcomes in VGI platforms. Geoforum, 124, 195–206. https://doi.org/10.1016/j.geoforum.2021.05.002
244. Aleksejeva, V., Lavrinenko, O., Betlej, A., & Danileviča, A. (2021). Increasing inequalities in what we do online: a longitudinal cross sectional analysis of Internet activities among the Dutch population 2010 to 2013 over gender, age, education, and income. Telematics and Informatics, 9, 332–345. https://doi.org/10.9770/jesi.2021.9.2(22)
245. Adam, I. O., & Alhassan, M. D. (2021). Bridging the global digital divide through digital inclusion: the role of ICT access and ICT use. Transforming Government: People, Process and Policy. https://doi.org/10.1108/tg-06-2020-0114
246. Freeman, J., Park, S., Middleton, C., & Allen, M. (2016). The importance of broadband for socioeconomic development: A perspective from Rural Australia. Australasian Journal of Information Systems, 20, 1–18. https://doi.org/10.3127/ajis.v20i0.1192
247. Ziemba, P., & Becker, B. (2019). Analysis of the digital divide using fuzzy forecasting. Symmetry, 11(166), 1–34. https://doi.org/10.3390/sym11020166
248. ITU. (2017). Measuring the information society. ITU International Telecommunication Union Place des Nations CH-1211, Geneva Switzerland. https://www.itu.int/en/ITU-D/Statistics/Documents/publications/misr2017/MISR2017_Volume1.pdf
249. Peres R, Muller E, Mahajan V (2010) Innovation diffusion and new product growth models: a critical review and research directions. Int J Res Mark 27:91–106. https://doi.org/10.1016/j.ijresmar.2009.12.012
250. Friemel, T.N.: The digital divide has grown old: determinants of a digital divide among seniors. New Media Soc. 18(2), 313–331 (2016)
251. UNESCO (2021a.a). One year into COVID: prioritizing education recovery to avoid a generational catastrophe: report of world's education ministers' online conference orgnanised on 29 March
252. UNESCO (2021b). Supporting learning recovery one year after into COVID-19: the global education coalition in action
253. UNESCO (2021c). Global education monitoring report 2021c, Central and Eastern Europe, Caucasus and Central Asia– Inclusion and education: All Means All
254. Government of India, (2020a) Press Information Bureau, Ministry of Human Resource Development, March 21 https://pib.gov.in/newsite/PrintRelease.aspx?relid=200579

255. Government of India, (2020b). Release of 'PRAGYATA guidelines on digital education', press information bureau, July 14 https://www.education.gov.in/sites/upload_files/mhrd/files/PR_PRAGYATA_0.pdf
256. Government of Tamil Nadu (2020) School education department, guidelines issued on 29-07-2020 https://cms.tn.gov.in/sites/default/files/go/sedu_e_65_2020.pdf
257. Blanch-Hartigan D, Chawla N, Moser RP, Rutten LJF, Hesse BW, Arora NK. Trends in cancer survivors' experience of patient-centered communication: results from the health information National Trends Survey (HINTS). J Cancer Surviv. 2016;10(6):1067–77
258. Weiner JP. Doctor-patient communication in the e-health era. Isr J Health Policy Res. 2012;1(1):33
259. Chou WS, Liu B, Post S, Hesse B. Health-related internet use among cancer survivors: data from the health information National Trends Survey, 2003–2008. J Cancer Surviv. 2011;5(3):263–70
260. Dobransky, K., Hargittai, E.: Unrealized potential: exploring the digital disability divide. Poetics 58, 18–28 (2016). https://doi.org/10.1016/j.poetic.2016.08.003
261. Goldfarb, A., & Prince, J. (2008). Internet adoption and usage patterns are different: Implications for the digital divide. Information Economics and Policy, 20(1), 2–15
262. McFadden, D.: The measurement of urban travel demand. J. Public Econ. 3, 303–328 (1974)
263. McFadden, D.: Econometrics models of probabilistic choice. In: Manski, C., McFadden, D. (eds.) Structural Analysis of Discrete Data, pp. 198–272. MIT Press, Cambridge, MA (1981)
264. Oaxaca, R.L., Ransom, M.R.: Identification in detailed wage decompositions. Rev. Econ. Stat. 81, 154–157 (1999)
265. Hafkin, N.: Gender issues in ICT policy in developing countries: an overview. Paper presented at the United Nations Division for the Advancement of Women Expert Group Meeting on "Information and Communication Technologies and Their Impact On and Use as an Instrument for the Advancement and Empowerment of Women", Seoul, Republic of Korea. http://www.un.org/womenwatch/daw/egm/ict2002/. Accessed 11–14 November 2002
266. Brennan, T. (2016) The post-internet order broadband sector: Lessons from the pre-open internet order experience. Review of Industrial Organization (forthcoming)
267. National Telecommunications and Information Administration. (1999). Falling through the net: Defining the digital divide. Retrieved from https://www.ntia.doc.gov/legacy/ntiahome/fttn99/FTTN.pdf
268. Economides, N., & Hermalin, B. E. (2012). The economics of network neutrality. The Rand Journal of Economics, 43(4), 602–629
269. Reggiani, C., Valletti, T. (2016). Net neutrality and innovation at the core and at the edge. International Journal of Industrial Organization, 63, 16–27
270. Krämer, J., & Wiewiorra, L. (2012). Network neutrality and congestion sensitive content providers: Implications for content variety, broadband investment, and regulation. Information Systems Research, 23(4), 1303–1321
271. Hermalin, B. E., & Katz, M. L. (2007). The economics of product-line restrictions with an application to the network neutrality debate. Information Economics and Policy, 19(2), 215–248
272. Anderson MP, Andrew; Jiang, Jingjing; Kumar, Madhumita. 10% of Americans don't use the internet | Pew Research Center. 2019
273. Blake KD, Chou W-YS, Prestin A, Hesse BW. Cancer prevention and control in the changing communication landscape. J Natl Cancer Inst Monographs. 2013;2013(47):131–2. https://doi.org/10.1093/jncimonographs/lgt032
274. Ofran Y, Paltiel O, Pelleg D, Rowe JM, Yom-Tov E. Patterns of information-seeking for cancer on the internet: an analysis of real world data. PLoS One. 2012;7(9):e45921. https://doi.org/10.1371/journal.pone.0045921
275. Zhang H, Wheldon C, Dunn AG, Tao C, Huo J, Zhang R, et al. Mining Twitter to assess the determinants of health behavior toward human papillomavirus vaccination in the United States. J Am Med Inform Assoc. 2020;27(2):225–35. https://doi.org/10.1093/jamia/ocz191

276. Miller LMS, Bell RA. Online health information seeking:the influence of age, information trustworthiness, and search challenges. Journal of Aging and Health. 2012;24(3):525–41. https://doi.org/10.1177/0898264311428167
277. Ginossar T. Predictors of online cancer prevention information seeking among patients and caregivers across the digital divide: a cross-sectional. Correlational Study JMIR Cancer. 2016;2(1):e2. https://doi.org/10.2196/cancer.5108
278. Aunt Bertha., Connecting people and programs. https://Www.auntbertha.com. Accessed on 9 Jun 2021
279. Kulldorff, M., A spatial scan statistic. Communications in statistics-theory and methods. 26(6):1481-1469, 1997
280. Dixon, B. E., Kharrazi, H., Lehmann, H. P., Public health and epidemiology informatics: recent research and trends in the United States. Yearb. Med. Inform. 10(1):199-206, 2015
281. Gamache, R., Kharrazi, H., Weiner, J. P., Public and population health informatics: the bridging of big data to benefit communities. Yearb. Med. Inform. 27(1):199-206, 2018
282. Hatef, E., Weiner, J. P., Kharrazi, H., A public health perspective on using electronic health records to address social determinants of health: the potential for a national system of local community. Int. J. Med. Inform. 124:86-89, 2019
283. Federal Communications Commission., Emergency Broadband Benefit. https://www.fcc.gov/broadbandbenefit. Accessed on 10 Sept 2021
284. Oliner, S., & Sichel, D. (2000). The resurgence of growth in the late 1990s: Is the information technology the story? The Journal of Economic Perspectives,14(4), 3–22
285. Jorgenson, D., & Stiroh, K. (2000). Raising the speed limit: U.S. economic growth in the information age. Brookings Papers on Economic Activity, 2000(1), 125–235
286. Schreyer, P. (2000). The contribution of information and communication technology to output growth: A study of the G7 countries. OECD Science, Technology and Industry Working Papers
287. Basu, S., Fernald, J., Oulton, N., & Srinivasan, S. (2003). The case of the missing productivity growth: or, does information technology explain why productivity accelerated in the United States but not the United Kingdom? NBER Working Paper No. 10010
288. Cardona, M., Kretschmer, T., & Strobel, T. (2013). ICT and productivity: Conclusions from the empirical literature. Information Economics and Policy,25(3), 109–125
289. Department of HHS, Eliminating minority health disparities, HHS Fact Sheet, July 12, 2004. HHS
290. Brailer, D. The decade of health information technology. HHS Report, July 21, 2004
291. Brailer, D., The decade of health information technology. HHS Report, July 21, 2004
292. Masi, C. M., Suarez-Balcazar, Y., Cassey, M. Z., et al., Internet access and empowerment: A community-based health initiative. J. Gen. Intern. Med. 18(7):525–530, 2003
293. Carr, N. (2008, July/August). Is Google making us stupid? The Atlantic. http://www.theatlantic.com/magazine/archive/2008/07/is-google-making-us-stupid/306868/
294. Carr, N. (2010). The shallows: What the Internet is doing to our brains. New York: W. W. Norton
295. Moran, M., Seaman, J., & Tinti-Kane, H. (2011). Teaching, learning, and sharing: How today's higher education faculty use social media. Babson Survey Research Group. Retrieved from https://files.eric.ed.gov/fulltext/ED535130.pdf
296. Prensky, M. (2001b). Digital natives, digital immigrants, part II: Do they really think differently? On the Horizon, 9(6), 1–9
297. Waycott, J., Bennett, S., Kennedy, G., Dalgarno, B., & Gray, K. (2010). Digital divides? Student and staff perceptions of information and communication technologies. Computers and Education, 54(4), 1202–1211
298. Prensky, M. (2001). Digital natives, digital immigrants. On the Horizon, 9(5), 1–6. https://doi.org/10.1108/10748120110424816
299. Bauerlein, M. (2008). The dumbest generation: How the digital age stupefies young Americans and jeopardizes our future (or, don't trust anyone under 30). London: Penguin Books

300. Pew Research Center. (2011). The digital revolution and higher education. Retrieved March 2017 from http://www.pewsocialtrends.org/2011/08/28/the-digital-revolution-and-higher-education/
301. Taipale, S. (2016). Synchronicity matters: Defining the characteristics of digital generations. Information, Communication & Society, 19, 80–94
302. Ajman, H., & Hartshorne, R. (2008). Investigating faculty decisions to adopt Web 2.0 technologies: Theory and empirical tests. The Internet and Higher Education, 11, 71–80. https://doi.org/10.1016/j.iheduc.2008.05.002
303. Pasek, J., More, E., & Hargittai, E. (2009). Facebook and academic performance: Reconciling a media sensation with data. First Monday, 14(5). Retrieved from http://firstmonday.org/ojs/index.php/fm/article/view/2498/2181
304. Criar, B. (2013, November). The period is pissed: When did our plainest punctuation mark become so aggressive? The New Republic. Retrieved July 10, 2015, from http://www.newrepublic.com/article/115726/period-our-simplest-punctuation-mark-has-become-sign-anger
305. Crystal, D. (2008). Txtng: The gr8 db8. Oxford: Oxford University Press
306. Riordan, M. A. (2016). Appear happier: Text with emojis. Presentation at the 57th annual meeting of the Psychonomic Society, Boston, MA
307. Irwin A (1997) Citizen science: a study of people, expertise and sustainable development. Sci Technol Hum Values 22(4):525–527
308. Hand E (2010) Citizen science: people power. Nature 466(7307):685–687
309. Kirch W, Menne B, Bertollini R (2005) Extreme weather events and public health responses. Ed. Springer-Verlag, ISBN 3-540-24417-4, 303 pp
310. WMO (2014) Implementation plan of the global framework for climate services. Access 14 May 14 2016. http://www.wmo.int/gfcs
311. The Joint Center Report, April 2012. Minorities, mobile broadband and the management of chronic diseases. Available at: http://jointcenter.org/sites/default/files/Minorities%20Mobile%20Broadband%20and%20the%20Management%20of%20Chronic%20Diseases_0.pdf. Accessed August 9, 2016
312. Lyles CR, Allen JY, Poole D, Tieu L, Kanter MH, Garrido T. "I want to keep the personal relationship with my doctor": understanding barriers to portal use among African Americans and Latinos. J Med Internet Res. 2016;18:e263
313. Tieu L, Sarkar U, Schillinger D, Ralston JD, Ratanawongsa N, Pasick R, et al. Barriers and facilitators to online portal use among patients and caregivers in a safety net health care system: a qualitative study. J Med Internet Res. 2015;17:e275
314. Schickedanz A, Huang D, Lopez A, Cheung E, Lyles CR, Bodenheimer T, et al. Access, interest, and attitudes toward electronic communication for health care among patients in the medical safety net. J Gen Intern Med. 2013;28:914–20
315. Moreno G, Lin EH, Chang E, Johnson RL, Berthoud H, Solomon CC, et al. Disparities in the use of internet and telephone medication refills among linguistically diverse patients. J Gen Intern Med. 2016;31:282–8
316. Chang, B. L., Bakken, S., Brown, S. S., Houston, T. K., Kreps, G. L., Kukafka, R., et al. (2004). Bridging the digital divide: reaching vulnerable populations. Journal of the American Medical Informatics Association, 11, 448–457
317. Gell NM, Rosenberg DE, Demiris G, LaCroix AZ, Patel KV. Patterns of technology use among older adults with and without disabilities. The Gerontologist. 2013. https://doi.org/10.1093/geront/gnt166
318. Kaelber DC, Jha AK, Johnston D, Middleton B, Bates DW. A research agenda for personal health records (PHRs). J Am Med Inform Assoc. 2008;15(6):729–736
319. Weinstein RS, Lopez AM. Health literacy and connected health. Health Aff. 2014;33(6):1103–1104
320. Sarkar U, Karter AJ, Liu JY, et al. The literacy divide: health literacy and the use of an Internet-based patient portal in an integrated health system—results from the Diabetes Study of Northern California (DISTANCE). J Health Commun. 2010;15(sup2):183–196

321. Heeringa SG, Connor JH. Technical description of the Health and Retirement Survey sample design. Ann Arbor: University of Michigan; 1995
322. Ofstedal M, Weir D, Chen K, Wagner J. Updates to HRS sample weights. Ann Arbor: University of Michigan; 2011
323. Wasserman S, Faust K (1994) Social network analysis: methods and applications. In: Granovetter M (ed) Social networks, vol 8. Cambridge University Press, Cambridge
324. E-Government Fact Sheet (2018), E-Government in the UK, European Commission, Available at: https://joinup.ec.europa.eu/sites/default/files/inline-files/eGovernment_in_United_Kingdom_2018_0.pdf, Last accessed on 29/12/2018
325. Kaplan, D. (2005). E-Inclusion: New challenges and policy recommendations. Brussels: e-Europe Advisory Group
326. Mancinelli, E. (2008) 'E-inclusion in the information society', Information Society. In Theory to Political Practice: Course book. Budapest: Gondolt-Új Mandátum
327. Helsper, E. (2008). Digital inclusion: An analysis of social disadvantage and the information society. London: Communities and Local Government
328. Tam, C., Santos, D., & Oliveira, T. (2018). Exploring the influential factors of continuance intention to use mobile Apps: Extending the expectation confirmation model. Information Systems Frontiers. https://doi.org/10.1007/s10796-018-9864-5
329. Ajzen, I. (1991). The theory of planned behaviour. Organizational Behavior and Human Decision Processes, 50, 179–211. https://doi.org/10.1016/0749-5978(91)90020-T
330. Rogers, E. M. (1995). Diffusion of innovations (4th ed.). New York: Free Press
331. Taylor, S., & Todd, P. (1995). Assessing IT usage: The role of prior experience. MIS Quarterly, 19, 561–570
332. Creswell, J. W. (2008). Research design: Qualitative, quantitative, and mixed methods approaches. Thousand Oaks: Sage
333. Bryman, A., & Bell, E. (2011). Business Research Methods (3rd ed.). New York: Oxford University Press
334. Tabachnick, B.G. and Fidell, L.S. (2001) 'Using multivariate statistics'
335. Comrey, A., & Lee, H. (1992). A first course in factor analysis. Hillsdale: Lawrence Erlbaum Associates
336. Schaupp, L. C., Carter, L., & McBride, M. E. (2010). E-file adoption: A study of US taxpayers' intentions. Computers in Human Behavior, 26(4), 636–644
337. Horst, M., Kuttschreuter, M., & Gutteling, J. M. (2007). Perceived usefulness, personal experiences, risk perception and trust as determinants of adoption of e-government services in the Netherlands. Computers in Human Behavior, 23(4), 1838–1852
338. Hair, J. F., Tatham, R. L., Anderson, R. E., & Black, W. (2006). Multivariate data analysis. Upper Saddle River: Pearson Prentice Hall
339. Wang, Y. S., & Shih, Y. W. (2008). Why do people use information kiosk? A validation of the unified theory of acceptance and use of technology. Government Information Quarterly, 26, 158–165
340. Eubanks V (2018) Automating inequality: how high-tech tools profile, police, and punish the poor. Picador St. Martin's Press, New York
341. Mittelstadt BD, Allo P, Taddeo M, Wachter S, Floridi L (2016) The ethics of algorithms: mapping the debate. Big Data Soc 3(2):1–21. https://doi.org/10.1177/2053951716679679
342. Aizenberg E, van den Hoven J (2020) Designing for human rights in AI. Big Data Soc. https://doi.org/10.1177/2053951720949566
343. Cruz TM (2020) Perils of data-driven equity: safety-net care and big data's elusive grasp on health inequality. Big Data Soc. https://doi.org/10.1177/2053951720928097
344. Wachter S, Mittelstadt B, Russell C (2021) Bias preservation in machine learning: the legality of fairness metrics under EU non-discrimination law. West Virginia Law Review, Vol 123, issue 3, pp 1–51
345. Floridi L, Cowls J, King TC, Taddeo M (2020) How to design ai for social good: seven essential factors. Sci Eng Ethics 26(3):1771–1796. https://doi.org/10.1007/s11948-020-00213-5

346. Chu CH, Nyrup R, Leslie K, Shi J, Bianchi A, Lyn A et al (2022) Digital ageism: challenges and opportunities in artificial intelligence for older adults. Gerontologist. https://doi.org/10.1093/geront/gnab167
347. Rosales A, Fernández-Ardèvol M (2020) Ageism in the era of digital platforms. Convergence 26(5–6):1074–1087. https://doi.org/10.1177/1354856520930905
348. Smith, G. 2009. Democratic innovations: designing institutions for citizen participation (theories of institutional design). Cambridge: Cambridge University Press
349. Escher, T. 2013. Mobilisierung zu politischer Partizipation durch das Internet: Erwartungen, Erkenntnisse und Herausforderungen der Forschung. Analyse & Kritik 35(2):449–476
350. Putnam, R. D. (2001). Bowling alone: The collapse and revival of American community. New York: Simon and Schuster
351. Moya, J. C. (2005). Immigrants and associations: A global and historical perspective. Journal of Ethnic and Migration Studies, 31(5), 833–864
352. Toral, G. (2010). Las asociaciones de inmigrantes como sociedad civil: Un análisis tridimensional. Revista Española de Investigaciones Sociológicas, 132(1), 105–130
353. Zmerli, S., & Newton, R. (2007). Networking among voluntary associations: Segmented or integrated? In W. A. Maloney & S. Roßteutscher (Eds.), Social capital and associations in European democracies: A comparative analysis. New York: Routledge
354. Allievi, S., & Nielsen, J. S. (2003). Muslim networks and transnational communities in and across Europe. Leiden and Boston: Brill
355. Oiarzabal, P. J. (2012). Diaspora Basques and online social networks: An analysis of users of Basque institutional diaspora groups on Facebook. Journal of Ethnic and Migration Studies, 38(9), 1469–1485
356. Oiarzabal, P. J. (2013). The Basque diaspora webscape. Identity, nation and homeland, 1990s–2010s. Reno: Center for Basque Studies, University of Nevada
357. Álvarez de los Mozos, F. J. (2013). Capital social de las asociaciones de inmigrantes: Asociaciones bolivianas, colombianas, ecuatorianas y peruanas en Barcelona. Bilbao: Bilbao, Madrid y Valencia, Publicaciones de la Universidad de Deusto
358. Vancea, M., & Boso, À. (2014). Connected migrants? Four methodological challenges for the analysis of ICT use through survey data. Migraciones Internacionales, 7(3), 43–72
359. Bouncken, R., & Barwinski, R. (2020). Shared digital identity and rich knowledge ties in global 3d printing—a drizzle in the clouds? Global Strategy Journal, 11(1), 81–108. https://doi.org/10.1002/gsj.1370
360. Ritter, T., & Pedersen, C. L. (2020). Digitization capability and the digitalization of business models in business-to-business firms: Past, present, and future. Industrial Marketing Management, 86, 180–190. https://doi.org/10.1016/j.indmarman.2019.11.019
361. OECD. (2019c). Unpacking E-commerce: Business models, trends and policies. OECD Publishing. Retrieved 20 Jan 2021 from https://doi.org/10.1787/23561431-en
362. Laudon, K. C., & Traver, C. G. (2018). E-commerce 2018: Business, technology, society. Pearson
363. OECD, WTO. (2017). Aid for trade at a glance 2017: Promoting trade, inclusiveness and connectivity for sustainable development. WTO/OECD Publishing. https://doi.org/10.1787/aid_glance-2017-en
364. Scupola, A. (2003). The adoption of internet commerce by SMEs in the south of Italy: An environmental, technological and organizational perspective. Journal of Global Information Technology Management, 6(1), 52–71. https://doi.org/10.1080/1097198x.2003.10856343
365. Cui, L., Zhang, C., Zhang, C., & Huang, L. (2006). Exploring e-government impact on shanghai firms' informatization process. Electronic Markets, 16(4), 312–328. https://doi.org/10.1080/10196780600999734
366. Papanastassiou, M., Pearce, R., & Zanfei, A. (2019). Changing perspectives on the internationalization of R&D and innovation by multinational enterprises: A review of the literature. Journal of International Business Studies, 51(4), 623–664. https://doi.org/10.1057/s41267-019-00258-0

367. Clegg, J. (2019). From the editor: International business policy: What it is, and what it is not. Journal of International Business Policy, 2(2), 111–118. https://doi.org/10.1057/s42214-019-00025-w
368. Jean, R.-J.B., & Tan, D. (2019). The effect of institutional capabilities on E-business firms' international performance. Management International Review, 59(4), 593–616. https://doi.org/10.1007/s11575-019-00389-4
369. OECD. (2018). Economic outlook for southeast Asia, China and India 2018 - update: Promoting opportunities in E-commerce. OECD Publishing. Retrieved on 20 Jan 2021 from https://doi.org/10.1787/9789264302990-en
370. Neeraj, R. S. (2019). Trade rules for the digital economy: Charting new waters at the WTO. World Trade Review, 18, S121–S141. https://doi.org/10.1017/s1474745618000423
371. Ahen, F. (2021). International mega-corruption inc: The structural violence against sustainable development. Critical Perspectives International Business Advance Online Publication. https://doi.org/10.1108/cpoib-04-2018-0035
372. Buckley, P. J., Doh, J. P., & Benischke, M. H. (2017). Towards a renaissance in international business research? Big questions, grand challenges, and the future of ib scholarship. Journal of International Business Studies, 48(9), 1045–1064. https://doi.org/10.1057/s41267-017-0102-z
373. Sinkovics, N., Sinkovics, R. R., & Archie-Acheampong, J. (2021a). The business responsibility matrix: A diagnostic tool to aid the design of better interventions for achieving the sdgs. Multinational Business Review, 29(1), 1–20. https://doi.org/10.1108/MBR-07-2020-0154
374. Sinkovics, N., Sinkovics, R. R., & Archie-Acheampong, J. (2021b). Small- and medium-sized enterprises and sustainable development: In the shadows of large lead firms in global value chains. Journal of International Business Policy, 4(1), 80–101. https://doi.org/10.1057/s42214-020-00089-z
375. McCormick, M., & Fernhaber, S. A. (2018). Are growth expectations being met? Implications for the internationalization of micro-sized ventures. Small Business Economics, 50(3), 591–605. https://doi.org/10.1007/s11187-017-9909-z
376. Heang JF, Khan HU (2015) The role of internet marketing in the development of agricultural industry: a case study of China. J Internet Commer 14:65–113
377. Hong YA, Zhou Z, Fang Y, Shi L (2017) The digital divide and health disparities in China: evidence from a national survey and policy implications. J Med Internet Res 19:e317
378. Li L, Du K, Zhang W, Mao JY (2019) Poverty alleviation through government-led e-commerce development in rural China: an activity theory perspective. Inf Syst J 29:914–952
379. Wei H (2020) E-commerce companies lend a helping hand to farmers via product sales. China Daily. http://www.chinadaily.com.cn/m/beijing/zhongguancun/2020-01/02/content_37531070.htm. Accessed Oct 1, 2020
380. Maruping LM, Bala H, Venkatesh V, Brown SA (2017) Going beyond intention: Integrating behavioral expectation into the unified theory of acceptance and use of technology. J Assoc Inf Sci Technol 68:623–637
381. Tripathi, S.: Does higher economic growth reduce poverty and increase inequality? Evidence from urban India. Indian J. Hum. Dev. 7, 109–137 (2013)
382. Taylor J, Gibson LK (2016) Digitisation, digital interaction and social media: embedded barriers to democratic heritage. Int J of Herit Stu 23(5):408–420. https://doi.org/10.1080/13527258.2016.1171245
383. Holcombe-James I (2021) Digital access, skills, and dollars: applying a framework to digitzl exclusion in cultural trends. Cult Trends. https://doi.org/10.1080/09548963.2021.1972282 (online first)
384. Mihelj S, Leguina A, Downey J (2019) Culture is digital: Cultural participation, diversity and the digital divide. New Media Soc 21(7):1465–1485. https://doi.org/10.1177/1461444818822816
385. Thomas J, Barraket J, Wilson C K, Holcombe-James I, Kennedy J, Rennie E, Ewing S, MacDonald T (2020) Measuring Australia's digital divide: The Australian Digital Inclusion Index 2020. RMIT and Swinburne University of Technology, Melbourne, for Telstra

386. Hargittai E, Dobransky K (2017) Old dogs, new clicks: digital inequality in skills and uses among older adults. Can J Commun 42(2):195–212. https://doi.org/10.22230/cjc2017v42n2a3176
387. Dewhurst B, Sumption K (2012) AMOL ten years on: a legacy of working beyond museum walls. In: Burton Jones K, Marty PF (eds) Museum informatics: people, information and technology in museums. Routledge, London, pp 124–130
388. Flinn A (2011) Archival activism: Independent and community-led archives, radical public history and the heritage professions. InterActions: UCLA J of Educ and Inf Studies 7(2). https://escholarship.org/uc/item/9pt2490x
389. Hurley G (2016) Community archives, community clouds: enabling digital preservation for small archives. Archivaria 81(May):129–150
390. Schumacher J, Thomas L M, VandeCreek D, Erdman S, Hancks J, Haykal A, Miner M, Prud'homme P-A, Spalenka D (2014) From theory to action: "Good enough" digital preservation solutions for under-resourced cultural heritage institutions. https://commons.lib.niu.edu/bitstream/handle/10843/13610/FromTheoryToAction_POWRR_WhitePaper.pdf?sequence=1&isAllowed=y. Acessed 31 Aug 2021
391. Stevens M, Flinn A, Shepherd E (2010) New frameworks for community engagement in the archive sector: from handing over to handing on. Int J of Herit Stu 16(1–2):59–76. https://doi.org/10.1080/13527250903441770
392. Pelzer P, Goodspeed R, te Brömmelstroet M (2015) Facilitating PSS workshops: a conceptual framework and findings from interviews with facilitators. In: Geertman S, Ferreira J Jr, Goodspeed R, Stillwell J (eds) Planning support systems and smart cities. Springer, Cham, pp 355–369
393. O'Cathain A, Croot L, Duncan E, Rousseau N, Sworn K, Turner KM, et al. Guidance on how to develop complex interventions to improve health and healthcare. BMJ Open. 2019;9:e029954. https://doi.org/10.1136/bmjopen-2019-029954
394. NHS Health Research Authority. Public involvement in a pandemic: Lessons from the UK COVID-19 public involvement matching service. https://s3.eu-west-2.amazonaws.com/www.hra.nhs.uk/media/documents/8948_Public_Involvement_in_Pandemic_Research_Report_V9_-_Accessible.pdf. Accessed 24 Jan 2022
395. Cadel L, Marcinow M, Sandercock J, Dowedoff P, Guilcher SJT, Maybee A, et al. A scoping review of patient engagement activities during COVID-19: More consultation, less partnership. PLoS ONE. 2021;16: e0257880. https://doi.org/10.1371/journal.pone.0257880
396. Rasburn M, Crosbie H, Tonkinson A, Chandler D, Dhanji T, Habgood S, O'Brien S. Innovative patient involvement during covid-19: keeping patients at the heart of HTA. Front Med Technol. 2021;3: 793119. https://doi.org/10.3389/fmedt.2021.793119
397. Denegri S, Starling B. COVID-19 and patient engagement in health research: What have we learned? CMAJ. 2021;193:E1048–9. https://doi.org/10.1503/cmaj.210998
398. Lampa E, Sonnentheil B, Tökés A, Warner G. What has the COVID-19 pandemic taught us about conducting patient and public involvement remotely? Insights from a series of digital meeting observations. Res Involv Engagem. 2021;7:73. https://doi.org/10.1186/s40900-021-00315-9
399. Brighton LJ, Pask S, Benalia H, Bailey S, Sumerfield M, Witt J, et al. Taking patient and public involvement online: qualitative evaluation of an online forum for palliative care and rehabilitation research. Res Involv Engagem. 2018;4:14. https://doi.org/10.1186/s40900-018-0097-z
400. National Coordinating Centre for Public Engagemenet. Online Engagement: A guide to creating and running virtual meetings and events. https://www.publicengagement.ac.uk/sites/default/files/publication/creating_and_running_virtual_events_-_april_2020_v1.pdf. Accessed 24 Jan 2022
401. NEA (2013). Research report, How a nation engages with art: Highlights from the 2012 Survey of Public Participation in the Arts. Washington D.C.: National Endowment for the Arts

402. Favaro, D., & Frateschi, C. (2007). A discrete choice model of consumption of cultural goods: The case of music. Journal of Cultural Economics, 31(3), 205–234
403. Prieto-Rodríguez, J., & Fernández-Blanco, V. (2000). Are popular and classical music listeners the same people? Journal of Cultural Economics, 24(2), 147–164
404. Sandelowski M, Leeman J (2012) Writing usable qualitative health research findings. Qual Health Res 22(10):1404–1413. https://doi.org/10.1177/1049732312450368
405. Davidsson P, Thoresson A (2017) Svenskarna och internet 2017. https://www.iis.se/fakta/svenskarna-och-internet-2017/. Accessed 6 Apr 2018
406. Regeringskansliet (2017) För ett hållbart digitaliserat Sverige—en digitaliseringsstrategi (N2017/03643/D). http://www.regeringen.se/informationsmaterial/2017/05/for-ett-hallbart-digitaliserat-sverige—en-digitaliseringsstrategi/. Accessed 30 Nov 2018
407. World Health Assembly (2018) Seventy-first world health assembly—agenda item 12.4 digital health (WHA71.7). http://apps.who.int/gb/ebwha/pdf_files/WHA71/A71_R7-en.pdf?ua=1. Accessed 7 Jan 2019
408. World Health Organization (2015) World report on ageing and health. WHO Press, Geneva. http://apps.who.int/iris/bitstream/10665/186463/1/9789240694811_eng.pdf?ua=1. Accessed 21 Aug 2017
409. Ihm J, Hsieh YP (2015) The implications of information and communication technology use for the social well-being of older adults Information. Commun Soc 18:1123–1138. https://doi.org/10.1080/1369118x.2015.1019912
410. Dahlin-Ivanoff S, Hultberg J (2006) Understanding the multiple realities of everyday life: basic assumptions in focus group methodology. Scand J Occup Ther 13:125–132
411. Doody O, Slevin E, Taggart L (2013a) Focus group interviews in nursing research: part 1. Brit J Nurs 22(1):16–19
412. Stewart DW, Shamdasani PN, Rook DW (2007) Focus groups: theory and practice. Sage, Thousand Oaks
413. Doody O, Slevin E, Taggart L (2013b) Focus group interviews. Part 3: analysis. Brit J Nurs 22(5):266–269
414. Davidsson P, Palm M, Mandre ÅM (2018) Svenskarna och internet 2018. https://2018.svenskarnaochinternet.se. Accessed 8 Oct 2018
415. Arora, A. K., & Srinivasan, R. (2020). Impact of pandemic COVID-19 on the teaching–learning process: A study of higher education teachers. Prabandhan: Indian Journal of Management, 13(4), 43–56
416. Barış, M. F., & Çankaya, P. (2016). Akademik personelin uzaktan eğitim hakkındaki görüşleri [Opinions of academic staff about distance education]. Journal of Human Sciences, 13(1), 399–413
417. Chen, G. D., Ou, K. L., Liu, C. C., & Liu, B. J. (2001). Intervention and strategy analysis for web group- learning. Journal of Computer Assisted Learning, 17(1), 58–71
418. Jin, S. H. (2005). Analyzing student-student and student-instructor interactıon through multiple communıcatıon tools in web-based learning. International Journal of Instructional Media, 32(1), 59–67
419. Kaya, Z., & Önder, H. (2002). İnternet yoluyla öğretimde ergonomi [Ergonomics in Internet teaching]. The Turkish Online Journal of Educational Technology, 1(1), 48–54
420. Fry, K. (2001). E-learning markets and providers: Some issues and prospects. Education and Training, 43(4/5), 233–239. https://doi.org/10.1108/EUM0000000005484
421. Wang, C. H., Shannon, D. M., & Ross, M. E. (2013). Students' characteristics, self-regulated learning, technology self-efficacy, and course outcomes in online learning. Distance Education, 34(3), 302–323
422. Wilde, N., & Hsu, A. (2019). The influence of general self-efficacy on the interpretation of vicarious experience information within online learning. International Journal of Educational Technology in Higher Education, 16(1), 1–20
423. Bailey, D., & Lee, A. (2020). Learning from experience in the midst of covid-19: Benefits, challenges, and strategies in online teaching. Call-Ej, 21(2), 176–196

424. den Brok, P., Brekelmans, M., & Wubbels, T. (2006). Multilevel issues in research using students' perceptions of learning environments: The case of the questionnaire on teacher interaction. Learning Environments Research, 9, 199–213
425. Ida Rafidah Haji Dzolkefflie. (1993). A study on silence in response to teachers' questioning in the classroom discourse of three lectures in Universiti Brunei Darussalam. Unpublished master's dissertation. Universiti Brunei Darussalam
426. Larking, L. (1996). Bilingualism through the classroom: Strategies and practices in Brunei Darussalam. Journal of Multilingual and Multicultural Development, 17(2), 296–311
427. Sinclair, J. M., & Coulthard, R. M. (1975). Towards an analysis of discourse: The English used by teachers and pupils. Oxford University Press
428. Weiner, M. (2020). Online college classes are disturbingly quiet. The Boston Globe. https://www.bostonglobe.com/2020/09/18/opinion/online-college-classes-are-disturbingly-quiet/
429. Carter, R. A., Rice, M., Yang, S., & Jackson, H. A. (2020). Self-regulated learning in online learning environments: Strategies for remote learning. Information and Learning Science, 121(5–6), 311–319. https://doi.org/10.1108/ILS-04-2020-0114
430. Guiller, J., Durndell, A., & Ross, A. (2008). Peer interaction and critical thinking: Face-to-face or online discussion? Learning and Instruction, 18, 187–200
431. Jurewitsch, B. (2012). A Meta-analytic and qualitative review of online versus face-to-face problem-based learning. Journal of Distance Education, 26(2), 1–6
432. Kim, I.-H., Anderson, R. C., Nguyen-Jahiel, K., & Archodidou, A. (2007). Discourse patterns during children's collaborative online discussions. The Journal of the Learning Sciences, 16(3), 333–370
433. Nguyen, T. (2015). The effectiveness of online learning: Beyond no significant difference and future horizons. MERLOT Journal of Online Learning and Teaching, 11(2), 309–319. https://jolt.merlot.org/Vol11no2/Nguyen_0615.pdf
434. Hsu, S. H., Chang, J. W., & Lee, C. C. (2013). Designing attractive gamification features for collaborative storytelling websites. Cyberpsychology, Behavior, and Social Networking, 16(6), 428–435
435. Liu, C. C., Chen, W. C., Lin, H. M., & Huang, Y. Y. (2017a). A remix-oriented approach to promoting student engagement in a long-term participatory learning program. Computers & Education, 110, 1–15
436. GNM press office. (2009, June 23). MPs' expenses: The Guardian launches major crowdsourcing experiment. The Guardian, http://www.theguardian.com/gnm-press-office/crowdsourcing-mps-expenses. Accessed 1 November 2015
437. Kitchin, Rob (2014). The data revolution: Big data, open data, data infrastructures and their consequences. London: Sage
438. Benkler, Yochai (2011). The Penguin and the Leviathan: How Cooperation Triumphs over Self-Interest. New York: Crown Business
439. Putnam, Robert (2000). Bowling Alone: The collapse and revival of American community. New York: Simon and Schuster
440. Fogg, B.J. (2009). A behavior model for persuasive design. In Proceedings of the 4th international Conference on Persuasive Technology, (p. 40). ACM
441. Bouchart, Marianne (2012). Crowdsourcing data at the Guardian datablog, In J. Gray, L. Bounegru, & L. Chambers (eds.), Data Journalism Handbook. http://datajournalismhandbook.org/1.0/en/getting_data_5.html. accessed 1 November 2015
442. Indiparambil, J. J. (2017). An empirical study on the detrimental effects of employee surveillance in India. International Journal of Research in Computer Application & Management, 7(12), 48–51
443. West, J. P., & Bowman, J. S. (2016). Electronic surveillance at work: an ethical analysis. Administration and Society, 48(5), 628–651
444. Holland, P. J., Cooper, B., & Hecker, R. (2015). Electronic monitoring and surveillance in the workplace: the effects on trust in management, and the moderating role of occupational type. Personnel Review, 44(1), 161–175

445. Alge, B. J., & Hansen, S. D. (2014). Workplace monitoring and surveillance research since 1984: a review and agenda. In M. D. Coovert & L. F. Thompson (Eds.), The psychology of workplace technology (pp. 209–237). New York: Routledge
446. Trevino, L. K., Weaver, G. R., & Reynolds, S. J. (2006). Behavioural ethics in organizations: a review. Journal of Management, 32(6), 951–990
447. Lyon, D. (2003a). Surveillance after September 11. Cambridge: Polity Press
448. Lyon, D. (2003b). Surveillance as social sorting: computer codes and mobile bodies. In D. Lyon (Ed.), Surveillance as social sorting: privacy, risk, and digital discrimination (pp. 13–32). London: Routledge
449. Alge, B. J. (2001). Effects of computer surveillance on perceptions of privacy and procedural justice. Journal of Applied Psychology, 86(4), 797–804
450. Lyon, D. (2001a). Surveillance society: monitoring everyday life. Buckingham: Open University Press
451. Rawlins, B. L. (2008). Measuring the relationship between organizational transparency and employee trust. The Public Relations Journal, 2(2), 1–21
452. Ciocchetti, C. A. (2011). The eavesdropping employer: a twenty-first century framework for employee monitoring. American Business Law Journal, 48(2), 285–369
453. Sewell, G., & Barker, J. R. (2006). Coercion versus care: using irony to make sense of organizational surveillance. Academy of Management Review, 31(4), 934–961
454. Ball, K. (2010). Workplace surveillance: an overview. Labour History, 51(1), 87–106
455. Zweig, D. (2005). Beyond privacy and fairness concerns: examining psychological boundary violations as a consequence of electronic performance monitoring. In J. Weckert (Ed.), Electronic monitoring in the workplace: controversies and solutions (pp. 102–122). Hershey: Idea Group Publishing
456. Rule, J., & Brantely, P. (1992). Computerized surveillance in the workplace: forms and distributions. Sociological Forum, 7(3), 405–423
457. D'Urso, S. C. (2006). Who's watching us at work? Toward a structural-perceptual model of electronic monitoring and surveillance in organizations. Communication Theory, 16, 281–303
458. Lyon, D. (2007c). Surveillance, security and social sorting: emerging research priorities. International Criminal Justice Review, 17(3), 161–170
459. Lyon, D. (2001b). Facing the future: seeking ethics for everyday surveillance. Ethics and Information Technology, 3(3), 171–181
460. Stanton, J. M., & Stam, K. R. (2003). Information technology, privacy, and power within organizations: a view from boundary theory and social exchange perspectives. Surveillance and Society, 1(2), 152–190
461. Nielson, I. K., Jex, S. M., & Adams, G. A. (2000). Development and validation of scores on a two-dimensional workplace friendship scale. Educational and Psychological Measurement, 60, 628–643
462. Katz Y. Technology and innovation in Israel: advancing competitive position in a global environment. Open J Political Sci. 2018;8(4):536–46
463. Baram G, Ben-Israel I. The academic reserve: Israel's fast track to high-tech success. Israel Studies Review. 2019;34(2):75–91
464. Mays, D., Cremeens, J., Usdan, S., Martin, R., Arriola, K. and Bernhardt, J. (2010) The feasibility of assessing alcohol use among college students using wireless mobile devices: Implications for health education and behavioural research. Health Education Journal 69 (3): 311–320
465. Laakso, E.-L., Armstrong, K. and User, W. (2011) Cyber-management of people with chronic disease: A potential solution to eHealth challenges. Health Education Journal 5 (3): 1–8
466. Haraway, D. (1988) A manifesto for cyborgs: Science, technology and socialist feminism in the 1980s. In: E. Weed (ed.) Coming to Terms: Feminism, Theory and Practice. New York: Routledge, pp. 173–204
467. Lupton, D. (1995a) The embodied computer/user. Body & Society 1 (3–4): 97–112

468. Tomas, D. (1995) Feedback and cybernetics: Reimaging the body in the age of the cyborg. Body & Society 1 (3–4): 21–43
469. Freund, P. (2004) Civilised bodies redux: Seams in the cyborg. Social Theory & Health 2 (3): 273–289
470. Shildrick, M. (2010) Some reflections on the socio-cultural and bioscientific limits of bodily integrity. Body & Society 16 (3): 11–22
471. Nettleton, S. and Burrows, R. (2003) E-scaped medicine? Information, reflexivity and health. Critical Social Policy 23 (2): 165–185
472. Nettleton, S. (2004) The emergence of e-scaped medicine? Sociology 38 (4): 661–679
473. Haggerty, K. D., & Ericson, R. V. (2000). The surveillant assemblage. British Journal of Sociology, 51(4), 605–22
474. Lyon, D. (2007) Surveillance Studies: An Overview. Cambridge, UK: Polity Press
475. Lyon, D. (2010) Surveillance, power and everyday life. In: P. Kalantzis-Cope and K. Gherab-Martin (eds.) Emerging Digital Spaces in Contemporary Society. Houndsmills, UK: Palgrave Macmillan, pp. 107–120
476. Bennett, C.J. (2011) In defence of privacy: The concept and the regime. Surveillance & Society 8 (4): 485–496
477. Carrion M, Madoff LC. ProMED-mail: 22 years of digital surveillance of emerging infectious diseases. Int Health. 2017;9(3):177–83
478. O'Shea J. Digital disease detection: a systematic review of event-based internet biosurveillance systems. Int J Med Inform. 2017;101:15–22. https://doi.org/10.1016/j.ijmedinf.2017.01.019
479. Freifeld CC, Mandl KD, Reis BY, Brownstein JS. Healthmap: global infectious disease monitoring through automated classification and visualization of internet media reports. J Am Med Inform Assoc. 2008;15(2):150–7
480. Rolland C, Lazarus C, Giese C, Monate B, Travert AS, Salomon J. Early detection of public health emergencies of international concern through undiagnosed disease reports in ProMED-mail. Emerg Infect Dis. 2020;26(2):336–9
481. Abbood A, Ullrich A, Busche R, Ghozzi S. EventEpi-A natural language processing framework for event-based surveillance. PLoS Comput Biol. 2020;16:1–16. https://doi.org/10.1371/journal.pcbi.1008277
482. Dellanzo A, Cotik V, Ochoa-Luna J. A corpus for outbreak detection of diseases prevalent in Latin America. In: Proceedings of the 24th conference on computational natural language learning. Association for Computational Linguistics; 2020. pp 543–51. https://doi.org/10.18653/v1/2020.conll-1.44. https://aclanthology.org/2020.conll-1.44
483. Grubaugh, N. D., Hodcroft, E. B., Fauver, J. R., Phelan, A. L. & Cevik, M. Public health actions to control new SARS-CoV-2 variants. Cell https://doi.org/10.1016/j.cell.2021.01.044 (2021)
484. Abadie, A. Using synthetic controls: Feasibility, data requirements, and methodological aspects. J. Econ. Lit. 20, 20 (2020)
485. Cavallo, E., Galiani, S., Noy, I. & Pantano, J. Catastrophic natural disasters and economic growth. Rev. Econ. Stat. 95, 1549–1561 (2013). 10.1162/REST_a_00413
486. Mitze, T., Kosfeld, R., Rode, J. & Wälde, K. Face masks considerably reduce COVID-19 cases in Germany. Proc. Natl. Acad. Sci. 117, 32293–32301. https://doi.org/10.1073/pnas.2015954117 (2020)
487. Friedson, A. I., McNichols, D., Sabia, J. J. & Dave, D. Shelter-in-place orders and public health: Evidence from California during the COVID-19 pandemic. J. Policy Anal. Manage. 40, 258–283. https://doi.org/10.1002/pam.22267 (2021)
488. Cho, S.-W.S. Quantifying the impact of nonpharmaceutical interventions during the COVID-19 outbreak: The case of Sweden. Econom. J. 23, 323–344. https://doi.org/10.1093/ectj/utaa025 (2020)
489. Cassini, A. et al. Attributable deaths and disability-adjusted life-years caused by infections with antibiotic-resistant bacteria in the EU and the European Economic Area in 2015: a

population-level modelling analysis. Lancet. Infect. Dis 19, 56–66. https://doi.org/10.1016/s1473-3099(18)30605-4 (2019)

490. Laxminarayan, R. et al. Antibiotic resistance-the need for global solutions. Lancet. Infect. Dis 13, 1057–1098. https://doi.org/10.1016/s1473-3099(13)70318-9 (2013)
491. Wozniak, T. M., Bailey, E. J. & Graves, N. Health and economic burden of antimicrobial-resistant infections in Australian hospitals: a population-based model. Infect. Control Hosp. Epidemiol. 40, 320–327. https://doi.org/10.1017/ice.2019.2 (2019)
492. Bowen, A. C., Daveson, K., Anderson, L. & Tong, S. Y. An urgent need for antimicrobial stewardship in Indigenous rural and remote primary health care. Med. J. Aust. https://doi.org/10.5694/mja2.50216 (2019)
493. Australian Commission on Safety and Quality in Health Care. Australian Passive Antimicrobial Resistance Surveillance (APAS), https://www.safetyandquality.gov.au/antimicrobial-use-and-resistance-in-australia/apas/
494. Turnidge, J. D. & Meleady, K. T. Antimicrobial use and resistance in Australia (AURA) surveillance system: coordinating national data on antimicrobial use and resistance for Australia. Aust. Health Rev. Public. Aust. Hospital Assoc. 42, 272–276. https://doi.org/10.1071/ah16238 (2018)
495. Australian Commission on Safety and Quality in Health Care. AURA 2019: third Australian report on antimicrobial use and resistance in human health. Sydney: ACSQHC; 2019, https://consultations.health.gov.au/ohpd-health-protection-policy-branch/consultation-on-next-amr-strategy/user_uploads/aura-2019-report.pdf (2019)
496. Australian Department of Health. National Antimicrobial Resistance Strategy 2015–2019, https://www.health.gov.au/internet/main/publishing.nsf/Content/1803C433C71415CACA257C8400121B1F/$File/amr-strategy-2015-2019.pdf (2015)
497. Stevens, C. L., Ralph, A., McLeod, J. E. & McDonald, M. I. Community-acquired methicillin-resistant Staphylococcus aureus in Central Australia. Commun. Dis. Intell. Q. Rep. 30, 462–466 (2006)
498. Poncet, Patrick; Ripert, Blandine Fractured space: a geographical reflection on the digital divide. GeoJournal (2007). https://doi.org/10.1007/s10708-007-9050-7
499. Choudrie, Jyoti; Zamani, Efpraxia; Obuekwe, Chike Bridging the Digital Divide in Ethnic Minority Older Adults: an Organisational Qualitative Study. Information Systems Frontiers (2021). https://doi.org/10.1007/s10796-021-10126-8
500. Choudrie, Jyoti; Pheeraphuttranghkoon, Sutee; Davari, Soheil The Digital Divide and Older Adult Population Adoption, Use and Diffusion of Mobile Phones: a Quantitative Study. Information Systems Frontiers (2018). https://doi.org/10.1007/s10796-018-9875-2
501. Carmichael, Alex; Rice, Mark; Sloan, David; Gregor, Peter Digital switchover or digital divide: a prognosis for usable and accessible interactive digital television in the UK. Universal Access in the Information Society (2005). https://doi.org/10.1007/s10209-005-0004-x
502. Sanders, Cynthia K.; Scanlon, Edward The Digital Divide Is a Human Rights Issue: Advancing Social Inclusion Through Social Work Advocacy. Journal of Human Rights and Social Work (2021). https://doi.org/10.1007/s41134-020-00147-9
503. Huang, Shu-Chin; Cox, John Lew Establishing a social entrepreneurial system to bridge the digital divide for the poor: a case study for Taiwan. Universal Access in the Information Society (2014). https://doi.org/10.1007/s10209-014-0379-7
504. Yelland, Nicola; Neal, Greg Aligning digital and social inclusion: A study of disadvantaged students and computer access. Education and Information Technologies (2012). https://doi.org/10.1007/s10639-012-9223-y
505. Rye, Ståle Angen Exploring the gap of the digital divide. GeoJournal (2008). https://doi.org/10.1007/s10708-008-9154-8
506. Thonipara, Anita; Sternberg, Rolf; Proeger, Till; Haefner, Lukas Digital divide, craft firms' websites and urban-rural disparities—empirical evidence from a web-scraping approach. Review of Regional Research (2022). https://doi.org/10.1007/s10037-022-00170-5

507. Lybeck, Robin; Koiranen, Ilkka; Koivula, Aki From digital divide to digital capital: the role of education and digital skills in social media participation. Universal Access in the Information Society (2023). https://doi.org/10.1007/s10209-022-00961-0
508. Cleary, Paul F.; Pierce, Glenn; Trauth, Eileen M. Closing the digital divide: understanding racial, ethnic, social class, gender and geographic disparities in Internet use among school age children in the United States. Universal Access in the Information Society (2005). https://doi.org/10.1007/s10209-005-0001-0
509. Devisakti, A.; Muftahu, Muhammad; Xiaoling, Hu Digital divide among B40 students in Malaysian higher education institutions. Education and Information Technologies (2023). https://doi.org/10.1007/s10639-023-11847-w
510. Gaved, Mark B.; Mulholland, Paul Networking communities from the bottom up: grassroots approaches to overcoming the digital divide. AI & SOCIETY (2010). https://doi.org/10.1007/s00146-009-0259-5
511. Otioma, Chuks; Madureira, Ana Mafalda; Martinez, Javier Spatial analysis of urban digital divide in Kigali, Rwanda. GeoJournal (2018). https://doi.org/10.1007/s10708-018-9882-3
512. Badran, Mona Farid Young people and the digital divide in Egypt: an empirical study. Eurasian Economic Review (2014). https://doi.org/10.1007/s40822-014-0008-z
513. Vimalkumar, M; Singh, Jang Bahadur; Sharma, Sujeet Kumar Exploring the Multi-Level Digital Divide in Mobile Phone Adoption: A Comparison of Developing Nations. Information Systems Frontiers (2020). https://doi.org/10.1007/s10796-020-10032-5
514. Hellemans, Johan; Willems, Kim; Brengman, Malaika Covid-19 and mobile payment in Belgium: Closing the digital divide or just for the young, social, and impulsive?. Electronic Commerce Research (2022). https://doi.org/10.1007/s10660-022-09655-4
515. Elena-Bucea, Anca; Cruz-Jesus, Frederico; Oliveira, Tiago; Coelho, Pedro Simões Assessing the Role of Age, Education, Gender and Income on the Digital Divide: Evidence for the European Union. Information Systems Frontiers (2020). https://doi.org/10.1007/s10796-020-10012-9
516. Busch, Thorsten Capabilities in, capabilities out: overcoming digital divides by promoting corporate citizenship and fair ICT. Ethics and Information Technology (2010). https://doi.org/10.1007/s10676-010-9261-3
517. Blažič, Borka Jerman; Blažič, Andrej Jerman Overcoming the digital divide with a modern approach to learning digital skills for the elderly adults. Education and Information Technologies (2019). https://doi.org/10.1007/s10639-019-09961-9
518. Krizanovic Cik, Visnja; Zagar, Drago; Grgic, Kresimir A framework for optimal techno-economic assessment of broadband access solutions and digital inclusion of rural population in global information society. Universal Access in the Information Society (2017). https://doi.org/10.1007/s10209-017-0560-x
519. Koiranen, Ilkka; Keipi, Teo; Koivula, Aki; Räsänen, Pekka Changing patterns of social media use? A population-level study of Finland. Universal Access in the Information Society (2019). https://doi.org/10.1007/s10209-019-00654-1
520. Picatoste, Xose; Mesquita, Anabela; González-Laxe, Fernando Gender wage gap, quality of earnings and gender digital divide in the European context. Empirica (2022). https://doi.org/10.1007/s10663-022-09555-8
521. Zhang, Bingqian; Peng, Guochao; Liu, Caihua; Zhang, Zuopeng Justin; Jasimuddin, Sajjad M. Adaptation behaviour in using one-stop smart governance apps: an exploratory study between digital immigrants and digital natives. Electronic Markets (2022). https://doi.org/10.1007/s12525-022-00538-y
522. Kumm, Aubrey J.; Viljoen, Marisa; de Vries, Petrus J. The Digital Divide in Technologies for Autism: Feasibility Considerations for Low- and Middle-Income Countries. Journal of Autism and Developmental Disorders (2021). https://doi.org/10.1007/s10803-021-05084-8
523. Khan, Naurin Farooq; Ikram, Naveed; Saleem, Sumera Effects of socioeconomic and digital inequalities on cybersecurity in a developing country. Security Journal (2023). https://doi.org/10.1057/s41284-023-00375-4

524. Chang, She-I; Yen, David C.; Chang, I-Cheng; Chou, Jung-Chu Study of the digital divide evaluation model for government agencies–a Taiwanese local government's perspective. Information Systems Frontiers (2011). https://doi.org/10.1007/s10796-011-9297-x
525. Qadikolaei, Mohsen Roohani; Zali, Nader; Soltani, Ali Spatiotemporal investigation of the digital divide, the case study of Iranian Provinces. Environment, Development and Sustainability (2022). https://doi.org/10.1007/s10668-022-02738-0
526. Huxhold, Oliver; Hees, Elena; Webster, Noah J. Towards bridging the grey digital divide: changes in internet access and its predictors from 2002 to 2014 in Germany. European Journal of Ageing (2020). https://doi.org/10.1007/s10433-020-00552-z
527. Jafar, K.; Ananthpur, Kripa; Venkatachalam, L. Digital divide and access to online education: new evidence from Tamil Nadu, India. Journal of Social and Economic Development (2023). https://doi.org/10.1007/s40847-023-00236-1
528. Jiang, Shaohai; Hong, Y. Alicia; Liu, Piper Liping Trends of online patient-provider communication among cancer survivors from 2008 to 2017: a digital divide perspective. Journal of Cancer Survivorship (2019). https://doi.org/10.1007/s11764-019-00742-4
529. Ngunan, Achakpa-Ikyo Patience; Regina, Ogaba-Egba Digital divide: A Gap in interpersonal communication amongst nigerian family relationship. GSTF Journal on Media & Communications (2016). https://doi.org/10.7603/s40874-016-0003-y
530. Johansson, Stefan; Gulliksen, Jan; Gustavsson, Catharina Disability digital divide: the use of the internet, smartphones, computers and tablets among people with disabilities in Sweden. Universal Access in the Information Society (2020). https://doi.org/10.1007/s10209-020-00714-x
531. Jiang, Wun-Ji; Luh, Yir-Hueih Gender digital divide in a patriarchal society: what can we learn from Blinder–Oaxaca decomposition?. Quality & Quantity (2016). https://doi.org/10.1007/s11135-016-0409-z
532. Connolly, Michelle; Lee, Clement; Tan, Renhao The Digital Divide and Other Economic Considerations for Network Neutrality. Review of Industrial Organization (2016). https://doi.org/10.1007/s11151-016-9554-8
533. Fareed, Naleef; Swoboda, Christine M.; Jonnalagadda, Pallavi; Huerta, Timothy R. Persistent digital divide in health-related internet use among cancer survivors: findings from the Health Information National Trends Survey, 2003–2018. Journal of Cancer Survivorship (2020). https://doi.org/10.1007/s11764-020-00913-8
534. Hatef, Elham; Ma, Xiaomeng; Shaikh, Yahya; Kharrazi, Hadi; Weiner, Jonathan P.; Gaskin, Darrell J. Internet Access, Social Risk Factors, and Web-Based Social Support Seeking Behavior: Assessing Correlates of the "Digital Divide" Across Neighborhoods in The State of Maryland. Journal of Medical Systems (2021). https://doi.org/10.1007/s10916-021-01769-w
535. Hawash, Ronia; Lang, Guenter Does the digital gap matter? Estimating the impact of ICT on productivity in developing countries. Eurasian Economic Review (2019). https://doi.org/10.1007/s40822-019-00133-1
536. Lorence, Daniel P.; Park, Heeyoung; Fox, Susannah Racial disparities in health information access: resilience of the digital divide. Journal of Medical Systems (2006). https://doi.org/10.1007/s10916-005-9003-y
537. Riordan, Monica A.; Kreuz, Roger J.; Blair, Alyssa N. The digital divide: conveying subtlety in online communication. Journal of Computers in Education (2018). https://doi.org/10.1007/s40692-018-0100-6
538. Fdez-Arroyabe, Pablo; Lecha Estela, Luis; Schimt, Falko Digital divide, biometeorological data infrastructures and human vulnerability definition. International Journal of Biometeorology (2017). https://doi.org/10.1007/s00484-017-1398-x
539. Casillas, Alejandra; Moreno, Gerardo; Grotts, Jonathan; Tseng, Chi-Hong; Morales, Leo S. A Digital Language Divide? The Relationship between Internet Medication Refills and Medication Adherence among Limited English Proficient (LEP) Patients. Journal of Racial and Ethnic Health Disparities (2018). https://doi.org/10.1007/s40615-018-0487-9

540. Levy, Helen; Janke, Alexander T.; Langa, Kenneth M. Health Literacy and the Digital Divide Among Older Americans. Journal of General Internal Medicine (2014). https://doi.org/10.1007/s11606-014-3069-5
541. James, Jeffrey The Digital Divide Across All Citizens of the World: A New Concept. Social Indicators Research (2007). https://doi.org/10.1007/s11205-007-9156-9
542. de Brito, Silvana Rossy; da Silva, Aleksandra do Socorro; Martins, Dalton Lopes; Vijaykumar, Nandamudi Lankalapalli; da Rocha, Cláudio Alex Jorge; Costa, João Crisóstomo Weyl Albuquerque; Francês, Carlos Renato Lisboa Employing online social networks to monitor and evaluate training of digital inclusion agents. Social Network Analysis and Mining (2013). https://doi.org/10.1007/s13278-012-0093-5
543. Al-Muwil, Ahlam; Weerakkody, Vishanth; El-haddadeh, Ramzi; Dwivedi, Yogesh Balancing Digital-By-Default with Inclusion: A Study of the Factors Influencing E-Inclusion in the UK. Information Systems Frontiers (2019). https://doi.org/10.1007/s10796-019-09914-0
544. Stypinska, Justyna AI ageism: a critical roadmap for studying age discrimination and exclusion in digitalized societies. AI & SOCIETY (2022). https://doi.org/10.1007/s00146-022-01553-5
545. Rottinghaus, Bastian; Escher, Tobias Mechanisms for inclusion and exclusion through digital political participation: Evidence from a comparative study of online consultations in three German cities. Zeitschrift für Politikwissenschaft (2020). https://doi.org/10.1007/s41358-020-00222-7
546. Uzcanga, Catalina; Oiarzabal, Pedro J. Associations of Migrants in Spain: An Enquiry into Their Digital Inclusion in the "Network Society" in the 2010s. VOLUNTAS: International Journal of Voluntary and Nonprofit Organizations (2019). https://doi.org/10.1007/s11266-019-00162-z
547. Ahi, Alan A.; Sinkovics, Noemi; Sinkovics, Rudolf R. E-commerce Policy and the Global Economy: A Path to More Inclusive Development?. Management International Review (2022). https://doi.org/10.1007/s11575-022-00490-1
548. Han, Honghua; Xiong, Jason; Zhao, Kexin Digital inclusion in social media marketing adoption: the role of product suitability in the agriculture sector. Information Systems and e-Business Management (2021). https://doi.org/10.1007/s10257-021-00522-7
549. Aslam, Aribah; Naveed, Amjad; Shabbir, Ghulam Is it an institution, digital or social inclusion that matters for inclusive growth? A panel data analysis. Quality & Quantity (2020). https://doi.org/10.1007/s11135-020-01008-3
550. Holcombe-James, Indigo 'I'm fired up now!': digital cataloguing, community archives, and unintended opportunities for individual and archival digital inclusion. Archival Science (2022). https://doi.org/10.1007/s10502-021-09380-1
551. Hofmann, Mathias; Münster, Sander; Noennig, Jörg Rainer A Theoretical Framework for the Evaluation of Massive Digital Participation Systems in Urban Planning. Journal of Geovisualization and Spatial Analysis (2019). https://doi.org/10.1007/s41651-019-0040-3
552. Engler, Jennifer; Brosse, Franziska; Dinh, Truc Sophia; Klein, Astrid-Alexandra; Brueckle, Maria-Sophie; Petermann, Jenny; Muth, Christiane; Mergenthal, Karola; van den Akker, Marjan; Voigt, Karen Digital participatory workshops with patients and health professionals to develop an intervention for the management of polypharmacy: results from a mixed-methods evaluation and methodological conclusions. Research Involvement and Engagement (2022). https://doi.org/10.1186/s40900-022-00387-1
553. Ateca-Amestoy, Victoria; Castiglione, Concetta Live and digital engagement with the visual arts. Journal of Cultural Economics (2022). https://doi.org/10.1007/s10824-022-09466-3
554. Fischl, Caroline; Lindelöf, Nina; Lindgren, Helena; Nilsson, Ingeborg Older adults' perceptions of contexts surrounding their social participation in a digitalized society—an exploration in rural communities in Northern Sweden. European Journal of Ageing (2020). https://doi.org/10.1007/s10433-020-00558-7
555. Ho, Debbie G. E.; Sa'adi, Munawwarah; He, Deyuan; Hoon, Chang-Yau Silence over the wire: student verbal participation and the virtual classroom in the digital era. Asia Pacific Education Review (2023). https://doi.org/10.1007/s12564-023-09834-4

556. Liu, Chen-Chung; Yang, Chen-Yu; Chao, Po-Yao A longitudinal analysis of student participation in a digital collaborative storytelling activity. Educational technology research and development (2019). https://doi.org/10.1007/s11423-019-09666-3
557. Handler, Reinhard A.; Ferrer Conill, Raul Open Data, Crowdsourcing and Game Mechanics. A case study on civic participation in the digital age. Computer Supported Cooperative Work (CSCW) (2016). https://doi.org/10.1007/s10606-016-9250-0
558. Indiparambil, Jijo James Privacy and beyond: socio-ethical concerns of 'on-the-job' surveillance. Asian Journal of Business Ethics (2019). https://doi.org/10.1007/s13520-019-00089-0
559. Keshet, Yael Fear of panoptic surveillance: using digital technology to control the COVID-19 epidemic. Israel Journal of Health Policy Research (2020). https://doi.org/10.1186/s13584-020-00429-7
560. Lupton, Deborah M-health and health promotion: The digital cyborg and surveillance society. Social Theory & Health (2012). https://doi.org/10.1057/sth.2012.6
561. Dellanzo, Antonella; Cotik, Viviana; Lozano Barriga, Daniel Yunior; Mollapaza Apaza, Jonathan Jimmy; Palomino, Daniel; Schiaffino, Fernando; Yanque Aliaga, Alexander; Ochoa-Luna, José Digital surveillance in Latin American diseases outbreaks: information extraction from a novel Spanish corpus. BMC Bioinformatics (2022). https://doi.org/10.1186/s12859-022-05094-y
562. Mitze, Timo; Rode, Johannes Early-stage spatial disease surveillance of novel SARS-CoV-2 variants of concern in Germany with crowdsourced data. Scientific Reports (2022). https://doi.org/10.1038/s41598-021-04573-1
563. Wozniak, T. M.; Cuningham, W.; Buchanan, S.; Coulter, S.; Baird, R. W.; Nimmo, G. R.; Blyth, C. C.; Tong, S. Y. C.; Currie, B. J.; Ralph, A. P. Geospatial epidemiology of Staphylococcus aureus in a tropical setting: an enabling digital surveillance platform. Scientific Reports (2020). https://doi.org/10.1038/s41598-020-69312-4
564. Amin, Heba Y.; Downey, Anthony Contesting post-digital futures: drone warfare and the geopolitics of aerial surveillance in the middle east. Digital War (2020). https://doi.org/10.1057/s42984-020-00021-y

Ethnicity, Race, and Identity in the Digital Age

5

Introduction

Digital technologies have had disruptive effects on self, self-identity, community, race, and ethnicity imaginations. Where the conception of the digital communities have moved beyond the immediate geographic locations to group of networked persons across the geographies, race and ethnicity identification as such have also become sharper as evident from the increased identification of people with similar physical or racial attributes.

Digital technologies like cameras associated with phones and laptops have become dominant sources of viewing the body and creating self-perceptions. Social media like Instagram, Facebook (now Meta), Twitter (now X), etc., work as mirrors to validate the image of others through digital hyperconnectivity. The like and dislike of the images, texts, and other forms of communication leads to either happiness or sadness and many times lead to psychological depression or adverse consequences to mental health. The construction of the digital self on social networking sites is often through the photos of the body through filters, and impacts emotions, and self-esteem among the users. Social media and platform interactions now also shape personal relationships and determine the social, economic, and political consequences and their spatial relations.

Many social media sites have now become sites of body surveillance and body shaming. These are not the true selves that are represented by the media spaces but the selves through the body. This is why the use of filters and photo editing has become so common among the users of social media before posting photos. However, there are also cases where individuals face 'digitized dysmorphia', specially females of their bodies and this also results in adverse self-esteem and personal consequences.

A. Shaban (ed.), *Digital Geographies—Theory, Space, and Communities*,
https://doi.org/10.1007/978-981-97-4734-4_5

Studies show that among the younger generations and specially among adolescents the construction of the digital self and looking through the same the real self has become narcissistic. These age-groups individuals are found to be more sensitive to their self-images and fall into the trap of what is known as 'you're Nothing but your body' on digital platforms.

How far the digital identity is able to overcome non-digital or real social identity remains a question, but certain social identities of age, gender, religion, race and ethnicity often are propagated by users in the same way as in the real world. Therefore, it is said that some of identities may be also 'born digital or fossilised digitally', and may lead to 'digital immortality'.

Identity stereotyping and circulating the same through digital technologies have become very pervasive in many societies. India is a good example of the same. The women from a religious group were assigned different types of identities, and they were traded on different websites as if they were available for sexual pleasure. The police had to act on the same to bring those involved to justice. However, there are many other websites where such morphing and trading continue. The digital has become a complex system like the ocean, and effectively governing the same has become an issue.

The papers reviewed in this chapter are again theoretically and empirically rich and offer readers a wide range of ideas and methods. The reviewed papers are divided into 4 sections. The first section deals with papers on digital expressions, perspectives and relationships, while the second section focuses on papers which relate to digital identities, their construction and propagation. The third section deals with the papers related to physical or body aspects of the digital self, and the last section surveys the papers related to the psychological aspects of the digital self.

A paper titled 'The Effect of Multimodal Emotional Expression on Responses to a Digital Human during a Self-Disclosure Conversation: a Computational Analysis of User Language' in the first section makes the case that the inclusion of a face and an emotion in the voice may enhance emotional responses to digital humans, while another paper on 'Born digital or fossilized digitally? How born-digital data systems continue the legacy of social violence towards LGBTQI+ communities: a case study of experiences in the Republic of Ireland' finds that digital mediums do allow freedom to express one's gender identity.

A study in part 2 titled 'Discussing human values in digital immortality: towards a value-oriented perspective' makes the case that we are on the verge of being able to use digital immortality to immortalize our thoughts, judgments, and actions.

According to the paper in section three of this chapter, 'The Impact of Social Media on Body Image Perceptions and Bodily Practices Among Gay, Bisexual, and Other Men Who Have Sex with Men: A Critical Review of the Literature and Extension of Theory', eating disorders and body image disturbance in men of sexual minorities are caused by a wide range of interrelated biological, psychological, social, and cultural determinants. In contrast, a different study titled 'From filters to fillers: an active inference approach to body image distortion in the selfie era' demonstrates that impaired self-modelling is unable to reduce interpersonal uncertainty during identity exploration, which eventually degenerates into uncontrollable epistemic habits that isolate the body image from corrective sensory evidence. Additionally, in this fresh situation, increased interpersonal unpredictability may hasten the establishment of body dysmorphic disorder and associated body image distortions, especially when selfie filters are used.

A paper on '"Why don't I look like her?" How adolescent girls view social media and its connection to body image' claims that adolescent girls are more likely than boys to experience mental health issues as a result of using social media, and that sexual objectification through photos may encourage adolescent females to believe that their value is based on beauty. The paper on 'Body Surveillance on Instagram: Examining the Role of Selfie Feedback Investment in Young Adult Women's Body Image Concerns' finds that young adult female Instagram users who value viewer reactions to their selfies are more likely to display disordered eating attitudes and intentions.

Another paper 'Adolescents' Body Shame and Social Networking Sites: The Mediating Effect of Body Image Control in Photos' finds that attention paid to online physical appearance seems to trigger some body-related behaviours (such as body image monitoring) that may be connected to self-objectification and problematic social networking site (SNS) use, and body shame strongly predicts body image control in SNS photos, indirectly leading to both male and female adolescent body shame.

A paper in section four on 'Digital hyperconnectivity and the self' finds that The self has been colonized by hyperconnectivity and forced to serve technological and social institutions, while another paper on 'How photo editing in social media shapes self-perceived attractiveness and self-esteem via self-objectification and physical appearance comparisons' argues that Self-perceived beauty and self-esteem are negatively correlated with photo editing behaviour for social media, which is mediated through self-objectification and physical appearance comparisons.

Machine generated summaries

Disclaimer: The summaries in this chapter were generated from Springer Nature publications using extractive AI auto-summarization: An extraction-based summarizer aims to identify the most important sentences of a text using an algorithm and uses those original sentences to create the auto-summary (unlike generative AI). As the constituted sentences are machine selected, they may not fully reflect the body of the work, so we strongly advise that the original content is read and cited. The auto generated summaries were curated by the editor to meet Springer Nature publication standards. To cite this content, please refer to the original papers.

Machine generated keywords: identity, image, body, body image, adolescent, social medium, appearance, photo, selfie, sns, girl, medium, self, medium use, behavior

5.1 Digital Expressions, Perspectives and Relationships

Machine generated keywords: emotional, treatment, identity, human, embody, expression, relationship, gender, voice, face, sex, male, academic, agent, case study

5.1.1 The Effect of Multimodal Emotional Expression on Responses to a Digital Human during a Self-disclosure Conversation: A Computational Analysis of User Language [480]

This is a machine-generated summary of:

Loveys, Kate; Sagar, Mark; Broadbent, Elizabeth: The Effect of Multimodal Emotional Expression on Responses to a Digital Human during a Self-Disclosure Conversation: a Computational Analysis of User Language [480]

Published in: Journal of Medical Systems (2020)

Link to original: https://doi.org/10.1007/s10916-020-01624-4

If you want to cite the papers, please refer to the original.

For technical reasons we could not place the page where the original quote is coming from.

Abstract-Summary

"Emotional expressiveness can improve social closeness in human relationships, especially for females."

"It is unknown whether multimodal emotional expression improves relationships with digital humans."

"The digital humans varied in modality richness (face, no face) and emotional expression (emotional voice, neutral voice; emotional face, neutral face)."

"A series of three-way ANOVA and ANCOVA were conducted to evaluate the effect of digital human face type, voice type, and participant gender on emotional content in participant language."

"A digital human with no face was associated with more first-person singular pronoun use than a neutral face and an emotional face digital human."

"A digital human with no face and a neutral voice received more general negative emotion language than a digital human with no face and an emotional voice."

"Findings suggest the presence of a face and emotion in the voice may improve emotional responses to digital humans."

Introduction

"For digital humans to be effective virtual therapists or companions, it is important that they can develop close relationships with patients."

"Emotional expressiveness has shown to improve social closeness in human relationships, especially for females [1]."

"This is because research has shown that women value self-disclosure and emotional expressiveness in close relationships, whereas men report closeness to be based more on shared interests and activities [2, 3]."

"While emotional expressiveness may promote closeness in human relationships, it is unclear whether this technique applies to relationships with digital humans."

"Digital humans have the capacity to engage in multimodal emotional expression during conversations."

"Multimodal emotional expression may help to increase the salience of a digital human's emotional responses during interactions and ultimately, improve its relationships with users."

"Multimodal emotional expression has yet to be tested with digital humans in close relationships."

"This study investigated the effect of multimodal emotional expression on emotional responses to a digital human."

Methods

"Participants were recruited for an experimental study looking at social closeness with a digital human."

"A total of 13 participants data were excluded from analyses for the following reasons: the digital human program would not start ($n = 1$), there was an error processing the audiovisual recording ($n = 10$), or the participant identified as gender non-binary and this group was not of sufficient size for adequate statistical power in gender analyses ($n = 2$)."

"Participants completed the conversation in private with a digital human."

"After the conversation was finished, the researcher returned to the room and participants completed a second questionnaire on their relationship with the digital human and psychological variables (stress, emotional state, social support, loneliness)."

"A series of three-way factorial analyses of variance (ANOVA) or analyses of covariance (ANCOVA) were conducted where appropriate to analyse the effect of face type, voice type, and participant gender on emotional responses to a digital human."

Results

"A three-way factorial ANCOVA was conducted to investigate the effect of face type, voice type, and gender on use of first-person singular pronouns during the conversation."

"A three-way factorial ANOVA including face type, voice type, and gender revealed no significant main effects on the use of negative emotion words (all $Fs < .762$)."

"There was a significant two-way interaction effect between face type and voice type ($F(2, 173) = 3.311$, $p = .039$)."

"A three-way factorial ANOVA examined the effect of face type, voice type, and gender on positive emotion language during the interaction."

Discussion

"Emotional responses to a digital human were improved by the presence of a face and an emotional voice."

"Results show that a digital human with no face and a neutral voice generated a more negative emotional response from users."

"While there is limited research on digital humans with which to make comparisons, these results are in keeping with findings from the ECA literature which indicate that embodied agents with an emotional voice (compared to a neutral voice) and a face (compared to speech or text only) are responded to more positively by users [4, 5]."

"This study also found no gender differences in responses to an emotional face and voice."

"Digital humans intended for use as virtual therapists or companions should be designed with a face and an emotional voice."

5.1.2 Personal Relationships in the Digital Age: Three Female Academics' Qualitative Research [481]

This is a machine-generated summary of:

Capecchi, Saveria: Personal relationships in the digital age: three female academics' qualitative research [481]

Published in: Quality & Quantity (2017)

Link to original: https://doi.org/10.1007/s11135-017-0544-1

If you want to cite the papers, please refer to the original.

For technical reasons we could not place the page where the original quote is coming from.

Abstract-Summary

"This essay deals with the theme of personal relationships on social media, linked to the theme of online identity, bringing together the research and reflections of some of the most important academics of the internet."

"It is interesting to see how the three academics place not only the quality of the relationship, but also the quality of research at the centre of their reflections, demonstrating their passion for the work which is part of their lives."

Introduction

"Sherry Turkle, Nancy Baym and danah boyd are well-known American academics who have something in common: they have dedicated many years of their lives (more than 30, 20 and 10 years respectively) to the study of online relationships through digital ethnography, and they continue to do so."

"Social Networks facilitate contact but run the risk of losing all the meaning and emotional intensity of human relationships, reducing them hasty exchanges rather than looking after the quality of the relationship, which needs reciprocal listening, empathy and non-verbal communication."

"Regarding relationships conducted on the web, Nancy Baym and danah boyd are more optimistic than Turkle, claiming that newer generations are able to manage their online/offline communication."

"A perspective that takes into account the many factors and complexity of the technology-society relationship, ascribing primary importance to the social processes, but agrees that the direction of influence is, at the very least, two-way."

The nineties: life on the screen

"In the field of Internet Studies at that time they started to explore the interactive potentials offered by the net such as the possibility to create, within the virtual communities, meaningful relationships among users, who for the most part did not know each other: relationships that are similar to those conducted face to face (Reinghold [6]; Turkle [7]; Markham [8]; Hine [9]; Baym [10])."

"To demonstrating the creation of a real virtual community, Baym highlighted that communicating with unknown people on the net was very natural, so much so that the participants claim that it felt like "talking", rather than writing messages."

"If the internet was conceived as a "place" where people meet, have conversations, build relationships and have new experiences in the first stage of Internet Studies, for the most part opposed to offline life, Turkle and Baym's research shows that the "two worlds" are both experienced as "real lives" at a psychological level."

The 2000s: life on and off the screen

"Baym states that the majority of online relationships are made up of weak ties (Granovetter [11]), but every so often it is possible to transform an online friendship into a close bond or a "real" friendship with face to face meetings, as happened to her with her Swedish friend Markus."

"The difference with offline relationships, Baym claims, is that on the net people tend to idealise the other person who they are communicating with, a lot more than happens offline."

"Turkle claims that, while communication technology such as smartphones and social network make interpersonal relationships easier, at the same time they lead to less human contact, diluting its nature and scope and making us feel emotionally alone."

"More and more in offline situations, both private and public (with family or in a café), even when individuals are physically present, their attention is elsewhere, each of them busy multitasking, connecting online with others far away."

Conclusions

"It is interesting to note how all three of the researchers have an attentive overview of the subtleties, dual aspects and the bond that is getting tighter and tighter between humans and technology, online and offline and public and private."

"This is where the charm and wealth of quality qualitative research lies: a large amount of data, gathered over the years with qualitative methods such as participant observation, online and offline interviews and conversations from which, rather than getting a purely impressionistic image that is not very scientific and very subjective, comes a set of information that give us order, classifying behavior and related themes, based on methodological rigor and a unparalleled passion for research (Baym [12]; Markham and Baym [13])."

"Their reflections lead to the questioning of what it means to establish a quality relationship: not hiding emotions, empathy, the sharing of moods that emoticons cannot translate; the conversation, based on listening and paying attention to who is speaking, identified as the essence of interpersonal relationships."

5.1.3 Born Digital or Fossilised Digitally? How Born Digital Data Systems Continue the Legacy of Social Violence Towards LGBTQI + Communities: A Case Study of Experiences in the Republic of Ireland [482]

This is a machine-generated summary of:

Donnelly, Noeleen; Stapleton, Larry; O'Mahoney, Jennifer: Born digital or fossilised digitally? How born digital data systems continue the legacy of social violence towards LGBTQI + communities: a case study of experiences in the Republic of Ireland [482]

Published in: AI & SOCIETY (2022)

Link to original: https://doi.org/10.1007/s00146-021-01374-y

If you want to cite the papers, please refer to the original.

For technical reasons we could not place the page where the original quote is coming from.

Abstract-Summary
"The AI and Society discourse has previously drawn attention to the ways that digital systems embody the values of the technology development community from which they emerge through the development and deployment process."

"Research shows how this effect leads to a particular treatment of gender in computer systems development, a treatment which lags far behind the rich understanding of gender that social studies scholarship reveals and people across society experience."

"Many people do not relate to the narrow binary gender options of male or female, and many people express their gender identity in much richer ways than the sex/gender binary female/woman and male/man Boolean terms will allow."

"We ask: are "born-digital" gendered datasets in digital systems experienced as marginalising by those who express their identity beyond the male/female binary?"

"To answer this universal question, this paper presents the findings of an empirical case study of people in Ireland with diverse gender identities and expressions, and their experiences with public data systems and new technologies."

"In spite of great social changes in Ireland which have led to constitutional change in favour of LGBTQI + people, born-digital systems were experienced by respondents as embodying socio-cultural values which were no longer accepted in society at large."

Introduction
"The case study was introduced in Ireland to explore the lived experiences and the perceptions of persons with technology in relation to their gender, where their gender identity and/or gender expression is outside of the sex/gender binary categories female/woman and male/man in Irish society."

"There has not been any research specifically focused on non-binary gender identities in Ireland and the experience with data collection systems and technology."

"The aims and objectives of this research are to examine the existing literature on gender and digital technology through the secondary research and to answer the following research questions through the primary research using a case study in Ireland: (RQ1) For people within the LGBTQI + community: a. What is the diversity of gender identities and gender expression?"

Literature review

"Categorising a persons' gender into the cisgender sex/gender binary female/woman or male/man (Feinberg [15]) in digital data systems, without considering the diversity of gender identities and gender expressions, marginalises anybody outside of these restrictive binary gender categories."

"To this, the criteria to make a machine, or programme a computer, to perform like a human, is normalised into the binary sex/gender identity categories of male/man and female/woman in society."

"Although machine learning is about the flexibility of the computer to reprogram itself without human intervention, the foundational algorithms on which AI technology is built utilises hardcoded data biases such as the gender binaries of male and female which are the object of this study."

"This research explores the experiences of digital data systems, social media and new smart technologies among gender diverse persons within the LGBTQI + community in Ireland."

Research design

"Gender Identity: the data related to the gender identity of the respondents was collected on the survey using a mandatory question containing a list of options where multiple selections could be made."

"The participants were recruited via the last survey question that gave the researchers contact details or an option to leave a number if they would like to participate in the focus group discussion."

"Gender Identity: within the focus group, the participants expressed their gender identity through the focus group discussion rather than being asked the question directly to avoid the assumption that everybody identifies with a gender."

"The participants in both the focus group and the questionnaire survey, were guaranteed confidentiality of the data collected and the option of remaining anonymous and all participants and respondents in this research were made aware of what the research is for and how the research will be utilized."

Findings and discussion

"To this, all respondents indicating a gender identity as a man or a woman and also transgender, agreed that online social spaces enabled them to express their gender, however, they answered "disagree" or "unsure" to the questions about online social spaces being inclusive of all genders and LGBTQI + people, you can be you in online social spaces and you feel safe in online social spaces."

"Most persons identifying with a non-binary gender identity do not agree that, or are unsure if, online social spaces are inclusive of all genders and LGBTQI + people."

"To this, persons identifying with a non-binary gender identity do not believe that online social spaces enable them to express their gender identity in comparison to persons identifying with binary genders including trans-binary."

"All participants in this research indicating a trans-binary gender identity, agreed that online social spaces enable them to express their gender."

Conclusion

"This paper contributes to the recent research (Hicks [16]) that digital data systems used for public services, are experienced by members of the LGBTQI + community as embodying restricted notions of a sex/gender binary."

"Through the experience of a transgender woman in this study, it is evident that there is cultural inclusion and understanding towards the LGBTQI + community from the staff, but not all of the time, and it is the digital data systems for which staff must use that are reinforcing the restrictive heteronormative gender binary assumptions long after legislation and society has moved on."

"The findings in this research also support Hick's [16] findings and the notion that today's arena of big data analytics, machine learning, automation and other artificial intelligence technologies are built on top of legacy systems with deep rooted mandatory heteronormative gender binary data structures where people identifying outside of the binary sex/gender categories of male/man or female/woman are rejected and treated as an anomaly in digital society."

5.2 Digital Identity

Machine generated keywords: identity, security, identification, infrastructure, construction, financial, customer, build, text, user, digital, management, twitter, actor, self

5.2.1 Fandom and Identity Construction: An Analysis of Thai Fans' Engagement with Twitter [483]

This is a machine-generated summary of:

Smutradontri, Pitchapa; Gadavanij, Savitri: Fandom and identity construction: an analysis of Thai fans' engagement with Twitter [483]

Published in: Humanities and Social Sciences Communications (2020)

Link to original: https://doi.org/10.1057/s41599-020-00653-1

If you want to cite the papers, please refer to the original.

For technical reasons we could not place the page where the original quote is coming from.

Abstract-Summary

"Fans are dedicated, and creating fan text (i.e., text production made by fans about their object of fandom) is one way of showing dedication and passion to the fan base."

"This article explores how Thai fans engage with fan text on popular social media platforms such as Twitter, and how fandom relates to identity construction among Thai fans who are online media users."

"This article discusses fans' shared lexicon called 'fan talk,' and how fans position themselves as relatives and friends of the source texts."

Introduction

"Influenced by de Certeau's work, the early scholars of fan studies also viewed fandom as a power-struggle against the media industries."

"Fandom studies tend to increasingly focus on fan culture as a space to explore fans' gender and sexual identities."

"Rooted in Stuart Hall's notions of encoding/decoding, fan studies see media audiences as members who engage in different degrees of semiotic productivity, in which they produce meanings and pleasures that pertain to their social situation (Fiske, [17])."

"Unlike the Western fan studies which focus on the power resistance of mainstream media by fans and the promotion of participatory culture, Yano (2004, as cited in Chin, [18]) proposed that East Asian fandom is tied to the concept of 'intimacy' between fans and their object of fandom, which "impels individuals to act in ways that go beyond the bounds of self to seek greater communion with the object of their adoration" (p. 44)."

Methods and analytical framework

"The unit of analysis is obtained from fan tweets, which is a kind of fan post on the social media platform Twitter."

"The size of the unit of analysis is 100 tweets (25 fan tweets per each source text)."

"According to Bengtsson [19], "there are no established criteria when using content analysis for the size of a unit of analysis, neither the number of informants or objects to study, nor the number of pages based on the informants' own written text or transcribed data" (p. 10)."

"To answer research question two, semiotic analysis is used as analytical framework in exploring identities in narrative texts such as fan tweets."

"As fans are always involved in creating, sending, receiving, and interpreting fan texts with multicultural elements, Berger [20] suggested that "what semiotics and semiology do is provide us with more refined and sophisticated ways of interpreting these messages-and of sending them."

Data analysis and results

"The data collected of fan tweets could be categorized into five types: hypothetical interpretation, fan arts, narrative of a personal anecdote regarding the source text, expression of personal opinions and feelings, and fan parody."

"Panwink #Linhoon #linhoon] This tweet started out by stating that he/she did not speak from the shipper's (fans who 'ship' a certain couple of the source text) point of view, or try to view it in romantic way."

"This type of fan tweets is statements or narrative of an anecdote regarding the source text."

"Most of the selected fan tweets (43 tweets out of 100 tweets) are expressions of emotions and personal opinions toward source text pictures, videos, or personal fan experiences."

"Unlike narratives of anecdotes, this type of tweet offers insight into the fans' emotion and feeling towards the source text."

Discussion

"Answering the first research question, the result shows that there are five ways Thai fans engaged with fan tweets: hypothetical interpretation, fan art, narrative of an anecdote regarding the source text, expression of personal opinions and feelings, and fan parody."

"In their creation of fan tweets, there are mainly two ways that Thai fans handled the differences of their own culture and the source text."

"Thai fans obtained some of the source text's cultural elements and integrated it in their fan tweets."

"Thai fans also integrated their own cultural elements with the source text in their fan texts."

"The result suggests that some fans included Thai cultures into their fan tweets."

"From the humorous nature of fan tweets, the expressive feature, and the space for identity play, it seems like one of the reasons for its popularity could be that fan tweets is an easier way that fans could integrate themselves with their source text than other types of fan texts."

Conclusion

"The first question asks 'How is the source text portrayed and rewritten in fan tweets?'"

"Research question one explores how Thai fans engage with (in other words, portray and rewrite) fan text in popular social media platforms such as Twitter."

"The second question asks 'how does fandom relate to identity construction among Thai fan online media users?'"

"This study has investigated fan tweets, which are not often mentioned in fan studies but are as popular as any other types of fan texts and how fans engage with them."

5.2.2 Discussing Human Values in Digital Immortality: Towards a Value-Oriented Perspective [484]

This is a machine-generated summary of:

Galvão, Vinícius Ferreira; Maciel, Cristiano; Pereira, Roberto; Gasparini, Isabela; Viterbo, José; Bicharra Garcia, Ana Cristina: Discussing human values in digital immortality: towards a value-oriented perspective [484]

Published in: Journal of the Brazilian Computer Society (2021)

Link to original: https://doi.org/10.1186/s13173-021-00121-x

If you want to cite the papers, please refer to the original.

For technical reasons we could not place the page where the original quote is coming from.

Abstract-Summary

"These traces are the digital legacy individuals build all through their lives."

"Advances in artificial intelligence have fed our dream to build artificial agents trained with these digital traces to behave similarly to a deceased person, and individuals are facing the possibility of immortalizing their ideas, reasoning and behavior."

"How do people perceive the possibility of letting digital avatars take care of their digital legacy?"

"These findings can serve as preliminary thoughts to inform system design, from the very early stage of development, that preserve the digital legacy while respecting the human needs and values concerning the delicate emotional moment that death provides."

Introduction
"The possibility of interacting with artificial profiles created from the deceased data raises natural issues and concerns through achieving digital immortality."

"Among the questions raised in the episode is the issue: "What human values related to technology are involved in digital immortality?""

"There has been a growing interest in the possibilities of creating digital immortality, either discussing the technological aspects [21–23], the philosophical issues [24], the legal implications [25, 26], the economic impact [27], the psychological changes on people [28], and the physical aspects of it [29]."

"Investigation, we conducted a series of focus group interviews to obtain the opinions of volunteers regarding technologies that promise digital immortality through digital memorials, software or hardware."

"According to the volunteers' opinions and the discussions emerged during the interviews, this paper presents an analysis based on the human values mapped on social software by Pereira and Baranauskas [30] responding to the focus of this study and offering a starting point for a value-oriented reflection on digital immortality."

Digital immortality
"In the present, companies like Facebook and Google offer options to users to destine their legacies, after confirmation of their deaths, to an heir or even complete transformation of their digital accounts into digital memorials [31]."

"Any information left by people on the Web is part of their digital legacy immortalized on the Internet."

"Given the possibility of immortalizing a person and allowing posthumous interactions with their data, system developers are faced with complex issues, both philosophical, technical, and social, with those predicted by Galvão and others [32]: When manipulating a person's data, one can be changing the "digital self," that is, his digital identity."

"As a continuation of the studies by Galvão and others [33], there are some complex issues concerning digital immortality: the differences when dealing with famous deceased versus non-famous deceased individuals, explicitly defined public data versus private data, rights and duties related to data preservation versus deletion, technological advances versus cultural traditions, and the reasons for desiring and justifying the use of digital immortalization."

Design values and value-sensitive design
"While digital immortality is not widely used or formally recognized as a social software, the values and approaches taken by the previously mentioned authors are considered in this study in order to understand what are users' key concerns about its usage, how it may be presented, how it operates, etc Having a better grasp of what users find interesting or problematic about digital immortality is

an important step in order to guide future technologies in grasping its limitations for immortality, what users expect, and what needs to be carefully approached before, during and after development of solutions that deal with such a sensitive and mostly unknown topic."

Existential HCI

"Associated with issues such as design values and digital immortality, another important factor is the existential nature of these tools."

"Kaptelinin [34] argues that according to the evolution of technology, HCI must also cover issues of existential nature (what the author calls the existential HCI)."

"While the impacts of AI technology for digital immortality are still largely unknown, through the existential IHC, Kaptelinin [34] recommends the prioritization of the following aspects of study: 1 The focus on existence in general, instead of pragmatic tasks; 2 The search to understand how people confront their own mortality; 3 The need for space for existential decisions (culture, creation, religion, personality and relationships and among other personal factors of an individual affect his or Her decisions during life); 4 The unpredictability of existential experiences."

"According to the author, by taking these factors into account, it is possible to recognize and understand what intimately human experiences can be treated appropriately through technology."

Research methodology

"The meeting consisted of the following activities: 1 Reception of volunteer participants: when they arrived at the place on the scheduled day and time, they were welcomed and each member received a pen, paper for notes, and the consent form to read and sign 2 A brief introduction to the research topic; and 3 Beginning of the focus group with a question aimed at identifying which members have already had the experience of dealing with death in the digital world and whether they had knowledge about ways or tools for treating it."

Findings

"From participants' responses regarding digital immortality, we can see informal and formal values were a priority, permeating their discussions."

"Discussions revealed participants perceive digital immortality as a way to keep loved ones virtually present for future interaction, a way to deal with grieving pain, to keep the memory of deceased loved ones alive, and also a new way to learn from the data of deceased people interactively."

"All participants agreed that, in order to use digital immortality systems, it is necessary the system to be flexible, capable of changing over time, respect the memory of deceased ones, allow interactions between the virtualized dead and the living, and especially that it aligns and respect for human values, both of its virtualized entities and its users."

Final considerations

"While it is not yet possible to understand all the impacts that digital immortalization will bring, this study allowed us to observe how important systems that deal with sensitive subjects, such as death, must consider human values in their development."

"The participants' opinions indicate that digital immortality systems should be prepared for future possibilities of interaction that respect the following characteristics: The object: the form from which the digital immortality system will take place must allow interactions that respect the relationship of users with their loved ones; The expression of values, such as emotion and affection: the feelings that users may be dealing with related to death; and The identity: since both, users and the deceased, share memories of their relationship."

"For digital immortality to reach its full potential, in a way that respects human sensitivities at the time of death, the discussion should go beyond the human-computer interaction issues abridging other areas of knowledge to consider the needs of users, family and loved ones."

5.2.3 The Identity Challenge in Finance: From Analogue Identity to Digitized Identification to Digital KYC Utilities [485]

This is a machine-generated summary of:

Arner, Douglas W.; Zetzsche, Dirk A.; Buckley, Ross P.; Barberis, Janos N.: The Identity Challenge in Finance: From Analogue Identity to Digitized Identification to Digital KYC Utilities [485]

Published in: European Business Organization Law Review (2019)
Link to original: https://doi.org/10.1007/s40804-019-00135-1

If you want to cite the papers, please refer to the original.

For technical reasons we could not place the page where the original quote is coming from.

Abstract-Summary

"Identity is fundamental in finance: to protect against fraud and crime, to fulfil know-your-customer obligations, and to ensure market integrity."

"Identification and know-your-customer rules can be major barriers to accessing financial services."

"This paper considers the various requirements for identification in the financial sector and the evolving nature of identity."

Introduction

"Whilst any and all may be used for authentication purposes (e.g. accessing social media or a banking account), they are not universally compatible nor acceptable for all purposes."

"While this number is lower for e-commerce platforms which process payments (false accounts come into existence once a payment is accepted enabling goods to be sent to whomever) e-commerce providers are usually unperturbed by unverifiable customer identification."

"While the acceptability of varying forms of online identification and identities can be solved by changing regulatory requirements, the cross-compatibility (legally or technically) of someone's online identity raises the issue of the necessity to develop a new regulatory framework or technical utility design."

"At a time when TechFins are making inroads into the financial services industry, a tension emerges as to which is the most effective method of customer identification."

The Identity Challenge in Finance

"The financial services sector supports economic growth and development through allocating financial resources, providing investment opportunities, and managing risks."

"Financial regulation seeks to promote these functions through minimizing the frequency and severity of financial shocks (i.e. financial stability), enhancing access to financial services (i.e. financial inclusion), and promoting market integrity."

"For an international financial centre such as Hong Kong, Luxembourg, London, or New York, competitiveness derives from balancing these objectives and providing the necessary infrastructure for financial markets to function well."

"CDD underpins how customer needs are understood, which is essential to providing appropriate financial services."

"These requirements restrict access to financial services, in particular for small and medium-sized enterprises (SMEs)."

"Identification and identity pose particular challenges in the context of finance because of the sometimes conflicting regulatory framework objectives of transparency, privacy, financial integrity, revenue collection, financial inclusion, economic growth, and financial stability."

The Evolution of Identification and Identity: Analogue to Digitalized to Digital

"A number of principles have been nominated which could guide the treatment of dynamic digital identities in light of these risks, such as the user control principle (i.e. that individuals can choose whether, and to what extent, to disclose their information), the transparency principle (i.e. that data processors must justify why they are collecting data), and the data minimisation principle (i.e. that only the minimum data required for a certain purpose be collected)."

"Until the trade-off between data provided and services rendered is made clear, and data protection frameworks are actively enforced, or information technology solutions to privacy concerns become globally implemented, the acceptance of digital identity as a standard for financial services will be delayed."

"They provide that individuals should have control over their digital identities and personal data, and argue that new digital technologies allow for both individual privacy protection and government access to consumer identification."

Solving the Identity Challenge: Designing Infrastructure for Digital Identification

"Rather than introducing a pan-European ID card system, which would double the work for Member States, the eIDASR seeks to ensure that people and businesses can use their own national eIDs to access public services in other EU countries where eIDs are available."

"By mandating liability on Member States as well as the eID provider for meeting certain identification obligations (i.e. including that the person identification data uniquely represents the individual and online authentication is available), the eIDASR creates trust in the eIDASR-based cross-border identification."

"This would not only go a long way in bringing in the 1.7 billion unbanked individuals around the world into the financial system, it would also build a Base ID system over time that is highly efficient and effective—and even more so if the same and additional data are linked to the Base ID authority in other sectors (e.g. payments to and from government, health registration cards etc)."

From Digital Identification to Digital Identity and KYC Hubs

"Base ID, particularly when linked electronically with other golden source data (e.g. tax and address information), provides the basis for a simple eKYC system which can dramatically reduce account opening costs while addressing concerns regarding market integrity (e.g. AML concerns)."

"If only the largest institutions underwrite the KYC utility, the argument for joint liability lies in incentivizing all members to invest in the maintenance and development of the utility."

"How to address these risks requires careful thinking that takes into account legal factors (e.g. property rights, liability, competition and antitrust concerns, and data privacy rules) together with non-legal factors (e.g. the technology used—with blockchain a natural candidate), cybersecurity risks, and the need to build a networked infrastructure to which hundreds, if not thousands, of entities can be linked."

Conclusion: Addressing the Identity Problem in Finance

"In addressing the identity problem in finance, this paper identified and considered two aspects which must be addressed strategically: digital identification infrastructure and eKYC infrastructure."

"100% digital identification is neither feasible nor likely in the short term and aiming at 100% coverage will either increase the risk of disruption or delay synergies from sector-wide eKYC systems for the foreseeable future."

"A sector-wide eKYC solution could first aim at digital identification of licensed domestic financial intermediaries, then include locally incorporated companies, then be utilized for non-face-to-face onboarding of individuals and finally address the challenges posed by foreign institutions."

"To focus on only identification, and ignore sector-specific needs and use cases, misses many of the potential opportunities an eKYC system could provide."

"Only when this is achieved will financial intermediaries benefit from the full potential of a sector-wide eKYC system."

5.2.4 Digital Identity Infrastructures: A Critical Approach of Self-sovereign Identity [486]

This is a machine-generated summary of:

Giannopoulou, Alexandra: Digital Identity Infrastructures: a Critical Approach of Self-Sovereign Identity [486]

Published in: Digital Society (2023)

Link to original: https://doi.org/10.1007/s44206-023-00049-z

If you want to cite the papers, please refer to the original.

For technical reasons we could not place the page where the original quote is coming from.

Abstract-Summary

"This paper sets out to uncover the principles, technological design ideas, and underlying guiding ideologies that are attached to self-sovereign identity infrastructures, carrying the promise of user-centricity, self-sovereignty, and individual empowerment."

"Considering the flourishing of digital identity markets, and the subsequent institutional interest on a European level in the techno-social promises that this identity architecture carries, this paper explores how the implementation of EU-wide self-sovereign identity shifts the already existing historical power balances in the construction of identity infrastructures."

"We argue that the European-wide adoption of self-sovereign ideals in identity construction does not address the shortcomings that identity and identification have historically faced and that instead of citizen empowerment, it puts individuals (a category broader than citizens) in a rather vulnerabilized position."

Introduction

"Control over digital identity and its ensuing infrastructures is key."

"We define digital identity infrastructures, as systems that construct, control, and commodify (facets of) user identities."

"8–9), digital identity corresponds to systems of identification of individuals, as well as to systems of authentication that modulate access rights and authorize the performance of pre-specified actions or predetermined access to services."

"This paper uses the above understanding of digital identity when discussing how it is implemented through self-sovereign technological infrastructures and corresponding ideologies."

"Capturing digital identity as an information transaction, the implementation of self-sovereign identity involves employing appropriate technological tools that attempt to maintain privacy, data protection, and security of the identification or information transfer process as these concepts are understood by the self-sovereign identity enthusiasts."

"Their implementation in self-sovereign identity schemes aims to transpose these features by treating digital (self-sovereign) identity as a set of identification credential transactions that can be described as an architectural problem."

Understanding Self-Sovereign Identity

"There are many promises attached to the evolution of identification online towards self-sovereign identity, and the potential it presents: "the SSI paradigm shift is also deeper than just a technology shift—it is a shift in the underlying infrastructure and power dynamics of the Internet itself" (Preukschat & Reed [35])."

"From a technological perspective, self-sovereign identity constitutes a technological architecture built to by design avoid the risks inherent in the current model of digital identification (Hoepman [36])."

"If these technological elements related to the creation of digital identities exist independently of self-sovereign identity, it is the rise of blockchain that has, it seems, succeeded in creating a revival for the latter."

"Against this backdrop, self-sovereign identity is appearing as a simple identity technological artifact, a digital solution aspiring both to formalize the individualization of access to computer networks and to digitally recreate the relationships that (in)form individual identity."

(Digital) Identity in Context

"In both the physical and digital realms, we construct our identity through the selective self-disclosure of our traces and markers."

"The digital revelation of these different identities is assessed by both actors in the process based on the context, the necessary minimum level of trust, and the individual control over the revealed information."

"On a policy level, digital identity has been defined as "a collection of electronically captured and stored identity attributes that uniquely describe a person within a given context and are used for electronic transactions" (World Bank Group, GSMA and Secure Identity Alliance [37])."

"Khatchatourov describes the double essence of the concept: "Digital identity can therefore have two complementary meanings, which precisely constitute the crux of the problematic of this domain: identification of the user and their actions in the digital environment and the effects of digital technology on the construction of identity understood as a relationship to oneself, to others and the public space" ([38], p.24)."

Shaping Public Digital Identity Infrastructures

"Online platforms, services, and digital technologies, which have the capacity to authenticate its users, can also use this function to collect, analyze the digital traces their users leave on their services and use that to build categories, and assign identities to those users."

"In France, since the first attempts to systematize digital identity for e-administration purposes, the concern over ensuring the accountability of the government materialized through the creation of the French Data Protection Authority (CNIL)."

"This process can be best understood as an ecosystem, one which is co-determined by the identity information at hand, the actors involved, the technologies and governance or design architectures which form the end digital artifact qualified as digital identity."

"The establishment of a (digital) identity system is certainly a state responsibility, as part of its sovereign power."

The Adoption of Self-Sovereign Ideals in Digital Identity Policy-Making

"Among the solutions for which blockchain pilot projects have been launched, self-sovereign identity technological principles are introduced both in the revision of the European identity regulatory framework and as a practical architectural design for the creation of European-wide digital identity infrastructures."

"The project called European self-sovereign identity framework (eSSIF) is being developed to constitute a "generic and interoperable self-sovereign identity framework (SSI), defining the necessary specifications and building the services and supporting capacities which will allow citizens to create, control and use their own digital identity (including identification, authentication and many other types of identity-related information) without having to depend on a single centralized authority"."

"If we take the example of the pan-European infrastructure providing a blockchain service (EBSI), its success will be based on the trust that citizens have in the providers of the services and on the latter's ability to engage the responsibility of the European institutions responsible for this infrastructure."

Conclusion

"The emphasis put in decentralizing the technical system of digital identity production, expression, and validation is but only one part of the process of claiming back control over our understanding of selective self-revelations vis-à-vis the state or private actors."

"The principles that guide self-sovereign identity, especially the user-centricity and individual empowerment run the risk of creating increased accountability of individuals in control of their information and a correlative disempowerment of other powerful actors involved in the identification processes, including public ones."

"Exercising its sovereign power in (legal) identity creation, validation, and authentication, the state is also engaging in a rather complex network of identification relationships under any of the capacities: in the performance of e-government services, in the cross-border digital identification of its citizens, and in the collaboration between the public and the private sector for infrastructural support and creation of digital identity management systems."

5.2.5 A Model of Digital Identity for Better Information Security in E-Learning Systems [487]

This is a machine-generated summary of:

Korać, Dragan; Damjanović, Boris; Simić, Dejan: A model of digital identity for better information security in e-learning systems [487]

Published in: The Journal of Supercomputing (2021)

Link to original: https://doi.org/10.1007/s11227-021-03981-4

If you want to cite the papers, please refer to the original.

For technical reasons we could not place the page where the original quote is coming from.

Abstract-Summary

"The trend of rapid development of information technologies is creating new challenges in information security such as security management of identities in e-learning systems."

"In order to cope with these challenges, this work has developed a model of digital identity (MDI) intended primarily for e-learning systems."

"Two main objectives are identified for development of the MDI, firstly, to contribute to raising of information security awareness and improving information security behavior, and secondly to enable the bridging of security gaps in the interaction between student/administrator."

"Practical application of this model is made in Moodle platform through the developed software application in the form of the security tool as a plugin for filtering inactive users (IUs)."

"This work should have a profound positive impact on further academic research and additionally it should contribute to a better understanding of challenges in information security that emerge with the development of the contemporary e-learning systems."

Introduction

"This platform offers a wide variety of opportunities with regard to storing a lot of sensitive information, like grades, tests, and students' private information, etc As such, Moodle satisfies many user priorities, like security, usability, accessibility, pricing, complexity, privacy, convenience (SUAPCPC)."

"Mobile technologies significantly improve student's way of learning but contribute to an increased exposure of personal information, which can lead to identity theft."

"Moodle LMS as an extremely large system contains many other problems of the loop holes, such as computing and accessing student grades and other personal information, creating and managing activities."

"An additional aim of this work is to obtain a clearer picture of DI concept: Does the model provide a lens through which student can gain deeper understanding of how to create a better security management of identities in e-learning systems?"

Related work

"Their research is focused on user self-awareness problems as a very important factor in creation of multiple identities in a system."

"Besides security challenges, there are a lot of works such as [39–44], in which identity issues related to others user factors such as privacy, convenience, or pricing, are only briefly touched upon as a relevant subproblem rather than the main topic."

"There are a lot of works such as [45–48], whose primary focus includes the most important security issues in e-learning platforms, such as the Moodle platform."

"Besides the security of personal information, Manzo 2017 [49] focused on the problems in the process of communication during the use of Moodle platform."

"That in all the previous works, research programs are unanimous in emphasizing the importance and role of information security issue in educational researches, where each one focuses on specific issues or looks at the problem from a particular perspective."

Proposed DI model

"The specifics of this model are the processes of planning and feedback."

"The strength of feedback processes is that they connect all processes, while the planning process pervades through all processes in DI lifecycle."

"Because of importance of planning and feedback processes in the MDI, they are described in more details below."

"The planning process involves a clear user decision of how DI needs to be created, maintained and used, and when it will be deleted."

"The planning process needs to motivate users to apply positive security behavior that require users to pay special attention and responsibility towards DI through its whole lifecycle, especially in the frontier points."

"The planning process is a place where user's need for a DI is recognized but, more importantly for this process, it is a place which recognizes user's need to revoke DI."

Workflow for DI diagram

"The sharing process is completed and DI is ready to use."

"If the process of the DI use is not based on one but multiple systems, then a SSO subprocess is required."

"In that case, the DI lifecycle is the shortest and does not require application of the updating process."

"The first frontier point A indicates the place in the DI lifecycle, where the users make an omission by not performing DI revocation."

"The DI activation is inevitably followed by its redistribution to the user/entity and repository."

"If there is no need for the DI, a revocation process follows."

"If there is a need for reuse of the same identity or choosing another DI for the same purpose, it is necessary to repeat the whole previous cycle."

"With this process user/entity is informed about DI revocation."

Development of the security tool as plugin for filtering IUs and criteria for filtering

"When it comes IUs, it is necessary to highlight that the misuse of IUs presents a serious threat to a system because an inactive identity is neither under control of a student nor an administrator."

"Without credentials, it is not possible to establish the connection between student/system (administrator)."

"Process, the software application in the form of a security tool as a plugin for filtering IUs is developed."

"Feedback process is used to refer to a situation in which a student and an administrator have a real mutual connection with each other."

"After determination of the key criteria, an administrator makes the choice of criteria for IUs filtering."

"The administrator has a choice to perform the process of criteria filtering individual/any or multiple/all."

"The security tool gives administrators the ability to filter IUs according to time."

Security tool implementation in Moodle LMS

"After the setup of the security tool and its implementation in Moodle LMS, the system for searching and selecting IUs is implemented in two countries (Bosnia and Herzegovina—BiH, and Serbia)."

"Testing, two types of filtering are applied: (1) process of filtering by the courses for subjects such as Informatics as a joint course, Internet Programming and e-banking as specific courses, and (2) process of filtering by the time (period of inactivity) selected by number of weeks of 500, 50 and 12."

"From the obtained results, it is evident that the technique of searching and selecting of the IUs presents a precise security tool for better security management in e-learning environments."

"This real-world scenario has validated that this work has not only made contributions to the theoretical study for better preservation of consistency of DIs in e-learning environments, but it is also beneficial to practical areas, such as a security engineering during searching and selecting of the IUs according to different filtering criteria."

Discussion

"Because of that, it is important to highlight that every student in the system has to reach the frontier point A. In that way, the student will have much more control over his personal information resulting in the raising of information security awareness and improving information security behavior as well."

"In that way, it gives an explanation of how a student can create more identities in one system and how it is possible to make potential IUs i.e., isolated "pockets.""

"These frontier points can be understood as "landmarks" for students, which will be enabled by themselves to raise information security awareness and to improve information security behavior in e-learning systems."

"To that, students need to be more careful in the context of the situational awareness, i.e., their timely perception, understanding and reaction to potential threats and risks of identity theft."

Conclusions and future work

"In that regard, this paper has given an answer to an underlying aim of this research: the MDI provides raising information security awareness and improving information security behavior for all participants, and it enables the bridging of security gaps in interaction between student/administrator."

"This work has given the answer to an additional aim of this research: through implementation of DI lifecycle, this model provides a new lens through which a student can gain deeper understanding of how to create better security management of identities in e-learning systems."

"As a direction for future research, one could highlight the development of a new security tool in m-learning environments, with which it will be possible to give an adequate answer to challenges in information security that follow the development of the contemporary and future e-learning systems."

5.2.6 The Digital Self and Customer Loyalty: From Theory to Virtual Reality [488]

This is a machine-generated summary of:

Bischoff, Jared; Berezan, Orie; Scardicchio, Lorenzo: The digital self and customer loyalty: from theory to virtual reality [488]

Published in: Journal of Marketing Analytics (2019)

Link to original: https://doi.org/10.1057/s41270-019-00065-4

If you want to cite the papers, please refer to the original.

For technical reasons we could not place the page where the original quote is coming from.

Abstract-Summary

"This work aims to bring Belk's (J Consum Res 15(2):139–168 [50]) seminal theory of self-development into present day by connecting its concepts with those of customer loyalty formation, highlighting how a modernization of this theory explains the means by which present-day customers are forming attachments to digital goods and the brands and companies associated with them."

"Through analysis of both historical and current literature regarding self-development, self-expression, and object attachment, we explore how consumers form connections to digital possessions, and how this process has evolved alongside the growing technological integration of daily life."

"Our intent is to expand upon existing loyalty literature by explaining how attachment is formed with digital objects, and highlight both the challenges and opportunities this creates in loyalty marketing analytics."

The self

"In order for a tangible to be part of the extended self, it is necessary to build a relationship with it through passion, authority, or personalization (Belk [50]). Passion is a catalyst that represents the willingness and dedication one puts forth to build a relationship with a tangible."

"The more authority the individual attains over the object, the more of the individual's self is invested within it, denoting that it is a growing extension of self (Belk [50]). Personalization involves building or adding to an item as a form of expressing one's identity through it."

"Sharing or giving an item that represents the extended self is a form of giving part of the self to another person (Belk [50])."

"Belk himself addresses this by pointing out the similarities between physical objects and digital content in terms of extended-self formulation, as both still contain passion, authority, and personalization [51]."

Creating and expressing the self

"Autonomy, in particular, is a need more easily satisfied in the digital sphere (rather than simply via social media) as the level of personalization and customization online is significant—from user-created content to the virtual avatars mentioned in the previous section."

"In today's digital era, however, users have much more control over their perceived self-image and presentation through online social networks (Boyd [52])."

"Social media has transformed this linear identity presentation into a publicized expression of self through the use of profile homepages."

"Online social networks give users the ability to reflect on their content prior to posting, managing their impressions through asynchronous online interactions—a level of control over users' self-presentation that previously did not exist in face-to-face interactions."

"The definition of self is evolving with the widespread growth of social media, as more forms of expression and creativity results in new ways of manifesting identities via different social media platforms."

Digitizing the self

"By placing these expressions of self in the digital sphere, individuals are now capable of actively managing their impressions by performing consistent and repeated public behaviors on social media to communicate their ideal identity (Spero and Stone [53])."

"Adolescents, who are prevalent online social media users, are viewed as the generation that commenced the reshaping of self-identity, self-presentation, and impression management in the twenty-first century (Doster [54])."

"The advantage of social media technology in this context is that it allows users to associate themselves with digital images, products, brands, idolized influencers, and other resources as means of self-expression (Boyd [52])."

"This generation fulfills its SDT need for relatedness by using social media to help shape their self-identities (and impact their sense of well-being) through behaviors such as posting, liking, and sharing (Raab and others [55]; Berezan and others [56])."

The digital self in marketing

"Because the average consumer can now reach a potentially large number of other users via social media, businesses have the opportunity to create marketing allies simply by allowing loyal customers to share their experiences with others."

"Attention should be paid to the relationship between the consumer's digital self and the product or business in question—evaluating not just how much they might share positive eWOM with their own social networks, but other means by which consumers display their attachments to these digital possessions."

"eWOM is likely to remain one metric, but relies upon consumers being 'outgoing' enough to share their experiences in such open form; some users may display their appreciation for a product, brand, or business in ways that are not so easily recognizable or categorized."

Conclusion

"By examining the ways in which self-formation occurs in the digital sphere, we aim to shed new light on how consumers form relationships with brands, products, and businesses."

"These metrics can then be utilized by businesses to accurately determine the effectiveness of new marketing campaigns, initiatives, and strategies that occur in the digital sphere."

"The addition of new metrics in marketing analytics research would provide a more accurate means of measuring the nuances of the increasingly digitized consumer self, their object attachment to digital possessions, and how they interact and communicate with other consumers and businesses in the digital sphere."

[Section 6]

"From here, the effects of digital and social media technologies on the process of self-development and self-expression are analyzed and discussed, with particular attention paid to the ways in which these traditional methods have changed as a result of the growing influence of the digital sphere."

"We offer suggestions on revisions to existing marketing analytics to account for the changes discussed prior, in order to better account for the effects of digital self-formation on inspiring positive eWOM and customer loyalty."

"Review and synthesis of relevant literature on the self and social media, we aim to illustrate how self-development and expression have changed from the pre-digital era, how an understanding of these changes can be put into practice in the field of loyalty formation, and the future research avenues that can be explored in order to produce metrics that can account for these changes and their effects on eWOM and loyalty formation."

5.2.7 Public Sector Engagement with Online Identity Management [489]

This is a machine-generated summary of:

Barnard-Wills, D.; Ashenden, D.: Public sector engagement with online identity management [489]

Published in: Identity in the Information Society (2010)

Link to original: https://doi.org/10.1007/s12394-010-0079-2

If you want to cite the papers, please refer to the original.

For technical reasons we could not place the page where the original quote is coming from.

Abstract-Summary

"The individual management of online identity, as part of a wider politics of personal information, privacy, and dataveillance, is an area where public policy is developing and where the public sector attempts to intervene."

"This paper attempts to understand the strategies and methods through which the UK government and public sector is engaging in online identity management."

"Communication theory suggested that the assumption of individual responsibility, alongside linguistic distortions created by this way of understanding the problematic of identity management, complicate and fundamentally limit engagement activity."

Introduction

"This paper attempts to understand how the UK government and public sector is engaging in the field of online identity management and the form this engagement takes."

"This theoretical approach to the study of government provides a number of perspectives that cast light upon public sector engagement efforts in general, and upon identity management in particular; particularly that government is best conceived of as an assemblage of actors, oriented towards shaping conduct of populations."

"It is hoped that a research strategy including the multiplication of narratives of personal privacy, including online identity management, and presentation of those narratives through multiple vectors might produce alternative ways of conceptualising online identity management that could be incorporated into public sector engagement efforts in a way that satisfies the three demands of verisimilitude, public sector needs, and accessible and effective forms of communication for the diverse variety of the population."

Online identity management

"'Online identity management' is understood as any attempt to consciously manage or control the "growing mass of information about ourselves and our social or business transactions and relationships that exists in digital form whether stored within commercial or government databases or scattered on the Web in blogs and social networking sites"(Evans-Pughe [57])."

"Identity management differs from concepts of 'impression management' in that it deals with the security of personal information."

"Online identity management involves not just the 'presentation of self' as discussed in the Goffman-inspired sociological literature (Ellison and others [58]) but also a (potential) response to a range of threats arising from the proliferation of personal information online."

"The Information Assurance Advisory Council defines identity assurance as: "A framework of technical, management, policy and regulatory initiatives aimed at preserving the confidentiality, integrity and privacy of identity related data, as well as the availability of information infrastructures and supporting identity management systems." (Information Assurance Advisory Council [59]). '''

"Online Identity Management' focuses upon the individual level."

Governmentality

"Engagement efforts and strategies around online identity management are therefore conceptualised as a particular strategy of government intervention, and a way to influence the conduct of governed populations."

"In demonstrating the need to protect themselves from online identity fraud or exploitation, this governmental guidance becomes one of a number of: "Instances of contriving practices of liberty in which the responsibilities for risk minimisation become a feature of the choices that are made by individuals, households and communities as the consumers, clients and users of services." (Dean [60]: 214) This suggests attention to the way that OIM engagement constructs and positions the individual as primarily responsible for managing their online risk."

"Whist it provides a strong basis for analysing the intentions and reasons for the engagements efforts on the part of the public sector, the analytics of government theory cannot, by itself comment on the effectiveness of these efforts, or the extent to which the guidance – the strategies and tactics of identity management – becomes taken up by various groups in society and effectively made part of their conduct."

Public sector engagement efforts

"ICO's education and influencing communication objectives related to identity management are: to 'maintain awareness of rights among individuals, encourage good practice, and maintain 'confidence in organisations' handling of information (among individuals)'(2009:4)."

"The high profile of recent data losses has drawn attention to the importance of safeguarding people's personal information, but it has also damaged public confidence that personal information is well protected." (Ibid:19) A report for the EU Commission examined the methods used by national data protection supervisory authorities across Europe (including ICO), to promote personal data protection (European Commission [61])."

"It does not provide any way of moving beyond this, of understanding what such a control would resemble, what individuals want to do with their personal information, or what processes and practices of identity management they would feel capable of, or even of ways in which such information could be better communicated."

Problems of engagement

"We can see from this example how important it is that we understand how individuals make sense of what they are being told about identity management."

"Applying this to online identity management, we can see a drive to present active citizens with a range of strategies and techniques to manage their 'identity risk' as part of their daily conduct, whilst the national identity card has been discussed in terms that suggest that it plays an interventionist role in identity management – protecting those parts of the population that are unable to protect themselves, for lack of resources or technological skills (Byrne [62])."

"With identity management these centres of authority are the broad assemblages of government, and the architecture of online identity management is predominantly presented as a technological fait accomplis and engagement with these models is only legitimation after the fact."

Moving forward with engagement

"Our approach towards engagement uses a theoretical framework derived from the discourse theory work of Glynos and Howarth, which emphases the importance of starting with contextualised self-interpretations of subjects involved in a discursive space (Glynos and Howarth [63])."

"Glynos and Howarth set out a discursive research strategy and theoretical approach based upon post-structural presuppositions and ontological framework, a development of Laclau and Mouffe's work in political discourse analysis (Laclau and Mouffe [64])."

"Taking inspiration from hermeneutic approaches, all meaningful social science explanations must pass through the contextualised self-interpretations of subjects."

"This contingency is highlighted by any moment where a subject's mode of being (which is generally habitual, and non-reflexive) is disrupted; when identities and existing discourses fail to make sense of the social space."

"Contingency occurs when identity practices and their supporting discourses fail to work, or to make sense of the world."

Conclusions
"This paper has examined contemporary UK public sector engagement efforts around the complex socio-technological issue of online identity management."

"Drawing upon governmentality analytics suggested attention to the plurality of actors with shared regimes of practice and fields of visibility, as well as to the extent to which individual actors are made responsible for their own identity management."

"Communication theory suggested that this assumption, alongside linguistic distortions created by this way of understanding the problematic of identity management, complicate and fundamentally limit engagement activity."

"Whist much of this is a practical issue, we have set out the first principles of an appropriate engagement strategy for online identity management; understanding (so as to respond better to and incorporate) existing public understandings of the space through a inductive qualitative research strategy."

5.2.8 Schumpeterian Entrepreneurial Digital Identity and Funding from Venture Capital Firms [490]

This is a machine-generated summary of:

Block, Jörn H.; Fisch, Christian; Diegel, Walter: Schumpeterian entrepreneurial digital identity and funding from venture capital firms [490]

Published in: The Journal of Technology Transfer (2022)

Link to original: https://doi.org/10.1007/s10961-022-09973-7

If you want to cite the papers, please refer to the original.

For technical reasons we could not place the page where the original quote is coming from.

Abstract-Summary
"From the perspective of digital identity, we examine the relationship between a Schumpeterian digital identity and venture capital (VC) funding."

"Because the VC industry celebrates innovative and visionary entrepreneurship, we posit that a founder's digital identity as a Schumpeterian-type entrepreneur influences the venture's chances of receiving VC funding."

"While some dimensions of Schumpeterian entrepreneurship have a positive relationship with the acquisition of resources from VC firms (entrepreneurial vision and optimism), other dimensions seem to have no (uncertainty tolerance and rationality) or even a decreasing (achievement motivation) effect."

"Our study contributes to research on Schumpeterian entrepreneurship, the financing of technology transfer, and the link between entrepreneurial digital identity and entrepreneurial finance."

"From a practical perspective, the results of our study demonstrate the limits of VC with regard to the financing of technology transfer and highlight the need for public funding through governmental VC or agencies for (disruptive) innovation."

Introduction

"Combining a digital identity perspective (e.g., Fisch & Block [65]; Fischer & Reuber [66]; Smith and others [67]) with Schumpeter's theory of entrepreneurship (Schumpeter [68]), our study investigates the relationship between founders' social media identities and resource acquisition from VC firms."

"While some dimensions of a Schumpeterian entrepreneurial digital identity increase the chances of resource acquisition (e.g., entrepreneurial vision and optimism) from VC firms, other dimensions seem to have no (e.g., uncertainty tolerance or rationality) or even decreasing effects (e.g., achievement motive)."

"Our study also contributes to the literature on the entrepreneur-related funding criteria of VC firms, which represents an established research stream in entrepreneurial finance (e.g., Block and others [69–71]; Franke and others [72])."

"For projects in which an entrepreneur shows a strong achievement motivation (e.g., in academic entrepreneurship, Hossinger and others, [73]), funding through governmental VC and/or agencies for (disruptive) innovation could be more suitable."

Schumpeterian entrepreneurship and entrepreneurial (digital) identity

"By exploiting these opportunities, entrepreneurs bring innovation into the market and destroy existing market equilibria (e.g., Audretsch & Link, [74]; Hébert & Link, [75, 76])."

"This literature argues and empirically documents that the benefits of entrepreneurship, such as economic development and job growth, are not generated by all types of new ventures but only by a small number of young and innovative high-growth ventures (e.g., Aldrich & Ruef, [77]; Henrekson & Johansson, [78]), many of which are backed by VC (e.g., Kortum & Lerner, [79]; Lerner & Nanda, [80])."

"An important source of financing for Schumpeterian entrepreneurs in the science or research stage is governmental money, which may come in the form of governmental venture capital (e.g., Audretsch and others, [81]; Colombo and others, [82]; Masiak and others, [83]; Minola and others, [84]) or funding from innovation agencies such as the Defense Advanced Research Projects Agency (DARPA) (Bonvillian & Van Atta, [85]; Colatat, [86]; DARPA, [87])."

"Exploring entrepreneurs' digital identities has gained momentum in recent entrepreneurship research because an entrepreneur's digital presence is increasingly important in the management of networks and business success (e.g., Fischer & Reuber, [66]; Nambisan, [88]; Smith and others, [67])."

Hypotheses about schumpeterian entrepreneurial digital identity and VC funding

"Such entrepreneurs are rare, and VC firms look for such rare cases of visionary entrepreneurship (e.g., Galbraith and others, [89]; Sudek, [90])."

"VC firms know from their experience with past investments that successful innovation is rare and that many entrepreneurs fail to achieve entrepreneurial success (Tian & Wang, [91])."

"We posit that VC firms will try to avoid investments in such self-centered entrepreneurs and propose the following hypothesis: An entrepreneurial digital identity that displays a strong degree of self-centrism is negatively associated with VC funding."

"These arguments, we posit that VC firms are skeptical about entrepreneurs who display a strong degree of power motivation."

"The following hypothesis should apply: An entrepreneurial digital identity that displays a strong degree of power motivation is negatively associated with VC funding."

"Arguments, we posit the following: An entrepreneurial digital identity that displays a strong achievement motivation is negatively associated with VC funding."

Data and variables

"To obtain reliable results from the text analysis software Linguistic Inquiry and Word Count (LIWC), we exclude all cases in which the founder tweeted less than 50 words in the examined period (i.e., all tweets that were posted from the first tweet to the date of the first funding or, in case that no funding occurred, to the last recorded tweet), resulting in a final sample of 3586 ventures founded by 3313 individuals."

"These variables capture founder characteristics, the characteristics of founders' use of Twitter, venture characteristics, and funding characteristics."

"To rule out any year-specific effects, for example, financial crises or cyclical movements of the economy (e.g., Block & Sandner, [92]; Gompers and others, [93]), we include a set of dummy variables that capture the venture's founding year as well as the year in which the venture received its first round of funding."

Results

"have a significantly positive effect on the probability of acquiring a second funding round."

"has a significantly negative effect on the probability of receiving a second round of funding."

"A stronger use of future-oriented language has a significant positive effect on the probability of receiving a second round of funding."

"Regarding the degree of confidence displayed, we find a significantly positive effect on the probability of a second round of funding."

"Because a second funding round could potentially occur after our sampling period, the dependent variable used in our logit models is right-censored."

"While the display of "entrepreneurial vision" (H1a), "optimism" (H1b), and "rationality" (H2a) significantly decrease the time to a second round of funding, the display of "self-centrism" (H2c) and of an "achievement motive" (H3b) increase the time to a second round of funding."

Discussion and conclusion

"Future research could revisit the reliability of our findings by constructing digital identities based on information gathered from more personal social media sites such as Facebook (e.g., Kosinski and others, [94]), more professional networks such as LinkedIn (Piva & Rossi-Lamastra, [95]), or other Web 2.0 platforms such as Reddit (Corbet and others, [96])."

"Future research could analyze the effects of Schumpeterian entrepreneurial digital identities on other types of funders or more fully explore the heterogeneity within the group of VCs."

"An additional promising area of future research would involve an investigation into the role of Schumpeterian founder identities in a non-US cultural and historical context, in which a different entrepreneurial image and view of entrepreneurship prevail (Pahnke & Welter, [97])."

"To the categorization of Fauchart and Gruber [98], future research could also focus on Kirznerian entrepreneurship and investigate whether a Kirnerian entrepreneurial identity has an influence on resource acquisition from VC firms or other providers of entrepreneurial finance."

5.2.9 Transnational Digital Identity as an Instrument for Global Digital Citizenship: The Case of Estonia's E-Residency [491]

This is a machine-generated summary of:

Tammpuu, Piia; Masso, Anu: Transnational Digital Identity as an Instrument for Global Digital Citizenship: The Case of Estonia's E-Residency [491]

Published in: Information Systems Frontiers (2019)

Link to original: https://doi.org/10.1007/s10796-019-09908-y

If you want to cite the papers, please refer to the original.

For technical reasons we could not place the page where the original quote is coming from.

Abstract-Summary
"We explain the adoption of e-residency by analyzing how individual motives to apply for a digital identity are affected by both individual-level socio-demographic characteristics and macro-level characteristics measuring digital and economic development in applicant's country of origin."

"The findings suggest that individual motives to adopt e-residency vary depending on both the citizenship of applicants and the level of e-government development in the country of origin."

"Although attracting more citizens from digitally advanced countries, individual motives indicate that e-residency can compensate certain digital disadvantage to citizens of countries with lower levels of e-government development."

Introduction
"This paper seeks to extend existing research on digital identity systems beyond territorially bound national (or subnational) contexts by exploring the social implications of Estonia's e-residency programme, the first transnational digital identity scheme implemented by the Estonian government, for global digital citizenship."

"The aim of "electronic residency" or "virtual residency" enabled by this programme is to give non-residents of the country, independent of their citizenship and place of residence, remote access to Estonia's well-advanced digital infrastructure and e-services via government-supported digital identity issued in the form of a smart identity card (the e-resident's ID)."

"We therefore seek to explore these assumptions by analyzing the overall adoption of e-residency and the ways in which individual motives to apply for the digital identity offered by the e-residency programme vary depending on applicant's citizenship and the level of e-government development in the applicant's country of origin."

Digital Identity Systems as Mechanisms of Digital Inclusion and Enablement
"Most nations, both in developed and developing countries, have been developing their digital identity systems as part of their e-government initiatives, the scope and functionality of these systems is highly varied (Goodstadt and others [99]; Kubicek and Noack [100]; Sullivan [101]; World Bank Group [102, 103])."

"The variances between national digital identity systems thus reflect and are part of broader international disparities of e-government development among countries and regions (Warf [104, 105]; World Bank Group [102]; Zhao and others [106])."

"To a range of scope and functionality, government-supported digital identity schemes also vary in terms of adoption and usage."

"Drawing on these assumptions, researchers have argued that as long as there are other methods and means available for digital authentication, for example such as those offered by banks, government-supported digital identity systems do not necessarily provide sufficient additional value for citizens."

"In cases where government-supported digital identities have been promoted as secure and efficient means for digital authentication for private sector services as well, for example Internet banking, adoption and usage rates may consequently be higher (Kubicek and Noack [100])."

Estonia's E-Residency as a Transnational Digital Identity Scheme

"Established on a well-functioning national digital identity system (Martens [107]; Vassil [108]), Estonia's e-residency is the first transnational digital identity scheme that extends a government-supported digital identity system beyond a territorially bound national society by providing a transnational digital identity in the form of a smart identity card (the e-resident's ID) to non-residents of the country independent of their nationality (citizenship) and place of residence."

"Given Estonia's membership in the EU, the e-residency programme is being promoted as providing digital access not only to Estonia's digital society and e-services but also to the broader EU digital market, particularly in the context of the recently enforced European Regulation on Electronic Identification and Trust Services (eIDAS) (Godoy and Heal [109])."

"Although Estonia's e-residency programme may have a "compensation effect" for citizens and entrepreneurs from digitally less developed countries and regions, as some authors argue (Godoy and Heal [109]; Kotka and others [110]), it may actually come to serve those who already have more digital opportunities, as earlier studies on e-government adoption suggest (Taipale [111]; Warf [104])."

Research Questions and Methodological Considerations

"What is the spread of applicants' individual motives to apply for e-residency as a transnational digital identity across individual and country-level characteristics?"

"More specifically: 1.1 How does the applicant's citizenship, compared to other individual-level characteristics, characterize the adoption of e-residency across motive types?"

"1.2 How does the level of e-government development in the applicant's country of origin, compared to other macro-level characteristics, characterize the adoption of e-residency across motive types?"

"The multilevel regression modelling enables to take into consideration both individual-level structural variables as well as country-level contextual characteristics in order to explain the variations in individual motives to adopt Estonian e-residency."

"The individual motives to adopt e-residency as an instrument that enables digitally mediated virtual mobility and digital access to the EU presumably vary, depending on the applicant's citizenship status."

Results

"In the case of the first three motive types – "business activity or interests related to Estonia", being a "fan of e-residency", and interested in "location-independent international business" – male applicants were most often represented (see negative sign of regression coefficients), whereas female applicants mostly expressed "other motives not directly related to business"."

"Of macro-level variables included in the regression analysis, the level of e-government development (EDGI-level) was statistically significant for explaining the first two motive types: the applicants applying for e-residency because of having "business activity or interests related to Estonia" are more often from the countries with high or medium level of e-government development, whereas "fans of e-residency" are more often citizens of countries with very high level of e-government development."

Discussion

"As the vast majority of applicants of e-residency are citizens of countries with relatively higher levels of digital and economic development, the e-residency programme tends to further increase the digital opportunities of citizens and entrepreneurs who are digitally better positioned by their own national governments rather than advance those with less digital opportunities, as suggested by some authors (Godoy and Heal [109]; Kotka and others [110])."

"As existing studies on e-government adoption suggest, the socio-geographic context and the differences in the levels of digital and economic development between applicants' countries of origin not only condition the access but also the need for certain e-government solutions (Taipale [111])."

"In case of citizens from digitally less developed countries, the adoption is more frequently driven by instrumental, usage-related needs, which highlights the potential of e-residency to compensate certain digital disadvantage related to a lower level of e-government development, whereas in case of applicants from digitally more developed countries the motives rather indicate a more general, non-instrumental interest in e-residency."

Conclusions

"We sought to examine the ways in which both individual-level socio-demographic and country-level socio-economic variables that measure digital and economic disparities between applicants' countries of origin explain the variations in individual motives to adopt the digital identity offered by the Estonian e-residency programme, using a multilevel modelling regression method."

"While citizens from countries with very high level of e-government development express more frequently a non-instrumental and a more general interest in e-residency, the motives of citizens from countries with relatively lower levels of e-government development rather relate to the instrumental advantages of the Estonian e-residency programme."

"We can conclude based on our findings that despite providing more frequently an advantage to citizens from wealthier and digitally more developed countries, the individual motives for joining the e-residency programme still indicate that the transnational digital identity can compensate certain digital disadvantage in case of citizens and entrepreneurs from countries with relatively lower levels of e-government and economic development."

Limitations and Future Research

"The limitations of the current study mainly relate to the administrative register-based data used for the empirical analysis."

"The limited individual-level data and the pre-defined standardized motives included in the individual application form of e-residency provide only partial insights into the needs and incentives underlying the individual motives to apply for Estonian e-residency, which sets limits for the analytical and methodological research design."

"Combined with other (administrative) data describing the actual use of the e-resident ID-s (the "digital footprints" of e-residents), the information included in the individual applications would make possible to analyze the differentiated use of e-residency as shaped both by individual factors and different social, economic, and digital contexts."

"A qualitative analysis of e-residents' individual motives to apply for Estonian e-residency would make possible a more in-depth study of the different dimensions of digital citizenship and mobile residency as enabled by the transnational digital identity offered by the Estonian e-residency programme."

5.2.10 Pre-service Educators Developing a Digital Identity [492]

This is a machine-generated summary of:

O'Byrne, W. Ian; Hunter-Doniger, Tracey: Pre-Service Educators Developing a Digital Identity [492]

Published in: TechTrends (2021)

Link to original: https://doi.org/10.1007/s11528-021-00617-5

If you want to cite the papers, please refer to the original.

For technical reasons we could not place the page where the original quote is coming from.

Abstract-Summary

"This article discusses a study on preservice education students and the implementation of digital identities through self-constructed websites in a technology course."

"Researchers investigated the process of helping educators create a create a domain of their own while they consider the role of technology and digital identities in their future classrooms."

"Data revealed themes that focused around building the website and building a digital identity."

Introduction

"As educators in networked spaces, we utilize a variety of digital texts and tools to create and curate representations of our presented identities."

"To address these challenges, we believe that students need to develop and maintain a domain of one's own, one online space that students build up from Pre-K through higher education to allow for documentation of learning over time (O'Byrne & Pytash, [112])."

"Educators and students can create and document learning over time, and when they move on to another physical institution, their digital work can follow them."

"There is a need for digital spaces in which educators can actively create, build, and modify content as they create representations of their identities."

"How do students in undergraduate and graduate-level educational programs view the role of digital content construction in the context of social and connected online learning environments?"

Moving from Theory to Construction of Digital Identities

"In effect, when an individual attempts to create meaning in the world, the associated activities can be viewed as collective socio-collaborative acts of meaning that impact their educational and social identities."

"While they intended for portfolios to provide an authentic assessment of students' learning, they also found that a portfolio system had many benefits, including teachers acting as collaborative colleagues, rather than isolated dispensers of student grades."

"As Clark and others [113] note, "For some theorists, learning itself is seen as the construction of identities as individuals take up and take on different social practices in different contexts with different social communities at different times.""

"The work of compiling a portfolio yields the possibility for students to begin constructing a sense of their identity."

"Students are engaged in media production, constructing websites, and showcasing digital evidence of their learning."

Setting & Participants

"The class is required in the sequence of courses as students are enrolled in a pre-service teacher education program."

"This study focused on the development of a digital portfolio to archive student work product across the program, and the impact of owning this domain on the creation of this space."

"The instructional model used to guide pre-service teachers as they use the Internet identifies a continuum of three stages that move learners from consumers to producers of digital content."

"The third stage of this sequence involves having learners construct or create digital content."

"There are many parallels between online content construction and the writing process as learners plan, generate, organize, compose, and revise digital work product."

"Throughout the semester, students were given two to three weeks to work their way across several elements of building their digital identity, embodied in a WordPress website."

Methodology

"To address the research focus, we analyzed qualitative data to identify connections between the data and the research questions (Thomas, [114])."

"We collected data from three different sources to allow for triangulation of findings (Denzin, [115]): (a) student work product on their website, (b) focus group interviews, and (c) researcher notes."

"Qualitative data were collected throughout a 13-week semester course in digital literacies in education and analyzed to answer the research questions."

"This study was triangulated through the use of multiple methods and sources of data collection such as observations, field notes, and the websites the students created."

"This process insured the recursive nature of data collection and analysis necessary in naturalistic qualitative inquiry (Patton, [116])."

"Analysis of patterns and themes in qualitative data allowed the dynamics of change to be more evident and permitted us to better understand how participants' decisions during the study."

Findings

"At that same time, Student #2 had gone beyond the aesthetic concerns the rest of her classmates had and was beginning to envision what she wanted to reveal as her digital identity as well as how she wanted to present that to her intended audience."

"As students worked their way through the learning curve necessary to build in WordPress, they gradually saw opportunities to utilize this digital space in their program and other coursework, as well as other aspects of their identity."

"Although, once the aesthetics were acceptable to the student, they did not fully consider their digital identity, even though the assignment was to create the website and digital identity concurrently."

"As students continued to develop their website, they were regularly engaged in a discussion to consider the impact on their digital identity and issues of privacy or security."

Discussion

"The length and timing of the study may have provided challenges as we examined possible changing of student dispositions and use of aspects of digital literacy and pedagogy in classroom instruction."

"This research highlights the need for all individuals, especially educators, need opportunities to learn how to develop agency and ownership over their digital identity."

"To engage in digital literacy practices, individuals need to consider opportunities to move learners from consumers to producers of digital content."

"Through the findings in this study, we believe there is a continuum of stages that exist as we help students become the digitally literate citizens that they need to be in the future (NCTE, [117]; O'Byrne, [118])."

"As we have indicated throughout this study, identity construction through the development of digital content can only happen if educators embed opportunities to build and utilize these new and digital literacies in instruction."

Conclusion

"Despite the transformative possibilities associated with the inclusion of the Internet and related networked technologies in instruction, relatively little is known about the regular use of these technologies in our daily lives."

"For students in particular, their ability to best utilize these digital and web literacies in work is central to their collective future."

"This means that students need to leave our schools with the skills necessary to not only read, write, and participate on the web."

"As technologies connect our global marketplace, students need to identify opportunities to empower themselves as true natives of these digital spaces."

5.3 Digital Self-physical Aspects

Machine generated keywords: body image, image, body, adolescent, appearance, photo, sns, peer, girl, ideal, networking site, social networking, social medium, young, post

5.3.1 The Impact of Social Media on Body Image Perceptions and Bodily Practices among Gay, Bisexual, and Other Men Who Have Sex with Men: A Critical Review of the Literature and Extension of Theory [493]

This is a machine-generated summary of:

Filice, Eric; Raffoul, Amanda; Meyer, Samantha B.; Neiterman, Elena: The Impact of Social Media on Body Image Perceptions and Bodily Practices among Gay, Bisexual, and Other Men Who Have Sex with Men: A Critical Review of the Literature and Extension of Theory [493]

Published in: Sex Roles (2019)
Link to original: https://doi.org/10.1007/s11199-019-01063-7

If you want to cite the papers, please refer to the original.

For technical reasons we could not place the page where the original quote is coming from.

Abstract-Summary
"Little research has been done to investigate whether social media differentially influence gay, bisexual, and other men who have sex with men, who are known to be at increased risk of body image disorders compared to their heterosexual counterparts."

"The current paper aims to address this gap by surveying and extending existing theory, using a critical review methodology, to derive a provisional theoretical model that explains how social media influence body image and weight and shape control behaviours of sexual minority men in particular."

"Our proposed model serves as an extension to the transactional model of social media and body image concerns, which includes additions to individual vulnerability factors (perceived self-discrepancy, gender nonconformity, minority stress) and psychosocial mediating processes (sexual objectification, sociocultural processes, online disinhibition)."

Body Image among GBMSM
"Studies assessing aggregate disordered eating symptoms have found that sexual minority men and women (i.e., those who identify as gay, lesbian, bisexual, two-spirit, queer or otherwise not heterosexual) report greater levels of eating disorder symptomatology than their heterosexual counterparts, although disparities across men are more consistent (Calzo and others [119])."

"Purging behaviour (e.g., self-induced vomiting or laxative misuse for the purpose of weight loss) is reported 3 to 16 times more frequently by GBMSM than by heterosexual men (Matthews-Ewald and others [120]; Watson and others [121])."

"GBMSM are also more likely to report fasting (i.e., not eating for extended periods of time) (Watson and others [121]), dieting to lose weight (Matthews-Ewald and others [120]), and using diet pills for weight loss (Austin and others [122]; Matthews-Ewald and others [120]; Watson and others [121]) than heterosexual men, but findings concerning disparities in the use of dietary supplements (e.g., creatine) or drugs (e.g., anabolic steroids) for muscle building are inconclusive."

Use of Social Media

"Most (94%) individuals ages 18–24 years-old report some form of social media use, and a majority of users for several platforms (e.g., Facebook, Snapchat, Instagram) visit the site at least once daily (Smith and Anderson [123])."

"GBMSM may use social media to an even greater extent than the general population."

"Although data describing patterns in 2SLGBTQ+ (Two-spirit, lesbian, gay, bisexual, trans, queer, and other diverse sexual and gender identities) populations' media engagement are scant, one study has found that compared with heterosexual respondents, a greater proportion of lesbians, gay men, and bisexuals reported having a social networking account (Seidenberg and others [124])."

"GSN app use is remarkably widespread among GBMSM; a recent study of 3105 MSM (men seeking sex with men) from diverse ethno-racial and socioeconomic backgrounds revealed that 78.2% of the sample reported some or frequent use of general dating websites and apps."

Current Gaps in the Literature

"Despite social media's pervasiveness and central role in the social activity of GBMSM in particular, little research has been done to investigate their impact on body image disturbance and other adverse health outcomes that disproportionately affect this population."

"We only identified one study that investigated the links among social media, body image, and weight and shape control behaviours in GBMSM (Griffiths and others [125])."

"In the current paper, we aim to address this gap by surveying and extending existing theory, using a critical review methodology, to derive a provisional model detailing the impact of GSN apps (and social media more broadly) on body image-related outcomes and applying specifically to GBMSM."

"To that end, we seeks to answer the following questions: What pre-existing theories, models, and frameworks may be used to explain the influence of social media on body image and weight and shape control behaviours among GBMSM?"

Class I Theories—Body Image Disturbance Etiology

"Heatherton and Baumeister [126] expound on this notion with escape theory, which posits that when individuals are faced with an aversive self-awareness, such as an awareness of a discrepancy between one's actual and ideal body shape, they will engage in behaviours that help them cope and/or escape this awareness (e.g., drug use, self-harm)."

"To providing a useful framework for understanding the psychological processes underpinning body image disturbance and disordered weight and shape control behaviours, the constructs of self-discrepancy theory imply that standards of attractiveness are not fixed, rather that body dissatisfaction is determined in relation to standards that vary historically, cross-culturally, and individually (Vartanian [127])."

"The relatively fewer studies conducted among men suggest that although the model holds overall (Martins and others [128]; Wiseman and Moradi [129]), the degree of significance among its various proposed interrelations (e.g., among sexual objectification, self-surveillance, body shame, interoreceptive deficits, and disordered eating) may vary across gender and sexual orientation."

Class II Theories—Sexual Orientation-Based Differences in Body Image

"The perils of sexual objectification hypothesis, articulated and tested by Siever [130], posits that the similar vulnerability to body dissatisfaction and disordered eating seen among heterosexual women and gay men is attributable to their shared experience of being sexual objectified by men."

"A compelling inference may be drawn from the masculinity hypothesis in light of evidence demonstrating increased rates of gender nonconformity among sexual minority men (Bailey and Zucker [131]; Rieger and others [132]): Sexual orientation-based differences in body-image related pathology may be due to differences in adherence to gender norms."

"If it is, in fact, gender conformity (i.e., for men, deviance or adherence to hegemonic masculinity) that is associated with body image disturbance, rather than the presence of femininity, it would be inappropriate to suggest femininity itself is predisposed toward body image disturbance and that its prevalence among gay men explains why their rates of disordered eating approach women's."

Class III Theories—Impact of Social Media on Body Image

"Using a multipartite transactional model of social media and body image concerns, Perloff [133] contends that social media do not unidirectionally cause body image disturbance, but in the presence of individual predisposing factors and when mediated by certain psychosocial processes, they can have deleterious effects."

"Some pathways have yet to be verified, such as the interactive effects of perfectionism and centrality of appearance to self-worth on social media use and the mediational effects of transportation and normative influences on social media use and body image."

"Disinhibited online behaviour could be seen as a variable that mediates the association between social media use, sociocultural perspectives, and body image-related outcomes—if community integration is associated with body image disturbance among GBMSM by way of appearance conversations, one may speculate that disinhibited online behaviour resulting in appearance conversations accounts partly for the association between social media use and body image disturbance in GBMSM."

Developing an Integrated Model of Social Media and Body Image for GBMSM

"We leave the remaining pathways in Perloff's [133] transactional model intact: Individuals' pursuit of validation drives them to use social media, which through mediating processes results in adverse body-image related outcomes and disordered weight and shape control behaviours."

"Recent articles from 2SLGBTQ+ popular media suggest that the body positive movement has expanded to capture the interest of GBMSM, who are becoming increasingly cognizant of their shared vulnerability to the harmful effects of comparisons made with others who depict normative body image ideals on social media (Baker [134]; Feldman [135]; Villarreal [136])."

"Purposes, the main points we emphasize here are that (a) various intersecting subject positions can conceivably modify the vulnerability factors and mediating processes pertinent to how social media impact GBMSM's body image and (b) quantitative research has inadequately explored these intersections specifically among GBMSM."

Hypothesis Verification and Other Directions for Future Research

"None have hitherto been linked directly to social media use in quantitative studies, thus presenting opportunities for future research."

"Although appearance conversations have been shown to be related to body image disturbance, and social media use encourages disinhibited behaviour, it remains unclear if and how community involvement, sense of community, and online disinhibition are related to appearance conversations."

"Foremost, only one known study (Griffiths and others [125]) has tested whether the link between social media use and body image disturbance is generalizable to GBMSM."

"It is still unclear if the relationship between social media use and body image disturbance varies between social media platforms and, if so, to what platform-specific features those differences can be attributed."

"Future research should extend beyond comparing users to non-users and investigate associations between body image disturbance in GBMSM and app-specific behaviours, such as frequency and duration of use, representations in profile pictures, profile description content, intentions for app use, and number of apps used simultaneously."

Practice Implications

"Messaging that advocates curbing overall social media and dating app use could be well-received by heterosexual women, but GBMSM may find this directive to reflect a lack of cultural sensitivity and/or competence, given internet-mediated communication has served a unique historical role for this group in circumventing the social, cultural, and political barriers to forming connections in public spaces."

"The notion that social media use does not uniformly result in body image issues, underscored by the myriad vulnerability factors and mediating/protective processes identified in our model, also points to the potential ineffectiveness of this generic messaging."

"Any attempt to reduce social media use among all GBMSM would require overlooking that some are at minimal risk of body image issues and, indeed, could even benefit from protective factors like body-positive content."

Conclusion

"The present paper is the first known to offer a theoretical framework detailing how social media and other information technologies influence body image and weight and shape control behaviours in GBMSM."

"Our findings make clear that body image disturbance and eating pathology among sexual minority men result from a broad and complex array of interrelated biological, psychological, social, and cultural determinants."

"Uncovering a simple causal mechanism increasingly seems idealistic, especially because matters are further complicated by digital technologies and sexuality, but a more thorough understanding of the many risk factors at play will be required for effective treatment and prevention strategies targeted toward this vulnerable population."

[Section 11]

"A growing body of literature suggests that interactive media-based web technologies, such as social networking sites, have a measurable impact on users' perceptions of and attitudes toward their own bodies, as well as resultant weight and shape control behaviours (Holland and Tiggemann [137]; Mingoia and others [138])."

"However, most research has focused on girls and young, heterosexual White women, and little work has been done to understand how social media influence other populations."

"Our focus is on extending pre-existing theory to elucidate how social media may differentially impact gay, bisexual, and other men who have sex with men (GBMSM) who are at risk of body image-related issues."

5.3.2 From Filters to Fillers: An Active Inference Approach to Body Image Distortion in the Selfie Era [494]

This is a machine-generated summary of:

Tremblay, Simon C.; Essafi Tremblay, Safae; Poirier, Pierre: From filters to fillers: an active inference approach to body image distortion in the selfie era [494]

Published in: AI & SOCIETY (2020)

Link to original: https://doi.org/10.1007/s00146-020-01015-w

If you want to cite the papers, please refer to the original.

For technical reasons we could not place the page where the original quote is coming from.

Abstract-Summary
"We discuss how body dysmorphic disorder and related body image distortions could arise, using the conceptual resources provided by the active inference framework."

"We suggest that these disorders involve dysfunctional self-modelling which entails maladaptive internalization of sociocultural preferences during adolescent identity formation."

"Identity formation is hereby described as cycles of interpersonal active inference that arbitrate between identity exploration and commitment."

"We propose that impaired self-modelling is unable to reduce interpersonal uncertainty during identity exploration, which, over time, degenerates into uncontrollable epistemic habits that isolate the body image from corrective sensory evidence."

"We conclude that heightened interpersonal uncertainty in this novel context could precipitate the onset of body dysmorphic disorder and related body image distortions, particularly when selfie filters are involved."

Introduction
"Given the following explanatory model, it was perhaps to be expected that this democratization of filters would have an effect on the body image and self-esteem of heavy users of photo-sharing social networks, especially among the younger generations."

"While data has yet to link the use of selfie filtering apps such as Snapchat (and many others, e.g. Camera+, FaceTune, Front-Facing Flash, and dozens more) with the rise of this particular form of body image distortion, given how recent the phenomenon is, cosmetic medicine as well as mental health professionals have started warning against what they believe is an increasing role of such applications in the way we perceive ourselves."

"We outline a computational model whose dysfunction may explain BDD and predict SD as well as other, yet unknown, forms of body image distortions and self-related disorders that could emerge given further advances in AI."

Distinction between body representations and the self-model theory
"Unlike Anderson, she does not consider body image as being distinct from these two representations (i.e. body structural description and body lexico-semantic representation)."

"We will return to the distinction between types of body representations when we integrate our proposal with the self-model theory of subjectivity and active inference framework (Limanowski and Blankenburg [139]; Metzinger [140]; Limanowski and Friston [141])."

"This first-order representation of the intentionality relation generates the experience of a strong first-person perspective (strong 1PP or attentional 1PP) capable of attentional agency (i.e. deliberate control of attention), which accounts for the subjectivity of the self-model (Blanke and Metzinger [142])."

"Metzinger's account is somewhat compatible with Gallagher's distinction between body representations, the body schema being a global representation of the body (non-conscious functional self-model) that dynamically achieves second-order embodiment and the body image being a local representation of the body (conscious PSM) that realizes third-order embodiment (Metzinger [143], pp."

Self-consciousness as active inference

"The Active Inference Framework, in fact, extends hierarchical predictive coding by including action selection and planning, that is, the minimization of the expected free energy (or uncertainty) of alternative action policies (i.e. beliefs about action sequences that specify state-transitions and associated expected outcomes) (Friston and others [144])."

"Within this Bayesian framework, the neurocomputational process that builds the SMT's model (based on predictive processing and active inference) can be understood as a self-evidencing process, where the brain actively maximizes the evidence for the hypothesis it entertains in favor of its own existence by minimizing a proxy for (an upper bound on) surprisal (i.e. variational free energy, or its formal equivalent under some assumptions: precision-weighted prediction errors), where expected surprisal corresponds to uncertainty (Hohwy [145])."

Distorted self-models

"In active inference, identity formation is self-evidencing associated with motivated goal-directed planning in the sociocultural environment (i.e. interpersonal planning), a process that depends on shared intentionality (Tomasello and others [146]) and on subsequent synchronization of mental states through shared narratives (i.e. shared prior beliefs)."

"In the right sociocultural context, impaired epistemic agency (i.e. attentional selection) may lead to the internalization of overly precise prior beliefs about the negative body image (i.e. uncertain interpersonal outcome) in the higher levels of the self-model."

"According to our model, this process is achieved through cycles of social active inference that adaptively internalize (through perceptual inference) the relevant sociocultural norms as self priors (i.e. narratives) that serve to guide actions toward self-congruent goals (through active inference)."

Conclusion

"We began by showing that BDD and other body image distortions can be better understood as self-related disorders that primarily disrupt aspects of personal identity."

"We subsequently situated the hierarchical self-model within the active inference framework, describing conscious personal identity as the temporally thick, EAM-mediated process of interpersonal active inference."

"Since BDD generally arises during adolescence, we discussed how the identity formation process can be disrupted when EAM functioning is impaired, entailing the maladaptive internalization of sociocultural standards and the subsequent apparition of body image distortions."

"An impaired EAM is unable to adequately minimize interpersonal uncertainty (of salient/preferred interpersonal outcome or self-representation) during identity exploration, generating maladaptive identity commitments that exceedingly depend on conformity to sociocultural standards."

"With the progressive distortion of the body image through visual normalization, the automatization of epistemic behaviors is particularly problematic."

5.3.3 "Digitized Dysmorphia" of the Female Body: The Re/Disfigurement of the Image [495]

This is a machine-generated summary of:

Coy-Dibley, Isabelle: "Digitized Dysmorphia" of the female body: the re/disfigurement of the image [495]

Published in: Humanities and Social Sciences Communications (2016)

Link to original: https://doi.org/10.1057/palcomms.2016.40

If you want to cite the papers, please refer to the original.

For technical reasons we could not place the page where the original quote is coming from.

Abstract-Summary

"What has the digitization of female appearances, the altering of female bodies not only in media but individually among the general public through the use of apps, done to human sexual interactions and senses of self?"

"Particularly in Western culture, we can digitally alter ourselves through Photoshop and apps such as Perfect365, producing our own notions of normativity."

"How do these online ideal images affect women's relationships to their material bodies, do these digital images provide the freedom to express a self that would otherwise be overlooked or simply accentuate the disparity between female bodies and the images women feel they must embody?"

"In positing the new term of "Digitised Dysmorphia", I question whether transforming oneself into a virtual ideal, manipulating the image, is an act of dissidence, or a demonstration of how fully regulated we are by social norms, whereby digital modifications simply enable us to reach normative ideals."

"Perhaps this enablement irrevocably changes definitions of "normativity" as we alter our relations to the body and image."

"This article offers an initial exploration into how this digitized dsymorphia affects our perception of self individually and as a collective society."

Introduction

"As the age of digitized beauty progresses, alongside a growing personal capacity to technologically modify our own images as many in Western society, and other societies, now can, there is a greater need for discussion as to how this is affecting women's relationship to their bodies."

"This article examines women's relationship to media and cultural representations of female beauty standards and how this has filtered into the presentation of self via digitally modified images, ultimately positing that this disjuncture between body and image produces a socially collective "Digitised Dysmorphia"."

"As an introductory paper intended to initially conceptualize the term "digitised dysmorphia", I will explore its relation to image-editing technology, body theory and the way in which the image materialises in place of the body to question how the digitization of one's image, an "ideal" version of an individual's appearance, has further skewed what is considered a "standard" body, particularly a standard female body."

Body dysmorphic disorder and the female body

"I believe digitized dysmorphia is generated by societal pressures to look a certain way, and is currently directed more towards women's than men's bodies to meet such criteria and standards."

"Whilst society has normalized the preoccupation women perceivably have over their looks, the similar effect that increased visibility of the male body has had upon men suggests a highly influential link for both men and women between bodyimage and societal pressures upon this image—slowly body-image satisfaction or dissatisfaction is becoming more than just a women's issue."

"I use BDD as a comparison to digitized dysmorphia because of the way in which findings from mild cases have reflected how societal pressures currently impact women in greater ways than men, suggesting that women generally show more signs of body dissatisfaction, preoccupation with appearance matters and a social body dysmorphia of what the female body should look like."

The pressure to be beautiful: the modified image and feminine Masquerade

"After considering the public dimension of the body and the industries that shape the societal constructs of what the body should supposedly look like, this section explores the position of femininity and this societal pressure to attain certain beauty standards in relation to the technology that allows users to digitally alter their images."

"Susan Bordo (2003) notes that digital modification of images means that we are being educated to shift our perception of what a normal woman's body looks like, so that we see our own bodies as wanting because they do not match an unrealistic, polished, slimmed, and smoothed ideal … ([147]: 25) Consequently, these types of femininity and supposed social standards of beauty are in themselves skewed, no longer based upon anatomical, material bodies, but upon the way in which the female body can be and has been manipulated by technology itself."

Agency and the "interpretative flexibility" of technology

"Digitized dysmorphia is in itself a form of negative dysmorphia that has arisen due to the harmful societal pressures placed upon women's bodies to look a certain way and industries have capitalized upon this, manufacturing technology to assist this desire to attain beauty ideals."

"Having said this, perhaps the counter-argument is that although digitized dysmorphia is a largely negative phenomenon, shaped by prejudice and the belief that one can only attain such ideals of beauty through modifying the image of oneself, there is a certain liberation, a relief of sorts in that these technologies, while the perpetrators of such ideals, enable a person to finally embody these previously unattainable social standards and their own internal ideals of self-image."

"It is questionable as to whether the "interpretative flexibility" of the technology used within the visual manifestation of digitized dysmorphia, the modified images and the dysmorphic view of self they generate provide the freedom to play with or escape labels that have been societally inscribed upon material female bodies."

"Digitized Dysmorphia" and the online and offline representations of self

"Not only does digitized dysmorphia exclusively metamorphose based on the choices made within the editing process of an individual's digital image, but through the way in which these modifications may influence, and ultimately alter, an individual's online representation of self and the way people interact with them."

"In the context of online dating sites, chat rooms and social networking sites that exclusively allow or can be used predominantly with still-images and text, 2D interaction allows users to choose to maintain an unreadable and unverifiable level of self, which is perhaps restricted and ultimately less trustworthy than the physical self and the material body language this offers."

"Let us once again consider how the reality of the material body can affect the online self-image presented by individuals and the reasons one may decide to digitally modify their image, dysmorphically skewing their perception of self."

5.3.4 The Perfect Storm: A Developmental–Sociocultural Framework for the Role of Social Media in Adolescent Girls' Body Image Concerns and Mental Health [496]

This is a machine-generated summary of:

Choukas-Bradley, Sophia; Roberts, Savannah R.; Maheux, Anne J.; Nesi, Jacqueline: The Perfect Storm: A Developmental–Sociocultural Framework for the Role of Social Media in Adolescent Girls' Body Image Concerns and Mental Health [496]

Published in: Clinical Child and Family Psychology Review (2022)

Link to original: https://doi.org/10.1007/s10567-022-00404-5

If you want to cite the papers, please refer to the original.

For technical reasons we could not place the page where the original quote is coming from.

Abstract-Summary

"In this theoretical review paper, we provide a developmental–sociocultural framework for the role of social media (SM) in adolescent girls' body image concerns, and in turn, depressive symptoms and disordered eating."

"We propose that the features of SM (e.g., idealized images of peers, quantifiable feedback) intersect with adolescent developmental factors (e.g., salience of peer relationships) and sociocultural gender socialization processes (e.g., societal over-emphasis on girls' and women's physical appearance) to create the "perfect storm" for exacerbating girls' body image concerns."

"In the context of proposing this framework, we provide empirical evidence for how SM may increase adolescent girls' body image concerns through heightening their focus on (1) other people's physical appearance (e.g., through exposure to idealized images of peers, celebrities, and SM influencers; quantifiable indicators of approval); and (2) their own appearance (e.g., through appearance-related SM consciousness; exposure to idealized self-images; encouraging over-valuing of appearance; and peer approval of photos/videos)."

Introduction

"The field has lacked a unifying theoretical framework for understanding the exacerbation of body image concerns as a potential mechanism underlying associations between adolescent girls' SM use and mental health."

"We propose that the features of SM (e.g., idealized images of peers, quantifiable feedback) intersect with broader developmental and social processes in adolescence (e.g., salience of peer feedback and social status, heightened self-consciousness in the form of the "imaginary audience") and sociocultural gender socialization processes (e.g., emphasis on girls' and women's physical appearance) to create the "perfect storm" for exacerbating girls' body image concerns, and for some girls, worsening their mental health."

"Prior theoretical reviews and empirical papers have proposed specific pathways through which SM use may increase body image concerns, yet existing publications have focused on disordered eating without addressing depressive symptoms and/or have not focused specifically on adolescent developmental considerations (see Perloff, [133]; Rodgers & Melioli, [148]; Rodgers, [149]; Rodgers and others, [150])."

Adolescent Girls' Body Image Concerns, Depressive Symptoms, and Disordered Eating

"Although body image concerns begin in childhood for girls, they spike at the transition to adolescence (Rodgers and others, [151])."

"U.S. estimates suggest that approximately 81% of adolescent girls and 55–67% of adolescent boys experience at least some dissatisfaction with their bodies (Lawler & Nixon, [152]; Neumark-Sztainer and others, [153])."

"The increase in body dissatisfaction in adolescence is likely caused by numerous developmental transitions, including biological, social, and psychological changes (Markey, [154])."

"Depressive symptoms increase markedly during adolescence, especially among girls (Maughan and others, [155]), and rates of depression among adolescents have increased in recent years (Geiger & Davis, [156])."

"Longitudinal studies with adolescents provide robust empirical support for the link between body dissatisfaction and subsequent disordered eating and depressive symptoms, particularly among girls (Ferreiro and others, [157]; Sharpe and others, [158]; Stice and others, [159])."

Theories Regarding Gender-Related Sociocultural Appearance Pressures

"Both theories emphasized how Western cultural contexts socialize young women to over-value physical attractiveness; to devote substantial cognitive, emotional, and financial resources toward attempts to achieve cultural beauty standards; and to experience body image concerns, shame, and distress when these standards cannot be achieved (Fredrickson & Roberts, [160]; McKinley & Hyde, [161])."

"In the more than two decades since these sociocultural theories related to physical appearance concerns were published, Western beauty standards for women and girls have continued to prioritize thinness, while evolving to also encompass toned muscles and curviness (e.g., Deighton-Smith & Bell, [162])."

"While scholars have increasingly called for attention to culturally specific appearance ideals and greater diversity in research on body image and disordered eating (e.g., Beccia and others, [163]; Burke and others, [164]; McEntee and others, [165]; Watson and others, [166]; Wilfred & Lundgren, [167]), the majority of studies cited in the current paper relied on predominantly White samples from

Western nations, reflective of the current research literature on social media use and body image among young women."

Theories Regarding Features of Adolescence as a Developmental Period
"The adolescent developmental period is marked by increased salience of social norms and importance of peer feedback for an individual's self-evaluation and identity development (Brechwald & Prinstein, [168])."

"In the context of these complex biological, cognitive, and interpersonal changes, an appearance culture develops during adolescence that dictates normative appearance-related behaviors and reinforces beauty ideals, in which peers frequently discuss physical attractiveness and how to improve it, ultimately encouraging social comparison and exacerbating body dissatisfaction (Jones and others, [169])."

"During adolescence, physical attractiveness becomes important for social status among girls (Mayeux & Kleiser, [170])."

"Adolescent girls are socialized to prioritize physical appearance and are highly attuned to peers, social comparisons may be particularly impactful during this developmental stage."

Theories Regarding the Features of Social Media
"Theoretical perspectives on SM have increasingly moved away from considerations of specific platforms and instead have emphasized an "affordances" approach—i.e., considering specific facets of SM that impact individuals' experiences (boyd, [171]; Granic and others, [155]; Moreno & Uhls, [172])."

"Building on these principles, psychology and media effects scholars have argued that SM sites are made up of unique features (e.g., "24/7" availability, publicness) that impact, or even transform, adolescents' social experiences and developmental processes (boyd, [171]; Nesi and others, [173]; Subrahmanyam & Šmahel, [174]; Valkenburg & Peter, [175])."

"The framework describes how specific features of SM may intersect with developmental characteristics of adolescence to transform interpersonal relationships."

"Permanence, for example, is a feature of SM describing the permanent accessibility of content, which draws on prior discussions of persistence (boyd, [171]) and retrievability (Valkenburg & Peter, [175]) and may increase adolescents' attention to their appearance in SM photos."

"We revisit the seven features described in the transformation framework and discuss their important role in adolescent girls' body image."

Body Image Concerns as a Key Mechanism Linking Social Media Use with Mental Health
"Rather, our goal was to propose a theoretical framework, and to highlight key findings from the research literature on adolescents' social media use, appearance concerns, and mental health in service of this framework."

"These searches were designed to identify papers that focused on adolescence or emerging adulthood, social media use, appearance-related concerns and processes, and indicators of well-being or mental health symptoms related to depression or disordered eating."

"We prioritized incorporating studies that included participants in the early and middle adolescent developmental periods (roughly aged 11–17) and which used longitudinal or experimental designs."

"We included studies with late adolescent or emerging adult samples (roughly ages 18–25) when the findings were especially relevant and/or when rigorous empirical evidence from younger samples was lacking (e.g., rigorously controlled experimental studies were more common in college samples than in younger adolescent samples)."

Social Media and the Physical Appearance of Others

"Consistent with the tripartite influence model of body image (Thompson and others, [176]), and Perloff's [133] and Rodgers' [149] theoretical frameworks for SM use and body dissatisfaction, the two primary processes by which exposure to others' images may affect girls' body image and mental health include social comparison, whereby adolescents compare themselves to the physical appearance of others, and internalization of appearance ideals, whereby the prevailing sociocultural norms and values relating to physical appearance are conveyed to SM users through attractive images."

"By engaging in social comparison with these idealized images of one's friends and peers, adolescent girls may increase the perceived discrepancy between their ideal and actual appearance, resulting in body dissatisfaction (e.g., Scully and others, [177])."

"In an Australian sample, SM use was associated with adolescent girls' body dissatisfaction via social comparison, thin-ideal internalization, and internalization of appearance ideals (Rodgers and others, [150])."

Social Media and One's Own Physical Appearance

"SM offers the ideal platform for adolescents to fastidiously monitor and evaluate their own attractiveness, engaging in self-presentation strategies to increase the likelihood that peers will provide positive appearance-based feedback."

"Whether or not girls become preoccupied with their appearance on SM, the frequent taking, editing, and posting of images of the self (or "selfies") may directly impact body image."

"Among both adolescent boys and girls, sexy self-presentation on SM is associated with appearance conversations on SM, wanting attention on SM, and actually receiving more attention in the form of likes and followers/friends (Trekels and others, [178])."

"A recent narrative review of research on selfie practices found that adolescents' body image and feelings of self-worth were associated with how many "likes" they received on a photo (McLean and others, [179])."

Summary of Theoretical and Empirical Evidence

"SM likely exacerbates the emphasis on peers' physical appearance and invites upward social comparisons through its presentation of idealized and edited images, with quantifiable indicators of approval, of both known peers and glamorous celebrities and influencers."

"SM may also increase girls' focus on their own appearance, by heightening their appearance-related SM consciousness, magnifying the perceived value of physical appearance, and increasing the focus on quantifiable indicators of approval."

"While some may be less susceptible to appearance-related processes on SM, others may be especially affected if they have underlying vulnerabilities, such as heightened sensitivity to peer feedback or tendency to engage in social comparison."

"SM's availability 24/7 presents unprecedented access to one's own and other people's images—and the feedback those images receive—potentially creating an ever-present appearance culture."

Limitations and Future Directions

"Research suggests that sexual minority adolescents use SM in different ways and experience unique challenges compared to heterosexual youth (e.g., identity development and management in online communities; McConnell and others, [180]), but research has yet to examine how SM use may impact sexual minority adolescents' body image."

"No published work has examined the role of SM use in body image concerns among gender minority youth, although research has begun to explore gender minority adolescents' unique body image concerns more broadly (e.g., Romito and others, [181])."

"Through SM, Black adolescent girls may be exposed to more images of Black women than are available in mainstream mass media, and these images may depict a broader range of beauty ideals (e.g., related to body size and shape, hair, skin tone)."

Clinical Implications

"We recommend the following as preliminary guidelines for clinicians' work with adolescent girls: (1) If the adolescent client is struggling with body image concerns, the clinician might assess whether she is engaging with SM in ways that may exacerbate these concerns, such as through upward social comparisons, editing selfies, or following beauty-oriented influencer content; (2) if yes, the clinician might try using techniques from evidence-based approaches to treating other problematic behaviors, such as motivational interviewing, urge-surfing, and cognitive restructuring approaches (e.g., Bowen & Marlatt, [182]; Harris and others, [183]; Naar-King, [184]; Sudhir, [185]), to encourage the client to spend more time on SM or offline activities that improve, rather than worsen, her body image and mental health."

"School- and community-based preventive interventions can educate girls regarding SM-related risks for body image concerns, by increasing media skepticism, awareness of advertisers' motives, and rejection of unrealistic appearance ideals (e.g., Gordon and others, [186]; McLean and others, [187])."

Conclusions

"Our developmental–sociocultural framework integrates theoretical perspectives from across psychological science and other disciplines, exploring how developmental, sociocultural, and SM-specific factors influence adolescent girls' body image and subsequent mental health."

"We identify and elaborate on these broad theoretical factors, as well as more specific SM experiences related to physical appearance."

"The framework emphasizes a focus on examining body image concerns as a mechanism underlying associations between SM use and adolescent girls' mental health concerns."

"We highlight a new avenue for future research on adolescent girls' social media use and mental health, which recognizes the central role of body image."

5.3.5 "Why Don't I Look Like Her?" How Adolescent Girls View Social Media and Its Connection to Body Image [497]

This is a machine-generated summary of:

Papageorgiou, Alana; Fisher, Colleen; Cross, Donna: "Why don't I look like her?" How adolescent girls view social media and its connection to body image [497]

Published in: BMC Women's Health (2022)

Link to original: https://doi.org/10.1186/s12905-022-01845-4

If you want to cite the papers, please refer to the original.

For technical reasons we could not place the page where the original quote is coming from.

Abstract-Summary

"Adolescent girls appear more vulnerable to experiencing mental health difficulties from social media use than boys."

"Sexual objectification through images may reinforce to adolescent girls that their value is based on their appearance."

"This study explored how sexualized images typically found on social media might influence adolescent girls' mental health, in positive and/or negative ways."

"Participants identified body image as a major concern, reporting negative appearance comparisons when viewing images on social media."

"Comparisons also influenced adolescent girls' efforts to change their appearance and seek validation on social media."

"The findings of this study have important implications for professionals working with adolescent girls and for the development of health promotion programs addressing social media use and body image concerns."

Background

"Time spent on the Internet has been associated with increased body dissatisfaction among adolescent girls, with the interaction allowed by social media and appearance-focused content influencing body image concerns through negative social comparisons and peer normative processes [193–197]."

"There are few qualitative studies exploring the influence of sexualized images on social media or the role of social media use in body image development from the perspective of adolescent girls themselves."

"To the best of the authors' knowledge, there are no qualitative studies exploring adolescent girls' perceptions of the influence of sexualized images on social media on their mental health, or body image, as referred to in the present study."

"The aim of this study was to explore how sexualized images of females' bodies typically found on social media might influence adolescent girls' mental health, in positive and/or negative ways."

Methods

"A purposive sample of twenty-four adolescent girls aged 14–17 years (Grades 9–11) was recruited for the study from the Student Edge (an Australian student membership organization) youth research panel (n = 13, 54.17%), non-government schools (n = 6, 25%) and through snowball sampling techniques (n = 5, 20.83%) in Perth, Western Australia between 2016 and 2018."

"Girls were recruited through snowball sampling methods, with those who participated in the study asked to distribute project information to other girls aged between 14 and 17 years."

"Participants were also asked for their opinions about the sexualization of girls through images on social media, and in what ways this could be positive or negative."

"Prior to data collection, the interview protocol was pilot tested with a convenience sample of two adolescent girls aged between 16 and 17 years to provide feedback on question development and types of responses received, as well as the skills of the interviewer."

Results

"It's almost like a visual effect I guess if one particular group of teenage girls follow celebrities or whatever, begin to follow those sort of things [appearance-focused and popular attractive female Instagram accounts] and all people follow them, their friends, it [trying to look like the females in those accounts] becomes more of an expectation. (Brooke, 16 years) While encouraged to use third person disclosures during interviews, participants reported they made negative appearance comparisons when viewing images on social media."

Discussion

"Consistent with previous research, the influence of social media on adolescent girls' body image was perceived as negative by the participants in this study [198–202]."

"Adolescent girls in this study identified the importance of peers in relation to making appearance-based comparisons, with differences in the comparisons made to peers or celebrities, suggesting body image may be more negatively influenced by viewing images of peers on social media."

"While the present study did not collect data about participants' own body image, findings support the importance of girls' varying levels of body image when developing interventions aimed at reducing negative appearance comparisons when using social media."

"The findings of this study suggest that social media can have a negative influence on girls' body image through negative appearance comparisons when viewing images on social media, exacerbating appearance-based concerns and body dissatisfaction."

Conclusion

"This study provides some insight into the influence of social media on adolescent girls' body image from the perspective of girls in Perth, Western Australia."

"Further research should engage with adolescent girls to identify and investigate the impact of strategies to prevent and counteract negative body image related to social media utilizing an ecological approach to encompass all aspects of girls' lives."

5.3.6 "It Just Sends the Message That You're Nothing But Your Body" a Qualitative Exploration of Adolescent Girls' Perceptions of Sexualized Images on Social Media [498]

This is a machine-generated summary of:

Papageorgiou, Alana; Fisher, Colleen; Crossa, Donna: "It just Sends the Message that you're Nothing but your body" a Qualitative Exploration of Adolescent Girls' Perceptions of Sexualized Images on Social Media [498]

Published in: Sexuality & Culture (2022)

Link to original: https://doi.org/10.1007/s12119-022-10022-6

If you want to cite the papers, please refer to the original.

For technical reasons we could not place the page where the original quote is coming from.

Abstract-Summary

"This qualitative study used in-depth interviews to explore adolescent girls' perceptions of sexualized images they typically find when using social media."

"Twenty-four participants aged 14–17 years described sexualized images of females as normalized on social media."

"The interplay between gendered and social norms that endorsed and rewarded girls for posting sexualized images was seen to influence an expectation for girls to conform with their peers and post such images of themselves."

"There were tensions between whether girls' sharing of sexualized images of themselves on social media would be interpreted as a display of confidence (socially acceptable) or attention seeking (socially unacceptable)."

Introduction

"The sexualization of adolescent girls on social media has raised concerns among the community and researchers alike (Cookingham & Ryan, [203]; Hartas, [204]; Perloff, [133]; Skowronski and others, [205])."

"A prominent source in which adolescent girls are subjected to the sexualization of females, as described in the APA's definition, is within their use of social media."

"Of these adolescents, 70% of girls use social media every day compared to 56% of boys (Rideout & Robb, [206])."

"The high frequency in which adolescent girls use social media combined with the high prevalence of sexualized images of females on these platforms warrant further exploration of girls' perceptions of such images and how these might influence their mental health."

Sexualization on Social Media and Influences on Adolescent Girls' Mental Health

"Two mechanisms that may play a role in the relationship between sexualized images on social media and adolescent girls' mental health are that of self-objectification and social norms."

"While not focused on sexualized images specifically, research has also explored the relationship between adolescent girls' and young women's 'selfie' (self-taken images) behaviors on social media, self-objectification, and negative body image and disordered eating (Cohen and others, [207]; Wang and others, [208])."

"Self-objectification, therefore, may help to explain how sexualized images on social media influence adolescent girls' mental health."

"The influence of sexualized images on social media on adolescent girls' mental health is likely to involve an interplay between self-objectification and social norms, both which occur within broader sociocultural discourses related to the sexualization of females."

"This paper is part of a broader study exploring how sexualized images of females on social media might influence adolescent girls' mental health, in positive and/or negative ways."

Method

"This study, informed by Objectification Theory (Fredrickson & Roberts, [160]) and Social Cognitive Theory (Bandura, [209]), utilized a generic qualitative research design using one-on-one in-depth interviews to better understand how adolescent girls discern and interpret sexualized images typically found when using social media."

"To recruit participants through Student Edge, eligible members (adolescent girls aged 14–17 years in the capital city of an Australian state) were emailed an explanation of the study together with a screening survey to confirm their suitability to participate."

"The names and contact details of those who completed the screening survey (n = 45) were collated by Student Edge and provided to the first author who then contacted adolescent girls to arrange interviews."

"Written informed consent was obtained from both participating adolescent girls and their parent/guardian prior to or at the time of interviews."

Results

"Time spent using social media varied among participants from more than four hours to less than two hours per day on a weekday."

"Social media use of most participants increased to more than four hours per day on the weekend."

"Participants described an 'interplay between gendered and social norms', 'personal value based on appearance', 'tensions between confidence and attention seeking' in relation to sexualized images of females typically found in their use of social media."

"Participants also referred to the 'responsibility' placed on girls to protect themselves and prevent any risks associated with sharing such images."

Sexualization – as Understood by Girls

"They discussed sexualization as including a focus on females' bodies and referred to examples of imagery where girls are portrayed in sexually suggestive poses; I think of posters for perfume."

"It just makes them think that they're not equal to men and that they are only here for sex. (Julia, 14 years) Three of the four images shown in interviews were perceived by participants as sexualized."

"The images perceived as sexualized included a selfie of Gigi Hadid wearing a robe, open to show her cleavage in a push up bra and midriff, while pouting her lips; a selfie of Kendall Jenner in a bedroom wearing a bikini, focused on her stomach and bikini bottoms in a sexually suggestive pose; and a photo of Kylie Jenner holding a dog while posing on a couch in underpants and a tight t-shirt while pouting her lips."

"The image not perceived by the participating girls as sexualized was of Selena Gomez, taken from the ground on an upward angle to focus on her body while leaning on a balcony."

The Normalization of Sexualized Images of Females on Instagram

"Through participants' discussions about sexualized images of females on social media, and Instagram specifically, it became apparent that girls perceived such images as normalized on this platform."

"They're nothing like too scandalous or out of the norm for these days of what people are posting. (Grace, 15 years) Participants perceived that most girls would not be able to identify such images as sexualized or recognize the potential impacts to their wellbeing because of how normalized they are; I think people don't even pick up on it [sexualization of girls on social media]."

"I guess a lot of girls post them [sexualized photos] and it's just like a regular thing to be scrolling through your feed and finding things like that of people my age. (Ava, 14 years)"

Interplay Between Gendered and Social Norms
"I think people just post those pictures to boost their self-confidence sometimes. (Lucy, 14 years) While the quotes above illustrate the role of peers in contributing to girls' posting norms on Instagram, the role of female celebrities and the sexualized images they frequently post of themselves in influencing girls to post similar images was also apparent in participant discussions."

"They need to be like this person, skinny like this person, or perfect like this person, when in reality, it's not who they are. (Amelia, 16 years) I know sometimes I'm going through Instagram, seeing all these girls with a lot of makeup and everything makes me feel more obliged, I suppose, to wear makeup and more self-conscious about myself. (Abbey, 17 years) I think we view it [sexualized images of girls] quite a lot."

Tension Between Confidence and Attention Seeking
"Girls who shared sexualized photos of themselves on Instagram were considered by participants as confident, and they aspired to have the confidence to do this themselves; They [other girls] must be very confident to post themselves on social media in that way. (Josephine, 14 years) It's like you go through girls' Instagrams and they'll have a lot of photos kind of like that [images shown of celebrities posing provocatively], them posing, [and you think] "Oh, do I have to post stuff like that?""

"but then seeing that [image shown of celebrity posing provocatively], I'm thinking "Well, she kind of wants something." (Tahlia, 16 years) If you're doing it [posting a revealing photo of yourself on social media] to show that you're confident in your body, I reckon it's good, but it depends how explicit the photo is."

Responsibility
"Participants discussed the education and advice given to girls from adults and school in relation to potential risks and harms associated with viewing and/or sharing sexualized images on social media."

"For those who had, the onus was again on girls to protect themselves to reduce any potential negative consequences; The other week, we were talking about that [sexualization on social media], like, how girls can sexualize themselves and they were just saying, "Be careful what you're doing.""

"We were just talking about how people use their body in the wrong way online. (Madeleine, 14 years) Some participants, also believed that the onus was on girls for any negative consequences experienced as a result of sharing sexualized photos on social media; And if they [girls] want to sexualise themselves, I guess that's their option, that they've chosen, and then they have to deal with the repercussions if it goes south."

Discussion

"Findings from this study showed that adolescent girls perceived sexualized images of females on social media as normalized, reflecting an interplay between gendered and social norms of how girls should behave on social media and that such images promoted personal value based on appearance."

"It is also important to consider the social norms related to posting sexualized images on social media among girls' peers when targeting self-objectification in interventions aimed at addressing social media use among adolescent girls, to normalize challenging and critiquing sexualized images (van Oosten, [210])."

"Understanding the interplay between the gendered and social norms that adolescent girls encounter on social media where sexualized images are rewarded and place an expectation on girls to conform and behave this way themselves, and personal value based on appearance and self-objectification, can assist families, schools and youth mental health providers to support adolescent girls."

Limitations

"Despite not necessarily being able to apply their knowledge when using social media, girls in this study appeared to have high levels of knowledge in relation to gendered representations and the sexualization of females on social media."

"Age and more years of experience using social media may have influenced the study participants' interest in issues relation to social media and thus their interest in participating in the study leading to a sampling bias."

"Participating girls in this study all used the same social media platform most frequently (Instagram), and thus their responses during interviews were only discussed in relation to this type of social media."

"Given the important influence of peers highlighted in this study, findings may differ when girls discuss their perceptions of sexualized images on social media with other girls who are a similar age."

Conclusion

"They described how gendered and social norms contribute to the normalization of sexualized images on social media and an expectation for girls to conform and post sexualized images of themselves to gain approval and acceptance from their peers."

"The findings highlight the complexities of how adolescent girls navigate and interpret the sexualization of females both on social media and in society more broadly, and how they may be conflicted between their awareness of the impacts of sexualization and gendered and social norms and their ability and confidence to reject or challenge these notions to conform with their peers."

"When considering how sexualized images on social media might then influence adolescent girls' mental health, it is crucial to recognize the social norms and rules related to sexualized ideals of femininity within broader sociocultural discourses that influence how girls interpret and engage with such images and their self-efficacy in challenging these."

5.3.7 Body Surveillance on Instagram: Examining the Role of Selfie Feedback Investment in Young Adult Women's Body Image Concerns [499]

This is a machine-generated summary of:

Butkowski, Chelsea P.; Dixon, Travis L.; Weeks, Kristopher: Body Surveillance on Instagram: Examining the Role of Selfie Feedback Investment in Young Adult Women's Body Image Concerns [499]

Published in: Sex Roles (2019)

Link to original: https://doi.org/10.1007/s11199-018-0993-6

If you want to cite the papers, please refer to the original.

For technical reasons we could not place the page where the original quote is coming from.

Abstract-Summary

"We employ objectification theory to explore how value placed on selfie feedback among young women relates to markers of body image disturbance, including body dissatisfaction, drive for thinness, and bulimia action tendencies."

"We conducted an online survey of 177 English-speaking young adult (18–30 years-old) female Instagram users who were recruited through MTurk and who post selfies to examine the relationship between investment in selfie feedback and body image disturbance."

"We found that women who reported higher investment in selfie feedback were more likely to express body dissatisfaction and drive for thinness through the indirect influence of body surveillance, but this effect did not extend to bulimic tendencies."

"Our findings suggest that young adult female Instagram users who value audience responses to their selfies are more likely to exhibit disordered eating attitudes and intentions."

Selfie Feedback Investment

"The total number of likes, positive comments, and negative comments a post receives can also serve as powerful tools for self-assessment and validation."

"Despite these recent findings, little research has explored the effects of social media feedback on users' posting behaviors and self-concepts, particularly in regard to body image and surveillance."

"Translating the foundation of this interpersonal research to social media contexts, Ramsey and Horan [211] found that individuals who posted more highly self-sexualized images on Facebook and Instagram generally received more likes and followers than those who posted fewer objectifying photos on both platforms."

"Photos of people, including selfies, tend to receive more likes and comments on social media websites such as Instagram than images without people do (Bakhshi and others [212])."

"Selfies are appearance-oriented self-representations, and the feedback they earn can be interpreted as affirmations of attractiveness and social success."

Body Surveillance

"Objectification theory (Fredrickson and Roberts [160]) provides a framework for studying observed relationships between exposure to objectification of women in popular culture and disordered eating."

"Fredrickson and Roberts [160] suggest that repeated exposure to sexual objectification of female bodies in mass media or social interaction can lead women to self-objectify by "[internalizing] an observer's perspective of the self" (p. 177)."

"Researchers have begun employing objectification theory to study social media contexts."

"Social media has the potential to serve as a particularly effective tool for the body surveillance behaviors detailed by objectification theory."

"Existing applications of objectification theory to social media are limited and focus primarily on users' social media exposure and posting behaviors as opposed to the effects of the feedback responses these posts generate from audiences of friends, family, and unknown others (Cougar Hall and others [213]; Fardouly and others [214]; Rose and others [215])."

Body Image Disturbance

"Examined markers include body dissatisfaction, drive for thinness, and bulimia action tendencies (Garner and others [216]; Grabe and others [217])."

"Bulimia action tendencies is a distinctly behavioral component of body image disturbance that reflects proclivities toward excessive overeating followed by purging or vomiting."

"With regard to social media photos, Meier and Gray [218] suggest that individuals who dedicate their Facebook use to photo-sharing are more likely to experience body dissatisfaction and drive for thinness than do those who prioritize other aspects of Facebook use."

"With regard to selfies specifically, Cohen and others [207] found that investment in selfie sharing activities on social media was correlated with body dissatisfaction and bulimia action tendencies."

Other Factors

"African Americans, Latino/as, and Asians post online more frequently than Whites do, and they tend to use social media for different purposes, such as forming close connections with niche communities (Correa and Jeong [219])."

"African Americans may also present themselves differently in social media images than Whites do (Kapidzic and Herring [220])."

"Several studies have demonstrated that media exposure affects Women and Girls of Color differently than White women and girls, particularly with regard to body image."

"Milkie [221] found that Black adolescent girls did not experience significant negative body image effects from exposure to thin-ideal mass media images due to their tendency to favor depictions of White women."

"Because the Participants of Color did not view these White models as sufficiently similar to themselves or their peers, they did not internalize the idealized images in relation to their own bodies."

The Present Study

"We test the relationship between young adult women's investment in the likes and comments they receive on Instagram selfies and their body image disturbance expressed through symptomology-focused markers of body dissatisfaction, drive for thinness, and bulimia action tendencies."

"Investment in selfie feedback will be positively associated with (a) body dissatisfaction (perceptual), (b) drive for thinness (attitudinal), and (c) bulimia action tendencies (behavioral) (Hypothesis 1)."

"Body surveillance will mediate the relationship between investment in selfie feedback and (a) body dissatisfaction (perceptual), (b) drive for thinness (attitudinal), and (c) bulimia action tendencies (behavioral) (Hypothesis 2)."

Method

"Using 7-point semantic differential scales, participants rated the degree that "other people's feedback (comments and likes) to selfies I post on Instagram" is unimportant/important, useless/useful, uninteresting/interesting, and boring/entertaining ($\alpha = .86$, $M = 4.45$, $SD = 1.23$)."

"The nine-item body dissatisfaction subscale was utilized to assess participants' perceptions of their bodies as too fat or large ($\alpha = .86$, $M = 3.98$, $SD = 1.34$)."

"The seven-item bulimia action tendencies subscale was utilized to assess participant likelihood to enact disordered eating behaviors such as binge eating and self-induced vomiting that typify bulimia ($\alpha = .88$, $M = 3.49$, $SD = 1.56$)."

"Participants selected a response from half-hour increments ranging between 30 min and more than 20 h. The mean response was roughly 3 h. Participants also reported the frequency with which they posted selfies to Instagram during the last 3 months on a seven-item scale ranging from 0 (Never) to 7 (Multiple times per day)."

Results

"Selfie feedback investment was not positively correlated with body dissatisfaction to produce a significant nonmediated total effect as in c-path ($b = .11$, $SE = .07$), $t(170) = 1.57$, $p = .117$, or a significant mediated direct effect in c'-path ($b = -.01$, $SE = .07$), $t(170) = -.08$, $p = .938$."

"Of the five control variables, BMI exerted a significant positive effect ($b = .10$, $SE = .13$), $t(170) = 7.91$, $p < .001$, on b-path between body surveillance and body dissatisfaction."

"When the mediating influence of body surveillance was accounted for, however, the direct effect of selfie feedback investment on drive for thinness became nonsignificant in c'-path ($b = .06$, $SE = .09$), $t(170) = .68$, $p = .500$."

"BMI was the only control variable to affect the mediation, explaining variance in the b-path relationship between body surveillance and drive for thinness ($b = .07$, $SE = .02$), $t(169) = 3.97$, $p < .001$."

Discussion

"We found that participants' investment in selfie feedback was positively related to only the perceptual (body dissatisfaction) and attitudinal (drive for thinness) dimensions of body image disturbance through the mediating influence of body surveillance tendencies."

"Although Garner and colleagues (Garner and others [216]) associate body dissatisfaction and drive for thinness with clinical cases of anorexia nervosa, investment in selfie feedback was not significantly related to our behavioral body disturbance outcome (bulimia action tendencies) even when body surveillance was employed as a mediator."

"In determining how individuals value social media feedback on selfies differently based on their distinct characteristics, it would be beneficial to examine significant correlates of investment in selfie feedback and body surveillance beyond those measured in our survey."

"The present study demonstrates that attitudes toward social media feedback received on selfies are related to body image disturbance in young women."

[Section 9]

"In examining the relationship between selfie feedback investment and body image, we contribute to a developing body of research regarding online media use behaviors and young women's health (Karsay and others [222])."

"To examine this phenomenon, we conducted a survey of English-speaking young adult female Instagram users who post selfies with regard to their role as social media content creators, including their investment in selfie feedback, body surveillance tendencies, and body image disturbance."

"It adds to investigations of the technological affordances that may be behind observed relationships between social media use and body image disturbance for young women and girls and addresses a gap in the objectification theory literature."

"Young women's attitudes toward feedback on appearance-oriented social media posts, such as selfies, are important to studying its potential effects on well-being."

5.3.8 #fitspiration: A Comparison of the Sport-Related Social Media Usage and Its Impact on Body Image in Young Adults [500]

This is a machine-generated summary of:

Klier, Kristina; Rommerskirchen, Tessa; Brixius, Klara: #fitspiration: a comparison of the sport-related social media usage and its impact on body image in young adults [500]

Published in: BMC Psychology (2022)

Link to original: https://doi.org/10.1186/s40359-022-01027-9

If you want to cite the papers, please refer to the original.

For technical reasons we could not place the page where the original quote is coming from.

Abstract-Summary

"Following and posting sport-related content on social media is wide-spread among young people."

"Little is known about the interdependence between sport-related social media use and the thereby perceived personal body image."

"We conducted an online survey (N = 285) to examine how social media influences the sport-related body image."

"Social media are frequently used for sport (n = 136, 47.7%)."

"Resistance training correlated significantly with several motives of sport-related use of social media, and thus, represents the strong online presence of athletic sports."

"Regarding the perception of body image, it was found that the group of rejecting (negative) body image significantly correlated with the emulation of social media mediated sport-related beauty and body ideals ($r = 0.63$, $p = 0.001$), as well as with increased body dissatisfaction when viewing sport-related posts on social media ($r = 0.590$, $p = 0.001$)."

Introduction
"Social media offer platforms for presenting socially standardized body ideals [223]."

"Fitspiration-related content such as 'before and after' images, sugar-free or low-carb food, body tuning workouts and so on are intended to show the ideal of health and fitness."

"Studies indicate that the influence of the internet and especially of social media to convey body ideals is higher today than that of television and print media [224]."

"Social media are also responsible for shaping specific body ideals."

"A current guiding theme is the slim and athletic body: The Athletic Ideal, a body ideal, which emerged from a new understanding respectively a new level of perfection prevailing on social media as well as in wide-spread fitness magazines."

"The present study aimed to investigate the sport-related use of social media and the resulting influence on body image in a young, healthy sample."

Methods
"For investigation, we used the standardized body image questionnaire (FKB-20) by Clement and Löwe [225] and added further questions about participants' sportive background and (sport-related) social media usage behavior."

"The fourth section was related to the overarching themes of sport, social media and body image and were questioned by using a 5-point Likert scale (1 = does not apply at all, 5 = fully applies)."

"I actively look for confirmation of my own body image on social media."

"Further, we calculated chi-square and Cramer's V for normally and ordinally scaled data in order to examine the relationship between sport, social media and body image."

Results
"The tendency to emulate ideals of beauty and body shape on social media could only be significantly predicted by doing resistance training ($p < 0.001$)."

"The rejecting body image and the emulation of sport-related ideals of beauty and body shape correlated significantly ($p = 0.001$) showing a strong effect ($r = 0.63$)."

"The RBI showed a strong positive correlation with dissatisfaction towards the own body when viewing sport-related social media posts ($r = 0.60$, $p = 0.001$), and with the feeling of social pressure regarding the perfect body image on social media ($r = 0.50$, $p = 0.001$)."

"Also the correlation of the RBI and the comparison with the body ideals on social media resulted in a moderate to strong relation ($r = 0.48$, $p = 0.001$)."

Discussion

"The study aimed to examine the relationship between the sport-related use of social media and its influence on body image."

"Especially individuals engaging in athletic sports seem be more influenced by social media which in turn also correlates with less self-consciousness regarding their body image."

"The present study was also able to provide an informative insight into the sport-related use of social media."

"Our results showed a significant correlation between the engagement in resistance training and the motive for sport-related use of social media."

"We could not find any correlations between the sport-related use of social media and the motive of self-staging."

"The motives for sport-related use of social media mainly aim for the optimization of the body and allow the assumption that the participants may be concerned with conforming to an ideal of beauty and body shape."

Limitations

"In our study, we mostly related to photo- and video-based sporting social media content as they play a special role for the representation of ideals of beauty and body shape."

"The questioned target group might be a limiting factor itself as its acquisition via local sports clubs and Instagram per se addressed individuals who are sportively active and have an affinity to social media."

"It can be assumed that hereby the sport-related use of social media automatically gains more importance than it probably might in other target groups."

"The direction of the connections between the sport-related use of social media and its influence on the individual's body image emerged only superficially in our study, and needs to be examined more comprehensively."

Conclusion and future directions

"The findings of the present study reveal the important role as well as the influence of social media on sport-related use and the resulting impact on young person's body image."

"Social media do have the potential via sport-related content to motivate people to do sports, but at the same time, also some negative effects can be observed that should be prevented."

"Our results show that the influence of social media on the sport-related body image varies between person."

"The attempt to achieve the given ideal of beauty is associated with effort and social pressure."

"Tailored research is needed to assess the impact of sport-related use of social media more differentiated, and to establish, for example, suitable awareness interventions and user guidelines."

5.3.9 Instagram as a Digital Mirror: The Effects of Instagram Likes and Disclaimer Labels on Self-awareness, Body Dissatisfaction, and Social Physique Anxiety Among Young Italian Women [501]

This is a machine-generated summary of:

Di Gesto, Cristian; Matera, Camilla; Policardo, Giulia Rosa; Nerini, Amanda: Instagram As A Digital Mirror: The Effects of Instagram Likes and Disclaimer Labels on Self-awareness, Body Dissatisfaction, and Social Physique Anxiety Among Young Italian Women [501]

Published in: Current Psychology (2022)

Link to original: https://doi.org/10.1007/s12144-021-02675-7

If you want to cite the papers, please refer to the original.

For technical reasons we could not place the page where the original quote is coming from.

Abstract-Summary

"This study examined the effects of being appreciated on Instagram (numbers of Likes) and knowing whether or not the physical appearance of a person depicted in an Instagram image has been altered (Disclaimer)—on women's self-awareness, body dissatisfaction, and social physique anxiety, both independently and in combination."

"The study used a 2×2 between-participants design in which the number of Likes (Low vs. High) and the image Disclaimer (Presence vs. Absence of information about digital enhancement) were manipulated."

"Findings showed that participants exposed to an Instagram image with many Likes reported higher body dissatisfaction and social physique anxiety compared to those with few Likes."

"When many Likes were associated with an Instagram image, the presence of a Disclaimer was found to reduce women's self-awareness, lowering levels of attentional focus."

Introduction

"Through the present study, we aimed to extend the literature on the impact that specific information accompanying images on Instagram can have on women's body image, by examining experimentally the combined effect of the number of Likes and the presence of a Disclaimer."

"It is our contention that the presence of a Disclaimer can attenuate the negative effect that Likes attached to an Instagram image can have on women's body image:

whereas the numbers of Likes might increase women's self-awareness, body dissatisfaction, and social physique anxiety by means of comparison with an ideal self, the knowledge that the image is unrealistic might reduce these negative effects."

"Hypothesis 2 (H2): The presence of a Disclaimer related to an Instagram image will attenuate the effect that the number of Likes has on women's self-awareness, body dissatisfaction, and social physique anxiety."

Material and methods

"The study used a 2 × 2 between-participants design in which the number of Likes (Low vs. High) and the image Disclaimer (Presence vs. Absence) were manipulated."

"To ensure that the Low and the High Likes were perceived as such, before showing the experimental stimulus, participants were told that they were going to see an image posted on Instagram that received either a few (Low Likes condition) or many (High Likes condition) Likes."

"Participants were asked to give a series of sociodemographic details—Body Mass Index (BMI), use of Instagram, and internalization of thin ideals—to investigate whether there were differences between the experimental conditions with respect to these variables."

"We asked the participants to take part in a study on body image and Instagram use."

Results

"Participants in the High Likes condition reported viewing significantly more Likes ($M = 5.70$, $SD = .09$) than those in the Low Likes condition ($M = 1.98$, $SD = .09$, $F_{(1, 200)} = 802.66$, $p < .001$, $\eta_p^2 = .80$)."

"Participants also reported that the physical appearance of the person depicted in the image was more likely to have been altered in the Presence of the Disclaimer condition ($M = 2.97$, $SD = .12$) than in the Absence of the Disclaimer condition ($M = 1.56$, $SD = .12$, $F_{(1, 200)} = 70.37$, $p < .001$, $\eta_p^2 = .26$)."

"The ANCOVA with self-awareness as the dependent variable revealed no significant effect for either Likes ($F_{(1, 200)} = 1.26$, $p = .26$) or Disclaimer ($F_{(1, 200)} = .26$, $p = .61$)."

"There was a significant Likes × Disclaimer interaction effect for this variable ($F_{(1, 200)} = 4.03$, $p < .05$, $\eta_p^2 = .02$)."

Discussion

"The main aim of the present study was to examine experimentally the effects (if any) of two factors—(a) being appreciated on Instagram (expressed in the numbers of Likes attached to an image) and (b) knowing whether or not the physical appearance of a person depicted in an Instagram image has been altered or modified—on women's self-awareness, body dissatisfaction, and social physique anxiety, both independently and in combination."

"Our findings also support H2 to a certain degree: the presence of an Instagram image Disclaimer attenuated the effect that the number of Likes had on women's self-awareness."

"Future studies are needed to test the effectiveness of these programs in mitigating the deleterious effects that some specific Instagram features (i.e., number of likes) can have on young women's physical appearance, and body dissatisfaction and social physique anxiety in particular."

5.3.10 Body Shame and Problematic Social Networking Sites Use: The Mediating Effect of Perfectionistic Self-presentation Style and Body Image Control in Photos [502]

This is a machine-generated summary of:

Fioravanti, Giulia; Bocci Benucci, Sara; Vinciarelli, Viola; Casale, Silvia: Body shame and problematic social networking sites use: the mediating effect of perfectionistic self-presentation style and body image control in photos [502]

Published in: Current Psychology (2023)

Link to original: https://doi.org/10.1007/s12144-023-04644-8

If you want to cite the papers, please refer to the original.

For technical reasons we could not place the page where the original quote is coming from.

Abstract-Summary

"Previous studies have shown that body shame and body image control in photos are associated with Problematic Social Networking Sites Use."

"The current study aims to build upon previous evidence by examining the role of perfectionistic self-presentation in the relationship between body shame, on the one hand, and body image control in photos and Problematic Social Networking Sites Use, on the other hand."

"The hypothesized serial mediation effect of perfectionistic self-presentation style and body image control in photos in the association between body shame and Problematic Social Networking Sites Use was significant (indirect effect = 0.05; 95% CI: [0.02, 0.07] for young women; indirect effect = 0.02; 95% CI: [0.002, 0.04] for young men)."

"The gratification of the need for self-presentation through body image control in Social Networks photos activities could in turn lead to Problematic Social Networking Sites Use."

Introduction

"Building upon previous evidence, the current study focuses the attention on the potential role of perfectionistic-self-presentation as a behavioral strategy that might explain how body shame leads to body image control in photos which, in turn, leads to Problematic Social Networking Sites Use."

"Previous studies found that body shame is positively associated with body image control in photos, which, in turn, is associated with Problematic Social Networking Sites Use (Gioia and others, [226])."

"The goal of the present study was to provide an integration of these two research lines by testing a model that proposes that body shame generates the need to hide perceived imperfections (i.e., a perfectionistic self-presentation style), which leads to body image control in photos, which, in turn, leads to Problematic Social Networking Sites Use."

Materials and methods

"A web link directed the participants to the study website, and if they consented to participate, they were asked to answer some demographic information, questions about their use of Social Networks and to complete self-report questionnaires."

"This subscale is composed of eight items and measures shame over one's body that extends to the self as a result of a perceived failure to achieve cultural ideals and social standards of appearance."

"Answers are rated on a seven-point Likert scale, from 1 (strongly disagree) to 7 (strongly agree), and the scale comprises 3 subscales: Perfectionistic Self-Promotion (e.g., "I try always to present a picture of perfection") which contains 10 items that reflect the need to appear perfect to others; Nondisplay of Imperfection (e.g., "It would be awful if I made a fool of myself in front of others"), which includes 10 items that assess the need to avoid appearing imperfect to others."

"Higher scores indicate greater Problematic Social Networking Sites Use."

Results

"Perfectionistic self-presentation style mediated the relation between body shame and Problematic Social Networking Sites Use (indirect effect = 0.11; 95% CI: [0.05, 0.18]) and body image control in photos mediated this relationship (indirect effect = 0.09; 95% CI: [0.06, 0.14])."

"The hypothesized serial mediation effect of perfectionistic self-presentation style and body image control in photos on the association between body shame and Problematic Social Networking Sites Use was significant (indirect effect = 0.05; 95% CI: [0.02, 0.07])."

"The hypothesized serial mediation effect of perfectionistic self-presentation style and body image control in photos on the association between body shame and Problematic Social Networking Sites Use was significant (indirect effect = 0.02; 95% CI: [0.002, 0.04])."

Discussion

"The study tested a serial mediation model to explore the role of perfectionistic self-presentation style and body image control in photos as mediators in the relationship between body shame and Problematic Social Networking Sites Use."

"The serial mediating role of perfectionistic self-presentation style and body image control in photos in the association between body shame and Problematic Social Networking Sites Use emerged, and this path was confirmed among both young men and women."

"The current study contributes to the existing debate about the psychological correlates of Problematic Social Networking Sites Use, by (i) confirming the association between body shame, body image control in photos and problematic social networking and (ii) evidencing for the first time the role of perfectionistic self-presentation style as a motivational factor involved in the relationship between body shame and Problematic Social Networking Sites Use."

Conclusion

"The present study contributes to the understanding of the possible predictors of Problematic Social Networking Sites Use and demonstrates novel findings in the field of self-objectification research."

"The previously unexplored role of perfectionistic self-presentation style in the association between self-objectification dimensions (i.e., body shame and body monitoring) and Problematic Social Networking Sites Use was confirmed in the current study."

"The gratification of the need for self-presentation through body image control in social networking sites photo activities could in turn lead to Problematic Social Networking Sites Use."

5.3.11 Adolescents' Body Shame and Social Networking Sites: The Mediating Effect of Body Image Control in Photos [503]

This is a machine-generated summary of:

Gioia, Francesca; Griffiths, Mark D.; Boursier, Valentina: Adolescents' Body Shame and Social Networking Sites: The Mediating Effect of Body Image Control in Photos [503]

Published in: Sex Roles (2020)

Link to original: https://doi.org/10.1007/s11199-020-01142-0

If you want to cite the papers, please refer to the original.

For technical reasons we could not place the page where the original quote is coming from.

Abstract-Summary

"Attention directed toward online physical appearance appears to trigger some body-related behaviors (e.g., body image monitoring) which potentially are related to self-objectification and problematic SNS use."

"The present study evaluated the previously unexplored predictive role of body shame on SNS use, testing the mediating effect of body image control in photos on SNS."

"Results only partially confirmed the direct effect of body shame on problematic SNS use."

"Body shame strongly predicted body image control in SNS photos, leading indirectly to both male and female adolescents' problematic social media use."

Self-Objectification Experiences and Social Networking

"Scholars' findings highlighted how men and male adolescents are increasingly facing self-objectification experiences and becoming involved in body image-related activities (Daniel and Bridges [227]; Dakanalis and others [228, 229]; Holland and Tiggemann [137]; Karsay and others [222]; Manago and others [230]; Moradi [231]; Moradi and Huang [232]; Vandenbosch and Eggermont [233])."

"Ubiquitous online social media are increasingly replacing traditional mass media, and the use of online platforms (including the creation of content and peer interactions) seems to provide a novel and highly accessible medium for male and female individuals' socializing with self-objectification experiences and objectified body consciousness (Caso and others [234]; Cohen and others [207]; De Vries and Peter [235]; Manago and others [230])."

"A more recent study showed the predictive role of older Chinese adolescents' body-related conversations on SNSs on body shame via body surveillance (Wang and others [236])."

"On SNS profiles, individuals appear to literally look at themselves from an observer's perspective (Fardouly and others [237]), confirming a clear interrelationship between social networking and self-objectification."

"Only one known previous study has explored the association between self-objectification issues and problematic social networking."

The Current Study

"How individuals interact with their body image prior to posting on SNSs, often dealing with experiences of shame, appears a matter of interest for body image-related issues and problematic social networking (Cohen and others [207])."

"In the present study we evaluated the main and indirect effects of body shame and body image control in photos on adolescents' problematic SNS use."

"We predicted that body shame would be positively associated with problematic SNS use and that body image control in photos would significantly mediate this relationship (Hypothesis 1)."

"Higher body shame is expected to be positively associated with greater body image in photo monitoring, which in turn would be related to greater problematic social networking (i.e., preference for online social interactions, mood regulation, cognitive preoccupation, compulsive SNS use, and negative outcomes)."

Method

"For the purposes of the present study we used the full continuous measure of BICP-R. The 15-item GPIUS2 (Caplan [238]) assessed the degree of generalized problematic internet use assessing five constructs: preference for online social interactions (e.g., "I prefer communicating with people online rather than face-to-face"), mood regulation (e.g., "I have used the internet to make myself feel better when I was down"), cognitive preoccupation (e.g., "I think obsessively about going online when I am offline"), compulsive internet use (e.g., "I have difficulty controlling the amount of time I spend online"), and negative outcomes (e.g., "My internet use has created problems for me in my life")."

Results

"Female adolescents reported higher mean scores in OBCS body shame, BICP selfie-related factors, BICP positive body image factors, BICP negative body image factors, PSNSU mood regulation, PSNSU cognitive preoccupation, and PSNSU compulsive SNS use."

"In the male sample, body shame had a significant direct effect on BICP, PSNSU mood regulation, PSNSU cognitive preoccupation, and PSNSU negative outcomes."

"BICP was significantly and positively associated with PSNSU preference for online social interactions, PSNSU mood regulation, PSNSU cognitive preoccupation, PSNSU compulsive SNS use, and PSNSU negative outcomes though, in the female sample, body shame had a significant direct effect on BICP, PSNSU mood regulation, and PSNSU negative outcomes."

"In the female sample, there were more significant and stronger direct effects of BICP on PSNSU preference for online social interactions, PSNSU mood regulation, PSNSU cognitive preoccupation, PSNSU compulsive SNS use, and PSNSU negative outcomes."

Discussion

"The study surveyed a sample of Italian adolescents and tested a mediation model to explore the predictive role of body shame on problematic SNS use via the body image control in photos across male and female subsamples."

"Similar to Boursier and Manna [239] who found a positive correlation between body appearance control in photos and adolescents' problematic internet use, the findings of the present study showed the strong and positive co-occurrence of all dimensions of problematic SNS use and control on self-photos online and offline, especially among the female sample."

"According to Casale and Fioravanti [240], individuals who feel body shame might display problematic SNS use responding to their need to control self-presentation and facilitate social acceptance."

"The present study showed the unexplored effect of body shame and body image control in photos on problematic SNS use."

Conclusion

"The previously unexplored association between body shame and problematic social networking was confirmed in the present study."

"Adolescents who feel ashamed of their bodies because of the perceived discrepancy between their appearance and culturally promoted standard appear to be engaged in strategies of control over their body image in photos."

"The present findings showed that body shame and body image control in photos strongly influence problematic social networking, contributing to both the self-objectification research field and the ongoing debate on possible predictors of problematic SNS use."

"Despite the higher relevance of the relationship between body shame and problematic SNS use among female adolescents, similar findings among male adolescents might suggest that they are increasingly becoming engaged in self-objectification experiences."

[Section 7]

"Social networking sites (SNSs) are constantly available, virtual communities that allow users to create individual private or public profiles, interact with "offline" friends, meet other people sharing common interests, view others' activities, comment on others' activities, "like" others' activities, and share various forms of content (Balakrishnan and Griffiths [241]; Boursier and others [242]; Boyd and Ellison [243]; Cohen and others [207]; Holland and Tiggemann [137]; Kuss and Griffiths [244, 245]; Perloff [133]; Tiggemann and Slater [246])."

"SNS use allows users to (a) avoid face-to-face difficulties, (b) provide greater control over informational disclosure, and (c) be strategic in managing their own self-presentation, often through visual content (e.g., pictures, videos, stories)."

"It is likely that disembodied SNS environments and photo-based activities allow individuals who are dissatisfied with their appearance to create, construct, and edit their best self-presentation, overinvesting in their online body image (Boursier and Manna [247]; Casale and Fioravanti [240]; Cohen and others [207]; Fox and Rooney [248]; Lonergan and others [249]; Manago and others [230]; McLean and others [250])."

"Body image-related activities and strategies of self-presentation (such as appearance monitoring) might lead to problematic SNS use (Boursier and Manna [239]; Boursier and others [251]; Cohen and others [207])."

5.3.12 The Psychology of Social Networking: The Challenges of Social Networking for Fame-Valuing Teens' Body Image [504]

This is a machine-generated summary of:

Te'eni-Harari, Tali; Eyal, Keren: The Psychology of Social Networking: the Challenges of Social Networking for Fame-Valuing Teens' Body Image [504]

Published in: Philosophia (2017)

Link to original: https://doi.org/10.1007/s11406-016-9806-8

If you want to cite the papers, please refer to the original.

For technical reasons we could not place the page where the original quote is coming from.

Abstract-Summary

"The article argues that youth's exposure to thin-idealizing content posted by their adored celebrities on interactive and highly engaging social networking sites poses potential challenges for these young people."

"The article suggests that even ostensibly "innocent" SNS content that presents thin-idealizing images and messages about the body may place pressures on youth and encourage a preoocupation with and risky behaviors associated with body image even unintentionally."

"The article further recognizes that netusers, readers, and social networking sites all share in the responsibility to minimize the risks posed to youth in this context."

Social Networking Sites and Adolescent Development

"Adolescents spend about 9 h a day with entertainment media; 58% of teens use social networking sites (SNSs), averaging 2 h of use a day (Common Sense Media [252])."

"Social networking likely appeals to teens due to technical reasons, such as its accessibility on mobile technologies, but also because of the large range of interpersonal communication opportunities that it offers: to stay connected with others, make new friends, exchange ideas, and access information easily, blurring geographical boundaries and connecting people around the globe, mostly free from parental supervision (Dror & Gershon; Wiederhold [253])."

"Social networking seems to satisfy several adolescent developmental needs."

"One manner through which social networking responds to adolescents' fame valuation is through providing them with opportunities to "befriend" and follow famous celebrities on SNSs (Greenwood [254]; Soukup [255])."

The Mediated Thin-Ideal, Youth Body Image, and Social Networking Sites

"The concern is that SNS content, much like thin-idealizing messages in traditional media, promotes unhealthy body standards that may lead to negative self-perceptions without providing youth with the tools needed to critically assess this content and deal with the pressures it places on them as it sets "almost impossible standards for both women and men" (Tiggemann & Zaccardo, p. 6)."

"Adolescents' involvement with thin-ideal supportive messages through social networking may place youth at risk for negative body perceptions."

"The study suggests that even ostensibly "innocent" SNS content that presents thin-idealizing images and messages about the body may place pressures on youth and encourage a preoccupation with and risky behaviors associated with body image even unintentionally."

"Teens' preoccupation with appearances and body image, coupled with the above-mentioned high value they attribute to fame and public recognition (Uhls and Greenfield [256]), may drive them to seek social networking content that endorses both fame and the thin-ideal without providing critical tools with which to make healthy decisions."

Explaining the Link between Social Networking Sites, Youth Fame Valuation, and Body Image: Theoretical Perspectives

"Social comparison processes highly characterize adolescence, also relative to media models and, specifically, in the context of body-related comparisons (Clay and others [257]; Eyal and Te'eni-Harari [258])."

"Within the broad context of U&G theorizing, Perloff [133] suggested that social media, and among them SNSs, are an especially pertinent arena within which to explore media uses, gratifications, and body image effects."

"He offers a helpful theoretical framework incorporating the basic components of U&G logic and linking together social media and its gratifications, personal factors (i.e., low self-esteem, depression, and thin-ideal internalization) and other mediating psychological processes (e.g., identification and social comparison with mediated personae), and body image effects (i.e., body dissatisfaction, adverse moods, and, ultimately, eating disorders)."

"According to this model, social media's unique characteristics of enhanced interactivity, heavy reliance on the visual, and incorporation of peer communication all join together to create opportunities for negative body perceptions."

Conclusion

"We contend that youth's exposure to thin-idealizing content posted by their adored celebrities on interactive and highly engaging social networking sites poses potential challenges for these young, developing individuals."

"We argue that concern should be directed also at seemingly mundane online content which can potentially impact young people: the presentation of extreme and unhealthy body ideals – specifically, the thin-ideal – by adored celebrities in SNSs."

"In light of the dangers posed to adolescents by their exposure to thin-idealizing content posted by celebrities on social networks, there is clearly a need to protect youth from such online content."

"Indeed, it may be timely for the SNSs themselves to work alongside their users, generally, and celebrity users, specifically, toward the creation of a set of agreed-upon guidelines or codes for responsible content posting and sharing specifically with guidelines about thin-idealizing and digitally manipulated content."

[Section 5]

"Social networking sites (SNSs) serve as a central platform for adolescent online activity and engender both positive and potentially harmful outcomes for youth."

"The current article argues that there is a potential meaningful link between social networking sites and youth's body image perceptions; following Cohen-Almagor's [259] argument about the moral and social responsibility of different agents to online content and its consequences, the article further recognizes that netusers, readers, and social networking sites all share in the responsibility to minimize the risks posed to youth in this context."

"The article begins by reviewing youth use of SNSs and the ways in which these networks respond to the developmental needs of adolescents, continues by addressing the mediated thin ideal and its potentially dangerous presence and promotion within SNSs, and suggests theoretical perspectives that shed light on the uses and effects of SNSs on adolescents."

5.3.13 Body Image Comparisons on Social Networking Sites and Chinese Female College Students' Restrained Eating: The Roles of Body Shame, Body Appreciation, and Body Mass Index [505]

This is a machine-generated summary of:

Yao, Liangshuang; Niu, Gengfeng; Sun, Xiaojun: Body Image Comparisons on Social Networking Sites and Chinese Female College Students' Restrained Eating: The Roles of Body Shame, Body Appreciation, and Body Mass Index [505]

Published in: Sex Roles (2020)

Link to original: https://doi.org/10.1007/s11199-020-01179-1

If you want to cite the papers, please refer to the original.

For technical reasons we could not place the page where the original quote is coming from.

Abstract-Summary

"Based on social comparison, sociocultural, and objectification theories, the present study aimed to investigate the association between body image comparisons on SNS and restrained eating, as well as the mediating role of body shame and the moderating roles of body appreciation and body mass index (BMI) among Chinese young adult women."

"A sample of 567 Chinese college women were recruited to complete a questionnaire assessing body image comparisons on SNS, body shame, body appreciation, restrained eating, and information about weight and height."

"Body appreciation and BMI each moderated the association between body shame and restrained eating as well as the association between body image comparisons on SNS and restrained eating."

Body Image Comparisons on SNS and Restrained Eating

"To address this gap, in the present study we investigated whether body appreciation could moderate the relationship between body image comparisons on SNS and restrained eating as well as the mediating role of body shame among Chinese college women."

"We hypothesized that body appreciation would moderate the associations between body image comparisons on SNS and body shame (Hypothesis 3a), between body shame and restrained eating (Hypothesis 3b), and between body image comparisons on SNS and restrained eating (Hypothesis 3c); specifically, we predicted that these associations would be attenuated among individuals with higher body appreciation."

"Drawing on social comparison, sociocultural, and objectification theories, we investigated the relationship between body image comparisons on SNS and restrained eating among Chinese college women, as well as the mediating role of body shame and the moderating roles of body appreciation and BMI."

Method

"This scale has been successfully used with Chinese samples and was found to have good construct validity and internal consistency with a Cronbach's alpha of .89 among adolescents (Wei and others [260])."

"The Cronbach's alpha for the scale was .91."

"The measure has been successfully used with Chinese samples and was found to have good construct validity and internal consistency with a Cronbach's alpha of .82 among undergraduate students (Jackson and Chen [261]; Wang and others [236])."

"The BMI of participants ranged from 15.21 to 28.93 ($M = 19.83$, $SD = 2.03$)."

"Participants rated each of the 10 items on a 5-point scale from 1 (never) to 5 (very often)."

"This scale has been used with Chinese samples and was found to have good construct validity and internal consistency with a Cronbach's alpha of .86 among college women (Niu and others [262])."

Results

"Regression analysis held body image comparisons on SNS as the independent variable, restrained eating as the dependent variable, body shame as the mediator, and both body appreciation and BMI as moderators."

"The interaction between body image comparisons on SNS and body appreciation on restrained eating also was not significant ($b = .04$, $p = .52$), indicating that the relationship between body image comparisons on SNS and restrained eating was not moderated by body appreciation."

"Hypothesis 3c was not supported in that high levels of body image comparisons on SNS was related to more restrained eating independent of women's levels of body appreciation."

"The interaction between body image comparisons on SNS and BMI on restrained eating was also significant ($b = .07$, $p < .001$), showing that the relationship between body image comparisons on SNS and restrained eating was moderated by BMI."

Discussion

"We found the relationship between body shame and restrained eating was stronger among women with lower BMI, which is inconsistent with our hypothesis; however, the relationship between body image comparisons on SNS and restrained eating was stronger among women with higher BMI, which is consistent with our hypothesis."

"The present study revealed the mediating role of body shame in the association between body image comparisons on SNS and restrained eating, indicating that body shame is a proximal factor affecting disordered eating behaviors."

"Consistent with our hypothesis, the association between body image comparisons on SNS and restrained eating was stronger among women with higher BMI."

"The present study constructed a mixed model with body shame as a mediator, and body appreciation and BMI as moderators, to clarify the association between body image comparisons on SNS and restrained eating."

Conclusions

"The present study revealed the positive relationship between body image comparisons on SNS and restrained eating among Chinese college women, as well as the mediating role of body shame in this association."

"The findings indicate that body shame is of great value for explaining the relationship between body image comparisons on SNS and restrained eating."

"The findings also revealed that body appreciation can buffer the relationship between body shame and restrained eating, which has some implications for prevention interventions."

"The moderating effect of BMI indicated that both thinner and heavier (within a normal weight range) young women are vulnerable to developing restrained eating, but through different mechanisms."

[Section 6]

"Non-clinical restrained eating behavior adopted by worldwide young women in pursuit of slim and thin bodies has gradually been more and more popular."

"A survey of 850 Chinese college women also found that restrained eating is one of the most common means they use to lose weight (Wang and others [263])."

"Relevant studies found that consumption of media, especially sources that purvey images of ideal slender women, is one of the robust factors contributing to restrained eating (Mclean and others [264])."

"Relevant studies found that SNS use was associated with negative body image and disordered eating (Morgan and others [265]; Venus and others [266])."

"In the present study we aimed to investigate the association between body image comparisons on SNS and restrained eating among Chinese college women, as well as identity mediating and moderating mechanisms in this association."

5.4 Digital Self-psychological Aspects

Machine generated keywords: adolescent, selfie, behavior, social medium, medium use, medium, identity, appearance, personality, sns, selfesteem, self, scale, photo, selfpresentation

5.4.1 Digital Hyperconnectivity and the Self [506]

This is a machine-generated summary of:
Brubaker, Rogers: Digital hyperconnectivity and the self [506]
Published in: Theory and Society (2020)
Link to original: https://doi.org/10.1007/s11186-020-09405-1

If you want to cite the papers, please refer to the original.

For technical reasons we could not place the page where the original quote is coming from.

Abstract-Summary
"To recasting social interaction, culture, economics, and politics, it has profoundly transformed the self."

"It has created new ways of being and constructing a self, but also new ways of being constructed as a self from the outside, new ways of being configured, represented, and governed as a self by sociotechnical systems."

"Subsequent sections show how digital hyperconnectivity has engendered new ways of objectifying, quantifying, producing, and regulating the self—considered both as active, reflexive practices and as systemic, data- and algorithm-driven processes."

"I conclude by reflecting on the broader implications of contemporary modes of governing the self and by underscoring the ways in which hyperconnectivity has colonized the territories of the self, conscripting the self into the service of techno-social systems."

Exploring the self
"In Sherry Turkle's influential account, the mutual anonymity characteristic of these contexts allows people to adopt and enact digital personas of their own choosing, to "express multiple and often unexplored aspects of the self, to play with their identity and to try out new ones" ([267], p. 12)."

"Some play with multiple online identities that may have little or no relation to offline selves."

"Participants in multiple virtual worlds were "pioneers" whose "experience of constructing selves in cyberspace will become increasingly important" in a post-modern "culture of simulation" in which identity is "multiple, heterogeneous, and fragmented" (Turkle [268], p. 157)."

"Even if experimentation and play with multiple virtual identities in contexts of mutual anonymity are less common than Turkle suggested, the web is more than ever an expansive space of self-exploration and self-formation—and not just for participants in virtual worlds, but for everyone."

Emancipating the self

"If hyperconnectivity has failed to emancipate or uncouple the self from the body, it has nonetheless contributed powerfully—for better or worse—to emancipating the self from family and local community."

"While emancipation from the body is a postmodernist dream, emancipation from family and community is a core narrative of sociological modernity: the long-running story of the progressive erosion of familial and communal control over the means of socialization in an increasingly literate, mobile, urban, and media-saturated world."

"While the smartphone—as a private anywhere/anytime portal to the internet—is a powerfully individualizing technology that circumvents the dense network of communal surveillance that is so central to the reproduction of ultra-orthodox or other conservative religious communities (Deutsch [269], p. 8), it (and hyperconnectivity more generally) can also be powerfully collectivizing, calling into being counter-publics and alternative communities."

"In both individualizing and collectivizing ways, hyperconnectivity works to emancipate the self from family and local community."

Objectifying the self

"The digital residues used for these purposes—which dwarf those that we ourselves use—remain invisible to us: they are fed into vast and opaque systems of data aggregation and analysis that enable the algorithmic objectification of the self—a theme I return to in the next section.) Verbal and visual practices of representing the self as an object are of course in no way new."

"We routinely create objectified digital representations of the self, sometimes through deliberate, self-conscious self-work, sometimes unselfconsciously; and once these representations are in place, we see the self from the outside, as an object."

"Digital hyperconnectivity thus proliferates objectified representations of the self; it creates new ways of seeing ourselves from the point of view of others; and it generalizes, democratizes, routinizes, and quantifies self-objectification by holding up to us a ubiquitous digital mirror that turns the digital gaze back on ourselves."

Quantifying the self

"Built-in integration with social media platforms allows self-tracking data to be readily shared with others, thereby harnessing an additional, social layer of motivation, also evident in the "gamification" of self-tracking."

"Whatever its value to the individual, self-tracking data are potentially valuable to insurance companies (who can set rates for auto, health, or life insurance based on individualized rather than aggregate data and penalize those unwilling to share personal data); to employers (who can encourage or even require employees to use fitness-monitoring apps in an effort to improve productivity, reduce healthcare costs, or discipline workers); and to retailers (who can use such data to target customers more effectively)."

"The numbers I have been discussing so far enter into our experience in conspicuously visible ways: prominently displayed by our self-tracking reports and social media feeds, the numbers are inescapable."

Producing the self

"In the landscape of digital hyperconnectivity, self-entrepreneurship comes to mean not only improving the corporeal self as a factor of production, adapted to competition in the labor market, but producing the digital self as an object of consumption, adapted to competition for attention in the social media market."

"The influencer is a person who has cultivated a large social media following not by virtue of pre-existing celebrity status but by producing and enacting a digital self that succeeds in engaging—and being consumed by—his or (more often) her followers."

"In a context in which the cultural obligation to produce the self as a distinctive, authentic individual is increasingly difficult to fulfill, the burdensome work of individualizing the self is turned over increasingly to algorithms, which after all know us and our uniqueness better than we know ourselves."

Regulating the self

"Having discussed new forms of measuring, monitoring, and tracking bodily and mental processes under the heading of quantifying the self, I turn here to new ways of intervening in and regulating these processes, focusing especially on the regulation of mood and emotion."

"ASMR videos regulate the self in a manner that stands in sharp contrast to the prevailing physiological, cognitive, and emotional rhythms of digital hyperconnectivity."

"ASMR videos are a technology of self-regulation that responds to characteristic ways in which the self is dysregulated, physiologically and emotionally, in an ecology of hyperconnectivity."

"ASMR videos are part of a much broader phenomenon of the use of online resources—music, videos, games, social media feeds, and so on—for emotional and physiological self-regulation."

"In a context of hyperconnectivity, we all draw on online resources to modulate, alter, or regulate our moods and emotions."

Governing the self

"Beyond introducing specific new technologies of the self, digital hyperconnectivity has expanded the ways in which people actively govern themselves in a more general way that deepens and intensifies neoliberal forms of governmentality."

"I discussed above one such affinity: the ways in which hyperconnectivity has enabled—and rendered quasi-obligatory—new forms of self-entrepreneurship, new ways of producing a digital "self of value.""

"Returning to Foucault's distinction, if the neoliberal self is formed through technologies of the self, the post-neoliberal self is formed through technologies of power, "which determine the conduct of individuals and submit them to certain ends or domination.""

"Digital hyperconnectivity has enmeshed individuals in an ever-tighter web of algorithmic processes that objectify, quantify, produce, regulate, and govern the self."

"The reflexive practices of self-making have been increasingly overlaid by algorithmic process that objectify, quantify, produce, regulate, and govern the self from the outside."

[Section 8]
"This article explores the last of these transformations: the ways in which digital hyperconnectivity has transformed the self."

"Digital hyperconnectivity, in sum, has created new ways of being and constructing a self, but it has also created new ways of being constructed as a self from the outside, new ways of being configured, represented, and governed as a self by sociotechnical systems."

"How has digital hyperconnectivity transformed the social formation of the self?"

"Subsequent sections show how digital hyperconnectivity has engendered new ways of objectifying, quantifying, producing, and regulating the self—considered both as active, reflexive practices and as systemic, data- and algorithm-driven processes."

"The new practices of objectifying, quantifying, producing, regulating, and governing the self are likewise complex and ambivalent: they can be construed in ways that highlight the active, reflexive moment or the systemic, cybernetic, disciplinary moment."

5.4.2 Streaming Your Identity: Navigating the Presentation of Gender and Sexuality Through Live Streaming [507]

This is a machine-generated summary of:

Freeman, Guo; Wohn, Donghee Yvette: Streaming your Identity: Navigating the Presentation of Gender and Sexuality through Live Streaming [507]

Published in: Computer Supported Cooperative Work (CSCW) (2020)

Link to original: https://doi.org/10.1007/s10606-020-09386-w

If you want to cite the papers, please refer to the original.

For technical reasons we could not place the page where the original quote is coming from.

Abstract-Summary

"We endeavor to contribute towards this research agenda in yet another new media context -- live streaming -- by analyzing female and LGBTQ streamers' practices to present and manage their gender identity and sexual identity."

"Our findings highlight streamers' gender representation and sexual representation as a demonstration of controlling their own bodies, an awareness of the audiences and the resistance to their expectations, and an exhibition of the affordances and power structure of the specific online social space."

"We extend existing studies on live streaming by exploring the understudied gender identity and sexual identity aspect of the streaming practices."

"We add nuance to the existing HCI/CSCW studies on gender and sexuality by investigating a highly dynamic, interactive, and multilayered self-presentation mechanism emerging in live streaming and point to the need for potential new lenses to analyze technology-supported identity construction."

Introduction

"There is a growing interest in HCI and CSCW to explore more nuanced presentations of identity emerging in online social spaces that have not been thoroughly studied, in hopes of 1) shedding light on how these nuanced presentations extend our current understandings of gender and sexuality that are mediated, constructed, and shaped by technology; and 2) exploring how emerging online social spaces can further help online users portray, enact, and experience their online identities in novel ways."

"We endeavor to contribute towards the above-mentioned growing research agenda in yet another new media context-- live streaming-- by analyzing female and LGBTQ streamers' practices to present and manage their gender identity and sexual identity."

Background

"Two strands of research are shaping this study: the presentation of gender and sexuality in online social spaces; and live streaming as a new social interaction space."

"The other lens, visibility, especially attends to marginalized users' (e.g., LGBTQ) practices and challenges of presenting gender identity and sexual identity in online social spaces."

"Our goal of this paper is to contribute towards addressing these two limitations by investigating how users present, experience, and affirm their gender identity and sexual identity through live streaming, a new social interaction space that affords and supports high-fidelity and multidimensional physical presence by real time audio, video, and textual interaction."

"Despite live streaming has been investigated as a form of cultural production (Gray [270]; Pellicone and Ahn [271]), a part of digital economy (Johnson [272]; Wohn and Freeman [273]; Wohn and others [274]), an engagement with cultural heritage (Lu and others [275]), and affective labor and performance (Woodcock and Johnson [276]), very little study explores how this emerging social space both affords people's (especially female and LGBTQ streamers') perceptions, experiences, and interpretations of their gender identity and sexual identity and introduces new challenges in this process."

Methodology

"For this study we especially focused on Twitch streamers who were self-identified as women and/or LGBTQ."

"We attended TwitchCon in person, a convention for Twitch streamers, to directly recruit participants."

"We used keywords search (including words such as "Twitch, female, woman, streamer, gay, black, Asian, harassment, lesbian, trans") on Twitter to find people who identified as being a female and/or LGBTQ live streamer in their Twitter profile and/or had Tweeted about being harassed during streaming."

"Inspired by phenomenology (Eatough and Smith [277]), we then used an empirical, in-depth qualitative analysis of the collected data to explore streamers' self-presentation on live streaming, especially with regard to their gender and/or sexuality."

Findings

"Compared to cis female streamers, LGBTQ streamers especially highlighted how presenting their gender and sexuality through live streaming was an official 'ritual' to announce and acknowledge their non-binary or non-traditional identity to others."

"According to this quote, live stream has become an official way for some LGBTQ streamers to reveal their gender identity and sexual identity to the public (i.e., chose to show their face and voice to all viewers)."

"They will go viral. (Amy, cis female, 30, Latino, bisexual) Collectively, these two quotes highlight female streamers' proactive self-reflections on presenting gender in order to build a counter-culture against sexism and misogyny in the live streaming community."

"For female and LGBTQ streamers involved in this study, presenting and promoting their gender identity and sexual identity as activism and empowerment was an unexpected but emergent outcome of their streaming practices."

Discussion

"To answer our RQs, our findings have shown: 1) female and LGBTQ streamers in this study tended to adopt a less audience-oriented but more self-driven approach to present their gender and sexuality, highlighting the negotiation of gender and sexuality in the context of content creation, live streaming as an official 'ritual' to acknowledge gender and sexuality in public, and advocating gender and sexuality as an online activism (RQ1); 2) they used unique strategies to support this self-driven approach for self-presentation, involving manipulating webcams and microphones and managing clothing and appearance (RQ2); and 3) they often faced various challenges in their identity practices, including the mismatch between viewers' expectations and streamers' aspiration as well as tensions between live streaming Identity and offline limitations (RQ3)."

Conclusions

"The increasingly popular live stream platforms not only create new forms of social interaction experiences and human relationships but also shape how people perceive and define themselves online."

"We have highlighted streamers' self-presentations with regard to gender and/or sexuality as emerging in a highly dynamic, interactive, and multilayered self-presentation mechanism of content, viewers, streamers, and subculture; demonstrated by controlling their own bodies; featuring an awareness of diverse audiences and the resistance to their expectations; and supported by the affordances and power structure of the specific online social space."

"To answer the research questions we raised at the beginning of this paper, live streaming, as a new and novel social interaction space, mediates a more self-driven approach of self-presentation by highlighting the importance of affirmation and empowerment in this process."

5.4.3 "We're Continually Comparing Ourselves to Something": Navigating Body Image, Media, and Social Media Ideals at the Nexus of Appearance, Health, and Wellness [508]

This is a machine-generated summary of:

Monks, Helen; Costello, Leesa; Dare, Julie; Reid Boyd, Elizabeth: "We're Continually Comparing Ourselves to Something": Navigating Body Image, Media, and Social Media Ideals at the Nexus of Appearance, Health, and Wellness [508]

Published in: Sex Roles (2020)

Link to original: https://doi.org/10.1007/s11199-020-01162-w

If you want to cite the papers, please refer to the original.

For technical reasons we could not place the page where the original quote is coming from.

Abstract-Summary

"In modern Western society, women commonly encounter media that conflates appearance and health."

"The present study describes how a group of women perceive appearance, health, and wellness ideals promoted through media and social media as well as how they attempted to challenge and negotiate these."

"Thematic analysis of the data revealed multiple overarching themes, including the women's recognition of unrealistic ideals in the media, the potential for harmful social comparisons, their desire to achieve the appearance of health and wellness, and the use of various strategies to modify their exposure to such media and uphold a positive body image."

"Some participants were challenged by media claiming a more "holistic" orientation in which ideals of appearance, health, and wellness are closely intertwined."

The Media and Body Image

"Media ideals can become part of women's own belief system through a process of internalisation (i.e., cognitive endorsement) (Calogero and others [278])."

"Self-objectification can ensue, whereby women learn to view themselves as an object valued primarily for its appearance (Calogero and others [278]; Fredrickson and Roberts [160])."

"By adopting an observer's perspective of the body and consistently anticipating others' evaluation, women and girls feel compelled to engage in vigilant processes of self-surveillance and body monitoring, leading to appearance anxiety and body shame, and resulting in serious mental health risks (Fredrickson and Roberts [160])."

The Conflation of Appearance and Health in the Media

"Ironically, women's health magazines are just as likely to promote health behaviour to their readers for appearance reasons as they are for health reasons (Aubrey [279])."

"Other research suggests the editorial content common to popular women's health and fitness magazines features an overemphasis on appearance compared to health (Willis and Knobloch-Westerwick [280])."

"The framing of health advice primarily as a matter of appearance can provoke women's feelings of shame, likely because they perceive themselves as failing to live up to the standard of strict health-related behaviors (Aubrey [279])."

"Even those magazines that lay claim to a holistic orientation (e.g., with the goal of improving mind, body, and spirit) are likely to privilege women's physical appearance over their physical and emotional health (Aubrey [279])."

Social Media Influences on Body Image

"Specific types of social media use, such as engaging in photo activities on Facebook and following appearance-focused accounts on Instagram, are associated with thin-ideal internalisation and body surveillance among young women (Cohen and others [281])."

"Exposure to thin-ideal Instagram images has been associated with increased body dissatisfaction among young adult women (Tiggemann and others [282])."

"Highly visual social media is associated with body image concerns and internalising symptoms among both male and female adolescents (Marengo and others [283])."

"The recent phenomenon of posting inspirational health and fitness messages and images (i.e., "Fitspiration") via social media reflects shifting cultural trends constructing the thin, athletic, ultra-fit body as an aesthetic ideal for which women should strive (Bozsik and others [284]; Homan [285]; Homan and others [286])."

Holistic Health, Media, and Body Image

"A qualitative approach (Agar [287]; Taylor and Francis [288]) can illuminate an understanding of the intricacies of women's interactions with media and social media and the overlapping ideals of appearance and health."

"Although some women perceive Fitspiration content in an optimistic way as a source of health advice and motivation to reach health and fitness goals as well as a benchmark against which to measure their success in striving toward an ideal representation of health (Goldstraw and Keegan [289]; Raggatt and others [290]), they can also be critical of the artificial nature of such content and the unrealistic ideal it promotes (Goldstraw and Keegan [289]; Raggatt and others [290])."

"We sought to further a nuanced understanding of how women engage with appearance and health ideals promoted through the media, including social media."

"Whereas previous qualitative research has explored women's perceptions of Fitspiration content on social media (Goldstraw and Keegan [289]; Raggatt and others [290]), we were particularly concerned with the nature of women's engagement with media and social media content that has a more holistic orientation, in which the closely intertwined ideals of appearance, health, and wellness may be even more difficult to discern and disentangle."

Method

"Our research took place in a broader ethnographic study that aimed to explore how women's participation in holistic health culture shaped their expressions of positive body image."

"This inclusion criterion was chosen to allow a deeper understanding of how women participating in holistic health culture engage with media conflating appearance, health, and wellness in their daily lives."

"After the interview, participants were asked if they would be willing to pass information on about the study to anyone they knew who was involved in holistic health and would be interested in being interviewed about their experiences."

"As the first author and primary researcher, the female interviewer could draw on personal experiences and engagement in holistic health culture over the past decade to enrich the process of interviewing."

"Although analytic concepts included themes emanating inductively from the data, in some cases these were informed by a priori issues from the research literature on positive body image and holistic health culture."

Results

"Then I was able to recognise that that was creating this pressure and this unrealistic kind of ideal of how I should be...you just see pictures of people at their best...there's just this almost unrealistic idea of health and wellness, of a particular way that you should look, and that's shoved in people's faces all day, every day (Alison) Social media presents a unique threat to body image, and the nature of social media may mean that appearance and health ideas which are promoted by contemporary health culture can be particularly difficult for these women to challenge."

Discussion

"Notwithstanding the limitations we mention, the current study makes an important contribution to the field, by illuminating the closely intertwining ideals of appearance and health as represented through holistic health culture and their implications for promoting positive body image among women."

"Future research could also confirm if, and how, the content of Fitspiration messages might differ from holistic health messages and determine whether women's exposure to these media ideals influences their body image in different ways."

"Future research could also investigate the specific elements of holistic health messages that appeal to women and how some elements of this culture could be appropriated in intervention strategies to promote positive body image."

"Then, positive body image intervention strategies could benefit from expanding their focus to incorporate critical health media literacy (Higgins and Begoray [291]) to enable women and girls to engage critically with media messages about health, particularly those messages pertaining to women's appearance."

Conclusions

"Our research aimed to provide a nuanced perspective on how women perceive appearance and health ideals on media and social media, as well as how they attempted to challenge and negotiate these ideals."

"Participants were critical of thin-ideal media, and they recognised the pressure for harmful social comparisons created by such unrealistic standards."

"It is fundamental that such interventions promote the skills to critically assess appearance ideals that are promoted as representing health and wellness."

"Future research focused on gaining deeper understandings of how holistic health ideals influence women's body image may assist in mitigating potential negative effects and promote adaptive features of positive body image that enhance women's health and well-being."

[Section 9]
"Media can have profound influence on body image, particularly among women (Grabe and others [217]; Hesse-Biber and others [292])."

"The overarching objective of the present study is to contribute to extant research on this topic by exploring how women engage with media and social media that promote appearance and health ideals."

"This study aimed to gain an understanding of how women perceive and respond to those appearance ideals that reflect contemporary 'wellness' trends."

"Positive body image emphasises acceptance and appreciation of the body, taking care of the body through healthy behaviours for the benefit of overall well-being rather than appearance-related reasons, as well as processing and responding to body-related messages (including media messages) in a body-protective way (Menzel and Levine [293]; Tylka [294]; Webb and others [295]; Wood-Barcalow and others [296])."

5.4.4 Digital Ethnography and the Social Dimension of Introspection: An Empirical Study in Two Colombian Schools [509]

This is a machine-generated summary of:

Rieken, Johannes; Garcia-Sanchez, Efraín; Trujillo, Mónica Pérez; Bear, Daniel: Digital Ethnography and the Social Dimension of Introspection: An Empirical Study in Two Colombian Schools [509]

Published in: Integrative Psychological and Behavioral Science (2015)

Link to original: https://doi.org/10.1007/s12124-015-9299-1

If you want to cite the papers, please refer to the original.

For technical reasons we could not place the page where the original quote is coming from.

Abstract-Summary
"We developed a teaching-led research project to empirically ground methodological reflection about digital ethnography."

"Drawing on Cordelois' collective ethnographic observation approach, fifteen emerging professionals (from a private general education university and a Police Academy in Bogota) collaborated in a method seminar on digital ethnography."

"They worked in cross-institutional research teams, each carrying SenseCams for 3 days."

"The research design enabled emerging professionals to introspect about what it is to be a member of their institution."

"Our research highlights the relevance of the social dimension of introspection for digital ethnography."

Introduction and Literature Review

"The discussion situates our research within the field of digital ethnography by comparing it to that of Cordelois [297] and argues that digital ethnography requires us to focus on the social dimension of introspection that emerges when the first-person perspective captured by SenseCams is negotiated within a group."

"Digitally capturing first person perspectives in ethnography allows access to real-time sequential analysis and self-confrontation, along with the ensuing implications for introspection."

"This gives both the participating introspecting subject and the inquiring interviewer equal access to the first-person perspective digital representations (SenseCam images), limiting the asymmetry of professional experience."

"It is both the participant and researcher together that introspect about digitally captured practices during self-confrontation."

"We need to be mindful of the above processes in digital ethnography when we accept that with this methodology the social dimension of introspection increases, as participants and researchers jointly negotiate the first-person individual perspective captured in digital ethnographic data."

Methodology

"In each organization, students were invited to participate in a Research Methods Seminar that involved taking part in an educational project."

"Following data collection, two researchers (authors Rieken & Garcia-Sanchez) met with each team of students, composed by two ECSAN students and one Los Andes student, and jointly conducted self-confrontation interviews."

"Researchers used the digital data as the focus of the interview and asked participants to summarize the content of the SenseCam material to identify the higher-level goals they pursued in the recorded activity."

"Presentation students were asked to highlight what insights they have gained from capturing and reviewing their material in terms of awareness about own practice, how their practice compares to that of their peers from the other participating institution, what recurring temporal patterns they noticed, and finally any observations about the effects of carrying the Sensecam."

Results

"The use of self-confrontation interviews helped the research team to identify differences between Los Andes and ECSAN students concerning the notion of time, the use of spaces, Information and Communication Technologies and its applications, and references for professional identity."

"One participant stated: "What I see is that [the Los Andes student] manages his time in a more individualistic way, he does his own activities that he had scheduled previously; here [at ECSAN] we don't have the ability to do that" (ECSAN student-5-Male)."

"When listening to his partner's account, a student from Los Andes replied, "What's interesting to me is the contrast [with students from ECSAN] regarding interaction with people [when wearing the camera], because I didn't feel in Los Andes that they had changed their behavior with me or because of me; I felt strange and I thought they were going to see me in a bad way, but after 15 min I forgot I was wearing the camera."

Discussion

"The difference in detail of the digital data-capturing device results in different levels of analysis."

"The clarity with which these differences in learning practices between the two institutions can be seen is precisely also a function of choosing SenseCams as digital capture devices for the research."

"The research design specifically sought to contrast two very different institutions, and the students were asked to look at the digital data and reflections of their co-participants in order to make sense of their own organizations as was illustrated in the finding section by the way participants from the two institutions would compare their study routines."

"This may suggest more generally that different professional practices may be better expressed in account format that follows a different logic (in our case that of goals vs. routines), which in turn may be indicative about the type of practice and normative set-up that is predominate in a given institution."

Conclusions

"Further, we illustrated that with digital ethnography the social dimension of introspective processes is increasingly relevant."

"This is because digital ethnographic data allows for self-confrontational interview settings where interviewees, researchers and other participants have equal access to the now digital, not mental, representations that are the object of introspection."

"Particularly, research that focuses on professional groups characterized by sharing and developing a body of professional knowledge may particularly benefit from increased sensitivity to the social dimension of introspection."

"While it is clear that further research comparing the effects of different social and individual analysis procedures is essential for the development of introspection in digital ethnography, we have identified some important methodological elements in this paper."

"We have reaffirmed the importance of Sensecam data, highlighted the social dimension that is always part of introspection, and explored the duality of both the danger of the social dimensions of introspection to bias research but also its potential to improve it."

5.4.5 Profiling the Self in Mobile Online Dating Apps: A Serial Picture Analysis [510]

This is a machine-generated summary of:

Degen, Johanna Lisa; Kleeberg-Niepage, Andrea: Profiling the Self in Mobile Online Dating Apps: a Serial Picture Analysis [510]

Published in: Human Arenas (2021)

Link to original: https://doi.org/10.1007/s42087-021-00195-1

If you want to cite the papers, please refer to the original.

For technical reasons we could not place the page where the original quote is coming from.

Abstract-Summary

"Profiles in the widely used phenomenon of mobile online dating applications are characteristically reduced to condensed information mostly containing one or a few pictures."

"While profile pictures in social media are omnipresent and some research has already focused on these pictures, especially selfies, there has been little attention with regards to the actual self-presentation when mobile online dating."

"We show the results of a reconstructive serial analysis of 524 mobile online dating profile pictures investigating how subjects present themselves in the context of a mobile online dating app."

"Contemplating on subject's good reasons for presenting the self as one of many and not as varied and unique when mobile online dating, we refer to the discourse of the private self (Gergen, The saturated self: Dilemmas of identity in contemporary life, Basic Books, New York, 1991; Rose, Governing the soul: Shaping of the private self, Free Association Books, London, 2006) and to (Holzkamp, 1983."

Finding a Match Based on a Single Picture: the Characteristics of Online Mobile Dating

"This does not deny a significance of the profile texts, yet the profile-pictures play a predominant and crucial role in the practice of mobile online dating."

"Despite the obvious importance of profile pictures in online dating apps and the general fact that pictures play a (more and more) important role in subjects' lives and experiences, psychological research so far does not address the significance of (these) pictures, but largely ignores them as independent data (Reavey, [298])."

"To bridge this gap and to explore how subjects visually present the self in the context of mobile an online dating, we conducted a qualitative serial analysis of 542 Tinder profile pictures."

"We will first shortly review the state of research in the field of mobile online dating and in profile pictures and discuss significant theoretical perspectives to grasp the specifics of visually presenting the self in this context."

Research on Mobile Online Dating and the Profile Pictures

"Psychological research has so far focused on personal traits and behavior regarding the degree of honesty when presenting the self or on specific traits and personality characteristics of dating-app users (Charpenter & McEwan, [299]; Musil and others, [300]), user's motivations (Timmermans & DeCaluwe, [301]; Ward, [302]) or the extent of disclosure (e.g., their face) (Fitzpatrick and others, [303]), accuracy of profile (-pictures) (Duguay, [304, 305]; Hancock & Toma, [306]) or potentially hidden information in the profile pictures in the form of conclusions about ones personality based on the presentation (Ward, [302])."

"While there is some research about profile pictures (e.g., on Facebook), little research regards profile pictures in the specific context of mobile online dating."

"The little research deals with profile pictures in the context of mobile online dating specifically subdivides into qualitative and quantitative approaches and their respective perspectives."

"Ranzini and Lutz [307] combine location-based real-time dating and personal traits and picture analysis in regard to narcissism, loneliness, and self-esteem with motives and real or deceptive self-presentation."

"One can thoroughly create and optimize a profile and pictures, meaning the self-presentation is a planned enactment with a decisive time advantage, but this leaves the user with the question to what degree the self should be optimized to avoid insecurity and disappointment when meeting in real life."

The Self and Its Profile Picture

"The self-presentation of subjects becomes more and more important either as a means of recognizability (like a Facebook profile picture), or to tell stories about one's life or current activities (Instagram, WhatsApp status)."

"The self-presentation in an online dating profile, being reduced from a complex process of verbal, para- and nonverbal behavior and activities (Goffman, [308]) to a few static characteristics, leaves the subject with limited options to prevent the self from setbacks."

"These theoretical considerations provide the framework for our serial analysis of profile pictures and for responding to the following questions: How do subjects shape—construct and reconstruct—the self in the specific context of dating app profiles (accelerated daily practice, continuous judgment by others, unseen competition, validation of the self, exposure, self-esteem), in the tension between authenticity, exposure, and subjectivity on the one hand and the implied social norms and disclosure to unknown others on the other hand?"

Methodological Framework and Methodical Procedure

"Such an understanding of pictures corresponds to the reconstructive research paradigm within which specifically Documentary Method developed a picture-analytical procedure to reveal the creators implicit (or atheoretical) knowledge contained in the pictures (e.g., Bohnsack, [309]; Bohnsack and others, [310])."

"For Mietzner and Pilarczyk both serial and single picture analysis are equally significant and do not have a strict order: "The aim of serial analyses is to detect continuous or discontinuous developments, anomalies and differences as well as to classify and interpret them." (p. 142, translated by the authors)."

"Mietzner and Pilarczyk [311] who consider a sample of 300 to 500 pictures as suitable for a serial analysis propose different strategies like arranging series of motives to register changes (e.g., over time, to record the elements of the picture design, common motives or topics) to eventually create types as well."

The Tinder Typology of Self-presentation: Being One of Many—countless Opportunities, Few Types

"The selfie type, characterized by self-control and limits: Description: A selfie is a picture taken by the person him- or herself where the angle, the proportion of head, shoulder, and arm show that photographer and photographed person are the same subject."

"There are different kinds of selfies coming with "direct eye contact," where the subject looks into the camera," "indirect eye contact," where the person is watching the self on the screen, "showing skin" where subjects are (seemingly) partly naked, "no eye contact" where subjects seemingly look away or have closed eyes, look towards the ground, and "mirror selfies," where the picture is taken in front of a mirror often showing the device and selfie "with context reference," where the picture contains explicit information about the surrounding, for instance one person taking a selfie with an elephant."

Profile Pictures (in Mobile Online Dating Apps) at the Intersection Between Social Norms and Subjective Sovereignty

"We can infer that subjects (visual) self-presentation in mobile online dating apps relies on implicit knowledge about social norms, unwritten rules and (assumed) expectations of an anonymous, imagined audience."

"The prescribed logic of self-presentation in online dating app profiles seems to reduce the uniqueness of the subject to a uniqueness of many."

"This is reasonable as presenting the self on the app exposes a person to a liberal logic in Roses terms [312]."

"Inner psychological processes, such as validation and adjustment of the self, in this specific online-dating context is thus negotiated with imagined others, possibly idealized others creating a digital projective surface for supposedly social yet mostly projective self-processes."

"Being resistive towards the social norms of appropriateness and attractiveness in an (online) dating context might be a sign of generalized agency (Holzkamp, [313]), for instance of overcoming restrictive social rules and thereby widening one's scope of action."

Ethics and Limitations

"In the conduct of research, we collected data taking the role of gender into account, presuming differences, aiming for a balanced sample."

"Gender differences shown in former research for instance regarding sport equipment did not present great relevance for the findings."

"The reconstructive serial analysis is focused and limited to the intersubjective orientation that again functions as a storyline for behavior—the inherent logic and social rules—it cannot and does not aim for a reconstruction of the individual expectation, motives, or aspired impression/intent nor does it contribute to the individual perception of these forms of self-presentation."

"When building typologies, the researcher is biased to search for minimum and maximum contrast, resulting in typicality of self-presentation; this does not imply that there is not individuality in the self-presentation and individual nuances."

Conclusion and Outlook: Coming to an Understanding of Selves When Mobile Online Dating

"Serial picture analysis in the predominantly visual context of mobile online dating gives valuable insights into implicit social norms and respective subjective referencing in the presentations of the selves, beyond the seemingly obvious."

"The great majority of users orient the self-presentation strongly on these social norms and do not exhaust the vast possibilities of variety, presenting contrary motives or create unique and individual profiles."

"This goes far beyond the discourse on self-presentation in mobile online dating where picture analysis so far mostly focuses on quantitative analysis of presented motives, colors, and objects and mere speculations about users' motifs."

"This contributes to the research of online dating by analyzing the actual self-presentation enriching the understanding in the sphere between motives, (high) hopes and expectations and the actual behavior."

"What does the practice of common and somehow unspecific self-presentation mean for user's ideas of intimate relationships?"

5.4.6 How Photo Editing in Social Media Shapes Self-perceived Attractiveness and Self-esteem via Self-objectification and Physical Appearance Comparisons [511]

This is a machine-generated summary of:

Ozimek, Phillip; Lainas, Semina; Bierhoff, Hans-Werner; Rohmann, Elke: How photo editing in social media shapes self-perceived attractiveness and self-esteem via self-objectification and physical appearance comparisons [511]

Published in: BMC Psychology (2023)

Link to original: https://doi.org/10.1186/s40359-023-01143-0

If you want to cite the papers, please refer to the original.

For technical reasons we could not place the page where the original quote is coming from.

Abstract-Summary

"As photo editing behavior to enhance one?s appearance in photos becomes more and more prevalent on social network sites (SNSs), potential risks are increasingly discussed as well."

"The purpose of this study is to examine the relationship between photo editing behavior, self-objectification, physical appearance comparisons, self-perceived attractiveness, and self-esteem."

"A parallel-sequential multiple mediation model was conducted to examine the relationship between photo editing behavior and self-esteem considering multiple mediators."

"The results indicate that photo editing behavior is negatively related to self-perceived attractiveness and self-esteem mediated via self-objectification and physical appearance comparisons."

Introduction

"Social media including social media sites constitute a new context of social interaction which is contrasted with face-to-face interaction and digital communication media like email and video conferencing [314]."

"Photo editing behavior increases the options available for self-presentation on social network sites and constitutes a significant restriction on verifiability of the accuracy of profile photos."

"The importance of photo editing behavior in the context of social network sites is high."

"According to a survey of the Renfrew Center Foundation [315], 50% of SNS users edit their photos before posting them to Social Network Sites (SNSs)."

"Whereas almost all previous studies have investigated the impact of photo editing behavior solely on body image, this study refers to self-perceived attractiveness in general by including the body and face as part of self-perceived attractiveness."

"Almost all studies on photo editing have only included women, as they are more likely to engage in photo editing behavior [316] and experience higher pressure to conform to the cultural beauty ideal [317]."

Theoretical background

"The first hypothesis states: Photo editing behavior is negatively correlated with self-perceived attractiveness in terms of appearance."

"The nature of photo editing behavior activates feelings of self-objectification [12; 14] and physical appearance comparisons."

"Photo editing behavior leads to self-objectification [12; 14; 20; 32] and physical appearance comparisons [318]."

"Physical appearance comparisons are likely to trigger photo editing behavior to compensate for one's optical flaws (cf., the occurrence of objective facial attractiveness; [319])."

"Since physical appearance comparisons [320, 321] and self-objectification [218] are negatively correlated with body image and self-perceived attractiveness, it is reasonable to assume that the two constructs are mediators of the relationship between photo editing behavior and self-perceived attractiveness."

"Photo editing behavior is associated with higher self-objectification and more physical appearance comparisons, that result in lower self-perceived attractiveness, which, in turn, implies lower self-esteem."

Method

"Demographic variables, photo editing behavior, self-objectification, physical appearance comparison, self-perceived attractiveness and self-esteem, and Instagram activity were obtained."

"The questionnaire consists of 38 items rated on a 5-point Likert scale from 1 = "never" to 5 = "very often", based on the two factors: Active (27 items; e.g., "I post pictures.") and Passive (11 items; e.g., "I look at the photos of other users.")."

"Participants exhibited moderately high ratings of self-objectification, $M_{Factor\_1} = 2.92$, $SD_{Factor\_1} = 0.88$, $M_{Factor\_2} = 1.70$, $SD_{Factor\_2} = 0.62$, $M_{SOBBS\_Total} = 2.31$, $SD_{SOBBS\_Total} = 0.66$, with highest ratings on the first factor (internalizing an observer's perspective of the body) and comparatively low ratings on the second factor (equating the body to who one is as a person and valuing physical appearance above other attributes)."

Results

"In correspondence with expectations significant negative correlations were found between self-perceived attractiveness and self-objectification, r_s (401) = – 0.569 $p < .001$, as well as on corresponding subscales, referring to internalizing an observer's perspective of the body, r_s (401) = – 0.462, $p < .001$, and referring to equating the body to who one is as a person and valuing physical appearance above other attributes), r_s (401) = – 0.477, $p < .001$), respectively."

"This hypothesis was confirmed because a significant negative correlation was found between photo editing behavior and self-perceived attractiveness in terms of appearance, r_s (401) = – 0.146, $p < .01$."

"The statistical mediation model summarized in H5 proposed that photo editing behavior is associated with higher self-objectification and more physical appearance comparisons, and that both mediators are associated with a lower self-perceived attractiveness, which, in turn, is associated with lower self-esteem."

Discussion

"Consistent with other studies [322, 323], the results regarding H1 indicate that photo editing behavior is associated with lower self-perceived attractiveness in terms of appearance."

"The confirmation of H2b indicates that higher sores on the photo editing scale are associated with more intense immersion in physical appearance comparisons including the construction of self-other contrasts with respect to good looks."

"The findings generally supported hypothesis H3 that physical appearance comparisons are negatively associated with self-perceived attractiveness in terms of the subscale Appearance of the BES."

"The results indicate that mediation via self-objectification and via physical appearance comparisons occupy central switching points in the model which are both associated with self-perceived attractiveness."

"The link between photo editing behavior and self-esteem was sequentially mediated via self-objectification, physical appearance comparisons, and self-perceived attractiveness."

"Individuals who engage in photo editing behavior more often perform physical attractiveness comparisons with others sand self-objectify more frequently."

Limitations and future research

"In future studies the results of online and offline measurement of the assessment of photo editing variables including personality variables like self-esteem should be compared with each other in order to increase the generalizabilitv of results."

"Although we determined a low score on photo editing behavior in our study, we found robust results confirming our hypotheses."

"Based on sample characteristics (i.e., age, gender, and participants's photo editing behavior), some points of criticism with respect to the generalizability of our data arise."

"The list of potential mediators between photo editing behavior and self-esteem includes variables that were not considered in our research design."

"Future research might investigate the outcome of photo editing behavior in contexts like dating platforms."

"As especially adolescents are vulnerable in terms of self-esteem and appearance-based self-worth, further research should also be conducted on the impact of photo editing behavior on this vulnerable target group."

5.4.7 Adolescents' Selfie-taking and Selfie-editing: A Revision of the Photo Manipulation Scale and a Moderated Mediation Model [512]

This is a machine-generated summary of:

Gioia, Francesca; McLean, Siân; Griffiths, Mark D.; Boursier, Valentina: Adolescents' selfie-taking and selfie-editing: A revision of the photo manipulation scale and a moderated mediation model [512]

Published in: Current Psychology (2021)

Link to original: https://doi.org/10.1007/s12144-021-01702-x

If you want to cite the papers, please refer to the original.

For technical reasons we could not place the page where the original quote is coming from.

Abstract-Summary

"Few studies focused on body image control in photos (BICP) and PM, and no studies evaluated the influence of selfie-expectancies on photo-taking and photo-editing."

"Study1 psychometrically evaluated the PM scale (N = 1353)."

"Study2 evaluated the mediating role of BICP and the moderating role of gender in the relationship between selfie-expectancies and PM (N = 453)."

"BICP mediated the relationship between selfie-expectancies and PM and being male significantly affected the relationship between the variables."

Introduction

"The taking and sharing photos by individuals may appear to be straightforward and innocuous activities, selfie behavior represents a nuanced, complex, and multidimensional phenomenon reflecting selfies intentional creation, and comprises multiple actions including taking (such as preparation and posing), modifying (such as editing and filtering), and posting onto SNSs (McLean and others, [179])."

"According to previous studies (Boursier & Manna, [239]; Bij de Vaate and others, [324]; McLean and others, [250]; Pelosi and others, [325]), photo investment and control over body image in pictures represent adolescents' and emerging adults' attempt to manage concerns about their own self-portrait quality and how they portray themselves, following specific strategies in taking and choosing selfies before sharing online."

"Previous studies confirm that selfie practice is a widespread daily activity and the increased social media use for sharing personal photos make adolescent body image an extremely contemporary issue (Boursier & Manna, [239]; Franchina & Lo Coco, [326]; Pelosi and others, [325])."

"Selfie behavior appears to be a complex phenomenon that comprises not only selfie-sharing on social media but also taking and editing personal photos following specific control and manipulation strategies."

Study 1

"The first study tested the psychometric properties of the Italian version of the 10-item Photo Manipulation Scale (PMS; McLean and others, [250]) among a large sample of Italian adolescent boys and girls."

"The aim of the first study was to evaluate the factor structure of a revised version of PMS (McLean and others, [250]) among a sample of Italian adolescents."

"The revised eight-item PMS showed good Cronbach's α value (.80), lower but comparable to the Cronbach's α value (.85) in the original study of McLean and others [250]."

"Due to the changed length of the scale, the Spearman-Brown coefficients for the PMS and the two-item facial image manipulation subscale were tested, showing a good reliability (.78 and .75, respectively)."

"The factors of the 8-item PMS (photo filter use, body image modification, and facial image manipulation) significantly and positively correlated with one another."

Study 2

"Positive correlations of large effect size were observed among boys between selfie expectancy variables self-confidence and self-presentation and body image control in photo variables selfie-related factors and privacy filters, and the photo manipulation factor, filter use."

"The 1000 bias-corrected bootstrapped estimates showed a significant indirect effect of selfie-expectancies on photo manipulation via body image control in photos (β = .219; SE = .039; Bootstrap 95% CI [.15, .308]) and for conditional direct effects of selfie-expectancies on photo manipulation bootstrapping estimates confirmed the significant gender effect on the relationship between the variables."

"In this second study, a moderated mediation model was cross-sectionally tested in which body image control in photos mediated the association between expectancies underlying selfie behavior and photo manipulation, and gender moderated this relationship in a sample of Italian adolescents."

"Concerning the gender-related differences, selfie-expectancies, body image control in photos, and photo manipulation mean scores highlighted significant differences between male and female adolescents."

General Discussion and Conclusions

"In the second study, gender differences between selfie expectancies, body image control in photos, and photo manipulation were tested, and correlations between variables were examined."

"The tested mediation model demonstrated that selfie-expectancies were both directly and indirectly (via body image control in photos) positively associated with photo manipulation."

"It is likely that males' expectations concerning selfie-taking and selfie-sharing (mainly negative or related to the sexual component of the selfies) directly predicted their photo-editing strategies, especially to manipulate their own body image."

"A possible interpretation of the current findings could be that girls engage in a consistent level of photo manipulation regardless of their selfie expectancies and body image control in photo behaviors, whereas for boys, photo manipulation may occur only under circumstances where they are more highly concerned about the outcomes of their selfie behaviors, such as presenting an ideal appearance."

Practice Implications

"These findings might enhance the psychologists' and clinicians' attention toward teenagers' digital self-images on social media (King, [327]), helping them to better counsel adolescents about the integration and mentalization of a changing body."

"The association between online body image-related issues and self-objectification experiences deserves clinicians and health workers attention suggesting a focus on the outcomes and mechanisms that underlie selfie practices (Veldhuis and others, [328]), thus supporting the adolescents' identity construction processes and making them aware of the risk of digitally modified body standards internalization."

"Media literacy programs might be useful tools to educate teenagers about own real body image, about culturally and peer-to-peer promoted body standards, and about their selfie-sharing on social media (Fardouly and others, [329]; McLean and others, [187])."

Limitations and Future Directions

"The novel findings of the present study provide evidence of the constant evolution of the social media landscape, with many platforms being replaced by new ones which are increasingly focused on visual content, especially among adolescents (Gioia & Boursier, [330]; Feltman & Szymanski, [331])."

"Some limitations of the present studies need to be addressed when interpreting the findings."

"The present studies only explored a small number of variables in relation to the complex phenomenon of selfie-behavior."

"Photo manipulation and editing could be explored in association with self-objectification experiences, body image-related issues, and other online creative or problematic activities."

"The present studies' findings provided some novel and previously unreported issues."

"And unexpectedly, the present findings provided novel and unreported gender differences that highlight modifications concerning appearance-related issues and online activities carried out by both boys and girls."

5.4.8 The Role of Identity in Digital Consumer Behavior: A Conceptual Model and Research Propositions Based on Gender [513]

This is a machine-generated summary of:

Rogova, Natalia; Matta, Shashi: The role of identity in digital consumer behavior: A conceptual model and research propositions based on gender [513]

Published in: AMS Review (2022)

Link to original: https://doi.org/10.1007/s13162-022-00237-z

If you want to cite the papers, please refer to the original.

For technical reasons we could not place the page where the original quote is coming from.

Abstract-Summary

"This article aims to understand these digital consumer behaviors by focusing on identity strength and the identity signaling phenomenon."

"Further, we propose how digital and offline identity signaling behaviors are intertwined, and discuss the online behaviors of trans consumers."

"Our conceptual work highlights the unique features of digital identity signaling behaviors as well as the complexity of identities, including gender, and provides useful insights for researchers and marketers."

Introduction

"We develop a conceptual model, and several research propositions, to discuss the influence of both consumers' internal identity-related characteristics and external factors that they face in the online environment on their digital identity signaling behaviors."

"Digital content generation (e.g., creating social media posts) is one of the major and novel forms of identity signaling that has not been explored by consumer researchers in depth (Gal, [332])."

"Digital behaviors such as posting on social media signal consumer identity (Grewal and others, [333]; Jensen Schau & Gilly, [334])."

"Taking the example of gender identity, consumers could signal their gender online by actively supporting or, on the contrary, criticizing gendered brands on their social media, following gendered brands' social media pages and communities, sharing posts created by gendered brands and related influencers, etc Consumers' reactions and discussions of gendered brands and/or their gender-related content are often very active and controversial."

What is (gender) identity?

"Gender identity is one part of a consumer's self-concept (Wood & Eagly, [335])."

"Wood and Eagly [335] define the personality traits approach as an older tradition, which started to gain popularity in the 1970s, when Bem [336] and Spence and others [337] proposed that gender identity consists of two dimensions of gender-stereotypical personality traits—communal traits as a symbol of femininity, and agentic traits as a symbol of masculinity."

"Gender identity based on self-categorization can be defined as a "descriptive or prescriptive categorization of oneself as female or male, along with the importance of this categorization for one's self-definition" (Wood & Eagly, [335], p. 464)."

"The personality tradition to defining gender relates to personal identity and its gendered nature, whereas the self-categorization tradition is focused on individuals' self-labeling with a particular social group (women/men/genderqueer etc) along with its strength, and the meaning of identity that individuals derive from that (Wood & Eagly, [335])."

(Gender) identity in consumer behavior research

"Situational salience cues include, for instance, symbols related to in-group or out-group, whereas chronic salience is often influenced by identification strength, which reinforces the association between the self and a particular identity (Reed and others, [338])."

"Connecting these findings to the identity salience principle and the approaches to gender identity, we conclude that many studies on gender as social identity conceptualize gender as a self-category (a man vs. a woman) and investigate its influence on evaluations of gender-related in-group and out-group associated products and brands, that can prime salience of gender identity situationally."

"Findings in consumer psychology on gender as social identity might indicate that for one group (in that case, men) it is more important than for others to act congruently with their identity and signal it to the self and the others, which one might want to transfer to the digital domain as well."

Our conceptual framework and research propositions

"The moderation effect of gender role beliefs for strongly identified men also affects the relationship between their desire to signal identity and their consumer behavior, but it is less pronounced than for women."

"When exploring cis men's digital identity signaling behaviors, including strong criticism and hatred toward non-stereotypical gendered brands, one might use the Gender Role Conflict Scale (O'Neil and others, [339]) developed to assess how tightly a man follows a traditional gender male role, whereas this measure will not be applicable for women or mixed-gender consumer sample."

"Men with weaker gender identities and especially egalitarian gender role beliefs could want to distance themselves from their gender group by criticizing their behaviors, whereas men with stronger gender identities would be more likely to keep signaling their gender and masculinity by further supporting the hashtag

#LikeABoy or even engaging in consumer-to-consumer interaction online to derogate the outgroup, in that case, either women or a group of weaker identified men who criticized their behaviors."

Transgender and non-binary consumers
"Taking into consideration our model and research propositions, we thus hypothesize that gender role beliefs of strongly identified trans people who prefer to self-identify in binary terms of men and women, would affect how they want to signal their gender in their digital and offline consumer behaviors, i.e., P2a can apply to transgender consumers too."

"Rather, we think that their gender self-category might give more insights on how they signal gender, i.e., whether they identify as genderblended, agender/postgender, etc In general, we believe that our conceptual model is broad enough to explain the online and offline behaviors of non-binary consumers, but more research in this field is needed to understand identity signaling behaviors of different identities within the non-binary."

Conclusion
"We conceptualize the role of identity in digital consumer behaviors using identity strength as well as identity signaling theory."

"We classify consumer digital behaviors by the level of intensity, i.e., their effortfulness and visibility, to illustrate a continuum of actions that consumers can undertake online to signal their identities."

"We also separate digital behavior intensity and digital behavior content, i.e., what exactly consumers decide to share and post online, and show that different identity-related constructs might predict these behaviors."

"We further contribute to a more nuanced understanding of the role of gender identity in digital and offline consumer behaviors."

"We contend that other identities can also be analyzed multidimensionally in relation to digital identity signaling behaviors, i.e., not only through identity label and identity strength, but also with the focus on identity meaning, which is rarely looked at in consumer psychology (Wooten & Rank-Christman, [340])."

5.4.9 Consumer Self-concept and Digitalization: What Does This Mean for Brands? [514]

This is a machine-generated summary of:

Bartoli, Chiara: Consumer self-concept and digitalization: what does this mean for brands? [514]

Published in: Italian Journal of Marketing (2022)

Link to original: https://doi.org/10.1007/s43039-022-00059-8

If you want to cite the papers, please refer to the original.

For technical reasons we could not place the page where the original quote is coming from.

Abstract-Summary

"This study aims to contribute to the discourse on the impact of digitalization on the consumer self-concept."

"The objective is to provide a broad understanding of the impact that digitally enabled consumption environments have on the consumer self-concept and to unveil how this relates to brands."

"The study also has managerial implications pertaining to the relationship between the consumer self-concept and brands in digital spaces involving consumer–brand relationships, brand values, seamless positioning, and personalization."

Introduction

"Despite acknowledgment of the relevance of the interplay between the self-concept and digitalization for firms, there is a lack of studies that capture the point of view of businesses on this topic and that provide a broad overview of the implications for brands of the evolving consumer self-concept in digitally enabled consumption environments."

"In light of this gap, the objective of this study is to contribute to the theoretical discourse on the impact of digitalization on the consumer self-concept by taking the viewpoint of firms to understand what happens to the consumer self-concept in hybrid consumption environments and unveil what this means for brands."

"The study also shows the understanding of managers' awareness of how digitalization is impacting the consumer self-concept and the changes this demands to strategies for articulating brand personality and approaches to consumer data management."

Background

"The ideal and real self-concepts intervene in the creation of emotional connections between consumers and brands (MacInnis & Folkes, [341]; Malär and others, [342])."

"The resources provided by brands in digital environments are integrated in consumers' digital narratives of self-expression (Sung and others, [343]) and function as tools to build a coherent narration of the self that contributes to identity construction (Halliday, [344]) and to the transformation of the self into a consumption object in the form of the "branded self" (Hearn, [345])."

"The digital objects of Industry 4.0 are also seen as natural extensions of the consumer self-concept and prosthetic enablers of individuals (Belk, [346])."

"These studies have not interpreted how the consumer self-concept relates to brands in such new digitally led physical environments."

"As the impact of digitalization on the self-concept has been addressed in relation to consumer behavior, there is a lack of empirical contributions that consider marketing practitioners' cognition about the topic."

"This study answers the following research question: What does the interaction of the consumer's self-concept with digitally enabled environments mean for brands?"

Methodology

"Purposive sampling was employed to select the final sample of managers to interview (Robinson, [347])."

"A key informant approach was adopted for the panel selection (Robson & Foster, [348]), as the informants identified for interviews were recognized as having a privileged point of view in the investigated domain—that is, digitally enabled brand environments."

"The request for an interview was submitted to the selected marketing managers via email and LinkedIn."

"In the first phase, interviewees were provided with an initial stimulus (for interviews via phone and email) or a set of slides (for interviews via Meet and Zoom) with a list of themes that had emerged from a previous SLR about the digital self (Bartoli and others, [349])."

"In phase two, informants were asked three guiding questions about the topic: (a) their general views on the five topics shared, (b) their thoughts about emerging aspects relating to the consumer self-concept and its relationship with technology, and (c) the implications for brands resulting from the impacts of digitalization on the consumer self-concept."

Findings

"Practitioners recognized that this activity of individuals presenting the self in digital environments is independent from the phenomenon of consumption, but it nevertheless has a strong impact on them, as consumers ask brands and products to be part of their narratives and they act as brands: There's obviously a different role of the customer who has the power to narrate a brand, and that's an important element in the relationship, as a consumer interacting with brands and digitalization also makes them ambassador and spokesperson of the brand."

"Managers' narratives further highlighted consumers' different approaches to digital, increasing the need for brands to improve the understanding of how technology impacts the construction of consumer–brand connections and how this varies among generations and types of consumers."

"The discourse of practitioners regarding the implications for brands arising from the new centrality of the consumer self-concept with digitization focused on four aspects: seamless positioning, consumer–brand relationship, brand values, and personalization."

Discussion and conclusions
"This study investigated the impact of digitalization on the consumer self-concept and its implications for brands by adopting the marketing managers' lens."

"This emphasizes the need for a renewed alliance between business and research (Arnould and others, [350]) to define relevant topics to the advancement of marketing knowledge (Kohli & Haenlein, [351]), such as the hybridization of experience, with the new forms of phygital and metaverse, and its impact on the consumer's self-concept, and how digitization changes the relationship between consumers and brands."

"Concerning the brand implications resulting from the focus on consumer self-concept that digitalization has added to the consumer–brand relationship, the study highlights that the brand identity must correspond to consumers' expectations in terms of values and the transparency of messages."

"In digitally enabled environments, brands should also show their ability to express their own sociality in their relationships with consumers."

Limitations and further research
"To compensate for these limitations and corroborate the results, future research could extend the survey panel by including practitioners in other business fields by distinguishing between B2B and B2C markets and by comparing views among business fields."

"It would also be relevant to consider the perspectives of other relevant actors on this subject, such as academics, to extend the research outlook on the topic."

"Future investigations should employ cross-country analysis and even investigate how managers' understandings can vary with respect to culture and age."

"This suggests the need for future research employing quantitative techniques to collect data, such as online surveys or experimental design, to test the relationship between digitalization and consumer self-concept and their effects on the individual variables related to the brand that were highlighted in the present study."

5.4.10 Digital Media as a Medium for Adolescent Identity Development [515]

This is a machine-generated summary of:

Sebre, Sandra B.; Miltuze, Anika: Digital Media as a Medium for Adolescent Identity Development [515]

Published in: Technology, Knowledge and Learning (2021)

Link to original: https://doi.org/10.1007/s10758-021-09499-1

If you want to cite the papers, please refer to the original.

For technical reasons we could not place the page where the original quote is coming from.

Abstract-Summary
"The aim of the present study was to examine the interrelationships between adolescent media activity, adaptive or maladaptive cognitions related to media use, and identity development processing styles."

"The results showed negative associations between informational identity processing style and time spent gaming, as well as negative associations with time spent communicating via digital media."

"Informational identity processing style was positively associated with adaptive media-related cognitions, such as schoolwork benefits from media use."

"Normative identity processing style was negatively associated with time spend using media for schoolwork or for communicating."

"Diffuse—avoidant identity processing style was positively associated with time spent gaming, and positively associated with ratings of preoccupation with being on media."

Introduction
"In light of the ubiquitous nature of digital media in the lives of many adolescents, and in light of the undeniable advantages and benefits provided by digital literacy, it is somewhat surprising that there have been relatively few studies examining the associations between digital media use and adolescent identity development."

"Digital media, which involves digitally encoded information on a digital electronic device with a screen (Smith [352]), provides a medium which opens up a vast virtual horizon within which the adolescent is free to engage in his or her identity explorations."

"The various purposes of digital media use, such as information gathering, social networking and gaming, each can potentially provide opportunities for the adolescent to develop aspects of his or her identity."

Adolescent Identity Development
"These are referred to as identity processing styles and they include the following: diffuse-avoidant processing style; informational processing style; normative identity processing style."

"According to Berzonsky, individuals with the diffuse-avoidant identity processing style will try to avoid making identity-related decisions as long as possible, they will procrastinate whenever the opportunity allows."

"Individuals with an informational identity processing style will intentionally seek out information which may be self-relevant, they will explore and evaluate relevant information before committing themselves to certain beliefs or values."

"Berzonsky and Kinney [353] have shown that university students with diffuse-avoidant identity processing style reported higher ratings of depression symptoms, and this association was mediated by self-regulation ability."

"Diffuse-avoidant identity processing style has been shown to be associated with alcohol and substance abuse, delinquent behavior, low self-esteem, procrastination and anxiety (e.g. Courey & Pare [354]; Soenens, Berzonsky, Dunkel, Papini & Vansteenkiste [355])."

Digital Media as a Medium for Adolescent Identity Development

"Advocates emphasize that digital media provides a medium whereby the adolescent can be more creative, and can access vast amounts of information which may assist in identity development and identity construction."

"Social networking provides opportunity for self-presentation, an important aspect of identity development, with the possibility of immediate feedback."

"Feedback from peers in the form or "likes" and "followers" may contribute to the adolescent's online identity and sense of self-esteem in a positive or negative manner."

"Several studies analyzing messages conveyed by YouTubers and participants of online fandom communities, have found these digital sources to provide meaningful sites for the exploration and self-acceptance of sexual and gender identity (McInroy and Craig [356])."

"Information that one receives or the interaction one engages with via digital media is only one potentially contributing aspects of adolescent identity development, in conjunction with influences from parents, peers, broader sociocultural and socioeconomic aspects."

Preoccupation with the Internet and Precipitating Factors

"Studies with adolescents have consistently shown that the amount of time spent in online gaming is an important risk factor for the development of problematic internet use (e.g. Ko, Yen, Yen, Lin and Yang [357])."

"Several studies have demonstrated the independent effect of maladaptive cognitions as predictors of problematic internet use (e.g. Sebre, Miltuze and Limonovs [358]; Zhang, Li and Li [359])."

"In relation to identity development, this potentially bidirectional process implies the following: the adolescent's identity processing style may affect identity-seeking behavior in general, including the specifics of his or her internet use; at the same time, the content and form of internet use may encourage and strengthen the adolescent's identity processing style."

"This cross-sectional study does not allow for exploration of causality, but rather provides opportunity to examine the interrelationships between adolescent digital media activity, adaptive or maladaptive cognitions related to media use, and identity processing styles as presented by Berzonsky [360]."

Method

"For purposes of the present study we conducted an initial pilot study in order to ascertain if young adolescents (ages 11–14 years old) would be able to comprehend the meaning of the items."

"Zero-order correlational analyses were conducted to examine bivariate associations among the adolescents' ratings of identity processing styles (informational, normative, diffuse-avoidant), amount of time spent on digital media activity (for schoolwork, communicating and gaming) and cognitions related to digital media activity (useful information, schoolwork benefits and preoccupation)."

"A series of three regression analyses ("stepwise") were conducted, each with one of the identity processing styles (informational, normative, and diffuse-avoidant) as the dependent variable, and as independent variables: the amount of time spent on digital media activity (for schoolwork, communicating, gaming) and cognitions related to digital media use (useful information, schoolwork benefits and preoccupation)."

Results

"Normative identity style was negatively correlated with ratings of "digital media provides useful information" ($r = -0.14$, $p < 0.01$)."

"Informational identity style ratings were shown to be predicted by ratings of the cognition that "digital media provides useful information" ($\beta = 0.16$, $p = 0.006$), and positively correlated with ratings of "my schoolwork benefits from my use of digital media" ($\beta = 0.14$, $p = 0.017$), as well as by less time spent communicating via digital media use ($\beta = -0.16$, $p = 0.003$)."

"Normative identity processing style ratings were shown to be negatively predicted by time using digital media for schoolwork ($\beta = -0.12$, $p = 0.024$), and negatively predicted by age and gender ($\beta = -0.30$, $p = 0.000$, $\beta = -0.10$, $p = 0.046$, respectively)."

Discussion

"The goal of this study was to examine these interrelationships, including media-related cognitions, and specifically in regard to identity processing styles (informational, normative and diffuse-avoidant), as delineated by Berzonsky [360]."

"The results that adolescents with informational processing style are relatively less prone to spend time engaged in digital social communication—e.g. texting, chatting, sending Instagram messages—implies that such activity may not be particularly conducive to identity development, especially if it is not personally meaningful."

"This extensive engagement and preoccupation with gaming may support and facilitate the diffuse-avoidant processing style as well as identity uncertainty, as the results of this study imply."

"Adolescents participating in this study with high normative identity processing style ratings indicated that they are not prone to spend much time on using the computer or other forms of digital media for school purposes."

Conclusions and Practical Implications

"The results of this study highlight the interrelationships between identity processing styles, time spend engaged with various forms of media activity and media-related cognitions."

"Adolescents with informational identity processing style reported media-related cognitions and expectations that digital media will provide useful information and that schoolwork will benefit."

"Adolescents with diffuse-avoidant study did not endorse such adaptive cognitions, but rather reported that they spend greater amounts of time gaming, and tend to be preoccupied with media use."

5.4.11 From Internet Access to Internet Skills: Digital Inequality Among Older Adults [516]

This is a machine-generated summary of:

Hargittai, Eszter; Piper, Anne Marie; Morris, Meredith Ringel: From internet access to internet skills: digital inequality among older adults [516]

Published in: Universal Access in the Information Society (2018)

Link to original: https://doi.org/10.1007/s10209-018-0617-5

If you want to cite the papers, please refer to the original.

For technical reasons we could not place the page where the original quote is coming from.

Abstract-Summary

"Although much research examines the factors that affect technology adoption and use, less is known about how older adults as a group differ in their ability to use the Internet."

"We analyze survey data about older American adults' Internet skills to examine whether skills differ in this group and if they do, what explains differential online abilities."

"The results suggest that attempts to achieve a knowledgeable older adult population regarding Internet use must take into account these users' socioeconomic background and available access points."

Introduction

"While it is important to understand who among older adults is not online, digital inequality scholarship has also emphasized the need to examine differences among people even after they have become users [361]."

"While much of the digital inequality literature examines the population as a whole, the present paper contributes a more nuanced understanding of how older adults vary in their Web-use skills."

"Our analyses suggest that even among a group of online older adults, there is significant variation in skills, and this is linked to people's education, income, and autonomy of use, i.e., the freedom to use the technology when and where one wants."

"Understanding differences in Internet skills among online older adults has implications for future policy, training programs, and technology design."

Related work

"Internet use among U.S. older adults drops off significantly for people aged 75 and older [362]."

"While prior work makes clear what factors matter for Internet adoption among older demographics, limited research has examined whether and how they relate to older adults' Internet skills, that is, their ability to use the Internet effectively and efficiently [363]."

"Known diversity among older adults' technology adoption and use, there have been few studies that systematically examine differences among older adults in terms of the skills required to make full use of the Internet."

"As an exception, based on a national survey of older adults conducted in 2009 in the United States, Hargittai and Dobransky [363] found that the "oldest old" (age 80 and older) and individuals with less than college education had lower levels of Internet skills than younger users of mature ages (55–64)."

Method

"The survey respondents were 505 people from the U.S. aged 60 and older whom we reached through the Cint's OpinionHUB, an online survey panel service also used by others in the literature [364, 365]."

"Cint contacts potential participants through telephone, face-to-face interactions, email, social media, and banner ads."

"We asked participants their age (open field), gender (two options), their highest level of education (six categories), and their income (nine categories)."

"We recoded age into the following categories: 60–64, 65–69, 70–74, and 75–80 (80 was the age of the oldest participants in the study)."

"Regarding Internet experiences, we asked respondents "which of the following categories of devices you typically use to access the Internet (check all that apply)" to assess autonomy of use, i.e., the freedom to use the technology when and where one wants."

Findings

"The age group 60–64 has an average of 1.9 locations of access to the Internet, which is a statistically significant difference compared to all other age groups that have fewer locations of access."

"Being in the most highly educated group and having more income are the most important factors explaining who has more freedom to use the Internet when and where they want."

"Both education and income are positively related to general Internet skills."

"Participants in the highest income bracket have the highest skills averaging 3.5."

"Running the model without the autonomy of use measure (not shown), those with higher income are the most skilled."

"It is through the possibility of having more freedom to use the Internet where and when one wishes that income influences Internet skills."

"The highest income bracket group is again the savviest with a 4.0 average skill score."

Discussion

"While prior work has established that Internet adoption varies considerably among the older population depending on age, income, and educational level [362, 366–369], the findings presented here show that similar factors influence older adults' Internet skill levels."

"Drawing on the theory of digital inequality [370, 371], our analysis calls attention to the importance of differences in Internet skills among older adults once they have gone online."

"Our analysis, along with prior work [363], demonstrates that issues of digital inequality exist among older adults who are active online, and highlights a need for future work examining differences in online participation based on age, socioeconomic status, and even disability status."

"Technologies and training programs designed for addressing digital inequality should also focus on helping these older adults engage in capital-enhancing activities, such as new systems for staying socially connected without needing a home computer or smart mobile device (e.g., [372])."

Conclusion

"The goals of this paper are two-fold in addressing the Internet skills of older adults."

"This paper tested whether online older adults differ when it comes to Web know-how and found that skills vary both within and across age groups, even among the oldest in society."

"Digital inequality literature suggests that people from less-privileged backgrounds have lower Internet skills than those from more advantaged backgrounds, a proposition heretofore rarely tested for lack of appropriate datasets about older adults."

5.4.12 Investigating the Occurrence of Selfie-based Emotional Contagion Over Social Network [517]

This is a machine-generated summary of:

Sasaki, Wataru; Nishiyama, Yuuki; Okoshi, Tadashi; Nakazawa, Jin: Investigating the occurrence of selfie-based emotional contagion over social network [517]

Published in: Social Network Analysis and Mining (2021)

Link to original: https://doi.org/10.1007/s13278-020-00712-0

If you want to cite the papers, please refer to the original.

For technical reasons we could not place the page where the original quote is coming from.

Abstract-Summary

"What occurs under this propagation of emotion is called "emotional contagion," a phenomenon wherein through perception, people experience the same emotion expressed by someone when communicating with them."

"It is not yet clear if emotional contagion among people occurs through selfie photographs posted on the social network media."

"We implemented "SmileWave," the social networking system for investigating selfie-based emotional contagion."

"Our in-the-wild user studies with 38 participants for 2 weeks revealed the occurrence of selfie-based emotional contagion over the social network, based on the results that the users' smile degree improved (15% on average) when the user looked at posted selfie photographs."

Introduction
"We intend to reveal which emotional contagion occurs on mobile online social media, particularly focusing on selfie photographs in which a user is smiling as communication media to propagate happiness among people."

"We also found that there is a higher effect on selfie-based emotional contagion when the user looks at selfie photographs with a higher smile degree."

""SmileWave," the first system for revealing the occurrence of selfie-based emotional contagion on social network and conduct in-the-wild user study with 38 participants for 2 weeks."

"We confirm that the first stage of selfie-based emotional contagion occurs in social network since the users' smile degree improved 15% on average when the user browses posted selfie photographs."

"We find a slight tendency that there is a higher effect on selfie-based emotional contagion when the users view selfie photographs with higher smile degree."

Emotional contagion
"In the early research, emotional contagion was considered to occur only through limited types of human behavior, such as gesture, pose, and facial expression (Condon and Ogston [373])."

"Emotional contagion is composed of a two-stage process (Hatfield et al. [374])."

"Emotional contagion is confirmed to occur in real-world social network."

"We focus on emotional contagion over online social network."

"Along with that, research works on emotional contagion of online social network have also appeared."

"Kramer showed that emotional contagion occurs via text-based computer-mediated communication, without face-to-face interaction through social network (Kramer et al. [373])."

"These two literature works focus on the emotional contagion on the text-based social network."

"There is a research regarding emotional contagion of an image social network."

"Our work is at the same viewpoint in terms of focusing on the emotional contagion over the image social network."

SmileWave
"Since we cannot easily and accurately evaluate the effect of emotional contagion solely by smile selfie images on such environments, we decided to implement our original social network "SmileWave" in order to enable accurate evaluation with the real users."

"We first asked 10 participants to take 11 photographs each with different smile degrees starting from 0% and increasing by 10% until they reached 100%, all of which is based on the user's subjectivity."

"Since it is already known to validate that the change duration of facial expression is immediately (Ekman [373]), we conducted the small-scale experiment using SmileWave to test how quickly the smile degree reached the maximum and whether the system can detect every frame of user's smile degree."

"We selected 1 second as the maximum duration (maximum number of images = 15) after which the user started changing his/her facial expression; we also analyzed how long it takes for the smile degree to reach a maximum."

In-the-wild evaluation

"OBJ3: We want to evaluate if there is any difference in viewing user's reactive smile degree according to multiple different sets of a series of posted photographs ordered differently based on the smile degrees."

"The analysis also revealed that user's smile degree increased by 15% on average in the reactive facial expressions when viewing other user's posted photograph."

"After viewing five photographs with the Random strategy, users' average smile degree in posted selfies was 31.4% (standard deviation 20.0)."

"We discuss that, in the case of photographs with higher smile degree, the rate of user's selfie-based emotional contagion increases when the time to some extent has passed."

"After viewing five photographs with the HighToLow strategy, users' average smile degree in their posted selfies was 21.6% (standard deviation 16.2)."

"After viewing five photographs with the WithHistory strategy, users' average smile degree in their posted selfies was 31.8% (standard deviation 22.4)."

Discussion

"We thus concluded that the first stage of smile emotional contagion does occur in a selfie-based social network."

"We mainly focus on the change of smile degree for "individual" user in order to confirm the occurrence of selfie-based emotional contagion over online communication."

"It can be seen that selfie-based emotional contagion occurred regardless of the scale of this experiment since it was investigated from an individual point of view."

"When we further investigate the online emotional contagion over social network such as how to spread the effect of emotional contagion among people and how users' positions in social network affect their influential strength, we need to conduct longer-term and larger-scale user study in order to generalize the results."

"Existing studies in the past have mentioned emotional contagion as a property of all human beings, regardless of culture, age, and gender (Hatfield et al. [374])."

Conclusion

"We proposed SmileWave, a novel social network that shares selfie photographs among users and detects the change of smile degree while the users browse others' selfie photographs, in order to evaluate the occurrence of facial mimicry."

"We also found a slight tendency that there is an effect on selfie-based emotional contagion by showing smile selfies with higher smile degree."

"We observed that there is a stronger effect on selfie-based emotional contagion after he/she viewed the posted selfie photographs ordered from the one with the lower smile degree to the one with the higher degree."

"Our extensive evaluation is the first work that deeply investigated and revealed the effects of selfie-based emotional contagion over the social network."

5.4.13 Mobile Social Media Use and Self-identity Among Chinese Adolescents: The Mediating Effect of Friendship Quality and the Moderating Role of Gender [518]

This is a machine-generated summary of:

Wang, Wei; Qian, Guoying; Wang, Xingchao; Lei, Li; Hu, Qian; Chen, Jing; Jiang, Suo: Mobile social media use and self-identity among Chinese adolescents: The mediating effect of friendship quality and the moderating role of gender [518]

Published in: Current Psychology (2019)

Link to original: https://doi.org/10.1007/s12144-019-00397-5

If you want to cite the papers, please refer to the original.

For technical reasons we could not place the page where the original quote is coming from.

Abstract-Summary

"The current study investigated the effect of mobile social media use on adolescents' self-identity, and explored the mediating role of friendship quality and the moderating role of gender."

"Results indicated that mobile social media used was positively correlated with adolescents' self-identity and friendship quality."

"Friendship quality mediated the relationship between mobile social media use and adolescents' self-identity."

"Adolescents' mobile social media use was associated with a higher quality of friendship, which, in turn, contributed to their formation of self-identity."

Introduction

"Our study is guided by the idea that the Internet has provided a range of new contexts, such as social networking sites, instant messaging platforms, mobile social media (hereafter, "MSM"), and so on, within which adolescents can present and explore their self-identities."

"Social media use has been found to be closely related to the formation of adolescents' self-identity (Matsuba [375]; Yang and others [376])."

"The self-identity control theory (Kerpelman and others [377]) offers a useful approach for understanding how friendship quality may affect adolescents' self-identity."

"Based on existing theories and previous studies, the following hypothesis was proposed: H1: MSMU would be positively related to adolescents' self-identity and this association was expected to be explained by friendship quality."

"The present study examined the effect of MSMU on self-identity among Chinese adolescents and explored whether MSMU and self-identity were indirectly linked via friendship quality."

Materials and Methods

"Among the 560 adolescents, 87 reported that they never used mobile social media, so these participants were excluded from the data analysis."

"It consists of 15 items (e.g., I post my feelings and moods on mobile social media) and includes the following three dimensions: (1) interpersonal communication and presentation behavior (Cronbach's $\alpha = .67$), (2) information acquisition and sharing behavior (Cronbach's $\alpha = .64$), and (3) having fun and recreation behavior (Cronbach's $\alpha = .88$)."

"Items were assessed on a 5-point Likert scale that ranged from 1 (never) to 5 (all at time)."

"We used a 12-item ego identity scale (Tan and others 378) to measure participants' global evaluation of their self-identity."

"Responses to all items were summed to produce a composite score, with higher scores indicating higher levels of self-identity."

"Items were assessed on a 5-point Likert scale that ranged from 1 (strongly disagree) to 5 (strongly agree)."

Results

"We found that MSMU was positively correlated with friendship quality ($r = .29$, $p < .001$) and self-identity ($r = .12$, $p = .008$)."

"Model 4 was adopted to test the mediating effect of friendship quality on the relation between MSMU and self-identity."

"MSMU was significantly associated with friendship quality ($\beta = .29$, $p < .001$), and friendship quality was significantly associated with self-identity ($\beta = .12$, $p = .02$)."

"Friendship quality fully mediated the relation between MSMU and self-identity."

"The indirect effect of MSMU on self-identity mediated by friendship quality was significant, ab = .035, SE = .02, 95% CI [.006, .065]."

"The 95% confidence interval did not consist of zero, signifying that MSMU exerted a significant indirect effect on self-identity via friendship quality."

Discussion

"The current study proposed a mediated moderation model to test whether MSMU was indirectly related to self-identity via friendship quality as well as whether gender moderated the effects of MSMU on friendship quality and self-identity."

"We focused on two factors that may have significant contributions to the formation of self-identity during adolescence, that is, MSMU and friendship quality."

"Consistent with these ideas, we found that MSMU and friendship quality both had positive correlations with adolescents' self-identity, which was in line with previous studies (e.g., Lei and Wang [379]; Loughlin [380])."

"Our study found that friendship quality mediated the association between MSMU and adolescents' self-identity."

"Gender indirectly moderates the effect of MSMU on self-identity by moderating the relation between MSMU and friendship quality."

5.4.14 The Association Between Self-reported Depressive Symptoms and the Use of Social Networking Sites (SNS): A Meta-analysis [519]

This is a machine-generated summary of:

Vahedi, Zahra; Zannella, Lesley: The association between self-reported depressive symptoms and the use of social networking sites (SNS): A meta-analysis [519]

Published in: Current Psychology (2019)

Link to original: https://doi.org/10.1007/s12144-019-0150-6

If you want to cite the papers, please refer to the original.

For technical reasons we could not place the page where the original quote is coming from.

Abstract-Summary

"A recent proliferation of research regarding SNS use pertains to its association with self-reported depressive symptoms."

"The current meta-analysis evaluates this literature by assessing the relationship between SNS use and self-reported depressive symptoms."

"Based on 55 independent samples (N = 80, 533), an effect size of r = .165 was obtained, indicating a small, positive, significant association between self-reported symptoms of depression and SNS use."

"The type of SNS use was identified to be a significant moderator of this relationship."

Social Networking Sites: Definition, Users, and Motivations

"SNS use often involves the customization of a public or private "profile," with the aim of interacting with one's social group in an online community."

"Although UGT has generally been employed to examine the use of more traditional types of media (i.e. newspapers), researchers have recently applied it to newer forms of communication."

"Previous literature has generally identified two main motivations for SNS use: 1) maintaining existing social connections; and 2) developing new social connections? (e.g., Tosun [381]; Reich and others [382])."

"General entertainment has also been found to be a motivation for SNS use (e.g., Al-Menayes [383]; Hunt and others [384])."

"Despite these motivations, however, the use of SNS's has been associated with various detrimental consequences, including depressive symptoms (e.g., Frison and Eggermont [385]; Lee [386]; Nesi and Prinstein [387])."

Depression: Diagnosis and Etiology

"Rates of depression generally decrease over the lifespan and – starting from adolescence – females are twice as likely as males to be diagnosed with depression (Galambos and others [388])."

"An individual is diagnosed with clinical depression through a structured or semi-structured interview conducted by a trained clinician."

"To structured interviews, several self-report rating scales also are used to screen the severity of depressive symptoms."

"These scales are generally not meant to be used as diagnostic tools, but rather are usually implemented in a) research settings, to assess the severity of depressive symptoms in a sample of participants; b) clinical settings, to screen for and identify individuals at-risk for depression; and c) clinical settings, for individuals who have already been diagnosed with depression, in order to track their symptoms in response to treatment."

SNS Use and Depression: Negative Quality and Problematic Use

"Previous literature (e.g., Davila and others [389]; Feinstein and others [390]) has suggested that the quality of SNS use – that is whether an individual experiences the activity as positive or negative in nature – is a greater predictor of depressive symptoms than the frequency of SNS use."

"Negative quality SNS use has not been concretely or consistently defined in the literature; rather, it has instead been used as an umbrella term to encompass a variety of incidents on SNSs including detrimental social comparisons (Feinstein and others [390]), receiving unwanted friend-requests from previous romantic partners (Tsai and others [391]), and cyberbullying (Best and others [392])."

"Several authors, concede that clear theoretical distinctions between these two concepts – negative quality and problematic SNS use – have not been clearly defined (Bányai and others [393]; Baker and Algorta [394])."

"Although researchers acknowledge theoretical differences between SNS use that is negative in quality, and SNS use that is problematic, consistent definitions differentiating these two concepts have not been delineated in the literature."

Rationale for Current Meta-Analysis

"Huang [395] recently conducted a meta-analysis to investigate the relationship between time spent on SNSs and various constructs of psychological well-being, including depression."

"A weak relationship between depression and SNS use was identified ($r = .11$; $k = 24$), and no significant moderators were obtained."

"Although Huang's [395] meta-analysis presents an indication of the relationship between time spent on SNSs and depression, it does not provide a comprehensive assessment of the relationship between specific types of SNS use and depressive symptoms."

"An inclusive meta-analysis that investigates the type of SNS use as a moderator is the next step in contributing to the existing literature."

Hypotheses

"Based on the existing literature, we hypothesized the following: H1: There will be a positive relationship between overall social networking site use and self-reported depressive symptoms."

"Although mixed results do exist in literature, existing reviews (e.g., Baker and Algorta [394]; Huang [395]) do indicate that overall, increased SNS use is associated with increased symptoms of depressive."

"H2: Problematic – or negative quality – SNS use will be more strongly associated with increased depressive symptoms than general SNS use."

"The type of SNS use – that is, whether it is problematic or not – will be a significant moderator of the relationship between SNS use and self-reported depressive symptoms."

"H3: The relationship between SNS use and self-reported depressive symptoms will be weakened as the mean age of the participants increases."

Method

"Studies must have included the participants' self-reported measure of SNS use."

"Studies where SNS use was included as a subscale of a larger scale – for example, one that investigated the various types of internet use – were included if the unique statistic between the SNS use subscale and depression was provided."

"Studies where depressive symptoms were assessed as a subscale of a larger scale (i.e., the severe depression subscale of the GHQ-28) were included if the appropriate statistic between the depression subscale and SNS use was provided."

"Studies must have included a measure of participants' levels of SNS use."

"Studies often assessed multiple aspects of SNS use; in order to ensure only one effect size was obtained from each sample, the extraction of effect sizes was prioritized as follows: a) general frequency of SNS use; b) addictive SNS use; c) SNS use that was negative in quality; and d) SNS use that was positive in quality."

Results

"Studies that investigated problematic or negative quality SNS use obtained a stronger effect size than those that assessed general SNS use ($r = .27$ vs. $r = .11$, respectively)."

"The results indicate a similar effect for problematic ($r = .26$) and negative quality use ($r = .30$), with significant heterogeneity still attained; furthermore, an analysis comparing only the studies assessing problematic/addictive SNS use ($k = 11$) and negative quality SNS use ($k = 8$) obtained an insignificant heterogeneity Q-value statistic, $Q = .394$, $p = .53$, indicating no significant moderating effect between these two types of SNS use."

"These results suggest that there is no significant difference between studies assessing problematic SNS use and negative quality SNS use; therefore, these two groups of studies were combined into one subgroup to achieve a more well-powered moderator analysis."

Discussion

"The obtained results suggest a small, positive association between SNS use and self-reported symptoms of depression, with the quality of social networking as a significant moderator of this relationship."

"Although a heterogeneous group of studies comprised the subgroup of negative quality use, our results suggest that a medium-sized association ($r = .27$) exists between SNS use that pertains to one or more of these negative activities, and depressive symptoms."

"Combined with the insignificant moderating effect of daily time spent on SNSs, this result supports the proposition by various researchers (Davila and others [389]; Feinstein and others [390]), who argue that the quality of SNS use is a stronger associate of depressive symptoms than the quantity of use."

"Given these results, the only claim that can be made confidently is that there is a positive association between self-reported SNS use and depressive symptoms, and that the causal direction of this relationship is currently unknown."

Conclusions

"The current meta-analysis provides a comprehensive review of the literature on social networking site use and self-reported depressive symptoms."

"The detection of a significant moderator provides a key area of future research to further delineate the differing associations between the various types of SNS use and depression."

"With the ubiquity of social networking sites in modern-day society, it is imperative that the psychological impact of its use is rigorously investigated."

"It is the intention of this review to provide a framework for understanding self-reported depressive symptoms associated with SNS use, and highlight the various questions that remain open to future investigation."

[Section 10]

"The use of SNSs has been associated with various social and psychological benefits, such as enhanced feelings of subjective well-being (e.g., Kim and Lee [396]), there has also been a surge in research documenting the adverse effects, such as psychological distress (Radovic and others [397])."

"A sizable proportion of this research is devoted to identifying the relationship between depression and SNS use (e.g., Frison and Eggermont [385]; Lee [386]; Nesi and Prinstein [387]; Woods and Scott [398])."

"Given depression is a serious mental illness, with varying psychological, cognitive, somatic, and social effects, researchers have thoroughly investigated the cross-sectional relationship between SNS use and depression Despite the proliferation of this research, the findings remain inconsistent; for instance, some studies have found a positive relationship (e.g., Landoll and others [399]), others have identified a negative relationship (e.g., Park and others [400]), and some indicate no significant relationship at all (e.g., Banjanin and others [401])."

5.4.15 Problematic Social Media Use, Personality Traits, and Mental Health Among Bangladeshi University Students [520]

This is a machine-generated summary of:

Ahmed, Oli; Sultana, Tasnim; Alam, Najifa; Griffiths, Mark D.; Hiramoni, Fatema Akhter: Problematic Social Media Use, Personality Traits, and Mental Health Among Bangladeshi University Students [520]

Published in: Journal of Technology in Behavioral Science (2022)

Link to original: https://doi.org/10.1007/s41347-021-00235-1

If you want to cite the papers, please refer to the original.

For technical reasons we could not place the page where the original quote is coming from.

Abstract-Summary

"Problematic social media use (PSMU) has received increasing attention among mental health professionals."

"There are a few studies that have explored the relationships between PSMU, personality traits, and mental health."

"There is scarcity of information related to usage duration, usage purpose of different social media, and prevalence rates of the risk of PSMU in Bangladesh."

"Results also showed (i) low conscientiousness (i.e., being careless, impulsive, disorganized) was significantly associated with PSMU; (ii) low agreeableness (i.e., cold, unfriendly), low conscientiousness, and high social media use were predictors of PSMU; and (iii) low conscientiousness, high neuroticism, and PSMU were significantly associated with self-rated poor mental health."

Introduction

"Existing literature also suggests individual differences among those who use social media and that one of the key determinants appears to be personality factors (Kircaburun and others, [402])."

"Excessive social media use and problematic social media use (PSMU) can negatively affect users' mental health (Chou & Edge, [403]; Rahman & Ahmed, [404]; Valkenburg and others, [405])."

"Individuals with higher conscientiousness engage in social media use to maintain existing relationships, and individuals lower in this trait utilize social media for news and information (Choi and others, [406])."

"Kircaburun and others [402] found that higher problematic social media users were more likely female, introverted, neurotic, non-conscientious, and agreeable."

"Although the majority of studies found a positive association between time spent on social media and depression, a few studies reported non-significant association between frequency of social media use and depression."

"The present study explored the relationship between personality traits, social media use, PSMU, and mental health among young Bangladeshi individuals."

Method

"Confirmatory factor analysis from the present study data indicated that the scale had a good model fit ($\chi^2/df = 1.93$, GFI = 0.98, CFI = 0.98, TLI = 0.96, SRMR = 0.03, RMSEA = 0.05)."

"The inter-item correlations of the scale in the present study were sufficient (0.29–0.65) (Pallant, [407])."

"Confirmatory factor analysis in the present study indicated that the scale had good model fit ($\chi^2/df = 1.92$, GFI = 0.97, CFI = 0.97, TLI = 0.94, SRMR = 0.06, RMSEA = 0.05)."

"Confirmatory factor analysis in the present study data indicated that the scale had acceptable model fit ($\chi^2/df = 2.46$, GFI = 0.90, CFI = 0.91, TLI = 0.90, SRMR = 0.06, RMSEA = 0.07)."

Results

"The obtained score on Breusch-Pagan test (9.347, $p < 0.252$) and Koenker test (10.099, $p < 0.170$) indicated homoskedasticity of the data."

"A large proportion of users who used social media more than 4 h daily (18.9%) were classed as problematic social media users ($\chi^2 = 20.056$, $p < 0.001$) than those who used social media less than 4 h daily (4.0%), a difference that was significant (odds ratio = 5.607, 95% CI [2.456, 12.800], $p < 0.001$)."

"Among personality traits, only low conscientiousness ($\beta = -0.19$, $p < 0.01$) significantly predicted social media usage duration."

"Low agreeableness, low conscientiousness, and high social media usage duration were significant predictors of PSMU (agreeableness: $\beta = -0.14$, $p < 0.01$; conscientiousness: $\beta = -0.22$, $p < 0.001$; social media usage duration: $\beta = 0.33$, $p < 0.001$)."

Discussion

"The present study explored the role of personality traits (extroversion, openness, agreeableness, neuroticism, and conscientiousness), social media usage duration, and problematic social media use (PSMU) on social media users' mental health."

"Path analysis, for testing hypothesized model, explored the direction of influence of personality traits, social media usage duration, and PSMU on mental health."

"Extroversion and openness had non-significant influence on social media usage duration, PSMU, and mental health."

"Findings also suggested that low conscientiousness was strong and negative predictor of social media usage duration, PSMU, and self-rated poor mental health."

"Findings here showed that social media use duration was significantly associated with PSMU (as found in other studies, such as Rahman & Ahmed, [404]) but not to self-rated poor mental health."

"The present study's findings also showed that PSMU was a strong positive predictor of self-rated poor mental health."

Recommendations and Conclusion

"Personality traits, especially low agreeableness and low conscientiousness, affect problematic social media use that, in turn, affect users' mental health."

"Awareness programs and workshops need implementing in Bangladeshi universities to raise awareness among young users about responsible use of social media."

"Counseling and behavior modification therapy could also be provided for problematic social media users and to counter the negative effects of PSMU and facilitate improved mental health."

"Further in-depth research should be undertaken to explore how personality traits impact problematic social media use and further delineate what factors underly this association in impacting social media users' mental health."

5.4.16 Narcissism and Selfie-posting Behavior: The Mediating Role of Body Satisfaction and the Moderating Role of Attitude Toward Selfie-posting Behavior [521]

This is a machine-generated summary of:

Wang, Yuhui; Xie, Xiaochun; Wang, Xingchao; Wang, Pengcheng; Nie, Jia; Lei, Li: Narcissism and selfie-posting behavior: the mediating role of body satisfaction and the moderating role of attitude toward selfie-posting behavior [521]

Published in: Current Psychology (2018)
Link to original: https://doi.org/10.1007/s12144-018-9795-9

If you want to cite the papers, please refer to the original.

For technical reasons we could not place the page where the original quote is coming from.

Abstract-Summary
"The present study aimed to examine whether body satisfaction mediated the relation between narcissism and selfie-posting behavior, and whether this mediating process was moderated by attitude toward selfie-posting behavior."

"The results indicated that body satisfaction mediated the association between narcissism and selfie-posting behavior."

"Attitude toward selfie-posting behavior moderated the relation between body satisfaction and selfie-posting behavior."

"Findings of the present study highlight the significance of identifying the mechanisms that moderate the mediated paths between narcissism and selfie-posting behavior."

Introduction
"Given the pervasiveness of selfie-posting, a number of studies explored the factors (e.g., self-esteem, Big Five personality, and narcissism) that may drive this behavior (Kim and Chock [409]; McCain and others [410]; Qiu and others [411]; Sorokowska and others [412])."

"Factors, narcissism is regarded as an important predictor that can influence posting selfies on SNSs (Barry and others [413]; Fox and Rooney [414]; Kim and others [415]; Kim and Chock [409]; Lee and Sung [416]; McCain and others [410]; Sorokowski and others [417]; Sung and others [418]; Weiser [419])."

"Previous studies focused primarily on the direct relation between narcissism and selfie-posting behavior, little is known about the mediating and moderating mechanisms underlying this relation."

"The present study aimed to explore how and when narcissism relates to selfie-posting behavior."

Narcissism and Selfie-Posting Behavior
"Since selfie-posting behavior allows individuals to selectively present attractive and self-promoting photos of themselves (Chua and Chang [420]; Diefenbach and Christoforakos [421]; Fox and Rooney [414]; Fox and Vendemia [422]), it is feasible that such behavior would be more common in those with high narcissism."

"According to the dynamic self-regulatory processing model of narcissism, narcissists attempt to gain positive feedback from other people by regulating their behavior, which further leads to the ultimate goal of maintaining their grandiose and positive self-views (Morf and Rhodewalt [423, 424])."

"The act of selecting attractive selfies and posting them on SNSs may function as a self-regulatory behavior through which narcissistic individuals attempt to meet their self-regulatory objectives—enhancing their grandiose and positive self-views (McCain and others [410]; Weiser [419])."

"Previous empirical studies support the positive relation between narcissism and selfie-posting behavior."

"These suggestions indicate that individuals' body satisfaction may mediate the relation between narcissism and selfie-posting behavior."

"The present study aims to examine the possible mediating role of body satisfaction between narcissism and selfie-posting behavior."

Body Satisfaction as a Mediator

"Body satisfaction, as a intrapersonal variable (Wright and others [425]), may mediate the association between narcissism and selfie-posting behavior (interpersonal self-regulatory behavior)."

"Individuals with high narcissism may have cognitive and affective evaluation about their bodies (e.g., satisfied with their bodies) during the intrapersonal process, which will underlie their external self-regulatory behavior (e.g., post selfies on SNSs) in the interpersonal process."

"Empirical research has demonstrated that narcissism has a positive relation to body satisfaction (Bleske-Rechek and others [426]; Chan and Tsang [427]; Lipowska and Lipowski [428])."

"Specific to selfie-posting behavior, body satisfaction also plays a predicting role."

"Empirical research also showed that body image satisfaction has a significant and positive correlation to posting photos of oneself (Fox and Vendemia [422]; Ridgway and Clayton [429])."

"These findings highlight that individuals who have higher levels of body satisfaction post more selfies compared to their counterparts."

"Body satisfaction would mediate the positive association between narcissism and selfie-posting behavior."

Attitude toward Selfie-Posting Behavior as a Moderator

"Heterogeneity of outcomes may originate from individual characteristics that moderate the effect of body satisfaction on selfie-posting behavior, for example, attitude toward selfie-posting behavior."

"Attitude toward selfie-posting behavior may moderate the relation between body satisfaction and selfie-posting behavior."

"In light of the dynamic self-regulatory processing model of narcissism (Morf and Rhodewalt [423, 424]), body satisfaction and attitude toward selfie-posting behavior, as two intrapersonal processing variables, may interact with each other and influence selfie-posting behavior together."

"Positive attitude toward selfie-posting behavior may enhance the predicting role of body satisfaction on selfie-posting behavior; in contrast, negative attitude may weaken the effect of body satisfaction on selfie-posting behavior."

"To our knowledge, no previous study has examined whether attitude toward selfie-posting behavior moderate the relation between body satisfaction and selfie-posting behavior."

"The effect of body satisfaction on selfie-posting behavior would be stronger for individuals who have positive attitude toward selfie-posting behavior than those who have negative attitude toward selfie-posting behavior."

The Present Study

"The purposes of the current study were twofold."

"The present study examined whether body satisfaction would mediate the relation between narcissism and selfie-posting behavior."

"We explored whether the relation between body satisfaction and selfie-posting behavior would be moderated by attitude toward selfie-posting behavior."

Methods

"The initial sample consisted of 390 undergraduate students who had ever posted their own selfies on SNSs from a university in China."

"Body satisfaction was assessed using Body Areas Satisfaction Scale (BASS), a subscale of Multidimensional Body-Self Relations Questionnaire (MBSRQ) (Cash [430])."

"BASS uses a fully anchored, 5-point response scale ranging from 1 = very dissatisfied to 5 = very satisfied."

"Higher scores on the BASS indicate higher satisfaction with the body."

"Attitude toward selfie-posting behavior was measured along a seven-point (1–7) semantic differential scale (Lee and Sung [416])."

"These eight items were averaged to form an attitude toward the selfie-posting behavior index."

"Higher scores indicate more positive attitude."

"Selfie-posting behavior was assessed using Sung and colleagues' method by asking "How often would you say that you post selfies on SNSs?" (Sung and others [418])."

"Higher scores indicate higher selfie-posting frequencies."

Results

"Results showed that narcissism, body satisfaction, and attitude toward selfie-posting behavior were significantly correlated with selfie-posting behavior positively."

"Narcissism and attitude toward selfie-posting behavior were positively associated with body satisfaction."

"Shown in model 3, body satisfaction positively predicted selfie-posting behavior significantly ($b = .16$, $p < .01$), whereas the direct effect of narcissism on selfie-posting behavior was not significant ($b = .10$, $p > .05$)."

"Simple slope tests indicated that for individuals who had positive attitude toward selfie-posting behavior, higher body satisfaction were associated with more selfie-posting behavior, $b_{simple} = .23$, $p < .01$."

Discussion

"The present study formulated a moderated mediation model to examine whether narcissism would be indirectly related to selfie-posting behavior through body satisfaction and whether the association between body satisfaction and selfie-posting behavior was moderated by attitude toward selfie-posting behavior."

"Consistent with our hypothesis, we found that body satisfaction mediates the relationship between narcissism and selfie-posting behavior."

"Given the relation between personality traits and body image (Allen and Walter [431]; Davis and others [432]; Fortes and others [433]), the present study innovatively integrated narcissism and body satisfaction together to examine selfie-posting behavior and tested the mediating role of body satisfaction in the relation between narcissism and selfie-posting behavior."

"The predicting role of body satisfaction on selfie-posting behavior may be explained by the impact of intrapersonal process on interpersonal self-regulatory strategy noted in the dynamic self-regulatory processing model of narcissism (Morf and Rhodewalt [423, 424])."

5.4.17 Analyzing the Links Between Problematic Social Media Use, Dark Triad Traits, and Self-esteem [522]

This is a machine-generated summary of:

Kircaburun, Kagan; Demetrovics, Zsolt; Tosuntaş, Şule Betül: Analyzing the Links Between Problematic Social Media Use, Dark Triad Traits, and Self-esteem [522]

Published in: International Journal of Mental Health and Addiction (2018)

Link to original: https://doi.org/10.1007/s11469-018-9900-1

If you want to cite the papers, please refer to the original.

For technical reasons we could not place the page where the original quote is coming from.

Abstract-Summary

"The purpose of this study was to examine the association of problematic social media use (PSMU) with Dark Triad (DT) traits (i.e., Machiavellianism, psychopathy, and narcissism) and self-esteem."

"A total of 827 university students [495 (60%) women and 332 (40%) men, $M_{age} = 20.36$, SD = 1.47] filled out a questionnaire that included the Social Media Disorder Scale, Dark Triad Dirty Dozen Scale, and Single Item Self-esteem Scale."

"The direct effects of DT traits and self-esteem on PSMU and the indirect effect of narcissism on PSMU through self-esteem were investigated by applying structural equation modeling."

"Results indicated that, when controlling for gender and age, Machiavellianism and narcissism had small significant direct effects on PSMU, while the partial indirect link of narcissism via self-esteem was negligible."

PSMU and Dark Triad

"Individuals with emotional and psychological dysfunctions are more likely to be problematic social media users (Andreassen and others [434]; van den Eijnden and others [435]); therefore, we hypothesized that PSMU will be positively associated with DT traits (i.e., narcissism, psychopathy, and Machiavellianism)."

"It is known that problematic use of social media differs according to various individual characteristics such as self-esteem and narcissism (Andreassen and others [436]; Carmody [437])."

"When aforementioned studies are all considered together, it is observed that the relationships between DT traits, self-esteem and social media use, and other antisocial behaviors have been investigated in previous researches; however, their direct and indirect effects on PSMU have not been examined."

"Based on the UGT and I-PACE, which suggest that demographics and personality features are important individual differences regarding uses and gratifications of problematic use of technology (Brand and others [438]; Katz and others [439]; Rubin [440]), this research investigates the direct effects of DT traits and self-esteem on PSMU and indirect effects of narcissism on PSMU through self-esteem."

Method

"After the scale adaptation, 827 students from a public university in Turkey, aged between 17 and 24 ($M_{age} = 20.36$, SD = 1.47), have participated in the study."

"The Social Media Disorder Scale was developed by van den Eijnden and others [435] based on the Internet gaming disorder criteria determined in DSM-5 and was adapted to Turkish by the researchers of this study."

"Confirmatory factor analysis (CFA) produced good fit values with the Turkish form (χ^2/df = 1.63, root mean square error of approximation (RMSEA) = .06 CI 90% [.02, .09], standardized root mean square residual (SRMR) = .05, comparative fit index (CFI) = .94, normed fit ındex (NFI) = .86, goodness-of-fit index (GFI) = .96, adjusted goodness-of-fit ındex (AGFI) = .92)."

"The Single Item Self-esteem Scale was developed by Robins and others [441] and was adapted to Turkish by the researchers of this study."

Results

"Multivariate analysis of covariance (MANCOVA) showed that disordered social media users had higher scores on all DT traits compared with the non-disordered users, self-esteem did not differ across groups, and gender was a significant covariate ($F_{(4, 820)} = 13.85$; $p < .001$; $\eta^2 = .06$)."

"According to the findings, the direct effects to PSMU of narcissism ($\beta_S = .30$, $p < .001$, 95% CI [.19, .41]), Machiavellianism ($\beta_S = .23$, $p < .05$, 95% CI [.02, .45]), and self-esteem ($\beta_S = -.09$, $p < .05$, 95% CI [− .17, − .01]) were significant, and that of psychopathy ($\beta_S = -.10$, $p > .05$, 95% CI [− .30, .09]) was non-significant."

"Further, the indirect effect of narcissism to PSMU through self-esteem ($\beta_S = -.01$, $p <.05$, 95% CI [− .02, .00]) was significant; however, the effect size was negligible."

Discussion

"The present study analyzed the relations between Dark Triad traits, self-esteem, and problematic social media use."

"These results are in accordance with recent studies indicating that DT traits are related to time spent on social networking sites (Fox and Rooney [414]), and with the theoretical frameworks which suggest that problematic social media use is differentiated according to personality differences (Brand and others [438]; Ryan and others [442])."

"Previous studies have put forward that narcissism was positively related to number of friends (it may be deceptive and exaggerated) and higher time spent on social media (Eşkisu and others [443])."

"Individuals with lower self-esteem may want to compensate their need for real-life social relations or perhaps to simply occupy themselves by using virtual platforms such as social media excessively in order to avoid problematic relations."

"Individuals may be using social media for higher self-esteem through gathering likes."

"Future studies should examine the mediating role of social media use motives between DT traits and PSMU."

[Section 5]

"Researches that have been carried out after the widespread use and expansion of social media sites revealed that each social media site is being used for different purposes."

"The use of social media becomes uncontrolled and compulsive in terms of the situations of addicted individuals (Griffiths [444])."

"Uses and gratifications theory (UGT) indicates that individuals' media use motives and intensity are differentiated regarding their social and psychological needs, individual differences such as personality features, media use motives, and outcome expectations (Katz and others [439]; Rubin [440])."

"The UGT framework has been used by many researchers in order to investigate the relationships between media use motives, individual differences, and problematic use of technology (Floros and Siomos [445]; Huang [446])."

"Self-esteem is another personality difference that should be taken into account when considering social media use and addiction."

5.4.18 Selfie Image Super-resolution Using an Implicit Prior Learned from Self-examples [523]

This is a machine-generated summary of:

Hans, W. Jino; Venkateswaran, N.: Selfie image super-resolution using an implicit prior learned from self-examples [523]

Published in: Cluster Computing (2018)

Link to original: https://doi.org/10.1007/s10586-018-2466-7

If you want to cite the papers, please refer to the original.

For technical reasons we could not place the page where the original quote is coming from.

Abstract-Summary

"Selfie image is a self-captured photograph of oneself using the front camera of a smartphone."

"a high-resolution primary camera in the rear and a low-resolution secondary camera in the front of the smartphone."

"Selfie images are captured using low-resolution front camera and hence the spatial resolution of these images will be very much limited."

"We propose an efficient approach to improve the spatial resolution of selfie images by exploiting the self-similarity across various scales of selfie images using a self-example based super-resolution algorithm."

"The super-resolution algorithm is formulated by learning a local regression from in-place self-example patches extracted from various scales of the given selfie image."

"In-place matching ensures that image patches extracted from different scales are spatially close and hence will have the same high-frequency details."

Introduction

"Selfie image is a self-captured photograph of one-self using the front camera of a smartphone."

"As the front camera will have limited pixel sensor elements compared with rear camera, selfie image will be typically a low-resolution image."

"It is significant to super-resolute the selfie images."

"If the training images are not relevant to the test selfie image, it will learn false high-frequency details and the reconstructed image will look unnatural with displeasing artifacts."

"To avoid the counterfeit high-frequency details in the reconstructed selfie image, self-examples are used for training."

"We propose an efficient approach to improve the spatial resolution of selfie images by exploiting the self-similarity across various scales using a self-example based super-resolution algorithm."

"The super-resolution algorithm is formulated by learning a local regression from in-place self-example patches extracted from various scales of the given selfie image [447]."

Selfie image super-resolution using self-examples

"The selfie image super-resolution algorithm dealt in this paper is based on in-place self-examples and a local regression."

"The in-place self-similarity [447] between various scales of the input test selfie image is used for learning a regression function to map the LR image patches with its corresponding HR counterpart."

"The self-examples should be in-place with same spatial location at various scale of the image."

"Self-example based SR approaches produce more visually pleasing super-resolved images as they do not rely on any external training examples in the reconstruction process."

"The succeeding section gives further details of the in-place self-examples and regression learning."

"Self-example based SR algorithms rely only on the internal statistics of the input test image to learn the HF details."

"The self-example patch-pairs are used as an implicit prior to super-resolve the selfie image."

Results and discussions

"SSIM value (typically close to 1) is a correlative measure, which gives the similarity between the super-resolved image and the ground truth [448]."

"The proposed approach super-resolves the LR image effectively."

"To have a clear understanding on it, the test selfie image #3 is super-resolved and a small portion from the face is magnified and is compared with other algorithms."

"This approach generates more sharp images than other algorithms."

"It is observed from the zoomed-in image that a few high-frequency details are visualized in the facial region of the image."

"The eyelash in the zoomed image looks natural and also the freckles in the face region are distinct."

"This is due to the fact that the implicit prior is learned from the given test image itself and hence the super-resolved from distortions."

"SSIM features the correlation index between the input and the super-resolved image."

Conclusion

"An efficient implicit prior learned from self-examples to super-resolve low-resolution selfie images is presented."

"The proposed SR algorithm establishes a relation between the low and high-resolution image by learning an implicit prior from self-examples."

"The relation between self-example image patches is established by a local regression approximated by Taylor's series and the regression co-efficients are estimated by a dictionary based approach."

5.4.19 The Relationship Between Self-objectification and Social Avoidance Among Chinese Middle Adolescent Girls: The Mediating Role of Appearance Comparison and Self-esteem [524]

This is a machine-generated summary of:

Guo, Qinglong; Wu, Mingxia: The relationship between self-objectification and social avoidance among Chinese middle adolescent girls: The mediating role of appearance comparison and self-esteem [524]

Published in: Current Psychology (2021)

Link to original: https://doi.org/10.1007/s12144-021-01705-8

If you want to cite the papers, please refer to the original.

For technical reasons we could not place the page where the original quote is coming from.

Abstract-Summary

"We investigated self-objectification, appearance comparison, self-esteem, and social avoidance among 262 Chinese middle adolescent girls ($M_{age} = 15.97$ years, ranging from 14 to 17) living in Guangdong province, China."

"The results showed a positive correlation between self-objectification and social avoidance, which was further mediated by appearance comparison and self-esteem."

"Girls with high levels of self-objectification were more likely to compare their appearance to others, which in turn, was linked to lower self-esteem, and subsequently correlated with greater social avoidance."

"Interventions addressing social avoidance among adolescent girls may include reducing self-objectification, decreasing appearance comparison, and improving self-esteem."

Self-Objectification and Social Avoidance

"According to cognitive models of SAD, individuals may construct a mental representation of their appearance and behavior from an observer's perspective (that is, the supposed audience) and feel anxiety and fear regarding social situations when this mental representation is related to negative evaluation from the audience (Rapee & Heimberg, [449])."

"According to objectification theory, self-objectification increases the possibility of negative evaluation of physical appearance from an observer's perspective (Fredrickson & Roberts, [160]), and may increase social avoidance in this way."

"Self-objectifying girls tend to feel higher anxiety and shame about their bodily appearance from an observer's perspective (Dakanalis and others, [229]; Tiggemann & Slater, [450])."

Appearance Comparison as a Potential Mediator

"Appearance comparison is the application of social comparison (Festinger, [451]) to physical appearance, and occurs when individuals evaluate their physical attractiveness by comparing their appearance to others (Schaefer & Thompson, [452])."

"Self-objectifying women habitually monitor their physical appearance, which may prompt them to compare their appearance to others, and previous empirical studies demonstrated that women engaged in self-objectification were more likely to make appearance comparisons with others (Lindner and others, [453]; Tylka & Sabik, [454]; Yang and others, [455])."

"Based on previous studies, we proposed that self-objectification may be linked to greater social avoidance through increased appearance comparison (Hypothesis 2)."

Self-Esteem as another Potential Mediator

"Adolescent girls have lower self-esteem than adolescent boys (von Soest and others, [456]; Zuckerman and others, [457])."

"Adolescent girls tend to exhibit lower self-esteem when they perceive actual–ideal weight discrepancy (Argyrides & Sivitanides, [458])."

"Lower self-esteem in early adolescence predicted greater avoidance motivation and fewer social connections (Masselink and others, [459])."

"Chinese adolescents with lower self-esteem tended to have greater social anxiety (Pan and others, [460])."

The Relation between Appearance Comparison and Self-Esteem

"This finding indicates that comparing appearance to the norms of a social group may cause people to strive toward maintaining ideal body standards, thereby decreasing self-esteem."

"Adolescent girls' comparison of their physical appearance to that of others was related to lower self-esteem (Shroff & Thompson, [461])."

"Self-objectification may be associated with greater social avoidance via increased appearance comparison and decreased self-esteem (Hypothesis 4)."

The Present Study

"This study investigated the relationship between self-objectification and social avoidance in Chinese culture, which is useful for understanding the adverse consequences of self-objectification on interpersonal well-being in different cultures."

"Previous studies demonstrated that mid-adolescent girls showed greater self-objectification than early adolescent girls (Harrison & Fredrickson, [462]) and exhibited greater SAD symptoms than early and late adolescent girls (Nelemans and others, [463])."

"Middle adolescent girls have a higher level of social avoidance compared to early adolescent girls in situations where they are observed by others (Sumter and others, [464])."

"The present study investigated how self-objectification predicts social avoidance through appearance comparison and self-esteem among Chinese middle adolescent girls."

Methods

"A total of 262 Chinese high school girls in southern China voluntarily participated in this study."

"The Female Questionnaire of Trait Self-Objectification (FQSO, Wu & Lang, [465]) was used to measure the extent participants experienced female self-objectification within a Chinese cultural context."

"The FQSO showed excellent test-retest reliability and criterion-related validity in a sample of Chinese college females (Wu & Lang, [465])."

"We used the Chinese version of the Rosenberg Self-esteem Scale (Rosenberg, [466]), which was edited by Wang and others [467]."

"The scale consists of 10 items (e.g., "I feel that I am a person of worth, at least on an equal plane with others"), and is rated on a 4-point Likert scale ranging from 1 ("strongly disagree") to 4 ("strongly agree") with higher total scores indicating higher self-esteem."

"The Chinese version showed satisfying internal consistency reliability, test-retest reliability and criterion-related validity in Chinese college and high school students (Peng and others, [468])."

Results

"Self-objectification was related to greater social avoidance through increased appearance comparison (indirect effect = .087, 95% CI [.041, .143]), which supports Hypothesis 2."

"The mediating role of self-esteem in the relationship between self-objectification and social avoidance was also significant (indirect effect = .112, 95% CI [.058, .169])."

"A serial mediation model between self-objectification and social avoidance through appearance comparison and self-esteem (Hypothesis 4) was verified (indirect effect = .038, 95% CI [.014, .066]); thus, self-objectification was associated with greater appearance comparison, which was linked to lower self-esteem ($\beta = -.19$, $p < .01$, 95% CI [−.319, −.065]), and lower self-esteem was subsequently related to greater social avoidance ($\beta = -.43$, $p < .001$, 95% CI [−.538, −.323])."

Discussion

"Self-objectification was related to increased appearance comparison which, in turn, was related to lower self-esteem, which was subsequently linked to greater social avoidance (supporting Hypothesis 4)."

"Girls with higher levels of self-objectification have a higher tendency to make appearance comparisons with others which, in turn, is linked to greater social avoidance."

"Building on previous research, the present study tested the mechanism underlying self-objectifying adolescent girls' tendency to have lower self-esteem, which increases their propensity to avoid social situations."

"Considering that physical development increases the risk of girls' self-objectification (Fredrickson & Roberts, [160]) and social avoidance (Carter and others, [469]), future studies may be able to obtain novel findings on the relationship between self-objectification and social avoidance after taking puberty development during the whole period of adolescence into consideration."

"Based on our findings, future interventions addressing social avoidance among adolescent girls could begin by reducing their self-objectification, which may decrease their tendency to compare their appearance to others', which may subsequently improve their self-esteem."

Conclusion

"The results of the current study supported the relation between Chinese adolescent girls' self-objectification and social avoidance and helped to elucidate the processes that link the two."

"The data supported our hypotheses that Chinese adolescent girls with higher self-objectification are more likely to avoid social situations, and that this relation is mediated by appearance comparison and global self-esteem."

"These findings extend the applicability of objectification theory to adolescent girls' interpersonal distress and help us understand why girls' physical changes during adolescence increase their social avoidance, from the perspective of objectification theory, in a non-Western cultural context."

[Section 10]

"Chinese adolescents with social avoidance are at an increased risk of peer problems (Ding and others, [470]) and social adjustment difficulties (Sang and others, [471])."

"The development of girls' secondary sex characteristics (e.g., breast development) may play an important role in increasing social avoidance (Carter and others, [469])."

"A key question emerges: why do adolescent girls' bodily changes increase their social avoidance?"

"Girls' pubertal development increases their risk of self-objectification (Fredrickson & Roberts, [160]), which may increase their social avoidance."

5.4.20 Mother–Daughter Negative Body Talk as a Moderator Between Body Surveillance and Body Shame in Adolescent Girls [525]

This is a machine-generated summary of:

Domoff, Sarah E.; Tan, Cin Cin; Chow, Chong Man: Mother–daughter negative body talk as a moderator between body surveillance and body shame in adolescent girls [525]

Published in: Eating and Weight Disorders - Studies on Anorexia, Bulimia and Obesity (2020)

Link to original: https://doi.org/10.1007/s40519-020-00925-4

If you want to cite the papers, please refer to the original.

For technical reasons we could not place the page where the original quote is coming from.

Abstract-Summary

"For both mothers and daughters, individuals who had higher body surveillance reported higher body shame."

"Negative body talk emerged as a significant moderator: girls with higher body surveillance experienced greater body shame when they engaged in more negative body talk with their mothers."

"The findings highlight the need for clinicians to address mothers' own surveillance and body shame, as well as negative body talk between mothers and daughters, in interventions that seek to reduce the impact of objectification on body shame in adolescents."

Introduction

"Across multiple levels of a young person's socialization, adolescents learn that women are judged and valued on their appearance and their appeal to the male gaze [160, 232]."

"Although some research has examined these consequences of experiencing objectification (e.g., heightened body surveillance, or frequent self-monitoring of appearance and ensuring that one meets cultural ideals, [161]) in adolescent samples, we still know little about social or relational risk factors that may be modifiable to reduce the risks associated with objectification."

"Identifying these exacerbating influences is critical for clinical interventions seeking to buffer young women from experiencing body shame."

Mother–adolescent relationship and objectification

"Prior research has suggested that mothers' modeling of gender-related behaviors and attitudes influences their daughters' internalization of objectification."

"One study examined the links between media consumption and young girls' self-sexualization and found that mothers' own self-objectification moderated the relation between media consumption and young girls' self-sexualization [472]."

"Mothers' modeling of self-objectification may increase young girls' risk of internalizing gendered behaviors and attitudes."

"Katz-Wise and others [473] investigated the influence of mother–adolescent relationship quality and mothers' own body surveillance and body shame as predictors of adolescents' body surveillance and body image."

"How mothers and daughters communicate about body size may be a factor contributing to daughters' body surveillance and body shame."

"Given the role that mothers play in implicitly influencing daughters' gendered behaviors and attitudes via modeling, it is likely that more explicit communication about weight and appearance could also impact aspects of young women's body image."

Mother–daughter negative body talk

"Related to the outcomes of self-objectification, Arroyo and Andersen [474] found in the same sample of young adult women and mothers, that fat talk predicted bulimic tendencies in the adult daughters, again supporting that mother–daughter negative body talk may be a risk factor for the consequences of self-objectification, at least in adult women."

"Only one study [475] has examined the impact of negative body talk between mothers and their younger adolescent daughters."

"Chow and Tan (2017) found that daughters reported greater disordered eating when they and their mothers had higher levels of negative body talk [475]."

"Given the interactive effect that mother–daughter negative body talk has on disordered eating in younger adolescents [475], and that it is a contributing factor to body surveillance and body dissatisfaction in adult daughters, it is likely that mother–daughter negative body talk likewise accounts for experiences of body surveillance and body shame in younger adolescents."

Current study

"Based on the Social Learning Theory, we hypothesize the following to be examined in our cross-sectional study: Mothers' body surveillance and body shame will positively associate with adolescents' body surveillance and body shame, respectively."

"Negative body talk will moderate the associations between body surveillance and body shame for both adolescents and mothers, such that stronger associations between body surveillance and body shame will occur for mothers and adolescents higher in negative body talk."

Method

"Mothers and daughters completed the Objectified Body Consciousness Scale [161] to capture their body shame (8 items)."

"To capture negative body talk, mothers and daughters were asked to complete the Body Talk Scale [476]."

"Mothers' and daughters' reports of negative body talk were significantly related, $r = 0.38$, $p < 0.01$."

"Because body talk reflects a dyadic construct, both mothers' and daughters' reports were averaged to form a composite score of overall body talk in the relationship."

"This model also estimated the direct association between negative body talk and both members' body shame, along with two interaction terms: (1) negative body talk with mothers' body surveillance and (2) negative body talk with daughters' body surveillance."

"These interaction effects examined whether body talk would moderate the actor and partner effects of body surveillance and body shame, for both mothers and daughters (Hypothesis 2)."

Results

"Actor effect showed that mothers who had higher body surveillance reported higher body shame."

"Actor effect showed that daughters who had higher body surveillance also reported higher body shame."

"Partner effect showed that mothers' body surveillance was not significantly related to their daughters' body shame."

"Partner effect showed that daughters' body surveillance was not significantly related to their mothers' body shame."

"The direct association between negative body talk was not related to mothers' and daughters' reports of body shame."

"The association between daughters' body surveillance and their own body shame was significantly moderated by negative body talk (moderated actor effect; Hypothesis 2)."

Discussion

"This study extended previous work using a dyadic design by examining the relational risk factors, particularly negative body talk in mothers and their daughters, in moderating the link between mothers' and adolescents' body surveillance and body shame."

"The hierarchical nature of the mother–child relationship may explain for the lack of partner effect of adolescents' body surveillance on mothers' body shame, such that adults may be less susceptible to change in response to children's behaviors and attitudes."

"Our finding is also consistent with prior work that found that mothers' body surveillance was not associated adolescents' body shame [473]."

"Consistent with the Social Learning Theory that postulates that learning involves behavioral (e.g., reinforcement) and cognitive factors (e.g., motivation) [477], we found that negative body talk interacted with body surveillance to predict body shame in adolescent girls."

Limitation and future directions

"Another limitation was that this study only focused on the negative aspect of body talk but did not consider the content of this talk."

"Future research should clarify the content of this negative body talk (thin-ideal related versus age-related)."

"A recent study showed that positive body talk was not associated with negative outcomes, and that this type of weight communication is protective against pathological eating, especially among individuals with lower weight status [478]."

"Future research could investigate the differential influence of types of body talk on adolescent health and mental health outcomes."

"Future work should consider examining similar research questions with male participants, as some work suggests that adolescent boys also experience similar body image issues, although it could manifest differently (e.g., men/boys are being pressured to be masculine or muscular [479])."

Conclusions

"We demonstrated that when girls have high body surveillance, negative body talk between them and their mothers increases the level of body shame."

"To reduce body shame among girls with high body surveillance, mothers may be encouraged to minimize their expression of body dissatisfaction with their daughters."

"Clinicians and practitioners may consider the role of mothers' own surveillance and body shame, as well as the existence of negative body talk within dyads when designing interventions."

What is already known on this subject?

"Prior research has examined consequences of experiencing objectification (e.g., heightened body surveillance, or frequent self-monitoring of appearance) in adolescent samples."

"Little is known about social or relational risk factors that may be modifiable to reduce the risks associated with objectification."

"Identifying these exacerbating influences is critical for clinical interventions seeking to buffer young women from experiencing body shame."

What your study adds?

"We demonstrated that when girls have high body surveillance, negative body talk between mothers and daughters increases the level of body shame."

"To reduce body shame among girls with high body surveillance, mothers may be encouraged to minimize their expression of body dissatisfaction with their daughters."

"These results have practical implications for clinicians and practitioners who work with adolescents with body image issues."

Bibliography

1. Burleson BR (2003). The experience and effects of emotional support: what the study of cultural and gender differences can tell us about close relationships, emotion, and interpersonal communication. Personal Relationships 10:1–23. https://doi.org/10.1111/1475-6811.00033.
2. Parks MR, Floyd K (1996). Meanings for closeness and intimacy in friendship. Journal of Social and Personal Relationships 13:85–107. https://doi.org/10.1177/0265407596131005.
3. Floyd K (1995). Gender and closeness among friends and siblings. The Journal of Psychology 129:193–202. https://doi.org/10.1080/00223980.1995.9914958.
4. Lisetti C, Amini R, Yasavur U, Rishe N (2013). I can help you change! An empathic virtual agent delivers behavior change health interventions. ACM Transactions on Management Information Systems (TMIS), 4:1–28. https://doi.org/10.1145/2544103.
5. Creed C, Beale R (2012). User interactions with an affective nutritional coach. Interacting with Computers 24:339–350. https://doi.org/10.1016/j.intcom.2012.05.004.
6. Reinghold, H.: The Virtual Community. Finding Connection in a Computerized World. Addison-Wesley Longman Publishing, Boston (1993)
7. Turkle, S.: Life on the Screen. Identity in the Age of the Internet. Simon and Schuster, New York (1995)
8. Markham, A.: Life Online: Researching Real Experience in Virtual Space. AltaMira Press, Walnut Creek (1998)
9. Hine, C. (2000). Virtual ethnographies. London: Sage Publications Ltd.
10. Baym, K.N.: Tune in, Log on. Soaps, Fandom, and Online Community. Sage, Thousand Oaks (2000)
11. Granovetter M (1973) The strength of weak ties. Am J Sociol 78(6):1
12. Baym, K.N.: Finding the quality in qualitative research. In: Silver, D., Massanari, A. (eds.) Critical Cyberculture Studies, pp. 79–87. New York University Press, New York (2006)
13. Markham, A., Baym, K.N.: Internet Inquiry: Conversation About Method. Sage, Thousand Oaks (2009)
14. Bray F (2007) Gender and technology. Annu Rev Anthropol 36(1):37–53
15. Feinberg L (1996) Transgender warriors, 1st edn. Beacon Press, Boston
16. Hicks M (2019) Hacking the Cis-tem. transgender citizens and the early digital state. IEEE Ann Hist Comput 41(1):20–33. https://doi.org/10.1109/MAHC.2019.2897667
17. Fiske J (1992) The cultural economy of fandom. In: Lewis LA (ed.) The adoring audience: fan culture and popular media. Routledge, London, pp. 30–49
18. Chin B (2007) Beyond Kung-Fu and violence: locating East Asian cinema fandom. In: Gray J, Sandvoss C, Harrington CL (eds) Fandom: identities and communities in a mediated world. NYU Press, New York, pp. 210–219
19. Bengtsson M (2016) How to plan and perform a qualitative study using content analysis. NursingPlus Open 2:8–14. https://doi.org/10.1016/j.npls.2016.01.001
20. Berger AA (1995) Cultural criticism: a primer of key concepts. SAGE Publications Ltd, London

21. Baldi V, Rodrigues RPC (2016) Fantasmas digitais e relações swipe: a cultura infocomunicacional na era das proximidades automatizadas. Ciência da Informação 45(2):52–61.
22. Bell G, Gray J (2000) Digital immortality. Microsoft Research, San Francisco, CA.
23. Paul-Choudhury S (2011) Digital legacy: the fate of your online soul. New Sci 210(2809):41–43.
24. Grover D (1989) Posthumous harm. Philos Q (1950-) 39(156):334–353.
25. ULC (2015) Fiduciary access to digital assets act, revised, uniform law commission. www.uniformlaws.org/committees/community-home?CommunityKey=f7237fc4-74c2-4728-81c6-b39a91ecdf22 (Accessed on 03/22/2020).
26. Sperling D (2008) Posthumous interests: legal and ethical perspectives. Cambridge law, medicine and ethics. Cambridge University Press, Cambridge [UK] ; New York.
27. Öhman C, Floridi L (2017) The political economy of death in the age of information: a critical approach to the digital afterlife industry. Minds Mach 27(4):639–662.
28. Huberman J (2017) Dearly departed: communicating with the dead in the digital age. Soc Anal 61(3):91–107.
29. Campbell AV, Willis M (2005) They stole my baby's soul: narratives of embodiment and loss. Med Humanit 31(2):101–104.
30. Pereira R, Baranauskas MCC, da Silva SRP (2010) Softwares sociais: uma visão orientada a valores In: Proceedings of the IX Symposium on Human Factors in Computing Systems, 149–158.
31. Maciel C (2011) Issues of the social web interaction project faced with afterlife digital legacy In: Proceedings of the 10th Brazilian Symposium on Human Factors in Computing Systems and the 5th Latin American Conference on Human-Computer Interaction, 3–12.
32. Galvão VF, Maciel C, Pereira R, Gasparini I, Viterbo J (2019) Talking about digital immortality: a value-oriented discussion In: Proceedings of the 18th Brazilian Symposium on Human Factors in Computing Systems, 1–8.
33. Galvão VF, Maciel C, Pereira VC, Garcia ACB, Pereira R, Viterbo J (2021) Posthumous data at stake: an overview of digital immortality issues In: Proceedings of the XX Brazilian Symposium on Human Factors in Computing Systems IHC '21. Association for Computing Machinery, New York, USA. https://doi.org/10.1145/3472301.3484358.
34. Kaptelinin V (2016) Making the case for an existential perspective in HCI research on mortality and death In: Proceedings of the 2016 CHI Conference Extended Abstracts on Human Factors in Computing Systems, 352–364.
35. Preukschat, A., & Reed, D. (2021). Self-sovereign Identity, MEAP.
36. Hoepman, J.-H. (2021). Privacy is hard and seven other myths: Achieving privacy through careful design. The MIT Press.
37. World Bank Group, GSMA and Secure Identity Alliance. (2016). Digital identity: Towards shared principles for public and private sector cooperation. Washington, DC: World Bank Group.
38. Khatchatourov, A. (2019). Digital identities in tension. Between autonomy and control. Wiley.
39. Tabrizchi H, Kuchaki Rafsanjani M (2020) A survey on security challenges in cloud computing: issues, threats, and solutions. J Supercomput 76:9493–9532
40. Chen Y, Chen J (2021) A secure three-factor-based authentication with key agreement protocol for e-Health clouds. J Supercomput 77:3359–3380
41. Dhasarathan C, Kumar M, Srivastava AK et al (2021) A bio-inspired privacy-preserving framework for healthcare systems. J Supercomput. https://doi.org/10.1007/s11227-021-037 20-9
42. Akremi A, Rouached M (2021) A comprehensive and holistic knowledge model for cloud privacy protection. J Supercomput. https://doi.org/10.1007/s11227-020-03594-3
43. Shi Rh, Zhong H, Zhang S (2015) Comments on two schemes of identity-based user authentication and key agreement for mobile client-server networks. J Supercomput 71:4015–4018
44. Jafri R, Arabnia H (2009) A survey of face recognition techniques. J Inf Process Syst 5(2):41–68

45. Miguel J, Caballé S, Xhafa F (2017) Security for e-learning. In: Intelligent data analysis for e-learning: enhancing security and trustworthiness in online learning systems. Academic Press, pp 7–23
46. Ueda H, Nakamura M (2017) Deployment of multilanguage security awareness education online course by federated moodle in Japan. In: Computer Software and Applications Conference (COMPSAC), 2017 IEEE 41st Annual, vol 2. IEEE, pp 49–52
47. Hernandez JCG, Chavez MAL (2008) Moodle security vulnerabilities. In: 2008 5th International Conference on Electrical Engineering, Computing Science and Automatic Control. IEEE, pp 352–357
48. Korać D, Damjanović B, Simić D (2020) Information security in M-learning systems: challenges and threats of using cookies. In: 19th International symposium infoteh-Jahorina 2020, pp 1–6
49. Manzo M (2017) A model for users behavior analysis and forecasting in Moodle. J e-Learn Knowl Soc 13(2):129–139
50. Belk, R.W. 1988. Possessions and the extended self. Journal of Consumer Research 15 (2): 139–168.
51. Belk, R.W. 2013. Extended self in a digital world. Journal of Consumer Research 40 (3): 477–500.
52. Boyd, D. 2007. Why youth (heart) social network sites: The role of networked publics in teenage social life. InMacArthur Foundation series on digital learning—Youth, identity and digital media volume, ed. D. Buckingham. Cambridge: MIT Press.
53. Spero, I., and M. Stone. 2004. Agents of change: How young consumers are changing the world of marketing. Qualitative Market Research 7 (2): 153–159.
54. Doster, L. 2013. Millennial teens design and redesign themselves in online social networks. Journal of Consumer Behaviour 12: 267–279.
55. Raab, C., O. Berezan, A.S. Krishen, and S. Tanford. 2016. What's in a word? Building program loyalty through social media communication. Cornell Hospitality Quarterly 57 (2): 138–149.
56. Berezan, O., A.S. Krishen, S. Agarwal, and P. Kachroo. 2018. The pursuit of virtual happiness: Exploring the social media experience across generations. Journal of Business Research 89: 455–461.
57. Evans-Pughe C. Engineering digital identity. Eng Technol. 2008;3(10):16–18. 7–20th June.
58. Ellison N, Heino R, Gibbs J. Managing impression online: self presentation processes in online dating environments. J Comput-Mediat Commun. Available online: http://jcmc.indiana.edu/vol11/issue2/ellison.html (accessed 25/07/10); 2006; 11:2. Article 2.
59. Information Assurance Advisory Council. IAAC Position Paper on Identity Assurance (IdA): towards a policy framework for electronic identity. Available online: http://www.iaac.org.uk/Portals/0/identity_management_paper_v1-7.pdf ; 18th October 2005.
60. Dean M. Governmentality: power and rule in modern society. 2nd ed. London, Thousand Oaks, Delhi: Sage; 2010.
61. European Commission. Final report: evaluation of the means used by national data protection supervisory authorities in the promotion of personal data protection. Directorate General Justice, Freedom & Security. Available online: http://ec.europa.eu/justice/policies/privacy/docs/studies/final_report_kantor_management_consultants.pdf. Accessed 19/07/1010; 2007.
62. Byrne L. Securing our identity: a 21st century public good: speech by Liam Byrne MP, the Minister of State for Immigration, Citizenship & Nationality, to Chatham House. http://press.homeoffice.gov.uk/Speeches/sc-identity-21st-century, 19th June 2007.
63. Glynos J, Howarth D. Logics of critical explanation in social and political theory. London: Taylor & Francis; 2007.
64. Laclau E, Mouffe C. Hegemony and socialist strategy. London & New York: Verso; 1995.
65. Fisch, C., & Block, J. H. (2021). How does entrepreneurial failure change an entrepreneur's digital identity? Evidence from Twitter data. Journal of Business Venturing, 36(1), 106015.

66. Fischer, E., & Reuber, A. R. (2014). Online entrepreneurial communication: Mitigating uncertainty and increasing differentiation via Twitter. Journal of Business Venturing, 29(4), 565–583.
67. Smith, C., Smith, J. B., & Shaw, E. (2017). Embracing digital networks: Entrepreneurs' social capital online. Journal of Business Venturing, 32(1), 18–34.
68. Schumpeter, J. A. (1934). The Theory of Economic Development: An Inquiry into Profits, Capital, Credit, Interest, and the Business Cycle. Harvard University Press.
69. Block, J. H., De Vries, G., Schumann, J. H., & Sandner, P. (2014). Trademarks and venture capital valuation. Journal of Business Venturing, 29(4), 525–542.
70. Block, J. H., Fisch, C. O., Obschonka, M., & Sandner, P. G. (2019a). A personality perspective on business angel syndication. Journal of Banking & Finance, 100, 306–327.
71. Block, J., Fisch, C., Vismara, S., & Andres, R. (2019b). Private equity investment criteria: An experimental conjoint analysis of venture capital, business angels, and family offices. Journal of Corporate Finance, 58, 329–352.
72. Franke, N., Gruber, M., Harhoff, D., & Henkel, J. (2008). Venture capitalists' evaluations of start-up teams: Trade-offs, knock-out criteria, and the impact of VC experience. Entrepreneurship Theory and Practice, 32(3), 459–483.
73. Hossinger, S., Block, J., Chen, X., & Werner, A. (2021). Venture creation patterns in academic entrepreneurship: the role of founder motivations. The Journal of Technology Transfer, forthcoming.
74. Audretsch, D. B., & Link, A. N. (2012). Entrepreneurship and innovation: Public policy frameworks. The Journal of Technology Transfer, 37(1), 1–17.
75. Hébert, R. F., & Link, A. N. (1989). In search of the meaning of entrepreneurship. Small Business Economics, 1(1), 39–49.
76. Hébert, R. F., & Link, A. N. (2006). The entrepreneur as innovator. The Journal of Technology Transfer, 31(5), 589–597.
77. Aldrich, H. E., & Ruef, M. (2018). Unicorns, gazelles, and other distractions on the way to understanding real entrepreneurship in the United States. Academy of Management Perspectives, 32(4), 458–472.
78. Henrekson, M., & Johansson, D. (2010). Gazelles as job creators: A survey and interpretation of the evidence. Small Business Economics, 35(2), 227–244.
79. Kortum, S., & Lerner, J. (2000). Assessing the contribution of venture capital to innovation. The RAND Journal of Economics, 31(4), 674–692.
80. Lerner, J., & Nanda, R. (2020). Venture capital's role in financing innovation: What we know and how much we still need to learn. Journal of Economic Perspectives, 34(3), 237–261.
81. Audretsch, D. B., Lehmann, E. E., Paleari, S., & Vismara, S. (2016). Entrepreneurial finance and technology transfer. The Journal of Technology Transfer, 41(1), 1–9.
82. Colombo, M. G., Cumming, D. J., & Vismara, S. (2016). Governmental venture capital for innovative young firms. The Journal of Technology Transfer, 41(1), 10–24.
83. Masiak, C., Fisch, C., & Block, J. (2020). In which regions do governmental, independent, and corporate venture capital firms invest? An empirical investigation across 402 German regions. In: Moritz, A., Block, J., Golla, S., Werner, A. (Eds.). Contemporary Developments in Entrepreneurial Finance, 201–227.
84. Minola, T., Vismara, S., & Hahn, D. (2017). Screening model for the support of governmental venture capital. The Journal of Technology Transfer, 42(1), 59–77.
85. Bonvillian, W. B., & Van Atta, R. (2011). ARPA-E and DARPA: Applying the DARPA model to energy innovation. The Journal of Technology Transfer, 36(5), 469–513.
86. Colatat, P. (2015). An organizational perspective to funding science: Collaborator novelty at DARPA. Research Policy, 44(4), 874–887.
87. DARPA. (2008). DARPA: 50 years of bridging the gap. Faircount LLC.
88. Nambisan, S. (2017). Digital entrepreneurship: Toward a digital technology perspective of entrepreneurship. Entrepreneurship Theory and Practice, 41(6), 1029–1055.

89. Galbraith, C. S., McKinney, B. C., DeNoble, A. F., & Ehrlich, S. B. (2014). The impact of presentation form, entrepreneurial passion, and perceived preparedness on obtaining grant funding. Journal of Business and Technical Communication, 28(2), 222–248.
90. Sudek, R. (2006). Angel investment criteria. Journal of Small Business Strategy, 17(2), 89–104.
91. Tian, X., & Wang, T. Y. (2014). Tolerance for failure and corporate innovation. The Review of Financial Studies, 27(1), 211–255.
92. Block, J., & Sandner, P. (2009). What is the effect of the financial crisis on venture capital financing? Empirical evidence from US Internet start-ups. Venture Capital, 11(4), 295–309.
93. Gompers, P., Kovner, A., Lerner, J., & Scharfstein, D. (2008). Venture capital investment cycles: The impact of public markets. Journal of Financial Economics, 87(1), 1–23.
94. Kosinski, Michal, David Stillwell and Thore Graepel. 2013. Private traits and attributes are predictable from digital records of human behavior. PNAS 15:5802–5805.
95. Piva, E., & Rossi-Lamastra, C. (2018). Human capital signals and entrepreneurs' success in equity crowdfunding. Small Business Economics, 51(3), 667–686.
96. Corbet, S., Hou, Y., Hu, Y., & Oxley, L. (2022). We Reddit in a forum: The influence of message boards on firm stability. Review of Corporate Finance, 2(1), 151–190.
97. Pahnke, A., & Welter, W. (2019). The German Mittelstand: Antithesis to Silicon Valley entrepreneurship? Small Business Economics, 52(2), 345–358.
98. Fauchart, E., & Gruber, M. (2011). Darwinians, communitarians, and missionaries: The role of founder identity in entrepreneurship. Academy of Management Journal, 54(5), 935–957.
99. Goodstadt, L. F., Connolly, R., & Bannister, F. (2015). The Hong Kong e-identity card: Examining the reasons for its success when other cards continue to struggle. Information Systems Management, 32(1), 72–80.
100. Kubicek, H., & Noack, T. (2010). Different countries-different paths extended comparison of the introduction of eIDs in eight European countries. Identity in the Information Society, 3(1), 235–245.
101. Sullivan, C. (2018). Digital identity – from emergent legal concept to new reality. Computer Law & Security Review, 34, 723–731.
102. World Bank. (2016). World development report 2016: Digital dividends. Washington, DC: World Bank.
103. World Bank Group (2017). ID4D identification for development. 2017 annual report. Washington DC: The World Bank Group. http://pubdocs.worldbank.org/en/561391513609113984/2017-ID4D-Annual-Report.pdf. Accessed 12 February 2019.
104. Warf, B. (2014a). Asian geographies of e-government. Eurasian Geography and Economics, 55(1), 94–110.
105. Warf, B. (2014b). Geographies of e-government in Latin America and the Caribbean. Journal of Latin American Geography, 13(1), 169–185.
106. Zhao, F., Collier, A., & Deng, H. (2014). A multidimensional and integrative approach to study global digital divide and e-government development. Information Technology & People, 27(1), 38–62. https://doi.org/10.1108/ITP-01-2013-0022.
107. Martens, T. (2010). Electronic identity management in Estonia between market and state governance. Identity in the Information Society, 3(1), 213–233.
108. Vassil, K. (2015). Estonian e-government ecosystem: Foundation, applications, outcomes. Washington DC: The World Bank Group. http://pubdocs.worldbank.org/en/165711456838073531/WDR16-BP-Estonian-eGov-ecosystem-Vassil.pdf. Accessed 12 February 2019.
109. Godoy, D.G.S., & Heal, A. (2016). Trade in the digital age: Can e-residency be an enabler for Asia-Pacific developing countries? Trade Insights Issue No 17. United Nations Economic and Social Commission for Asia and the Pacific. United Nations Trade, Investment and Innovation Division. https://www.unescap.org/publications/trade-digital-age-can-e-residency-be-enabler-asia-pacific-developing-countries-escap. Accessed 12 February 2019.
110. Kotka, T., Ivan Vargas Alvarez del Castillo, C., & Korjus, K. (2015). Estonian e-residency: redefining the nation-state in the digital era. Cyber Studies Programme Working Paper Series

3: 2015. University of Oxford. https://www.politics.ox.ac.uk/materials/publications/14883/workingpaperno3kotkavargaskorjus.pdf. Accessed 12 February 2019.
111. Taipale, S. (2013). The use of e-government services and the internet: The role of socio-demographic, economic and geographical predictors. Telecommunications Policy, 37(4–5), 413–422. https://doi.org/10.1016/j.telpol.2012.05.005.
112. O'Byrne, W. I., & Pytash, K. E. (2017). Becoming literate digitally in a digitally literate environment of their own. Journal of Adolescent & Adult Literacy, 60(5), 499–504. https://doi.org/10.1002/jaal.59.
113. Clark, C., Chow-Hoy, T. K., Herter, R. J., & Moss, P. A. (2001). Portfolios as sites of learning: Reconceptualizing the connections to motivation and engagement. Journal of Literacy Research, 33(2), 211–241.
114. Thomas, D. R. (2006). A general inductive approach for analyzing qualitative evaluation data. American Journal of Evaluation, 27(2), 237–246.
115. Denzin, N. K. (1978). The research act: A theoretical introduction to sociological methods. New York: McGraw-Hill.
116. Patton, M. Q. (2002). Two decades of developments in qualitative inquiry: A personal, experiential perspective. Qualitative Social Work, 1(3), 261–283.
117. National Council of Teachers of English, NCTE (2019) Definition of Literacy in a Digital Age. https://ncte.org/statement/nctes- definition-literacy-digital-age/.
118. O'Byrne, W. I. (2019). Educate, Empower, Advocate: Amplifying marginalized voices in a digital society. Contemporary Issues in Technology and Teacher Education. Retrieved online at https://www.citejournal.org/volume-19/issue-4-19/english-language-arts/educate-empower-advocate-amplifying-marginalized-voices-in-a-digital-society/
119. Calzo, J. P., Blashill, A. J., Brown, T. A., & Argenal, R. L. (2017). Eating disorders and disordered weight and shape control behaviors in sexual minority populations. Current Psychiatry Reports, 19(49), 1–10. https://doi.org/10.1007/s11920-017-0801-y.
120. Matthews-Ewald, M. R., Zullig, K. J., & Ward, R. M. (2014). Sexual orientation and disordered eating behaviors among self-identified male and female college students. Eating Behaviors, 15(3), 441–444. https://doi.org/10.1016/j.eatbeh.2014.05.002.
121. Watson, R. J., Adjei, J., Saewyc, E., Homma, Y., & Goodenow, C. (2017). Trends and disparities in disordered eating among heterosexual and sexual minority adolescents. International Journal of Eating Disorders, 50, 22–31. https://doi.org/10.1002/eat.22576.
122. Austin, S. B., Nelson, L. A., Birkett, M. A., Calzo, J. P., & Everett, B. (2013). Eating disorder symptoms and obesity at the intersections of gender, ethnicity, and sexual orientation in US high school students. American Journal of Public Health, 103(2), e16–e22. https://doi.org/10.2105/AJPH.2012.301150.
123. Smith, A., & Anderson, M. (2018). Social media use in 2018. In Pew research center Retrieved July 29, 2018 from http://assets.pewresearch.org/wp-content/uploads/sites/14/2018/03/01105133/PI_2018.03.01_Social-Media_FINAL.pdf.
124. Seidenberg, A. B., Jo, C. L., Ribisl, K. M., Lee, J. G. L., Buchting, F. O., Kim, Y., … Emery, S. L. (2017). A national study of social media, television, radio, and internet usage of adults by sexual orientation and smoking status: Implications for campaign design. International Journal of Environmental Research and Public Health, 14(4), 1–14. https://doi.org/10.3390/ijerph14040450.
125. Griffiths, S., Murray, S. B., Krug, I., & McLean, S. A. (2018). The contribution of social media to body dissatisfaction, eating disorder symptoms, and anabolic steroid use among sexual minority men. Cyberpsychology, Behavior and Social Networking, 21(3), 149–156. https://doi.org/10.1089/cyber.2017.0375.
126. Heatherton, T. F., & Baumeister, R. F. (1991). Binge eating as escape from self-awareness. Psychological Bulletin, 110(1), 86–108. https://doi.org/10.1037//0033-2909.110.1.86.
127. Vartanian, L.R. (2012). Self-discrepancy theory and body image. In T. F. Cash (Ed.), Encyclopedia of body image and human appearance Vol. 2, (pp. 711–717). https://doi.org/10.1016/C2010-1-66177-9

128. Martins, Y., Tiggeman, M., & Kirkbride, A. (2007). Those speedos become them: The role of self-objectification in gay and heterosexual men's body image. Personality and Social Psychology Bulletin, 33(5), 634–647. https://doi.org/10.1177/0146167206297403.
129. Wiseman, M. C., & Moradi, B. (2010). Body image and eating disorder symptoms in sexual minority men: A test and extension of sexual objectification theory. Journal of Counseling Psychology, 57(2), 154–166. https://doi.org/10.1037/a0018937.
130. Siever, M. D. (1994). Sexual orientation and gender as factors in socioculturally acquired vulnerability to body dissatisfaction and eating disorders. Journal of Consulting and Clinical Psychology, 62(2), 252–260. https://doi.org/10.1037/0022-006X.62.2.252.
131. Bailey, J. M., & Zucker, K. J. (1995). Childhood sex-typed behavior and sexual orientation: A conceptual analysis and quantitative review. Developmental Psychology, 31(1), 43–55. https://doi.org/10.1037/0012-1649.31.1.43.
132. Rieger, G., Linsenmeier, J. A. W., Gygax, L., & Bailey, J. M. (2008). Sexual orientation and childhood gender nonconformity: Evidence from home videos. Developmental Psychology, 44(1), 46–58. https://doi.org/10.1037/0012-1649.44.1.46.
133. Perloff, R. M. (2014). Social media effects on young women's body image concerns: Theoretical perspectives and an agenda for research. Sex Roles, 71(11–12), 363–377. https://doi.org/10.1007/s11199-014-0384-6.
134. Baker, N. (2019). Meet the totally hot body positive guys on Instagram. Retrieved May 10, 2019 from https://www.thegayuk.com/meet-the-totally-hot-body-positive-guys-on-instagram/.
135. Feldman, J. (2019). How this LGBTQ advocate is using nudity to bring men into the body image conversation. Retrieved May 10, 2019 from https://www.huffpost.com/entry/nude-body-positivity-gay-advocate_n_5c37952ce4b0c469d76c40f0?guccounter=1&guce_referrer=aHR0cHM6Ly93d3cuZ29vZ2xlLmNvbS8&guce_referrer_sig=AQAAAJjoD-S_iYXdX_Pqt6Pr2F0EKvCGW_0lDjmisitBiwLOdFr0GJv-Ar0o9LGjV7dbuqt4mWCH5kKnABnGRivpBph6CLoWiPAK1Sbo-FNWYYYS_l8AU9rrhPY0StbbZtpNAsoPB1Y2A94B_wx6isOnHfxOXOuuKGuDkwmAqbjzpmjY.
136. Villarreal, D. (2019). 5 ways gay men can promote body positivity besides blocking Instagram "thirst traps." Retrieved May 10, 2019 from https://www.lgbtqnation.com/2019/02/5-things-gay-men-need-beyond-blaming-instagram-thirst-traps-negative-body-image/.
137. Holland, G., & Tiggemann, M. (2016). A systematic review of the impact of the use of social networking sites on body image and disordered eating outcomes. Body Image, 17, 100–110. https://doi.org/10.1016/j.bodyim.2016.02.008.
138. Mingoia, J., Hutchinson, A. D., Wilson, C., & Gleaves, D. H. (2017). The relationship between social networking site use and the internalizations of a thin ideal in females: A meta-analytic review. Frontiers in Psychology, 7(8), 1–10. https://doi.org/10.3389/fpsyg.2017.01351.
139. Limanowski J, Blankenburg F (2013) Minimal self-models and the free energy principle. Front Hum Neurosci 7(547):1–12
140. Metzinger T (2017) The problem of mental action: predictive control without sensory sheets. In: Metzinger T, Wiese W (eds) Philosophy and predictive processing. MIND Group, Frankfurt
141. Limanowski J, Friston K (2018) 'Seeing the dark': Grounding phenomenal transparency and opacity in precision estimation for active inference. Front Psychol 9(643):1–9
142. Blanke O, Metzinger T (2009) Full-body illusions and minimal phenomenal selfhood. Trends Cogn Sci 13(1):7–13
143. Metzinger T (2004) Being no one: the self-model theory of subjectivity. MIT Press, Cambridge
144. Friston KJ, Rosch R, Parr T, Price C, Bowman H (2018) Deep temporal models and active inference. Neurosci Biobehav Rev 90:486–501
145. Hohwy J (2016) The self-evidencing brain. Noûs 50(2):259–285
146. Tomasello M, Carpenter M, Call J, Behne T, Moll H (2005) Understanding and sharing intentions: the origins of cultural cognition. Behav Brain Sci 28(5):675–691

147. Grogan S (2008) Body Image: Understanding Body Dissatisfaction in Men, Women and Children, 2nd edn, Routledge: London.
148. Rodgers, R. F., & Melioli, T. (2016). The relationship between body image concerns, eating disorders and internet use, part I: A review of empirical support. Adolescent Research Review, 1, 95–119. https://doi.org/10.1007/s40894-015-0016-6
149. Rodgers, R. F. (2016). The relationship between body image concerns, eating disorders and internet use, part II: An integrated theoretical model. Adolescent Research Review, 1, 121–137. https://doi.org/10.1007/s40894-015-0017-5
150. Rodgers, R. F., Slater, A., Gordon, C. S., McLean, S. A., Jarman, H. K., & Paxton, S. J. (2020). A biopsychosocial model of social media use and body image concerns, disordered eating, and muscle-building behaviors among adolescent girls and boys. Journal of Youth and Adolescence, 49(2), 399–409. https://doi.org/10.1007/s10964-019-01190-0
151. Rodgers, R. F., Paxton, S. J., & McLean, S. A. (2014). A biopsychosocial model of body image concerns and disordered eating in early adolescent girls. Journal of Youth and Adolescence, 43(5), 814–823. https://doi.org/10.1007/s10964-013-0013-7
152. Lawler, M., & Nixon, E. (2011). Body dissatisfaction among adolescent boys and girls: The effects of body mass, peer appearance culture and internalization of appearance ideals. Journal of Youth and Adolescence, 40(1), 59–71. https://doi.org/10.1007/s10964-009-9500-2
153. Neumark-Sztainer, D., Paxton, S. J., Hannan, P. J., Haines, J., & Story, M. (2006). Does body satisfaction matter? Five-year longitudinal associations between body satisfaction and health behaviors in adolescent females and males. Journal of Adolescent Health, 39(2), 244–251. https://doi.org/10.1016/j.jadohealth.2005.12.001
154. Markey, C. N. (2010). Invited commentary: Why body image is important to adolescent development. Journal of Youth and Adolescence, 39, 1387–1391. https://doi.org/10.1007/s10964-010-9510-0
155. Granic, I., Morita, H., & Scholten, H. (2020). Beyond screen time: Identity development in the digital age. Psychological Inquiry, 31(3), 195–223. https://doi.org/10.1080/1047840X.2020.1820214
156. Geiger, A. W., & Davis, L. (2019, July 12). A growing number of American teenagers—particularly girls—are facing depression. Pew Research Center. Retrieved from https://www.pewresearch.org/fact-tank/2019/07/12/
157. Ferreiro, F., Seoane, G., & Senra, C. (2014). Toward understanding the role of body dissatisfaction in the gender differences in depressive symptoms and disordered eating: A longitudinal study during adolescence. Journal of Adolescence, 37(1), 73–84. https://doi.org/10.1016/j.adolescence.2013.10.013
158. Sharpe, H., Patalay, P., Choo, T.-H., Wall, M., Mason, S. M., Goldschmidt, A. B., & Neumark-Sztainer, D. (2018). Bidirectional associations between body dissatisfaction and depressive symptoms from adolescence through early adulthood. Development and Psychopathology, 30(4), 1447–1458. https://doi.org/10.1017/S0954579417001663
159. Stice, E., Marti, C. N., & Durant, S. (2011). Risk factors for onset of eating disorders: Evidence of multiple risk pathways from an 8-year prospective study. Behaviour Research and Therapy, 49(10), 622–627. https://doi.org/10.1016/j.brat.2011.06.009
160. Fredrickson, B. L., & Roberts, T. A. (1997). Objectification theory: Toward understanding women's lived experiences and mental health risks. Psychology of Women Quarterly, 21, 173–206.
161. McKinley, N. M., & Hyde, J. S. (1996). The objectified body consciousness scale. Psychology of Women Quarterly, 20, 181–215.
162. Deighton-Smith, N., & Bell, B. T. (2018). Objectifying fitness: A content and thematic analysis of #fitspiration images on social media. Psychology of Popular Media Culture, 7(4), 467–483. https://doi.org/10.1037/ppm0000143
163. Beccia, A. L., Baek, J., Jesdale, W. M., Austin, S. B., Forrester, S., Curtin, C., & Lapane, K. L. (2019). Risk of disordered eating at the intersection of gender and racial/ethnic identity among US high school students. Eating Behaviors, 34, 101299. https://doi.org/10.1016/j.eatbeh.2019.05.002

164. Burke, N. L., Schaefer, L. M., Hazzard, V. M., & Rodgers, R. F. (2020). Where identities converge: The importance of intersectionality in eating disorders research. International Journal of Eating Disorders, 53(10), 1605–1609. https://doi.org/10.1002/eat.23371
165. McEntee, M. L., Serier, K. N., Smith, J. M., & Smith, J. E. (2020). The sum is greater than its parts: Intersectionality and measurement validity of the Eating Disorder Examination Questionnaire (EDE-Q) in Latinx undergraduates in the United States. Sex Roles, 84, 102–111. https://doi.org/10.1007/s11199-020-01149-7
166. Watson, L. B., Lewis, J. A., & Moody, A. T. (2019). A sociocultural examination of body image among Black women. Body Image, 31, 280–287. https://doi.org/10.1016/j.bodyim.2019.03.008
167. Wilfred, S. A., & Lundgren, J. D. (2021). The double consciousness body image scale: A body image assessment centering the experiences of Black women. International Journal of Eating Disorders, 54, 1771–1781. https://doi.org/10.1002/eat.23581
168. Brechwald, W. A., & Prinstein, M. J. (2011). Beyond homophily: A decade of advances in understanding peer influence processes. Journal of Research on Adolescence, 21(1), 166–179. https://doi.org/10.1111/j.1532-7795.2010.00721.x
169. Jones, D. C., Vigfusdottir, T. H., & Lee, Y. (2004). Body image and the appearance culture among adolescent girls and boys: An examination of friend conversations, peer criticism, appearance magazines, and the internalization of appearance ideals. Journal of Adolescent Research, 19(3), 323–339. https://doi.org/10.1177/0743558403258847
170. Mayeux, L., & Kleiser, M. (2020). A gender prototypicality theory of adolescent peer popularity. Adolescent Research Review, 5(3), 295–306. https://doi.org/10.1007/s40894-019-00123-z
171. boyd, D. (2010). Social network sites as networked publics: Affordances, dynamics, and implications. In Z. Papacharissi (Ed.), A networked self (pp. 47–66). Routledge.
172. Moreno, M. A., & Uhls, Y. T. (2019). Applying an affordances approach and a developmental lens to approach adolescent social media use. Digital Health, 5, 2055207619826678. https://doi.org/10.1177/2055207619826678
173. Nesi, J., Choukas-Bradley, S., & Prinstein, M. J. (2018a). Transformation of adolescent peer relations in the social media context: Part 1—A theoretical framework and application to dyadic peer relationships. Clinical Child and Family Psychology Review, 21(3), 267–294. https://doi.org/10.1007/s10567-018-0261-x
174. Subrahmanyam, K., & Šmahel, D. (2011). Digital youth: The role of media in development. Springer. https://doi.org/10.1007/978-1-4419-6278-2
175. Valkenburg, P. M., & Peter, J. (2011). Online communication among adolescents: An integrated model of its attraction, opportunities, and risks. Journal of Adolescent Health, 48(2), 121–127. https://doi.org/10.1016/j.jadohealth.2010.08.020
176. Thompson, J. K., Heinberg, L. J., Altabe, M., & Tantleff-Dunn, S. (1999). Exacting beauty: Theory, assessment, and treatment of body image disturbance. American Psychological Association.
177. Scully, M., Swords, L., & Nixon, E. (2020). Social comparisons on social media: Online appearance-related activity and body dissatisfaction in adolescent girls. Irish Journal of Psychological Medicine. https://doi.org/10.1017/ipm.2020.93
178. Trekels, J., Ward, L. M., & Eggermont, S. (2018). I "like" the way you look: How appearance-focused and overall Facebook use contribute to adolescents' self-sexualization. Computers in Human Behavior, 81, 198–208. https://doi.org/10.1016/j.chb.2017.12.020
179. McLean, S. A., Jarman, H. K., & Rodgers, R. F. (2019). How do "selfies" impact adolescents' well-being and body confidence? A narrative review. Psychology Research and Behavior Management, 12, 513–521. https://doi.org/10.2147/PRBM.S177834
180. McConnell, E. A., Clifford, A., Korpak, A. K., Phillips, G., & Birkett, M. (2017). Identity, victimization, and support: Facebook experiences and mental health among LGBTQ youth. Computers in Human Behavior, 76, 237–244. https://doi.org/10.1016/j.chb.2017.07.026
181. Romito, M., Salk, R. H., Thoma, B. C., Romito, M., Levine, M. D., & Choukas-Bradley, S. (2021). Exploring transgender adolescents' body image concerns and disordered eating:

Semi-structured interviews with nine gender minority youth. Body Image, 37, 50–62. https://doi.org/10.1016/j.bodyim.2021.01.008
182. Bowen, S., & Marlatt, A. (2009). Surfing the urge: Brief mindfulness-based intervention for college student smokers. Psychology of Addictive Behaviors, 23(4), 666–671. https://doi.org/10.1037/a0017127
183. Harris, J. S., Stewart, D. G., & Stanton, B. C. (2017). Urge surfing as aftercare in adolescent alcohol use: A randomized control trial. Mindfulness, 8(1), 144–149. https://doi.org/10.1007/s12671-016-0588-7
184. Naar-King, S. (2011). Motivational interviewing in adolescent treatment. Canadian Journal of Psychiatry, 56, 651–657. https://doi.org/10.1177/070674371105601103
185. Sudhir, P. M. (2015). Cognitive behaviour therapy with adolescents. In M. Mehta & R. Sagar (Eds.), A practical approach to cognitive behaviour therapy for adolescents (pp. 21–42). Springer Science + Business Media.
186. Gordon, C. S., Rodgers, R. F., Slater, A. E., McLean, S. A., Jarman, H. K., & Paxton, S. J. (2020). A cluster randomized controlled trial of the SoMe social media literacy body image and wellbeing program for adolescent boys and girls: Study protocol. Body Image, 33, 27–37. https://doi.org/10.1016/j.bodyim.2020.02.003
187. McLean, S. A., Paxton, S. J., & Wertheim, E. H. (2016). Does media literacy mitigate risk for reduced body satisfaction following exposure to thin-ideal media? Journal of Youth and Adolescence, 45(8), 1678–1695. https://doi.org/10.1007/s10964-016-0440-3
188. Vandenbosch L, Eggermont S. Understanding sexual objectification: a comprehensive approach toward media exposure and girls' internalization of beauty ideals, self-objectification and body surveillance. J Commun. 2012;62:869–87.
189. Roberts T, Gettman J. Mere exposure: gender differences in the negative effects of priming a state of self-objectification. Sex Roles. 2004;51:17–27.
190. Tolman D, Impett E, Tracy A, Michael A. Looking good, sounding good: femininity ideology and adolescent girls' mental health. Psychol Women Q. 2006;30:85–95.
191. Cohen R, Newton-John T, Slater A. The relationship between Facebook and Instagram appearance-focused activities and body image concerns in young women. Body Image. 2017;23:183–7.
192. de Lenne O, Vandenbosch L, Eggermont S, Karsay K, Trekels J. Picture-perfect lives on social media: a crossnational study on the role of media ideals in adolescent well-being. Media Psychol. 2020;23(1):52–78.
193. Burnette CB, Kwitowski MA, Mazzeo SE. "I don't need people to tell me I'm pretty on social media:" a qualitative study of social media and body image in early adolescent girls. Body Image. 2017;23:114–25.
194. Tiggemann M, Miller J. The Internet and adolescent girls' weight satisfaction and drive for thinness. Sex Roles. 2010;63(1):79–90.
195. Rodgers R, Melioli T. The relationship between body image concerns, eating disorders and internet use, Part I: A review of empirical support. Adolesc Res Rev. 2016;1(2):95–119.
196. Chang L, Li P, Loh R, Chua T. A study of Singaporean adolescent girls' selfie practices, peer appearance comparisons, and body esteem on Instagram. Body Image. 2019;29:90–9.
197. Faelens L, Hoorelbeke K, Cambier R, van Put J, Van de Putte E, De Raedt R, et al. The relationship between Instagram use and indicators of mental health: a systematic review. Comput Hum Behav Rep. 2021;4:100121.
198. Tiggemann M, Slater A. NetGirls: the Internet, Facebook and body image concern in adolescent girls. Int J Eat Disord. 2013;46:630–3.
199. Kleemans M, Daalmans S, Carbaat I, Anschutz D. Picture perfect: the direct effect of manipulated Instagram photos on body image in adolescent girls. Media Psychol. 2016;21(1):93–110.
200. Salomon I, Spears BC. The selfie generation: examining the relationship between social media use and early adolescent body image. J Early Adolesc. 2019;39(4):539–60.
201. Mahon C, Hevey D. Processing body image on social media: Gender differences in adolescent boys' and girls' agency and active coping. Front Psychol. 2021;12:1–11.

202. Marengo D, Longobardi C, Fabris M, Settanni M. Highly-visual social media and internalizing symptoms in adolescence: the mediating role of body image concerns. Comput Hum Behav. 2018;28:63–9.
203. Cookingham, L., & Ryan, G. (2015). The impact of social media on the sexual and social wellness of adolescents. Journal of Pediatric and Adolescent Gynecology, 28, 2–5
204. Hartas, D. (2021). The social context of adolescent mental health and wellbeing: Parents, friends and social media. Research Papers in Education, 36(5), 542–560. https://doi.org/10.1080/02671522.2019.1697734
205. Skowronski, M., Busching, R., & Krahé, B. (2020). Predicting adolescents' self-objectification from sexualized video game and Instagram use: A longitudinal study. Sex Roles. https://doi.org/10.1007/s11199-020-01187-1
206. Rideout, V., & Robb, M. B. (2019). The common sense census: Media use by tweens & teens, 2019. Common Sense Media. https://www.commonsensemedia.org/research/the-common-sense-census-media-use-by-tweens-and-teens-2019
207. Cohen, R., Newton-John, T., & Slater, A. (2018). 'Selfie'-objectification: The role of selfies in self-objectification and disordered eating in young women. Computers in Human Behavior, 79, 68–74. https://doi.org/10.1016/j.chb.2017.10.027.
208. Wang, Y., Xie, X., Fardouly, J., Vartanian, L. R., & Lei, L. (2021a). The longitudinal and reciprocal relationships between selfie-related behaviors and self-objectification and appearance concerns among adolescents. New Media & Society, 23(1), 56–77. https://doi.org/10.1177/1461444819894346
209. Bandura, A. (2001). Social cognitive theory of mass communications. In J. Bryant, & D. Zillman (Eds.), Media effects: Advances in theory and research (pp. 121–153). Erlbaum
210. van Oosten, J. M. F. (2021). Adolescent girls' use of social media for challenging sexualization. Gender Technology and Development, 25(1), 22–42
211. Ramsey, L. R., & Horan, A. L. (2018). Picture this: Women's self-sexualization in photos on social media. Personality and Individual Differences, 133, 85–90
212. Bakhshi, S., Shamma, D. A., & Gilbert, E. (2014). Faces engage us: Photos with faces attract more likes and comments on Instagram. Paper presented at the meeting of CHI, Toronto, Canada. https://doi.org/10.1145/2556288.2557403
213. Cougar Hall, P., West, J. H., & McIntyre, E. (2012). Female self-sexualization in Myspace.com personal profile photographs. Sexuality and Culture, 16, 1–16. https://doi.org/10.1007/s12119-011-9095-0.
214. Fardouly, J., Willburger, B. K., & Vartanian, L. R. (2018). Instagram use and young women's body image concerns and self-objectification: Testing mediational pathways. New Media and Society, 20. Advance online publication. https://doi.org/10.1177/1461444817694499.
215. Rose, J., Mackey-Kallis, S., Shyles, L., Barry, K., Biagini, D., Hart, C., … Jack, L. (2012). Face it: The impact of gender on social media images. Communication Quarterly, 60, 588–607. https://doi.org/10.1080/01463373.2012.725005.
216. Garner, D. M., Olmstead, M. P., & Polivy, J. (1983). Development and validation of a multidimensional eating disorder inventory for anorexia nervosa and bulimia. International Journal of Eating Disorders, 2, 15–34. https://doi.org/10.1002/1098-108X(198321)2:2<15::AID-EAT2260020203>3.0.CO;2-6.
217. Grabe, S., Ward, L. M., & Hyde, J. S. (2008). The role of the media in body image concerns among women: A meta-analysis of experimental and correlational studies. Psychological Bulletin, 134(3), 460–476. https://doi.org/10.1037/0033-2909.134.3.460.
218. Meier, E. P., & Gray, J. (2014). Facebook photo activity associated with body image disturbance in adolescent girls. Cyberpsychology, Behavior and Social Networking, 17(4), 199–206. https://doi.org/10.1089/cyber.2013.0305.
219. Correa, T., & Jeong, S. H. (2011). Race and online content creation: Why minorities are actively participating in the web. Information, Communication & Society, 14, 638–659. https://doi.org/10.1080/1369118X.2010.514355.

220. Kapidzic, S., & Herring, S. C. (2015). Race, gender, and self-presentation in teen profile photographs. New Media & Society, 17(6), 958–976. https://doi.org/10.1177/1461444813520301
221. Milkie, M. A. (1999). Social comparisons, reflected appraisals, and mass media: The impact of pervasive beauty images on black and white girls' self-concepts. Social Psychology Quarterly, 62, 190–210. https://doi.org/10.2307/2695857.
222. Karsay, K., Knoll, J., & Matthes, J. (2018). Sexualizing media use and self-objectification: A meta-analysis. Psychology of Women Quarterly, 42(1), 9–28
223. Vuong AT, Jarman HK, Doley JR, McLean SA. Social media use and body dissatisfaction in adolescents: the moderating role of thin- and muscular-ideal internalisation. Int J Environ Res Public Health. 2021;18(24):13222. https://doi.org/10.3390/ijerph182413222.
224. Cohen R, Blaszczynski A. Comparative effects of Facebook and conventional media on body image dissatisfaction. J Eat Disord. 2015;3:23. https://doi.org/10.1186/s40337-015-0061-3.
225. Clement U, Löwe B. Der" Fragebogen zum Körperbild (FKB-20)". Literaturüberblick, Beschreibung und Prüfung eines Messinstrumentes. Diagnostica. 1996;42(4):352–76.
226. Gioia, F., Griffiths, M. D., & Boursier, V. (2020). Adolescents' body shame and social networking Sites: The mediating effect of body image control in photos. Sex Roles, 83(11–12), 773–785. https://doi.org/10.1007/s11199-020-01142-0
227. Daniel, S., & Bridges, S. K. (2010). The drive for muscularity in men: Media influences and objectification theory. Body Image, 7(1), 32–38. https://doi.org/10.1016/j.bodyim.2009.08.003.
228. Dakanalis, A., Di Mattei, V. E., Prunas, A., Riva, G., Sarno, L., Volpato, C., … Zanetti, M. A. (2012). Il corpo oggettivato: Media, benessere psicofisico e differenze di genere [The objectified body: Media, psychophysical wellbeing, and gender differences]. Psicologia sociale, 7(2), 261–284.
229. Dakanalis, A., Carrà, G., Calogero, R., Fida, R., Clerici, M., Zanetti, M. A., & Riva, G. (2015). The developmental effects of media-ideal internalization and self-objectification processes on adolescents' negative body-feelings, dietary restraint, and binge eating. European Child & Adolescent Psychiatry, 24(8), 997–1010. https://doi.org/10.1007/s00787-014-0649-1
230. Manago, A. M., Ward, L. M., Lemm, K. M., Reed, L., & Seabrook, R. (2015). Facebook involvement, objectified body consciousness, body shame, and sexual assertiveness in college women and men. Sex Roles, 72(1–2), 1–14. https://doi.org/10.1007/s11199-014-0441-1.
231. Moradi, B. (2010). Addressing gender and cultural diversity in body image: Objectification theory as a framework for integrating theories and grounding research. Sex Roles, 63(1–2), 138–148. https://doi.org/10.1007/s11199-010-9824-0.
232. Moradi, B., & Huang, Y. (2008). Objectification theory and psychology of women: A decade of advances and future directions. Psychology of Women Quarterly, 32, 377–398. https://doi.org/10.1111/j.1471-6402.2008.00452.x.
233. Vandenbosch, L., & Eggermont, S. (2013). Sexualization of adolescent boys: Media exposure and boys' internalization of appearance ideals, self-objectification, and body surveillance. Men and Masculinities, 16(3), 283–306. https://doi.org/10.1177/1097184X13477866
234. Caso, D., Fabbricatore, R., Muti, F., & Starace, C. (2019). Sessualizzazione e oggettivazione femminile su Instagram: Il ruolo delle influencer. Psicologia Sociale, 14(3), 441–463. https://doi.org/10.1482/94944
235. de Vries, D., & Peter, J. (2013). Women on display: The effect of portraying the self online in women's self-objectification. Computers in Human Behavior, 29, 1483–1489. https://doi.org/10.1016/j.chb.2013.01.015.
236. Wang, Y., Wang, X., Yang, J., Zeng, P., & Lei, L. (2019a). Body talk on social networking sites, body surveillance, and body shame among young adults: The roles of self-compassion and gender. Sex Roles, 1–12. Advance online publication. https://doi.org/10.1007/s11199-019-01084-2.
237. Fardouly, J., Diedrichs, P. C., Vartanian, L. R., & Halliwell, E. (2015). The mediating role of appearance comparisons in the relationship between media usage and self-objectification

in young women. Psychology of Women Quarterly, 39(4), 447–457. https://doi.org/10.1177/0361684315581841.
238. Caplan, S. E. (2010). Theory and measurement of generalized problematic internet use: A two-step approach. Computers in Human Behavior, 26(5), 1089–1097. https://doi.org/10.1016/j.chb.2010.03.012.
239. Boursier, V., & Manna, V. (2019). Relational body identities: Body image control through self-portraits –a revision of the body image control in photos questionnaire. In R. T. Gopalan (Ed.), Intimacy and developing personal relationships in the virtual world (pp. 40–63). Hershey, PA: IGI Global. https://doi.org/10.4018/978-1-5225-4047-2.ch003.
240. Casale, S., & Fioravanti, G. (2017). Shame experiences and problematic social networking sites use: An unexplored association. Clinical Neuropsychiatry, 14(1), 44–48. https://flore.unifi.it/retrieve/handle/2158/1074075/212427/2017_shame_PIU_Clin_Neur.pdf.
241. Balakrishnan, J., & Griffiths, M. D. (2017). Social media addiction: What is the role of content in YouTube? Journal of Behavioral Addictions, 6, 364–377. https://doi.org/10.1556/2006.6.2017.058.
242. Boursier, V., Manna, V., Gioia, F., Coppola, F., & Venosa, N. (2018). Cyber-moms facing motherhood: Holding functions and regressive movements in parenting websites. In S. Sekalala & B. C. Niezgoda (Eds.), Global perspectives on health communication in the age of social media (pp. 29–58). Hershey, PA: IGI Global. https://doi.org/10.4018/978-1-5225-3716-8.ch002.
243. Boyd, D., & Ellison, N. B. (2007). Social network sites: definition, history, and scholarship. Journal of Computer-Mediated Communication, 13(1), 210–230.
244. Kuss, D. J., & Griffiths, M. D. (2011b). Online social networking and addiction – A review of the psychological literature. International Journal of Environmental Research and Public Health, 8(9), 3528–3552. https://doi.org/10.3390/ijerph8093528.
245. Kuss, D. J., & Griffiths, M. D. (2017). Social networking sites and addiction: Ten lessons learned. International Journal of Environmental Research and Public Health, 14(3), 1–17. https://doi.org/10.3390/ijerph14030311
246. Tiggemann, M., & Slater, A. (2017). Facebook and body image concern in adolescent girls: A prospective study. International Journal of Eating Disorders, 50(1), 80–83. https://doi.org/10.1002/eat.22640
247. Boursier, V., & Manna, V. (2018b). Selfie expectancies among adolescents. Construction and validation of an instrument to assess expectancies toward selfies among boys and girls. Frontiers in Psychology, 9, 839. https://doi.org/10.3389/fpsyg.2018.00839.
248. Fox, J., & Rooney, M. C. (2015). The Dark Triad and trait self-objectification as predictors of men's use and self-presentation behaviors on social networking sites. Personality and Individual Differences, 76, 161–165. https://doi.org/10.1016/J.PAID.2014.12.017
249. Lonergan, A. R., Bussey, K., Mond, J., Brown, O., Giffiths, S., Muray, S. B., … Mitchison, D. (2019). Me, my selfie, and I: The relationship between editing and posting selfies and body dissatisfaction in men and women. Body Image, 28, 39–43. https://doi.org/10.1016/j.bodyim.2018.12.001.
250. McLean, S. A., Paxton, S. J., Wertheim, E. H., & Masters, J. (2015). Photoshopping the selfie: Self-photo editing and photo investment are associated with body dissatisfaction in adolescent girls. The International Journal of Eating Disorders, 48(8), 1132–1140. https://doi.org/10.1002/eat.22449
251. Boursier, V., Gioia, F., & Griffiths, M. D. (2020). Objectified body consciousness, body image control in photos, and problematic social networking: The role of Appearance Control Beliefs. Frontiers in Psychology, 11, 147. https://doi.org/10.3389/fpsyg.2020.00147
252. Common Sense Media. (2015). Common Sense census: Media use by tweens and teens. Retrieved online: https://www.commonsensemedia.org/sites/default/files/uploads/research/census_researchreport.pdf
253. Wiederhold, B. K. (2012). As parents invade Facebook, teens tweet more. Cyberpsychology, Behavior, & Social Networking, 15, 385–385. https://doi.org/10.1089/cyber.2012.1554.

254. Greenwood, D. N. (2013). Fame, Faceboko, and twitter: how attitudes about fame predict frequency and nature of social media use. Psychology of Popular Media Culture, 2, 222–236. https://doi.org/10.1037/ppm0000013.
255. Soukup, C. (2006). Hitching a ride on a star: celebrity, fandom, and identification on the world wide web. Southern Communication Journal, 71, 319–337. https://doi.org/10.1080/10417940601000410.
256. Uhls, Y. T., & Greenfield, P. M. (2012). The value of fame: preadolescent perceptions of popular media and their relationship to future aspirations. Developmental Psychology, 48, 315–326. https://doi.org/10.1037/a0026369.
257. Clay, D., Vignoles, V. L., & Dittmar, H. (2005). Body image and self-esteem among adolescent girls: testing the influence of sociocultural factors. Journal of Research on Adolescence, 15, 451–477. https://doi.org/10.1111/j.1532-7795.2005.00107.x.
258. Eyal, K., & Te'eni-Harari, T. (2013). Explaining the relationship between media exposure and early adolescents' body image perceptions: the role of favorite characters. Journal of Media Psychology, 25, 129–141. https://doi.org/10.1027/1864-1105/a000094.
259. Cohen-Almagor, R. (2015). Confronting the internet's dark side: moral and social responsibility on the free highway. Washington, DC: Woodrow Wilson Center Press.
260. Wei, Q., Sun, X. J., Lian, S. L., & Song, Y. H. (2017). The effect of using social networking sites on body image satisfaction: The mediating role of body image comparison and the moderating role of self-objectification. Journal of Psychological Science, 40(4), 920–926. https://doi.org/10.16719/j.cnki.1671-6981.20170422 (In Chinese).
261. Jackson, T., & Chen, H. (2015). Features of objectified body consciousness and sociocultural perspectives as risk factors for disordered eating among late-adolescent women and men. Journal of Counseling Psychology, 62(4), 741–752. https://doi.org/10.1037/cou0000096.
262. Niu, G., Sun, L., Liu, Q., Chai, H., Sun, X., & Zhou, Z. (2020). Selfie-posting and young adult women's restrained eating: The role of commentary on appearance and self-objectification. Sex Roles, 82(3), 232–240. https://doi.org/10.1007/s11199-019-01045-9.
263. Wang, X. Y., Qi, Q. R., Lu, X. C., & Sun, S. X. (2015). Perception of obesity and weight reduction behaviors among undergraduates. Chinese Journal of School Health, 36(1), 54–60. https://doi.org/10.16835/j.cnki.1000-9817.2015.01.016 (In Chinese).
264. Mclean, S. A., Paxton, S. J., & Wertheim, E. H. (2016). The role of media literacy in body dissatisfaction and disordered eating: A systematic review. Body Image, 19, 9–23. https://doi.org/10.1016/j.bodyim.2016.08.002.
265. Morgan, M., et al. (2015). Facebook use and disordered eating in college-aged women. Journal of Adolescent Health, 57(2), 157–163. https://doi.org/10.1016/j.jadohealth.2015.04.026.
266. Venus, J. S., Ehri, R., & Aziz, M. (2018). Dieting 2.0!: Moderating effects of Instagrammers' body image and Instafame on other Instagrammers' dieting intention. Computers in Human Behavior, 87, 224–237. https://doi.org/10.1016/j.chb.2018.06.001.
267. Turkle, S. (1995). Life on the screen: Identity in the age of the internet. New York: Simon & Schuster.
268. Turkle, S. (1996). Parallel lives: Working on identity in virtual space. In D. Grodin & T. R. Lindlof (Eds.), Constructing the self in a mediated world (pp. 157–175). Thousand Oaks: SAGE.
269. Deutsch, N. (2009). The forbidden fork, the cell phone holocaust, and other Haredi encounters with technology. Contemporary Jewry, 29(1), 3–19. https://doi.org/10.1007/s12397-008-9002-7.
270. Gray, Kishonna L (2017). They're just too urban: Black gamers streaming on Twitch. Digital Sociologies, vol. 1, pp. 355–368.
271. Pellicone, Anthony J; and June Ahn (2017). The Game of performing play: Understanding streaming as cultural production. In CHI 2017: Proceedings of the 2017 CHI Conference on Human Factors in Computing Systems. Denver, CO, USA, 6 – 11 May 2017. New York: ACM Press, pp. 4863–4874.

272. Johnson, Mark R (2019). Inclusion and exclusion in the digital economy: Disability and mental health as a live streamer on Twitch. tv. Information, Communication & Society, vol. 22, no. 4, pp. 506–520.
273. Wohn, Donghee Yvette; and Guo Freeman (2020). Live Streaming, Playing, and Money Spending Behaviors in eSports. Games and Culture, vol. 15, no. 1, January 2020, pp. 73–88.
274. Wohn, Donghee Yvette; Guo Freeman; and Caitlin McLaughlin (2018). Explaining viewers' emotional, instrumental, and financial support provision for live streamers. In CHI 2018: Proceedings of the 2018 CHI Conference on Human Factors in Computing Systems. Montréal, Canada, 21 -26 April 2018. New York: ACM Press, article 474, https://doi.org/10.1145/3173574.3174048
275. Lu, Zhicong; Michelle Annett; Mingming Fan; and Daniel Wigdor (2019). I feel it is my responsibility to stream: Streaming and engaging with intangible cultural heritage through livestreaming. In CHI 2019: Proceedings of the 2019 CHI Conference on Human Factors in Computing Systems. Glasgow, UK, 4–9 May 2019. New York: ACM Press, article 229.
276. Woodcock, Jamie; and Mark R Johnson (2019). The affective labor and performance of live streaming on Twitch. tv. Television & New Media, vol. 20, no. 8, December 2019, pp. 813–823.
277. Eatough, Virginia; and Jonathan A. Smith (2008). Interpretative phenomenological analysis. In C. Willig; and W. S. Rogers (eds.): The Sage Handbook of Qualitative Research in Psychology, London: Sage, pp. 194–211.
278. Calogero, R. M., Davis, W. N., & Thompson, J. K. (2005). The role of self-objectification in the experience of women with eating disorders. Sex Roles, 52, 43–50. https://doi.org/10.1007/s11199-005-1192-9.
279. Aubrey, J. S. (2010). Looking good versus feeling good: An investigation of media frames of health advice and their effects on women's body-related self-perceptions. Sex Roles, 63(1), 50–63. https://doi.org/10.1007/s11199-010-9768-4.
280. Willis, L. E., & Knobloch-Westerwick, S. (2014). Weighing women down: Messages on weight loss and body shaping in editorial content in popular women's health and fitness magazines. Health Communication, 29(4), 323–331. https://doi.org/10.1080/10410236.2012.755602.
281. Cohen, R., Newton-John, T., & Slater, A. (2017). The relationship between Facebook and Instagram appearance-focused activities and body image concerns in young women. Body Image, 23, 183–187. https://doi.org/10.1016/j.bodyim.2017.10.002.
282. Tiggemann, M., Hayden, S., Brown, Z., & Veldhuis, J. (2018). The effect of Instagram "likes" on women's social comparison and body dissatisfaction. Body Image, 26, 90–97. https://doi.org/10.1016/j.bodyim.2018.07.002
283. Marengo, D., Longobardi, C., Fabris, M. A., & Settanni, M. (2018). Highly-visual social media and internalizing symptoms in adolescence: The mediating role of body image concerns. Computers in Human Behavior, 82, 63–69. https://doi.org/10.1016/j.chb.2018.01.003
284. Bozsik, F., Whisenhunt, B. L., Hudson, D. L., Bennett, B., & Lundgren, J. D. (2018). Thin is in? Think again: The rising importance of muscularity in the thin ideal female body. Sex Roles, 79(9–10), 609–615. https://doi.org/10.1007/s11199-017-0886-0.
285. Homan K. Athletic-ideal and thin-ideal internalization as prospective predictors of body dissatisfaction, dieting, and compulsive exercise. Body Image. 2010;7(3):240–5. https://doi.org/10.1016/j.bodyim.2010.02.004.
286. Homan, K., McHugh, E., Wells, D., Watson, C., & King, C. (2012). The effect of viewing ultra-fit images on college women's body dissatisfaction. Body Image, 9(1), 50–56. https://doi.org/10.1016/j.bodyim.2011.07.006.
287. Agar, M. (2013). The lively science. Minneapolis: Hillcrest Publishing Group.
288. Taylor, B., & Francis, K. (2013). Qualitative research in the health sciences: Methodologies, methods and processes. London: Routledge.
289. Goldstraw, D., & Keegan, B. J. (2016). Instagram's 'Fitspiration' trend and its effect on young women's self-esteem. Proceedings of the 29th Bled eConference (p. 190–198). University of Maribor. Bled, Slovenia. Retrieved from http://aisel.aisnet.org/bled2016/35.

290. Raggatt M, Wright CJC, Carrotte E, et al. "I aspire to look and feel healthy like the posts convey": engagement with fitness inspiration on social media and perceptions of its influence on health and wellbeing. BMC Public Health. 2018;18(1):1002. https://doi.org/10.1186/s12889-018-5930-7.
291. Higgins, J. W., & Begoray, D. (2012). Exploring the borderlands between media and health: Conceptualizing 'critical media health literacy'. Journal of Media Literacy Education, 4, 136–148. https://doi.org/10.1007/SpringerReference_223477.
292. Hesse-Biber, S., Leavy, P., Quinn, C. E., & Zoino, J. (2006). The mass marketing of disordered eating and eating disorders: The social psychology of women, thinness and culture. Women's Studies International Forum, 29(2), 208–224. https://doi.org/10.1016/j.wsif.2006.03.007.
293. Menzel, J. E., & Levine, M. P. (2011). Embodying experiences and the promotion of positive body image: The example of competitive athletics. In R. Calogero, S. Tantleff-Dunn, & J. Thompson (Eds.), Self-objectification in women: Causes, consequences, and counteractions (pp. 163–186). Washington, DC: American Psychological Association. https://doi.org/10.1037/12304-008.
294. Tylka, T. (2011). Positive psychology perspectives on body image. In T. F. Cash & L. Smolak (Eds.), Body image: A handbook of science, practice, and prevention (2nd ed., pp. 56–64). New York: Guilford Press.
295. Webb, J. B., Wood-Barcalow, N. L., & Tylka, T. L. (2015). Assessing positive body image: Contemporary approaches and future directions. Body Image, 14(2), 130–145. https://doi.org/10.1016/j.bodyim.2015.03.010
296. Wood-Barcalow, N. L., Tylka, T. L., & Augustus-Horvath, C. L. (2010). "But I like my body": Positive body image characteristics and a holistic model for young-adult women. Body Image, 7(2), 106–116. https://doi.org/10.1016/j.bodyim.2010.01.001.
297. Cordelois, A. (2010). Using digital technology for collective ethnographic observation: an experiment on 'coming home'. Social Science Information Sur Les Sciences Sociales, 49(3), 445–463.
298. Reavey, P. (2016). Visual methods in psychology. Using and interpreting images in qualitative research. Abingdon: Routledge.
299. Carpenter, C., & McEwan, B. (2016). The players of micro-dating: Individual and gender differences in goal orientations toward micro-dating apps. First Monday, 21(5).
300. Musil, B., Preglej, A., Ropert, T., Klasinc, L., & Babic, N. C. (2017). What is seen is who you are: Are cues in selfie pictures related to personality characteristics? Frontiers in Psychology, 8(1), 1–13.
301. Timmermans, E., & De Caluwé, E. (2017). Development and validation of the Tinder Motives Scale (TMS). Computers in Human Behavior, 70, 341–350.
302. Ward, J. (2016). Swiping, Matching, Chatting. Self-Presentation and Self-Disclosure on Mobile Dating Apps. In Human IT: Journal for Information Technology Studies as a Human Science, 13(2), 81–95.
303. Fitzpatrick, C., Birnholtz, J., & Brubaker, J. R. (2015). Social and Personal Disclosure in a Location-Based Real Time Dating App. In 48th Hawaii International Conference on System Sciences, (pp. 1983–1992). Kauai.
304. Duguay, S. (2016a). Dressing up Tinderella: Interrogating authenticity claims on the mobile dating app Tinder. Information, Communication & Society, 20(3), 351–367.
305. Duguay, S. (2016b). Lesbian, gay, bisexual, trans, and queer visibility through selfies: Comparing platform mediators across Ruby Rose's Instagram and Vine presence. Social Media + Society. https://doi.org/10.1177/2056305116641975
306. Hancock, J. T., Toma, C., & Ellison, N. B. (2007). The truth about lying in online dating profiles. In Proceedings of the 2007 Conference on Human Factors in Computing Systems, CHI 2007, San Jose, California, USA, April 28 - May 3, 2007.
307. Ranzini, G., & Lutz, C. (2017). Love at first swipe? Explaining Tinder self-presentation and motives. Mobile Media & Communication, 5(1), 80–101.
308. Goffman, E. (1959). Wir alle spielen Theater. New York: Doubleday.

309. Bohnsack, R. (2010). The interpretation of pictures and the documentary method. In R. Bohnsack, N. Pfaff, & W. Weller, (Eds.), Qualitative analysis and documentary method in international educational research, 9(3), 267–292.
310. Bohnsack, R., Michel, B., & Przyborski, A. (2015). Dokumentarische Bildinterpretation. Methodologie und Forschungspraxis. Opladen: Budrich.
311. Pilarczyk, U., & Mietzner, U. (2008). Das reflektierte Bild. Die seriell-ikonografische Fotoanalyse in den Erziehungs- und Sozialwissenschaften (Vol. 1.). Bad Heilbrunn: Klinkhardt.
312. Rose, N. (2006). Governing the soul: Shaping of the private self. London: Free Association Books.
313. Holzkamp, K. (1983). Grundlagen der Psychologie. Frankfurt a.M.: Campus.
314. McFarland LA, Ployhart RE. Social media: a contextual framework to guide research and practice. J Appl Psychol. 2015;100(6):1653–77.
315. Renfrew Center Foundation. (2014). Afraid to be your selfie? Survey reveals most people photoshop their images Retrieved March 13, 2022 from http://renfrewcenter.com/news/afraid-be-your-selfie-survey-reveals-most-people-photoshop-their-images.
316. Othman S, Lyons T, Cohn JE, Shokri T, Bloom JD. The influence of photo editing applications on patients seeking facial plastic surgery services. Aesthetic Surg J. 2021;41(3):101–10.
317. Myers, T. A., & Crowther, J. H. (2009). Social comparison as a predictor of body dissatisfaction: A meta-analytic review. Journal of Abnormal Psychology, 118(4), 683–698. https://doi.org/10.1037/a0016763.
318. Chae, J. (2017). Virtual makeover: Selfie-taking and social media use increase selfie-editing frequency through social comparison. Computers in Human Behavior, 66, 370–376. https://doi.org/10.1016/j.chb.2016.10.007.
319. Kanavakis G, Halazonetis D, Katsaros C, Gkantidis N. (2021) Facial shape affects self-perceived facial attractiveness. PLoS ONE 16(2): e0245557. https://doi.org/10.1371/journal
320. Lee M, Lee HH. Social media photo activity, internalization, appearance comparison, and body satisfaction: the moderating role of photo-editing behavior. Comput Hum Behav. 2021;114:106–579.
321. O'Brien KS, Caputi P, Minto R, Peoples G, Hooper C, Kell S, et al. Upward and downward physical appearance comparisons: development of scales and examination of predictive qualities. Body Image. 2009;6(3):201–6.
322. Cohen, R., Newton-John, T., & Slater, A. (2020). The case for body positivity on social media: Perspectives on current advances and future directions. Journal of Health Psychology. https://doi.org/10.1177/1359105320912450
323. Tiggemann, M., Anderberg, I., & Brown, Z. (2020). Uploading your best self: Selfie editing and body dissatisfaction. Body Image, 33, 175–182. https://doi.org/10.1016/j.bodyim.2020.03.002
324. Bij de Vaate, A. J. N. B., Veldhuis, J., Alleva, J. M., Konijn, E. A., & van Hugten, C. H. (2018). Show your best self (ie): An exploratory study on selfie-related motivations and behavior in emerging adulthood. Telematics and Informatics, 35(5), 1392–1407. https://doi.org/10.1016/j.tele.2018.03.010.
325. Pelosi, A., Zorzi, G., & Corsano, P. (2014). The body image control in photos questionnaire (BICP). BPA-Applied Psychology Bulletin (Bollettino di Psicologia Applicata), 62(269), 42–52.
326. Franchina, V., & Lo Coco, G. (2018). The influence of social media use on body image concerns. International Journal of Psychoanalysis and Education, 10(1), 5–14. File:///C:/users/franc/downloads/218-661-1-PB.Pdf.
327. King, V. (2016). "If you show your real face, you'll lose 10 000 followers"– The gaze of the other and transformations of shame in digitalized relationships. CM Komunikacija i Mediji, 11(38), 71–90. https://doi.org/10.5937/comman12-11504.
328. Veldhuis, J., Alleva, J. M., Bij de Vaate, A. J., Keijer, M., & Konijn, E. A. (2020). Me, my selfie, and I: The relations between selfie behaviors, body image, self-objectification, and self-esteem in young women. Psychology of Popular Media, 9(1), 3. https://doi.org/10.1037/ppm0000206

329. Fardouly, J., Diedrichs, P. C., Vartanian, L. R., & Halliwell, E. (2015). Social comparisons on social media: The impact of Facebook on young women's body image concerns and mood. Body Image, 13, 38–45. https://doi.org/10.1016/j.bodyim.2014.12.002.
330. Gioia, F., & Boursier, V. (2019a). Emotion dysregulation and adolescents' preference for online social interactions: The moderating role of gender. Advance online publication. http://ceur-ws.org/Vol-2524/paper23.pdf
331. Feltman, C. E., & Szymanski, D. M. (2018). Instagram use and self-objectification: The roles of internalization, comparison, appearance commentary, and feminism. Sex Roles, 78(5–6), 311–324. https://doi.org/10.1007/s11199-017-0796-1
332. Gal, D. (2015). Identity-signaling behavior. In M. I. Norton, D. D. Rucker, & C. Lamberton (Eds.), The Cambridge handbook of consumer psychology (pp. 257–281). Cambridge University Press.
333. Grewal, L., Stephen, A. T., & Coleman, N. V. (2019). When posting about products on social media backfires: The negative effects of consumer identity signaling on product interest. Journal of Marketing Research, 56(2), 197–210.
334. Jensen Schau, H., & Gilly, M. C. (2003). We are what we post? Self-presentation in personal web space. Journal of Consumer Research, 30(3), 385–404.
335. Wood, W., & Eagly, A. H. (2015). Two traditions of research on gender identity. Sex Roles, 73(11), 461–473.
336. Bem, S. L. (1974). The measurement of psychological androgyny. Journal of Consulting and Clinical Psychology, 42(2), 155–162.
337. Spence, J. T., Helmreich, R., & Stapp, J. (1975). Ratings of self and peers on sex role attributes and their relation to self-esteem and conceptions of masculinity and femininity. Journal of Personality and Social Psychology, 32(1), 29–39.
338. Reed, I. I., Forehand, A., Puntoni, M. R., S., & Warlop, L. (2012). Identity-based consumer behavior. International Journal of Research in Marketing, 29(4), 310–321.
339. O'Neil, J., Helms, B., Gable, R., David, L., & Wrightsman, L. (1986). Gender-role conflict scale: College men's fear of femininity. Sex Roles, 14, 335–350.
340. Wooten, D. B., & Rank-Christman, T. (2022). The significance and meaning of racial identity in consumer research: A review and call for research. Consumer Psychology Review, 5(1), 19–32.
341. MacInnis, D. J., & Folkes, V. S. (2017). Humanizing brands: When brands seem to be like me, part of me, and in a relationship with me. Journal of Consumer Psychology, 27 No(3), 355–374
342. Malär, L., Krohmer, H., Hoyer, W. D., & Nyffenegger, B. (2011). Emotional brand attachment and brand personality: The relative importance of the actual and the ideal self. Journal of Marketing, 75 No(4), 35–52
343. Sung, Y., Kim, E., & Choi, S. M. (2018). "#Me and brands: understanding brand-selfie posters on social media", International Journal of Advertising, Vol. 0487, available at: https://doi.org/10.1080/02650487.2017.1368859
344. Halliday, S. (2016). User-generated content about brands. Journal of Business Research, 69 No(1), 137–144
345. Hearn, A. (2008). 'Meat, mask, burden': Probing the contours of the branded 'self.' Journal of Consumer Culture, 8(2), 197–217. https://doi.org/10.1177/1469540508090086.
346. Belk, R. (2014). Digital consumption and the extended self. Journal of Marketing Management, 30 No, 11–12
347. Robinson, R. S. (2014). "Purposive Sampling", edited by Michalos, A.C.Encyclopedia of Quality of Life and Well-Being Research, Springer Netherlands, Dordrecht
348. Robson, S., & Foster, A. (1989). Qualitative Research in Action, edited by Butterworth-Heinemann Ltd, Stoughton, Hodder, London
349. Bartoli, C., Nosi, C., Mattiacci, A., & Sfodera, F. (2022) Consumer self-concept in the cyberspace: how digitization has shaped the way we self-disclose to others. Journal of Strategic Marketing https://doi.org/10.1080/0965254X.2022.2056501

350. Arnould, E., Press, M., Salminen, E., & Tillotson, J. S. (2019). Consumer culture theory: Development, critique, application and prospects. Foundations and Trends in Accounting, 12 No(2), 80–166
351. Kohli, A. K., & Haenlein, M. (2021). Factors affecting the study of important marketing issues: Implications and recommendations. International Journal of Research in Marketing, Vol. 38 No(1), 1–11
352. Smith, R. (2013). What is digital media? Centre for Digital Media. https://thecdm.ca/news/what-is-digital-media
353. Berzonsky, M. D., & Kinney, A. (2019). Identity processing style and depression: The mediational role of experiential avoidance and self-regulation. Identity, 19(2), 83–97.
354. Courey, M., & Pare, P. P. (2016). A closer look at the relationship between low self-control and delinquency: The effects of identity styles. Crime and Delinquency, 62, 368–396.
355. Soenens, B., Berzonsky, M. D., Dunkel, C. S., Papini, D. R., & Vansteenkiste, M. (2011). Are all identity commitments created equally? The importance of motives for commitment for late adolescents' personal adjustment. International Journal of Behavioural Development, 35, 358–369.
356. McInroy, L. B., & Craig, S. L. (2020). "It's like a safe haven fantasy world": Online fandom communities and the identity development activities of sexual and gender minority youth. Psychology of Popular Media, 9(2), 236–246.
357. Ko, C. H., Yen, J. Y., Yen, C. F., Lin, H. C., & Yang, M. J. (2007). Factors predictive for incidence and remission of internet addiction in young adolescents: A prospective study. Cyberpsychologt & Behavior, 10(4), 545–551.
358. Sebre, S. B., Miltuze, A., & Limonovs, M. (2020). Integrating adolescent problematic internet use risk factors: Hyperactivity, inconsistent parenting, and maladaptive cognitions. Journal of Child and Family Studies, 29, 2000–2009.
359. Zhang, H., Li, D., & Li, X. (2015). Temperament and problematic internet use in adolescents: A moderated mediation model of maladaptive cognition and parenting styles. Journal of Child and Family Studies, 24, 1886–1897. https://doi.org/10.1007/s10826-014-9990-8.
360. Berzonsky, M. D. (2011). A social-cognitive perspective on identity construction. In S. J. Schwartz, K. Luyckx, & V. Vignoles (Eds.), Handbook of identity theory and research (Vol. 1, pp. 55–76). New York: Springer.
361. DiMaggio, P., Hargittai, E., Celeste, C., Schafer, S.: Digital Inequality: From Unequal Access to Differentiated Use. In: K. Neckerman (ed.) Social Inequality, pp. 355–400. Russell Sage Foundation, New York (2004). https://www.russellsage.org/research/reports/dimaggio
362. Smith, A.: Older adults and technology use: adoption is increasing but many seniors remain isolated from digital life. Pew Research Center (2014). http://www.pewinternet.org/2014/04/03/older-adults-andtechnology-use/
363. Hargittai E, Dobransky K (2017) Old dogs, new clicks: digital inequality in skills and uses among older adults. Can J Commun 42(2):195–212. https://doi.org/10.22230/cjc.2017v42n2a3176
364. De Choudhury, M., Morris, M.R., White, R.W.: Seeking and Sharing Health Information Online: Comparing Search Engines and Social Media. In: Proceedings of the SIGCHI Conference on Human Factors in Computing Systems, CHI '14, pp. 1365–1376. ACM, New York, NY, USA (2014). https://doi.org/10.1145/2556288.2557214
365. Zyskowski, K., Morris, M.R., Bigham, J.P., Gray, M.L., Kane, S.K.: Accessible Crowdwork?: Understanding the Value in and Challenge of Microtask Employment for People with Disabilities. In: Proceedings of the 18th ACM Conference on Computer Supported Cooperative Work & Social Computing, CSCW '15, pp. 1682–1693. ACM, New York, NY, USA (2015). https://doi.org/10.1145/2675133.2675158
366. Carpenter, B. D., & Buday, S. (2007). Computer use among older adults in a naturally occurring retirement community. Computers in Human Behavior, 23(6), 3012–3024. https://doi.org/10.1016/j.chb.2006.08.015.

367. Cresci, M.K., Yarandi, H.N., Morrell, R.W.: Pro-Nets Versus no-nets: differences in urban older adults' predilections for internet use. Educ. Gerontol. 36(6), 500–520 (2010). https://doi.org/10.1080/03601270903212476
368. Friemel, T.N.: The digital divide has grown old: determinants of a digital divide among seniors. New Media Soc. 18(2), 313–331 (2016)
369. Perrin, A., Duggan, M.: Americans' Internet Access: 2000–2015. Tech. rep., Pew Research Center (2015). http://www.pewinternet.org/2015/06/26/americans-internet-access-2000-2015/
370. Hargittai, E.: The Digital Reproduction of Inequality. In: Grusky, D. (ed.) Social Stratification, pp. 939–944. Westview Press, Boulder (2008)
371. van Dijk, J.: The Deepening Divide. Sage Publications Inc, Thousand Oaks (2005)
372. Brewer, R., Garcia, R.C., Schwaba, T., Gergle, D., Piper, A.M.: Exploring traditional phones as an e-mail interface for older adults. ACM Trans. Access. Comput. 8(2), 6:1–6:20 (2016). https://doi.org/10.1145/2839303
373. Kelly JR, Barsade SG (2001) Mood and emotions in small groups and work teams. Organ Behav Hum Decis Process 86(1):99–130
374. Easterlin RA (2003) Explaining happiness. Proc Natl Acad Sci 100(19):11176–11183
375. Matsuba, M. K. (2006). Searching for self and relationships online. CyberPsychology and Behavior, 9(3), 275–284.
376. Yang, C. C., Holden, S. M., & Carter, M. D. K. (2017). Emerging adults' social media self-presentation and identity development at college transition: Mindfulness as a moderator. Journal of Applied Developmental Psychology, 52, 212–221.
377. Kerpelman, J. L., Pittman, J. F., & Lamke, L. K. (1997). Toward a microprocess perspective on adolescent identity development: An identity control theory approach. Journal of Adolescent Research, 12(3), 325–346.
378. Tan, A. L., Kendis, R. J., Fine, J. T., & Porac, J. (1977). A short measure of Eriksonian ego identity. Journal of Personality Assessment, 41(3), 279–284.
379. Lei, L., & Wang, W. (2015). The relationship between the use of mobile social media and friendship quality in adolescents. Studies of Psychology and Behavior, 13(5), 664–670.
380. Loughlin, A. (2014) Relationship status and its effect on interpersonal relationships, self-identity and life satisfaction (Bachelor's thesis). Department of Psychology, Dublin Business School, Ireland. Retrieved from https://pdfs.semanticscholar.org/42e7/78fcf97649ea248e761e0728f202b1374d88.pdf. Accessed September 2017.
381. Tosun, L. P. (2012). Motives for Facebook use and expressing "true self" on the Internet. Computers in Human Behavior, 28, 1510–1517.
382. Reich, S. M., Subrahmanyam, K., & Espinoza, G. (2012). Friending, IMing, and hanging out face-to-face: Overlap in adolescents' offline and online social networks. Developmental Psychology, 48, 356–368.
383. Al-Menayes, J. J. (2015). Motivations for using social media: An exploratory factor analysis. International Journal of Psychological Sciences, 7(1), 43–50.
384. Hunt, D., Atkin, D., & Krishnan, A. (2012). The influence of computer-mediated communication apprehension on Motives for Facebook use. Journal of Broadcasting & Electronic Media, 56(2), 187–202.
385. *Frison, E., & Eggermont, S. (2016). Exploring the relationships between different types of Facebook use, perceived online social support, and adolescents' depressed mood. Social Science Computer Review, 34(2), 153–171.
386. *Lee, S. Y. (2014). How do people compare themselves with others on social network sites?: The case of Facebook. Computers in Human Behavior, 32, 253–260.
387. Nesi, J., & Prinstein, M. J. (2015). Using social media for social comparison and feedback-seeking: Gender and popularity moderate associations with depressive symptoms. Journal of Abnormal Child Psychology, 43, 1427–1438.
388. Galambos, N. L., Leadbeater, B. J., & Barker, E. T. (2004). Gender differences in and risk factors for depression in adolescence: A 4-year longitudinal study. International Journal of Behavior Development, 28, 16–25.

389. *Davila, J., Hershenberg, R., Feinstein, B. A., Gorman, K., Bhatia, V., & Starr, L. R. (2012). Frequency and quality of social networking among young adults: Associations with depressive symptoms, rumination, and corumination. Psychology of Popular Media Culture, 1(2), 72–86.
390. *Feinstein, B. A., Hershenberg, R., Bhatia, V., Latack, J. A., Meuwly, N., & Davila, J. (2013). Negative social comparison on Facebook and depressive symptoms: Rumination as a mechanism. Psychology of Popular Media Culture, 2, 161–170.
391. *Tsai, C.-W., Shen, P.-D., & Chiang, Y.-C. (2015). Meeting ex-partners on Facebook: Users' anxiety and severity of depression. Behaviour & Information Technology, 34(7), 668–677.
392. Best, P., Manktelow, R., & Taylor, B. (2014). Online communication, social media and adolescent wellbeing: A systematic narrative review. Children and Youth Services Review, 41, 27–36.
393. *Bányai, F., Zsila, A., Klraly, O., Maraz, A., Elekes, Z., Griffiths, M. A., Andreassen, C. S., & Demetrovics, Z. (2017). Problematic social media use: Results from a large-scale nationally representative adolescent sample. PLoS One, 12(1), e0169839.
394. Baker, D. A., & Algorta, G. P. (2016). The relationship between online social networking and depression: A systematic review of quantitative studies. Cyberpsychology, Behavior and Social Networking, 19, 638–648.
395. Huang, C. (2017). Time spent on social network sites and psychological well-being: A meta-analysis. Cyberpsychology, Behavior and Social Networking, 20(6), 346–354.
396. Kim, J., & Lee, J.-E. R. (2011). The Facebook paths to happiness: Effects of the number of Facebook friends and self-presentation on subjective well-being. Cyberpsychology, Behavior and Social Networking, 14, 359–364.
397. Radovic, A., Gmelin, T., Stein, B. D., & Miller, E. (2017). Depressed adolescents' positive and negative use of social media. Journal of Adolescence, 55, 5–15.
398. *Woods, H. C., & Scott, H. (2016). #Sleepyteens: Social media use in adolescence is associated with poor sleep quality, anxiety, depression and low self-esteem. Journal of Adolescence, 51, 41–49.
399. *Landoll, R. R., La Greca, A. M., & Lai, B. S. (2013). Aversive peer experiences on social networking sites: Development of the Social Networking-Peer Experiences Questionnaire (SN-PEQ). Journal of Research on Adolescence, 23(4), 695–705.
400. *Park, S., Lee, S. W., Kwak, J., Cha, M., & Jeong, B. (2013). Activities on Facebook reveal the depressive state of users. Journal of Medical Internet Research, 15(10), e217.
401. *Banjanin, N., Banjanin, N., Dimitrijevic, I., & Pantic, I. (2015). Relationship between internet use and depression: Focus on physiological mood oscillations, social networking and online addictive behavior. Computers in Human Behavior, 43, 308–312.
402. Kircaburun, K., Alhabash, S., Tosuntaş, ŞB., & Griffiths, M. D. (2019). Uses and gratifications of problematic social media use among university students: A simultaneous examination of the Big Five of personality traits, social media platforms, and social media use motives. International Journal of Mental Health and Addiction, 17, 336–345. https://doi.org/10.1007/s11469-018-9940-6
403. Chou, H. -T. G., & Edge, N. (2012). "They are happier and having better lives than I am": The impact of using Facebook on perceptions of others' lives. Cyberpsychology, Behavior, and Social Networking, 15, 117–121. https://doi.org/10.1089/cyber.2011.0324
404. Rahman, M., & Ahmed, O. (2018). Facebook use, Facebook addiction, and mental health of Chittagong university students. Bulgarian Journal of Science and Education Policy, 12(2), 345–358.
405. Valkenburg, P. M., Peter, J., & Schouten, A. P. (2006). Friend networking sites and their relationship to adolescents' well-being and social self-esteem. Cyberpsychology & Behavior, 9, 584–590. https://doi.org/10.1089/cpb.2006.9.584.
406. Choi, T. R., Sung, Y., Lee, J. -A., & Choi, S. M. (2017). Get behind my selfies: The Big Five traits and social networking behaviors through selfies. Personality and Individual Differences, 109, 98–101. https://doi.org/10.1016/j.paid.2016.12.057

407. Pallant, J. (2016). SPSS survival manual: A step by step guide to data analysis using IBM SPSS. McGraw Hill Education.
408. Hayes, A. F. (2013). Introduction to mediation, moderation, and conditional process analysis: A regression-based approach. New York: Guilford Press.
409. Kim, J. W., & Chock, T. M. (2017). Personality traits and psychological motivations predicting selfie posting behaviors on social networking sites. Telematics and Informatics, 34(5), 560–571.
410. McCain, J. L., Borg, Z. G., Rothenberg, A. H., Churillo, K. M., Weiler, P., & Campbell, W. K. (2016). Personality and selfies: Narcissism and the dark triad. Computers in Human Behavior, 64, 126–133.
411. Qiu, L., Lu, J., Yang, S., Qu, W., & Zhu, T. (2015). What does your selfie say about you? Computers in Human Behavior, 52, 443–449.
412. Sorokowska, A., Oleszkiewicz, A., Frackowiak, T., Pisanski, K., Chmiel, A., & Sorokowski, P. (2016). Selfies and personality: Who posts self-portrait photographs? Personality and Individual Differences, 90, 119–123.
413. Barry, C. T., Doucette, H., Loflin, D. C., Rivera-Hudson, N., & Herrington, L. L. (2017). "Let me take a selfie": Associations between self-photography, narcissism, and self-esteem. Psychology of Popular Media Culture, 6(1), 48–60.
414. Fox, J., & Rooney, M. C. (2015). The dark triad and trait self-objectification as predictors of men's use and self-presentation behaviors on social networking sites. Personality and Individual Differences, 76, 161–165.
415. Kim, E., Lee, J.-A., Sung, Y., & Choi, S. M. (2016). Predicting selfie-posting behavior on social networking sites: An extension of theory of planned behavior. Computers in Human Behavior, 62, 116–123.
416. Lee, J. A., & Sung, Y. (2016). Hide-and-seek: Narcissism and "selfie"-related behavior. Cyberpsychology, Behavior and Social Networking, 19(5), 347–351.
417. Sorokowski, Sorokowska, A., Oleszkiewicz, A., Frackowiak, T., Huk, A., & Pisanski, K. (2015). Selfie posting behaviors are associated with narcissism among men. Personality and Individual Differences, 85, 123–127.
418. Sung, Y., Lee, J.-A., Kim, E., & Choi, S. M. (2016). Why we post selfies: Understanding motivations for posting pictures of oneself. Personality and Individual Differences, 97, 260–265.
419. Weiser, E. B. (2015). #Me: Narcissism and its facets as predictors of selfie-posting frequency. Personality and Individual Differences, 86, 477–481.
420. Chua, T. H. H., & Chang, L. (2016). Follow me and like my beautiful selfies: Singapore teenage girls' engagement in self-presentation and peer comparison on social media. Computers in Human Behavior, 55, 190–197.
421. Diefenbach, S., & Christoforakos, L. (2017). The selfie paradox: Nobody seems to like them yet everyone has reasons to take them. An exploration of psychological functions of selfies in self-presentation. Frontiers in Psychology, 8, 7.
422. Fox, J., & Vendemia, M. A. (2016). Selective self-presentation and social comparison through photographs on social networking sites. Cyberpsychology, Behavior and Social Networking, 19(10), 593–600.
423. Morf, C. C., & Rhodewalt, F. (2001a). Expanding the dynamic self-regulatory processing model of narcissism: Research directions for the future. Psychological Inquiry, 12(4), 243–251.
424. Morf, C. C., & Rhodewalt, F. (2001b). Unraveling the paradoxes of narcissism: A dynamic self-regulatory processing model. Psychological Inquiry, 12(4), 177–196.
425. Wright, P. J., Tokunaga, R. S., Kraus, A., & Klann, E. (2017). Pornography consumption and satisfaction: A meta-analysis. Human Communication Research, 43(3), 315–343.
426. Bleske-Rechek, A., Remiker, M. W., & Baker, J. P. (2008). Narcissistic men and women think they are so hot--but they are not. Personality and Individual Differences, 45(5), 420–424.

427. Chan, L. S., & Tsang, H. W. E. (2014). "Hey, look at my body!": An exploratory study of body display on Facebook among Hong Kong young adults. International Journal of Interactive Communication Systems & Technologies, 4(1), 31–46.
428. Lipowska, M., & Lipowski, M. (2015). Narcissism as a moderator of satisfaction with body image in young women with extreme underweight and obesity. PLoS One, 10(5), e0126724.
429. Ridgway, J. L., & Clayton, R. B. (2016). Instagram unfiltered: Exploring associations of body image satisfaction, Instagram# selfie posting, and negative romantic relationship outcomes. Cyberpsychology, Behavior and Social Networking, 19(1), 2–7.
430. Cash, T. F. (2000). Multidimensional Body–Self Relations Questionnaire (MBSRQ). Washington, D.C: American Psychological Association and Oxford University Press.
431. Allen, M. S., & Walter, E. E. (2016). Personality and body image: A systematic review. Body Image, 19, 79–88.
432. Davis, C., Dionne, M., & Shuster, B. (2001). Physical and psychological correlates of appearance orientation. Personality and Individual Differences, 30(1), 21–30.
433. Fortes, L. D. S., Neves, C. M., Almeida, S. S., Ferreira, M. E. C., Fortes, L. D. S., Neves, C. M., … Ferreira, M. E. C. (2015). Disordered eating, body dissatisfaction, perfectionism, and mood state in female adolescents. Revista de Nutrição, 28(4), 371–383.
434. Andreassen, S. C., Billieux, J., Griffiths, M. D., Kuss, D. J., Demetrovics, Z., Mazzoni, E., & Pallesen, S. (2016). The relationship between addictive use of social media and video games and symptoms of psychiatric disorders: a large-scale cross-sectional study. Psychology of Addictive Behaviors, 30(2), 252–262.
435. van den Eijnden, R. J., Lemmens, J. S., & Valkenburg, P. M. (2016). The social media disorder scale. Computers in Human Behavior, 61, 478–487.
436. Andreassen, C. S., Pallesen, S., & Griffiths, M. D. (2017). The relationship between addictive use of social media, narcissism, and self-esteem: findings from a large national survey. Addictive Behaviors, 64, 287–293.
437. Carmody, C. L. (2012). Internet addiction: just Facebook me! The role of social networking sites in internet addiction. Computer Technology and Application, 3(3), 262–267.
438. Brand, M., Young, K. S., Laier, C., Wölfling, K., & Potenza, M. N. (2016). Integrating psychological and neurobiological considerations regarding the development and maintenance of specific Internet-use disorders: an Interaction of Person-Affect-Cognition-Execution (I-PACE) model. Neuroscience and Biobehavioral Reviews, 71, 252–266.
439. Katz, E., Blumler, J. G., & Gurevitch, M. (1973). Uses and gratifications research. The Public Opinion Quarterly, 37(4), 509–523.
440. Rubin, A. M. (1993). Audience activity and media use. Communications Monographs, 60(1), 98–105.
441. Robins, R. W., Hendin, H. M., & Trzesniewski, K. H. (2001). Measuring global self-esteem: construct validation of a single-item measure and the Rosenberg Self-Esteem Scale. Personality and Social Psychology Bulletin, 27(2), 151–161.
442. Ryan, T., Chester, A., Reece, J., & Xenos, S. (2014). The uses and abuses of Facebook: a review of Facebook addiction. Journal of Behavioral Addictions, 3, 133–148.
443. Eşkisu, M., Hoşoğlu, R., & Rasmussen, K. (2017). An investigation of the relationship between Facebook usage, Big Five, self-esteem and narcissism. Computers in Human Behavior, 69, 294–301.
444. Griffiths, M. D. (2010). The role of context in online gaming excess and addiction: some case study evidence. International Journal of Mental Health and Addiction, 8(1), 119–125.
445. Floros, G., & Siomos, K. (2013). The relationship between optimal parenting, Internet addiction and motives for social networking in adolescence. Psychiatry Research, 209, 529–534.
446. Huang, H. (2012). Social media addiction among adolescents in urban China: an examination of sociopsychological traits, uses and gratifications, academic performance, and social capital. (Doctoral thesis).
447. Yang, J., Lin, Z., Cohen, S.: Fast image super-resolution based on in-place example regression. In: Computer Vision and Pattern Recognition (CVPR), 2013 IEEE Conference on, pp. 1059–1066 (2013)

448. Wang, Z., Bovik, A.C., Sheikh, H.R., Simoncelli, E.P.: Image quality assessment: from error visibility to structural similarity. IEEE Trans. Image Process. 13(4), 600–612 (2004)
449. Rapee, R. M., & Heimberg, R. G. (1997). A cognitive-behavioral model of anxiety in social phobia. Behaviour Research and Therapy, 35(8), 741–756. https://doi.org/10.1016/S0005-7967(97)00022-3.
450. Tiggemann, M., & Slater, A. (2015). The role of self-objectification in the mental health of early adolescent girls: Predictors and consequences. Journal of Pediatric Psychology, 40(7), 704–711. https://doi.org/10.1093/jpepsy/jsv021.
451. Festinger, L. (1954). A theory of social comparison processes. Human Relations, 7(2), 117–140. https://doi.org/10.1177/001872675400700202.
452. Schaefer LM, Thompson JK. The development and validation of the physical appearance comparison scale-revised (PACS-R). Eat Behav. 2014;15(2):209–17.
453. Lindner, D., Tantleff-Dunn, S., & Jentsch, F. (2012). Social comparison and the 'circle of objectification'. Sex Roles, 67(3), 222–235. https://doi.org/10.1007/s11199-012-0175-x.
454. Tylka, T. L., & Sabik, N. J. (2010). Integrating social comparison theory and self-esteem within objectification theory to predict women's disordered eating. Sex Roles, 63(1–2), 18–31. https://doi.org/10.1007/s11199-010-9785-3.
455. Yang, J., Fardouly, J., Wang, Y., & Shi, W. (2020). Selfie-viewing and facial dissatisfaction among emerging adults: A moderated mediation model of appearance comparisons and self-objectification. International Journal of Environmental Research and Public Health, 17(2), 672. https://doi.org/10.3390/ijerph17020672.
456. von Soest, T., Wichstrøm, L., & Kvalem, I. L. (2016). The development of global and domain-specific self-esteem from age 13 to 31. Journal of Personality and Social Psychology, 110(4), 592–608. https://doi.org/10.1037/pspp0000060.
457. Zuckerman, M., Li, C., & Hall, J. A. (2016). When men and women differ in self-esteem and when they don't: A meta-analysis. Journal of Research in Personality, 64, 34–51. https://doi.org/10.1016/j.jrp.2016.07.007.
458. Argyrides, M., & Sivitanides, M. (2017). Body image, self-esteem, media, disordered eating and actual ideal weight discrepancy: Findings in Cyprus. The European Journal of Counselling Psychology, 6(1), 12. https://doi.org/10.5964/ejcop.v5i2.109.
459. Masselink, M., Van Roekel, E., & Oldehinkel, A. J. (2018). Self-esteem in early adolescence as predictor of depressive symptoms in late adolescence and early adulthood: The mediating role of motivational and social factors. Journal of Youth and Adolescence, 47(5), 932–946. https://doi.org/10.1007/s10964-017-0727-z.
460. Pan, Z., Zhang, D., Hu, T., & Pan, Y. (2018). The relationship between psychological Suzhi and social anxiety among Chinese adolescents: The mediating role of self-esteem and sense of security. Child and Adolescent Psychiatry and Mental Health, 12(1), 50. https://doi.org/10.1186/s13034-018-0255-y.
461. Shroff, H., & Thompson, J. K. (2006). Peer influences, body-image dissatisfaction, eating dysfunction and self-esteem in adolescent girls. Journal of Health Psychology, 11(4), 533–551. https://doi.org/10.1177/1359105306065015
462. Harrison, K., & Fredrickson, B. L. (2003). Women's sports media, self-objectification, and mental health in black and white adolescent females. Journal of Communication, 53(2), 216–232. https://doi.org/10.1111/j.1460-2466.2003.tb02587.x.
463. Nelemans, S. A., Hale, W. W., Branje, S. J. T., Raaijmakers, Q. A. W., Frijns, T., van Lier, P. A. C., & Meeus, W. H. J. (2014). Heterogeneity in development of adolescent anxiety disorder symptoms in an 8-year longitudinal community study. Development and Psychopathology, 26(1), 181–202. https://doi.org/10.1017/S0954579413000503.
464. Sumter, S. R., Bokhorst, C. L., & Westenberg, P. M. (2009). Social fears during adolescence: Is there an increase in distress and avoidance? Journal of Anxiety Disorders, 23(7), 897–903. https://doi.org/10.1016/j.janxdis.2009.05.004.
465. Wu, M., & Lang, M. (2019). Female questionnaire of trait self-objectification: Initial development and validation in China. Sex Roles, 80(11), 758–769. https://doi.org/10.1007/s11199-018-0972-y.

466. Rosenberg, M. (1965). Society and the adolescent self-image. Princeton University Press.
467. Wang, P., Gao, H., Xu, J. Y., Huang, J. J., & Wang, C. J. (1998). 自尊量表信效度研究 [the Rosenberg self-esteem scale: The development and validation in China]. Shangdong Archives of Psychiatry. 11 (4): 22, 31–32.
468. Peng, C., Fan, X., & Li, L. (2003). 社交回避与苦恼量表在学生群体中的信效度研究 [The validity and reliability of social avoidance and distress scale in Chinese students]. Chinese Journal of Clinical Psychology, 11(4), 279–281.
469. Carter, R., Jaccard, J., Silverman, W. K., & Pina, A. A. (2009). Pubertal timing and its link to behavioral and emotional problems among 'at-risk' African American adolescent girls. Journal of Adolescence, 32(3), 467–481. https://doi.org/10.1016/j.adolescence.2008.07.005.
470. Ding, X., Coplan, R. J., Deng, X., Ooi, L. L., Li, D., & Sang, B. (2019). Sad, scared, or rejected? A short-term longitudinal study of the predictors of social avoidance in Chinese children. Journal of Abnormal Child Psychology, 47(7), 1265–1276. https://doi.org/10.1007/s10802-018-0476-9.
471. Sang, B., Ding, X., Coplan, R. J., Liu, J., Pan, T., & Feng, X. (2016). Assessment and implications of social avoidance in Chinese early adolescents. The Journal of Early Adolescence, 38(4), 554–573. https://doi.org/10.1177/0272431616678988.
472. Starr CR, Ferguson GM (2012) Sexy dolls, sexy grade-schoolers? Media & maternal influences on young girls' self-sexualization. Sex Roles 67:463–476. https://doi.org/10.1007/s11199-012-0183-x
473. Katz-Wise SL, Budge SL, Lindberg SM, Hyde JS (2013) Individuation or identification? Self-objectification and the mother–adolescent relationship. Psychol Women Q 37:366–380. https://doi.org/10.1177/0361684312468425
474. Arroyo, A., & Andersen, K. K. (2016). The relationship between mother-daughter self-objectification: Identifying direct, indirect, and conditional direct effects. Sex Roles, 74(5), 231–241. https://doi.org/10.1007/s11199-015-0554-1.
475. Chow C, Tan C (2017) The role of fat talk in eating pathology and depressive symptoms among mother–daughter dyads. Body image 24:36–43. https://doi.org/10.1016/j.bodyim.2017.11.003
476. Rudiger JA, Winstead BA (2013) Body talk and body-related co-rumination: associations with body image, eating attitudes, and psychological adjustment. Body Image 10:462–471. https://doi.org/10.1016/j.bodyim.2013.07.010
477. Bussey K, Bandura A (1999) Social cognitive theory of gender development and differentiation. Psychol Rev 106:676
478. Hart E, Chow CM, Tan CC (2017) Body talk, weight status, and pathological eating behavior in romantic relationships. Appetite 117:135–142. https://doi.org/10.1016/j.appet.2017.06.012
479. Murray T, Lewis V (2014) Gender-role conflict and men's body satisfaction: the moderating role of age. Psychol Men Mascul 15:40. https://doi.org/10.1037/a0030959
480. Loveys, Kate; Sagar, Mark; Broadbent, Elizabeth The Effect of Multimodal Emotional Expression on Responses to a Digital Human during a Self-Disclosure Conversation: a Computational Analysis of User Language. Journal of Medical Systems (2020). https://doi.org/10.1007/s10916-020-01624-4
481. Capecchi, Saveria Personal relationships in the digital age: three female academics' qualitative research. Quality & Quantity (2017). https://doi.org/10.1007/s11135-017-0544-1
482. Donnelly, Noeleen; Stapleton, Larry; O'Mahoney, Jennifer Born digital or fossilised digitally? How born digital data systems continue the legacy of social violence towards LGBTQI + communities: a case study of experiences in the Republic of Ireland. AI & SOCIETY (2022). https://doi.org/10.1007/s00146-021-01374-y
483. Smutradontri, Pitchapa; Gadavanij, Savitri Fandom and identity construction: an analysis of Thai fans' engagement with Twitter. Humanities and Social Sciences Communications (2020). https://doi.org/10.1057/s41599-020-00653-1
484. Galvão, Vinícius Ferreira; Maciel, Cristiano; Pereira, Roberto; Gasparini, Isabela; Viterbo, José; Bicharra Garcia, Ana Cristina Discussing human values in digital immortality: towards

a value-oriented perspective. Journal of the Brazilian Computer Society (2021). https://doi.org/10.1186/s13173-021-00121-x
485. Arner, Douglas W.; Zetzsche, Dirk A.; Buckley, Ross P.; Barberis, Janos N. The Identity Challenge in Finance: From Analogue Identity to Digitized Identification to Digital KYC Utilities. European Business Organization Law Review (2019). https://doi.org/10.1007/s40804-019-00135-1
486. Giannopoulou, Alexandra Digital Identity Infrastructures: a Critical Approach of Self-Sovereign Identity. Digital Society (2023). https://doi.org/10.1007/s44206-023-00049-z
487. Korać, Dragan; Damjanović, Boris; Simić, Dejan A model of digital identity for better information security in e-learning systems. The Journal of Supercomputing (2021). https://doi.org/10.1007/s11227-021-03981-4
488. Bischoff, Jared; Berezan, Orie; Scardicchio, Lorenzo The digital self and customer loyalty: from theory to virtual reality. Journal of Marketing Analytics (2019). https://doi.org/10.1057/s41270-019-00065-4
489. Barnard-Wills, D.; Ashenden, D. Public sector engagement with online identity management. Identity in the Information Society (2010). https://doi.org/10.1007/s12394-010-0079-2
490. Block, Jörn H.; Fisch, Christian; Diegel, Walter Schumpeterian entrepreneurial digital identity and funding from venture capital firms. The Journal of Technology Transfer (2022). https://doi.org/10.1007/s10961-022-09973-7
491. Tammpuu, Piia; Masso, Anu Transnational Digital Identity as an Instrument for Global Digital Citizenship: The Case of Estonia's E-Residency. Information Systems Frontiers (2019). https://doi.org/10.1007/s10796-019-09908-y
492. O'Byrne, W. Ian; Hunter-Doniger, Tracey Pre-Service Educators Developing a Digital Identity. TechTrends (2021). https://doi.org/10.1007/s11528-021-00617-5
493. Filice, Eric; Raffoul, Amanda; Meyer, Samantha B.; Neiterman, Elena The Impact of Social Media on Body Image Perceptions and Bodily Practices among Gay, Bisexual, and Other Men Who Have Sex with Men: A Critical Review of the Literature and Extension of Theory. Sex Roles (2019). https://doi.org/10.1007/s11199-019-01063-7
494. Tremblay, Simon C.; Essafi Tremblay, Safae; Poirier, Pierre From filters to fillers: an active inference approach to body image distortion in the selfie era. AI & SOCIETY (2020). https://doi.org/10.1007/s00146-020-01015-w
495. Coy-Dibley, Isabelle "Digitized Dysmorphia" of the female body: the re/disfigurement of the image. Humanities and Social Sciences Communications (2016). https://doi.org/10.1057/palcomms.2016.40
496. Choukas-Bradley, Sophia; Roberts, Savannah R.; Maheux, Anne J.; Nesi, Jacqueline The Perfect Storm: A Developmental–Sociocultural Framework for the Role of Social Media in Adolescent Girls' Body Image Concerns and Mental Health. Clinical Child and Family Psychology Review (2022). https://doi.org/10.1007/s10567-022-00404-5
497. Papageorgiou, Alana; Fisher, Colleen; Cross, Donna "Why don't I look like her?" How adolescent girls view social media and its connection to body image. BMC Women's Health (2022). https://doi.org/10.1186/s12905-022-01845-4
498. Papageorgiou, Alana; Fisher, Colleen; Crossa, Donna "It just Sends the Message that you're Nothing but your body" a Qualitative Exploration of Adolescent Girls' Perceptions of Sexualized Images on Social Media. Sexuality & Culture (2022). https://doi.org/10.1007/s12119-022-10022-6
499. Butkowski, Chelsea P.; Dixon, Travis L.; Weeks, Kristopher Body Surveillance on Instagram: Examining the Role of Selfie Feedback Investment in Young Adult Women's Body Image Concerns. Sex Roles (2019). https://doi.org/10.1007/s11199-018-0993-6
500. Klier, Kristina; Rommerskirchen, Tessa; Brixius, Klara #fitspiration: a comparison of the sport-related social media usage and its impact on body image in young adults. BMC Psychology (2022). https://doi.org/10.1186/s40359-022-01027-9
501. Di Gesto, Cristian; Matera, Camilla; Policardo, Giulia Rosa; Nerini, Amanda Instagram As A Digital Mirror: The Effects of Instagram Likes and Disclaimer Labels on Self-awareness,

Body Dissatisfaction, and Social Physique Anxiety Among Young Italian Women. Current Psychology (2022). https://doi.org/10.1007/s12144-021-02675-7

502. Fioravanti, Giulia; Bocci Benucci, Sara; Vinciarelli, Viola; Casale, Silvia Body shame and problematic social networking sites use: the mediating effect of perfectionistic self-presentation style and body image control in photos. Current Psychology (2023). https://doi.org/10.1007/s12144-023-04644-8
503. Gioia, Francesca; Griffiths, Mark D.; Boursier, Valentina Adolescents' Body Shame and Social Networking Sites: The Mediating Effect of Body Image Control in Photos. Sex Roles (2020). https://doi.org/10.1007/s11199-020-01142-0
504. Te'eni-Harari, Tali; Eyal, Keren The Psychology of Social Networking: the Challenges of Social Networking for Fame-Valuing Teens' Body Image. Philosophia (2017). https://doi.org/10.1007/s11406-016-9806-8
505. Yao, Liangshuang; Niu, Gengfeng; Sun, Xiaojun Body Image Comparisons on Social Networking Sites and Chinese Female College Students' Restrained Eating: The Roles of Body Shame, Body Appreciation, and Body Mass Index. Sex Roles (2020). https://doi.org/10.1007/s11199-020-01179-1
506. Brubaker, Rogers Digital hyperconnectivity and the self. Theory and Society (2020). https://doi.org/10.1007/s11186-020-09405-1
507. Freeman, Guo; Wohn, Donghee Yvette Streaming your Identity: Navigating the Presentation of Gender and Sexuality through Live Streaming. Computer Supported Cooperative Work (CSCW) (2020). https://doi.org/10.1007/s10606-020-09386-w
508. Monks, Helen; Costello, Leesa; Dare, Julie; Reid Boyd, Elizabeth "We're Continually Comparing Ourselves to Something": Navigating Body Image, Media, and Social Media Ideals at the Nexus of Appearance, Health, and Wellness. Sex Roles (2020). https://doi.org/10.1007/s11199-020-01162-w
509. Rieken, Johannes; Garcia-Sanchez, Efraín; Trujillo, Mónica Pérez; Bear, Daniel Digital Ethnography and the Social Dimension of Introspection: An Empirical Study in Two Colombian Schools. Integrative Psychological and Behavioral Science (2015). https://doi.org/10.1007/s12124-015-9299-1
510. Degen, Johanna Lisa; Kleeberg-Niepage, Andrea Profiling the Self in Mobile Online Dating Apps: a Serial Picture Analysis. Human Arenas (2021). https://doi.org/10.1007/s42087-021-00195-1
511. Ozimek, Phillip; Lainas, Semina; Bierhoff, Hans-Werner; Rohmann, Elke How photo editing in social media shapes self-perceived attractiveness and self-esteem via self-objectification and physical appearance comparisons. BMC Psychology (2023). https://doi.org/10.1186/s40359-023-01143-0
512. Gioia, Francesca; McLean, Siân; Griffiths, Mark D.; Boursier, Valentina Adolescents' selfie-taking and selfie-editing: A revision of the photo manipulation scale and a moderated mediation model. Current Psychology (2021). https://doi.org/10.1007/s12144-021-01702-x
513. Rogova, Natalia; Matta, Shashi The role of identity in digital consumer behavior: A conceptual model and research propositions based on gender. AMS Review (2022). https://doi.org/10.1007/s13162-022-00237-z
514. Bartoli, Chiara Consumer self-concept and digitalization: what does this mean for brands?. Italian Journal of Marketing (2022). https://doi.org/10.1007/s43039-022-00059-8
515. Sebre, Sandra B.; Miltuze, Anika Digital Media as a Medium for Adolescent Identity Development. Technology, Knowledge and Learning (2021). https://doi.org/10.1007/s10758-021-09499-1
516. Hargittai, Eszter; Piper, Anne Marie; Morris, Meredith Ringel From internet access to internet skills: digital inequality among older adults. Universal Access in the Information Society (2018). https://doi.org/10.1007/s10209-018-0617-5
517. Sasaki, Wataru; Nishiyama, Yuuki; Okoshi, Tadashi; Nakazawa, Jin Investigating the occurrence of selfie-based emotional contagion over social network. Social Network Analysis and Mining (2021). https://doi.org/10.1007/s13278-020-00712-0

518. Wang, Wei; Qian, Guoying; Wang, Xingchao; Lei, Li; Hu, Qian; Chen, Jing; Jiang, Suo Mobile social media use and self-identity among Chinese adolescents: The mediating effect of friendship quality and the moderating role of gender. Current Psychology (2019). https://doi.org/10.1007/s12144-019-00397-5
519. Vahedi, Zahra; Zannella, Lesley The association between self-reported depressive symptoms and the use of social networking sites (SNS): A meta-analysis. Current Psychology (2019). https://doi.org/10.1007/s12144-019-0150-6
520. Ahmed, Oli; Sultana, Tasnim; Alam, Najifa; Griffiths, Mark D.; Hiramoni, Fatema Akhter Problematic Social Media Use, Personality Traits, and Mental Health Among Bangladeshi University Students. Journal of Technology in Behavioral Science (2022). https://doi.org/10.1007/s41347-021-00235-1
521. Wang, Yuhui; Xie, Xiaochun; Wang, Xingchao; Wang, Pengcheng; Nie, Jia; Lei, Li Narcissism and selfie-posting behavior: the mediating role of body satisfaction and the moderating role of attitude toward selfie-posting behavior. Current Psychology (2018). https://doi.org/10.1007/s12144-018-9795-9
522. Kircaburun, Kagan; Demetrovics, Zsolt; Tosuntaş, Şule Betül Analyzing the Links Between Problematic Social Media Use, Dark Triad Traits, and Self-esteem. International Journal of Mental Health and Addiction (2018). https://doi.org/10.1007/s11469-018-9900-1
523. Hans, W. Jino; Venkateswaran, N. Selfie image super-resolution using an implicit prior learned from self-examples. Cluster Computing (2018). https://doi.org/10.1007/s10586-018-2466-7
524. Guo, Qinglong; Wu, Mingxia The relationship between self-objectification and social avoidance among Chinese middle adolescent girls: The mediating role of appearance comparison and self-esteem. Current Psychology (2021). https://doi.org/10.1007/s12144-021-01705-8
525. Domoff, Sarah E.; Tan, Cin Cin; Chow, Chong Man Mother–daughter negative body talk as a moderator between body surveillance and body shame in adolescent girls. Eating and Weight Disorders - Studies on Anorexia, Bulimia and Obesity (2020). https://doi.org/10.1007/s40519-020-00925-4

Feminism and Digital Spaces

6

Introduction

Like any technology, digital technologies also have both positive and negative gender impacts determined by the intersectionalities of women. The digital divide and inequalities impact a large share of women and other non-male genders across societies mainly because their social positions are often subordinate to the males. The algorithms that are developed for various programmes and schemes also bring forth and highlight women and other non-male gender identities having consequences for the inclusion of such identities. Women face more body-shaming and the language they use also comes into sharp inquiry in social media. Therefore, even on the digital social platforms women remain cautious in their expressions and their posting of text or images is often done after enormous back-and-forth decisions by women themselves. Additionally, cyberspace of the digital sex industry has developed which threatens even to drag the women non-willing to participate in the same by morphing their images by certain actors. The social consequence of morphing often results in shaming, abandoning, and social violence.

Notwithstanding some adverse use of social media and digital platforms against women and the gendered digital divides, digital technologies are also proving liberating force for women and other genders. They are now able to form their own alliances of like minded to shame the patriarchal and unjust behaviours (one such group was MeToo), participate in the labour market through digital technologies, attain skills and education from digital platforms, receive digital medicines, express the opinions and participate in governance processes, etc.

The digital revolution has created the possibility of 'cyberfeminism', whereby women can fight for their rights and protest online despite being in different locations, as their geographical mobility has often been culturally compromised and

A. Shaban (ed.), *Digital Geographies—Theory, Space, and Communities*,
https://doi.org/10.1007/978-981-97-4734-4_6

need permission from men to travel. By writing from scattered locations, they have been able to generate enough force to impact civil society groups and policymakers. For instance, the 'Me Too' movement started in the USA, got support from women from different parts of the world and it became a worldwide movement. This movement had several consequences for the women's liberation. First, it propagated their rights of harassment-free and exploitation-free existence. Second, it impacted the policy and law enforcement agencies to devise appropriate regulations to protect the rights of women. Third, it also made women conscious to report the cases which earlier used to be hidden. This was all done through Twitter (now X), and other social media platforms. Further, in India, the 'Nirbhaya' movement, which emerged from the infamous rape and murder of a young women in a bus in Delhi, was largely created through the digital platforms. This led to long protests by civil society groups and feminist groups for appropriate laws and regulations. This also resulted in a big change in the Government in Delhi, as people voted for an alternative political party which they thought could better safeguard their rights.

However, digitalization has also opened the possibility of body shaming or disgracing women by morphing their images or through other methods. The online sex industry is a billion-dollar industry and is running because of the active compromise between the companies owning the digital systems, the entrepreneurs who want to make money by disgracing women, and the lack of active engagement of the law-enforcing agencies. In many countries, although several such websites have been banned, one also finds that messages and videos could be circulated from peer-to-peer.

A range of studies related to digitalization and women have been reviewed in this chapter. The chapter is divided into three sections. The first section deals with the review of the gendered self in the digital while section two reviews the papers on inequality and gender justice in the digital age. The last section presents a review of papers related to the feminism in digital age.

According to research titled 'My iPhone Changed My Life: How Digital Technologies Can Facilitate Women's Consumption of Online Sexually Explicit Materials', digital technologies continue to alter how women can access and consume pornographic content online. According to the paper, women can use digital technology to access pornography online in ways that will enable them to satisfy their sexual needs, embrace and explore their sexual selves, form romantic connections, and normalize their sexual wants.

Another paper, 'Gender Inclusivity of India's Digital Financial Revolution for Attainment of SDGs: Macro Achievements and the Micro Experiences of Targeted Initiatives', makes the case that women's efforts to combat poverty and gender inequality as well as work toward achieving the Sustainable Development Goals (SDGs) of the UN have been aided by digital financial inclusion. The study also shows that while India has made considerable strides in increasing digital financial inclusion, it has had difficulty achieving gender parity, particularly within programs specifically focused on money and intended to increase gender inclusivity.

According to a paper titled 'Piety as a Concept Underpinning Muslim Women's Online Discussions of Marriage and Professional Career', the Islamic influence sphere has grown, and the Internet has evolved into a forum for the exchange of unique understandings of Islam between believing women (and men) around the world, in both the geographical and social senses of the word. Another study, 'In a World Where You Can Be Anyone: An Investigation into the Gendered Social Practices of Pakistani Facebook Users', contends that Pakistan's religious, class, cultural, and traditional discourses interact in complex ways to shape, negotiate, and express gender identities on Facebook.

According to a study titled 'Meanings of Bodily and Sexual Expression in Youth Sexting Culture: Young Women's Negotiation of Gendered Risks and Harms', youth sexting is a gendered phenomenon in which young men can use it to gain social capital while young women are subject to harassment and social shaming.

In section three, a paper on 'Feminist data studies: using digital methods for ethical, reflexive and situated socio-cultural research' argues that the present data-logical turn emphasizes the digital datafication of daily life, expanding algorithmic processing, and emphasizing data as an emerging regime of power and knowledge. The study asks for moral, anti-oppressive digital data-driven social science and humanities research in this feminist and postcolonial intervention into gender, race, and geography- blind 'big data' ideology. It makes an argument for reflexive data research that can result from the combination of feminist and postcolonial science studies with care ethics values.

Another paper 'Perpetuating Gender Inequality via the Internet? An Analysis of Women's Presence in Spanish Online Newspapers' finds that the inequalities in gender reporting because female journalists frequently cover more women's news than their male counterparts. There is evidence that women are still underrepresented, stereotyped, and discriminated against in internet news, which reinforces gender inequality.

Machine generated summaries

Disclaimer: The summaries in this chapter were generated from Springer Nature publications using extractive AI auto-summarization: An extraction-based summarizer aims to identify the most important sentences of a text using an algorithm and uses those original sentences to create the auto-summary (unlike generative AI). As the constituted sentences are machine selected, they may not fully reflect the body of the work, so we strongly advise that the original content is read and cited. The auto generated summaries were curated by the editor to meet Springer Nature publication standards. To cite this content, please refer to the original papers.

Machine generated keywords: woman, feminist, gender, women, man, sexual, sexuality, inclusion, young, girl, power, practice, student, forum, job

6.1 Gendered Self in Digital Age

Machine generated keywords: woman, man, gender, women, sexual, girl, feminist, young, sexuality, culture, inclusion, stress, online, partner, job

6.1.1 "Becoming a Sugar Baby Will Change Your Life. Let's Talk About How": Sugar Dating Advice on Tumblr [200]

This is a machine-generated summary of:

Ellis, Lauren D.; Patterson, Callie L.; Walters, Andrew Schrack: "Becoming a Sugar Baby Will Change Your Life. Let's Talk About How": Sugar Dating Advice on Tumblr [200]

Published in: Sexuality & Culture (2022)

Link to original: https://doi.org/10.1007/s12119-022-10024-4

If you want to cite the papers, please refer to the original.

For technical reasons we could not place the page where the original quote is coming from.

Abstract-Summary

"Sugar dating crosses the boundaries of not only what is considered a traditional relationship."

"The current study focused on advice sharing within Tumblr communities of sugar babies, including sugar-dating specific language."

"Tumblr, a popular blogging website, houses a prolific subculture for sugar dating, wherein blogs provide information/advice to potential or novice sugar babies about how to navigate the sugar experience."

"These themes include sugar dating advice, relationship maintenance once a sugar relationship begins, and what to expect in a sugar dating relationship."

Introduction

"For some persons, sugar dating is understood as a liberating and agentic orchestration of sexuality and relational performance, while others consider the exchange-based relationship component of many sugar relationships as akin to sex work (Lam, [1]; Motyl, [2]; Nayar, [3])."

"Sugar babies recognize the importance of compatibility and connection in a sugar dating relationship and represent themselves as able to provide a connection above and beyond sexual exchanges only."

"Some assert that sugar dating lends sugar babies a vector of power within the relationship in the same ways that sexual agency may be enacted in some arenas of sex work (Cordero, [4]; Nayar, [3])."

"Despite arguments that sugar relationships are, in some ways, different than work that is more commonly recognized as sex work, sugar dating often is discussed in the context of prostitution (Green, [5]; Kuate-Defo, [6]; Lam, [1]; Miller, [7]; Motyl, [2]; Rakic, [8])."

Method

"Tumblr data are freely available to anyone who accesses the internet-based site and the proposed research was limited to posted content."

"The researchers identified broad posts discussing sugar culture by initially using the hashtags sugar baby and sugar dating."

"After recognizing that the most popular Tumblr posts were those focused on advice for sugar babies, or individuals looking to become sugar babies, the top posts were utilized to identify more specific hashtags focused on sugar dating advice."

"The hashtags used to identify posts about advice included hashtags such as sugar bowl, sugaring, sugar dating, sugar dating advice, sugaring tips, sugar dating tips, how to become a sugar baby, sugar help, sugar advice, and sb tips."

"Posts were then discarded if they did not include advice for sugar babies."

"Researchers began by separately reading each post pulled for analysis and individually making memos about prevalent themes within the posts."

Results

"Tumblr posts which included advice about autonomy within the sugar bowl were often characterized by personal narratives that described how the sugar baby authoring the post asserted autonomy in sugar dating situations, and how sugar babies can enact and maintain control over the arrangement with a sugar parent."

"One user suggested that online platforms would not lend success in finding an ideal sugar parent nor making adequate money and suggested that sugar babies who use dating sites should re-evaluate their reasoning."

"Strategies for relationship maintenance included suggestions for activities a sugar baby could do with a sugar parent to stimulate conversation and strengthen commitment: It's just a fact, in the beginning especially if you are new to sugaring, you're going to be searching for things to talk about with your sugar daddy."

"In a post describing the misconceptions about sugar dating, this user suggests that sugar babies should save their money and that they have the ability to look the part without overspending."

Discussion

"We have no way of knowing why some posts were more frequent (e.g., dating advice, insistence on forms of safer sex if the relationship with a sugar daddy involves sexual activity) than other topics."

"We do wonder – given how freely sugar babies talked about the value of transactional relationships – if the women who made these Tumblr posts see sugaring relationships, which are exchange-based, as still dissimilar from the exchanged-based commodities of what, largely, is referred to sex work."

"We do not know the range of relationships the sugar babies whose posts were included in this study have, the nature of those relationships, or their satisfaction with them."

"In what appears to be a forum for the exchange of ideas and the provision of advice, sugar babies offered unadulterated opinions on their choices for involvement in the sugar bowl and in dates or relationships that might evolve from interacting in sugar culture."

6.1.2 Using Focus Groups to Understand the Dynamics of Mothers and Daughters on Social Networking Sites [201]

This is a machine-generated summary of:

Santarossa, Sara; Woodruff, Sarah J.: Using Focus Groups to Understand the Dynamics of Mothers and Daughters on Social Networking Sites [201]

Published in: Journal of Child and Family Studies (2020)

Link to original: https://doi.org/10.1007/s10826-020-01700-w

If you want to cite the papers, please refer to the original.

For technical reasons we could not place the page where the original quote is coming from.

Abstract-Summary

"The mother/daughter relationship has been deemed critical for the positive development of self-esteem, body image, and eating behaviors for adolescent girls."

"As maternal modeling exists offline between mothers and adolescent daughters, a similar influence could be exercised by the fast evolution of digital culture, such as social networking sites (SNSs), thus making it imperative that online mother/daughter relationship be considered and investigated."

"Sixteen mothers of girls born in 2003–2007 (M_{age} of their daughters = 12.78, SD = 1.31) participated in one of four focus group discussions."

"Results suggest that further research is needed to aid in the development of tools that will help foster positive online mother/daughter relationships."

Methods Participants

"Based on study intent (e.g., data was not yoked between mother and daughter), and to avoid ethical concerns related to dyads (e.g., setting up undue influence on the daughter to participate), mothers and girls were not recruited together (i.e., as dyads)."

"Mother focus groups were grouped by the age of their daughter, similarly, among girls, focus groups were constructed based on age cohorts."

"Eight focus groups were conducted (four with mothers, ranging from 3–5 participants; four with girls, ranging from 4–9 participants), and all focus groups were stratified based on the age of the mother's daughter or the girls (i.e., 11–12 and 13–14 years)."

Results

"When it came to talking about filtering or editing photographs before posting on SNSs, some mothers discussed how their daughters and/or girls their daughter's age should not be using a filter if it is to "edit who they are," but it is ok if they are "using cool apps to look neat—you know like puppy dog ears, pig nose, etc " (Mother 5, daughters 11, 13, and 14 years)."

"The type of content a mother should want to display on their SNSs was discussed, as it could influence how their daughter would behave: "Because you know they have a child and like if they starting posting bad stuff it may run in the family, and you know they need to set a good example." (Girl 8, 12 years); "But, if a mom would do it [asking to be rated or graded online], I would be like, you're supposed to look up to them."

Discussion

"The mothers' discourse within the current study included the need to consider the long term consequences of the digital footprint they were creating on SNSs, specifically in terms of the quantity of SNS posts and inappropriate or self-objectifying content that may indirectly influence their daughters in the future."

"In the current study, offline conversation based on SNSs appeared to be content driven, particularly when online behaviors of the daughter were deemed to be risky by the mother."

"Our findings, around appearance-based feedback (e.g., compliments), concur with previous evidence concerning the associations between the ability of SNSs to create an environment for social comparison and contribute to the transmission of beauty ideals (Holland and Tiggemann [11]; Tiggemann and Slater [12]) and further emphasize the importance of exploring the mother/daughter relationship in the online world."

Conclusion

"Future research should continue to use a constructivist approach and interventions should be created to teach mothers that identities may be created through social interactions (verbal and otherwise), thus teaching their daughter how their bodies are perceived by others and the social significance of that perception."

"The results of the current study may also have important implications for body image and disordered eating prevention programs, suggesting that in addition to fostering positive body image in young girls, greater emphasis could be placed on discouraging negative modeling behaviors among mothers."

"Interventions encouraging mothers to model healthful and positive body esteem on their own SNSs and that discourage mothers from making appearance-based comments on their daughters SNS post, may be effective in preventing the onset of body dissatisfaction and eating disturbances during adolescence."

[Section 5]

"As parent modeling exists offline between mothers and adolescent daughters, a similar influence could be exercised in the online world, such as social networking sites (SNSs), thus making it imperative that online mother/daughter relationship be considered and investigated."

"Understanding the role of the mother/daughter relationship throughout thc lifespan of women, but especially for adolescents is critically important for if mothers serve as an important model for their daughters, then their daughters' behavior may be influenced by what they observe in their mothers."

"As SNSs are a primary media source and have created an online social environment it is important to investigate the influence of mothers on their daughters' psychosocial health in this online world."

"Gaining further understanding of the influence of mothers on daughters' SNSs beliefs (e.g., individuals' subjective estimates about whether a particular behavior will lead to particular consequences; Bandura [13, 14]), attitudes (e.g., determined by personal conceptions concerning a given object/behavior and thus creating a learned disposition to respond in a consistently favorable or unfavorable manner with respect to a given object/behavior; Fishbein and Ajzen [15]), subjective norms (e.g., the expectation of other significant persons' opinions and beliefs and the degree/social pressures to which an individual feels the motivation to comply; Fishbein and Ajzen [15]) and ultimately behaviors is important."

6.1.3 Piety as a Concept Underpinning Muslim Women's Online Discussions of Marriage and Professional Career [202]

This is a machine-generated summary of:

Piela, Anna: Piety as a concept underpinning Muslim women's online discussions of marriage and professional career [202]

Published in: Contemporary Islam (2011)

Link to original: https://doi.org/10.1007/s11562-011-0162-y

If you want to cite the papers, please refer to the original.

For technical reasons we could not place the page where the original quote is coming from.

Abstract-Summary
"I address piety as a concept shaping Muslim women's online discussions about gender roles, marriage and professional careers."

"I explore motivations and religious interpretations of three categories of participants in these discussions: egalitarians, for whom gender equality is a necessary component of piety (Barlas [16]); traditionalists, identified by other authors as Islamists (Karam [17]) or social conservatives (Gül and Gül 48:1–26, [18]; Mahmood [19]) and finally, holists, a group that cannot be mapped out on the political landscape by using the progressive–conservative binary (Badran, Agenda 50:41–57, [20]) and which exists and acts outside of it, neither subverting nor enacting norms of any dominant system, be it secular–liberal or patriarchal."

Introduction
"Piety is fostered online through interaction with Islamic knowledge and other Muslims; online groups such as discussed in this article are spaces where women can re-read religious texts and hear their own voices, as 'women's voices in traditional religions have been filtered through institutional restraints that have shaped, interpreted, translated, and even silenced them' (Griffin [21]: 195)."

"Bunt [22, 23] and Eickelman and Anderson [24] were amongst the first to identify and indicate possible implications of new channels of distribution of information; a growing body of Islamic knowledge online; new opportunities for interaction with religious authorities; and finally, interaction amongst Muslims in discussion groups."

"Bastani [25] noticed the use of discussion groups by Muslim women, but she focused on social interaction, and although she briefly mentioned religious discussions taking place there, she did not explore their importance for new interpretations of Islam."

Methodology
"The data, generated for the purpose of thematic analysis, were extracted from the archives of 12 women-only Islamic online discussion groups to which I was granted ethical access in 2006."

"It was my objective to identify women-only groups as I was interested in discussions of Islamic scriptures from women's perspective; I also wished to engage with women's priorities (as debated in their groups), bound to be different from those in mixed-gender or men-only groups."

"Online groups such as the ones I discuss here exemplify the appropriation of English by Muslims online who share religion but differ in cultural and linguistic terms."

"As I was interested in an active, dynamic negotiation of members' identities as Muslim women, I found the newsgroup discussions particularly useful in my investigation of this research question."

Marriage, sexuality and career: Muslim women's negotiations of a pious lifestyle
"Traditionalists offer a practical explanation again and suggest that the underlying reason for the requirement of women's sexual submission is biological: Sis i know that man is verry eager specially in bed, and yes thier libido is greater than women.. We women can be patient without man in our life but most man can't be patient because they are most greater in libido than women… (…) Because if our husband needs us and we dont come or obey them, then that man will find a way to get out his libido in his body (…) And we all know that Zinah (sexual intercourse to non of her wife or husband) is a great sin… Due to what is seen as men's greater libido and more powerful sexuality (Dunne [26]), traditionalist participants claim that it is a wise marital strategy to make oneself sexually available to the husband, so that his sexual desires are channelled towards the wife, not other women (which prevents him from committing zina, a sexual sin)."

Conclusions
"The sphere of Islamic influence has expanded, and the Internet has come to act as a platform for the exchange of personalised understandings of Islam between believing women (and men) globally, in both the geographic and social meaning of the word."

"or 'What are sexual rights of women in Islam?') based on reflecting on the principles of Islam intertwines with a pragmatic, down-to-earth way of understanding problematic issues."

"Regardless of each participant's point of view, discussants strive to use Islamic references in their contributions."

"Women debate issues that concern them and ask others for suggestions of the best course of action in particular situations; they also discuss theoretical problems and ways to address them from an Islamic perspective."

6.1.4 In a World Where You Can Be Anyone: An Investigation into the Gendered Social Practices of Pakistani Facebook Users [203]

This is a machine-generated summary of:

Salam, Rauha: In a World Where You can be Anyone: An Investigation into the Gendered Social Practices of Pakistani Facebook Users [203]

Published in: Gender Issues (2021)
Link to original: https://doi.org/10.1007/s12147-021-09289-0

If you want to cite the papers, please refer to the original.

For technical reasons we could not place the page where the original quote is coming from.

Abstract-Summary
"By employing thematic analysis and taking insights from theory of performativity and intersectionality, the findings of the interview data suggest that the formation, negotiation, and expression of gender identities on Facebook occurs through complex interplay between the discourses of religion, class, culture, and tradition."

Introduction
"Pakistani men's and women's Facebook experiences are also deeply connected to and influenced by the daily practices and social interactions of their everyday lives, rendering the role of social media in bringing about social change even more complicated and powerful (ibid)."

"By providing unfiltered access to information on issues like politics, gender and religion and enabling users to express their points of view, post potentially transgressive or contentious material and/or use these platforms to stir up discussion on various subjects, the role of social media in advancing societal transformation and empowering users in Pakistan has been much debated [27]."

"Examining Facebook users' practices in relation to the dominant gender ideology can illuminate the current social transformative potential of social media in Pakistan."

"I interpret Facebook users' stances on and subjective experiences of a patriarchal gender ideology in relation to the socio-cultural, political, religious and historical context of Pakistan."

Research Context: Gender Ideology in Pakistan

"Most of the problems plaguing Pakistan, such as the rise of the religious right, ethnic and sectarian violence, the politicisation of religion, and, in particular, the (re)instatement of misogynist and conservative attitudes towards women are rooted in Zia-ul-Haque's policies."

"By promoting this gender ideology, Zia-ul-Haque aimed to demonstrate that Pakistan was not only implementing a major "Islamic" reform but had also distanced itself from Western influences and discarded modernity."

"Zia-ul-Haque's gender policies were actively resisted by women's movements in Pakistan and by NGOs concerned with women rights [28]."

"The notion of feminism as a western-derived liberation discourse and its supporters as "English-educated, westernized, upper middle-class women with imported ideas, having no link to grassroot realities" [29]: p.22), is an example of Zia-ul-Haque's legacy that continues to dominate Pakistan's collective consciousness."

"This vigilantism, perceptions of women activists as western agents, and the public/private gender divide are evident in social media interactions."

Conceptualising Gender

"Since social media offers a space where in their profiles people can construct their gender identities, consciously or unconsciously, by choosing what they want to share,gender identities are thus located in particular 'repeated acts' [30, 31], such as the kind of content people repeatedly share on their Facebook profile pages."

"These performative acts constitute individuals' gender identities in their cyber social sphere [32]."

"Using performativity theory to explain the construction of gender identities helps me to locate and understand my participants' gender identities in their broader socio-cultural and religious online and offline contexts."

"This means that users' digital gender identities may be considered as an extension of their physical or real-world identities (this is particularly significant in cases where my participants' offline or physical lives were affected by how they portrayed themselves in their social media accounts,see, e.g., excerpt 4)."

Gender and Social Media Studies

"In Pakistan, however, research on gender has overwhelmingly focused on the construction and representation of womanhood and on women's issues in the context of religion, law, politics, development, economics, education, health (including reproductive rights), sexuality, and the media [33, 34]."

"Recent studies on women and social media in the Pakistani context have also looked at online feminist resistance [35], the socio-economic empowerment of women via social media [36], the impact of social media on family dynamics [37], women's buying behaviour on social media [38] and the linguistic and semiotic construction of gender identities on Facebook [39]."

"The present study is important because unlike the dominant readings of Pakistani femininities and masculinities from a religious or political perspective, it offers an exploration of Pakistani femininities and masculinities as situated and performed at the intersection of various socio-cultural forces in a social media context, in this instance Facebook."

"It endeavours to fill a major research gap by bringing a South Asian-Pakistani perspective into the global and local debate on how men and women in patriarchal societies such as Pakistan negotiate multiple socio-cultural and religious pressures in constructing their gender identities on Facebook."

Research Design

"[39, 40], I investigated the role of visual and linguistic choices in Pakistani men's and women's construction of their gender identities on Facebook."

"The results of the present interview study on Pakistani men's and women's thoughts and evaluations regarding their own Facebook activities complement these earlier discourse analytic findings."

"The aim was to establish how and for what purposes the participants used Facebook and how, in their view, the prevailing socio-cultural and religious norms of Pakistani society influenced their Facebook posts."

"My aim was to find participants with whom I could establish good rapport and who would be available not only for interview but also for possible clarification during the data analysis."

"The analysis focuses in particular on the content of the interviews, that is, on gender and the construction of gender identities, as discussed by the participants."

Analysis of the Interview Data

"Discussing gender issues on Facebook is not the right forum because for me these issues are a private matter and discussing such things so publicly on social media will only create a rift between men and women in our society."

"One female interviewee, who actively makes and shares sarcastic videos about gender stereotyping and women issues in Pakistan on Facebook, described her experience: I have particularly received abuse and even rape threats from men."

"As the female interviewee in excerpt 5, he initially refers to his critics as people, but immediately specifies that men in particular attack him for talking about gender issues on social media."

"He further points out that many men agree with him (My most rewarding experience doing these videos on Facebook, Instagram and YouTube (on gender issues) has been the number of men who agree privately with what I share), claiming that his videos have made a difference and that men acknowledge the plight of women in Pakistani society."

Conclusion

"Applying thematic analysis to interviews and drawing on Butler's [30] theory of performativity and intersectionality, I investigated the views of twelve young, well-educated Pakistani men and women on Facebook as a platform for the discussion of gender issues."

"My analysis also revealed the strongly gendered nature of socio-cultural discourse (see, e.g., Excerpts 1, 2 & 5) in which demands for women's rights and autonomy were reductively framed as a demand for sexual freedom."

"Further, the interviews foregrounded that the traditional gendering of space (see Excerpt 2), that is, the relegation of women to the private domain, means the privatisation, and thereby silencing, of the woman question (Habiba and others [41])."

"For a collectivist patriarchal system to function in Pakistan, it is paramount that men and women adhere strictly to the established gender models."

"Marginalised women (and men) have benefitted from Facebook and for men it has become a space for learning about gender inequality in Pakistan."

6.1.5 Meanings of Bodily and Sexual Expression in Youth Sexting Culture: Young Women's Negotiation of Gendered Risks and Harms [204]

This is a machine-generated summary of:

Setty, Emily: Meanings of Bodily and Sexual Expression in Youth Sexting Culture: Young Women's Negotiation of Gendered Risks and Harms [204]

Published in: Sex Roles (2018)

Link to original: https://doi.org/10.1007/s11199-018-0957-x

If you want to cite the papers, please refer to the original.

For technical reasons we could not place the page where the original quote is coming from.

Abstract-Summary

"The present paper explores how young people construct gendered social meanings and cultural norms surrounding sexual and bodily expression in youth sexting culture."

"Previous research suggests youth sexting is a gendered phenomenon in which young men are able to seek social capital through sexting, whereas young women are subject to social shaming and harassment."

"Young people attributed agency and legitimacy to young men's sexual practices, whereas young women were disempowered, denied legitimacy, and tasked with managing gendered risks of harm in youth sexting culture."

"The accounts of two young women, who shared experiences sexting and social shaming, are presented to show some of the ways young women make sense of social meanings and cultural norms on individual and interpersonal levels."

Perceptions of Young Men and Women Who Sext

"Young women's sexual and bodily self-expression can be met with slut-shaming and abuse, whereas young men are lauded for engaging in sexting and can obtain social capital through the unauthorised distribution of images of young women in their peer group (Albury and others [43]; Coy and others [44]; Harris and others [45]; Phippen [46]; Ringrose and others [47]; Walker [48])."

"Risk and harm narratives dominate among young people, in which gender inequalities are taken-for-granted and underpin the victim-blaming of young women who engage in sexting and have their images distributed without their consent (Bond [49]; Jonsson and others [50])."

"They argue that young women seem to have little "[legitimate entitlement] to digitally mediate sexuality or express sexual desire" (p. 11); instead, they are tasked with managing the risk of unauthorised distribution by young men."

"Salter [51] found that these narratives are reproduced by young adult women who sext and experience unauthorised distribution."

Unwanted and Pressured Sexting

"Pressure to sext, particularly on young women by young men, is prevalent according to previous research (Albury and others [43]; Coy and others [44]; Crofts and others [52]; Harris and others [45]; Lippman and Campbell [53]; McGraw [54]; Phippen [46]; Ringrose and others [47]; Walker [48])."

"The construction of sexting as an accomplishment for young men can lead to pressure on them to adhere to the demands of compulsory masculine sexuality by obtaining and distributing the images of young women (Bailey and Mouna [55]; Ringrose and others [47])."

"Unwanted sexting through unsolicited image-sharing similarly affects both young men and women, but previous research suggests the meaning and experience of this is gendered."

"In her study of young adults, Burkett [56] found that men also experience unsolicited image-sharing, but these men did not construct it as threatening or report experiencing self-doubt or offense as the women did."

"Rather than seeing themselves as victimised, these young men constructed themselves as adhering to a form of etiquette toward young women by not sharing their images (Bond [49])."

Pleasurable Youth Sexting

"Surveys and qualitative research with young people suggest they sext for a variety of reasons, including to build and maintain relationships, to demonstrate intimate connection with others and for self-expression, as well as due to boredom and a desire for risk-taking (Albury and others [43]; Cooper and others [57]; Lenhart [58]; Walker [48])."

"Notwithstanding, research suggests that young women hold more negative perceptions of sexting than young men do, although these perceptions may relate as much to meanings surrounding the legitimacy of female sexual expression and the risks of such expression as to experiences of harm (Crofts and others [52]; Klettke and others [59]; Strassberg and others [60])."

"Burkett [56] found that young adult women, compared to young adult men, spoke less about sexting being fun, pleasurable, and spontaneous and more about it helping to build, maintain, and demonstrate intimacy and connection in relationship—suggesting women may conceive of sexting as a necessary and expected part of fulfilling intimate relationships."

The Present Study

"The aim of my research was to explore young people's practices and perceptions surrounding sexting to understand the social meanings and cultural norms underpinning their sexting practices, in particular for the purposes of the present article, with regard to gender and the implications—both harmful and beneficial—for young women."

"Three questions shaped the research and the analysis that underpins my work: (a) What are young people's practices and perceptions surrounding sexting, particularly regarding ethical digital practices in terms of privacy and consent?; (b) What are the underlying social meanings and cultural norms regarding gender that shape these practices and perceptions?; and (c) How are these meanings and value systems incorporated into young women's self-concepts, practices, and decision-making in sexting?"

"After outlining young people's gendered constructions of youth sexting and how these constructions underpin risk and harm for young women, I discuss how the young women navigated and negotiated gendered logics in their peer group and how their practices and perceptions were shaped by broader meanings and norms, even while, for some, they were resisting narratives of risk and shame."

Method

"In these interviews, I referred back to the group interview discussions and asked participants whether they had any personal experiences with what had been discussed and the conversation flowed depending upon their perspectives."

"Interested young people were told to go to the private room in the youth club at the time of the group interview that evening."

"In different ways, the gatekeepers at the schools and the youth clubs imposed barriers to conducting the research, but overall they were relatively flexible and content for me to explore sexting with participants in an honest and open way."

"Symbolic interactionism enabled me to explore how participants produced meaning in sexting in their local peer contexts and how they reflexively incorporated meaning into their self-concepts, decision-making, and practices."

"I compared between participants and research contexts (group and one-to-one interviews) to explore how participants constructed issues differently."

Results

"When discussing sexting, I observed the young women positioning (and thus legitimising) their practices as non-sexual by locating them within monogamous relationship contexts: I think with boyfriends and girlfriends... they respect each other, each other's bodies and stuff, and if you're like, hey, I think I kind of look nice, it's nice to have someone be like, oh yeah, you do look nice, like give you confidence, not necessarily in... a sexual way. (Rosie) I caught glimpses of more embodied, affective sexual desire, namely among Charlie, who identified as bisexual, and Skye-Rose, who identified as pan-sexual, who spoke one-to-one about their personal sexting experiences."

Discussion

"The elevation of male sexual desire and the riskiness of young men's sexual and bodily expression shaped young women's articulation of pleasurable sexting and the harmful practices perpetrated against them."

"Given the role of wider meanings and norms in harmful sexting, young women's perspectives and experiences should be interrogated for how they reflect and feed into broader social processes and cultural practices (Gill [61]; Schwalbe and others [62])."

"Challenging harm and empowering young women involves going beyond interpersonal dynamics and forms of resistance to deconstructing social norms and cultural practices at a wider level, but this necessarily entails interrogating the full range of practices of sexual and bodily self-expression within youth sexting culture."

"My research has revealed how youth sexting culture was shaped by gendered meanings regarding young women's lack of legitimacy and agency in sexual matters, as well as young men's sexuality and agency."

[Section 8]

"Young men, meanwhile, are presented as inherently unable to engage in sexting "ethically" and as likely to pressure and coerce young women to sext and, subsequently, to distribute young women's images around their peer group for social gain (Herriot and Hiseler [63])."

"The present article centralises the perspective of young people and examines the role of gender in their practices and perceptions surrounding sexting."

"I explore the implications of gendered norms and meanings for young women in youth sexting culture, and I discuss how young women navigated gendered sociocultural constraints on their bodily and sexual expression."

"Alternative narratives and practices that challenge the assumption that youth sexting is inherently harmful and, by association, the reification of young women as passive victims in sexting dynamics are present among some of those who report engaging in the practice (see Crofts and others [52])."

6.1.6 Work Values of Working Women in Israel: A Comparison of Haredi Women with Those from the Secular and Traditional Segments [205]

This is a machine-generated summary of:

Sharabi, Moshe; Kay, Avi: Work Values of Working Women in Israel: A Comparison of Haredi Women with Those from the Secular and Traditional Segments [205]

Published in: Contemporary Jewry (2023)

Link to original: https://doi.org/10.1007/s12397-023-09480-3

If you want to cite the papers, please refer to the original.

For technical reasons we could not place the page where the original quote is coming from.

Abstract-Summary

"The Haredi (or Ultra-Orthodox) sector is the fastest growing population in Israel, but the ability to study their working world is limited, and as far as we know, the work values of Haredi women (who are usually the main breadwinners) have not been studied to date."

"This unique study compares their work values with those of secular and traditional Jewish-Israeli women."

"The Meaning of Work (MOW) questionnaire was administered to 467 employed Jewish-Israeli women (309 Secular, 138 Traditional, and 120 Haredi), addressing values, attitudes, and aspirations at the workplace."

"The findings indicate that, while secular women ascribe more importance to individualistic values related to interesting work and variety than do traditionalist and Haredi women, no significant differences exist among the three groups concerning the interest in good pay, autonomy, interpersonal relations, or job security."

Introduction

"This research has not only theoretical value associated with the study of the juxtaposition of religion and work in general, but the study of the work lives of Haredi women also has tremendous practical value for those interested in Israeli society, for those working with or managing Israeli Haredi women, and for those entrusted with advancing economic policy in Israel."

"Due to the tremendous value placed on men engaging in life-long religious study, women in Israeli Haredi society are typically charged with being the primary breadwinner, while having their workplace endeavors downgraded in importance by community norms regarding traditional gender-related familial roles (e.g., Caplan [65])."

"This is the first study that examines the way in which work values and societal characteristics play out among the three largest groups of Israeli Jewish women: Secular, Traditional, and Haredi women."

The Israeli Context

"For the purpose of this study, we have chosen to examine the three largest populations within the Israel Jewish community: (1) those who define themselves as Secular, (2) those who define themselves as Traditional, and (3) those who define themselves as Haredim; in total they make up 89% of the Jewish population in Israel."

"The third largest but fastest growing sector is the Haredi sector, which currently makes up 17% of the Jewish population (and 13% of the total population) in Israel (Israel Central Bureau of Statistics [66])."

"Over the last few decades, Haredi society has been characterized by a lifestyle based on the idea that men should strive to remain in religious seminaries studying the holy books for as long as possible, while women should strive to work to enable that endeavor (Caplan [65]; Kay and Sharabi [67])."

Hypotheses

"This is particularly true for the Haredi society that, as an enclave society, has been far less exposed to mass media and Western secular culture (Kay and Levine [68]; Sharabi and Kay [69])."

"While secular Israeli Jewish society is closer to the individualistic pole of the spectrum, individualism is extremely frowned upon in religious circles, with individuals directed toward the acceptance of and allegiance to the community norms and needs (Cahaner and Shilhav 2012; Leon [70])."

"We expect that: The importance of interesting work, variety, learning new things, good pay, and autonomy will be highest among secular women and lowest among Haredi women."

"The importance of interpersonal relations and job security will be highest among Haredi women and lowest among secular women."

Method

"As was the case in previous studies (Sharabi and Harpaz [71, 72]), this sample was drawn on the basis of the specifications of the Israeli Central Bureau of Statistics [66] (1 year before the sample was executed) regarding Israeli Jewish working women and men (by gender, age, education, and more) representing the Jewish labor force (excluding those in the Haredi labor force, as was the case in the previous studies)."

"Since this study focuses on women, we excluded the data regarding men in the labor force and included women who were randomly selected on the basis of the previously noted specifications of the Israeli Central Bureau of Statistics."

"Due to our concern that Haredi women might not answer the question according to the instructions (to rank order all 11 items from the most important to the least important), we checked the data and found that the data (from filling in the question regarding the ranking of work goals) were as accurate as those who were interviewed."

"The total sample consists of 567 employed Israeli Jewish women: 309 secular, 138 traditional, and 120 Haredi."

Findings

"There are wider differences between Traditional and Haredi women (five work value differences) than between Traditional and Secular women (three work value differences)."

"Secular women had the highest need for interesting work, but traditional (and not Haredi women) have the lowest need."

"Coming from a more "collectivist" society, the importance of interpersonal relations was the highest among Haredi women, but it was the Traditional (and not the Secular) who expressed the lowest interest in this matter."

"In Hypothesis 4, Secular women expressed the least interest in convenient work hours than Haredi women."

"With that, it is negatively correlated with the importance of convenient hours ($\beta = -0.09$, $P < 0.05$), good pay ($\beta = -0.11$, $P < 0.05$), and work conditions ($\beta = -0.15$, $P < 0.01$)."

Discussion

"This is the first study to examine work-related values among Israeli Haredi women in comparison to their Secular and Traditional counterparts."

"The current study that focused on Israel's fastest growing community, Haredi Jews and is the first to examine work-related values of Israeli women through the prism of religiosity, adds to that literature."

"The relationships between economic and social development and values related to individualism and collectivism and socioeconomic status can also explain the differences in work values of Haredi and non-Haredi women as two distinguishable social classes."

"It is important to note that in recent years Israeli Haredi women are increasingly found in new and less traditional fields (i.e., finance, high tech, etc) as they may increasingly seek jobs offering opportunities for personal interest and growth (Freund et al. [73])."

6.1.7 Gender Inclusivity of India's Digital Financial Revolution for Attainment of SDGs: Macro Achievements and the Micro Experiences of Targeted Initiatives [206]

This is a machine-generated summary of:

Duvendack, Maren; Sonne, Lina; Garikipati, Supriya: Gender Inclusivity of India's Digital Financial Revolution for Attainment of SDGs: Macro Achievements and the Micro Experiences of Targeted Initiatives [206]

Published in: The European Journal of Development Research (2023)

Link to original: https://doi.org/10.1057/s41287-023-00585-x

If you want to cite the papers, please refer to the original.

For technical reasons we could not place the page where the original quote is coming from.

Abstract-Summary

"It has promoted digital financial inclusion to further its fight against poverty and gender inequality and support the attainment of UN's Sustainable Development Goals (SDGs)."

"We take stock of how India's digital financial revolution has affected financial transactions and services with a focus on gender inclusivity for the attainment of the SDGs."

"We propose a framework to understand the gender inclusivity of digital financial inclusion initiatives that connects the macro developments in the sector with the micro-level experiences of improving women's access and use of these services."

"Our findings suggest that India has made great advances in promoting digital financial inclusion but at the same time, the country has struggled to achieve gender parity even within specific finance-focused programmes designed to improve gender inclusivity."

Introduction

"India has in the last few years moved towards digital financial inclusion by promoting a national biometric identity programme together with an open access digital financial transactions infrastructure; this created the foundations upon which both public and private digital financial inclusion activities could be launched."

"We examine digital financial inclusion in relation to women in India and the particular challenges they may face in accessing and using digital financial services."

"Understanding the factors that may enable or constrain women in accessing digital financial services is important in order to enable gender-inclusive digital financial inclusion."

"The government of India has consciously laid down the foundations for a national digital finance infrastructure to enable the country to leapfrog to the next generation of digital financial inclusion."

"This suggests that women's access to digital financial services and thus digital financial inclusion, may be hindered by factors related to digital technology."

Digital Financial Inclusion and Gender Inclusivity: Experiences from the Global South

"Women are seen to be particularly affected by the downsides of DFS as outlined by Gammage et al [74] in their systematic review examining gender and digital financial inclusion."

"e.g., there is a gender gap in national identity cards which in many countries are needed to obtain mobile phones and or register for services, women are more likely to be illiterate which hampers their use of DFS apps which require certain levels of literacy; finally, women are less likely to own phones and are thus excluded from accessing and using DFS (Gammage and others [74]; Lee and others [75]; GSMA [76])."

"Above, the bulk of the evidence base centres on the Sub-Saharan experience with a focus on M-PESA not yet engaging much with women's access and use of mobile phones which is surprising given the more general discourse in relation to financial inclusion is very women's focused."

Analytical Framework and Study Methodology

"This section first develops a framework to better understand the factors affecting women's access to and use of digital financial inclusion initiatives, before outlining the methodology guiding the remainder of this paper."

"We adapt the framework to consider five different levels of factors within a system that influence women's ability to engage with digital financial inclusion."

"While the original framework is not gender-focussed, our framework focusses specifically on factors at various levels of the system that affect women's ability to access and use digital finance."

"This framework allows us to explore the various complex factors that impact women's access to and use of digital financial inclusion initiatives within the larger system."

"While we acknowledge the limitations of relying on such grey literature evidence, it allows us to consider up to date individual, organisational and community level factors influencing women's access to digital financial inclusion."

Policy and Infrastructure Enablers of Digital Financial Inclusion in India
"The three pillars that have spurred digital financial inclusion over the last decade are commonly referred to as the JAM trinity, an acronym representing Jan Dhan Yojana bank accounts, the Aadhaar ID programme, and Mobile technology penetration (Ravi [77])."

"Administered by the Reserve Bank of India (RBI), UPI works like a financial scaffolding that uses Aadhaar, mobile number, account number and address in a unified way to facilitate the next generation online immediate payments leveraging increase in smartphone adoption and app downloads, regional language interfaces, and improved access to internet and data."

"A second digital finance technology the government has invested in is the Aadhaar-enabled payment system (AePS), which enables agents such as the business correspondents discussed in the next section, to process account openings and basic banking services such as transfers, deposits and withdrawals, and undertaking Aadhaar-based biometric identification to verify customers and approve service requests on a portable device (Pinto and Arora [78])."

Case Study: The Business Correspondent Model and Women
"Women business correspondents essentially work as an extension of banks, providing services such as bank account applications and account openings, cash withdrawals and deposits, bank transfers, etc They also offer options for accessing government direct benefit transfers such as pensions, scholarships for school children, and payments under employment schemes such as the Mahatma Gandhi National Rural Employment Guarantee Act and the COVID relief scheme (Hindustan Times [79])."

"There was a 46% increase in the number of Bank Sakhis, or women business correspondents during April to June 2020 (Pinto and Arora [78], p. 10)."

"Female business correspondents are likely to have a role to play in enabling such proxy users to gain agency to use the bank accounts themselves, e.g., to ease the many restrictions that limit women to be full users which includes gendered restrictions on mobility outside of the home, a lack of financial literacy and limited knowledge about digital technologies (MSC [80])."

Discussion
"Putting in place digital enablers, referred to above as the JAM trinity (e.g., PMJDY, Aadhaar and Mobile technology penetration), allows female business correspondents to provide women with access to banking services."

"At the interpersonal and family level, it is clear that factors such as conditional access to and use of mobile phones as well as agency over one's finances plays an important role in enabling women to engage in digital finance."

"At the organisational level, we note that Bank Sakhis work through SHGs and NGOs,forming communities of women who can provide peer support not just with regard to their digital training and financial expertise, but also with regard to their social and cultural needs."

"At the level of the community, factors such as cultural and social norms towards the use of digital technology and women's access to finance are important enabling or constraining factors."

Concluding Comments

"Through a multi-level framework, we have connected the macro-level perspective of digital financial inclusion policy and infrastructure with the micro-level view of women's experiences of using finance through the BCM."

"This has implications for women's ability to access and use DFS, and by extension their scope for digital financial inclusion."

"We have also seen that the use of digital finance is gendered, and thus social norms need to be considered in designing interventions to enable women to fully participate in financial institutions and markets."

"We showed that having female business correspondents helped provide digital financial access for women that better suited their requirements."

"Government officials and practitioners have a role to play to find ways to overcome the technical challenges that still persist hampering the growth of digital financial inclusion for low-income women, while cultural barriers need to be broken down to facility true gender equality."

6.1.8 The Latent Profile of Self-control Among High School Students and Its Relationship with Gender and Cyberbullying [207]

This is a machine-generated summary of:

Wu, Wenzhi; Yu, Lei; Cao, Xiang; Guo, Zeyi; Long, Qing; Zhao, Xinling; You, Xu; Teng, Zhaowei; Zhang, Yunqiao; Zeng, Yong: The latent profile of self-control among high school students and its relationship with gender and cyberbullying [207]

Published in: Current Psychology (2022)

Link to original: https://doi.org/10.1007/s12144-022-03995-y

If you want to cite the papers, please refer to the original.

For technical reasons we could not place the page where the original quote is coming from.

Abstract-Summary

"Mechanisms by which self-control influences cyberbullying were unclear, it is worth to explore the latent profile of self-control among high school students and analyze the antecedents of latent self-control profiles and their impact and moderating effect on cyberbullying victimization and perpetration."

"The latent profile analysis was used to explore the latent profiles of self-control of high school students, and the multinomial logistic regression analysis of the R3STEP method was used to explore the different effects of demographic variables on the latent profiles of high school students' self-control."

"The BCH method was used to analyze the effect of latent profiles of self-control of high school students on the cyberbullying victimization and perpetration. (1) There are four latent profiles of self-control among high school students, namely "low-control type"(3.49%), "high-impulse type" (18.49%), "high-control type" (13.78%) and "balance type" (64.24%); (2) Female high school students are more likely to be low-control than male high school students; (3) Different profiles of self-control have a moderating effect on the impact of cyberbullying victimization on perpetration, and compared with balance type individuals, high-impulse type individuals are more likely to implement cyberbullying with the increase in the level of cyberbullying victimization."

Introduction

"Self-control is usually defined as an individual's ability to adjust their behavior, emotions, and other responses in a timely manner to achieve certain activity goals, which is an important indicator of one's early socialization, and an important cognitive factor in regulating individual assessment and decision-making when cyberbullying occurs (Moffitt and others, [81]; Xie and others, [82])."

"Strong individual control systems can help mitigate the negative effect of risk factors on the cyberbully, but researchers mainly focused on variables, that is, from different dimensions of self-control, to analyze different behavioral tendencies that may be triggered by different dimensions in the past (Yajun and others, [83])."

"Considering the current serious situation of cyberbullying among high school students in China (Dou and others, [84]; Wang & Ngai, [85]; Zhan and others, [86]), the present study proposes to analyze the latent structure of self-control of high school students based on the dual-system model of self-control through LPA, and to achieve accurate classification of self-control of high school students and understand the proportion of each latent profile in the whole according to their response patterns in various dimensions."

Methods

"The average age of participants was 15.5 years, SD = 0.57 years."

"Using a 5-point scale (1 = completely disagree, 5 = completely agree), the items are summed up to get a total score, with the higher the score representing the higher level of self-control."

"The Cronbach's α coefficients of impulse system and control system were 0.895 and 0.876, and six dimensions of impulsivity, distractibility, poor delay of gratification, problem solving, and future time perspective were 0.876, 0.9, 0.78, 0.668 and 0.875, respectively, and the Cronbach's α coefficient of the total scale was 0.939."

"Using a 5-point scale (0 = never, 4 = almost every day), items are added to obtain a total score, with higher scores indicating higher levels of individual cyberbullying victimization and perpetration."

Research procedures

"The latent profile analysis of self-control was conducted to seek a model that fits the optimal index (Nylund-Gibson and others, [87])."

"Used information criteria include LL (the model-based log-likelihood statistic), AIC (Akaike Information Criteria), BIC (Bayesian Information Criteria), and aBIC (Sample-Size Adjusted BIC); the commonly used classification criteria is Entropy, a standardized index for determining the accuracy of a model classification, with values between 0 and 1; the likelihood ratio test derivatives is used to compare two nested models (k profile models and k-1 profile models) commonly used metrics include LMR (Lo-Mendell Rubin Likelihood Ratio Test), BLRT (Bootstrap Likelihood Ratio Test) (Nylund-Gibson and others, [87]; Tofighi & Enders, [88])."

"In the third part, using the latent profile results obtained from the first part as the dependent variable, polynomial logistic regression (R3STEP command) was conducted to explore the effect of demographic variables (gender, location, only children or not) on the different profile of self-control (Asparouhov & Muthén, [89])."

Results

"Post-hoc multiple comparison analysis found that there were significant differences in the total scores and dimensions of self-control among different profiles of high school students, which indicated that the latent profile of self-control of high school students could distinguish the degree of self-control of high school students, and also showed that the latent profile was effective, supporting H1."

"In order to explore the influence of different types of self-control on high school students cyberbullying victimization and perpetration, BCH command was used to analyze the differences between cyberbullying victimization and perpetration of different self-control profiles."

"Using the BCH command, with the four latent profiles of self-control as moderator variables, high school students' cyberbullying victimization as an independent variable, and cyberbullying perpetration as a dependent variable, grouped regression analysis was performed."

Discussion

"By analyzing the relationship between different profiles of self-control and cyberbullying among high school students, this study further revealed the moderating effect of self-control between cyberbullying victimization and perpetration, that is, high-impulse type individuals are more likely to bully others in a similar way after being cybervictims."

"The results of the moderation effect test showed that the relationship between cyberbullying victimization and perpetration of high school students varied with the differences in self-control profiles, specifically, when experiencing cyberbullying victimization, high-impulse type individuals that had higher scores on the impulsive system and lower scores on the control system are more likely to develop cyberbullying behaviors than balance type individuals, in other words, those with high control system scores are less likely to commit criminal behavior in the face of violent situations (Brown, [90])."

6.1.9 "My iPhone Changed My Life": How Digital Technologies Can Enable Women's Consumption of Online Sexually Explicit Materials [208]

This is a machine-generated summary of:

McKeown, Janet K. L.; Parry, Diana C.; Penny Light, Tracy: "My iPhone Changed My Life": How Digital Technologies Can Enable Women's Consumption of Online Sexually Explicit Materials [208]

Published in: Sexuality & Culture (2017)

Link to original: https://doi.org/10.1007/s12119-017-9476-0

If you want to cite the papers, please refer to the original.

For technical reasons we could not place the page where the original quote is coming from.

Abstract-Summary

"Digital technologies continue to change the ways women can access and consume sexual materials online, including pornography."

"We consider the ways women who identify as porn consumers can use digital technologies to enable and enhance their consumption of these pornographic materials."

"Drawing on qualitative, in-depth interviews with 11 women, our findings illustrate the ways women can use digital technologies to consume pornography online in ways that help them to fulfill their sexual needs, embrace and explore their sexual selves, connect in sexual relationships, and normalize their sexual desires."

Introduction

"Smits [91] suggested women represent one of the fastest growing groups of online pornography consumers."

"Lending support to this suggestion, de Cadenet ([92], October 19) found 31% of women reported watching pornography every week and 90% reported accessing pornographic content online."

"These studies illustrate that there is a growing demand for online pornography, and women are making up a larger segment of the porn consumer-base (Penley and others [93])."

"Despite the rise in both access to and use of sexual content by women, to date, much of the research on pornography consumption continues to focus predominantly on men's consumption."

"Using a cyberfeminist lens, the purpose of this paper is to examine the ways women consume sexually explicit material online, focusing specifically on women's experiences who identify as online pornography consumers."

Literature Review

"This work was particularly influential in drawing attention to the expanding landscape of women's consumption of SEM and encouraging feminists to pay more attention to understanding the ways women access and consume pornography in their everyday lives, when previously, little to no research has considered women's experiences consuming SEM."

"Research remains limited as to how and why women adopt new digital technologies to consume pornography online, more specifically, and the implications of their consumption (Attwood [94]; Döring [95])."

"Although we recognize women can experience limitations when consuming pornography online, we have chosen to focus this paper on exploring the possibilities of women's consumption enabled through digital technologies."

"The research questions explored in this paper include: How do women use digital technologies to consume pornography online?"

"In what ways do digital technologies enable women's consumption of online pornography?"

Theoretical Orientation

"Richards [96] noted, because 'cyberfeminsm relies on plurality, fluidity, and ironic play' (p. 10), it is more accurately described in its plural form, cyberfeminism(s) (Daniels [97]; Richards [96] Despite resisting categorization, Daniels [97] noted the common thread apparent in all cyberfeminist work is an interest in understanding the relationship between gender and digital technologies."

"Richards [96] pointed out, simply exploring the ties between gender and digital technologies 'does not create a cyberfeminist' (p. 10)."

"She continued, "cyberfeminists engage in the nexus of gender and digital technologies in a manner that is subversive while playful, hopeful even if hostile" (p. 10)."

"With these ideas in mind, in this paper we utilize a cyberfeminist theoretical lens to examine how women can use digital technologies to enable and enhance their consumption of online pornography."

Methods

"Criterion based sampling, as a form of purposeful sampling, was used to recruit women who were open to speaking about their sexuality and who were also consumers of online SEM, including pornography."

"If women met the criteria, the study was open to them regardless of their racial and ethnic backgrounds, sexual identities, socio-economic statuses, religions, etc Participants were recruited through a sexual workshop centre in Toronto, Canada that is focused on providing a safe and comfortable environment for women to explore their sexuality."

"We selected these eleven specifically as these participants focused their interviews on their consumption of online pornography, both images and videos, as opposed to some other areas related to SEM consumption (e.g., explicit educational materials) explored in the larger study."

"The participant profile represented diversity in terms of sexual identities, relationship status, employment status, race and ethnicity, education levels, and household incomes."

Findings

"Hazel also spoke about how she consumed pornography online for escapism reasons: "I go to websites that I know and then any free porn website has links to other sites…and you can just sort of fall down a rabbit hole until you find something that you like.""

"Women also spoke about consuming online pornography to explore their desires and interests in sexual relationships, whether they were in a sexual relationship currently or not."

"Nora shared, her online consumption using digital technologies helped to "sort of inform me of what I do want within a relationship to a degree.""

"Women also spoke about how through their use of digital technologies to consume online pornography, their sexual desires and interests were socially validated and normalized."

"She also noted that she consumed online porn and talked about her consumption to help normalize sexual desires and interests for other women."

Discussion

"Of the paper, we draw on our findings to discuss women's experiences consuming online pornography through digital technologies."

"Our findings highlight how women from diverse backgrounds and identity factors can consume online pornography and that women use different digital technologies to do so."

"Our findings support this work, given that women in this study spoke about consuming online pornography to inform their sexual relationships and to improve their level of communication with their sexual partners."

"Connected to these ideas, the findings from this study illustrate the ways consuming online pornography through digital technologies can help women to normalize their sexual desires, and to build acceptance of any tensions they confront if their desires do not align with the beliefs and understandings they hold in their everyday life."

Conclusion

"Digital technologies continue to change the ways in which women are accessing and consuming sexual materials online, including pornography, yet, research remains limited as to how women are using digital technologies to enable their consumption of these materials."

"In this study women spoke about consuming these materials through digital technologies to explore their sexual identities, desires, and needs, independently as well as in sexual relationships."

"The findings also illustrate the ways digital technologies can help women who identify as porn consumers, consume pornography in ways that emphasize personal choice, freedom, and control."

"With these ideas in mind, we encourage other researchers to pay greater attention to examining the ways women do consume online pornography in their everyday lives using digital technologies, and the implications of their consumption."

6.1.10 Slut-Shaming on Facebook: Do Social Class or Clothing Affect Perceived Acceptability? [209]

This is a machine-generated summary of:

Papp, Leanna J.; Erchull, Mindy J.; Liss, Miriam; Waaland-Kreutzer, Lauren; Godfrey, Hester: Slut-shaming on Facebook: Do Social Class or Clothing Affect Perceived Acceptability? [209]

Published in: Gender Issues (2016)

Link to original: https://doi.org/10.1007/s12147-016-9180-7

If you want to cite the papers, please refer to the original.

For technical reasons we could not place the page where the original quote is coming from.

Abstract-Summary

"The goal of this study was to examine how the use of that term affects perceptions of both the "slut" and the "shamer" as well as how clothing choice and social class of the targets, as well as feminist identity of the participant, affect these judgments."

"We presented 186 self-identified feminist and non-feminist female college students with a Facebook interaction in which one woman calls another woman a "slut.""

"Participants generally had negative perceptions about the "shamer," but results showed that both social class and clothing had an effect on how women perceived both the "slut" and the "shamer.""

"Participants' feminist identity also played a role in shaping perceptions: self-identified feminists were more willing to spend time with the "slut" and found the "shamer" less justified in her actions than did non-feminist participants."

Introduction

"Although, in some contexts, men may be disparaged for their sexual behavior (e.g., be referred to as a "man-whore" or "man-slut"), women receive the bulk of "slut-shaming" [98–100]. ""

"One way in which young women navigate these complex social messages is through a woman-on-woman control of female sexuality in the form of "slut-shaming" [98]."

"Slut-shaming" may be precipitated by female jealousy, in that women want the attention the sexually promiscuous woman receives from men [101]."

"Wearing provocative clothing may be seen to justify "slut-shaming," we also hypothesized that a "slut-shamer" would be judged less harshly, and the "slut-shaming" would be seen as more justified, if the woman being shamed were dressed provocatively as opposed to conservatively. ""

"Feminist blogs coined the term "slut-shaming" in order to point out and problematize the fact that women tend to be valued and denigrated based on their sexuality [102]."

Method

"Participants over the age of 18 who were enrolled in college were recruited using the psychology department subject pool consisting of students."

"The posted study link, hosted through Surveygizmo.com, took participants to an online informed consent."

"Participants evaluated both the "slut" (Jessica) and the "shamer" through a semantic-differential measure of person perception [103]."

"This 15-item measure was used to assess people's willingness to participate in common social interactions for college students with varying degrees of closeness (e.g., "Would you study with this person?""

"Participants were asked the extent to which they believed Ashley's (the "shamer") comment was justified on a 6-point scale ranging from 1 (Not At All) to 6 (Very)."

"The participants were asked to rate the extent to which they believed the "shamer" was jealous, judgmental, and serious, and they responded on a 7-point scale ranging from 1 (Not At All) to 7 (Very Much So)."

Results

"The univariate results, $F(1, 177) = 15.59$, $p <.001$, $\eta =.08$, indicated that participants who identified as non-feminist reported wanting significantly more social distance from the "slut," regardless of attire or SES ($M = 3.48$, $SD = 0.86$), than did the self-identified feminists ($M = 2.94$, $SD = 0.94$)."

"With serious tone as a significant covariant, the "shamer" received the lowest evaluation when she commented on the photo of the conservatively dressed, high SES "slut" and the highest evaluation when she commented on the photo of the provocatively dressed, high SES "slut.""

"Follow-up analyses indicated no significant difference between feminists and non-feminists in their assessment of whether the comment was justified within the conservative attire condition, $F(1, 78) = 0.08$, $p =.77$, $\eta =.001$."

Discussion

"The purpose of this research was to build upon prior research of perceptions of "sluts" and "slut-shamers" on social media by varying the socioeconomic status and attire of our target "slut" and examining the role of feminist identity of the participant."

"The low SES, provocatively dressed "slut" may be perceived as behaving in line with what our participants expect from a low SES woman, whereas the high SES, provocatively dressed target is expected to behave with more 'class' [104]."

"We hypothesized that our participants would perceive the "slut-shamer" more negatively if she shamed a high SES/conservatively dressed "slut" than if she shamed a low SES/provocatively dressed "slut.""

"Further, while participants generally viewed "slut-shaming" as unacceptable, feminists indicated that the "shamer's" comment was less justified than did non-feminist participants when the target was dressed provocatively."

"It could be that feminists are more willing to socially affiliate with provocatively dressed women whether or not she has been "slut-shamed.""

6.1.11 Predictors of Occupational and Organizational Commitment in Information Technology: Exploring Gender Differences and Similarities [210]

This is a machine-generated summary of:

Major, Debra A.; Morganson, Valerie J.; Bolen, Heather M.: Predictors of Occupational and Organizational Commitment in Information Technology: Exploring Gender Differences and Similarities [210]

Published in: Journal of Business and Psychology (2012)

Link to original: https://doi.org/10.1007/s10869-012-9282-5

If you want to cite the papers, please refer to the original.

For technical reasons we could not place the page where the original quote is coming from.

Abstract-Summary

"This study examined predictors of occupational and organizational commitment outcomes among IT professionals and explored gender differences and similarities in the relative importance of predictors."

"Satisfaction with growth opportunities, job security, job stress, and work–family culture generally were related to commitment outcomes as expected, accounting for the greatest variance in organizational commitment."

"Relative weights for men and women differed in the prediction of occupational commitment, where growth satisfaction and work–family culture were weighted more strongly by women and job stress was more strongly weighted by men."

"For organizational commitment, the relative weights among predictors were similar for men and women, except that men weighted job security more strongly than women."

"Findings suggest that garnering commitment to the IT field requires a gender-specific approach, highlighting growth and work–family support for women and addressing job stress for men."

"Increasing organizational commitment for both men and women calls for an emphasis on opportunities for growth and development."

Introduction

"Although we could find no studies examining the relationship between occupational commitment and satisfaction with job security, meta-analytic evidence has demonstrated a negative relationship between job insecurity and organizational commitment ($r = -.36$; Sverke and others [106])."

"A positive relationship is expected between satisfaction with job security and both occupational and organizational commitment."

"Satisfaction with job security will be positively related to (a) occupational commitment and (b) organizational commitment."

"Research on women working in gender-incongruent roles (i.e., jobs sex-typed as masculine) shows no significant differences between women's and men's organizational commitment (Jacobs and others [107]; van der Velde and others [108])."

"We are interested in the extent to which the predictors of occupational and organizational commitment in IT (i.e., satisfaction with growth opportunities, satisfaction with job security, job stress, and work–family culture) are weighted similarly and differently by men and women."

Method

"An example item is, "If I could get another job outside of IT that paid the same amount, I would probably take it.""

"An example item is "I am proud to tell others that I am part of this organization.""

"An example item is "The amount of personal growth and development I get in doing my job.""

"An example item is "How secure things look for me in the future in this organization.""

"Job stress was measured using a shortened 6-item version of Bernas and Major's [109] 12-item scale."

"An example item is "It makes me tense to think about my job.""

"An example item is "Higher management in this organization encourages supervisors to be sensitive to employees' family and personal concerns.""

"Participants responded on a scale ranging from 1 (strongly disagree) to 7 (strongly agree)."

Results

"Of the hypothesized predictors, growth satisfaction (γ =.13, p <.001), job stress ($\gamma = -.12$, p <.001), and work–family culture (γ =.06, p <.01) all significantly predicted occupational commitment."

"The hypothesized predictors accounted for 18 % of the variance in occupational commitment beyond the control variables."

"Growth satisfaction (γ =.36, p <.001), satisfaction with job security (γ =.15, p <.001), and work–family culture (γ =.24, p <.001) were significant predictors for organizational commitment, supporting Hypotheses 1b, 2b, and 4b."

"We used relative weight analysis (RWA) to assess the proportion of variance in the commitment variables associated with each predictor for men and for women."

"Job stress ($z = -9.45$, p <.05) was relatively more important in predicting occupational commitment for men than for women."

Discussion

"It was the only gender difference that emerged in the RWA for organizational commitment with men weighting it more strongly than women."

"To ensure fulfillment of this provider role, it appears that men's commitment to a particular IT organization is more likely than women's to be driven by job security."

"Finding an organization with a family-friendly culture may be a more important driver of women's commitment to the IT profession than men's commitment to the field."

"The extant research on organizational and occupational commitment (Lee and others [110]; Mathieu and Zajac [111]), as well as the handful of studies that have examined gender issues in IT (King [112]; Sumner and Niederman [113]–2004), suggest that there are more similarities than differences between men and women in IT."

"Our findings demonstrate that in predicting commitment to organizations, opportunity for growth and work–family culture are factors rated equally highly by both men and women."

"Satisfaction with growth opportunities was the strongest predictor of both occupational and organizational commitment, particularly for women."

Limitations and Future Research

"Understanding occupational and organizational commitment in the IT workforce was the impetus in this study, which employed a large sample of IT professionals."

"A great deal of contemporary research breaks down organizational commitment into affective, continuance, and normative dimensions (Meyer and others [114])."

"From a practical perspective when considering IT and other STEM fields, affective and normative commitment may be most important in that research suggests that they are most strongly associated with turnover intentions (Meyer and others [114])."

"Results showed that predictors accounted for a greater percentage of variance in organizational commitment than in occupational commitment."

"Future research should investigate the extent to which the predictors investigated here account for greater variance in occupational commitment when they have an occupational referent (e.g., the IT career field as a whole)."

Conclusion

"This study contributes to the broader commitment literature by demonstrating that, at least in some industries, the magnitude of predictors for organizational and occupational commitment are not always as similar as one would assume based on the extant meta-analyses (Lee and others [110]; Mathieu and Zajac [111])."

"This study also demonstrates that, at least in the IT workforce and perhaps in other STEM occupations, there may be important gender differences in the relative importance of predictors of occupational commitment."

"In line with calls for research to examine gender in new ways (Ely and Padavic [115]) this study employed RWA, which factors out the collinearity between variables for purer comparisons of the relative strength with which predictors are associated with commitment across gender."

"Findings from this study are timely as are the associated implications for fostering occupational and organizational commitment."

6.1.12 Gender and Sexuality Online on Australian Muslim Forums [211]

This is a machine-generated summary of:

Marcotte, Roxanne D.: Gender and sexuality online on Australian Muslim forums [211]

Published in: Contemporary Islam (2009)

Link to original: https://doi.org/10.1007/s11562-009-0104-0

If you want to cite the papers, please refer to the original.

For technical reasons we could not place the page where the original quote is coming from.

Abstract-Summary

"The study of two online discussions on MuslimVillage forums—one of Australia's largest online Muslim communities—about polygamy and homosexuality will illustrate how online interaction within virtual Islamic environments provides both greater and lesser fluidity to e-Islamic normative discourses associated with gender and sexuality."

"Muslim forums provide opportunities for members to display a variety of views and opinions: on the one hand, they allow Muslims to post views that may challenge, contest, or even transgress Islamic gender and sexuality norms, while equally allowing members, on the other hand, to reaffirm more authoritative normative Islamic views."

Australia's Islamicyberspace: MuslimVillage forums

"E-bulletins, e-newsletters, online posting, e-discussion boards, Muslim forums, Facebook, MySpace, or its Muslim equivalent, Muxlim Space (http://spaces.muxlim.com), and blogs become new tools that make it easier for them to migrate online and stay in contact with one another in these new virtual social networked environments."

"MuslimVillage forums members are Australian converts, second and third generation Australian Muslims."

"MuslimVillage forums help sustain a virtual Australian Muslim community."

"MuslimVillage forums provide opportunities for Muslims to engage in non-vertical and non hierarchical exchange of information, including religious information, that is most relevant to them and the virtual community in which they participate and mold (Young [94, 116])."

"The ability to lurk and to become almost invisible on MuslimVillage forums provided an opportunity to observe the everyday online interaction of a number of Australian Muslims."

Gender and sexuality and the elusive normative in Australia's Islamicyberspace
"I would argue that, because of the characteristics of online virtual communities, the set of beliefs that members share with one another and that structure the manifestations of the e-Islamic normative discourses of MuslimVillage forums are more flexible, as members need to negotiate the construction of a shared set of beliefs for their virtual community."

"This may be one of the reasons why e-Islamic normative discourse on Muslim forums possesses such a high degree of elusiveness."

"MuslimVillage forums also promote e-Islamic norms of modesty by offering closed and private gender specific forums for the "brothers" and the "sisters.""

"Online Muslim forums and the social e-networks they sustain provide particular religiously value-laden understandings of gender or sexuality."

"Participating members and administrators continuously co-construct e-Islamic normativity with their consciously and/or unconsciously held assumptions about gender and sexuality and the manner these find their way into the e-Islamic discourse of the forum and shape it."

Constructing, contesting, and reaffirming E-Islamic normativity
"The polygamy debate provided members of MuslimVillage forums a virtual public space to post a variety of views and thus demonstrates the potential of Muslim forums to be virtual agoras, where Muslims are able to express more freely their personal views, opinions, understandings, concerns and questions about polygamy in Islam and in the context of Australian society than they would be in their real life Muslim communities."

"Graw2007 (Mar 28 2008, 04:08 PM) kept questioning those Islamic norms with examples of Muslim practices that contradict Islamic views on homosexuality."

"Graw2007, who (may or) may not be a Muslim, was thus able to entertain and present non e-Islamic views about sexuality and, in the process, challenge discursive e-Islamic normative views that the overwhelming majority of members on the forum held."

Gender and sexuality in Muslim virtual E-third space
"Online interaction equally reflects and reproduces social relations, gender biases, and religiously informed views about polygamy (both polygyny and polyandry) and homosexuality in Islam which all members of MuslimVillage forums bring to their discussions—but it can also be a means to question many of them."

"What might the fact that most of the authoritative voices that sought to reaffirm e-Islamic normative views on the homosexuality forum may have been Muslim men or that the views were more polarized than those on polygamy tell us about sexuality and gender on MuslimVillage forums?"

"Members of MuslimVillage forums can, in fact, be viewed as agents of a process of "localizing the global," as Australian Muslims draw from global "e-Islamic capital" and use innumerable hyperlinks and posts to make it available in Australia's Islamicyberspace."

"Members of MuslimVillage forums bring their own social, cultural, and religious views, opinions, prejudices, and preconceptions about gender and sexuality to their virtual Islamic community."

[Section 5]
"I will try to answer the following questions: Can discussions of controversial issues, such as polygamy and homosexuality, by members of such a large virtual Australian Muslim community shed some light on the construction and/or the contestation of normative views—what I call "e-Islamic normativity"?"

"Do online faith-based forums constitute virtual public spaces where traditional normative views about gender and sexuality are merely reproduced or can these virtual spaces provide Muslims with new venues to present and discuss different, alternative, or even transgresssive, normative views?"

"On account of the nature of online interaction, it can be argued that Muslims who participate in the virtual public spaces provided by MuslimVillage forums have greater freedom to challenge (even radically) Islamic normative views."

"This is not to say that members of MuslimVillage forums do not attempt to recreate, reiterate, and restate normative views online."

"I would like to suggest that e-Islamic normative discourses on Muslim forums possess a high degree of elusiveness."

6.2 Inequality and Gender Justice in Digital Age

Machine generated keywords: feminist, article, inequality, justice, inclusion, critical, commission, conflict, policy, child, language, equality, woman, politic, pandemic

6.2.1 The Language of Inclusion: Using Critical Corpus-Based Methods to Study the Presence and Representation of "Women, Children and Vulnerable Groups" in Liberia's Truth Commission [212]

This is a machine-generated summary of:

Evrard, Elke: The Language of Inclusion: Using Critical Corpus-Based Methods to Study the Presence and Representation of "Women, Children and Vulnerable Groups" in Liberia's Truth Commission [212]

Published in: Social Justice Research (2023)
Link to original: https://doi.org/10.1007/s11211-023-00411-z

If you want to cite the papers, please refer to the original.

For technical reasons we could not place the page where the original quote is coming from.

Abstract-Summary

"While inclusion, participation and victim-centredness have become catchwords in transitional justice discourse, this rhetoric has not necessarily enabled the articulation of more complex identities and experiences, or the pursuit of varied justice claims."

"The article combines expressive theories of justice with an innovative corpus-based methodology to critically examine how the Commission made visible, defined and construed these actors through its language of inclusion."

"This duality points to inherent tensions in the expressive messaging of TJ institutions, and rather locates the transformative potential of their inclusionary language in the strategic openings it affords for victims' groups, women and youth organizations in their broader trajectories towards justice and change."

A Normative Commitment to Inclusion

"Through promises and practices of inclusivity, TJ institutions acknowledge the rights, experiences and needs of those affected by conflict, endorse democratic ideals, and symbolize the state's responsiveness towards individuals and groups 'who might otherwise be excluded or marginalized from decision-making processes, such as women, youth, people with disabilities, and members of minority groups' (van der Merwe & Masiko, [117], p. 4)."

"Throughout this article, I trace this disconnect to inherent tensions in the expressive intentionality and messaging of TJ institutions—by pinpointing incongruencies between the promotion of inclusion as a vehicle for emancipation and democratization, and the implicit discursive representation of disenfranchised and conflict-affected actors 'included' in the language of these institutions."

"The article proceeds by conceptualizing inclusion as part of the expressive function of TJ mechanisms, and of Truth Commissions specifically, while arguing for the utility of a critical corpus-based approach to probe and evaluate this expressive function empirically."

The Expressive Functions of Inclusion

"Beyond increasing the legitimacy and efficacy of TJ processes, meaningful inclusion should also contribute to altering long-standing asymmetries in voice, access and power that delimit the capacity of conflict-affected and disenfranchised actors to exercise their agency as rightsholders and citizens (Bundschuh, [118]; Robins, [119])."

"Through meaningful inclusion, a Commission moreover demonstrates that victims and disenfranchised groups are part of the political community, that they are rightsholders and citizens who can access and make demands of the state (Bakiner, [120])."

"The participation of these actors as stakeholders and drivers of post-conflict justice processes, creates a new narrative about their role in society, signalling a point of rupture and transformation as the transitional state 'ritually inverts' the practices of political, legal and social exclusion that underpinned rights violations (Humphrey, [121], p. 173)."

A critical Corpus-Based Approach

"This study takes inspiration from CDA as a 'way of doing discourse analysis from a critical perspective' (Ibid, p. 273), unravelling dynamics of power by connecting micro-level patterns in text, to their discursive and social context (Fairclough, [122])."

"Corpus-based methods can arguably offset several core critiques levelled at traditional CDA, including the subjective selection of texts and features for analysis, the questionable reliability of findings derived from small numbers of text fragments, or the limited replicability of the qualitative approach (Widdowson, [123])."

"Drawing from multiple fields of computational text analysis, the article uses the programming environment R to combine word frequency, co-occurrence and sentiment analysis on a corpus annotated through natural language parsing."

"A final analysis uses dictionary-based sentiment analysis and Valence Shift Word Graphs to explore the distribution of positively and negatively charged words in texts about women and children."

Situating the Liberian Commission

"The 2003 Comprehensive Peace Agreement called for the creation of a Truth and Reconciliation Commission and in the TRC Act of 2005, the Commission was mandated to investigate violations and abuses committed between 1979 and 2003, identify main perpetrators, establish the historical root causes of conflict, investigate economic crimes, enable victims and perpetrators to share their experiences, and to compile its findings and recommendations in a public report."

"Article IV Section 4e of the TRC Act explicitly called for 'specific mechanisms and procedures to address the experiences of women, children and vulnerable groups, paying particular attention to gender-based violations, as well as to the issue of child soldiers [...].'"

"As defined in the TRC Act, these groups were largely subsumed under the banner women, children and (other) vulnerable groups and various efforts were undertaken to stimulate their involvement and visibility in the TRCL process."

Compiling and Describing the Corpus

"Following the termination of TRCL activities, the website now functions as a partial archive of its work, containing the sections about, news, videos, photos, hearings, reports and diaspora."

"The resulting corpus contains nearly 300 documents, which can be subsumed under three headers or sub-corpora: public hearing transcripts, various news items and the Commission's reports."

"According to these principles, a corpus of over a million words, reflecting different aspects and stages of the TRCL's work (hearings, press releases and updates, reports) and including different registers (spoken and written, vernacular and formal) can bring valuable insights into the language use of the TRCL and the discourses present therein."

"The website was mainly intended to target international (diaspora) audiences, whereas Liberians domestically could have been exposed to text or speech from this corpus through live attendance to the public hearings, broadcasts on radio and television, TRCL mobile video story-telling units, or excerpts cited in newspaper reporting (Weah, [124])."

Linguistic Analyses

"While permutation testing reveals no meaningful differences in the occurrence of terms man, soldier or brother between male and female statements, male witnesses are consistently less likely to reference terms signifying women or children (child, woman, boy, girl, sister)."

"Beyond actor-referring terms, the word frequency list for female witness statements also contains many words signifying different types of harms these women experienced or witnessed."

"A final analysis explores how the emotive word use of the TRCL further contributes to the evaluative context wherein women or children appear."

"The underlying calculation determines the percentage contribution of each emotion word to the observed difference in mean valence between a reference and comparison text, on the basis of (1) the word's numerical valence score relative to the reference text average and (2) its change in relative frequency between the reference and comparison text (Dodds & Danforth, [125])."

Discussion

"Other disenfranchised or minority groups are also far less visible in the language of the TRCL, obscuring their pertinence and turning the dyad of women and children into a placeholder or device to signify all vulnerable social identities."

"While it is crucial to recognise the pervasive use of sexual violence as weapon of conflict, equal attention should be afforded to the gendered impact and long-term consequences of displacement, deprivation and destruction on women's socio-economic rights, needs and agency, given the prevalence of these themes across their testimonies, and the likelihood for these harms to exacerbate pre-existing cleavages in access to property, livelihood opportunities, health and education services (Ní Aoláin, [126])."

"The TRCL's focus on children and young people also invited attention to social dimensions of harm that reflect ongoing damage to 'human relationships and connectedness' and which are often said to be side-lined by TJ initiatives (Sankey, [127], p. 9), including the breakdown of social, community and family structures and ties, and the high resulting number of minors living in the streets."

Limitations

"It is important to point out several limitations to the corpus-based approach, and of this study in particular."

"The choice for prior mark-up of the corpus precludes a more inductive approach (associated with the corpus-driven method), where 'linguistic constructs themselves emerge from analysis of a corpus', and that may have produced additional, diverging or unexpected results (Biber, [128], p. 159)."

"More fundamentally, a quantitative discussion of the corpus inevitably reiterates particular linguistic devices that are precisely the object of critique, such as the value-laden term vulnerable groups, the repeated co-referencing of women and children or the binary constructions of gender through the juxtaposition of man/woman or boy/girl."

"The choice for a macro-reading of the corpus, rather than a micro-analysis of selected testimonies, may also inadvertently gloss over the multiplicity of voices present in the corpus."

Conclusion

"This article adopted a critical corpus-based approach to investigate how and to what extent the TRCL's expressive commitment to inclusive justice, created openings or constraints for these and other disenfranchised and conflict-affected actors to voice their experiences and interests, subvert norms and systems that engender inequality, and assert themselves as protagonists in post-conflict change processes."

"The emancipatory potential of TJ's language of inclusion may rather lie in the way victims' groups, women and youth organizations strategically harness and transpose the expressive openings afforded by this language, to pursue their voice, rights and interests across multiple moments and spaces that make up their broader trajectory towards justice and change."

"The corpus-based approach offers a plethora of avenues for exploring this assertion, and for studying the ways in which conflict-affected and disenfranchised groups in post-conflict societies are impacted by, and in turn interact with, the ubiquitous rhetoric of inclusion and participation."

6.2.2 Understanding Digital Inequality: A Theoretical Kaleidoscope [213]

This is a machine-generated summary of:

Kuhn, Caroline; Khoo, Su-Ming; Czerniewicz, Laura; Lilley, Warren; Bute, Swati; Crean, Aisling; Abegglen, Sandra; Burns, Tom; Sinfield, Sandra; Jandrić, Petar; Knox, Jeremy; MacKenzie, Alison: Understanding Digital Inequality: A Theoretical Kaleidoscope [213]

Published in: Postdigital Science and Education (2023)

Link to original: https://doi.org/10.1007/s42438-023-00395-8

If you want to cite the papers, please refer to the original.

For technical reasons we could not place the page where the original quote is coming from.

Abstract-Summary

"The pandemic affected more than 1.5 billion students and youth, and the most vulnerable learners were hit hardest, making digital inequality in educational settings impossible to overlook."

"As a product of this collaboration, we propose a methodological toolkit: a theoretical kaleidoscope to examine and critique the constitutive elements and dimensions of digital inequalities."

"We argue that such a tool is helpful when a critical attitude to examine 'the ideology of digitalism', its concomitant inequalities, and the huge losses it entails for human flourishing seems urgent."

"We argue that the postdigital does not mean that the digital is over, rather that it has mutated into new power structures that are less evident but no less insidious as they continue to govern socio-technical infrastructures, geopolitics, and markets."

Introduction

"It became evident during the Covid-19 pandemic how precarious the nervous system of the networked society is as more than a third of all students globally were unable to access education, detrimentally affecting their present and future life (UNESCO [129]; Jandrić and others [130]) and further reinforcing already well-known historical inequalities."

"Traditional accounts of digital inequality have centred on the lack of access to ICTs, framing the discussion in terms of the 'haves' and 'have-nots' (see Light [131]; van Dijk [132, 133]; DiMaggio and Hargittai [134]), and have also concentrated on the technology as such, emphasising these devices as neutral within the contexts they are placed, overlooking the fact that they are impacted by greater socio-cultural constraints and users' agency, as well as new articulations of uneven power relations."

"Digital inequality is a supercomplex social phenomenon, stratified, and multidetermined, it is our contention that different ways of theorising can clarify a greater range of structural solutions to the social problem of inequities."

A Theoretical Kaleidoscope

"We use the metaphor of a kaleidoscope to describe the need for different theories, or sometimes, the intersection of multiple theories, to unpack and understand the complexity of digital inequality."

"Just like a kaleidoscope, as the reader turns to a new theory or the intersection of many, we aim to support them to appreciate and understand the nuanced nature of digital inequality that the theory or the intersection of them unveils as well as its relationship to previously discussed theories."

"For Jan van Dijk's resources appropriation theory, the problem of digital inequality starts with how people use digital media in their daily lives."

"We describe each theoretical lens in detail, exploring their advantages and sharing relevant examples where the theory has proved useful in shedding light on various aspects of digital inequality."

The Kaleidoscope of Theories to Study Digital Inequality

"By underscoring both the historical and cultural aspects of human activity with semiotic and physical tools, CHAT promotes a nuanced understanding of the powerful ways in which access to digital technologies promotes unequal relationships and how these tools may further perpetuate historical and social inequalities in their use and uptake."

"We are building on the premises presented in the 'Critical Realism' and the 'Human Development and Capability Theory' sections that put forward that digital inequality is multifaceted and part of larger social inequalities, positing that it is useful to think about what it means to have 'equal access', not just in terms of access to technology and broadband, but access to the academic and cultural capital that allows educators and students to use the digital for research, study, voice, and liberation."

Discussion

"A kaleidoscope can serve as an analytical tool to examine and critique the constitutive elements of digital inequalities using the intersection of different theories and finding generative connections to shed light on different dimensions of digital inequality so that we can decentre our attention from emerging technologies without dismissing them."

"In crafting this kaleidoscope using this particular intersection of theories, we are able to connect injustices that are not necessarily related only with the digital but which consequences affect the participation in the digital sphere."

"Through using collaborative writing, we have put together a toolkit for researchers who want to choose a kaleidoscopic approach to studying digital inequality."

"We have presented an overview of some theories and the possibility of finding generative intersections and connections to craft a powerful kaleidoscope that serves to inquire into the world of digital inequalities examining and critiquing its constitutive elements, its theoretical approaches, and its consequences for society more generally."

Conclusion

"We further argue, in line with Knox [135], that there is a need to broaden the scope of educational research on socio-technical systems within which the project of digital education is constituted."

"This requires mobilising the intersection of different theoretical perspectives through a kaleidoscopic approach that allow us to transcend the tendency to understand technology, in particular educational technology, in terms of the newest gadget and the prevailing idea of an outdated educational system that needs to constantly catch up with the latest trend in technology."

"We argue in line with Wakunuma for an urgent need to explore the aspect of power in light of the digital transformation of the Global South."

"We ought, as researchers, to transcend the idea of a tempting and seductive novelty and linear progress that seems to be implicit in the digital if our aim is to strengthen the social justice agenda in HE in the Global South or North, as both are in the midst of a severe crisis concerning digital injustices and its accompanying consequences for education."

Open Review 1: the Glitter and Gloom of Kaleidoscope Research (Petar Jandrić)
"Kaleidoscope research is a breath of fresh air in often-stale literature on research methods: highly creative, visually attractive, and above all pedagogical."

"Back to work, researchers' choice of approaches and theories in a kaleidoscope will always create a unique optic."

"Back to square one, the kaleidoscope approach to postdigital research does not escape the eternal dichotomy between inclusion and exclusion."

"With full recognition of issues arising from commensurability of research methods (see Jandrić [136]), the kaleidoscope pushes researchers towards more varied approaches–and that's a good thing."

"Displaying links and connections between approaches and methods that I never thought of, it helped me think of postdigital research in a new way."

"The other is the need to resist definitions of the kaleidoscope research method (see Jandrić and Ford [137]; Jandrić [138])."

"We should not get too infatuated by the glaze and glitter of kaleidoscopic research methods and we should neither be put off by their gloom(ier) sides."

Open Review 2: David Brewster's Kaleidoscope: Precision and 'Supercomplexity' (Jeremy Knox)
"As Brewster's Treatise on the Kaleidoscope [139] makes clear, the device required a meticulous positioning of three key elements: the reflectors, the object, and the eye of the viewer."

"If the theories outlined in this paper are the reflectors of the kaleidoscope, as suggested, they may need to be positioned and aligned in very precise and predefined orientations in order to produce the proposed beauty of a kaleidoscopic insight."

"The complex image one encounters by using the device is assumed to derive, not from the object itself, but from the array of reflections produced by reflectors."

"Interpreted thusly, the kaleidoscope does something fundamentally different from what is proposed in this paper, which is to assume 'supercomplexity' as intrinsic to the object itself (in this case digital inequality), and to draw on multiple theoretical frameworks to discern its intricacy, nuance, and convolution."

Open Review 3: Not the Observation of 'Beautiful' Forms, But How to Undertake a Theoretical Kaleidoscope of Inequality (Alison MacKenzie)
"I have often thought how valuable it would be to look at an issue of injustice from several theoretical or applied philosophical perspectives."

"Kaleidoscope' might, then, seem an odd choice of word to describe the authors' approach because inequality is not charming, it is not beautiful or innocent, and its form changes only insofar as technology or progress finds new ways for inequality to persist."

"We do need to understand inequality and injustice from multiple perspectives in one place."

"The standard wordcount does not permit good quality analysis of injustice and inequality from anything more than one perspective as academics have no taste for theoretical 'soups'."

"A single issue–digital inequalities–explored over eight chapters from eight perspectives, with a final chapter that concludes on the insights of digital inequality from a capabilities approach, critical realism, affective injustice, and so on."

6.2.3 Feminist Digital Diplomacy and Foreign Policy Change in Sweden [214]

This is a machine-generated summary of:

Aggestam, Karin; Rosamond, Annika Bergman; Hedling, Elsa: Feminist digital diplomacy and foreign policy change in Sweden [214]

Published in: Place Branding and Public Diplomacy (2021)

Link to original: https://doi.org/10.1057/s41254-021-00225-3

If you want to cite the papers, please refer to the original.

For technical reasons we could not place the page where the original quote is coming from.

Abstract-Summary

"This article analyses how the launch of Sweden's feminist foreign policy marked a change in Sweden's digital diplomatic efforts."

"It draws on three strands of research: digital diplomacy, foreign policy analysis (FPA) and feminist scholarship."

"Informed by FPA, the article explores the relevance of political leadership, bureaucratic agency and political context as drivers of policy change, and, specifically, Sweden's feminist digital diplomacy."

"The article provides an empirical case analysis of Sweden's foreign policy change and conduct of digital diplomacy during the period 2006–2020."

Introduction

"This article examines the ways in which Sweden has sought to promote feminist values in foreign policy practice through the conduct of digital diplomacy."

"This involves studying the ways in which digital diplomacy have developed and adapted to the emergence of feminist foreign policy as a method of achieving transformative gender justice globally."

"Sweden's communication of feminist foreign policy has been conducted through active use of digital diplomacy to project its external priorities and visions (Government Offices of Sweden [140])."

"Our study primarily explores the integration of feminist foreign policy into digital diplomacy rather than assessing the extent of Sweden's gender equality activities in social media (cf. Jezierska [141])."

"We outline the methodology employed to trace the development of Sweden's digital diplomacy narrative, before and after its adoption of a distinctively feminist foreign policy."

Digital diplomacy and foreign policy change

"We explore the significance of women's political leadership, their inclusion in bureaucratic politics as well as the prevalence of pro-gender norms in domestic society for the adoption of Sweden's feminist foreign policy and its distinct conduct of feminist digital diplomacy."

"The rise of feminist foreign policy has entailed change through both the employment of a pronounced feminist agenda, governing domestic and global politics, as well as a reorientation of diplomatic practice."

"Though FFP is located within Sweden's longstanding commitment to gender equality and justice globally, the use of digital technologies to communicate its contents and ambitions is innovative and reflects a significant change in Sweden's overall foreign policy narrative and practice."

"The distinctive features of a particular state's political culture, domestic politics, norms and national interests also affect the extent to which thoroughgoing foreign policy change is possible (Bergman Rosamond [142])."

Methodology

"Digital diplomacy, as discourse and practice, faced a major shift when an explicitly feminist foreign policy was adopted by the Swedish red-green coalition government in 2014, giving rise to changing endogenous and exogenous conditions for Sweden's external conduct."

"In analysing how feminist foreign policy became embedded in Sweden's digital diplomacy we draw on narrative methodology (Somers [143])."

"To illustrate such continuity and foreign policy change we analyse Sweden's digital diplomacy during the period 2006 to 2020."

"This enables us to make assessments of the distinct policy changes that were introduced as a consequence of the adoption of Sweden's feminist foreign policy in 2014, many of which were communicated through digital diplomatic channels."

"Our empirical analysis focuses on the development of Sweden's digital diplomacy both before and after the launch of Sweden's feminist foreign policy in 2014."

From digital promotion to feminist action: political leadership, bureaucratic agency and domestic politics

"We argue that domestic support for the new Swedish government in 2014 opened up for the launch of an innovative feminist platform for foreign policy, which in turn provided a window of opportunity for the Swedish MFA to broaden, and deepen its repertoire of digital diplomacy, adding a new chapter to its strategic self-narrative."

"The digital projection of Sweden's feminist foreign policy has helped to sustain the country's self-narrative as a state committed to the rights, resources and representation of women and girls globally."

"While the adoption of FFP is a significant foreign policy change, facilitated by political leadership and commitment as well as the MFA's active employment of digital campaigns and strategies, it is also a continuation of Sweden's domestic state feminist tradition and support for gender equality."

Conclusion: towards feminist digital diplomacy

"We have demonstrated that this turn in Swedish diplomacy can be intimately linked to the digital communication of FFP, in particular, Sweden's commitment to feminist transformations of global injustices."

"Sweden's active conduct of feminist digital diplomacy has both strengthened its feminist self-narration and offered new opportunities to communicate typically Swedish pro-gender values to large audiences."

"Our analysis shows that Sweden's foreign policy leadership and the bureaucratic agency of its MFA as well as a favourable domestic political context have been central to the incremental development of Sweden's feminist digital diplomacy."

"As our analysis has shown, the adoption of a feminist platform for foreign policy-making and embracing new diplomatic strategies have altered the trajectory of Sweden's digital diplomacy."

"Feminist digital diplomacy is evolving in an antagonistic world where the WPS agenda and the core pillars of Sweden's feminist foreign policy (rights, representation and resources) are increasingly contested and resisted."

6.3 Feminism in Digital Age

Machine generated keywords: feminist, woman, gender, power, sexuality, news, big, visual, ethical, relation, indicator, record, story, health, data

6.3.1 Feminist Adventures in Hypertext [215]

This is a machine-generated summary of:

Joyce, Rosemary A.; Tringham, Ruth E.: Feminist Adventures in Hypertext [215]

Published in: Journal of Archaeological Method and Theory (2007)

Link to original: https://doi.org/10.1007/s10816-007-9036-2

If you want to cite the papers, please refer to the original.

For technical reasons we could not place the page where the original quote is coming from.

Abstract-Summary

"Through the medium of hypermedia and hypertext (multilinear) narratives the complexities of the feminist practice of archaeology (including its multivocal interpretive process) can be grasped, enjoyed, and participated in by a non-archaeological audience more fluidly than in traditional linear texts."

Ruth Tringham on Visualization in The Feminist Practice Of Archaeology

"We have to ask, is that to be the aim of the use of computer-generated imagery in the production of images that use archaeological data?"

"Drawings that look like photographs (computer-generated imagery) can be imbued with the same power as photographs, as described by Berger, to offer a memory, an experience, even if it is more explicitly the product of interpretation."

"The Chimera Web project develops the theoretical implications of this computerized imagery for the construction of multiple realities of the past, all of which are illusions, but based on the material parameters that are founded empirically in the archaeological data."

"I have used computerized imagery not to reconstruct what I think Opovo "looked like" in its original form, but to construct multiple realities of Opovo's past, all of which are explicit illusions, but all of which are based on the material parameters that are founded empirically in the archaeological architectural and other data."

Rosemary Joyce on Multivocality, Authority, and Closure

"The final density of Sister Stories results from the operation of multiple individual authors."

"Even mastery is a relative thing to the extent that it is based on a chosen idea of what makes some work preferable to others...value is always a contextual element.... When we say Master we mean precisely a hierarchy of imposed value that claims domination and priority over not just another person's work, but the other person (Guyer [144])."

"Guyer [145] has already thought the underlying issues through in her own work on hypertext authoring and collaboration: The great cultural question of our time is how to accommodate our growing recognition of multiplicity.... It turns out that the boundaries between people, between groups of people, are permeable."

"By retelling the stories of Mexica women differently, foregrounding different narratives, in Sister Stories we implicitly critique the kind of singular authoritative representations of the past that are all too common in archaeology."

Ruth Tringham on Participation–Interactivity–Pedagogy

"These aims for The Chimera Web still hold true, but in the 7 years between writing about these aims and now in January 2006, several developments have taken place, not only in the technologies available, but in my own experience of putting hypermedia authoring into practice in my teaching and research, that have affected the way in which the interpretational hypermedia Chimera Web articulates in practice with the primary archaeological data."

"Of these developments, and as a result of my experience working with the Interactive University of UC Berkeley (Tringham [146]) as well as teaching courses that involve such access and re-purposing of digital data The Chimera Web has been re-created as an interpretive hypermedia guide to the archived data of the Opovo Archaeological Project."

Rosemary Joyce on Multivocality in Conventional Media of Representation

"Museum exhibits, of course are an authorized mode of presentation of stories about the past based on material traces."

"She lists some of the constraints on the construction of narratives by museum visitors, constraints that may be equally applicable to other forms of archaeological narrative (Perin [147]: 207–211): museum visitors display "resistance to assimilating the unfamiliar", showing a desire to conserve what they already know by fitting new information alongside it; "resonance with familiar subjects heightens interest" and allows visitors to reach satisfactory judgments about the authenticity of presentations about other cultures; visitors are interested in stories of "material and cultural minutiae"."

"Museum exhibits are the major genre in which non-specialists actively experience themselves as authors providing the coherence to the stories being told."

"Part of what these non-specialist visitors want from us is, paradoxically, authoritative statements: not to close off their role as co-authors, but to incorporate along with their own experience of the material things, into their own heterarchical story-telling."

A Look Back from the Future of Feminist Adventures in Hypermedia

"In a trajectory that parallels that of all the literature on hypertext and digital media, we now find a world where the limitations of networked technology no longer block any visually rich project based on access to extensive databases—at least for those in privileged enclaves in the developed world with direct Ethernet access or a high speed connection and plenty of processing power."

"What has not changed with the otherwise dramatic transformations that see both of us, once wary of the internet as a medium of interpretation, densely connected to it today, is our belief that what will distinguish good (responsible, ethical, multivocal, heterarchical) archaeological hypermedia in the future will continue to be the breadth of information to which we provide access, and the presence in our works of multiple voices and multiple scales of interpretation."

[Section 6]
"At the SAR conference we each presented hypertext works in progress, Ruth Tringham's Chimera Web, a multimedia opera, and Sister Stories, a dialogue in multiple voices (Joyce and others [148]), arguing that the new digital media offered unique opportunities for feminist archaeology to realize some of its deepest values."

"Consalvo ([149]: 115) suggests that, even given differences in access by race, class, ethnicity, and gender, networked digital media (the Internet or World Wide Web) "provides Western societies...a place where more individual voices can be heard.... These voices are largely absent from other media, due to concentration of ownership, economies of scale, and our structures of corporate capitalism.... Yet the Internet, for the moment, provides a place for more people to express ideas, concerns, and realities that may not be expressed elsewhere.""

6.3.2 Feminist Data Studies: Using Digital Methods for Ethical, Reflexive and Situated Socio-cultural Research [216]

This is a machine-generated summary of:

Leurs, Koen: feminist data studies: using digital methods for ethical, reflexive and situated socio-cultural research [216]

Published in: Feminist Review (2017)

Link to original: https://doi.org/10.1057/s41305-017-0043-1

If you want to cite the papers, please refer to the original.

For technical reasons we could not place the page where the original quote is coming from.

Abstract-Summary

"The current datalogical turn foregrounds the digital datafication of everyday life, increasing algorithmic processing and data as an emergent regime of power/knowledge."

"In this feminist and postcolonial intervention into gender-, race- and geography-blind 'big data' ideologies, I call for ethical, anti-oppressive digital data-driven research in the social sciences and humanities."

"I argue that a reflexive data scholarship can emerge from the reintegration of feminist and postcolonial science studies and ethics of care ideals."

"I align awareness of uneven relations of power and knowledge with the messy relation of dependency between human and non-human actors in data analysis."

"This framework productively recognises that digital data cannot be expected to speak for itself, that data do not emerge from a vacuum, and that isolated data patterns cannot be the end-goal of a situated and reflexive research endeavor."

"Data-driven research, in turn, shows the urgency for renewed feminist ethical reflection on how digital mediation impacts upon responsibility, intersectional power relations, human subjectivity and the autonomy of research participants over their own data."

Introduction

"This reflective article on data-driven ethnographic research is based on a two-year qualitative study with eighty-four young Londoners about their digital identities."

"Through the methods used, young people involved in the study not only decided for themselves what data they wanted to share, but more importantly, they also directed the course of the study as they became co-researchers of their own digital practices."

"I am certainly not claiming any novelty; like all research involving people, digital data-driven research involves complex power relations and mixed emotions in raising questions of ethics, voice and accountability."

"I draw on feminist and postcolonial science studies to explore urgent ethical questions including, what is required of a self-reflexive digital data researcher?"

"I ground my road map for an alternative data-analysis practice in empirical data from researching digital identities with young Londoners."

Aligning feminist internet studies with digital data-driven research

"While such scholars have critiqued big data 'to question how we think, see, and ultimately act on our knowledge' (Manovich, [150]), to think through the 'political consequences' of 'computational logics' (Berry, [151], p. 40) and to demand 'de-reification' (Couldry, [152]), these criticisms are neither especially feminist nor postcolonial, as they only marginally address (if at all) the specific people to whom their claims apply."

"Mirroring transcendental empiricist approaches, the 'data-rush' mentality (Mahrt and Scharkow, [153], p. 20) in the digital humanities and computational social science tends to naturalise the politics of knowledge production."

"Cultural geographers draw heavily on feminist literature to argue that data analysis should develop a 'situated, reflexive and contextually nuanced epistemology' (Kitchin, [154], p. 1)."

"How can we mobilise feminist and postcolonial philosophies of science to develop a situated, reflexive and more ethically-sound epistemology and praxis of data-driven research?"

"I draw on feminist and postcolonial ethics of care to reconsider data analytics for social-justice oriented research."

ethics of care and data analysis

"This literature offers ideal-types for ethical, caring, committed and responsible researchers to establish trusting and anti-oppressive relationships between scholars and those studied."

"A feminist and postcolonial ethics of care is value-based and recognises the dependencies, partiality, political commitments and personal involvements of researchers."

"Translated to the domain of research, feminist ethics confronts cultures of knowledge production, for example by focussing on concrete lived experiences, acknowledging dependencies, and assuming researchers are personally accountable and responsible for protecting research participants from risk or vulnerabilities (Edwards and Mauthner, [155]; Lomax, [156])."

"Those critiquing an ethics of care approach as merely validating the perspective of Western, white, middle-class feminists or for perpetuating paternalistic, neo-imperialistic and hierarchical assumption of caring are right insofar as (usually privileged) researchers get to decide who receives this care."

feminist data studies in practice

"Research participants like Jason were highly active on dedicated discussion boards and other atomised sites; however, they were non-users of mainstream platforms, resisting volunteering their data to profit-driven platforms and seeking to keep some control of their digital personae for specific intended audiences."

"This development raises questions about whether it is ethical to work with the data I have gathered, and it also means I cannot hold the promise I made to the research participants: that they would be able to study their own networks using TouchGraph at a later time."

"Although I do not want to claim that data elicitation technologies can allow researchers to bridge the power gap with research subjects, young people involved in the study did appreciate that they were in charge of analysing their own Facebook friendship visualisation."

conclusions

"In times of increasing digital datafication, the question arises: how can data researchers better ensure that human subjects in research maintain their rights to make decisions and have a say over their own digital data?"

"This article draws on feminist and postcolonial science studies to intervene in gender-, race- and geography-blind big-data ideologies and draws on feminist ethics of care to propose an alternative, research participant-centred data-analysis agenda."

"A feminist and postcolonial ethics of care draws our attention to digital data in relation to human meaning-making, context-specificity, dependencies and temptations, as well as benefits and harm."

"Data-driven research, in turn, shows the urgency for renewed feminist, critical race and postcolonial engagement and situated moral focus on how digital mediation impacts upon agency, relationality, responsibility, human subjectivity and the autonomy of research participants."

"Notwithstanding the celebratory data-rush mentality of most digital humanities and computational social science research, knowledge is not out there to be scraped."

6.3.3 Radical Pleasure: Feminist Digital Storytelling by, with, and for Women Living with HIV [217]

This is a machine-generated summary of:

Carter, Allison; Anam, Florence; Sanchez, Margarite; Roche, Juno; Wynne, S. T.; Stash, Just; Webster, Kath; Nicholson, Valerie; Patterson, Sophie; Kaida, Angela: Radical Pleasure: Feminist Digital Storytelling by, with, and for Women Living with HIV [217]

Published in: Archives of Sexual Behavior (2020)

Link to original: https://doi.org/10.1007/s10508-020-01822-8

If you want to cite the papers, please refer to the original.

For technical reasons we could not place the page where the original quote is coming from.

Abstract-Summary

"Despite the fact that HIV can be controlled with medication to undetectable levels where it cannot be passed on, stigmatization of women living with HIV persists."

"We situated our work within a participatory arts-based knowledge translation planning framework and used multiple data sources (Google Analytics, stories and comments on the website, team reflections over multiple meetings) to report on interim outcomes and impacts."

"In our first 1.5 years, we recruited and mentored 12 women living with HIV from around the world (Canada, Australia, New Zealand, Kenya, South Africa, Spain, Nigeria, and the U.S.) to write their own stories, with the support of a mentor/editor, as a way of regaining control of HIV narratives and asserting their right to have pleasurable, fulfilling, and safer sexual lives."

"Our social media community grew to 1600, and our website received approximately 300 visits per month, most by women (70%) and people aged 25–44 years (65%), from more than 50 cities globally, with shifts in use and demographics over time."

"We offer recommendations to others interested in using arts-based digital methods to advance social equity in sexual health."

Introduction

"Despite tremendous trauma and adversity, empowered narratives about sexuality, both inside and outside the context of risk prevention, have emerged—written by, with, and for women and other communities who are living with HIV, often by merging the worlds of art, activism, and (more recently) online technologies (AIDES, [158]; Becker, [159]; Caballero, [160]; Fratti, [161]; McClelland & Whitbread, [162]; Prevention Access Campaign, [163]; Whitbread, [164]; Whitbread, [165])."

"While strong, sex-positive activism is flourishing outside academic circles, negative, risk-focused approaches dominate published research about the sexuality of women who are living with HIV, particularly in quantitative work (see Carter and others [166] for a review)."

"In supporting women to share their stories of life and love with HIV, we sought to destabilize the stigmatized narrative and representation of women living with HIV globally and to show how women's sexual health is part of the larger social, economic, cultural, and political context of women's lives."

Method

"After further discussion about the purpose of our project, which is distinct from other institution-affiliated websites, we reached the agreement that ownership lies with the community: the women living with HIV who create stories on the platform."

"Prior to arriving at a decision around payment, we discussed the ethical tensions and logistical challenges of renumeration at length, such as the potential for coercion, the desire to share stories for self-healing and to help others rather than for money, the importance of acknowledging women's expertise and resourcing them for their work, the different writing skills on the team (e.g., some women might write a blog in 2 h, while others might take 8 h), the inequities in currency value between countries, whether to donate this money to a global organization for women living with HIV versus provide to individual writers, and how to send money from Canada to sites around the world and whether our institution would cover the transaction fees."

Results
"Across articles, comments highlighted the power of community storytelling: to connect emotionally to readers: "This was a beautiful empowering story that gave me chills and goosebumps;" to validate what women are feeling: "I felt that…100%…thank you for vocalizing what I feel;" to change attitudes toward dating and relationships: "As someone currently thinking of trying online dating again, this was timely;" to impact women's lives: "I think stories like these are going to have a huge impact on a lot of people and bring a positive change that is long overdue;" and to share information cross-culturally: "I love the story, the motivation and the three-date rule, but will it work in Nigeria?"

Discussion
"Extending the impact of sex research beyond the academic knowledge base to women living with HIV, service providers, policymakers, and the general public is important to help reduce HIV stigma and advance social equity in sexual health."

"We were also successful in creating a structure containing safety, support, and mentorship that enabled a group of women from across the globe with completely different lives to come together and form an identity in writing about sex, pleasure, and living with HIV."

"By involving researchers and women living with HIV as story editors/mentors and holding regular team meetings, we were able to offer guidance to women, promote information sharing and support, and help ensure scientific accuracy of the content, particularly blogs that addressed complicated issues such as HIV disclosure and the law This safe space enabled women to write engaging, creative, and diverse stories in order to help women living with HIV realize that their sexuality and well-being matters."

6.3.4 Feminist Online Interviewing: Engaging Issues of Power, Resistance and Reflexivity in Practice [218]

This is a machine-generated summary of:

Linabary, Jasmine R.; Hamel, Stephanie A.: feminist online interviewing: engaging issues of power, resistance and reflexivity in practice [218]

Published in: Feminist Review (2017)

Link to original: https://doi.org/10.1057/s41305-017-0041-3

If you want to cite the papers, please refer to the original.

For technical reasons we could not place the page where the original quote is coming from.

Abstract-Summary

"Drawing on feminist standpoint theory, 'reflexive email interviewing' is proposed as a method for feminist research."

"Through this case study, issues of power and resistance in the researcher/researched relationship and of participant reflexivity are interrogated."

"Criteria for reflexive email interviewing are proposed, including 1) strategies to interrogate and disrupt power hierarchies within the research process, 2) researcher reflexivity as a continuous part of the research process, and 3) continued invitations for participants to directly reflect on and respond to the research process."

feminist standpoint theory(ies)

"Within the research process, feminist scholars have long criticised traditional methodological approaches for reproducing unequal power relations and reinforcing the researcher's epistemic authority (Hesse-Biber, [167])."

"Feminist research then pays close attention to issues of power in the researcher/researched relationship, with goals of interrogating and disrupting unequal power relations as well as recognising participants' agency and resistance (Hesse-Biber, [167])."

"Feminist methodologies also call for critical attention to the social locations of the researcher and the participants in any research process."

"Reflexive knowledge building is another component of feminist research (Hesse-Biber and Piatelli, [168])."

"While in feminist standpoint research the terms perspectives and standpoints have often been used interchangeably, a standpoint is considered different from a perspective or experience, as it is achieved through critical reflection and dialogue and by struggling against dominant power structures (O'Brien Hallstein, [169]; Harding, [170])."

interviewing online

"Given the goals outlined above, existing literature on online interviewing, and specifically asynchronous email interviewing, suggests that this method can provide a means to uncover and engage women's experiences, disrupt power hierarchies in the research process, and produce a more reflective/reflexive interview than more traditional approaches."

"Email interviews are most comparable to offline practices of letter writing (Letherby and Zdrodowski, [171]), exchange diaries (White, [172]), or personal diary keeping (Pennebaker, Colder and Sharp, [173]), where researchers and participants engage in a prolonged, asynchronous information exchange."

"When using the email interview method, participants are engaging in a dialogue with the researcher over a prolonged period of time; they can proceed at their own rate and respond at a time that is convenient to them (James and Busher, [174])."

"Although conducting these interviews over an extended period of time allows for the collection of in-depth data and increased reflection, time can also be a disadvantage for online interviews, as there becomes a greater chance participants may drop out of the research project (James and Busher, [174])."

case study: women's voice and (dis)empowerment

"To expand upon the potentials of email interviewing as a feminist method, we will provide as an example email interviews conducted in the context of research on self-identified women's experiences with voice online."

"Our online method was selected with care, given existing critiques of mediated forms of interviewing which claim that the distancing of researcher and participant creates artificial contexts 'that strip away natural contexts of meaning from both questions and responses' (Mishler, [175], p. 121)."

"After initial invitations, including informed consent and the identification of an appropriate and secure email account for the study, the exchange began with pre-interview questions about the participants' backgrounds, as well as their internet access and use, to guide the interview and to allow us to be attuned to potential access issues for the interview itself."

power of 'the researcher'

"To account for our role in this, we have attempted to actively interrogate our positionality and the various power relations operating within the research process through discussions with other scholars and with the participants."

"Participants in some cases were surprised at this, noting that researchers do not usually do so, as a woman in Sri Lanka stated, 'THIS GENERALLY DOES NOT HAPPEN THAT ANY RESEARCHERS SHARE THEIR FINDINGS WITH THE RESPONDENTS—GRASS ROOT LEVEL UNLESS THEY SHARE WITH THEM WITH THE KEY INFORMANTS OR WITH THE ORGANIZATIONS/ COMMITTEES THEY INTERVIEW SO LET ME THANK YOU FOR ASKING OUR VIEWS AND IDEAS AND SENDING THEM YOUR FINDING TO YOUR RESPONDENTS' (text as written)."

"More implicitly, in her interactions with participants, Jasmine made a concerted effort to defer to and affirm the participants' preferences wherever possible, including in the platform for the interview, as indicated above, the number of questions or question sets exchanged in each response, and the format of the questions, as will be discussed next."

participant resistance and control

"In their email interviews, participants found ways to resist and subvert the power structure that is inherent in the research process, including ways that were medium-specific."

"The instructions requested that participants respond at the top of their email rather than below each question to provide a continuous record of their responses."

"Some participants entered responses below questions anyway."

"Participants also dictated the length of their responses, as email interviews allow."

"Despite these advantages and the agency that these women demonstrated, restraints to their participation based on their positionality and the research process became evident over the course of the email interviews."

"These issues with internet usage did not appear to be specific only to their participation in the interview, as their responses to questions about internet use indicated most used the internet almost daily for various purposes."

participant reflexivity

"Participants both reflected on their experiences and engaged in a reflexive process with the researchers through the process of reading and rereading their responses over an extended period of time."

"Participants specifically commented on their own about how the email interview had given them the opportunity to reflect."

"The opportunity to reflect at the end of the email interview on the entire record of the exchange resulted in many women assessing their own experiences, leading to crucial insights for this study and in some cases increased critical consciousness."

"Having the time and opportunity to reflect, afforded by the email interview method and the ease of engaging in an extended email dialogue between interview participant and researcher, made this experience perhaps more valuable for participants than a synchronous interview conducted face-to-face or digitally."

implications for a method of reflexive email interviewing

"Our own experience with email interviewing provides evidence of the ways in which it provides a means to start research from the perspectives of a geographically dispersed population of women, for both the participants and the researcher to subvert the traditional power dynamic in the research process, and to engage in a reflexive dialogue through the content and structure of email exchanges over time—all commitments of FST."

"Based on existing literature and our experiences in the case study, reflexive email interviewing is proposed to rely on the following criteria: 1) the researcher's use of explicit and implicit strategies to disrupt power hierarchies within the email interview process, 2) the researcher's engagement in reflexivity of her/his own positionality as a continuous part of the research process, and 3) continued invitations for participants to directly reflect on and respond to the research process, including the use of 'reflexive' questions that invite participants to ascribe meaning to their responses."

[Section 8]

"Given the expansion of the internet globally and increased interest in internet-based methods, more attention is needed to understand the implications of interviewing online for feminist research."

"Most discussions of online interviewing, including those in texts about feminist research, have tended to briefly highlight their benefits and drawbacks without an explicit articulation of how one might engage in feminist online interviewing."

"We respond to these calls by proposing reflexive email interviewing as a method for feminist research."

"We first discuss feminist standpoint theory as a theoretical framework, describing how it relates to existing approaches to online interviewing, and then offer a recent case study to examine this method in action."

6.3.5 Gender Bias and Conversational Agents: An Ethical Perspective on Social Robotics [219]

This is a machine-generated summary of:

Fossa, Fabio; Sucameli, Irene: Gender Bias and Conversational Agents: an ethical perspective on Social Robotics [219]

Published in: Science and Engineering Ethics (2022)

Link to original: https://doi.org/10.1007/s11948-022-00376-3

If you want to cite the papers, please refer to the original.

For technical reasons we could not place the page where the original quote is coming from.

Abstract-Summary

"The increase in the spread of conversational agents urgently requires to tackle the ethical issues linked to their design."

"The present paper discusses whether and to what extent it is ethically sound to intentionally trigger gender biases through the design of virtually embodied conversational agents."

"After outlining the complex dynamics involving social biases, social robots, and design, we evaluate the ethics of integrating gender cues in conversational agents, analysing four different approaches to the problem."

Introduction

"It is pivotal to ask to what extent it is ethically permissible to deliberately exploit pre-existing social biases in order to build products that successfully meet user expectations, blend in perfectly with their context of use, and maximize the perceived quality of interactions."

"This paper presents a discussion on whether it is ethically sound to leverage pre-existing biases in order to maximize the quality of human-machine interactions."

"Since such an issue potentially pertains to the entire field of social robotics and might apply to all kinds of bias, narrowing the scope of the analysis is both necessary and likely to be more productive in terms of theoretical insights and practical applications."

"The main question we consider is: Is it ethically permissible to align the design of ECAs to gender biases in order to improve interactions and maximize user satisfaction?"

Bias, Social Robots, and Ethics

"The cues that in human-human relations trigger biases that are conducive to effective interactions can be reproduced in the design of the artificial agent to which we want the corresponding human task to be delegated."

"According to this strategy, artificial agents should be specifically designed to trigger the same social biases that are triggered in the corresponding human-human interaction that is being automated."

"Gender biases triggered by design cues importantly influence user mental models and expectations."

"The bias alignment design strategy would suggest inserting some design cues that nudge users into projecting the male gender onto the technology."

"Through bias alignment, social biases can be leveraged to design better human-machine interactions."

"We are finally ready to pose our main research question: Is it ethically permissible to align the design of ECAs to gender biases in order to improve interactions and maximize user satisfaction?"

Ethics of Bias Alignment

"We will discuss the following answers: A1: It is ethically permissible to exploit widespread gender biases in order to maximize the quality of interactions between ECAs and users."

"A4: the inclusion of gender cues in ECA design should be used proactively to limit the spread of discriminatory bias and promote ethical attitudes."

"Let's start with discussing the affirmative answer to our question, which could be phrased as follows: It is ethically permissible to exploit widespread gender biases in order to maximize the quality of interactions between ECAs and users."

"it is ethically permissible to insert gender cues into ECA design as long as those cues do not spread a discriminatory vision of gender dynamics, reinforcing unethical social biases."

"The proactive approach is defined as follows: the inclusion of gender cues in ECA design should be used proactively to limit the spread of discriminatory bias and promote ethical attitudes."

Conclusions

"If complied with, this approach would make it possible to exploit the power of bias to our advantage without giving rise to ethically troubling feedback and lock-in effects, thus providing us with effective technologies and leaving the chances of correcting discriminatory social bias untouched."

"We are sympathetic to the idea of fighting discriminatory social bias by nudging people's beliefs and behaviours through the ethically informed design of ECAs and other technologies (A4)."

"A mild and entirely transparent approach, which would insist exclusively on widely acknowledged (e.g., legally sanctioned or institutionally opposed) discriminatory biases and would inform users about the embraced proactive design approach, might avoid some of the pitfalls usually associated with manipulation and paternalism."

6.3.6 Diversity via Datafication? Digital Patient Records and Citizenship for Sexuality and Gender Diverse People [220]

This is a machine-generated summary of:

Davis, Mark D. M.; Schermuly, Allegra; Smith, Anthony K. J.; Newman, Christy: Diversity via datafication? Digital patient records and citizenship for sexuality and gender diverse people [220]

Published in: BioSocieties (2022)

Link to original: https://doi.org/10.1057/s41292-022-00277-5

If you want to cite the papers, please refer to the original.

For technical reasons we could not place the page where the original quote is coming from.

Abstract-Summary

"In 2018, the Australian Government adopted an 'opt-out' strategy to increase participation in My Health Record (MHR), the national digital patient record system."

"LGBT community organisations warned that people with complex health needs should consider their privacy and legal situation when deciding whether or not to opt out of MHR."

"With reference to the health needs of sexuality and gender diverse people, we examine MHR's rights universalism, possessive individualism, and state-based rationalisation of health governance."

"It is a technological solution to state-based imperatives for health governance, an emphasis that does not serve minority communities or address health needs that attract stigma and prejudice."

"We suggest ways in which data medicine, of which MHR is but one example, can be made more relevant and effective for individuals and communities whose healthcare is poorly served by mainstream health systems."

My Health Record and the right to 'opt out'

"The current opt-in participation arrangements have not effectively encouraged use of My Health Records by individuals and their healthcare providers, and are creating a barrier to achieving the full benefits of the system for individuals. (Minister for Health [176]) This statement invoked the datafication of health as a universal human right, conferring on MHR potent powers over health outcomes that are difficult to refuse."

"The document went on to explain how consent to use of data would be undertaken in MHR and the related legal ramifications: Currently health information stored in medical files can only be accessed by a third party provided they gain court approval."

"Because MHR consent differs from approaches taken internationally (Mendelson [177]), it may lead to undesirable uses of data for individuals who travel between health systems."

Compulsory possessive individualism

"To MHR's quality as conferred citizenship technology, it also inscribes the notion that the database is owned by the citizen, something that is somewhat undermined by the co-existing expectation that citizens should participate in the curation of the individual record and how the data record is created by multiple parties."

"The collection of data – the record – became not the shared knowledge of a doctor and a patient in an intimate conversation, but the property of the patient, subject to, and under threat of violation by, malign external parties (Komesaroff and Kerridge [178], p. 1292)."

"Through its apparent focus on the record as 'belonging' to the patient, MHR decentres medical authority over healthcare, though as we have seen, there is also some basis for doubt that MHR is owned by anyone in particular."

Data medicine and reflexivity

"It is also important to recognise how sexuality and gender diverse people reflexively manage knowledge about themselves in clinical and online settings, in an effort to shape health outcomes for themselves while reducing stigma and mistaken assumptions on the part of people with whom they interact."

"Much can be learned, then, from how sexuality and gender diverse people already manage their lives in clinical settings and through digital media, practices that potentially intersect in the form of MHR and related digital health technologies."

"Such crafted self-reflexivity is important to understand because MHR has the potential to be an important source of knowledge about the healthcare needs of sexuality and gender diverse people with complex health needs."

"MHR is conceptualised as the technological solution through which health quality will be pursued, but there is little consideration of how MHR data assemblages in and of themselves might contribute to health and wellbeing, particularly for people with complex health needs and considering high risk disclosures and exposures."

Which citizenship?

"Against this backdrop, it becomes important to focus on how MHR can be optimised for sexuality and gender diverse people in ways that move beyond the conferral of citizenship through the right to opt out or by default."

"This framing of citizenship is concerned with the rights of individuals to freely experience sexual relationships without coercion and threat to health, freedom of expression and action, and the responsibility of economic and social institutions to support sexual citizenship."

"These other framings of citizenship also indicate how MHR and similar technologies may fall short of extending autonomy, control and protecting privacy in the interests of sexuality and gender diverse people in particular."

"Because these framings of citizenship help to problematise state conferred rights as they are articulated though data assemblages like MHR, they contribute to counterpublic dialogues on healthcare for sexuality and gender diverse people."

My Queer Health Record?

"A focus could be the creation of a shared and nuanced discourse on the benefits and drawbacks of data medicine for sexuality and gender diverse people and the invention of new kinds of health advocacy."

"Data medicine is only one example of health care challenges that raise deep and enduring tensions over identity, autonomy and harm for sexuality and gender diverse people."

"Like HIV for some sexuality and gender diverse people, data medicine will remain a feature of health care and is likely to shift and grow in importance."

"Too, MHR is likely to be only one important case of data medicine for sexuality and gender diverse people."

"It is likely then that MHR and data medicine in general will not succeed in terms of the interests and needs of sexuality and gender diverse people."

[Section 6]

"We examine Australia's My Health Record (MHR; Mendelson [177]) as a key component of data medicine with ramifications for sexuality and gender diverse people and their healthcare."

"Attention to the participation of sexuality and gender diverse people in MHR and similar technologies is significant for reasons of healthcare quality, right to privacy and the generation of knowledge about health that MHR is thought to be able to deliver."

"With reference to the healthcare needs of sexuality and gender diverse people, we seek in this paper to reflect on MHR as a significant instance of data medicine assemblages."

"By drawing, then, on different framings of citizenship—digital, biological, sexual—we consider how MHR could become a tool for generating insight into the relations of sexuality and gender diverse people with state-based electronic patient records and prospects for increased autonomy with regard to data medicine."

6.3.7 Perpetuating Gender Inequality via the Internet? An Analysis of Women's Presence in Spanish Online Newspapers [221]

This is a machine-generated summary of:

Mateos de Cabo, Ruth; Gimeno, Ricardo; Martínez, Miryam; López, Luis: Perpetuating Gender Inequality via the Internet? An Analysis of Women's Presence in Spanish Online Newspapers [221]

Published in: Sex Roles (2013)

Link to original: https://doi.org/10.1007/s11199-013-0331-y

If you want to cite the papers, please refer to the original.

For technical reasons we could not place the page where the original quote is coming from.

Abstract-Summary

"Our general objective was to investigate possible gender bias in Spanish online news."

"Analysis of article length and publication day showed that women appear more frequently in shorter news items and in the Sunday news, which we interpreted as indicators of male association with newsworthiness."

"We also found differences in gender reporting since female journalists tend to include more women in the news they report than their male peers."

"These results provide evidence that online newspapers continue to perpetuate underrepresentation, stereotyping, and discrimination of women in web news thereby reinforcing gender inequality."

Introduction

"We examine article sections, the reporter's gender relation with the use of female or male subjects, and story placement indicators (length and publication day) in order to explore possible gender bias in the online news."

"The assumption behind the existence of possible gender differences in reporting, according to the gender model of socialization, is that female reporters bring with them a different value system because of their gender differences in socialization, resulting in more diverse sources in the news and in the topics (this model is supported by several works, e.g.: Price and Wulff [179] in US broadcast networks and Rodgers and Thorson [180] in US daily newspapers)."

"Studies of both network television and newspapers in the United States in the 1990s (Armstrong [181]; Rodgers and Thorson [180]; Sutcliffe and others [182]; Zoch and Turk [183]) determined clear differences between female and male reporters in their selection of news sources."

"Many studies demonstrate that female reporters are routinely assigned to women's issues and softer news stories."

Method

"The second step, once we created the database, was to automatically identify male and female first names with a text analysis tool based on a natural language-processing tool (Schmid [184]) capable of identifying all the words in each sentence and their grammatical functions, namely, nouns, adjectives, verbs, and so on."

"The stored news items processed with the natural language-processing tool provided the word count of the news item, the number of male and female first names in the body copy, the reporter's gender, and the presence of a male or female first name in the headline."

"We obtained a Krippendorff alpha (which measures the degree of agreement between human experts and automatic coding) of 0.8138 for the proportion of female names, which implies that the automatic count of women mentioned in the news provides a data set that is sufficiently reliable for statistical analysis (Hayes and Krippendorff [185]; Krippendorff [186])."

Results

"To measure the reporter's gender (Hypothesis 2, that women and mixed male–female team reporters tend to include more women in their news stories), we introduced three dummy variables denoting whether the news item was bylined by a female journalist, a male journalist, or a mixed female–male team, respectively."

"In the cases of items with either a female journalist or a mixed female–male team, the proportion of women cited was higher than when the news was bylined by a man, corroborating Hypothesis 2."

"Many of the elements are closely related (e.g., women journalists report more often on "female" sections, such as society or people, and the number of news items in each section varies across weekdays), preventing a descriptive analysis from providing reliable conclusions since we are not always sure which variable is responsible for the differences found."

Discussion

"Female journalists seem to be gender complicit leading them to include more women in news items they byline than their male counterparts, even after controlling for news section."

"Since the 'glass ceiling' is a result of a systematic underestimation of women's abilities and discriminatory practices, it could be argued that male-dominated newsrooms would tend to consider women as less newsworthy than men, rewarding males with what has been traditionally considered as more prominent and prestigious news, i.e., the longer and weekday items."

"We offer evidence that, in the Spanish online newspapers women continue to be stereotyped in traditional "female" sections, marginalized to the less newsworthy news and discriminated by male reporters that tend to exclude them from the stories they write, perpetuating this way gender inequality in Spanish society despite women's advances in employment, politics, and education over the last 20 years."

6.3.8 Feminist Postdigital Inquiry in the Ruins of Pandemic Universities [222]

This is a machine-generated summary of:

Hurley, Zoë; Al-Ali, Khadija: Feminist Postdigital Inquiry in the Ruins of Pandemic Universities [222]

Published in: Postdigital Science and Education (2021)

Link to original: https://doi.org/10.1007/s42438-021-00254-4

If you want to cite the papers, please refer to the original.

For technical reasons we could not place the page where the original quote is coming from.

Abstract-Summary

"Some learners have been developing resistant practices to technology-driven learning, including being on mute and turning off cameras, but these silences, gaps and evasions are difficult to grasp through normative perspectives."

"We need alternate conceptions of learners' varying responses to technologies."

"This mashrabiyya structure is applied as a theoretical metaphor to consider Arab women learners' technological veiled affordances of filters, avatars and not replying."

"The mashrabiyya feminist postdigital framework develops unique inquiry into learners' subtle practices; the authors' self-reflexivity; and analysis of a (silent) email exchange and a Twitter avatar."

"Theorising suggests silences, invisibilities and disconnection are not necessarily a deficit but refractive responses enabling students and educators to stay below the radar."

Introduction

"While algorithmic data tallies up the numbers of suffering, redundancies, deaths and profit, the study asks: (1) How can we conceptualise Arab women's technological practices in the postdigital condition? (2) How have Arab women responded to the rapid technological shift to online delivery of education?"

"Arab women learners, at universities in Kuwait and the United Arab Emirates (UAE), are taken as a case to illustrate postdigital practices within 'scopic regimes' (Mirzoeff [188])."

"To consider the question of Arab women students' postdigital practices, the following folds of inquiry occur."

"We develop the mashrabiyya postdigital feminist framework, concerned with theorising Arab women learners' postdigital practices."

"Discussion of the findings helps to address the study's research questions, which consider Arab women students' postdigital practices within the pandemic university."

Literature Review: Pandemic Learning in the Ruins

"Roy does not advance a focus on Arab women students or how they might get ready for the 'fight', while the pandemic has turned the clock back on gender equality by several decades (Hurley [189]; Johnston [190])?"

"These issues align with the research questions of the inquiry which explore conceptualisations of Arab women's postdigital practices and their responses to online education within the domestic sphere."

"To consider further, how Arab women learners have been affected by Covid-19, next we discuss the Middle East North African (MENA) context."

"The varying experiences of Arab women during Covid-19 vary significantly across MENA."

"Exploring women's learning in Middle Eastern contexts from a feminist perspective is an important endeavour, particularly in view of Western stereotypes surrounding Muslim women as passive and needing to be unveiled, while Arab women do not necessarily view themselves in these terms (Ahmed [191]; Abu-Lughod [192])."

"We thus recognise that Arab women students' postdigital practices could occur in tandem and/or independently from Western pedagogic traditions and conceptions."

Oblique Gaze: Mashrabiyya Feminist Postdigital Framework

"Developing mashrabiyya feminist postdigital theoretical tools, to explore alternate scopic regimes, are required but do not merely involve side-stepping androcentric and ethnocentric philosophy."

"Mashrabiyya feminist postdigital inquiry is a conceptual process for considering social actors' hybrid practices."

"The analytic folds of the mashrabiyya feminist postdigital framework are explicit while inquiry considers discreet practices."

"How can mashrabiyya feminist postdigital inquiry address the research questions of the study which ask, (1) How can we conceptualise Arab women students' practices in the postdigital condition? (2) How have Arab women responded to the rapid technological shift to online delivery of education?"

"Mashrabiyya feminist postdigital inquiry considers researchers' perspectives as always informed by intersections of social class, gender, ethnicity, religion, etc This indicates that you cannot avoid researcher bias."

"As mashrabiyya feminist postdigital researchers, we realise these tricky questions of postdigital context as well as the intersections of offline/online sociocultural histories and phenomena."

Findings

"To address how the above mashrabiyya practices come together as a chain of meanings, for possible interpretation within the sociocultural context, we can observe in the email exchange that Khadija engaged in (unintentional) silent responses to Amal's nine textual emails."

"In relation to addressing the first research question of the study asking, how can we conceptualise Arab women students' practices in the postdigital condition, Khadija says that in this email incident, it is evident that Amal was engaged and part of a dialogical exchange with silence (of a silent/absent tutor) throughout."

"My silences were crucial and preceded Amal's leap into building the dialogue, presenting her demands, negotiating the solution the way she did. (Al-Ali [193]) In response to the second research question, asking how Arab women responded to the rapid technological shift to online delivery of education, application of mashrabiyya feminist postdigital nodes helps to articulate an understanding of the relational, sociocultural and postdigital practices of the emails."

Discussion
"It perceives of Arab women students' practices in postdigital terms, refractive silence and invisibility."

"The mashrabiyya feminist postdigital inquiry views Arab women students' practices as located within structures of power."

"She explains that social media, like the veil, has certain affordances that enable Arab women to be online while adhering to the contextual patriarchal bargain in the Middle East (Kandiyoti [194]), but within online settings, and according to liberal feminist research in particular, the dichotomy of voice and silence is pervasive, as is the decision of whether or not to show face online."

"Liberal feminist online theorising also seems biased and based on the preference of voice over silence, showing face and being seen as an assertion of women's agency, authority and presence."

"It also contests forcible unveiling, demasking and obligations for Arab women learners to show face online or to speak when they opt for silence."

Conclusion
"Mashrabiyya feminist postdigital inquiry is developed as an interpretive theoretical framework."

"It conceptualises Arab women students' practices in the postdigital condition and how Arab women have responded to the rapid technological shift to online delivery of education."

"Arab women's relative resistance, of silence and invisibilities, like Zoë's student use of the Twitter avatar, are entangled within intersectional practices."

"It appreciates Arab women learners' refractive postdigital practices."

"The mashrabiyya feminist postdigital framework could also be generalisable to other research agendas informed by social actors' discreet onto-epistemologies."

"Although we might like to cast off cultural baggage, to pass through the Covid-19 portal in a light, breezy yet combative manner, suggested by Roy [195], gendered practices are not necessarily disposable, but new theoretical tools, of mashrabiyya inquiry, refuse essentialist narratives of scopic regimes."

"Arab women learners have other insights into refractive practices of veils, silences and mashrabiyya to evade hegemony."

6.3.9 Beyond Tools: Technology as a Feminist Agenda [223]

This is a machine-generated summary of:

Ramilo, Chat Garcia: Beyond Tools: Technology as a feminist agenda [223]
Published in: Development (2006)
Link to original: https://doi.org/10.1057/palgrave.development.1100230

If you want to cite the papers, please refer to the original.

For technical reasons we could not place the page where the original quote is coming from.

Abstract-Summary
"Chat Garcia Ramilo argues strongly for a feminist agenda on technology."

Introduction
"Alarmingly absent or pitifully lacking from all processes and spaces that are constructing the frameworks, rules, structures, standards and tools of the new technologies."

"Disturbingly absent or still on the fringes of the discourses and debates that animate the feminist movement despite profound developments in new technologies that are changing our world in unalterable ways."

"These were the twin recurring themes in all the events that tackled new technologies at the AWID Forum, whether they were about genetics and eugenics, nanotechnology or information technologies."

The raging revolution of genetics
"What of the change that new technologies are already creating and are likely to wreck on women's bodies, identities and consciousness?"

"The handful of workshops dealing with technologies during the four days of collective reflection demonstrated why it is imperative for the women's movement to pay closer attention to new technologies and the corporate strategies that shape their development."

"Darling drew the line between the cutting edge and the bleeding edge of new technologies for women."

"She spoke of technological transformations in electronics that happens external to the human body, whereas the raging revolution in genetics is being played out in women's bodies and threatens to privatize all forms of life."

Information and communication technologies (ICTs)
"The term ICT has been used to encompass technological innovation and convergence in information and communication leading to the development of the so-called information or knowledge societies."

"The growth of ICT in the past decade is making the universal communication and flow of information possible and has been heralded as a powerful force in advancing equitable economic and social development, including gender equality."

"The rapid development of new technologies that have blurred the boundaries between information, communication and the various types of media have facilitated these tendencies, with mergers between different sectors of the communication economy creating a greater concentration of power."

"The shift towards the dominance of public policy that supports the concentration of the information and communications industries towards private enterprise have magnified the monopolistic power of new information and media moguls."

Trade, communication and pornography

"The muted response from feminists relegated the question about the very real and material consequences of creating a specialized space for the marketing of (predominantly) women's sexuality into one of lesser concern than the 'larger' political issue of Internet governance."

"Left as it is, with minimal intervention from women's movements and the structural dominance of masculinist discourse in both the fields of technology and sexually explicit content, our de-prioritization of this issue is augmenting unequal gender relations and the discursive construction of subjects based on highly contentious sexist, racist and related paradigms."

"It is without question that the dominant mass of pornography currently available through digital technologies is discriminatory toward women."

"Produced largely for the heterosexual male audience, images of women are shuttled across digital spaces as passive objects with minimal sexual agency."

Defining the feminist agenda on technology

"The connection that needs to be made is to shift the understanding of ICTs from a ghettoized area relevant only to those who are privileged enough to have technological access, to the larger framework of women's human rights, which includes violence against women."

"The other, less obvious agenda, is for the feminist movement to engage with technologies as a site of feminist political struggle."

"The challenge for the women's movement is to take control over the technologies that they use and ultimately influencing the decisions that direct their development."

"A number of feminists see the newly emerging cultural analyses of technology as a suitable framework for analysing gender and ICT relationships."

"This framework understands both technology and gender not as fixed and given, but as cultural processes which (like other cultural processes) are subject to 'negotiation, contestation, and, ultimately transformation'."

6.3.10 Measuring the Digital Inclusion of Women: A Poset-Based Approach to the Women in Digital Scoreboard [224]

This is a machine-generated summary of:

Damiani, Filippo; Rodríguez-Modroño, Paula: Measuring the digital inclusion of women: a poset-based approach to the women in digital scoreboard [224]

Published in: Quality & Quantity (2023)

Link to original: https://doi.org/10.1007/s11135-023-01666-z

If you want to cite the papers, please refer to the original.

For technical reasons we could not place the page where the original quote is coming from.

Abstract-Summary

"The participation of women in digital society is essential to achieving the objectives of Agenda 2030 and an essential component in the EU strategy for the digital transition."

"This article applies a poset-based approach to the European Women in Digital (WiD) Scoreboard, to examine the digital inclusion of women in EU Member States and in the UK."

"Our research contributes to better understanding the dynamics and the factors facilitating the digital inclusion of women in the EU-28 Member States, providing a clustering of EU countries into four performance groups depending on the digital inclusion of women."

Introduction

"Although the gender gap in IT skills and the use of digital devices is gradually levelling out across the European Union (EU), women are still behind men in the use of various ICT technologies at work (EIGE [196])."

"Gender equality needs to be considered as a primary objective in the EU strategy for the digital transition, incorporating the measurement of advances in digitalisation for women and men as an essential component of this strategy."

"The WiD index, part of the Digital Economy and Society Index (DESI), consists of twelve indicators to assess the performance of Member States in the areas of Internet use, Internet user skills as well as specialist skills and employment (European Commission [197])."

"Our findings about the significance of indicators, depending on the region, contribute to identifying areas where policy intervention continues to be needed, and to the design of more targeted and effective policies for integrating gender equality into the EU digital transition strategy."

From the digital gender divide to women's digital inclusion

"Research on the digital gender divide and women's digital inclusion can be divided into three main phases."

"This first-order digital gender gap focused on the lack of access to ICT."

"Despite the measures implemented to enhance women's digital skills and to increase the participation of women in the ICT workforce, studies show that gender disparities in digital proficiency are still more evident at the higher levels of specialisation."

"The third digital gender gap refers to this differentiated use of the most advanced ICT technologies and applications."

"The digital gender divide becomes even more pronounced when it comes to women as creators of technology or software developers."

"One of the main challenges is the development of gender-specific and gender-sensitive indicators and indices that provide insights into the breadth and depth of women's digital inclusion, since ICT-focused indices which include gender dimensions have a relatively short history (Brimacombe and Skuse [198])."

Materials and methods

"The data matrix considered in this study consists of 28 countries and 12 indicators; hence, the total number of observations is 336."

"We apply the partial order theory, or poset-based approach, a discipline associated with discrete mathematics, in which the objects of a data set, consisting of multiple indicators, are compared and ordered to obtain a ranking (Brüggemann and Patil [199])."

"In the analysis presented in this study, the first step consists in the identification of the Hasse diagram, which represents the relations between the 28 countries according to their scores considering all 12 indicators together."

"If we create a ranking by simply computing the unweighted arithmetic average, we obtain a different ranking, with country B better ranked than country C. In our analysis we use the LPOM to create the ranking considering first the entire dataset, and then we repeat the same process for each of the three dimensions of indicators."

Results

"The last positions of the ranking are occupied by Hungary and Slovakia: they both present incomparabilities with 19 countries, meaning that eight different countries have higher scores on all indicators compared to them."

"Hungary and Slovakia are in the last position of the poset ranking since their downset is 1, which means that every country has better scores in at least one indicator compared to Hungary and Slovakia, and their upset is 9; in fact, 8 countries perform better than Hungary and Slovakia in every indicator."

"As explained in the discussion above, Malta belongs to the small group of countries that cannot be compared with the rest, according to the poset-based approach theory, thanks above all to the positive results in relation to indicator 1.5, where Malta ranks second."

Discussion and conclusions

"We applied poset theory to analyse women's digital inclusion in the EU-28 countries using the data from the 12 indicators of the Women in Digital Scoreboard 2020."

"A ranking computed with this method can provide just an overview of the aggregate performance of a given country; whereas, the application of the poset-based approach permits to obtain a ranking which already takes into account the comparison between countries and their scores indicator by indicator, introducing the concept of incomparability in case the performances of two given countries are not coherent among the indicators considered (country A is better than country B in the indicator 1, but is worse in the indicator 2)."

"According to the poset-based approach, the leaders consist of those countries that achieve better results in relation to all indicators compared to at least three other countries."

"The United Kingdom is the country leading the ranking, since it achieves better scores on all the indicators with respect to eleven other countries; Ireland and Spain, the last two countries in the leaders' group, achieve better results on all indicators in comparison to three different countries."

Bibliography

1. Lam, R. (2020). Don't sugar it: An analysis of sugar dating culture on Reddit. In A closer look in unusual times, 97–108
2. Motyl, J. (2012). Trading sex for college tuition: How sugar daddy dating sites may be sugar coating prostitution. Penn State Law Review, 117, 927–957
3. Nayar, K. I. (2017). Sweetening the deal: Dating for compensation in the digital age. Journal of Gender Studies, 26(3), 335–346
4. Cordero, B. D. (2015). Sugar culture and SeekingArrangement.com participants: What it means to negotiate power and agency in sugar dating (Master's thesis, California State University, Sacramento, California). Retrieved from http://dspace.calstate.edu/handle/10211.3/159309
5. Green, S. P. (2016). What counts as prostitution? Bergen Journal of Criminal Law and Criminal Justice, 4(1), 65–101
6. Kuate-Defo, B. (2004). Young people's relationships with sugar daddies and sugar mummies: What do we know and what do we need to know? African Journal of Reproductive Health, 8(2), 13–37
7. Miller, A. (2011). Sugar dating: A new take on an old issue. Buffalo Journal of Gender Law & Social Policy, 33, 33–68

8. Rakic, V. (2020). Prostitutes, sex surrogates, and sugar babies. Sexuality & Culture, 24, 1207–1217
9. Barak-Brandes, S., & Lachover, E. (2015). Branding relations: mother–daughter discourse on beauty and body in an Israeli campaign by Dove. Communication, Culture & Critique, 9(3), 379–394. https://doi.org/10.1111/cccr.12111
10. Crotty, M. (1998). The foundations of social research: meaning and perspective in the research process. Thousand Oaks, CA: Sage Publications
11. Holland G, Tiggemann M."Strong beats skinny every time": disordered eating and compulsive exercise in women who post fitspiration on Instagram. Int J Eat Disord. 2017;50(1):76–9. https://doi.org/10.1002/eat.22559
12. Tiggemann, M., & Slater, A. (2014). NetTweens: The internet and body image concerns in preteenage girls. The Journal of Early Adolescence, 34(5), 606–620. https://doi.org/10.1177/0272431613501083
13. Bandura, A. (1977). Social learning theory. Englewood Cliffs, NJ: Prentice-Hall
14. Bandura, A. (1986). Social foundations of thought and action: a social cognitive theory. Englewood Cliffs, NJ: Prentice-Hall
15. Fishbein, M., & Ajzen, I. (1975). Belief attitude, intention and behavior: an introduction to theory and research. Reading, MA: Addison-Wesley
16. Barlas, A. (2006)."Believing women"in Islam: unreading patriarchal interpretations of the Qur'ān. Austin: University of Texas Press
17. Karam, A. (1998). Women, Islamisms and the State: contemporary feminisms in Egypt. Houndmills: Macmillan Press
18. Gül, S. S., & Gül, H. (2000). The question of women in Islamic revivalism in Turkey: a review of the Islamic Press. Current Sociology, 48(2), 1–26
19. Mahmood, S. (2005). Politics of piety: the Islamic revival and the feminist subject. Princeton: Princeton University Press
20. Badran, M. (2001). Locating feminisms: the collapse of secular and religious discourses in the Mashriq. Agenda, 50, 41–57
21. Griffin, W. (2004). The goddess net. In L. L. Dawson & D. E. Cowan (Eds.), Religion online: finding faith on the Internet (pp. 174–188). London: Routledge
22. Bunt, G. (2000). Virtually Islamic: computer-mediated communication and cyber Islamic environments. Lampeter: University of Wales Press
23. Bunt, G. (2003). Islam in the digital age: E-jihad, online Fatwas and cyber Islamic environments. London: Pluto Press
24. Eickelman, D., & Anderson, J. (1999). Redefning Muslim Publics. In D. Eickelman & J. Anderson (Eds.), New media in the Muslim world: the emerging public sphere (pp. 1–18). Bloomington: Indiana University Press
25. Bastani, S. (2001). Muslim women online. Arab World Geographer, 3(1), 41–74
26. Dunne, B. (1998). Power and sexuality in the Middle East. Middle East Report. 206. np. http://www.merip.org/mer/mer206/bruce.htm. Accessed on 7 April 2010
27. Schoemaker, E. (2016). Digital faith: Social media and the enactment of religious identity in Pakistan (Doctoral thesis). London School of Economics and Political Science
28. Shaheed, F. (2002). Imagined citizenship: Women state and politics in Pakistan. Shirkat Gah
29. Haeri, S. (2002). No shame for the Sun: Lives of professional Pakistani Women. Syracuse University Press
30. Butler, J. (1990). Gender Trouble: Feminism and the Subversion of Identity. Routledge
31. Litosseliti, L. (2006). Gender and language theory and practice (1st ed.). Routledge
32. Cover, R. (2012). Performing and undoing identity online: Social networking, identity theories and the incompatibility of online profiles and friendship regimes. Convergence: The International Journal of Research into New Media Technologies, 18(2), 177–193
33. Pillay, N. (2008). The portrayal of women in television advertisements on SABC3: A reflection on stereotypical representation (Master's thesis). University of the Witwatersrand, Johannesburg. Retrieved from https://genderlinks.org.za/gmdc/research/the-portrayal-of-

women-intelevision-advertisements-on-sabc3-a-reflection-on-stereotypical-representation-2011-08-01/

34. Raza, S. H., & Liaqat, M. (2016). Honor killing of women: A comparative study of the Pakistani print media. VFAST Transactions on Education and Social Sciences, 10 (1), https://vfast.org/journals/index.php/VTESS/article/view/419
35. Rehman, Z. (2017). Online feminist resistance in Pakistan. Sur International Journal on Human Rights, 14 (26), 151–159. https://jyu.finna.fi/PrimoRecord/pci.proquest2075503095
36. Zafar, M. Z., Toor, A., & Hussain, T. (2019). Social media as a conduit for women entrepreneurs in Pakistan. Pakistan Business Review, 20(4), 886–889. https://pbr.iobm.edu.pk/wp-content/uploads/2019/06/10-Article-1109.pdf
37. Ali, R., & Batool, S. (2015). Stereotypical identities: Discourse analysis of media images of women in Pakistan. Multidisciplinary Journal of Gender Studies, 4(1), 690–717. https://doi.org/10.4471/generos.2015.54
38. Nasir, S., Vel, P., & Mateen, H. (2012). Social media and buying behaviour of women in Pakistan towards the purchase of textile garments. Business Management Dynamics, 2(2), 61–69
39. Author, R. (2020). Compliance and resistance: An investigation into the construction of gender identities by Pakistani women on Facebook. Asian Journal of Women's Studies, 26(4), 503–527. https://doi.org/10.1080/12259276.2020.1854414
40. Author, R. (2021). Men will be Men?: Masculinities on display in the Facebook communication practices of Pakistani men. NORMA, 16(1), 38–56. https://doi.org/10.1080/18902138.2021.1875640
41. Habiba, U., Ali, R., & Ashfaq, A. (2016). From patriarchy to neopatriarchy: Experiences of women from Pakistan. International Journal of Humanities and Social Science, 6(3), http://ijhssnet.com/journals/Vol_6_No_3_March_2016/24.pdf
42. Ringrose, J., & Harvey, L. (2015)."BBM is like match.com": Social networking and the digital mediation of teen's sexual cultures? In J. Bailey & V. Steeves (Eds.), eGirls, eCitizens: Putting technology theory, policy & education into dialogue with girls' and young women's voices (pp. 199–227). Canada: University of Ottawa Press
43. Albury, K., Crawford, K., Bryon, P., & Matthews, B. (2013). Young people and sexting in Australia: Ethics, representation and the law, final report. Sydney: Australian Research Centre for Creative Industries and Innovation, University of New South Wales. Retrieved from: http://www.cci.edu.au/sites/default/files/Young_People_And_Sexting_Final.pdf
44. Coy, M., Kelly, L., Elvines, F., Garner, M., & Kanyeredzi, A. (2013)."Sex without consent, I suppose that is rape": How young people in England understand sexual consent. London: Office of the Children's Commissioner
45. Harris, A., Davidson, J., Letourneau, E., Paternite, C., & Miofsky, K. T. (2013). Building a preventative framework to address teen"sexting"behaviours. US Department of Justice. Retrieved from https://www.ncjrs.gov/pdffiles1/ojjdp/grants/244001.pdf. Accessed 10 Dec 2014
46. Phippen, A. (2012). Sexting: An exploration of practices, attitudes and influences. London: NSPCC and UK Safer Internet Centre. Retrieved from http://www.nspcc.org.uk/globalassets/documents/research-reports/sexting-report-2012
47. Ringrose, J., Gill, R., Livingstone, S., & Harvey, L. (2012). A qualitative study of children, young people and 'sexting': A report prepared for the NSPCC. London: NSPCC. Retrieved from https://www.nspcc.org.uk/globalassets/documents/researchreports/qualitative-study-children-young-people-sexting-report.pdf
48. Walker, S. J. (2012). Sexting and young people: A qualitative study (Unpublished masters thesis). University of Melbourne, Australia. Retrieved from https://minervaaccess.unimelb.edu.au/bitstream/handle/11343/37683/287493_SJW%20thesis%207-11-12.pdf?sequence=1
49. Bond, E. (2010). The mobile phone = bike shed? Children, sex and mobile phones. New Media and Society, 13(4), 587–604. https://doi.org/10.1177/1461444810377919

50. Jonsson, L. S., Bladh, M., & Priebe, G. (2015). Online sexual behaviours among Swedish youth: Associations to background factors, behaviours and abuse. European Child and Adolescent Psychiatry, 24(10), 1245–1260. https://doi.org/10.1007/s00787-015-0673-9
51. Salter, M. (2015). Privates in the online public: Sex(ting) and reputation on social media. New Media and Society, 18(11), 2723–2739. https://doi.org/10.1177/1461444815604133
52. Crofts, T., Lee, M., McGovern, A., & Milivojevic, S. (2015). Sexting and young people. Basingstoke and New York: Palgrave Macmillan
53. Lippman, J. R., & Campbell, S. W. (2014). Damned if you do, damned if you don't… if you're a girl: Relational and normative contexts of adolescent sexting in the United States. Journal of Children and Media, 8(4), 371–386. https://doi.org/10.1080/17482798.2014.923009
54. McGraw, J. M. (2013). Sexting unleashed: the social phenomenon according to youth sexters. Dissertations and Theses, Paper 1491. Retrieved from http://pdxscholar.library.pdx.edu/open_access_etds/1491
55. Bailey, J., & Mouna, H. (2011). The gendered dimensions of sexting: Assessing the applicability of Canada's child pornography provision. Canadian Journal of Women and the Law, 23(2), 405–441. https://doi.org/10.3138/cjwl.23.2.405
56. Burkett, M. (2015). Sex(t) talk: A qualitative analysis of young adults' negotiations of the pleasures and perils of sexting. Sexuality and Culture, 19(4), 835–863. https://doi.org/10.1007/s12119-015-9295-0
57. Cooper, K., Quayle, E., Jonsson, L., & Svedin, C. G. (2016). Adolescents and self-taken sexual images: A review of the literature. Computers in Human Behavior, 55(Part B), 706–716. https://doi.org/10.1016/j.chb.2015.10.003
58. Lenhart, A. (2009). Teens and sexting: How and why minor teens are sending sexually suggestive nude or nearly nude images via text messaging. Washington, DC: Pew Research Centre. Retrieved from http://pewresearch.org/assets/pdf/teens-andsexting.pdf
59. Klettke, B., Hallford, D. J., & Mellor, D. J. (2014). Sexting prevalence and correlates: A systematic literature review. Clinical Psychology Review, 34(1), 44–53. https://doi.org/10.1016/j.cpr.2013.10.007
60. Strassberg, D. S., Rullo, J. E., & Mackaronis, J. E. (2014). The sending and receiving of sexually explicit cell phone photos ("sexting") while in high school: One college's students' retrospective reports. Computers in Human Behaviour, 41, 177–183. https://doi.org/10.1016/j.chb.2014.09.008
61. Gill, R. (2007). Critical respect: The difficulties and dilemmas of agency and 'choice' for feminism (a reply to Duits and van Zoonen). European Journal of Women's Studies, 14(1), 69–80. https://doi.org/10.1177/1350506807072318
62. Schwalbe, M., Godwin, S., Holden, D., Schrock, D., Thompson, S., & Wolkomir, M. (2000). Generic processes in the reproduction of inequality: An interactionist analysis. Social Forces, 79(2), 419–452. https://doi.org/10.2307/2675505
63. Herriot, L., & Hiseler, L. E. (2015). Documentaries on the sexualization of girls: Examining slut-shaming, victim blaming and what's being left off screen. In E. Renold, J. Ringrose, & R. D. Egan (Eds.), Childhood, sexuality and sexualisation (pp. 289–304). Basingstoke: Palgrave Macmillan
64. MOW-International Research Team. 1987. The Meaning of Work: An International View. London: Academic Press
65. Caplan, Kimmy and Nurit Stadler, (Eds.). 2012. From survival to consolidation: change in Israeli Haredi society and its scholarly study, Van Leer Jerusalem Institute
66. Ji, Li., Guiyao Tang, Xinran Wang, Ming Yan, and Zhiqiang Liu. 2012. Collectivistic-HRM, firm strategy and firm performance: an empirical test. The International Journal of Human Resource Management 23: 190–203
67. Kemp, Linzi J., Susan R. Madsen, and J. James Davis. 2015. Women in business leadership: a comparative study of countries in the Gulf Arab states. International Journal of Cross Cultural Management 15: 215–233

68. Kay, Avi, Livia Levine, and Ephraim Shapiro. 2022. Reducing anxiety through workplace communication during COVID-19: who, what, when and how? International Journal of Business Communication. https://doi.org/10.1177/23294884221105580
69. Sharabi, Moshe, and Itzhak Harpaz. 2016. Impact of Generational Differences on Work Values in the Israeli Context. In Generational Differences in Work Values and Work Ethic: An International Perspective, ed. M. Sharabi, 19–41. Commack, N. Y.: Nova Science Publishers
70. Lyons, Sean T., Linda Duxbury, and Christopher Higgins. 2007. An empirical assessment of generational differences in basic human values. Psychological Reports 101: 339–352
71. Sharabi, Moshe, and Kay Avi. 2021. The relative centrality of life domains among secular, traditionalist and Ultra-Orthodox (Haredi) men in Israel. Community, Work and Family 24: 60–76
72. Sharabi, Moshe, and Itzhak Harpaz. 2011. Gender and the relative centrality of major life domains: changes over the course of time. Community, Work and Family 14: 57–62
73. Freund, Anat, Amit Zriker, and Ester Shor. 2019."All Glorious is the Princess within Her Chamber": ideological-spiritual fulfillment of Israeli ultra-Orthodox Jewish women in the changing employment environment. Journal of Religion and Spirituality in Social Work: Social Thought 38: 140–160
74. Gammage, S., Kes, A., Winograd, L., Sultana, N., Hiller, S., and Bourgault, S. 2017. Gender and digital financial inclusion: What do we know and what do we need to know?. ICRW
75. Lee, J., Morduch, J., Ravindran, S., and Shonchoy, A. 2021. Narrowing the Gender Gap in Mobile Banking, Working paper 2108, Florida International University, Department of Economics. https://economics.fiu.edu/research/working-papers/2021/2108/2108.pdf
76. GSMA. 2022. Connected women: The Mobile Gender Gap Report 2022. GSMA Report. GSMA – The Mobile Gender Gap Report 2022
77. Ravi, S. 2018. Is India ready to JAM? Brookings India, Impact Series No. 082018-3. https://www.brookings.edu/wp-content/uploads/2018/08/JAM-paper.pdf
78. Pinto, A.R., and Arora, A. 2021. Digital doorstep banking: Female banking agents lead digital financial inclusion through the pandemic and beyond. ADBI Working Paper Series. No. 1285. https://www.adb.org/publications/digital-doorstep-banking-female-banking-agents-lead-digital-financial-inclusion-through-the-pandemic-and-beyond
79. Hindustan Times. 2022. Women banking correspondents widen financial inclusion in rural Jharkhand. Hindustan Times. https://www.hindustantimes.com/cities/ranchi-news/women-banking-correspondents-widen-financial-inclusion-in-rural-jharkhand-101643041030346.html. Accessed 24 Jan 2022
80. MSC. 2019. The real story of women's financial inclusion in India. Microsave Consulting. https://www.microsave.net/wp-content/uploads/2020/01/191125_The-real-story-of-womens-financial-inclusion-in-India_Gender-research-report.pdf
81. Moffitt, T. E., Arseneault, L., Belsky, D., Dickson, N., Hancox, R. J., Harrington, H., Houts, R., Poulton, R., Roberts, B. W., & Ross, S. (2011). A gradient of childhood self-control predicts health, wealth, and public safety. Proceedings of the National Academy of Sciences, 108(7), 2693–2698. https://doi.org/10.1073/pnas.1010076108
82. Xie, D., Wang, L., Tao, T., Fan, C., & Gao, W. (2014). Validity and reliability of the Chinese version of the Dual-Mode of Self-Control Scale for adolescents. Chinese Mental Health Journal, 28(5), 386–391
83. Yajun, L., Fumei, C., Furong, L., & Yun, W. (2015). Effect of Cyber Victimization on Deviant Behavior in Adolescents: The Moderating Effect of Self-control. CHINESE JOURNAL OF CLINICAL PSYCHOLOGY, 23(5), 896–900
84. Dou, G., Xiang, Y., Sun, X., & Chen, L. (2020). Link between cyberbullying victimization and perpetration among undergraduates: Mediating effects of trait anger and moral disengagement. Psychology Research and Behavior Management, 13, 1269–1276. https://doi.org/10.2147/prbm.S286543
85. Wang, L., & Ngai, S.S.-Y. (2021). Cyberbullying Perpetration Among Chinese Adolescents: The Role of Power Imbalance, Fun-seeking Tendency, and Attitude Toward Cyberbullying.

Journal of Interpersonal Violence, Article 08862605211062988. https://doi.org/10.1177/08862605211062988
86. Zhan, J., Yang, Y., & Lian, R. (2022). The relationship between cyberbullying victimization and cyberbullying perpetration: The role of social responsibility. Frontiers in Psychiatry, 13, 995937–995937. https://doi.org/10.3389/fpsyt.2022.995937
87. Nylund-Gibson, K., Asparouhov, T., & Muthén, B. (2007). Deciding on the number of classes in latent class analysis and growth mixture modeling: A Monte Carlo simulation study. Structural Equation Modeling: A Multidisciplinary Journal, 14(4), 535–569. https://doi.org/10.1080/10705510701575396
88. Tofighi, D., & Enders, C. K. (2008). Identifying the correct number of classes in growth mixture models. Advances in Latent Variable Mixture Models, 2007(1), 317–341
89. Asparouhov, T., & Muthén, B. (2014a). Auxiliary variables in mixture modeling: Three-step approaches using Mplus. Structural Equation Modeling: A Multidisciplinary Journal, 21(3), 329–341. https://doi.org/10.1080/10705511.2014.915181
90. Brown, W. (2019). The influence of self-control on the impact of exposure to violence among youths. Victims & Offenders, 14(6), 692–711. https://doi.org/10.1080/15564886.2019.1630539
91. Smits, K. (2016). Applying political theory: Issues and debates (2nd ed.). New York: Palgrave
92. de Cadenet, A. (2015, October 19). More women watch (and enjoy) porn than you ever realized: A Marie Claire study. Marie Claire. Retrieved from http://www.marieclaire.com/sex-love/a16474/women-porn-habits-study/
93. Penley, C., Shimizu, C. P., Miller-Young, M., & Taormino, T. (2013). The feminist porn book: The politics of producing pleasure (pp. 9–20). New York City: The Feminist Press at CUNY
94. Attwood, F. (2005). What do people do with porn? Qualitative research into the consumption, use, and experience of pornography and other sexually explicit material. Sexuality and Culture, 9(2), 65–86
95. Döring, N. M. (2009). The Internet's impact on sexuality: A critical review of 15 years of research. Computers in Human Behavior, 25, 1089–1101
96. Richards, R. S. (2011)."I could have told you that wouldn't work": Cyberfeminist pedagogy in action. Feminist Teacher, 22(1), 5–22
97. Daniels, J. (2009). Rethinking cyberfeminism(s): Race, gender, and embodiment. WSQ Women's Studies Quarterly, 37(1–2), 101–124
98. Attwood, F. (2007). Sluts and riot grrrls; female identity and sexual agency. Journal of Gender Studies, 16, 233–247. https://doi.org/10.1080/09589230701562921
99. Bamberg, M. (2004). Form and functions of 'slut bashing' in male identity constructions in15-year-olds. Human Development, 47, 331–353. https://doi.org/10.1159/000081036
100. Ringrose, J., & Renold, E. (2012). Slut shaming, girl power and 'sexualisation': Thinking through the politics of the international SlutWalks with teen girls. Gender and Education, 24, 333–343. https://doi.org/10.1080/09540253.2011.645023
101. Baumeister, R. F., & Twenge, J. M. (2002). Cultural suppression of female sexuality. Review of General Psychology, 6, 166–203. https://doi.org/10.1037//1089-2680.6.2.166
102. Zeisler, A. (2013). New media, new feminism: Evolving feminist analysis and activism in print, and on the Web and beyond. In H. Thornham & E. Weissmann (Eds.), Renewing feminisms: Radical narratives, fantasies and futures in media studies (pp. 178–184). London: I.B. Tauris
103. Gentry, M. (1998). The sexual double standard: The influence of number of relationships and level of sexual activity on judgments of women and men. Psychology of Women Quarterly, 22, 505–511. https://doi.org/10.1111/j.1471-6402.1998.tb00173.x
104. Bettie, J. (2003). Women without class: Girls, race, and identity. Berkeley, CA: University of California Press
105. Hackman, J. R., & Oldham, G. R. (1975). Development of the job diagnostic survey. Journal of Applied Psychology, 60, 159–170. https://doi.org/10.1037/h0076546

106. Sverke, M., Hellgren, J., & Naswall, K. (2002). No security: A meta-analysis and review of job insecurity and its consequences. Journal of Occupational Health Psychology, 7, 242–264. https://doi.org/10.1037//1076-8998.7.3.242
107. Jacobs, P. A., Tytherleigh, M. Y., Webb, C., & Cooper, C. L. (2010). Breaking the mold: The impact of working in a gender-congruent versus gender-incongruent role on self-reported sources of stress, organizational commitment, and health in U.K. universities. International Journal of Stress Management, 17, 21–37. https://doi.org/10.1037/a0018026
108. van der Velde, M. E. G., Bossink, C. J. H., & Jansen, P. G. W. (2003). Gender differences in the influence of professional tenure on work attitudes. Sex Roles, 49, 153–162
109. Bernas, K. H., & Major, D. A. (2000). Contributors to stress resistance: Testing a model of women's work–family conflict. Psychology of Women Quarterly, 24, 170–178. 0361-6843/00
110. Lee, K., Carswell, J. J., & Allen, J. J. (2000). A meta-analytic review of occupational commitment: Relations with person- and work-related variables. Journal of Applied Psychology, 85, 799–811. https://doi.org/10.1037/0021-9010.85.5.799
111. Mathieu, J. E., & Zajac, D. M. (1990). A review and meta-analysis of the antecedents, correlates and consequences of organizational commitment. Psychological Bulletin, 108, 171–194. https://doi.org/10.1037/0033-2909.108.2.171
112. King, J. (2012). The No. 1 place to work in IT: USAA scores a three-peat. Computerworld. Retrieved July 10, 2012, from http://www.computerworld.com/s/article/9227900/The_No._1_Place_to_Work_in_IT_USAA
113. Sumner, M., & Niederman, F. (2003–2004). The impact of gender differences on job satisfaction, job turnover, and career experiences of information systems professionals. Journal of Computer Information Systems, Winter, 29–39
114. Meyer, J. P., Stanley, D. J., Herscovitch, L., & Topolnytsky, L. (2002). Affective, continuance, and normative commitment to the organization: A meta-analysis of antecedents, correlates, and consequences. Journal of Vocational Behavior, 61, 20–52. https://doi.org/10.1006/jvbe.2001.1842
115. Ely, R., & Padavic, I. (2007). A feminist analysis of organizational research on sex differences. Academy of Management Review, 32, 1121–1143. https://doi.org/10.5465/AMR.2007.26585842
116. Young, G. (2004). Reading and praying online: The continuity of religion online and online religion in internet Christianity. In L. Dawson & D. Cowan (Eds.), Religion online: Finding faith on the internet (pp. 93–106). New York: Routledge
117. van der Merwe, H. and Masiko, N. (2020) Addressing diversity and inclusion through transitional justice. https://www.africaportal.org/publications/addressing-diversity-and-inclusion-through-transitional-justice (accessed 15 September 2021)
118. Bundschuh, T. (2015). Enabling transitional justice, restoring capabilities: The imperative of participation and normative integrity. International Journal of Transitional Justice, 9(1), 10–32. https://doi.org/10.1093/ijtj/iju030
119. Robins, S. (2017). Failing victims? The limits of transitional justice in addressing the needs of victims of violations. Human Rights and International Legal Discourse, 11(1), 41–58
120. Bakiner, O. (2014). Truth commission impact: An assessment of how commissions influence politics and society. International Journal of Transitional Justice, 8(1), 6–30. https://doi.org/10.1093/ijtj/ijt025
121. Humphrey, M. (2003). From victim to victimhood: Truth commissions and trials as rituals of political transition and individual healing. The Australian Journal of Anthropology, 14(2), 171–187
122. Fairclough, N. (2003). Analysing discourse: Textual analysis for social research. Routledge
123. Widdowson, H. G. (2004). Text, Context, Pretext: Critical issues in Discourse Analysis. Blackwell Publishing
124. Weah, A. (2012). Hopes and uncertainties: Liberia's journey to end impunity. International Journal of Transitional Justice, 6(2), 331–343. https://doi.org/10.1093/ijtj/ijs007

125. Dodds, P. S., & Danforth, C. M. (2010). Measuring the happiness of large-scale written expression: Songs, blogs, and presidents. Journal of Happiness Studies, 11(4), 441–456. https://doi.org/10.1007/s10902-009-9150-9
126. Ní Aoláin, F. (2019). Transformative gender justice? In P. Gready & S. Robins (Eds.), From Transitional to Transformative Justice (pp. 150–171). Cambridge University Press
127. Sankey, D. (2016). 'Recognition of gendered experiences of harm at the extraordinary chambers in the courts of Cambodia: The promise and the Pitfalls', Feminist Legal Studies. Springer, Netherlands, 24(1), 7–27. https://doi.org/10.1007/s10691-016-9309-9
128. Biber, D. (2012). Corpus-based and corpus-driven analyses of language variation and use. In B. Heine & H. Narrog (Eds.), The Oxford Handbook of Linguistic Analysis (pp. 159–192). Oxford Academic
129. UNESCO. (2021). Education: from Disruption to Recovery. https://www.unesco.org/en/covid-19/education-response. Accessed 18 January 2023
130. Jandrić, P., Bozkurt, A., McKee, M., & Hayes, S. (2021b). Teaching in the Age of Covid-19 - A Longitudinal Study. Postdigital Science and Education, 3(3), 743-770. https://doi.org/10.1007/s42438-021-00252-6
131. Light, J. (2001). Rethinking the Digital Divide. Harvard Educational Review, 71(4), 709–734. https://doi.org/10.17763/haer.71.4.342x36742j2w4q82
132. Van Dijk, J. A. G. M. (1999). The Network Society: Social Aspects of New Media. London: Sage
133. van Dijk, J. (2006). Digital divide research, achievements and shortcomings. Poetics, 34, 221–235
134. DiMaggio, P., & Hargittai, E. (2001). From the 'Digital Divide' to 'Digital Inequality': Studying Internet Use as Penetration Increases. Working Paper No. 47. Princeton, NJ: Princeton University, School of Public and International Affairs, Centre for Arts and Cultural Policy Studies. https://econpapers.repec.org/paper/pricpanda/15.htm. Accessed 12 December 2022
135. Knox, J. (2019). What Does the 'Postdigital' Mean for Education? Three Critical Perspectives on the Digital, With Implications for Educational Research and Practice. Postdigital Science and Education, 1(2), 357–370. https://doi.org/10.1007/s42438-019-00045-y
136. Jandrić, P. (2021). The postdigital challenge of critical educational research. In C. Mathias (Ed.), The Handbook of Critical Theoretical Research Methods in Education (pp. 31-48). Abingdon and New York: Routledge
137. Jandrić, P., & Ford, D. (2022). Postdigital Ecopedagogies: Genealogies, Contradictions, and Possible Futures. Postdigital Science and Education, 4(3), 672-710. https://doi.org/10.1007/s42438-020-00207-3
138. Jandrić, P. (2022). History of the Postdigital: Invitation for Feedback. Postdigital Science and Education. https://doi.org/10.1007/s42438-022-00345-w
139. Brewster, D. (1819). Treatise on the Kaleidoscope. Edinburgh: Archibald Constable & Co
140. Government Offices of Sweden. 2019b. Handbook Sweden's feminist foreign policy. edited by Ministry of Foreign Affairs
141. Jezierska, Katarzyna. 2021. Incredibly loud and extremely silent: Feminist foreign policy on Twitter. Cooperation and Conflict 36: 177
142. Bergman-Rosamond, Annika. 2020. Swedish feminist foreign policy and"Gender Cosmopolitanism."Foreign Policy Analysis 16 (2): 217–235. https://doi.org/10.1093/fpa/orz025
143. Somers, Margaret R. 1994. The narrative constitution of identity: A relational and network approach. Theory and Society 23 (5): 605–649
144. Guyer, C. (1996b). Fretwork: ReForming me. Readerly/Writerly Texts, Spring
145. Guyer, C. (1996a). Along the Estuary. In S. Birkerts (Ed.), Tolstoy's dictaphone: Technology and the muse. Graywolf Forum 1. St. Paul, MN: Graywolf Press
146. Tringham, R. (2004). Interweaving digital narratives with dynamic archaeological databases for the public presentation of cultural heritage. In W. Börner & W. Stadtarcheologie (Eds.), Enter the past: The E-way into the four dimensions of Cultural heritage—CAA2003 (pp. 196–200). (full version on accompanying CD-ROM). Oxford, UK: Computer Applications and Quantitative Methods in Archaeology; Archeopress; BAR International Series 1227

147. Perin, C. (1992). The communicative circle: Museums as communities. In I. Karp, C. M. Kreamer, & S. D. Lavine (Eds.), Museums and communities: The politics of public culture (pp. 182–220). Washington DC: Smithsonian Institution Press
148. Joyce, R. A., Guyer, C., & Joyce, M. (2000). Sister stories. New York: New York University Press. Retrieved January 4, 2004, from http://www.nyupress.org/sisterstories/index.html
149. Consalvo, M. (2002). Selling the internet to women: The early years. In M. Consalvo & S. Paasonen (Eds.), Women and everyday uses of the Internet: Agency and identity (pp. 111–138). New York: Peter Lang
150. Manovich, L., 2017. Cultural analytics, social computing and digital humanities. In M. Schäfer and K. van Es, eds. The Datafied Society: Studying Culture Through Data. Amsterdam: Amsterdam University Press, pp. 55–68
151. Berry, D., 2014. Critical Theory and the Digital. London: Bloomsbury
152. Couldry, N., 2017. The myth of big data. In K. van Es and M. Schäfer, eds. The Datafied Society: Studying Culture Through Data. Amsterdam: Amsterdam University Press, pp. 55–68
153. Mahrt, M. and Scharkow, M., 2013. The value of big data in digital media research. Journal of Broadcasting & Electronic Media, 57(1), pp. 20–33
154. Kitchin, R. (2014). Big Data, new epistemologies and paradigm shifts. Big Data & society, 1(1), 1–12
155. Edwards, R. and Mauthner, M., 2012. Ethics and feminist research: theory and practice. In T. Miller, M. Birch, M. Mauthner and J. Jessop, eds. Ethics in Qualitative Research. London: Sage Publications Ltd, pp. 14–28
156. Lomax, H., 2015. Seen and heard? Families, Relationships and Societies, 4(3), pp. 493–502
157. Kukkonen, T., & Cooper, A. (2019). An arts-based knowledge translation (ABKT) planning framework for researchers. Evidence and Policy: A Journal of Research, Debate and Practice, 15(2), 293–311. https://doi.org/10.1332/174426417X15006249072134
158. AIDES. (2016). #Révélation. AIDES. Retrieved May 19 from http://www.aides.org/campagne/revelation
159. Becker, D. (2014). The dating game: Accepting the past and living in the present. Positive Living Newsletter: Girl Talk, 6–7
160. Caballero, M. (2016). Pleasure is a human right. Retrieved June 1 from http://www.hivequal.org/hiv-equal-online/feature-pleasure-is-a-human-right
161. Fratti, K. (2017). I am HIV positive. This is what it's like to date. Retrieved January 25 from http://www.redbookmag.com/love-sex/relationships/a48131/dating-with-hiv/
162. McClelland, A., & Whitbread, J. (2016). PosterVirus: Claiming sexual autonomy for people with HIV through collective action. In C. Kelly & M. Orsini (Eds.), Mobilizing metaphor: Art, culture, and disability activism in Canada (pp. 76–97). Vancouver: UBC Press
163. Prevention Access Campaign. (2017). Undetectable = untransmittable. Prevention Access Campaign. Retrieved May 19, 2017 from https://www.preventionaccess.org/undetectable
164. Whitbread, J. (2017). Love positive women. Retrieved February 3 from http://jessicawhitbread.com/project/love-positive-women/
165. Whitbread, J. L. (2016). Space dates. Retrieved July 29 from http://jessicawhitbread.com/project/space-dates/
166. Carter, A., Greene, S., Money, D., Sanchez, M., Webster, K., Nicholson, V., … Kaida, A. (2017). The problematization of sexuality among women living with HIV and a new feminist approach for understanding and enhancing women's sexual lives. Sex Roles, 77(11–12), 779–800. https://doi.org/10.1007/s11199-017-0826-z
167. Hesse-Biber, S.N., ed., 2014. Feminist Research Practice: A Primer. Thousand Oaks: Sage Publications, Inc.
168. Hesse-Biber, S.N. and Piatelli, D., 2012. The feminist practice of holistic reflexivity. In S.N. Hesse-Biber, ed. Handbook of Feminist Research: Theory and Praxis. Thousand Oaks: Sage Publications, Inc., pp. 557–582
169. O'Brien Hallstein, L., 2000. Where standpoint stands now: an introduction and commentary. Women's Studies in Communication, 23(1), pp. 1–15

170. Harding, S., 2004. The Feminist Standpoint Theory Reader: Intellectual and Political Controversies. New York: Routledge
171. Letherby, G. and Zdrodowski, D., 1995. 'Dear Researcher': the use of correspondence as a method within feminist qualitative research. Gender and Society, 9(5), pp. 576–593
172. White, M.I., 2003. Taking note of teen culture in Japan: dear diary, dear fieldworker. In T. Bestor, P. Steinhoff and V. Bestor, eds. Doing Fieldwork in Japan. Honolulu: University of Hawai'i Press, pp. 21–35
173. Pennebaker, J.W., Colder, M. and Sharp, L.K., 1990. Accelerating the coping process. Journal of Personality and Social Psychology, 58(3), pp. 528–537
174. James, N. and Busher, H., 2009. Online Interviewing. Thousand Oaks: Sage Publications Ltd
175. Mishler, E.G., 1986. Research Interviewing: Context and Narrative. Cambridge: Harvard University Press
176. Minister for Health. 2017. Explanatory Statement: Issued by Authority of the Minister for Health, My Health Records Act 2012, My Health Records (National Application) Rules 2017
177. Mendelson, D. 2018. The European Union General Data Protection Regulation (EU 2016/679) and the Australian My Health Record Scheme—A comparative study of consent to data processing provisions. JLM 26: 23
178. Komesaroff, P., and I. Kerridge. 2018. The My Health Record debate: Ethical and cultural issues. Internal Medicine Journal 48: 1291–1293
179. Price, C., & Wulff, S. (2005). Does sex make a difference? Job satisfaction of television network news correspondents. Women's Studies in Communication, 28, 207–234. https://doi.org/10.1080/07491409.2005.10162492
180. Rodgers, S., & Thorson, E. (2003). A socialization perspective on male and female reporting. International Journal of Communication, 53, 658–675. https://doi.org/10.1111/j.1460-2466.2003.tb02916.x
181. Armstrong, C. (2004). The influence of reporter gender on source selection in newspaper stories. Journalism and Mass Communication Quarterly, 81, 139–154. https://doi.org/10.1177/107769900408100110
182. Sutcliffe, J., Lee, M., & Soderlund, W. (2005). Women and crisis reporting: Coverage of television news coverage of political crises in the Caribbean. Harvard International Journal of Press/Politics, 10, 99–124. https://doi.org/10.1177/1081180X05279147
183. Zoch, L. M., & Turk, J. V. (1998). Women making news: Gender as a variable in source selection and use. Journalism and Mass Communication Quarterly, 75, 762–775. https://doi.org/10.1177/107769909807500410
184. Schmid, H. (1994). Probabilistic part-of-speech tagging using decision trees. In Proceedings of International Conference on New Methods in Language Processing (pp. 44–49). Manchester, UK
185. Hayes, A. F., & Krippendorff, K. (2007). Answering the call for a standard reliability measure for coding data. Communication Methods and Measures, 1, 77–89. https://doi.org/10.1080/19312450709336664
186. Krippendorff, K. (2004). Content analysis: An introduction to its methodology (2nd ed.). Thousand Oaks: Sage
187. Pfohl, S., Ayes, B., Turner, A., Amoo-Adare, E., Zecchin, M., & Borowski, M. et al. (2021). Simple, Dark, and Deep: Photographic Theorizations of As-Yet Schools. Postdigital Science And Education. https://doi.org/10.1007/s42438-021-00233-9
188. Mirzoeff, N. (2006). On visuality. Journal of Visual Culture, 5(1), 53–79. https://doi.org/10.1177/1470412906062285
189. Hurley, Z. (2020). Postdigital Feminism and Cultural Visual Regimes: Covid-19 at Women's Only University in the Gulf. Postdigital Science And Education, 2(3), 645-650. https://doi.org/10.1007/s42438-020-00134-3
190. Johnston, I. (2021). UK risks 'turning clock back' on gender equality in pandemic. Guardian, 9 February. https://www.theguardian.com/lifeandstyle/2021/feb/09/uk-risks-turning-clock-back-on-gender-equality-in-pandemic. Accessed 26 June 2021

191. Ahmed, L. (2011). A Quiet Revolution: The Veil's Resurgence, From the Middle East to America. New Haven, CN: Yale University Press
192. Abu-Lughod, L. (2005). Dialects of Women's Empowerment: The International Circuitry of the Arab Human Development Report 2005. International Journal of Middle East Studies, 41(1), 103a. https://doi.org/10.1017/s0020743808090521
193. Al-Ali, K. (2021). 'Silence', the Invisible Tool of a Dialogically Extended Mind: An Email Experience of a Kuwaiti Tutor in Higher Education. Studies in Technology Enhanced Learning. https://doi.org/10.21428/8c225f6e.514b8258
194. Kandiyoti, D. (1988). Bargaining with Patriarchy. Gender & Society, 2(3), 274–290. https://doi.org/10.1177/089124388002003004
195. Roy, A. (2020). The pandemic is a portal. Financial Times, 3 April. https://www.ft.com/content/10d8f5e8-74eb-11ea-95fefcd274e920ca. Accessed 26 June 2021
196. EIGE: Gender Equality Index 2020—Digitalisation and the future of work. (2020). https://eige.europa.eu/publications/gender-equality-index-2020-digitalisation-and-future-work
197. European Commission: Women in Digital Index 2020. Methodology. (2020b). https://comunic.ro/wp-content/uploads/2021/03/WomeninDigitalIndex2020-methodology.pdf
198. Brimacombe, T., Skuse, A.: Gender, ICTs, and indicators: Measuring Inequality and Change. Gend. Technol. Dev. 17(2), 131–157 (2013). https://doi.org/10.1177/0971852413488713
199. Brüggemann, R., Patil, G.P.: Ranking and Prioritization for Multi-Indicator Systems. Introduction to Partial Order Applications. Springer-Verlag, New York (2011). https://doi.org/10.1007/978-1-4419-8477-7
200. Ellis, Lauren D.; Patterson, Callie L.; Walters, Andrew Schrack"Becoming a Sugar Baby Will Change Your Life. Let's Talk About How": Sugar Dating Advice on Tumblr. Sexuality & Culture (2022). https://doi.org/10.1007/s12119-022-10024-4
201. Santarossa, Sara; Woodruff, Sarah J. Using Focus Groups to Understand the Dynamics of Mothers and Daughters on Social Networking Sites. Journal of Child and Family Studies (2020). https://doi.org/10.1007/s10826-020-01700-w
202. Piela, Anna Piety as a concept underpinning Muslim women's online discussions of marriage and professional career. Contemporary Islam (2011). https://doi.org/10.1007/s11562-011-0162-y
203. Salam, Rauha In a World Where You can be Anyone: An Investigation into the Gendered Social Practices of Pakistani Facebook Users. Gender Issues (2021). https://doi.org/10.1007/s12147-021-09289-0
204. Setty, Emily Meanings of Bodily and Sexual Expression in Youth Sexting Culture: Young Women's Negotiation of Gendered Risks and Harms. Sex Roles (2018). https://doi.org/10.1007/s11199-018-0957-x
205. Sharabi, Moshe; Kay, Avi Work Values of Working Women in Israel: A Comparison of Haredi Women with Those from the Secular and Traditional Segments. Contemporary Jewry (2023). https://doi.org/10.1007/s12397-023-09480-3
206. Duvendack, Maren; Sonne, Lina; Garikipati, Supriya Gender Inclusivity of India's Digital Financial Revolution for Attainment of SDGs: Macro Achievements and the Micro Experiences of Targeted Initiatives. The European Journal of Development Research (2023). https://doi.org/10.1057/s41287-023-00585-x
207. Wu, Wenzhi; Yu, Lei; Cao, Xiang; Guo, Zeyi; Long, Qing; Zhao, Xinling; You, Xu; Teng, Zhaowei; Zhang, Yunqiao; Zeng, Yong The latent profile of self-control among high school students and its relationship with gender and cyberbullying. Current Psychology (2022). https://doi.org/10.1007/s12144-022-03995-y
208. McKeown, Janet K. L.; Parry, Diana C.; Penny Light, Tracy"My iPhone Changed My Life": How Digital Technologies Can Enable Women's Consumption of Online Sexually Explicit Materials. Sexuality & Culture (2017). https://doi.org/10.1007/s12119-017-9476-0
209. Papp, Leanna J.; Erchull, Mindy J.; Liss, Miriam; Waaland-Kreutzer, Lauren; Godfrey, Hester Slut-shaming on Facebook: Do Social Class or Clothing Affect Perceived Acceptability?. Gender Issues (2016). https://doi.org/10.1007/s12147-016-9180-7

210. Major, Debra A.; Morganson, Valerie J.; Bolen, Heather M. Predictors of Occupational and Organizational Commitment in Information Technology: Exploring Gender Differences and Similarities. Journal of Business and Psychology (2012). https://doi.org/10.1007/s10869-012-9282-5
211. Marcotte, Roxanne D. Gender and sexuality online on Australian Muslim forums. Contemporary Islam (2009). https://doi.org/10.1007/s11562-009-0104-0
212. Evrard, Elke The Language of Inclusion: Using Critical Corpus-Based Methods to Study the Presence and Representation of"Women, Children and Vulnerable Groups"in Liberia's Truth Commission. Social Justice Research (2023). https://doi.org/10.1007/s11211-023-00411-z
213. Kuhn, Caroline; Khoo, Su-Ming; Czerniewicz, Laura; Lilley, Warren; Bute, Swati; Crean, Aisling; Abegglen, Sandra; Burns, Tom; Sinfield, Sandra; Jandrić, Petar; Knox, Jeremy; MacKenzie, Alison Understanding Digital Inequality: A Theoretical Kaleidoscope. Postdigital Science and Education (2023). https://doi.org/10.1007/s42438-023-00395-8
214. Aggestam, Karin; Rosamond, Annika Bergman; Hedling, Elsa Feminist digital diplomacy and foreign policy change in Sweden. Place Branding and Public Diplomacy (2021). https://doi.org/10.1057/s41254-021-00225-3
215. Joyce, Rosemary A.; Tringham, Ruth E. Feminist Adventures in Hypertext. Journal of Archaeological Method and Theory (2007). https://doi.org/10.1007/s10816-007-9036-2
216. Leurs, Koen feminist data studies: using digital methods for ethical, reflexive and situated socio-cultural research. Feminist Review (2017). https://doi.org/10.1057/s41305-017-0043-1
217. Carter, Allison; Anam, Florence; Sanchez, Margarite; Roche, Juno; Wynne, S. T.; Stash, Just; Webster, Kath; Nicholson, Valerie; Patterson, Sophie; Kaida, Angela Radical Pleasure: Feminist Digital Storytelling by, with, and for Women Living with HIV. Archives of Sexual Behavior (2020). https://doi.org/10.1007/s10508-020-01822-8
218. Linabary, Jasmine R.; Hamel, Stephanie A. feminist online interviewing: engaging issues of power, resistance and reflexivity in practice. Feminist Review (2017). https://doi.org/10.1057/s41305-017-0041-3
219. Fossa, Fabio; Sucameli, Irene Gender Bias and Conversational Agents: an ethical perspective on Social Robotics. Science and Engineering Ethics (2022). https://doi.org/10.1007/s11948-022-00376-3
220. Davis, Mark D. M.; Schermuly, Allegra; Smith, Anthony K. J.; Newman, Christy Diversity via datafication? Digital patient records and citizenship for sexuality and gender diverse people. BioSocieties (2022). https://doi.org/10.1057/s41292-022-00277-5
221. Mateos de Cabo, Ruth; Gimeno, Ricardo; Martínez, Miryam; López, Luis Perpetuating Gender Inequality via the Internet? An Analysis of Women's Presence in Spanish Online Newspapers. Sex Roles (2013). https://doi.org/10.1007/s11199-013-0331-y
222. Hurley, Zoë; Al-Ali, Khadija Feminist Postdigital Inquiry in the Ruins of Pandemic Universities. Postdigital Science and Education (2021). https://doi.org/10.1007/s42438-021-002 54-4
223. Ramilo, Chat Garcia Beyond Tools: Technology as a feminist agenda. Development (2006). https://doi.org/10.1057/palgrave.development.1100230
224. Damiani, Filippo; Rodríguez-Modroño, Paula Measuring the digital inclusion of women: a poset-based approach to the women in digital scoreboard. Quality & Quantity (2023). https://doi.org/10.1007/s11135-023-01666-z

Digital Rights, Digital Representation, and Digital Justice—Towards Digital Democracy and Freedom of Expression

7

Access to the internet has been advocated as human right or basic right as it mediates and leads to not only participation in the democratic process but also access to the basic services which a society or state has to offer. The state and corporate governance is also being shaped from the usual bureaucratic ones to that which works on the network of flow of information. Digitalization is also making possible not only the interaction of citizens with the state machinery in real-time and also economical but also leading to 'open governance', where the rules, processes, and deliveries or actions have become more transparent. 'Digital governance', and/or 'e-governance' is the term used to define the internet-mediated governance process.

Digitalization has in varied ways influenced governance and types of governments. It has had liberalize impacts on some of the countries as in Tunisia and other Arab countries; as demanded by the citizens for democratic governments against authoritarian ones; it has helped in overcoming the corruption and delays in the governance systems in several countries of the Global South; while without exception the rise of digital technologies have also led to increased surveillance of the citizens. The big and deep data, biometry, and development of advanced algorithms with the ability to discriminate people on their physical and social characteristics have also added to the settled and silent exclusions.

The revolution in social media and the ease with which it can be operated by citizens has deepened the assertation of the rights by the citizens, which has had its consequences on the voting behaviour of the citizens. Digital media have also led to the popping up of 'digital civil society', and 'digital activists', who through cyberspace are able to network with citizens and governments and influence both. Digital technologies have also assisted in the timely conduct of elections and coordination between different spatial units within individual countries. It has

A. Shaban (ed.), *Digital Geographies—Theory, Space, and Communities*,
https://doi.org/10.1007/978-981-97-4734-4_7

also helped in networking the countries in real-time which may have implications for the 'digital global democracy' and breaking down of the national walls.

Specific interest groups like women, sexual, religious, and racial minorities, environment activists, etc., using digital technologies are able to mainstream their points of view and negotiate with state or the public at large. The LGBTQ and green movements are some of such examples. Similarly, the repression of or violation of the rights of ethnic, religious, and racial minorities has got flip due to social media and e-participation and e-advocacy from across the countries. It has also made the repression and crime visible by states and non-state actors and helped in delivering justice by overcoming the transaction (spatial, temporal, and representational) costs.

Information is key to knowledge, and that in turn is key to exercising civil rights in today's democratic societies. The information which used to be earlier available in print forms is now turning into digital documents. The availability of the internet remains key to accessing those documents. However, many countries and regions where digital infrastructures are low-developed preclude groups and people residing in those areas from accessing the information. Many civil rights groups therefore have claimed that internet access should be considered as a human right or key to citizenship as it enables the citizens to exercise their rights. In several countries, including India, several NGOs and civil society groups work hard to provide internet access to the citizens. These NGOs either lay the digital infrastructure in those areas, as being done by Digital Empowerment Foundation (DEF), India, or use Jugad to modify the system such that they make the mobile and internet signals accessible to the people in remote, inaccessible or excluded areas.

The digital infrastructure not only includes the towers, cable, and bandwidth but also the personally owned mobile, software, etc. Many corporate groups and NGOs help people with community mobile phones or devices like Tabs which they can use for information. Such NGOs also provide digital literacy to the people to use the devices for information and also make them aware of their legal and citizenship rights. Especially, such digital initiatives which use various mediums from text, images, and cartoons to movies, help in the empowerment of women and marginalized groups. There are several civil society groups working in many countries which provide internet cafes in marginalized areas and help students and citizens fill out educational forms, apply for various documents, and apply for changes in documents or corrections in the same. Volunteers and related support are also provided by several civil society organizations to the citizens in these cafes. As such, these civil society groups play an important role in protecting the rights of citizens and helping them to avail themselves of the needed services from the public or private spheres.

The need for digitally empowering the citizens is more important as all the documentation and also statutes which used to be available in print forms earlier are shifting to digital forms. Further, many government and private facilities are now available only through digital applications. The economic cost of accessing digital infrastructure and acquiring digital literacy to exercise democratic and citizenship rights has put enormous burdens on marginalized social groups. Digitalization has created possibilities of not only social but also territorial injustices. It is, therefore, important that both the governments and the civil society groups need to make an effort to create a level playing field for all citizens so that all of them can exercise their rights and avail the goods that society at large has to offer.

The papers reviewed in this chapter deal with a variety of issues of digital rights, digital justice (smart digital courts), digital representations, freedom of expression, social media, algorithms of exclusion and inclusions, democratization, e-participation, hate and extremism, voting, and digital violence. The chapter is roughly divided into three sections. Section 7.1 deals with the papers on digital democracy, while Sect. 7.2 reviews the papers on digital justice. The last section discusses papers on digital equality.

To highlight the findings of a few papers, the very first paper in this chapter brings out the impact of social media political activists and voting patterns in Ghana, while another paper focuses on the effect of technology on the political and social aspects of Iran's e-government. In Sect. 7.2, a paper discusses the algorithms of the smart criminal justice system in Switzerland, while another paper deals with the online security and protection of civil rights. In Sect. 7.3, the first paper deals with digital inclusion and exclusion in German cities, while another paper highlights digital games and gender equality in Africa. The last paper of the chapter deals with accessible software development, which is key to the digital participation of citizens.

Machine generated summaries

Disclaimer: The summaries in this chapter were generated from Springer Nature publications using extractive AI auto-summarization: An extraction-based summarizer aims to identify the most important sentences of a text using an algorithm and uses those original sentences to create the auto-summary (unlike generative AI). As the constituted sentences are machine selected, they may not fully reflect the body of the work, so we strongly advise that the original content is read and cited. The auto generated summaries were curated by the editor to meet Springer Nature publication standards. To cite this content, please refer to the original papers.

Machine generated keywords: law, legal, government, justice, democracy, egovernment, citizen, democratic, political, public, civil, governance, gender, policy, right

7.1 Digital Democracy

Machine generated keywords: government, democracy, egovernment, democratic, political, citizen, public, governance, transparency, civil, social medium, ict, actor, platform, movement

7.1.1 The Impact of Social Media Political Activists on Voting Patterns [274]

This is a machine-generated summary of:

Kofi Frimpong, Adasa Nkrumah; Li, Ping; Nyame, Gabriel; Hossin, Md Altab: The Impact of Social Media Political Activists on Voting Patterns [274]

Published in: Political Behavior (2020)

Link to original: https://doi.org/10.1007/s11109-020-09632-3

If you want to cite the papers, please refer to the original.

For technical reasons we could not place the page where the original quote is coming from.

Abstract-Summary

"This paper investigates how social media affects general voting patterns."

"Unlike previous studies investigating whether citizens' use of social media affects political participation, this paper considers the connections that social media users have with political activists on social media, and how this connectedness influences general voting patterns, using data from Ghana."

"Structural equation modeling was used to test the conceptual model, which demonstrates that the connectedness with political and social media activists is significant and positively influences modifications in voting patterns."

"The relationship between connections with social media political activists and online political participation is significant, as indicated by a strong covariance observed in the model."

Introduction

"Studies show that the U.S. 2008 presidential election was one of the first political elections in which politicians used online social media (SM) as an essential platform for political activities (Smith [1])."

"Choi and others ([2]) studied social media heterogeneity and reported a significantly positive correlation with political participation."

"Their study did not consider the causal impact of SM political activists on people's general voting patterns."

"We attempt to address the connections that SM users have with political activists on these platforms, and how this connectedness influences general voting patterns."

"What impact does political affect have on the relationship between people's connections with social activists via social media, and people's voting patterns?"

"What impact does political disposition have on the relationship between people's connections with social activists via social media, and people's voting patterns?"

Literature Review

"The study did not consider the connections that people have with SM political activists and the combined effect on voting patterns."

"We investigate how online political participation in SM relates to voting patterns, as this relationship seems unexplored."

"The relationship between online political participation in SM concerning connections with political activists and the impact on voting patterns appears unclear."

"This leads to the present research question: is there a negative relationship between political participation in SM and people's voting patterns?"

"The relationships that SM users have with political activists, and how this affects the overall voting patterns, lead to the following hypothesis: Connectedness with social activists via social media positively influences voting patterns."

"This literature review revealed a positive relationship between online political participation and SM, but it did not examine connections with political activists or voting patterns."

About Ghana

"Analysis by Lindberg and Morrison ([3]) of the 1996 and 2000 general elections in Ghana found that four-fifths of Ghanaians voted consistently for the National Democratic Congress (NDC) and the New Patriotic Party (NPP) as the most popular political parties."

"Ichino and Nathan ([4]) studied the relationship between ethnic groups and voting in Ghana, but excluded SM, and concluded that people in a given ethnic group are less likely to vote for a political party of the same ethnic group."

"They further explained that the same people are more likely to vote for the political party of another ethnic group when they foresee that the local ethnic geography favors the other group."

"In Uganda, Carlson ([5]) explained that voters are more likely to vote for political candidates belonging to the same ethnic group."

"Lindberg and Morrison's ([6]) study shows that in Ghana, the majority of people consider neither ethnic voting nor clientelism."

Methodology

"The sample may be biased toward people interested in politics or the influence of SM on politics."

"Data collection considered respondents who had used them for more than six months before the 2016 general election in Ghana."

"Equally, we considered possible biases in all cases."

"We considered the fact that different respondents have (or may have) a different understanding of the level of influence and the possibility of providing different answers."

"We base this on our belief that questions with few and limited choices are likely to drive respondents to select loosely among restricted answers."

"As it was an online questionnaire, we made sure there was no possibility of skipping questions."

"Respondents might provide more desirable answers to the question: "I have adequate political knowledge," thus forcing measurement error."

Research Model

"The model also accounts for covariance between political affect and online political participation and offline political participation as subtypes of political involvement."

"Although offline political participation was relevant to explain political affect in the previous study, this is just a complementary factor to explain voting patterns, used in the model as a control variable."

"We deduce that political affect is a cause for the relationship between online civic participation (ONPP), the connections with social media political activists (CWSA), and the covariance between political affect and ONPP/CWSA."

"We consider the possibility of covariance between connections with social media political activists and political affect."

"We consider political affect, political disposition, and time online as moderating variables because their relationship to voting seems consciously clear in the academic literature (Pinkleton and others [7]; Alexander and Wood [8]; Hamlin [9]; Valentino and others [10]; Protevi [11]; Parker and Isbell [12]; Iyengar and others [13]; Gan and others [14])."

Results

"On the relationship between connections with social media political activists and online political participation (ONPP), the association is significant, as indicated by strong covariance observed in the adjusted model (estimate = 0.447 standard error = 0.074, $p < 0.001$)."

"These results show that political affect and online political participation did not have a significant effect on voting patterns for users of either of the two most populated platforms (Facebook and WhatsApp), in contrast to previous models."

"In this group, political affect showed an evident influence on voting patterns ($\beta = 0.145$, $p = 0.005$), online political participation ($\beta = 0.154$, $p = 0.005$), and connections with social media political activists ($\beta = 0.585$, $p = 0.005$)."

"The main observed difference from other groups is the significance of political affect's moderator effect on the influence of connections with social media political activists on voting patterns ($\beta = -0.187$, $p = 0.014$)."

Final Considerations

"We confirmed the first hypothesis that online political participation is positively related to Ghana's voting patterns, even though with a lower coefficient."

"They found that online news platforms (including SM) are positively related to higher political participation, but not voting patterns in the US, Nigeria, and Kenya."

"This relationship accounts for the positive association between online political participation and voting patterns."

"We build on existing studies, showing that political affect has a higher impact on men than women towards changing voting patterns (Gan and others 14; Bosnjak and others [15])."

"The moderator effect of political disposition on the relationship between connections with social media political activists and voting patterns is quite evident, as one's disposition, to some extent, is related to the somatic."

"The sample shows evidence of the influence of connections with social media political activists on voting patterns."

7.1.2 ICT-Driven Projects for Land Governance in Kenya: Disruption and E-Government Frameworks [275]

This is a machine-generated summary of:

Huggins, Christopher; Frosina, Natasha: ICT-driven projects for land governance in Kenya: disruption and e-government frameworks [275]

Published in: GeoJournal (2016)

Link to original: https://doi.org/10.1007/s10708-016-9710-6

If you want to cite the papers, please refer to the original.

For technical reasons we could not place the page where the original quote is coming from.

Abstract-Summary

"Critical academic research has yet to comprehensively identify conceptual linkages and tensions between information communication technologies (ICTs) and land governance projects."

"In order to make a contribution to these complex research fields, this article examines three Kenyan projects to illustrate different aspects of competing theoretical frameworks for ICT-based land rights projects."

"Drawing particularly on conceptual frameworks that emphasize the 'disruptive' potentials of ICTs, and frameworks based on e-government models, the article argues that these projects include both disruptive aspects, which work through applying pressure on the state, and more 'integrative' approaches which seek to build state capacity."

Introduction

"There is currently much academic and professional interest in emerging uses of information communication technologies (ICTs) for improving land governance (Moreri and others [16]; Lemmen and others [17]; Datta [18], Makau and others [19]; McLaren [20]; Hagen [21]), partly because of the potentials to make land governance monitoring more participatory."

"There is an increasingly large technical literature on ICTs in land administration, geospatial technologies in land governance, and related subjects, as well as academic journals devoted to these fields."

"Critical theoretical engagements with land governance rarely take ICT innovations as case studies."

"Some of the ICT for development (ICT4D) literature is based on communications theory, which has a critical sociological perspective, but rarely has a well-informed land governance focus."

Land governance and ICTs: brief overview

"Part of the interest in ICTs for land governance stems from the nature of key weaknesses in capacities of sub-Saharan African states to provide citizens with formalized land rights which include, first, the weakness in the geo-spatial data related to landholdings, and second, an inability to effectively link geo-spatial data on landholdings with text documents proving ownership (such as 'title deeds', attestations from local leaders, witness statements, etc.) (Byamugisha [22])."

"There are various ways in which ICT platforms can improve land governance, ranging from SMS-based monitoring of land administration processes (where automated systems require civil servants to document their activities by SMS), to informal GIS-enabled photographic documentation of landholdings to guard against land grabbing."

"Land records can be digitized and accessed through mobile phones, using SMS-based information services or the internet if smartphones are widely used (World Bank [23])."

Elements of a conceptual framework for ICTs in land governance

"Discussions of the benefits and risks of using ICTs for land governance are not informed by an explicit conceptual framework."

"Many governments remain nervous about the overall accuracy and dependability of ICT-based approaches, and are concerned about losing state control over the land registration and related activities."

"The e-government literature sometimes acknowledges that adoption of ICTs can increase control of systems by certain actors, resulting in 'concentration of power' (Sæbø [24], citing Schuppan [25])."

"The e-government literature has not been well integrated with discussions of ICT for land governance in less-developed countries, partly because effective land governance in LDCs requires significant structural, legal, administrative and other reforms prior to any e-government approach to land administration becoming effective."

"There are tensions when applying such models to land governance activities, as the 'prosumer' model tends to assume that information is an inherent 'good' (i.e. an end in itself) rather than an instrument for social change (i.e. a means to an end)."

Background to the case studies

"The size and population density of urban informal settlements in Kenya is a factor not only of economic inequalities but also a lack of comprehensive urban planning policies (Otiso [26]), weak municipal governments and problematic land tenure systems (Otiso and Owusu 2008), and corruption in the land sector."

"In places such as Kibera, youth gangs, such as the Kamkunji gang, comprised of ethnic Luos (Joireman and Vanderpoel 2011) are influential and limit the abilities of Chiefs to collude with landlords (ibid.) Politically connected bureaucrats and business people living outside of the informal settlements have irregularly accessed settlement land and properties, becoming slum landlords."

"After President Moi's regime ended in 2003, many former leaders of civil society organizations, with backgrounds in the human rights and democratization movements, became part of government agencies, and slum upgrading and land tenure reform moved up the state's agenda."

Background to ICT adoption in Kenya

"The Government of Kenya has invested heavily in data infrastructure and e-government services, including an Open Data Initiative."

"The government could consider providing full autonomy to the Kenya National Bureau of Statistics, increasing investment in data collection and capacity building, creating data standardization policies, and promoting open data."

"In terms of ICT for land governance, the Ministry of Land, Housing and Urban Development developed a National Land Information Management System (NLMIS) in 2009, based largely on digital technologies, but it has yet to be fully implemented (Makoro [27])."

"The state-managed Land Information for Informal Settlements (LIIS) project, initiated in 2005, used geo-spatial technologies to map informal settlements."

"Other ICT-related land sector initiatives overseen by the government, include digitization of topographic maps, establishment of an Automated Land Rent Information Database System, establishment of a national spatial data infrastructure (with support from JICA), and the establishment of a file-tracking system at the Ministry of Lands (Mbaria [28])."

Methodologies

"Following a desk-study of existing academic and policy literature on ICT platforms in Kenya, three projects were selected on the basis that they were well-established (and hence impacts and social dynamics were already visible), and covered different land tenure situations (informal settlement in urban areas, and public land in a more rural area)."

"We selected both urban and rural case studies to understand how ICT and land governance is operating and affecting rights in both settings."

"Information provided by project personnel (staff and volunteers) was triangulated with information gathered through interviews and focus group discussions (FGDs) with project participants (i.e. those using or 'benefiting' from the projects but not affiliated to them), members of locally-based organizations unaffiliated to the three case study projects, as well as secondary data (both grey and published)."

Case study one: Map Kibera

"It is paradoxical that Kibera residents make videos and upload them to the internet, permitting access by people outside of the settlement, whereas residents may never see the videos.(ibid.) This is a critique sometimes heard about the mapping project more generally; one former mapper told a researcher, 'only people outside Kibera and Kenya use services like Map Kibera."

"The local government used Map Kibera to help determine in which areas to install security lights and establish police posts (ibid.)."

"Map Kibera created special maps during the 2013 Presidential elections, which highlighted areas vulnerable to political violence, as well as police stations, security firms etc. The security maps were distributed to state security agencies, displayed in strategic locations, and placed online."

Case study two: GROOTS Kenya

"Recognizing that there existed no geo-spatial registry of public land in Kenya, GROOTS developed a participatory program to have women in Lari Sub-County map public land."

"They hoped the data could be used to pressure the local government to ensure that public land is productively used for social service delivery such as school, health centers, and conservation sites etc. GROOTS chose Lari Subcounty for this intervention because of their existing relationship with a local partner organization, Kijabe Environment Volunteers (KENVO) that gave them an 'entry point' to work with community members."

"To field mapping, GROOTS-Kenya used existing paper maps, testimony of former civil servants, and other informants to identify public land (WTV [29])."

"GROOTS argues that the maps discourage further grabbing of public land, as they have increased knowledge amongst community members of what existing land is public or private."

"Building on the success of the project, GROOTS Kenya has since produced a handbook outlining the step by step process of mapping public land for future replication."

Case study three: global land tenure network, Pamoja Trust, and partners
"The global land tenure network (GLTN) of UN-Habitat partnered with the local NGO Pamoja Trust and other actors to map the informal Mashimoni settlement in Nairobi."

"The Pamoja Trust partnered with UN-Habitat to use the STDM tool to map all structures within the settlement using GIS, identify the structure owners and tenants, the number of people living in each structure, and the location of social service delivery points."

"To complete the mapping, several community members, selected in conjunction with Muungano Wa Wanavijiji (the federation of slum dwellers) were trained by Pamoja Trust on how to use STDM and carry out the mapping survey (Makau and others [19])."

"Once compiled, the maps were validated at a community meeting and then presented to the local government, as part of the community's lobbying efforts to gain formal land ownership."

Conclusions
"The difficult politics of land ownership in informal urban settlements, where politicians and their close supporters sometimes own properties, and where political violence has often taken place, are part of the reason that the government has yet to transfer land in informal settlements, such as Kibera or Mashimoni, to community members."

"Unlike the other case studies in this article, Map Kibera didn't initially have any direct government involvement or support, but it was successful because it leveraged an already-high level of international awareness about Kibera, and existing networks of community activism."

"GLTN is therefore seeking to change the structural conditions of land administration in Kenya, and leveraging projects such as the Mashimoni mapping initiative to do that."

"Geo-spatial mapping of land and property rights provide multiple opportunities for improving land governance, though progress is often incremental and less visible than the web-pages of high-profile projects."

7.1.3 Social Media, Meet Old Politics: Preservation and Innovation in Colombian Presidential Elections, 2010–2018 [276]

This is a machine-generated summary of:

Torres-Echeverry, Nicolás: Social media, meet old politics: preservation and innovation in Colombian presidential elections, 2010–2018 [276]

Published in: Theory and Society (2023)

Link to original: https://doi.org/10.1007/s11186-022-09509-w

If you want to cite the papers, please refer to the original.

For technical reasons we could not place the page where the original quote is coming from.

Abstract-Summary

"This article develops a framework to analyze how political actors adopt social media in systems characterized by clientelism and populism, tracing the consequences and disruptive capabilities of the forms of social media adoption."

"The first locates actors' structural positions in the political system (internal/external) and their relationship with the mainstream media (allied/antagonistic)."

"In examining the structural positions and problem-solving stages of Colombian political actors, this article articulates three paths of adoption: habit preservation, internal innovation, and external innovation."

"External innovators develop practices that integrate physical spaces and online communication, displaying a disruptive potential for existing core practices and the political system."

"The framework and empirical case link and develop the literatures on clientelism and political communication."

Technological adoption in the literatures on clientelism and political communication

"Of understanding clientelism accounts for recent emphasis scholars have given to brokers within the structure (Auyero, [30]; Holland & Palmer-Rubin, [31]; Stokes and others, [32]) and helps parse the relationship between clientelism and technologies of communication."

"While relevant studies on clientelism and communication technologies exist, this literature has produced little reflection on their relationship, and even suggests three contradictory accounts."

"She does not theorize who is likely to be an early adopter of a new communication technology but makes clear the relevance of communication technologies (specifically, television) for non-clientelist competitors to engage in political branding."

"Contrary to the political science literature, studies in this line of research hint at the possible ways in which social media integrates with clientelist practices, without changing their fundamental logic and without weakening clientelism."

"Contradictory, all three approaches demonstrate a literature that agrees the relationship between communication technologies and clientelism is relevant."

A framework of elite adoption of new communication technologies

"The second step helps relate specific forms of social media adoption and specific problems actors face."

"Structural positions serve to assess the schemas associated with political work—stressing the habits tied to positions in the field that explain how actors make sense of it (Bourdieu, 33; Martin, [34]; Sewell, [35])."

"The structural position of actors prior to the introduction of technologies allows to define the problem situations, the political schemas, and the resources and practices available to actors when the context of action changes—in this case, the technological shift."

"Two structural dimensions are central to the question of adoption of communication technologies: the relation of the actors to the political system and their relationship to the mainstream media."

"In internal-antagonistic positions are actors whose political schemas are tied to the working of the political system but that have an antagonistic relation to the mainstream media."

Clientelism and populism in Colombia and the context framing social media adoption

"By the end of the 1900s, the country went through a period of institutional reform which included a new political constitution."

"It is in this last version of clientelism that actors found ways to effectively incorporate social media into the political repertoire."

"The fragmentation of the party system, the articulation of legal and illegal networks, and the rise of the fight against terrorism created structural factors that would allow another of the main political actors in this paper, Álvaro Uribe, to get elected to office in 2002."

"This turn created a political realignment expressed in the 2014 presidential election and the 2016 referendum that rejected the peace agreement."

"This realignment of political forces around war and peace proves fundamental to understand the recent changes in the political field, but it must be understood in tandem with the development of political practices, and in particular, the changing environment brought upon by the introduction of social media."

Data and methods

"Interview data served to grasp the understandings of politics and social media, and news archives served to trace practices over time and triangulate events that came up in interviews."

"This data source is useful both to trace the political actions politicians took during the period of study as well as their previous trajectories."

"With the help of a research assistant, I selected and read in full LSV entries that covered one of the following three topics: 1) clientelist practices (such as vote buying or exchange of political favors), 2) social media use, and 3) networks of alliances with family members, other politicians, and business associates."

"Twitter data from the 14 politicians provided an overview of the use of and engagement within this social media site over the entire period of study."

Three paths of social media adoption

"My findings show three ideal–typical paths that actors took in adopting social media."

"The first path, most prevalent and least creative, comes from actors that were in internal-alliance structural positions."

"The second path, less prevalent and more creative, came from actors who were in internal-antagonistic positions."

"They found a place both for clientelist practices as well as for social media ones."

"They incorporate new technologies of communication, innovate, but also preserve core practices in the system, granting it stability."

"They developed social media practices confronting problem situations that enable them to merge online and offline practices to trigger political activation, combining physical spaces with symbolic value in the political imagination with online practices."

"This path led to a creative development and integration of practices in interaction with the public that had high disruptive potential for existing core practices and the political system."

Habit preservation path

"From different places in the campaign, interviews brought up the emphasis Lleras placed on the political machinery and the understanding Lleras had of social media through the lens of a previous information environment that no longer existed."

"For Roberto, who was part of Vargas 2018 campaign, "[Lleras] did not pay enough attention to social media."

"To do the social media work for the 2018 campaign, Lleras hired one of the multiple companies that are owned by the five large world-wide advertising agencies or another equally large multinational, like the French mass media conglomerate Vivendi."

"Considering the events that marked Lleras' campaigns and the decisions to delegate the work on social media shows the projection of an old understanding of politics upon the new technology, and limitations that delegating the work had for creating the types of interactions that triggered engagement on social media."

Internal innovation path

"Pedro, a journalist who had worked for the major news media and as part of Uribe's 2002 campaign, explained: "[Early in the campaign,] the big news media (Caracol, RCN, El Tiempo, El Espectador) did not know who Álvaro Uribe was and, when they learned, they did not like him… So, we needed to think [of] alternatives to communicate." "

"From his position as a new "independent," Uribe operated with part of the political machinery of the two long-standing parties and the practices common to them, as well as the repertoire of the paramilitaries which also incorporated clientelist practices (Gutiérrez, [36]).”

“In both campaigns, Uribe’s communication practices and clientelist practices endured.”

“Uribe’s innovations transformed the power relations between the national elites and the local ones and employed social media to maintain political relevance after leaving office.”

[Section 8]

“Knowing Petro’s campaigns from within and having worked in political campaigns for three decades, Emilio explained to me, “Since [Petro] was the mayor of Bogotá, the candidate had an almost obsessive concern for social media.””

“The characteristics of this mobilization would bridge Petro’s use of social media in institutional politics and the form of social media use in social movements (Castells, 37; Tufekci, [38]).”

“Further, he understood that form of activation to respond to a specific decision by the campaign and Petro as mayor to dispute the political support of “the poor” against Uribe and his candidate Duque.”

“Social media was not only used to call people to the plaza—which the interviewees interpreted as being the route social movements had taken—but to “move the plaza online.” “

“Images and videos of the public square went online linked the emotions on Facebook and Twitter with the emotion in the plaza reinforced one another, activating people to mobilize for Petro.”

Discussion and conclusion

“In the second, they engage in iterative cycles of problem situations that demarcate their idiosyncratic forms of social media adoption, the combinatorial blends of practices that emerge, and their disruptive potential.”

“Internal innovators confront problem situations that lead them to find a place for clientelist practices as well as social media practices.”

“Their approach shows how clientelist practices and social media ones can be complementary.”

“Studying both types of actors may help answer relevant questions about the working of clientelism: How are brokers using social media?”

“Regarding clients, are their political schemas changing in any way because of the information they are accessing through social media?”

“Will social media contribute to end clientelist practices?”

“The provisional answer is no; those practices have shown great malleability in the process of political actors confronting problem situations and creatively responding.”

[Section 10]

"It studies elite actors' social media adoption, the consequences of its adoption over the stability of these political practices, and the disruptive potential of the forms of adoption over the political system."

"Despite the relevant advances in understanding the adoption of new communication technologies, particularly those tied to the internet by scholars of political communication (Epstein, [39]; Howard, [40]; Karpf, [41]; Stromer-Galley, [42]), the attention has been centered on the Global North and, most importantly, there is an absence of frameworks to explain variation in terms of the actors that adopt, the ways in which they do, and the consequences of each form of adoption."

"More concretely, this literature has emphasized technological development (Kreiss, [43]) and the interaction of different media (Chadwick, [44]), but offered no framework to understand the relationship of the political actors with the broader social environment, how that environment affects practices, and with what consequences."

"The framework foregrounds actors' structural relations to the political system and the mainstream media."

7.1.4 Technology, Small States and the Legitimacy of Digital Development: Combatting De-risking Through Blockchain-Based Re-risking? [277]

This is a machine-generated summary of:

Campbell-Verduyn, Malcolm; Rodima-Taylor, Daivi; Hütten, Moritz: Technology, small states and the legitimacy of digital development: combatting de-risking through blockchain-based re-risking? [277]

Published in: Journal of International Relations and Development (2020)

Link to original: https://doi.org/10.1057/s41268-020-00198-5

If you want to cite the papers, please refer to the original.

For technical reasons we could not place the page where the original quote is coming from.

Abstract-Summary

"As banks based in the Global North have moved to withdraw financial services from many emerging economies, de-risking has attracted attention for undermining financial inclusion and developmental efforts."

"Contributing to IR theorising of legitimacy and re-risking, this article illustrates how technology-based de-risking efforts that seek to attend to the perceptions of foreign financiers can undermine the legitimacy of financial inclusion projects."

"Contrasting unfolding blockchain-based financial inclusion initiatives in two regions of small states in the Eastern Caribbean and Eastern Europe, our analysis stresses the need for greater local participation and clearer distribution of benefits from finance and technology (fintech) centred forms of digital development."

Introduction: de-risking and the legitimacy of financial inclusion and digital development efforts

"How can blockchain-based initiatives address the exclusionary nature of de-risking without 're-risking' or undermining the legitimacy of digital development and financial inclusion efforts?"

"If technology-led digital development and financial inclusion efforts are to be perceived as legitimate, we argue, blockchain-based efforts to combat de-risking must facilitate the participation of, and benefits to, multiple publics."

"While heralded for overcoming foreign perceptions of informal local activities at the heart of de-risking, blockchain applications can perpetuate and extend the legitimacy challenges facing digital development and financial inclusion efforts."

"A second, contrasting case study of blockchain-based projects to combat de-risking unfolding in the Baltic states of Eastern Europe then reveals how wider participation of multiple publics can avoid 're-risking' the legitimacy of digital development and financial inclusion projects."

"These case studies reveal the greater involvement by an array of publics needed to minimise, though perhaps not entirely resolve, the paradox that technology-centered de-risking efforts can re-risk the legitimacy of digital development and financial inclusion strategies."

Legitimacy and multiple publics in blockchain-led de-risking

"Two subsections then outline the contributions and limits of blockchain technologies for generating wider participation and input legitimacy, as well as for ensuring broad benefits and output legitimacy across multiple transnational communities."

"We elaborate on the importance of spreading legitimacy more broadly across multiple publics in the context of blockchain applications for combatting de-risking after elaborating additional dimensions to our analysis of varying perceptions of rightfulness: input and output legitimacy."

"Blockchain applications can contribute to input legitimacy by enhancing the range of publics involved in decision-making."

"If only a narrow range of publics possessing sufficient technical knowledge and access are able to participate in decision-making, broader publics are unlikely to regard blockchain-based schemes as legitimate conditions for advancing digital development and financial inclusion."

"The application of blockchain technologies in promoting financial inclusion and overcoming de-risking problems can theoretically provide benefits to multiple publics."

De-risking and re-risking in blockchain-based digital development
"Blockchain applications intended to enhance financial inclusion and combat de-risking in the Eastern Caribbean can risk undermining the legitimacy of digital development if these projects are not explicitly accompanied by wider efforts to ensure that multiple publics have access, and the ability, to harness these complex and evolving technology-centered schemes."

"Despite a continued role of market actors in blockchain-based financial inclusion strategies, national governments have played significant roles in ensuring that digital development in the Baltic region are linked to wider digital infrastructures and education projects."

"By examining evolving blockchain-based initiatives in the Baltic sub-region we illustrate the importance of persistent roles for government in ensuring that multiple publics participate and benefit from digital financial inclusion projects."

"The involvement of governments in the blockchain-based projects across the Baltic region has broadened the range of publics involved in ways that risk the legitimacy of digital development and financial inclusion schemes much less."

Conclusion
"Reliance on technological solutions alone—whether blockchains or other emergent technologies—for combating de-risking can exacerbate socio-economic inequalities, leading a relatively small set of publics to participate and to benefit from digital development."

"As our second case study has shown, unfolding digital development efforts in the Baltic states of Eastern Europe have more productively involved a wider range of publics in blockchain-based projects."

"While there remained market-based processes in this region, blockchain-based de-risking was situated within digital development strategies that involved government funding of start-ups and digital education to ensure consumer protections from scams and privacy violations."

"It will be important to investigate how precisely Russian and Chinese state and non-state actors are employing blockchains and other emergent technologies in achieving development goals through more top-down and state- centralised governance models that are becoming particularly attractive in many of the regions of the Global South afflicted by de-risking (Mahrenbach [45]; Ngetich [46]; Vnesheconombank [47])."

7.1.5 E-Democracy and the Matter of Scale. Revisiting the Democratic Promises of the Internet in Terms of the Spatial Dimension [278]

This is a machine-generated summary of:

Kneuer, Marianne; Datts, Mario: E-democracy and the Matter of Scale. Revisiting the Democratic Promises of the Internet in Terms of the Spatial Dimension [278]

Published in: Politische Vierteljahresschrift (2020)
Link to original: https://doi.org/10.1007/s11615-020-00250-6

If you want to cite the papers, please refer to the original.

For technical reasons we could not place the page where the original quote is coming from.

Abstract-Summary
"We argue that the question of whether the Internet and digital tools can fulfill the democratic promises must be confronted in a more differentiated way, examining under which spatial conditions e-democracy is more likely to incentivize citizens to engage and to involve more citizens as well as marginalized citizens in public debates and decision making."

"The article introduces scale as a new element for a better theoretical understanding of e-democracy."

"The spatial perspective brings an innovative spin to the somewhat petrified positions on the mobilizing or reinforcing effects of digital tools as well as a new theoretical argument for the debate on e-democracy in general."

Introduction
"Understanding this technical connectivity as an enabling attribute for new and "better" democratic process, the spatial dimension adds an important piece to the puzzle of whether and how e-democracy tools really do improve overall democratic quality and to what extent."

"While theorizing on the correlation between size and democratic quality is certainly not a new issue for democracy studies (in particular, see Dahl and Tufte [48]), it is surprising that the recent research so far did not transfer this fundamental theoretical concern of scale to the concept of e-democracy."

"We argue that the question of whether the Internet and digital tools can fulfill the democratic promises must be confronted in a more differentiated way, examining under which spatial conditions e-democracy is more likely to incentivize citizens to engage and to involve more citizens as well as marginalized citizens in public debates and decision making."

The Internet, Democratic Promises, and Changing Cycles: Taking Stock of the Normative and the Empirical Debate

"The few empirical findings on transnational e-democracy projects indicate that digital tools are not capable of fulfilling any of the initial democratic promises on this level."

"Their findings show that the seven examined e-democracy projects on the local level mobilized many citizens to participate."

"Although the eight ICT projects on the regional level seemed to mobilize more citizens to participate in the process—with three projects mobilizing more than 100,000 citizens—one has to take into account that Peixoto and Fox ([49], p. 12) count the total number of participants."

"It seems fair to conclude that the subnational and especially the local e-democracy-projects were more successful in mobilizing potential participants than the three examined projects on the national level."

"Peixoto and Fox ([49]) show that e-democracy projects on the local level are able to mobilize more citizens than national projects and even as much as regional ones."

Theorizing on the Scale of E-Democracy

"The most important differentiation refers to the three levels of engagement: enabling (support for accessibility and understandability of information by the audience), engaging (consulting with a wider audience and supporting deliberative debate on policy issues), and empowering (active participation and bottom-up ideas to influence the political agenda; Macintosh [50], p. 3)."

"The discussion on place effects focuses on regional disparities in Internet access, which is one aspect of the popular concept of the digital divide (Mossberger et al. [51], p. 772)."

"While the aforementioned authors, above all Mossberger et al., provide important evidence for the "place matters" argument in relation to the matters of access to the Internet, our concept aims to expand the argument beyond the dimension of technological infrastructure, addressing online interaction on the basis of e-democracy tools."

Conclusion

"According to our overall argument, however, this testing should take into consideration the (vertical) comparison between the different levels in order to assess under which conditions of scale e-democracy has better chances of increasing citizens' participation and political actors' responsiveness, and thus democratic quality."

"Our theoretical considerations suggest that distance matters when it comes to e-democracy and that future studies on the effectiveness and impact of e-democracy are well advised to account for the level on which it is focused."

"By introducing the theoretical argument of the spatial dimension of e-democracy, we do not disregard other important factors influencing the effectiveness of e-democracy tools such as usability, privacy and the like."

"According to our claim of scale, we argue that the design of an e-democracy tool has to be tailored to the level where it will be applied."

7.1.6 E-Government in Kazakhstan: Challenges and Its Role to Development [279]

This is a machine-generated summary of:

Bhuiyan, Shahjahan H.: E-Government in Kazakhstan: Challenges and Its Role to Development [279]

Published in: Public Organization Review (2009)

Link to original: https://doi.org/10.1007/s11115-009-0087-6

If you want to cite the papers, please refer to the original.

For technical reasons we could not place the page where the original quote is coming from.

Abstract-Summary

"This paper critically examines the progress made in introducing and implementing e-government programs and policies in Kazakhstan."

"This study illustrates some international development experiences to understand the benefit of e-government."

"Such experiences may serve as policy guidelines to the successful implementation of e-government to ensure overall development in Kazakhstan."

Introduction

"In the United Nations system, the World Bank launched an e-government website, and in November 2002 its Information for Development Program released The E-Government Handbook for Developing Countries."

"In the midst of transitional challenges, the Government has chosen the introduction of e-government for the twin objectives of providing fast and quality access to public services and of improving public services' effectiveness through the widespread use of ICT in the public sector (World Bank [52])."

"In order to measure how far e-government initiatives have contributed to the improvement of relationships between politicians, bureaucrats and citizens in post-independence Kazakhstan, this paper: (i) analyzes the background to the introduction of e-government programs and policies; (ii) describes the overall development of e-government; (iii) investigates the challenges facing the implementation of e-government initiatives; and finally (iv) highlights its contribution to development."

What is E-government?

"The term e-government arises by analogy to the concepts and practices of electronic commerce applied to the public sector, referring to the delivery of government services to the public 'on-line' (typically over the Internet) or to the technological infrastructure required to deliver those services (Brown [53], p.242)."

"Like e-commerce, the popular interpretation of e-government is one that defines it exclusively as an Internet driven activity…to which it may be added "that improves citizen access to government information, services and expertise to ensure citizen participation in, and satisfaction with the government process…it is permanent commitment by government to improving the relationship between private citizen and the public through enhanced, cost-effective and efficient delivery of services, information and knowledge."

"The World Bank definition indicates the areas of operation of the concept and also lays down the broad benefits accruing out of this utilization of ICT to the field of governance, namely, to promote citizen empowerment, improve service delivery, strengthen accountability, increase transparency, or improve government efficiency (cited in Tandon [54], p.3)."

Kazakhstan: context

"Over the period 1998–2004, the population living below the poverty line in Kazakhstan declined significantly from 39% in 1998, to 20% in 2004 (Agarwal [55])."

"The long-term development strategy 'Kazakhstan 2030: Prosperity, security and improved living standards for all Kazakhs' was adopted in 1997."

"It identified seven priorities for the country's development: (i) National security, (ii) domestic stability and social cohesion, (iii) economic growth, (iv) health, education and welfare for the citizens of Kazakhstan, (v) energy resources, (vi) infrastructure, transport and communications, and (vii) a professional state."

"In 2007/8, the HDI for Kazakhstan is 0.794, which gives the country a rank of 73rd out of 177 countries (UNDP [56])."

"The backdrop painted a landscape that suggests the critical importance of implementation of e-government scheme in Kazakhstan with an aim to quick and quality access to public services, improving its effectiveness, combating corruption, poverty reduction through income generation, and thus building a breeding ground for development."

The state of E-government in Kazakhstan

"Kazakhstan's accomplishments in fostering e-government include (World Bank [52]): Recognition of e-government as a priority at the highest political level and the elaboration of an e-government strategy; Establishment of the Agency for Informatization and Communications (AIC) as an independent regulatory authority empowered to implement state ICT policy; Creation of government agency Web sites (32 out of 42 government agencies have their own Web sites); Development of a number of corporate networks and databases (e.g., integrated taxation, customs, pension information systems) by individual government agencies; and Enacting important legislations such as the laws on e-documents and e-signatures."

"Kazakhstan's e-government program incorporates a three-stage approach (World Bank [52]): 1st stage: creation of the basic components of e-government infrastructure, such as the governmental portal, a "payment gateway" providing a linkage with the banking system, national identification system, government-wide ICT network backbone infrastructure, creation of cross-agency information systems, provision of mainly informational and transactional e-government services, promoting Internet use among the citizens and bridging the digital divide."

The challenges

"Kazakhstan confronts with multifaceted challenges to the introduction of e-government, and some of them are described here."

"Kazakhstan e-government program receives strong support from the President and his office."

"The lack of political participation, however, compartmentalizes the development of e-governance."

"The emergence of e-governance has significantly changed the nature of the relationship between citizens and public servants (Haque [57])."

"A 2002 World Bank survey made an assessment of Kazakhstan's governance and service delivery."

"The survey suggests that general areas where Government can work on to improve the quality of public services through e-government."

"To this effect, several municipal governments in Germany have facilitated the development of high-speed network cables, fiber optics, and public access to improve digital economy (Hasse [58])."

"There is no denial that in most of the developing countries e-government programs suffer due to the lack of quality human resources."

E-government in the context of development

"The Government of Brazil also provides an e-participation platform that permits Members of Parliament and citizens to communicate through chat rooms, discussion forums and the service "Falm com Deputado" or "Talk to the MP"."

"E-Seva (electronic service) program was launched to effectively deliver public and social services to the citizens."

"No jurisdiction limits- any citizen in the twin cities can avail of the services at any of the 46 e-Seva service centers."

"The Government of India data shows that e-Seva centers offer the total of 132 G2C services from 16 government departments, and 15 B2C services from 10 business firms."

"Over and again, control of corruption and poverty reduction are two important development challenges that are being facilitated through e-government."

"Heeks ([59]) points out that the level of corruption in the public sector sharply decreases in countries where e-government exists."

Conclusion

"In order to keep pace with the articulated development goals, particularly to achieve Kazakhstan 2030, the GoK has started to move toward e-government paradigm to establish a citizen-centered, accountable and transparent government."

"E-government offers opportunities, though rudimentary at the present stage, to the government to improve service delivery across the country."

"Such experiences may serve as policy guidelines to the implementation of e-government in Kazakhstan, after careful evaluations to their acceptability in Kazakh society."

"An important challenge to e-government implementation in developing countries is the lack of financial resources."

7.1.7 Music and Democracy in America: Historical Perspectives on 'Democratization' in the Digital Age [280]

This is a machine-generated summary of:

Hanrahan, Nancy Weiss: Music and democracy in America: historical perspectives on 'democratization' in the digital age [280]

Published in: American Journal of Cultural Sociology (2021)

Link to original: https://doi.org/10.1057/s41290-021-00138-x

If you want to cite the papers, please refer to the original.

For technical reasons we could not place the page where the original quote is coming from.

Abstract-Summary

"What meanings of democracy are invoked in talk about democratization and music, and how does this discourse reflect struggles over democracy in our time?"

"Two historical instances in which musicians and critics in the US engaged in democratic controversies through music highlight the contingency of the linkage between network metaphors and liberal notions of the public sphere that underpin the current democratization discourse."

"Debates about American classical music in the mid-nineteenth century and jazz in the mid-twentieth illustrate that notions of democracy are tied to historical context and contested as different social issues and conflicts arise."

"I reflect on whether democracy might be decoupled from network structures and reconnected to historical struggles in this rethinking, or if different criteria are needed to speak to the question of why or how music matters."

Introduction

"It did not take long for the celebratory claim that digitalization would democratize the music world to be deflated, brought down from the realm of wishful fancy to confront the reality on the ground."

"The critique of digital democratization most often pursued by music scholars (e.g., Hesmondhalgh and Meier [60]; Carter and Rogers [61]) proceeds on the grounds that the promised democratic utopia has not been realized."

"The critique I wish to pursue here starts from a different premise; that is, the narrative of musical democratization has been problematic from the start because democracy is conceived as a technological rather than a historical achievement."

"While music scholars and commentators have relied on this de-historicized understanding of democratization to assess the promises and possibilities of music in the digital age, it is a notion that has lost much of its normative grounding."

The contemporary discourse of democratization in music

"Even the critique that suggests that the promise of democratization in music has not been realized operates largely within the terms of 'democracy' set by the discourse."

"The current discourse of democratization and music borrows from liberal notions of the public sphere and relies on the network as both a technical platform and a metaphor."

"The grounds for this faith are that networks appear to be horizontally rather than hierarchically organized, and they operate on a binary of inclusion and exclusion where inclusion is an inherent good."

"Democratic, as opposed to network, inclusion is not an operational parameter but contingent, situated, and principled."

"What I am suggesting is a move beyond the by now comfortable ways in which network metaphors and democratic principles seem to align, to recognizing the singular difference of historical struggle over the terms."

Nineteenth century classical music: democratic culture and national identity

"Because 'great' music was music that could produce common associations and sympathies, this was an appeal to a 'universal' aesthetic that transcended nationality: We too want, hope for and believe in, not an 'American system' in music but an American new era of musical Art: a new manifestation of musical genius which would be distinguished not by narrow nationality but by the universality, the generous humanity, the broad and glorious inspiration that shall make it the language of a brighter period of a whole human family redeemed and reconciled. (Dwight 1853, cited in Hatch [62], p. 582) This position sparked a debate that escalated throughout the 1850s with critics and composers situated in the very different musical culture of New York."

Jazz: contradictions of democracy and freedom
"Commentary about jazz engaged the same substantive issues about democratic citizenship that animated movements for civil rights and Black liberation: social and political recognition, racial equality, and freedom."

"The critical establishment in the 1940s to early 1960s was dominated by white men writing about and often promoting a historically Black music, while Black authors and jazz musicians became increasingly vocal about the politics of their work throughout the postwar period."

"Critics and musicians, Black and white, debated whether the increasing calls for Black political and artistic freedom were good for the music or were alienating audiences, were a form of protest or of spiritual transcendence, whether freedom was a personal ideal or foundational to creating art, or was necessarily tied to Black nationalism (Gennari [63], p. 255, see also Hersch [64])."

Democratization without democracy
"Historical context frames both the nature of the contestation in which music is engaged and the meanings of democracy in circulation."

"In many respects, what has been called democratization in the contemporary music world is more accurately a form of libertarianism that emphasizes individual empowerment and personal freedom, symbolized by the heroic (and gendered) DIY figure."

"This slippage between democratization and libertarianism in the current discourse shows how a functionalized conception of democracy, detached from the specific content of historical struggles, becomes detached in turn from normative democratic conceptions of emancipation, equality, or social justice."

"In lieu of accepting the terms of democratization that have been normalized in this discourse, I am suggesting that, if indeed democracy still is a basic normative principle, analysis and critique be framed in light of its contested meanings within the social conditions in which music is engaged."

Music, democracy, and the problem of normativity
"While new songs in support of Black Lives Matter, for example, are regularly produced, they have not generated a new musical aesthetic that is the subject of debate or criticism (as was the case with nineteenth century American classical music and with free jazz), as the politics are expressed primarily through lyrics."

"At its root, critique has to be grounded in some normative idea of why or how music matters, some positive vision of what music might offer against which research findings and observations can be measured."

"Rather, it is a way of understanding and a framework for articulating the possibilities music holds for enhancing human experience, social connection, or political expression in a specific historical period."

"It is time to think beyond democracy, particularly in its currently embattled state, time to undertake the project of coming to new understandings of how or why music matters."

7.1.8 Digital Campaigning as a Policy of Democracy Promotion: Applying Deliberative Theories of Democracy to Political Parties [281]

This is a machine-generated summary of:

Odzuck, Eva; Günther, Sophie: Digital campaigning as a policy of democracy promotion: Applying deliberative theories of democracy to political parties [281]

Published in: Zeitschrift für Politikwissenschaft (2021)

Link to original: https://doi.org/10.1007/s41358-021-00308-w

If you want to cite the papers, please refer to the original.

For technical reasons we could not place the page where the original quote is coming from.

Abstract-Summary

"Our article addresses this gap by drawing on recent research on the normative theory of political parties in the field of deliberative democratic theory."

"The deliberative theories of democracy proposed by Habermas and Rawls contain structural elements of a normative theory of the political party: the special status of political parties as mediators between background culture and the political forum, between the political system and the public sphere, and between the individual and the state, confers on them a central position as actors in in the public use of reason and deliberation."

"We point to the implications for the evaluation and design of digital political microtargeting that arise from the application of deliberative principles to political parties and consider the need they reveal for the ongoing development of detailed, nuanced normative theories of democracy."

The practice of digital political microtargeting in election campaigns and the gap in normative theories of democracy

"Because political parties are among the key actors in election campaigns as they take place in modern democracies, it makes sense to develop an outline for an ethos of digital campaigning through the lens of a normative theory of the political party as a deliberative actor."

"Political parties' mediating function, their dual role in representation, and their influence on public deliberation confer on them the obligation to conceive of digital campaigning not only as a tool for mobilizing voters and for winning votes, but also as part of a broader policy of promoting democracy, from which it follows that they should develop and follow an ethos for and in their digital campaigning."

"We will first outline principles of deliberation and public reason, before discussing the extent to which we may, or indeed need to, apply these principles to campaigning political parties, drawing on recent studies on political parties as actors of public deliberation in normative theories of democracy."

Digital political microtargeting as a policy of democracy promotion

"A second factor determining campaigning political parties' deliberative responsibility is their contribution to the "overlapping consensus" that, in Rawls's view, guarantees stability in diverse societies (see Bonotti 65, p. 3 f.): By acting as "loudspeakers", setting and structuring the political agenda, and advancing political programs, they add legitimacy to, and shore up a stable common ground for public deliberation in a democratic constitutional regime."

"Rawls emphasizes the need for a public political culture that supports and encourages such self-respect: "If citizens of a well-ordered society are to recognize one another as free and equal, basic institutions must educate them to this conception of themselves, as well as publicly exhibit and encourage this ideal of political justice. [...] Acquaintance with and participation in that public culture is one way citizens learn to conceive of themselves as free and equal [...]." (Rawls [66], p. 56) In our view, it is possible to reformulate Rawls's thesis of the social foundations of civic self-respect as a thesis of the sociotechnical foundations of civic self-respect (cf. Hoffmann [67]) and apply it, thus recontextualized, to the field of digital campaigning: Because the practice of digital political microtargeting is based on massive asymmetries of information (see Tufekci [68]) and power (cf. Bay [69], p. 1727) between party and voter, and among individual voters, it can impact a citizen's self-perception as someone with equal opportunities and rights to freely form her own political judgments."

Discussion and conclusion

"Increased public understanding of aspects of certain campaigning practices that are problematic from a normative perspective of deliberative democracy may cause some of these practices to lose their strategic appeal to political parties for fear that voters might act on their awareness and express their rejection of them in their voting decisions."

"We note a need for research to add further nuance and sharpness to deliberative theories of democracy to enable them to engage in analysis of the various remits and roles of political parties, including as they pertain to different points in time (such as before and after elections) and spheres of societal life (parliament, background culture, the public forum, intra-party communication)."

"The rich literature in the field of comparative political science around the roles and functions of political parties is likely to be of great help in producing a more realistic and context-sensitive account of political parties within theories of deliberative democracy and the normative demands we, as societies, can make of them in the context of election campaigns."

7.1.9 Digital Deceleration. Protest and Societal Irritation in the Internet Age [282]

This is a machine-generated summary of:

Mölders, Marc; Schrape, Jan-Felix: Digital Deceleration. Protest and Societal Irritation in the Internet Age [282]

Published in: Österreichische Zeitschrift für Soziologie (2019)

Link to original: https://doi.org/10.1007/s11614-019-00354-3

If you want to cite the papers, please refer to the original.

For technical reasons we could not place the page where the original quote is coming from.

Abstract-Summary

"Our article examines new potentials for systemic irritation in the Internet age and focuses on the processes of synchronization between media, politics, and law First, we discuss antipodal assessments of what is accelerated in the digital society: (a) the frequency and accessibility of attempts at "correcting society", or (b) the perceived need for societal rectification."

"Based on empirical case studies on social protest movements, investigative journalism, and online petition platforms, we subsequently elaborate our main thesis that the technical improvement of communication structures per se does not necessarily increase the likelihood of the addressed societal systems evaluating the correction queries."

"To irritate other societal contexts effectively, civil counter forces exercise self-control particularly regarding temporality, retarding their operations according to the operational speed of the addressed meaning systems."

Introduction

"From a systems-theoretical point of view, "correcting society" refers to how modern society deals with the societal consequences of functional differentiation such as global inequalities in wealth distribution or ecological problems (Luhmann 70, p. 124 ff."

"This knowledge stock does not prevent ever more civil counterforces from compiling ever more refined techniques for systemic irritation (Münte [71]; Bröckling [72]), i.e., from designing specific forms of irritation in communication that trigger the functional (or organizational) systems addressed to leave their established operational routines and to deal with the criticized aspects and conditions."

"4, we, therefore, argue that, while a central assumption of traditional differentiation theory still applies—namely that the autonomy of social systems renders direct interventions impossible—, the particular situational characteristics of irritation design make a fundamental difference in the effectiveness of queries for societal corrections and thus request detailed observation."

Digital Society and the Acceleration of (the Need for) Societal Correction

"We will introduce these stances more explicitly, resulting in a three-folded main theses: (1) Attempts at correcting society, aiming for resonance in functional meaning systems such as the mass media, politics, or the law, require evolving organizational efforts. (2) For social movements and other civil counterforces, this observation first and foremost implicates organizing communicational persistence. (3) This conclusion calls for context-sensitive communication strategies or a deliberate "irritation design" (Mölders [73]), including a continuous reflection of the specific timeframes of the addressed functional contexts and a controlled deceleration of publication efforts."

"Asked for the difference widely available digital technologies make in correcting society, proponents of the "digital society" stressed that digitally empowered citizens were no longer dependent on centralized distribution channels such as traditional mass media and that, therefore, even individual attempts could succeed in accelerating political or economic change in this day and age: "The network allows all citizens to change their relationship to the public sphere."

Countervailing Powers of Civil Society: Three Brief Case Studies

"The ICIJ, as a media organization, by no means aims at accelerating functional meaning systems such as politics or the law It instead decelerates its publication activities in an act of synchronizing the own operational pulse with the expected temporality of the addressed contexts."

"As with the Anti-ACTA demonstrations, the mass protests against TTIP and CETA were sustained by a variety of NGOs and initiatives; however, the initial nucleus for the considerable public attention to the movement in Germany emerged as a by-product of the upstream "market research" of the petition platform: As always, Campact tested the issue's potential for wide-spread support before initiating a broader campaign, in this case by contacting 10,000 randomly selected newsletter recipients."

Conclusion: Intersystemic synchronization, irritation design and digital deceleration

"To this effect, our empirical case studies emphasize the disparities in irritation capacities and the diversity in irritation design as significant factors in the effectiveness of collectively articulated impulses for social change: It makes a difference (1) whether the discontent is situationally revealed in volatile and algorithmically mediated counter-publics (e.g., along hashtags such as #MeToo) or whether social protest movements and NGOs implement collectively elaborated communication strategies; (2) to which degree civil counterforces, by carefully coordinating their means of publication, strive for a continual reflection of their position in the cross-societal description of a "common" present by the mass media; and (3) whether a protest message, or an impulse for systemic irritation, is intended for the general public at large or whether a more specific functional communication context in society is being addressed with respect to its temporal and structural logic.""

7.1.10 From Bureaucratic Administration to Network Administration? An Empirical Study on E-Government Focus on Catalonia [283]

This is a machine-generated summary of:

Welp, Yanina; Urgell, Ferran; Aibar, Eduard: From Bureaucratic Administration to Network Administration? An Empirical Study on E-Government Focus on Catalonia [283]

Published in: Public Organization Review (2007)
Link to original: https://doi.org/10.1007/s11115-007-0038-z

If you want to cite the papers, please refer to the original.

For technical reasons we could not place the page where the original quote is coming from.

Abstract-Summary

"This article aims to (1) analyse the extent to which there is a movement from a Weberian bureaucratic administration towards a new organisational model that we call network-administration and (2) consider whether the premises of New Public Management (NPM) theory are present in this process of transformation."

"Our initial conclusions indicate that while there is a change in Public Administration (PA), this change is incremental and has several nuances: a slight evolution towards network-administration is observed as currently existing alongside the traditional hierarchical and compartmentalised bureaucracy."

"We should add that there is also a new emphasis being placed on the relationship with the citizen as a service-user."

Introduction

"This alternative model centred on the proposal of New Public Management (NPM), which has been one of the most influential proposals in the context of post-bureaucratic theories."

"The NPM proposal is based on the private sector competency model and seeks to create smaller administrative structures that are decentralised and enable a degree of administrative flexibility."

"This model is basically characterised by (1) the connection of information systems and management procedures, and their ability to work together, both between units of the same administration and between distinct administrations (Hinnant and Moon 74; Dunleavy and Margetts [75]); (2) the tendency to adopt more flexible management methods, with flatter relationships than those that predominate in the Weberian style bureaucracy (Huse [76]; Mayntz [77, 78]; Rhodes [79]); (3) a noted orientation towards the citizen (Grandori [80]; Bimber [81]; Barber [82]; Fountain [83]); and (4) we associate the concept to some elements of modern governance (Mayntz [78]), particularly in what is referred to as the more co-operative way of governing, which includes the direct interaction of public and private actors in the public networks."

Methodological approach

"We monitored projects including the restructuring of citizen information databases, the implementation of a new manager of web contents, the design of a new portal for citizen attention, and the process of merging two citizen attention telephone services; (2) Interviews of actors involved in these processes, included records-managers and training staff, civil servants responsible for information system maintenance, politicians promoting the projects, and private sector actors involved in the design and implementation of the innovative projects. (It were done a hundred of interviews between September 2004 to December 2005) (3)

Access to documentation generated in relation to these initiatives and the channels of attention (proposals from different actors, indicators of use for the channels of attention, questionnaires and other types of user studies) Given our interest in evaluating the extent to which a transition process towards a new organisational structure can be observed, we have chosen to focus on a regional government."

Discussion

"Renate Mayntz ([84]: 92) notes that the horizontal co-operation with private sector actors and the establishment of policies in these networks highlights the problem of accountability and legitimacy of the private actors in the public policy networks do not generally have democratic legitimacy. (3) Citizen focused NPM adopted the metaphor of client-focused to redirect the relationship between the citizen and administration."

"Arguing against this it has been noted that there are differences of legitimacy: governments have the obligation to provide services to the public, and the relation between the public and their administrations is different from the relationship that exists with the private sector."

"The need to transform the administration is highlighted by both and involves the private sector in the design and implementation of public policies, and the reconsideration of the State–citizen relationship."

Empirical findings

"The System of Citizen Attention (SAC), the first database with transversal information, which combines data on organisational structures, procedures, support, etc. of the Catalonian Government, is fed with material from operators in each department who write the corresponding records, update the information, and so forth."

"The focus is on integrating information systems (implementing inter-operative technology, unifying criteria for the organisation of information, etc.) and improving citizen attention in all the channels, not just in the services provided by these new channels."

"3) Recovery of knowledge, privatisation of production: The current tendency in the Catalonian Government reveals a third scenario, with a strategy that has been formally outlined as being aimed at: (a) Recovering knowledge: the power accumulated by the private sector in the position of dominant supplier is seen as having the necessary know-how in management, such as control over the public information databases."

"The objective is transparency in management, which has the effect of control over government and control of citizens over their specific demands (information about the evolution of their administrative procedures, for example)."

Conclusions

"This exaggerates the complexity of the citizen's relationship with the administration, increasing the difficulties that result from administrative procedures:

difficulties in getting information, problems of information not either being up-to-date or valid, etc. The telephone service is completely different, with standardised attention by telephone for all the departments."

"Other authors affirm that certain characteristics of Weberian bureaucracy are incompatible with some aspects of the Network-Administration: that is, bureaucracies tend to identify the control of information (and secrecy) with an increase in power; along the same lines, there are no incentives for co-operation between administrative units to be able to capitalise their actions, which in turn strengthens accountability within the hierarchy rather than to the advantage of citizens; inertia which acts against innovation; and cooperation, which is suffocated because departments are used to competing for budget funds, something which goes along with the structural difficulties in undertaking joint-initiatives with shared resources (Jain [85], Fountain [86])."

7.1.11 Digital Civil Society: How Nigerian NGOs Utilize Social Media Platforms [284]

This is a machine-generated summary of:

Armstrong, Crystal; Butcher, Charity: Digital Civil Society: How Nigerian NGOs Utilize Social Media Platforms [284]

Published in: International Journal of Politics, Culture, and Society (2017)

Link to original: https://doi.org/10.1007/s10767-017-9268-4

If you want to cite the papers, please refer to the original.

For technical reasons we could not place the page where the original quote is coming from.

Abstract-Summary

"Building social capital requires constructing relationships between people, and social media will be most likely to accomplish this when it is used in an interactive way."

"Previous research has considered the issue of interactivity and social media usage in Western countries like the USA, but no research has focused on these issues within the developing world."

"Our paper explores the degree to which Nigerian NGOs are utilizing social media tools, such as Facebook and Twitter, in ways that are more interactive in nature, and thus in ways that are more likely to build robust and reciprocal relations."

Introduction

"Non-governmental organizations (NGOs), as part of a larger civil society, play a key role in providing the space for increased interactions among individuals within a community, and such interactions are essential for building strong communities and relationships within society."

"Our paper explores the degree to which Nigerian NGOs are utilizing social media platforms, such as Facebook and Twitter, in ways that are more interactive in nature and thus in ways that are more likely to build robust and reciprocal relations."

"Despite all this, however, many NGOs are seeking to connect with others via social media platforms, and understanding how those groups interact and seek to connect can provide us with foundational knowledge about how NGOs are engaging digitally with their audiences."

"Many countries within the developing world may have a special need for the relationships and connections that NGOs can help provide, it is important to assess how NGOs in these areas are utilizing social media."

Previous Literature

"Given the nature of social media which relies on user creation and interaction with content, there is enormous potential for NGOs to use these platforms to increase the reach of their organization, and more importantly, to encourage the type of relationships that will help communities and may help improve democracy and peace."

"Recent research has shown that organizations are reluctant to use more interactive technologies, such as Facebook and Twitter, and when they do use them, they tend to utilize them in passive (one-directional) ways."

"While there are likely differing reasons and circumstances for the degree and nature of NGO activity via social media platforms, examining them individually and with a more focused lens of analysis is a vital step in understanding the broad trends."

Interactivity

"Since this time, suggestions for improving interactivity with target audiences has been the focus of numerous studies for online interactivity of websites, social media platforms, governments, and more (see for example Lee ([87]), Mossberger and others ([88]), Sundar and others ([89]), and Teo and others ([90]))."

"In the case of humanitarian aid and emergency response, interactivity now includes information that is exchanged between stakeholders in local communities, as well as with humanitarian aid groups abroad."

"The inclusion of big data and distant volunteers in NGO crisis response efforts have also entered into notions of interactivity, though not without the criticism that these have the potential to reinforce power dynamics and inequality (see for example Burns ([91]), Crawford and Finn ([92]), Read and others ([93]), and Taylor and Broeders ([94]))."

"As some authors (Crawford and Finn [92]) have noted, the issue of privacy is particularly problematic during crisis situations, when people are sharing information during a time when they are still at risk."

NGO Usage of Technology and Social Media Platforms

"Waters and Feneley ([95]) look at both websites (i.e., one-directional communication) and social media (potentially more interactive platforms) use by The NonProfit Times top 100 organizations."

"They find that these organizations largely prefer to use their websites for cultivating relationships with stakeholders, and are less likely to fully embrace social media platforms."

"Of social work advocacy groups' websites, Edwards and Hoefer ([96]) find that these organizations rarely use social media for advocacy purposes, and when they did use these platforms for advocacy, they did so in ways that were not particularly effective."

"The authors also look at how these organizations are utilizing social media with an in-depth study of the use of Twitter by the organizations."

"The majority of tweets was used in a more passive way—merely providing information—and did not fully utilize the ability of social media to create a more interactive web environment."

NGOs and Nigeria

"Little research has attempted to look at how NGOs in countries outside of the West (where democracy is already flourishing) are utilizing social media."

"Given this gap in the research, our study is a first attempt to peek into the virtual NGO world within a democratizing state by looking at whether and how NGOs outside the West utilize social media."

"We consider the use of social media, especially Twitter, by NGOs in a place where it may have great consequences in terms of democratization."

"There are many countries that have attempted democratic transitions and are still struggling in their efforts; however, we have chosen to consider NGOs within Nigeria."

"A more detailed look at how NGOs are utilizing social media like Twitter, and thus possibly contributing to this civil society, is an important line of research."

Research Questions and Hypotheses

"Aside from the very basic question of whether NGOs in Nigeria are utilizing social media platforms, such as Twitter and Facebook, we are also interested in how these groups are using social media."

"Since previous research has shown that NGOs, at least in the USA, are not very likely to use Twitter in an interactive way, we also expect that most NGOs in Nigeria will use Twitter in a way that is more representative of one-way interaction: posting information about their group and activities but with little opportunity for (or expectation for) the "audience" to interact."

"H_6: Nigerian NGOs that have Twitter in addition to a traditional website or a Facebook page will be more likely to contribute interactive tweets than those NGOs that do not have a traditional website or Facebook."

Methodology and Data

"It was often difficult to distinguish between information being given (category 1) and when the NGO was interacting more with a specific person or group (category 5)."

"The first two categories discussed above represent one-way communication; NGOs can also present tweets that are more interactive in nature."

"This interaction has less potential of building a relationship between the audience and the NGO than an action that is highly interactive, but it is not merely one-way communication."

"There are tweets that we might consider to be highly interactive; where the NGO is attempting to create a back and forth with the audience."

"NGOs might not just post general calls to action, but might specifically interact with individuals or groups within Twitter."

"The NGO is attempting to reach beyond its general audience and respond to or interact with other NGOs and actors."

Analysis and Discussion

"The data support this hypothesis (hypothesis 3); around 29.1% of Nigerian NGOs in our dataset have websites, compared to only 18.4% with an active Facebook account and 7.6% with an active Twitter account."

"While we find support for hypothesis 5, we also find that the Nigerian NGOs in our sample are more interactive in their Twitter usage than we expected."

"Hypothesis 6 suggests that groups that do not have websites or Facebook pages might use Twitter for more one-way communications (in addition to more active communications) than those groups that have both mediums through which to communicate with their constituents."

"We find no support for hypothesis 6, which suggested that NGOs that have Twitter accounts in addition to a traditional website or Facebook would be more likely to use Twitter in a more interactive way."

Conclusions

"Of all, our findings are generally in agreement with previous research that suggests that NGOs often utilize social media, such as Twitter, in ways that do not fully capitalize on the interactive nature of these platforms."

"Contrary to previous studies, which focused on nonprofit organizations in the USA, our analysis of Nigerian NGOs demonstrated that these groups did use Twitter in a way that was more interactive than what might have been expected."

"Future research that focuses more specifically on some of these groups may be able to provide a more nuanced understanding of how and why Nigerian NGOs use Twitter in interactive ways and how successful these attempts at interaction have been."

"Interviews with these very active and interactive groups would likely provide some useful insights into their social media usage and whether their attempts at engaging with constituents have been successful."

7.1.12 Digital Pride on the Streets of the Internet: Facebook and Twitter Practices of the LGBTI Movement in Turkey [285]

This is a machine-generated summary of:

Caliskan, Ozgur: Digital Pride on the Streets of the Internet: Facebook and Twitter Practices of the LGBTI Movement in Turkey [285]

Published in: Sexuality & Culture (2021)

Link to original: https://doi.org/10.1007/s12119-021-09818-9

If you want to cite the papers, please refer to the original.

For technical reasons we could not place the page where the original quote is coming from.

Abstract-Summary

"This study investigates how the LGBTI movement changed utilizing social networking sites (SNS) and what kind of a communication culture has been established in the context of social movements of this transformation caused by SNS in Turkey."

"Communication practices that SNS provide to the LGBTI movement groups and use of these practices are interpreted and analyzed through collected data."

"This study has shown that the LGBTI movement in Turkey uses SNS as a space of action effectively with collective action, centralizes SNS with their actions and moves the new social movements' features to the Internet as a way of enhancing its exposure in the public sphere."

Introduction

"The recent studies on LGBTI identity and movement have focused on such topics; representation of LGBTI identity in the press and its association with women's movements, the treatment of LGBTI people as immigrants, the construction of gay identity in different classes, coming out as political action in the gay movement, the LGBT movement as KAOS GL group, images of masculinity and gay identity, the gay culture of life in Turkey presentation of gender in the context of political economy and new social movements."

"The areas of struggles which formed for the culture and identity groups that have caused formation of new social movements have been specified by many philosophers as peace, gender, ecology/environment, local/autonomy, sexual orientation, anti-globalization, alternative life, cultural/ethnic minority, youth and students, inside the third world, farmer-peasant, and psychedelic struggle areas (Heberle 97, pp."

"As all new social movements have relations with the Internet and SNS, the LGBTI movement uses these global communications networks both in the world and in Turkey."

Methods

"This observation aims to observe the Facebook and Twitter shares of the pages and reach conclusions about how the LGBTI movement uses social networking sites as a communication tool during the Pride Week event, which is the most public participation event."

"The activities of the groups that were selected through the unresponsive online observations were interpreted by observing their SNS activities during the Pride Week between June 24 and July 01, 2018 which is the only geographical annual public sphere event organized by the LGBTI movement in Turkey."

"The other postings of KAOS GL, which is prominent with its posts, about the Pride Week are as follows: the humor magazines that bring the Pride Week to the agenda, the messages and posts related to the Pride Week shared by publishers, politicians, associations outside the LGBTI movement, celebrities and various brands, Pera Film's Pride Week online film anthology, and the images related to Pride Week made by artists."

Measure

"The questionnaire was shared via communities with which semi-structured interviews were conducted and other accessible LGBTI pages on Facebook and Twitter and was kept active for two months to be answered between June and August 2018."

"Participants were asked if they shared pages' posts on their Facebook or Twitter pages."

"Participants were asked whether they participated in the public sphere activities of the pages they followed, and 105 (63.6 %) of the participants answered Yes, and 60 (36.4 %) answered No."

Discussion

"Although the page administrators say that there is not much difference between the public sphere and SNS, the followers of the pages think that the communities are more effective in the Internet environment."

"SNS have been able to create — for the LGBTI movement and its participants — a social communication space between the geographical public sphere and their private space."

"Considering that the page administrators have deleted posts and comments that include hate speech and that is for dating/friendship purposes and considering that a considerable number of followers commented on the posts, it is seen that there is also participating in the social space outside the public sphere."

"The majority of the followers who participated in the questionnaire did not use the anonymous identity feature provided by SNS, which is a favorable situation for a subject of a social movement who was also present outside the Internet."

Conclusions

"The voluntary and widespread posts about Pride Week, the diversity and abundance of the activities attended by the followers, and the importance attributed to the public sphere by the administrators confirm that they still consider the public sphere as a space of activism."

"The ideas of the administrators about the concept of slacktivism and the uncertainty of the followers reveal the dilemma they face between the public sphere and SNS."

"The reason for this is that it is not practical to find out the extent to which the people who did not participate in the questionnaire participated in public sphere events and the extent to which they carried out activism only through SNS."

"These raise the question of how actively people use SNS in an environment with so many pages and followers."

"The use of SNS by many other dominant identities, as is the case in the public sphere, pushes the LGBTI movement to fight to make itself visible."

Limitations and Future Perspective

"This study concluded that SNS are communication tools for new social movements and that social movements exhibit different practices with the communication tool they use."

"Page managers also emphasized that quantitative data are too much and qualitative data are insufficient, as they are generally reached for quantitative-oriented studies."

"It was tried to reach more page managers, but these pages were excluded from the study because no positive or negative feedback could be obtained."

"A study that can include these pages can also be made."

"In the interviews conducted in the study, it was observed that the page administrators were interested in SNS and offered many ideas."

"A larger number of studies, including only page administrators, can be put forward."

7.1.13 Public Access to Government Information in the Digital Age: Findings from China [286]

This is a machine-generated summary of:

Yu, Sun; Shengcui, Dang; Bin, Fang; Qiuyan, Zhao: Public Access to Government Information in the Digital Age: Findings from China [286]

Published in: Korean Social Science Journal (2014)
Link to original: https://doi.org/10.1007/s40483-014-0010-9

If you want to cite the papers, please refer to the original.

For technical reasons we could not place the page where the original quote is coming from.

Abstract-Summary
"The Chinese government makes great effort to publish administrative information and seek advice on good governance from the general public, but is nonetheless unable to meet citizens' legitimate demands for the right to know and participate."

"The findings showed that (a) netizens' heavy pressure rather than Chinese authorities' strong leadership pushes forward the disclosure of government information; (b) administrative information concerning the public agenda rather than the policy agenda best meets the needs of the general public; (c) commercial Weibo websites are a more satisfactory channel than official government websites for practical requests; and (d) as a unique institutional arrangement for examination and supervision, the annual report on government information disclosure should be standardized."

Introduction
"Regulations require that all administrative organs and organizations administering public affairs or providing public products and services, including schools, medical facilities, and utility companies, should proactively publish four types of administrative information: (a) that which concerns the vital interests of citizens, legal persons, or other organizations; (b) that which needs to be disseminated to or participated in by the general public; (c) that which shows the structure, function, and working procedures of any other matters relating to the administrative organ; and (d) that which should be disclosed by the administrative organ's initiative according to laws, regulations, and relevant state provisions."

"Citizens, legal persons, and other organizations may, according to the special needs of such matters as their own production, livelihood, and scientific and technological research, also file requests with departments of the State Council, local governments at all levels, and government departments at the county level and above to obtain relevant information."

"They offer four main viewpoints: (a) Chinese authorities only commit to open disclosure of administrative information, whereas what the general public really wants is democratic rights, therefore transparency without democracy is the fatal crux of China's problem (Horsely [98]; Zhou [99]; Zhou and others [100]; Fung and others [101]; Carter and Yanbin [102]; Tan [103]); (b) actual actions of all administrative organs and organizations lag behind their propaganda; however, citizens' awareness of their rights is rapidly growing due to the popularity of the Internet (Wang [104]); (c) Regulations in the legal sense can only restrict administrative organs and organizations, but other authorities, including the Communist Party apparatus, the People's Congress, the People's Political Consultative Congress, the People's Court, and the People's Procuratorate, are unrestricted (Zhou [105]; Sun [106]); and (d) Regulations gives administrative organs and organizations so much discretion that members of the general public tend to file an administrative lawsuit when their requests are rejected (Zhu and Yu [107])."

What is the driving force behind government information disclosure?
"Administrative organs and organizations are the biggest obstacle to government transparency for three reasons: (a) keeping secrets had always been the basic principle of public administration in China before Regulations; (b) there are conflicts between Regulations and other laws, including the State Secrets Law and the Archives Law; and (c) administrative organs and organizations have infinitely elastic discretion concerning work secrets and, therefore, could easily concoct excuses to avoid disclosure."

"Although some popular social networking sites, including Twitter and Facebook, are blocked in China because they may allow the posting of politically sensitive and anti-government content, Chinese native network platforms and applications, such as Weibo and WeChat, play an active and increasingly important role in government information disclosure."

"All administrative organs and organizations, especially governments at the local level, have been under unprecedented pressure for information disclosure."

What information does the general public really want?
"The core problem is that the general public is indifferent to the information published by administrative organs and organizations, but is intensely concerned about the hot issues publicized by grassroot opinion leaders through social media."

"The populace wants to take part in the decision-making process for issues concerning their daily food, clothing, shelter, and transportation, but Chinese authorities hope that the policy agenda will be made a matter of great concern for the general public via the media."

"When administrative organs and organizations publish information concerning the policy agenda, the typical response from the general public is indifference."

"In order to ensure the efficiency of public policy, all administrative organs and organizations must pay more attention to public and media agendas than to policy agenda when disclosing information."

Which information channel is the best?
"Shortly after the launch of government websites, public attention was overtaken by Web 2.0 tools and applications, which have had a profound impact on information consumption and on communication."

"Members of the general public obtain information and communicate with each other via social media websites, such as Weibo and Wechat."

"Information about the Wenzhou train crash and the rumblings of dissent in the days that followed was distributed first and expressed more forcefully on Weibo than on any government website."

"The government website is an important Internet-based platform, on which Chinese authorities attempt to publish information and communicate with the general public quickly and effectively, but it is technically not the best way—a reality that they have acknowledged."

"We think that the popular commercial Weibo platforms are the best choice as information channels."

What is wrong with the annual institutional report on government information disclosure?

"Regulations require that all administrative organs and organizations should publish their annual report on government information disclosure before March 31 each year."

"Only information actively disclosed by the provincial government office is covered by the annual reports of Jiangsu and Guangdong, because all sectors and all regions of these two provinces release independent annual reports."

"Other urgent issues include (a) the need to report on staffing and funding of government information disclosure, so as to estimate whether administrative organs and organizations regard information disclosure as a routine work; (b) the need to hold accountable administrative organs and organizations that do not prepare and release the annual report on time, or that include falsifications in their annual report; and (c) the need to establish a uniform online platform to present the annual reports of all regions from year to year in real time, thus making the reports conveniently accessible to the general public."

Conclusion

"Four principles were expounded in this paper: (a) netizens' heavy pressure rather than Chinese authorities' strong leadership push forward government information disclosure; (b) administrative information concerning the public agenda rather than the policy agenda best meets the needs of the general public; (c) commercial Weibo websites are a more satisfactory channel for practical requests than official government websites; and (d) as a unique institutional arrangement for examination and supervision, the annual report on government information disclosure should be standardized."

"Some issues still trouble Chinese authorities, primarily (a) how to handle the relationship between Internet governance and civil participation, (b) how to handle the relationship between governmental authorities and grassroot opinion leaders, and (c) how to handle the relationship between virtual space and real-world society."

"We have reason to expect that efforts accordant with the new round of administrative reform are likely to encourage the birth of a new relationship between government and society in China, according to the public agenda in the digital age."

7.1.14 The Conceptual Model for the Effect of Technology on the Political and Social Aspects of Iran's E-Government [287]

This is a machine-generated summary of:

Saghafi, Fatemeh; Zarei, Behrouz; Fadaei, Maryam: The conceptual model for the effect of technology on the political and social aspects of Iran's e-government [287]

Published in: Quality & Quantity (2015)

Link to original: https://doi.org/10.1007/s11135-015-0234-9

If you want to cite the papers, please refer to the original.

For technical reasons we could not place the page where the original quote is coming from.

Abstract-Summary

"Development in a country requires in-depth interdisciplinary studies and focusing on various economic, environmental, scientific, technological, political, and social aspects."

"The only conceptual model in Iran was used as the basis of study and was developed."

"The use of questionnaire, and correlation and regression analysis, the relationships in the conceptual model were developed and confirmed."

"Changing the attitude of people toward the government affects the way where trust in and satisfaction of the government is formed which in turn leads to their loyalty to the government and participation for promoting life quality that considered as the indices of sustainable development."

Introduction

"Achieving sustainable development in a country requires the conduction of interdisciplinary studies, and subsequently, developments in economic, environmental, science and technology, political, and social aspects."

"The fact that social indices are less developed reflects lengthy the discussions on the concept of social sustainability (The World Bank [108]), on the other hand, it might be the case that the relationship between these indices and sustainable development has not been considered well in scientific papers (Jaeger [109])."

"Regarding to the effect of participation and political development on security, economy, and social development, achieving them is one of the requirements of sustainable development."

"To express, developing is sustainable when it's enabled the rule of public political will by focusing on public participation."

"The present study starts with theoretical considerations about reviewing the studies and concepts on sustainable development and then, addresses the issue of social and political changes of e-government within the framework of good governance in public participation as the main index of sustainable development."

Research methodology

"Reviewing the extensive literature on various areas and summarizing their findings, the present study offers a conceptual model of the factors of e-government which exert influence upon sustainable development."

"The concepts of sustainable development are first reviewed the use of library and internet resources and their aspects are investigated."

"The review indicates that increasing of public participation can have a positive effect on reaching sustainable development."

"The review of literature revealed that e-government can facilitate the economic relations and leads to the high quality services."

"A questionnaire measuring the relations of the mode was developed to the validate model."

Literature review

"Considering the important role of Information and Communication Technology (ICT) in today's world, e-government is, therefore, viewed as a positive channel for enhancing trust in government by government accountability and the empowerment of its citizens (Kim and others [110]), it is considered as an opportunity for offering civil rights for people (Gunter [111]), and creates an opportunity for people to participate in the political life and challenges it (Gauld and others [112])."

"It is in line with the needs of citizens and government; i.e. using advanced tools, it becomes possible to have access to the specific information and political issues and involve citizens in a democratic process."

"e-Government's potential to increase transparency and combat the corruption in government administration has been gained popularity in the communities by e-government practitioners and researchers, such transparency can be achieved by providing citizens with more and direct access to the information regarding to the businesses of citizens concerned (Kim and others [110]), and discussed as a key to reducing of corruption and promoting good governance, and promoting transitions to democratic political systems and market economies."

Extraction of research conceptual model

"It is at this stage the government's transparency, answerability, and justice orientation are shown."

"It is hypothesized tin by this study that government's answerability and transparency have a direct effect on justice orientation."

"It is hypothesized that the decreasing digital divide through access to the transparent information is offered by the government that can affect to the government's justice orientation."

"Since these processes have occured in the body of government, one can investigate the effect of access on the information and decreasing the digital divide, government's transparency and answerability, and reduction of corruption and justice orientation on people's attitudes."

"The effect of these factors on individuals' attitudes and the impact of the resulting attitude on satisfaction and trust in the government were investigated in this study."

"This durable effect, if positive, the results in people's satisfaction and trust increasing their participation the government."

Findings and research conclusion

"To test research hypotheses, the effect of independent variables on dependent variables was investigated by the regression analysis."

"If the result of Kolmogorov–Smirnov test is a significant value over 0.05, the distribution will be normal or near normal."

"To test the research hypotheses (the effect of independent variable on dependent variables) the regression analysis was used."

"The value of F is calculated by SPSS and shows a significant linear relationship between the independent and dependent variables."

"Specifying the relationship between the independent and dependent variables with respecting to the significance level in result of the research hypothesis."

"Considering the F value of Fisher's test and its significance level, the regression analysis of the variables whose the significance level is less than 0.05, is significant, and for the variables whose significance level is more than 0.05 is not significant."

Conclusion

"The present study was conducted to offer a conceptual model about the participation of people in e-government on the one hand, and their participation in the political activities to promote the society and people's role in the government, on the other."

"The main question of the present study was "what are the indices affecting to sustainable development and how are they related to each other?" "

"Public participation, as one of the major indices in-progressing of the countries toward sustainable development, is the results of the people's satisfaction who trust in their government which is the outcome of their attitude toward the performance and policies of their leaders."

"To present a more comprehensive model of progress toward sustainable development, it is necessary to identify the other indices and their performance, such as economic, environmental, and other indices that might not be identified yet."

7.1.15 Empirical Assessment of the Role of Geospatial Technology in Delivering Governance and Strengthening Democracy in Nigeria [288]

This is a machine-generated summary of:

Agbaje, Ganiy I.; Bello, Innocent E.; Ojo, Adebayo Gbenga: Empirical assessment of the role of geospatial technology in delivering governance and strengthening democracy in Nigeria [288]

Published in: GeoJournal (2017)

Link to original: https://doi.org/10.1007/s10708-017-9797-4

If you want to cite the papers, please refer to the original.

For technical reasons we could not place the page where the original quote is coming from.

Abstract-Summary

"The paper examined the concept of development in line with the role of geospatial technology applications in governance and geospatial management in Nigeria."

"It also evaluates the previous efforts made by successive governments in managing governance process through geospatial technology initiatives using Lagos State as a proof-of-concept."

"It empirically demonstrates the role of monitoring spatial development initiatives using Landsat multi-dates satellite images 1985, 1990, 1995, 2000, 5005, 2010 and 2016 and Geographic Information System technique."

"The paper then identified, mapped and examined the settlement growth of Lagos using Ikorodu Division as an empirical proof-of-concept."

"The study reveals a geometric increase in uncoordinated human activities resulting from sporadic spatial development (builtup area) from 27.67 sq km (7.7%) in 1985–313.02 sq km (87.3%) spatial coverage in 2016."

"Based on the regression line, the study reveals a strong correlation between increasing settlement development and time."

Introduction

"Geospatial technology such as Geographic Information System (GIS) is able, to among other things, digitally generate or capture, store, retrieve, manipulate, analyze and present the results of georeferenced information in the forms of maps and supporting statistics necessary for good governance and democratic processes (Bello and Rilwani [113])."

"The geospatial information, in addition to other ancillary data, are important when planning and evaluating the level of development initiatives, especially those resulting from rural/urban spatial development as in the case of urban growth and coastal management in Warri and it environ in the Niger Delta region of Nigeria (Bello and Rilwani [113])."

"The justification for applying geospatial technologies also stems from the fact that it will allow for the assessment of the impacts and effects of emerging urban forms as well as formulation of policies, decisions and planning strategies needed in development and governance processes (Dekolo and Oduwoye [114])."

Literature review and conceptual issues

"Democratising governance through sustainable development initiatives that integrates geospatial technologies in mapping and monitoring rural/urban environment development processes become imperative (Guo and others [115]; Obanuso and Salihu [116])."

"This, to a large extent, makes Lagos State a good case study when discussing the role of geospatial technologies in delivering governance and strengthening democracy in Nigeria."

"A number of literatures and studies examining how various governments all over the world have, or can, use geospatial technologies in carrying out day-to-day governance responsibilities abound."

"Using the National Spatial Data Infrastructure (NSDI) for the production, distribution and accessing of geospatial data for national planning has also been regarded as a major mile stone in the delivery of good governance which most developing countries (like Nigeria) are still grappling with (Groot [117])."

"In Nigeria, application of geospatial technology includes border monitoring and management to conflict resolution using satellite-based sensors to acquire time series data (Bello and others [118]) and food security."

Research methodology

"Lagos State is the economic and financial capital of Nigeria."

"The choice of Lagos State in this study was informed by the mega city status it occupies and the rapid growth pattern of the rural areas."

"Using ArcGIS 10.2 software, the study further provided maps showing the landcover layers required for proper planning, community participation and delivery of good governance in Lagos State and by extension, Nigeria."

"In other to quantitatively ascertain the above, this study used multi-dates satellite images of different epochs to examine the spatial growth of Ikorodu as a proof-of-concept in generalising Lagos State."

"The study also used other secondary data especially Lagos State government statistics to examine the previous geoinformation efforts made so far in governing the State."

Results and discussions

"Lagos State Planning Information Centre (LASPIC) is a GIS Centre created for urban and regional planning purpose under the Ministry of Physical Planning and Urban Development (MPPUD) (Dekolo and Oduwaye [114])."

"Knowledge in 2011, Lagos State launched a N3bn Digital Mapping/ Geographic Information System (GIS) to solve the problem of the dearth of geospatial data for sustainable development planning."

"The LAGIS data infrastructure was expected to be integrated with the existing Electronic Data Management System (EDMS) in the Lands Directorate for accurate information about land ownership (Lagos State Government [119]; Dekolo and Oduwaye [114])."

"From inception, the objectives of the project include ensuring effective use and application of geospatial information in providing sustainable planning and development of the ever-growing Megacity."

"It is imperative to re-assert that managing mega cities (such as Lagos State) is a complex planning task requiring multi-dimensional urban information (Dekolo and Oduwaye [114]) and spatio-social indicators (Omofonmwan, [120])."

Conclusion

"These datasets include recent high resolution aerial photos/satellite images covering the country, recent large scale topographic maps, up-to-date official government statistics on citizen biometric and economies, boundary data at ward levels and home-grown Geo-ICT technologies for managing and delivery good governance to the people."

"Of the above findings, the establishment of Divisional GDI such as the proposed Ikorodu Division GDI (IKGDI) with adequate participation of the people in the community will, no doubt, ensure ownership and sustainability of infrastructure for informed decision making."

"When IKGDI with the sustaining policy is properly implemented in the state, it will facilitate rapid improvement of the economy including an efficient management of natural resources and environment."

"Its implementation and achievement can only be reached through the use of modern planning techniques and tools such as satellite remote sensing, GIS and integrated GDI for data production, dissemination, discovery and use by citizens and government at all levels of governance; thus building a virile institution and strong policy and polity."

7.1.16 The Christchurch Call: Insecurity, Democracy and Digital Media—Can It Really Counter Online Hate and Extremism? [289]

This is a machine-generated summary of:

Hoverd, William James; Salter, Leon; Veale, Kevin: The Christchurch Call: insecurity, democracy and digital media - can it really counter online hate and extremism? [289]

Published in: SN Social Sciences (2020)

Link to original: https://doi.org/10.1007/s43545-020-00008-2

Copyright of the summarized publication:

If you want to cite the papers, please refer to the original.

For technical reasons we could not place the page where the original quote is coming from.

Abstract-Summary

"We argue the existence of the Call helped change the conversation on the role played by online communication in hate, harassment and terrorism."

"The Christchurch Call is limited in its ability to counter online hate and extremism."

"It does not address the wicked problem of how to understand the social, communal and individual dynamics when the online expression of free speech turns to hate, and in turn, violence—together with the role platforms and states play in permitting and intervening in such digital interactions."

Introduction

"We set out to provide a cross-disciplinary analysis of the implications of the Call from three domains of knowledge and practice: (1) a national security perspective with a focus on online intelligence gathering and regulation, (2) a digital media focus with a tradition of investigation into online harm, and lastly (3) bringing that analysis into conversation with the critical study of liberal democracy."

"Through this cross-disciplinary analysis, we set out to answer the following three questions: (a) What are the national security goals and limitations of the Christchurch Call? (b) What are the practical implications of applying the Christchurch Call to online hate speech and extremism? (c) And, what are the implications of the Christchurch Call for the future of liberal democracy?"

"We argue that the Christchurch terror attack demonstrates how notions of national security are tied into broader transnational, collective and social/cultural challenges around the polarisation of politics and the creation of insecurity within online spaces."

The Christchurch terror attack and social media

"Prior to the attack, he allegedly announced his intentions on 8chan, and had earlier distributed his The Great Replacement manifesto' widely across a variety of internet platforms."

"An insight into the attacker's embedded social media extremist approach to violence can be found in his alleged post to 8Chan: Somewhere between 10 and 20 min before the first mosque was attacked, ******, logged on to the /pol/section of 8chan, an imageboard popular with the extreme right."

"I will carry out and [sic] attack against the invaders, and will even live stream the attack via facebook.'"

"In the attacker's alleged 8chan post, the author indicated their familiarity with social media, as well as the intent to use these platforms as a means to distribute their extremist views."

"The social media spread of the attacker's message through the manifesto and the live stream of the murders on Facebook was, for his purposes, extremely effective."

The Christchurch Call: a national security perspective

"We argue that the Christchurch Terror attack has clearly shown how notions of national security are tied into broader transnational, cultural and social challenges around the polarisation of politics and the creation of insecurity within online spaces."

"For nation states, the Christchurch Call has opportunistic implications for developing the surveillance and regulation of social media by intelligence apparatus through increased cooperation with social media companies."

"The challenge for social media companies and national security agencies is to discern which online shitposts are best understood as simply mischievous speech acts, which are best understood as harmful, and which represent potential physical threats in everyday life such as in the Christchurch attack."

"Aside from these enforcement issues, the broader question becomes how can democratic nation states' national security apparatus protect their populations against the potential incubation of such violent threats across social media, while still maintaining a free unregulated internet?"

The Christchurch Call: a digital media perspective

"For harassment campaigns more generally, the individuals who uncover personal information through deep research and circulate it to the community are in the primary tier, as are the people who directly harass targets either in-person, through physical proximity, or through media like phones (Campbell [121]; Dewey [122]; McKibben [123])."

"The Christchurch Call is focused on the most explicit expressions of terrorist intent expressed online, and thus it ignores the underlying disease of harassment communities in favour of its most obvious symptoms: exemplar first-tier individuals operating in physical spaces."

"The current aims and targets of the Christchurch Call either ignores or misunderstands the vital tiering dynamics underlying online hate and violence, and the role social media platforms play in profiting from and encouraging the communities from which primary-tier exemplar violence comes."

Where's democracy? The Christchurch Call and the turn to terror

"This section focuses on the absence of the subject of democracy from the Christchurch Call document, and the implications this may have for the exclusion of marginalised groups from public discourse due to digital hate."

"We argue that in the Christchurch Call document, and in a subsequent speech made by Facebook COO Sheryl Sandberg at the UN, there is a danger that terrorist becomes an empty signifier, enabling the positioning of the FAANG (Facebook, Apple, Amazon, Netflix, Google) as part of an us which excludes the very victims of the Mosque shootings and discourages further scrutiny of the relationship between their profit model and the proliferation of online hate."

"While the Christchurch Call document does make reference to "the operation of algorithms," there is a danger that debates have subsequently been allowed to shift onto FAANG's terms: the problem has been relabelled as a few bad eggs taking advantage of Facebook's drive to give everyone voice and enable a free marketplace of ideas [see also a recent speech by Mark Zuckerberg at Georgetown University (IANS [124])]."

Conclusion

"The Christchurch Call has helped change conversations about the current role played by the interactions of social dynamics and algorithms to produce digital hate, harassment and terrorism."

"The current aims and targets of the Christchurch Call either ignores or misunderstands the vital tiering dynamics within harassment communities and the algorithmic profiteering that encourages the communities from which primary-tier exemplar violence comes."

"The current agreement ignores the role that social media companies still play in fostering harassment communities from which first-tier exemplars (terrorists) might emerge."

"If nation states and social media companies genuinely wish to be seen as acting against the kinds of terrorist action and communication that the Christchurch Call stands against, then closer attention needs to focus on the capitalistic model that sustains harassment communities."

7.1.17 Implementing Open Government: A Qualitative Comparative Analysis of Digital Platforms in France, Italy and United Kingdom [290]

This is a machine-generated summary of:

De Blasio, Emiliana; Selva, Donatella: Implementing open government: a qualitative comparative analysis of digital platforms in France, Italy and United Kingdom [290]

Published in: Quality & Quantity (2018)

Link to original: https://doi.org/10.1007/s11135-018-0793-7

If you want to cite the papers, please refer to the original.

For technical reasons we could not place the page where the original quote is coming from.

Abstract-Summary

"We retrieved 979 open government platforms in France, Italy, and United Kingdom and analyzed them with an original set of indicators grouped in the three macro-areas of e-government, open data and transparency, participation and collaboration."

"Our study has methodological and policy implications: from the one hand, we provide the widest and deepest survey of open government platforms so far, contributing to the standardization of research methods on this field."

"From the other hand, we can observe that open government implementation is characterized by some typical traits, suggesting a trans-national convergence; this evidence is particularly prominent in the e-government and open data areas, whereas Participatory platforms are characterised by higher degrees of fragmentation both at the trans-national and at the intra-national levels."

"The analysis of open government platforms also reveals that, at the intra-national level, public administrations' approaches to digital government are not homogeneous, although at different degrees among the three countries involved in our study, suggesting the need of a better integration and coordination by central authorities."

Introduction

"The increasing importance of digital technology has dramatically changed the public administration landscape; in fact, the number of institutions publishing open government policies or developing digital platforms is expanding."

"The aim of this article is to compare how different national contexts are implementing open government policies with digital platforms."

"The work we outline here is part of a wider project on open government that enabled us to deepen our understanding of digital technologies, public policies, and the actors involved in the definition of national open government strategies across Europe."

"Because citizens and businesses are the end users of these platforms, the efficacy of a digital government project must be evaluated by considering whether and how a particular platform accomplishes its goals and affects the relationship between governments and their stakeholders."

"We describe the indicators of open government and how we applied them to digital platforms to evaluate their adherence to the specific ideal type."

The indicators of open government

"Open government can be conceived of as a policy agenda oriented towards transparency, participation and collaboration that relies on digital technologies to achieve its goals."

"Open government can be conceived of as a progressive stage model that moves from access to information (through transparency) to full collaborative governance, in which the power of decision-making is equally balanced among all the participants."

"Transparency, participation and collaboration have found fertile ground in digital platforms, where many actors can be involved in monitoring, participatory and collaborative practices; however, because open government platforms are a relatively new phenomenon, we could not rely on a consolidated body of literature and standardized methods of inquiry."

"The dataset has been subdivided into three sub-datasets according to specific macro-areas: e-government, open data and transparency, and participation and collaboration."

Research design

"The distribution of the categories of open government platforms across the countries led to several observations: the gap between the number of British platforms (483) and the number of platforms in the other two countries (298 in Italy and 198 in France) is noteworthy."

"We only considered ad hoc developed platforms: in the field of open data, for example, many institutions had released datasets—in fact, almost all local institutions had—but they had not set up a dedicated website."

"Those restrictions imply that the mere number of open data platforms in our sample does not describe how advanced a country is in this field, as the number of participatory platforms in our dataset does not reflect the whole picture of participatory practices in a specific national context; in effect, we are shifting the focus of attention to a more specific unit of analysis."

Coding procedure

"To analyse our sample of open government platforms, we adopted the method of content analysis, coding data retrieved from digital platforms as if they were plain text."

"These values describe two different conditions of an indicator (for example, 0 refers to closed code and 1 to open source) underlying an evaluation: in effect, we assumed that the presence of all positive conditions was the best possible result for the open government platform ideal type."

"Four contextual variables were further considered: the typology of the source code (proprietary or open), the financing source, and the presence of a legal device legitimizing the platform (i.e., a policy document, guidelines, or participatory pacts)."

"The first part of the codebook considers e-government indicators derived from international sources, which are organized into the variables of accessibility, usability, interactivity and collaboration, as shown in the previous section."

Qualitative comparative analysis of open government platforms

"The raw data matrix contains 979 cases, analysed considering a number of indicators that vary according to the macro-area: 24 for open data and transparency platforms; 27 for e-government platforms; and 32 for participatory and collaborative platforms."

"We ran the analysis on three datasets and considered as outcomes the indicators of E-SERVICE for e-government platforms, OPEN_DATA or TRANSPARENCY for open data and transparency platforms, and OBJECT for participatory and collaborative platforms."

"The last step consisted of an analysis of necessity for the combinations of indicators aimed at the selection of the combinations with values of necessity consistency higher than 0.9; those conditions are both sufficient and necessary for the outcome."

"For each macro-area and country, we use the values of coverage of sufficiency and consistency of necessity."

E-government platforms

"All the countries have a smaller number of platforms providing two-way interactivity (the consistency values are 0.81 in France, 0.8 in Italy and 0.65 in the UK); this evidence is even more meaningful if we consider that the other two indicators of interactivity, which are TRANSACTION and PERSONALIZATION, do not even appear in the list of necessary conditions."

"ANONYMITY is a distinctive feature of e-government platforms, though with some differences in consistency values across the three countries: 0.68 in France, 0.83 in Italy and 0.99 in the UK."

"French platforms correspond to five combinations of sufficient and necessary conditions: all of them have a common indicator in ONE_WAY interactivity; four combinations add another indicator, EASE_USE."

Open data and transparency platforms

"The indicator of OPEN_DATA obtains different values of consistency (0.85 in France and 0.75 in Italy) and is not even necessary in the UK (0.62)."

"With this indicator, DOWNLOAD, LICENSE and FILE_EXT are the most commonly used indicators to measure the quality of open data, and they obtain similar values of necessity consistency in France and Italy."

"With a necessity consistency value of 0.8, the presence of multiple sources indicates that France tends to concentrate public administration efforts in a few open data platforms."

"While this situation confirms the importance of OPEN_DATA and REUSE, as was observed in France and Italy, the absence of other sufficient and necessary indicators suggests that British platforms are much more diverse than in the other two countries, where a small set of indicators is consistently shared."

Participatory and collaborative platforms

"The fact that the indicator PERMANENCE is a necessary condition in most participatory and collaborative platforms suggests that public institutions try to engage in a long-term strategy of digital dialogue."

"The generic indicators necessary for participatory platforms in France were ORGANISED_INFO, which points to the presence of a news section or a blog where the institutions can publish news, events, comments, and so on, and POLICY_COMMUNICATION, which concerns the use of the platforms by public administrations to provide information about policy strategies, objectives and implementation."

"The absence of necessary combinations of conditions exceeding a value of 0.95 implies that the variability of platforms is difficult to synthesize in a few indicators, because the analysis cannot detect any completely adherent and robust explanations for the outcome participation in Italy."

Concluding remarks

"Discussed above, we outlined the countries' profiles in the three macro-areas of e-government, open data and transparency, and participation and collaboration by intersecting two axes, considering the degree of homogeneity or heterogeneity of the platforms at the intra-national level and at the trans-national level."

"France has a high degree of intra-national homogeneity in open data platforms but also shows a good level of integration in e-government and participatory platforms."

"Italy proves to be more homogeneous in the macro-area of participation, less homogeneous in open data, and decisively fragmented in e-government."

"The UK is more homogeneous in e-government and more fragmented in open data and participatory platforms."

"France and Italy converge in the realm of e-government, but they diverge with respect to open data and much more evidently with respect to participation; as already noted, the UK is the most divergent country of the three."

7.1.18 Usages of an E-Participation Platform by Legislators: Lessons from the French Parliament [291]

This is a machine-generated summary of:

Defacqz, Samuel; Dupuy, Claire: Usages of an E-participation platform by legislators: lessons from the French parliament [291]

Published in: French Politics (2021)

Link to original: https://doi.org/10.1057/s41253-021-00155-9

If you want to cite the papers, please refer to the original.

For technical reasons we could not place the page where the original quote is coming from.

Abstract-Summary

"It zeroes in on legislators who while key actors of representative democracy chose to be involved in an e-participation initiative."

"This article generates theoretical propositions on how they make use of e-participation platforms in their work as parliamentarians."

"Based on a qualitative analysis of interviews about the main e-participation platform in France, Parlement & Citoyens, the article shows that parliamentarians' usages of such participatory tools tend either toward a policy-oriented logic or a vote-seeking purpose."

Introduction

"This article asks how parliamentarians who decided to be involved in e-participation initiatives make use of participatory platforms in their work as parliamentarians."

"It also illuminates whether parliamentarians' usages of an e-participation platform supplement their representative role with a participatory orientation or mainly support their traditional role in representative democracies."

"Usages are defined as social practices that seize e-participation platforms as a set of opportunities in the work of parliamentarians and in the process of policy making, whether these opportunities are institutional or normative (based on Jacquot and Woll [125], p. 9)."

"We argue that parliamentarians' usages of e-participation platforms are structured along two main dimensions: The first dimension depicts the nature of their political work, that is, how they see and strategize about their role in the parliament; the second dimension refers to the normative orientation of their role toward a classic representative logic or toward its participatory supplement."

[Section 2]

"To the French context that magnifies the issues of elected politicians' usages of participatory tools, P&C as a case is also particularly relevant to draw theoretical propositions from its analysis."

"In that regard, P&C is a unique empirical case to investigate all types of parliamentarians' usages of e-participation platforms, regardless of the policy issue at stake, the specific context of e-participation, and whether these usages are related to conjectural or structural motives."

"As the purpose of this exploratory research is to generate theoretical propositions on legislators usages of e-participation platforms, we consider that the analysis of interviews about six parliamentarians (out of 17) with different profiles offers sufficient empirical traction to get access to a significant overview of the issue at stake."

The usages of an e-participation platform by French legislators
"We have the feeling that, it is unfair but... it's fascinating to see how even organized actors, politicized people, have a total distrust in the activity of elected officials, especially when actually we come to present them with a perfectly transparent, perfectly participative approach by saying: 'we work on this subject, we need you'. [...] It was necessary to build trust and these tools are essential to build this trust. (itw1) Overall, this thematic axis highlights the usage of P&C by legislators as a legitimizing instrument that garner and secure support to a policy proposal through the very involvement of stakeholders and citizens in the policy design."

Discussion: an analytical framework on legislators' usages of e-participation platforms
"We build a conceptual categorization of legislators' usages of P&C by relying on the usages identified by the thematic analysis and projecting them on a two-axis model that we derived from existing research on participation and parliamentary roles: first, is each usage about vote-seeking or is it policy-oriented?"

"The usages of P&C by French parliamentarians under scrutiny are leaning toward a vote-seeking logic rather than a policy-oriented logic as most of them stress some dimensions of the former and less of the latter."

"Other parliamentarians' usages of P&C are policy-oriented."

"As regards to its usage by the French legislators under scrutiny, Parlement & Citoyens falls short of fully fulfilling its promise to change policymaking through the inclusion of citizens' and stakeholders' inputs, as the platform usages tend to be oriented toward vote-seeking rather than toward policy objectives."

Conclusion
"Given the modest number of legislators under scrutiny, the objective of this article was not to draw general conclusions on why parliamentarians use participatory platforms but instead, build an analytical framework aiming at understanding further how central actors of representative democracy make use of these e-participation initiatives."

"Other parliamentarians' usage of the platform tend toward a vote-seeking logic as legislators may use the platform to strengthen their political reputation, make the parliamentary work more transparent to voters or position themselves as champions of participative democracy toward the electorate."

"French parliamentarians' usages of e-participation platforms can also be categorized based on normative underpinnings, tending toward either a representative democracy logic (strengthen political reputation, make the parliament more transparent, garner support to a policy or collect experiences and expertise) or a participatory democracy logic (engage in a collaborative policy-design or promote participative democracy as such)."

7.1.19 Digital Governance: An Assessment of Performance and Best Practices [292]

This is a machine-generated summary of:

Manoharan, Aroon P.; Melitski, James; Holzer, Marc: Digital Governance: An Assessment of Performance and Best Practices [292]

Published in: Public Organization Review (2022)

Link to original: https://doi.org/10.1007/s11115-021-00584-8

If you want to cite the papers, please refer to the original.

For technical reasons we could not place the page where the original quote is coming from.

Abstract-Summary

"E-government has evolved from providing static content and services to integrating user generated content and social media technologies."

"This article discusses the results of a global and comparative survey of e-government performance, based on an assessment of municipal government websites around the world."

"Along with a longitudinal assessment, the study identifies best practices, highlights key findings, and provides guidance for future research."

Introduction

"The diffusion of information and communication technologies (ICTs) is central to the study of e-government (Calista & Melitski, [126])."

"Open government and the improved quality of citizen participation are having a positive influence on public trust in government (Moon, [127])."

"From a public values framework, e-government creates value in several distinct areas."

"A second method for creating public value is derived by creating systems that improve democratic governance by engaging citizens, increasing transparency, and enhancing trust in government."

"The primary emphasis of smart cities and governments is the promotion of public value through the use of ICTs and internet-enabled devices like cameras and sensors to improve the quality of life (Manoharan & Mossey, [128])."

"A 2015–16 study of global municipalities (Holzer & Manoharan, [129]) determined that cities are at different stages of e-government adoption, with regard to their website privacy and security, usability, content, services, and citizen and social engagement."

Methodology

"Consistent with previous surveys conducted by the E-governance Institute, the 2018–19 survey identified the top 100 cities within the most wired nations based on data from the International Telecommunication Union (ITU), which is affiliated with the United Nations (UN)."

"Previous research consistently shows a positive relationship between population and e-government capacity (Manoharan, [130]; Moon, [131]; Zheng & Manoharan, [132])."

"Given the relationship between a city's e-government capacity and its population, the largest city by population was selected from these 100 nations."

"The survey used a comprehensive e-governance index of 86 measures, classified into five categories: 1) Privacy and Security; 2) Usability; 3) Content; 4) Services; and, 5) Citizen and Social Engagement."

Results and Discussion

"Seoul ranked first in the survey with an overall score of 84.07, increasing its score from the previous survey in 2015–16."

"Madrid was second with a score of 80.51, increasing its score from 69.24 in 2015–16."

"The highest ranked cities in each continent were Johannesburg (Africa), Seoul (Asia), Madrid (Europe), Toronto (North America), Auckland (Oceania), and Montevideo (South America)."

"Seoul was the top scoring city in Content and Services."

Results by E-Governance Categories

"For Privacy and Security, the top-ranked cities in 2018–19 were Madrid, Seoul, Yerevan, Bratislava, Singapore, Toronto, New York, Buenos Aires."

"The average score for all cities in this category was 6.16, an increase from a score of 5.55 in 2015–16."

"Lisbon came in fifth with a score of 15.93, and all cities scored an average of 12.40 in this category, a slight increase from 12.38 in 2015–16."

"The cities' average scores in the five e-government categories has also increased compared to 2015–16."

"The cities overall improved their average scores in all five categories, with the highest score in Usability and the lowest in Citizen and Social Engagement."

"Similar to the previous surveys, Seoul was the top performing city in Service Delivery, Content, and ranked second in Privacy and Security, third in Citizen and Social Engagement, and ninth in Usability."

Conclusion

"This research extends our knowledge of the capacity and performance of local governments to provide information, transact services and engage in participatory e-governance."

"The longitudinal analysis provides valuable comparative information demonstrating that e-government performance lagged in the years following the 2008 great recession before improving again in 2013–14."

"Future research needs to consider the impact of COVID-19 as a contextual factor in assessing the capacity of city e-government performance."

"This research indicates that in the 2018–19 survey, Privacy and Security and Citizen Social Engagement experienced the greatest gains."

"Melitski and Calista ([133]) further state that e-government capacity improves at a rate consistent with others technological innovations."

"E-government capacity is growing, but not at rates previously experienced, particularly among developed nations."

"Future research needs to continue evaluating e-government capacity and performance."

7.1.20 Digital Globalization and Democracy: The Territorial Walls Come Tumbling Down? [293]

This is a machine-generated summary of:

Youngs, Gillian: Digital Globalization and Democracy: The territorial walls come tumbling down? [293]

Published in: Development (2011)

Link to original: https://doi.org/10.1057/dev.2011.31

If you want to cite the papers, please refer to the original.

For technical reasons we could not place the page where the original quote is coming from.

Digital globalization

"Thinking in terms of digital globalization helps us to understand such technologically supported people power in grounded ways that avoid technological determinism (laying too much emphasis on the causal potential of technology), and emphasizes instead the intrinsic and dynamic linkages between technological processes and other social, political and economic ones, including crucially those related to identity, community and citizenship."

"Digital globalization shows us not only how global political, economic and cultural patterns have been changing the nature of international realities but also how information and communication technologies (ICTs) have been impacting on the ways we think, live, communicate and connect with other individuals and societies."

"Globalization and digital studies have both focused on time and space and time/space relations, and their growing significance as vectors for understanding what is happening in the world, whether in our intimate or civic lives and identities."

ICTS

"Globalization studies have demonstrated the thinning, if you like, of national territorial boundaries as economies have become increasingly interdependent across the world through trade, service, financial and investment networks, an interdependence starkly demonstrated by the current economic crises linked substantially to banking failures and high levels of government debt."

"In analytical terms, digital globalization takes the importance of time and space to a whole new level, especially with the mobile multimedia ICT revolution."

"As with globalization this concerns the effects of national territorial boundaries, which historically have acted container like for the exercise of sovereignty and authority within states and the political identities of citizens."

"States in their top-down exercise of vertical power (whether democratic and liberal or autocratic and oppressive in diverse forms) have historically had their security and identity heavily invested in the integrity of their boundaries defining their sovereign rights both over their citizens, as well as to act legitimately internationally."

New communications

"Many political and new media analysts including me have stressed the extent to which ICTs challenge the communicative character of political community as defined in sovereign state terms."

"This has concerned not only political action within specific states, but also the access of the wider world to news and direct testimony from within these countries about what is actually happening."

"Not only does this appear to be a factor in the wave like pro-democracy movement across a number of countries one after the other, but also in the grassroots multimedia witness testimony that has driven much of the mainstream media coverage and political responses internationally."

Neo-liberal challenges

"We might begin to think about the thinning of national boundaries in globalization becoming much more profound in the process of digital globalization, if these recent events are indicative of how people are prepared and anxious to use ICTs to mobilize and come together to press for peaceful democratic change."

"This point took me back to some conclusions I reached a few years ago: The time/space reconfigurations of the online world, its power to connect across the globe and to operate as an information-rich synchronous or asynchronous means of communication is enabling the building of temporary or longer-standing virtual communities ... Full access to this transcendent sphere of communication and action is now being incorporated into the framework of human rights."

"It is another dimension of what it means to be free as a society or individual, and to be able to participate fully in liberal and neoliberal terms. ([134]: 147)"

Freedom and information flows
"A related point that I explored in more depth was that the logic of free flow exchange embedded in ICTs was ideological in democratic terms as much as technological."

"Digital globalization does mean that the territorial walls of states are beginning to tumble down under the pressure of a growing global virtual community based on human rights and democratic principles."

"Key however is the grassroots horizontal nature of this pressure and there is much to celebrate in the peaceful role of ICTs as part of making that a real force for change."

7.2 Digital Justice

Machine generated keywords: law, justice, legal, privacy, protection, datum, big datum, data, right, analytic, balance, ethical, health, consumer, author

7.2.1 Societal and Ethical Issues of Digitization [294]

This is a machine-generated summary of:

Royakkers, Lambèr; Timmer, Jelte; Kool, Linda; van Est, Rinie: Societal and ethical issues of digitization [294]

Published in: Ethics and Information Technology (2018)

Link to original: https://doi.org/10.1007/s10676-018-9452-x

If you want to cite the papers, please refer to the original.

For technical reasons we could not place the page where the original quote is coming from.

Abstract-Summary
"We discuss the social and ethical issues that arise as a result of digitization based on six dominant technologies: Internet of Things, robotics, biometrics, persuasive technology, virtual & augmented reality, and digital platforms."

"We highlight the many developments in the digitizing society that appear to be at odds with six recurring themes revealing from our analysis of the scientific literature on the dominant technologies: privacy, autonomy, security, human dignity, justice, and balance of power."

"For other ethical issues concerning digitization such as discrimination, autonomy, human dignity and unequal balance of power, the supervision is not as well organized."

Introduction

"Digitization penetrates every aspect of our lives: the technology nestles itself in us (for example, through brain implants), between us (through social media like Facebook), knows more and more about us (via big data and techniques such as emotion recognition), and is continually learning to behave more like us (robots and software exhibit intelligent behaviour and can mimic emotions)."

"We describe what social and ethical issues arise when society becomes digitized on the basis of six dominant technologies: Internet-of-Things, robotics, biometrics, persuasive technology, virtual & augmented reality, and digital platforms."

"Many of the technologies we describe in this article are part of IoT: like the augmented-reality glasses which use the Internet to give users real-time additional information about their environment, or a biometric camera which can be linked to an online database to recognize faces."

Societal and ethical issues

"The extent to which a person has control over whether they submit the above data seems to be limited, as the collection of this information can be done remotely and covertly, for example, by inserting facial recognition technology in mannequins, without the knowledge of the person being observed (De Hert and Sprokkereef [135])."

"Examples include knowledge systems that make medical diagnoses based on a large amount of information, military robots that take life or death decisions using information from various sources, and the driver support systems that decide what speed we should drive on a particular stretch of road."

"One objection to persuasive technology is that users' actions have nothing more to do with ethics: they make no moral decisions but simply display controlled behaviour (Spahn [136])."

Conclusion

"In that digital world, the data which is processed and analysed forms the basis for people as well as automated systems to make decisions that subsequently have an impact on the physical world."

"Developments in the field of big data, smart algorithms based on artificial intelligence are indispensable elements of the technologies discussed above."

"We have already discussed various issues regarding privacy, and big data presents a specific challenge in this respect due to the re-use and potential combinations of different data sources."

"At European level, there has been an attempt to deal with big data issues by modifying the legislation."

"Unless government, industry, civil society and members of the public act now, there is a risk that while we are trying to get to grips with the new digital world, the frameworks to protect public values are meanwhile losing their relevance."

7.2.2 Will Big Data and Personalized Medicine Do the Gender Dimension Justice? [295]

This is a machine-generated summary of:

Carnevale, Antonio; Tangari, Emanuela A.; Iannone, Andrea; Sartini, Elena: Will Big Data and personalized medicine do the gender dimension justice? [295]

Published in: AI & SOCIETY (2021)

Link to original: https://doi.org/10.1007/s00146-021-01234-9

If you want to cite the papers, please refer to the original.

For technical reasons we could not place the page where the original quote is coming from.

Abstract-Summary

"Using Big Data poses several problems, especially in terms of representing the uniqueness of each patient and expressing the differences between individuals, primarily gender and sex differences."

"The first two sections of the paper provide a definition of "Big Data" and illustrate the uses of Big Data in medicine."

"The paper explores the struggle to represent exhaustively the uniqueness of the patient through Big Data is highlighted prior to a deeper investigation of the digital representation of gender in personalized medicine."

"The final part of the paper put forward a series of recommendations for better approaching the complexity of gender in medical and clinical research involving Big Data for the creation or enhancement of personalized medicine services."

Big Data: a definition still in progress

"Big Data is an expression that is used to refer not only to a gargantuan quantity but also to a mining process through technological means that have developed greatly over the last decade."

"Computer algorithms' ability to sort through and automatically identify in data recurring patterns, is the technological development that has made possible refining "the new oil." "

"Experts are realizing that referencing the volume of data and the velocity with which it must be analysed addresses aspects of a relationship, and thus of knowledge, that point to the value-based structure of society."

"In Askoff's hierarchy, as with the passage between data and information, the action of transitioning from knowledge to the next and final level, wisdom, implies the mediation of human sensibilities."

"Human beings lade data and information with ideological and socioeconomic values, creating knowledge and wisdom whose relevance depends on the expectations and the contextual needs of the humans involved."

Big Data in medicine: sources and potentiality

"One of the most important uses of Big Data in medicine regards the creation or enhancement of personalized and precision medical practices."

"Analyses of Big Data in pathology serve to support this practice by providing more apt information about each patient's unique characteristics."

"Beyond those provided by patients, data from a large variety of sources are utilized in the practice of personalized medicine."

"As Willems and others argue, these data contain the clinical data of patients, diseases, treatments, and outcomes, and may include "demographic details such as gender and age, presenting symptoms, family history, comorbidity, radiological data (such as CT, MRI, PET, US) as well as solid and liquid tissue-based analysis (such as histopathological diagnosis/features, immunohistochemistry, DNA/RNA sequencing experiments, blood analyses, and whole genome BAM files)" ([137], p. 9)."

The risks of a data-driven medicine in research and patient's representation

"The use of Big Data in medicine opens the doors to a series of complex problems, above all in moving from the descriptive analytics software, which allows information represented by the data to be displayed with graphs, tables, and other visual supports that highlight the underlying medical phenomena, to predictive analytics software, which through more or less transparently developed IT algorithms identify correlations between the provided medical data with an "if … then" logic (Cato and others [138]; Nicoletta and others [139])."

"After having briefly considered the potential effects of the use and correlation of biomedical Big Data, it is now interesting to acknowledge also that the entities that might benefit or profit from collecting "raw" biomedical Big Data or from their analyses, might be either private or public."

Gender and medicine

"After analysing the problems in the representation of the patient, in this part, we address the first crucial point of the article: what are the difficulties of representing the gender dimension in medicine?"

"The gender dimension, while opening up contingent and specific questions, poses a preliminary discriminant that is the consequence of the diagnostic paradigms that have arisen over the course of the history of medicine."

"Gender poses basic epistemic questions, such as those of "difference" and "intersectionality" (Crenshaw [140]; Gilligan [141])."

"There is an underlying gender discrimination that operates before medical research even produces its results."

"In medical practice and research, gender is understood as sexual dimorphism."

"Beyond pre-comprehensive discrimination, there is also contingent discrimination in medical research: the stringent requirements of clinical trials often mean that socioeconomic factors limit the representation of women therein."

Gender, Big Data, and personalized medicine

"The EU-funded project PERSIST (Patients-centered SurvivorShIp care plan after Cancer treatments based on Big Data and Artificial Intelligence technologies) is aimed at creating a system capable of improving treatments for cancer survivors and improving their health and well-being, through various phases and objectives that make use of Big Data models to facilitate decisions on the best treatments."

"Despite the explicit break with the past that Big Data epistemologies typically claim, many of the systemic biases around gender that characterised clinical trial data continue to be found in Big-Data-centred practices and technologies as well (Cirillo and others [142])."

"The inability to see gender equality as a possible driver of computer science research inevitably produces a dystopian image of data-driven technology as keen predictive analytics for approaching and solving all the risks and diseases of contemporary society."

Will Big Data and personalized medicine really do the gender dimension justice?

"Is the time to try to answer the question: will Big Data and personalized medicine really do the gender dimension justice?"

"Rather, the latter should be read and crossed with other analyses—to give examples: access to technological means that favour personalized medicine; social mobility that allows participation in clinical trials; equal representation in science."

"The road towards gender-sensitive sustainability of the use of Big Data in precision health care is still long."

"A policy-based recommendation could be to approach Big Data in medicine while being aware of the intrinsic risks of enacting technoscientific holism."

"On the basis of these preliminary considerations, at the end of the article, we provide some recommendations on how to approach a gender-sensitive use of Big Data in personalized and precision medicine."

Some gender-sensitive recommendations for Big Data in personalized medicine

"XAI is a relatively young field of research and its applications so far has not been particularly involved with sex and gender differences."

"An understanding of the contextuality of data (Ferryman and Pitcan [143]) is not only necessary to interpret them in meaningful ways, but also to increase sensibility for potential gender and other biases."

"In clinical trials, the underrepresentation of women was visible because the gender or sex of participants was recorded; missing women were, literally, missing bodies (Casper and Moore [144])."'

"The 'data bodies' that digital medicine operates with often have no explicit gender attribution anymore; when digital epidemiology analyses entries in online search engines or the movement of people throughout a city via their phone's geolocation, these data are often used without any information on gender."

"When medicine insists too much on personalisation, the individual uniqueness supersedes and resists group classifications, including gender."

Conclusions

"Gender biases challenge the quality of results yielded from Big Data analytics in precise and personalized medicine."

"The gender issues entail biomedical considerations, in an indirect way, since they help remind us that the definition of Big Data is still in progress and the specialist knowledge for analysing them is still in development."

"Gender biases persist in the context of biomedical Big Data in a direct way, creating new configurations of old problems."

"We have analysed the problems that characterise the generic relationship between medicine and gender; on the other, we have outlined the conceptual and practical obstacles that still make applying artificial intelligence algorithms to large streams of biomedical data contradictory and not very gender friendly."

7.2.3 Smart Criminal Justice: Exploring the Use of Algorithms in the Swiss Criminal Justice System [296]

This is a machine-generated summary of:

Simmler, Monika; Brunner, Simone; Canova, Giulia; Schedler, Kuno: Smart criminal justice: exploring the use of algorithms in the Swiss criminal justice system [296]

Published in: Artificial Intelligence and Law (2022)

Link to original: https://doi.org/10.1007/s10506-022-09310-1

If you want to cite the papers, please refer to the original.

For technical reasons we could not place the page where the original quote is coming from.

Abstract-Summary

"In the digital age, the use of advanced technology is becoming a new paradigm in police work, criminal justice, and the penal system."

"Algorithm-based applications are often deployed in this context, laying the groundwork for a 'smart criminal justice'."

"In this qualitative study based on 32 interviews with criminal justice and police officials, we explore the reasons why and extent to which such a smart criminal justice system has already been established in Switzerland, and the benefits perceived by users."

"Drawing upon this research, we address the spread, application, technical background, institutional implementation, and psychological aspects of the use of algorithms in the criminal justice system."

"We find that the Swiss criminal justice system is already significantly shaped by algorithms, a change motivated by political expectations and demands for efficiency."

"By discussing the main challenges to and issues with algorithm use in this field, this work lays the foundation for further research and debate regarding how to guarantee that 'smart' criminal justice is actually carried out smartly."

Introduction

"A strong tendency towards the establishment of a 'smart criminal justice' (i.e., the use of intelligent technology in the criminal justice system) is currently observable (Simmler and others [145])."

"It is evident that the criminal justice system is facing urgent challenges regarding its use of emerging technologies (cf. Završnik [146])."

"Research has yet to systematically deal with the implementation of advanced technology in the criminal justice system, in Switzerland and elsewhere."

"Against this background and in anticipation of the further unfolding of the digital age, this study explores the reasons why and extent to which such a smart criminal justice system has been established in Switzerland, as well as the benefits currently perceived by users."

"Based on these data, we argue that there is a need to thoroughly evaluate the effects of algorithm use in criminal justice, considering the legal foundations, as well as the accompanying psychological and institutional effects."

Smart criminal justice

"Smart criminal justice, a sub-category of smart government, can be defined as the use of technology in the criminal justice system based on algorithmic decision-making and the collection, analysis, and processing of big data (Simmler and others [145])."

"In the area of postcrime, when a crime has already been committed, we distinguish between the use of (3) advanced technology in police investigations and (4) the subsequent criminal proceedings."

"To forensic-psychiatric assessments during ongoing criminal proceedings or in the penal system, these predictive policing measures are intended to take effect before a crime is committed."

"Algorithmic risk assessment instruments can not only be used in ongoing criminal proceedings for forensic-psychiatric assessment, but also in the penal system (e.g., to determine the need for further examination of a convicted offender or decide on execution mode, probation, and early release)."

Methodology

"Methodologically, the study used qualitative guideline interviews to build upon the knowledge and experience of experts directly confronted with the implementation and use of algorithms."

"The study sample was composed of 25 interviews with 32 persons from 14 cantons."

"In order to capture (at least in outline) the use of algorithms in the 12 cantons where no representatives were personally interviewed, information was obtained by e-mail from 23 individuals."

"This overview of the algorithmic tools currently in use relied on this extended sample; the in-depth examination of the phenomenon, however, was based solely on the personal interviews, which were transcribed and qualitatively evaluated for this purpose."

"Although the interviews were carefully conducted to obtain additional information and cover as many algorithmic instruments as possible, it cannot be ruled out that the results are incomplete."

Results

"Three Swiss cantons use an algorithm-based predictive policing tool that supports the prevention of burglaries and relies on self-learning techniques."

"As the present study revealed, the majority of Swiss-German police corps currently use algorithm-based assessment instruments to support this process."

"Respondents unanimously felt that the tool itself was not decisive, since such algorithms do not propose what measures to take to prevent crimes."

"Some respondents stated that algorithmic tools could in no way replace human features, and abstract risk calculations could not fully guide the assessment of individual cases."

"The scepticism expressed was mostly related to the effectiveness or concrete benefit of the tools, and not based on socio-political aspects or a fundamental rejection of the use of algorithms in the criminal justice system."

"Respondents supporting the use of algorithms stated that the tools were a good way of incorporating the findings of current research into practice."

Discussion

"The use of algorithmic tools is often an expression of a trend generally observable in Western legal orders: a shift from the repressive function of criminal justice towards an increasingly preventive policing approach (Carvalho [147])."

"This study highlighted the psychological components accompanying the use of algorithms in the criminal justice system."

"Users of algorithmic tools in criminal justice should be aware of the interdependencies that can arise between individual assessments and algorithm-based decision-making, and pursue a better understanding of these psychological effects (and ways to handle them)."

"It is probable that the use of algorithms will further intensify in various areas and Switzerland will move towards a smart(er) criminal justice."

"Policymakers must actively promote such a debate regarding the potential and limitations of algorithm use in the criminal justice system."

Conclusion

"Drawing on a qualitative study, this research provides an overview of the algorithms used in the Swiss criminal justice system."

"Algorithms are gaining importance in criminal justice because they are part of the decision-making processes, even though they mainly serve as support for the humans in charge."

"As soon as algorithms are deployed and in any form become part of the criminal justice system, questions related to their implementation, effectiveness, legal foundation, and accountability arise."

"We argue that practice and research must face the challenging questions regarding what standards algorithm-based technology must meet and what an adequate and effective use of algorithms looks like."

"This study has shown that the emerging 'smart criminal justice' is promising, but factual, legal, and political limits will determine the boundaries of this development."

7.2.4 Blind-Sided by Privacy? Digital Contact Tracing, the Apple/Google API and Big Tech's Newfound Role as Global Health Policy Makers [297]

This is a machine-generated summary of:

Sharon, Tamar: Blind-sided by privacy? Digital contact tracing, the Apple/Google API and big tech's newfound role as global health policy makers [297]

Published in: Ethics and Information Technology (2020)

Link to original: https://doi.org/10.1007/s10676-020-09547-x

If you want to cite the papers, please refer to the original.

For technical reasons we could not place the page where the original quote is coming from.

Abstract-Summary

"Public debate has focused on the risks this technology poses to privacy, with advocates and experts sounding alarm bells about surveillance and mission creep reminiscent of the post 9/11 era."

"When Apple and Google launched their contact tracing API in April 2020, some of the world's leading privacy experts applauded this initiative for its privacy-preserving technical specifications."

"This article proposes to view the Apple/Google API in terms of a broader phenomenon whereby tech corporations are encroaching into ever new spheres of social life."

"These sphere transgressions raise numerous risks that are not captured by the focus on privacy harms."

"While privacy is certainly an important value, its centrality in the debate on digital contact tracing may blind us to these broader societal harms and unwittingly pave the way for ever more sphere transgressions."

Introduction: contact tracing and the automation of a public health practice
"Digital contact tracing seeks to address these limitations, by providing speed, scale and accuracy."

"While these limitations have dampened some of the initial enthusiasm around digital contact tracing as an easy solution to curbing the spread of the virus, the technology is still seen as an important complement in national post-lockdown strategies."

"More practical questions, one of the major points of contention in the implementation of digital contact tracing has been its potential to cause harm through privacy breaches (Ienca and Vayena [148]; McGee and others [149])."

"While privacy is an important intrinsic and instrumental value, the centrality that it has received in the debate on digital contact tracing, and arguably in other debates on digitalization, may blind us to these broader societal harms and unwittingly pave the way for ever more sphere transgressions."

Lessons learned: privacy takes central stage
"Beyond the pressing question of whether digital contact tracing will actually prove to be effective, public debate in many countries has focused on the risks contact tracing apps pose to privacy."

"Within the public discussions around digital contact tracing, these concerns have tended to be framed in terms of a trade-off between individual privacy rights and public health."

"Yet, the Apple/Google API has been presented as a response to some of the gravest privacy concerns voiced in the digital contact tracing debate."

"This alignment with privacy specifications led to widespread endorsement and applause for the Apple/Google initiative by some of the leading privacy experts in the contact tracing debate."

"Even if the Apple/Google contact tracing technology does get the privacy issue just right, what other trade-offs are involved in letting these companies contribute to the development and deployment of what might be the largest scale crisis management measure for the pandemic so far?"

The Googlization of pandemic response
"To answer this, it is helpful first to note that the development of digital contact tracing is only one of many ways that the large tech corporations, not just Apple and Google, but also Facebook, Amazon, Microsoft, other subsidiaries of Alphabet, as well as their Asian counterparts including Alibaba, Baidu Tencent and Huwaei, have contributed to addressing the threats posed by COVID-19."

"To providing key information and resources on COVID-19 on their respective platforms, many of these companies began developing COVID specific data collection, data sharing and data analysis tools, or have earmarked significant amounts of funds for COVID related research, early on during the pandemic."

"What this indicates is that the triage tool will soon be used for much more than just triage and testing, most likely for channeling users towards enrollment in trials, such as the serology study that Verily announced as the first of its Baseline COVID-19 Research Project initiatives (Verily [150])."

Sphere transgression: from digital goods to health and medicine
"The recent push in the health and medical sector to move towards more data-driven personalized medicine (Frohlich and others [151]) has turned digital expertise into a coveted good in the sphere of health, and has made tech companies attractive partners to collaborate with."

"As a number of public health experts have pointed out, there are good reasons to opt for a centralized approach that have to do with a thorough understanding of contact tracing as a practice involving the norms and skills discussed above, more than with privacy (Kelion [152], Leprince-Ringuet [153], Parker and others [154])."

"Sphere transgression, here the conversion of technical expertise into an advantage in the sphere of health and medicine, thus risks crowding out practices, norms and values that have been central to this sphere."

Sphere transgression: from digital goods to public services
"Decades of outsourcing and budget cuts, justified by a dominant political discourse that governments are inefficient, have significantly hampered governments' abilities to adequately provide health, housing, education, transportation, utilities and other essential services, including, in many countries, pandemic crisis response."

"Gratitude is to be expected in response to beneficence and charity, and many of the contributions by big tech to the pandemic response have been made as donations or through philanthropic foundations."

"The decimation of the public sector, and the role this plays in facilitating big tech's involvement in the political sphere's pandemic response, is characteristic of the US and the UK more than it is of continental Europe."

"Several reports on discussions that took place between government officials and Apple and Google portray another side of this story, pointing to a power play between sovereign states and the corporations, in which sovereign states had little say."

Conclusion
"This article has attempted to articulate what is at stake when two of the world's most powerful corporations move into the field of pandemic response management with a development like the Apple/Google API for digital contact tracing."

"Such sphere transgressions are not novel to the Apple/Google API case, nor are they limited to these two companies."

"Sphere transgression can happen in perfectly privacy-friendly ways, such that, as argued, the focus on how privacy-preserving the Apple/Google API is misses the point."

"Spheres clearly overlap at times, certainly during a pandemic: to be sure, digital contact tracing can be an important complement within a broad pandemic response strategy, and to be done properly it requires a combination of both epidemiological and technical expertise."

7.2.5 Hyperrealistic Jurisprudence: The Digital Age and the (Un)Certainty of Judge Analytics [298]

This is a machine-generated summary of:

Ferreira, Daniel Brantes; Gromova, Elizaveta A.: Hyperrealistic Jurisprudence: The Digital Age and the (Un)Certainty of Judge Analytics [298]

Published in: International Journal for the Semiotics of Law - Revue internationale de Sémiotique juridique (2023)

Link to original: https://doi.org/10.1007/s11196-023-10015-0

If you want to cite the papers, please refer to the original.

For technical reasons we could not place the page where the original quote is coming from.

Abstract-Summary

"This article is the first attempt to justify the "next" milestone in the development of legal realism: hyperrealism."

"The implications of digitalization have become the new fuel for the legal realist's jurisprudence prediction theory, that is, empirical research to predict the judge's or the court's decision."

"Based on a systemic, comparative, and interdisciplinary analysis that intermingles legal theory, data analytics and digital technologies, the article substantiates the concept of hyperrealism itself."

"It evaluates the advantages and disadvantages of its primary tool—judicial analytics."

Introduction

"Some authors claim that New Legal Realism is empirical research not limited to judicial decision-making (adding, for example, arbitration) and focusing on the judge's political ideology [155]."

"Others claim that New Legal realism tends to concentrate the empirical research in lower federal courts [156]."

"Some claim that the so-called new legal realism is just the contemporary use of empirical research which implicates the use of the same old paradigm of the 1930s realists who attempted to assess through social sciences what courts do [157, 158]."

"There is hyperrealism, which is the optimization of the old legal realism foundations by the new tools that give lawyers a complete HD picture of courts."

"We also applied interdisciplinary analysis that intermingles legal theory, arts (realism and hyperrealism), literature, data analytics and digital technologies to show the connection between semiotics, modern law, and technologies and to justify the need for specific regulation of judge analytics."

Realism and Hyperrealism in Arts

"The term realism came from the arts, and so is this paper's hypothesis which proposes a hyperrealist view of the jurisprudence due to the extensive use of technology as an attempt to predict its outputs."

"Realist artists attempt to be more accurate in their models' depictions—the term in art dates from 1826."

"Hyperrealism is the evolution of realism in arts and was born in the United States as a reaction to Pop Art and to the Minimalist art and Abstract Expressionism Movement in that country."

"As realism had the pre-realism artists, hyperrealism has realism as its predecessor."

"Hyperrealism intertwines technology and arts, and as artistic realism proved to apply to legal theory in the first half of the twentieth Century both in the United States and Scandinavia, hyperrealism is the twenty-first Century tendency by the ever-growing use of technology to analyze jurisprudence."

Legal Realism Doctrine

"Karl N. Llewellyn, the only American realist author who commented on Scandinavian legal realism, claimed that the Scandinavians were conducting a philosophical inquiry into the nature of law At the same time, the Americans attempted to develop legal technology [159, p. 125]."

"Roscoe Pound replies to Llewellyn in his Harvard Law Review 1931 paper The call for a realist jurisprudence [160] criticizing two main issues of the realism movement: the realist's persistence in applying empirical research and the use of psychology to analyze the judge's behavior."

"These would be the most notable contributions of the realistic movement to legal education and legal research: Interdisciplinarity (especially the integration of Law with sociology) proliferation of elective courses adoption of legal practice clinics application of empirical research to Law Improvement in teaching material (case books) progress of teaching techniques At the time, realists lacked the digital tools to reach the desired results by using empirical research applied to Law."

The Concept of Hyperrealism in the Age of Digital (Un)Certainty

"The advance of technologies [161] made it possible to resume discussions on legal realism."

"Speaking the language of artists, we can say that technologies can be the tools that help to make the picture of the court hyperreal."

"There is a technology based on Natural Language Processing (NLP) aimed at predicting the outcomes of the cases of the European Court of Human Rights (ECHtR) [162]."

"Data and technologies that can process it became the oil of the future society [163] and drivers for economic and social development."

"What makes us certain is that technologies are already integrated into our lives."

"We realize we are moving towards a digital future where technologies will play a much more crucial role [164, 165]."

"Technologies can significantly improve our well-being and living conditions."

"The potential of mentioned technologies is significant, and the possibilities of their use are almost limitless."

Judge Analytics as Hyperrealism's Primary Tool

"Existing judge analytics tools such as Context by LexisNexis, Supralegem.fr, Westlaw Edge, Litigation Analytics, Jurimetry and Predictice [166] can provide the following data (as its developers claim it): (a) the language, precedents and other judges that a particular judge finds the most compelling; (b) the "specific logic" that a judge tends to use when granting or denying a particular type of motion; (c) how likely a user's case is to prevail before a particular judge; (d) how long a particular judge takes to decide a particular type of motion; (e) how a particular judge tends to rule on a particular type of case, like a summary judgment motion; (f) how often a decision is confirmed or reversed (in whole or in part) on appeal; (g) outcome analysis by gender and race; (h) comparative function (statistics on a judge in comparison with other judges or a court average)."

Impact of Hyperrealism: Taking a HD Picture of Contemporary Law

"Modern technologies such as judge analytics show that personal attitudes and bias influence decision-making."

"If the latter has a narrow application sphere, hyperrealism applies to legal decision-making from a broader perspective."

"The concept of hyperrealism showed that the judge's personality plays a relevant role in decision-making."

"AI and other technologies can assist the judge in resolving disputes and support legal decision-making."

"It is too early to replace judges with AI or other technologies."

"In this case, the concept of hyperrealism and the use and regulation of technologies to make legal decision-making more transparent and accurate will help solve this replacement problem."

"The primary technological process is the discovery of the accumulated marked-up data sets of knowledge necessary to make decisions in crucial issues of legal practice."

"Companies and law firms increasingly seek data analysis to help make decisions and establish business and legal strategies."

Conclusion

"Digital technology is the primary tool of hyperrealism, which sheds new light on legal realism premises and confirms that, in some way, they were right in their empirical approach."

"Defined vulnerabilities of the AI-based judge analytic tools and recommendations for improvement can help design proper regulations on using AI-based analytical tools worldwide."

"Our findings help to understand and create better regulations for AI-based analytical tools from risk minimizing perspective."

"The results achieved in this research article are helpful for further research to promote trustworthy technologies and Artificial Intelligence based on respect for human rights in the digital transformation era."

"We recommend future research by studying the policy and experience of using AI-based analytical tools from a broader range of countries."

"Based on the achieved results, the authors plan to continue the research, focusing on confirming the idea of hyperrealism and creating effective models for regulating hyperrealistic tools."

7.2.6 Making Crime Visible in the Digital Age: The Ethnomethods of Data Policing [299]

This is a machine-generated summary of:

Buscariolli, André: Making Crime Visible in the Digital Age: The Ethnomethods of Data Policing [299]

Published in: Critical Criminology (2023)

Link to original: https://doi.org/10.1007/s10612-023-09683-y

If you want to cite the papers, please refer to the original.

For technical reasons we could not place the page where the original quote is coming from.

Abstract-Summary

"This article proposes that examining data policing using thick descriptive accounts of lived police work provides insights into police officers' judgments of what constitutes "reasonably objective" crime analysis."

"Drawing from ethnomethodology's contributions to police work and technological practices, I examine how the symbolic processes by which officers highlight criminologically relevant phenomena from seemingly decontextualized data points reproduce and reshape long-established methodical police work."

Introduction

"Despite reasonably warning about the potentially intrusive and prejudicial character of data-driven policing techniques and how they reproduce long-standing societal inequalities (Brayne, [167]; Eubanks, [168]; Ferguson, [169]; Osoba & Welser IV, [170]), criminal justice scholars have yet to engage with the epistemological bases by which officers make deductions from data, including concerning legal admissibility."

"This article proposes that examining data policing informed using thick descriptive accounts of lived police work provides insights into police officers' judgments of what constitutes "reasonably objective" crime analysis."

"The empirical analysis focuses on a case study where police officers explained in a public forum how they had used a suspect's GPS locational data to establish that his presence and movements in the city were consistent with burglary activities."

"Because officers render seemingly decontextualized data points observable-and-reportable (Garfinkel, [171]) evidence of criminal activity to an audience, this case study allows us to provide a thick description of the epistemological bases of the deductions they make from data."

Background

"To make deductions from data, officers articulate professionally based knowledge analogous to social scientists' qualitative inquiry, making "small" and "big" data policing epistemologically distinct and mutually constitutive practices (see Mair and others, [172])."

"Bittner's ([173]) classic article about policing skid-row discusses the relationship between methodical police work and law enforcement technologies; his insight grasps an essential epistemological dimension of officers' use of data."

"Bittner explains that police officers acquire and maintain knowledge about those socially structured features of everyday life as a response to the local contingencies of their work."

"What Bittner ([173]) terms "those socially structured features of everyday life" constitute ad hoc knowledge officers continuously develop, maintain, and adapt as a response to the local contingencies of their work."

"Officers' use of data builds upon, reproduces, and reshapes long-established methodical elements of police work through which they make criminologically relevant observations visible—namely, the ethnomethods of data policing."

Case Study

"Data come from a CompStat meeting held by a Police Department in the Washington D.C. metropolitan area on July 16, 2018."

"CompStat is a management and technological system that aims to optimize police departments' efficiency and transparency by gathering and analyzing crime-related statistics (Bratton, [174])."

"Nation-wide, CompStat meetings are leadership management processes where district commanders account for their efforts to reduce criminality."

"Because CompStat meetings aim at promoting transparency and accountability by presenting the department's efforts to reduce criminality, they constitute valuable material for ethnomethodological inquiry."

"Manning ([175]) explains that officers mobilize rationalizing forces during crime-analysis meetings, i.e., vocabularies of motive that justify and account for police work's effectiveness in its own institutional and organizational terms. "''

Steps of the Meeting

"Roughly half of the one-hour meeting discusses how police officers have used GPS location data from the suspect's cellphone to track down his location over time and establish that the presence of his phone signal in the city was consistent with burglary occurrences."

"Driving a rented car is not a violation of any sort; however, its presence occasioned "probable cause" for further investigation by the observing office in reflection of several factors, namely: the known transient character of the suspect whose cellphone data suggests they travel from out-of-city to commit burglaries; the time of both week and day being conducive or likely for burglaries to occur; the citizen report of an unknown vehicle with the same license plate, etc. The gestalt contexture produces the vehicle as suspicious instead of merely lost."

"Having established a reason for suspecting that Mr. Oak is responsible for the burglaries, the crime analyst returns on stage to explain how telephone data has contributed to the investigation.""

Discussion

"The reasonableness of officers' assertions about and through GPS location data depends on how they weigh professional expertise regarding offenders' habitual modes of activity and the public's commonsense presumed knowledge."

"The data requires interpretive work to bring potentially equivocal pieces of evidence into line with what the officers can account for assuming about a suspect's actions and motives."

"Having discussed how officers must fill data holes (Ferguson, [176]) between data points and the social order of the world they purportedly represent, the questions above suggest that the sergeant's account has its own interpretative gaps."

"Even though the social world officers deal with does not come "pre-coded or neatly packaged and self-assigned" (Benson, [177]), law enforcement agents still can use their ethnographic area knowledge to make crime visible through data points."

7.2.7 Development Justice in the Digital Paradigm: Agenda 2030 and Beyond [300]

This is a machine-generated summary of:

Gurumurthy, Anita; Chami, Nandini: Development Justice in the Digital Paradigm: Agenda 2030 and Beyond [300]

Published in: Development (2019)

Link to original: https://doi.org/10.1057/s41301-019-00203-6

If you want to cite the papers, please refer to the original.

For technical reasons we could not place the page where the original quote is coming from.

Abstract-Summary

"Digital and data technologies are not merely tools or enablers of development; they are the scaffolds of a new social paradigm."

"The intertwining of technological architectures and socio-economic structures presents key concerns for a transformative vision of development for people and planet alike."

Development in the Digital Paradigm

"The hopes and aspirations born with a young Internet of the late twentieth century seem to have considerably dissipated."

"The Internet-mediated techno paradigm of the early twenty-first century is emerging rather stridently as an instrument of exclusion and exploitation, belying the possibilities of its promised potential."

"This essay outlines the key dimensions of these trends, arguing why and how the digital paradigm needs to be urgently redirected for a radical and transformative vision of development 'for people and planet in the twenty-first century' (UN General Assembly [178], para 51)."

The New Capability Divide

"The bulk of those who are not connected are in the Least Developed Countries (LDCs)."

"They tend to be located '... in more remote, rural areas... [and] consist disproportionately of poorer, minority, less educated, and often female, members of society' (Broadband Commission [179]: 6)."

"Glance, this development is puzzling, considering the numerous supply-side advances in connectivity—whether the growth of international Internet bandwidth or the fall in broadband prices (ITU [180])."

"A closer examination reveals that while high-end broadband consumers are served well, poorer communities in remote and rural pockets are underserved—not just in the LDCs, but also in lower middle income countries with high levels of income inequality such as India, Brazil and China (Goldstein [181])."

"With the result, these populations are caught in an 'access trap' (Goldstein [181])—where non-availability of connectivity and poor demand feed off each other to create an exclusion loop."

The Network-Data Complex

"Growing on the network effect, transnational digital corporations have become gargantuan monopolies in an extremely short span of time They have stormed into every domain of social life, capitalizing on the might of data, creating and controlling the 'essential connections' for marketplace and social interactions (Iansiti and Lakhani [182])."

"In India, there are reports of start-ups in the financial technologies sector, commonly referred to as fintech firms, using up to 10,000 data points generated from device, social media, and browser fingerprinting of individuals to score their creditworthiness (Jain [183])."

"The system uses a complex algorithm that generates a trustworthiness score based on credit history, ability to fulfill contractual obligations, and problematically, behaviour and preference data mined from payment apps, e-comm platforms and social media."

"Although there is mounting concern about the role of social media in undermining democracy across the world, states in the Global South are yet to come up with effective policy measures in this regard."

The Invisible 'Expulsions'

"Digital technologies are likely to displace workers in routine jobs that involve a high percent of repetitive, codifiable tasks that can be easily replicated by computerized systems."

"In the on-demand economy (digitally based marketplaces) enabled by platformization, corporations are able to generate greater profits with a few high-skilled technology professionals, adopting new labour practices that result in casualization and precarity for the majority of the workforce (Scholz [184])."

"The shift to digital automation of jobs in both manufacturing and service sectors is likely to reinforce this trend of jobless growth."

"Economists predict that advancements in digital technologies are likely to increase automation of all economic activities, except those where wages are low enough to make automation uneconomic (Turner [185])."

"Manufacturers based in these countries are likely to make the switch from labour-intensive off-shoring factories to older industrial centres in the West, as AI-facilitated transition to Industry 4.0, shipping time, efficient local governance and maturity of digital infrastructure become more important considerations in factory siting decisions (Saxer [186])."

Ecological and Survival Costs

"The demand for data-intensive web services such as streaming videos or mining cryptocurrency has been linked to a rapid increase in energy demand."

"Emissions of global data centre web servers, such as those operated by Google and Facebook, are estimated to equate 2% of global greenhouse emissions, an amount equivalent to emissions generated by global aviation (Vaughan [187])."

"Bitcoin's 'mining' network uses more electricity in a year than the whole of Ireland (Hern [188])."

"The production of the material-technical infrastructure for the digital economy (microchips, cables, electricity) and creation of global digital commerce networks rests upon an unsustainable exploitation of natural and common property resources in the Global South."

The 2030 Agenda and the New Techno-paradigm

"The preferred route of 'multistakeholder partnerships (…) to facilitate development, transfer and dissemination of relevant technologies for the Sustainable Development Goals' (UN General Assembly [178]: para 70) furthers a simplistic, global-to-local solutionism, eliding the reality of power differentials between and among governments, corporations and citizens, and evading the question of accountability."

"In a 2018 workshop on digital trade and industrialization, organized by UNCTAD for developing countries, participants observed that:'Proposed rules (around cross-border data transfers, in trade talks) are all aimed at impeding independent and robust digital industrialisation by providing the way for foreign ownership and control over national data, digital intelligence and platform businesses in all sectors (including vital sectors like health, education, agriculture, manufacturing and indeed governance)."

The Way Forward: What Will it Take to Realize Development Justice in Digital Times?

"The Bangkok Civil Society Declaration: From Inclusive to Just Development imagines development justice as comprising five foundational dimensions (APWLD [189]):development of economies that enable dignified lives for all and promote the expansion of capabilities and well-being (economic justice)correction of the economic imbalance between the Global North and south (redistributive justice)elimination of all forms of discrimination, marginalization and exclusion (social justice)acknowledgment of historical responsibility by countries, and elites within countries, whose actions have contributed to the degradation of environmental resources and decimated livelihoods of the marginalized (environmental justice)right of all peoples to demand democratic and just governance structures for development decision-making (accountability)The challenge before us is to produce a cartography towards claiming each of these dimensions that is adequate to our digitalized and datafied existence."

"How norms may be rebuilt and institutions recast in the intertwining of network architectures and social structures is discussed below with respect to all five aspects of development justice."

Economic Justice

"State action on the following fronts will be central to such a new social contract:1.Ensuring universal and affordable access to the full Internet, through public provisioning of broadband infrastructure."

"For these efforts to succeed in the long term, new legal and regulatory mechanisms will be necessary, especially to restrain platform monopolies.3.Appropriating data as a collective resource and a public good, while respecting privacy."

"To secure citizen rights, such interventions need to adopt privacy-by-design principles, with special attention to collective data rights.4.Proactively responding to digital automation and platformization of work."

"Developing countries need to mobilize necessary investments in AI technologies for eliminating drudgery and indignity; commit to new models in the manufacturing sector promoting workers' share in ownership; provide workers' organizations and women's groups the wherewithal to set up and run on-demand work platforms; and provision new digital and educational architecture for mass skills development to ensure 'decent livelihoods for all' (Saxer [190])."

Re-distributive Justice

"As has been discussed, in the greed for data control, rich countries and their digital corporations have further entrenched the marketisation of development."

"They also need to call out the double standards of countries of the North in restricting the attempts of developing countries to independently develop data governance frameworks, whilst pushing data privacy standards that countries of the South may find difficult to implement, thus closing off the latter's market access."

"New age technology transfers through Smart City projects often tend to override the right of local governments to political self-determination, technicalizing public infrastructure and service provisioning, and transferring local data resources to foreign countries and companies."

"Digital industrialization policies can play a significant role in encouraging domestic digital enterprises and curtailing digital monopolies, and to set up enabling public infrastructure for data and digital intelligence relevant to local development priorities (Singh [191])."

Social Justice

"As governance gets completely restructured by the algorithmic turn, social justice outcomes will depend on algorithmic accountability—'the obligation to report, explain or justify algorithmic decision-making as well as mitigate any negative social impacts or potential harms' (Web Foundation [192])."

"As transnational digital platforms become new orchestrators of social and economic activity, they tend to exercise tremendous power over the emerging ecosystem comprising old and new actors—local firms, suppliers, producers, service providers, and customers/consumers."

"While their tactics to protect their competitive edge and maximize profits are often obfuscated from public scrutiny, they exercise considerable influence over local policy discourses, projecting themselves as the harbingers of opportunity (Cohen [193])."

"Policy frameworks in the Global South need to understand the nuances of value extraction in the platform economy in order to protect and promote local enterprises, informal sector workers and small producers, who tend to be negatively affected (James [194])."

Environmental Justice

"Environmental justice in current times must begin with curbing the adverse 'direct effects' on the environment triggered by the 'production, distribution, operation and disposal of digital technologies' (MacLean and others [195)."

"To improving energy and materials efficiency in manufacturing, transitioning to renewable energy sources for powering digital networks, and regulating e-waste management, the 'culture of obsolescence' promoted by the proprietary technology lobby must be challenged (Mobbs [196])."

"The 'indirect environmental effects' of the digital economy—unsustainable consumption cultures and the rise of planetary-scale distribution networks—need to be tackled as an important issue for cultural and economic self-determination."

"Developing countries must ensure that the expansion of digital commerce does not come with unsustainable environmental costs."

Accountable Governance for the Techno-paradigm
"An international treaty on human rights on the Internet that includes economic, social and cultural rights and the right to development must be adopted to rearticulate the non-negotiables for human rights in the twenty-first century (Gurumurthy and Chami [197])."

"Global governance frameworks must address this to safeguard the social and collective value of privacy, media freedoms, labour rights and economic sovereignty, especially in contexts of the Global South."

"Discussions on data will need to be decoupled from negotiations on trade, so that 'countries without the infrastructure and sophistication for collecting and managing their own data are not forced to trade their sovereignty' in exchange for participation in global markets (Gurumurthy and Chami [197])."

"The buzz around tech-innovation distracts us from the ways in which the dominant platform model of economic organization renders the real economy even more vulnerable to the caprices of capital markets, exacerbating the risk of exploitation, exclusion and expulsion for marginalized communities in the Global South."

7.2.8 Protecting Human Health and Security in Digital Europe: How to Deal with the "Privacy Paradox" ? [301]

This is a machine-generated summary of:

Büschel, Isabell; Mehdi, Rostane; Cammilleri, Anne; Marzouki, Yousri; Elger, Bernice: Protecting Human Health and Security in Digital Europe: How to Deal with the "Privacy Paradox" ? [301]

Published in: Science and Engineering Ethics (2014)

Link to original: https://doi.org/10.1007/s11948-013-9511-y

If you want to cite the papers, please refer to the original.

For technical reasons we could not place the page where the original quote is coming from.

Abstract-Summary
"The interdisciplinary approach is a unique feature and guarantees important new insights in the social, ethical and legal implications of these technologies for the individual and society as a whole."

"A special focus is put from the perspectives of psychology, medical ethics and European law on the contradiction between individuals' motivations for consented processing of personal data and their fears about unknown disclosure, transferal and sharing of personal data via information and communication technologies (named the "privacy paradox")."

"The authors point out the ethical and legal limitations inherent to the processing of personal data in a democratic society governed by the rule of law Finally, they seek to demonstrate that the impact of information and communication technology use on the individuals' well-being, the latter being closely correlated with a high level of fundamental rights protection in Europe, is a promising feature of the socalled "e-democracy" as a new way to collectively attribute meaning to large-scale online actions, motivations and ideas."

The Psychological Perspective

"A study on information disclosure and control on Facebook by adolescents and young adults shows that despite concerns for privacy induced by the disclosure of personal data, the survey population chose to disclose voluntarily personal data such as birthdays, e-mail addresses, profile pictures, pictures with friends, including pictures at parties (Christofies and others [198])."

"While UK adults and young people have expressed concerns about privacy, security, as well as control over access, use and misuse of personal data, they were in favour of using them for public health research and surveillance related to cancer (Luchenski and others [199])."

"A study on the use of mobile phones for increasing mental health and human well-being (mHealth) shows that mobile phone applications specifically associated with behavioural health including "developmental disorders, cognitive disorders, substance-related disorders as well as psychotic and mood disorders" respond to the shortcomings of conventional assessment procedures, namely difficulties in gathering information from the individual subject during the day (Luchenski and others [199])."

The Medical–Ethical Perspective

"The underlying idea is that patient information that is important in a medical emergency should be accessible by all doctors while other information that is not relevant, for example the fact that the patient is undergoing psychotherapy, is more strictly protected."

"Even more important in a pluralistic society governed by the harm principle (Mill [200]) are arguments that refer to the fact that patients who hide emergency information could cause harm to the health care system since expensive tests would need to be repeated by doctors although results might have already been available in the electronic records, and time would be lost in the diagnostic and therapeutic process, for example Intensive Care Unit (ICU) beds might be occupied longer than necessary which could harm other patients that are waiting for an ICU bed."

The European Law Perspective

"This is especially true since the revision of the Lisbon Treaty conferring a binding character on the closely related rights to the respect for private and family life and to the protection of personal data (Art 7 and 8 of the Charter of Fundamental Rights of the EU, European Court of Justice [201, 202]) and providing for a specific legal basis for adopting privacy laws independently from internal market aspects (Art 16 Treaty on the Functioning of the EU)."

"According to European law, the processing of data for other purposes than those for which they have initially been collected, as it risks violating the fundamental rights to privacy and data protection, is in principle prohibited."

[Section 4]
"Individuals fear threats to their personal autonomy and freedoms stemming from globalized data processing by governments and undertakings, while on the other hand they voluntarily proceed to the disclosure of personal data (e.g. by posting names, photographs, dates of birth, marital status on social networking sites, or medical data on health forums)."

"Law, while creating the conditions for the development of technological progress (e.g. via the freedom to conduct a business, the right to protection of intellectual property), also limits the use of ICT (e.g. through the rights to protection of private life, and of personal data)."

"In democratic societies, opinions diverge however as to the question whether the processing of personal data via ICT for the purposes of protecting human health and security is legitimate and which should be the legal rules governing it."

7.2.9 Fundamental Rights and the European Regulation of iConsumer Contracts [302]

This is a machine-generated summary of:

Mak, Chantal: Fundamental Rights and the European Regulation of iConsumer Contracts [302]

Published in: Journal of Consumer Policy (2008)

Link to original: https://doi.org/10.1007/s10603-008-9084-3

If you want to cite the papers, please refer to the original.

For technical reasons we could not place the page where the original quote is coming from.

Abstract-Summary
"This paper addresses the question of how fundamental rights affect European legislation and adjudication on contracts regarding digital information services (iConsumer contracts)."

"It is submitted that because of this double-faced nature, they can bring to the fore policy issues in contract legislation and case law Fundamental rights can thus play a role in evaluating the policy choices that are being made in the review of the acquis communautaire in the field of consumer law For iConsumer contracts, that means that the rights of consumers, authors, and suppliers of copyright-protected content affect the choice of rule-solutions on the European legislative level."

"Fundamental rights help define the various rule-solutions the Court can choose from and thus demarcate the law-making capacity of the judiciary."

Introduction

"An answer to the question of how policy choices regarding the protection of the iConsumer will turn out may be found by studying a development that is obtaining an equally prominent place in the further harmonisation of European contract law: the impact of fundamental rights on the regulation of (consumer) contracts (see also Mak [203])."

"The European legislature and the European judges have to strike a balance between these rights when formulating specific rules in the area of iConsumer contract law This paper addresses the question of how fundamental rights affect European legislation and adjudication concerning iConsumer contracts."

"It is submitted that fundamental rights argumentation brings into the open the policy questions underlying the regulation of iConsumer relations, which concern the balance between principles of autonomy and solidarity in these contracts."

Fundamental Rights and European (Consumer) Contract Law

"In Dutch law, rules similar to the German ones have been developed in case law determining the impact of fundamental rights on the colouring of general clauses of private law The Dutch Supreme Court, the Hoge Raad, has for instance held that the legality of a contractual waiver of the right to practice the Mensendieck method for improvement of posture should be assessed in light of the fundamental right to teach."

"Although the Dutch courts tend to lean towards a somewhat less dogmatic approach than their German counterparts, the rules established in these cases resemble the German theory of reciprocal effect or Wechselwirkung between fundamental rights and private law In English law, finally, the debate on effects of fundamental rights in contract cases has been rekindled by the Human Rights Act 1998 (HRA), which came into force on 2 October 2000 and aims to give further effect to ECHR rights in British law Should the Act also apply in relations between private parties?"

Legislating iConsumer Contracts

"The review process has recently resulted in a proposal by the European Commission for a Directive on Consumer Rights (hereafter: Draft Directive), which aims to revise four of the aforementioned Directives, namely, those on doorstep selling, unfair contract terms, distance contracts, and consumer sales and guarantees (Commission Proposal 204])."

"The review of the European consumer law acquis has important implications for contracts concerning digital information services (Commission [205], p. 3)."

"It is the subject of debate whether the protection offered by certain Directives on consumer sales of goods should be extended to other types of contracts, such as those concerning digital information services (Commission [205], p. 24)."

"An expansion of the scope of the Directives, on the other hand, wished for by consumer organisations and most academic respondents to the Green Paper, rather favours the iConsumer's freedom to receive information, rights to privacy and non-discrimination, thus opting for a further protection of the weaker contracting party."

Adjudicating iConsumer Contracts

"It confirms the aforementioned theory of "mediation" of fundamental rights between contract law and public policy, since the application of the CRC colours the ECJ's interpretation of public morality."

"In effect, the Court's application of fundamental rights shows similarities with the aforementioned German theory of Wechselwirkung: a reciprocal effect between a general clause of private law (public morality and public policy) and a fundamental right (Art 17 CRC)."

"It is remarkable that the ECJ has ruled quite differently in case law concerning the enforcement of EC law: insofar as Community law is concerned, prevalence in that case is given to the ECJ's interpretation of fundamental rights over the interpretation by a national Constitutional Court."

Conclusion

"A brief look into the European legislature's review of the acquis communautaire has shown that the policy choices made in the process of this revision relate to the outcome of this balancing process: an emphasis on consumer rights will, for instance, result in an expansion of the scope of application of the Consumer Sales and Distance Selling Directives, whereas a focus on copyright holder's rights points in the opposite direction."

"Fundamental rights can help the legislature and judiciary define these issues in questions concerning iConsumer contracts."

"In line with the Commission's aim to enhance the coherence of European Contract Law, further thought should therefore be given to the role of the European courts, in particular the ECJ, in the uniform interpretation of (iConsumer and consumer) contract law rules in light of fundamental rights."

7.2.10 Learning Analytics: at the Nexus of Big Data, Digital Innovation, and Social Justice in Education [303]

This is a machine-generated summary of:

Aguilar, Stephen J.: Learning Analytics: at the Nexus of Big Data, Digital Innovation, and Social Justice in Education [303]

Published in: TechTrends (2017)
Link to original: https://doi.org/10.1007/s11528-017-0226-9

If you want to cite the papers, please refer to the original.

For technical reasons we could not place the page where the original quote is coming from.

Abstract-Summary
"We are still designing educational experiences for the average student, and have room to improve."

"In so doing, learning analytics has the potential to contribute to more equitable and socially just educational outcomes for students who might otherwise be seen through the lens of the average student."

"These digital innovations will enable us to do away with a model of education that teaches toward the non-existent average student, replacing it with one that is more socially just—one that addresses the individual needs of every student."

Big Data and Learning Analytics
"In education, learning analytics has emerged as the discipline associated with analyzing and reporting big data, though work in this area is relatively new."

"Predicated on the availability of large amounts of student data (e.g., student demographic information from student information systems, student grades, and/or student behaviors in learning management systems), learning analytics is poised to transform the ways in which schools and universities leverage big data to address issues of retention and student success, thus moving away from the average student and towards addressing the needs of each student in a personalized and data-driven way."

"What distinguishes learning analytics from traditional modes of academic feedback (e.g., grades, or comments on an essay) is its capacity to extract data from user interactions with the learning environment, analyze it, and "load" it onto applications that are used by those invested in the learning process (Lonn and others [206])."

The Fallacy of the "Average"
"The protocol developed seemed simple and achievable: to build a cockpit for the average man (the use of "man" here is deliberate—women were not seen as potential pilots), the Air Force would measure over 140 physical dimensions of 4000 of pilots and take the average of those measurements."

"Daniels concluded that the average, for all its merits, was a woefully inadequate statistic to use in the design process: "While the use of average dimension is generally unsatisfactory even when only one dimension is being considered at a time, the inadequacy of the 'average man' method is compounded many times when more than one dimension is to be considered in a design problem." (Daniels [207] pg 2) Daniels highlights an insidious problem in design, namely, if you design for an "average" person, not only are you designing for no one in particular, you are, in fact, designing for no one at all."

Moving beyond the "Average" Student

"A defining characteristic of learning analytics is to utilize data in any form it comes in, and tie it to a particular student for the purpose of personalizing their learning experience."

"Related work has aggregated data from audio sources; video sources coded for hand and wrist movements; and electro-dermal activation devices (i.e., changes in the skin's resistance to small electrical current due to sweat), in order to distinguish between when students were using "principle based reasoning" versus "example based reasoning" in STEM contexts The latter was associated with better learning outcomes (Worsley and Blikstein [208])."

"When paired with other types of data (e.g., surveys, standardized test results, and demographic information), AWEs can help predict the quality of students' writing."

"Many of the insights derived through learning analytics techniques are operationalized through the use of "dashboards," which provide instructors with real-time information regarding the learning outcomes of their students (Diana and others [209]; Verbert and others [210]; West [211])."

Learning Analytics and Social Justice

"Regardless of the form they take, applications of learning analytics instantiate the goals of various stakeholders in their designs for the purpose of personalizing students' learning experiences."

"With personalization as a starting point, learning analytics has the potential to give students new tools to study themselves as learners and be more reflective of their study habits (e.g., Cutumisu and others [212]; Roll and Winne [213]), while also connecting them to institutional resources that may have otherwise remainded hidden."

"The learning analytics application can also suggest a staff person at writing center who the student can contact."

"In the classroom, the insights generated by learning analytics-based applications can help teachers truly differentiate and alleviate some of the burden from new or struggling teachers; both of which have been shown to play a role in underserved student populations (Gusky [214])."

"Learning analytics-based applications thus have the potential to help students reflect on their learning while also helping them to connect to resources they might have otherwise not realized were available."

Not a Silver Bullet

"While learning analytics techniques, practices, and applications can potentially change the way various stakeholders understand student learning, there is still more work to be done."

"Advancements in the collection, analysis, and integration of student data into various applications, for example, come with risks, and also pose ethical dilemmas that need to be understood and addressed as learning analytics-based educational technologies are implemented."

"As with other analytic techniques, learning analytics can suffer from a latent variable problem, i.e., an important dimension about student learning may exist, and not be captured."

"A perceived lack of privacy on the part of the students whose data is being used, for example, can undermine the legitimacy of learning analytics applications before they get off the ground (Hoel and Chen [215])."

"The resulting set of groups (or clusters) have the potential to wash out individual behaviors, thus replicating a key issue learning analytics is meant to address."

Conclusion

"In 2016 we are still designing educational experiences for the average student, and have room to improve."

"It is unfair and ineffective to expect teachers to personalize the experiences of each of their students, all at once, without help."

"With the aid of learning analytics-driven educational technologies, we can move closer to the personalized instruction method championed by Keller ([216])—learning analytics provides a way forward."

"However, these digital innovations will enable us to finally do away with a model of education that teaches toward the non-existent average student, replacing it with one that is more socially just and equitable; one that acknowledges and supports the individual needs of every student."

[Section 7]

"It is challenging to reach every student in a personalized way."

"Differentiation is challenging, and stakeholders often compromise and teach to the average student."

"Smaller class sizes speak to an intuitive characteristic of education; namely, that it is easier to teach and learn when instruction is personalized."

"Educational technologies have begun address this challenge by taking elements of personalized learning to scale."

"Learning analytics-driven technologies are capable of providing learners and instructors alike with insights into learning processes while also detecting students who struggle or become disengaged, making it less likely for those students to fall through the cracks."

"The unique capabilities of learning analytics-driven educational technologies speak to their potential to be a positive force for social change, and they are poised to have a central role in the reduction of social inequity because they represent a unique, scalable learning technology that can be used to address individual students' learning needs."

7.2.11 Online Security and the Protection of Civil Rights: A Legal Overview [304]

This is a machine-generated summary of:

Pagallo, Ugo: Online Security and the Protection of Civil Rights: A Legal Overview [304]

Published in: Philosophy & Technology (2013)

Link to original: https://doi.org/10.1007/s13347-013-0119-6

If you want to cite the papers, please refer to the original.

For technical reasons we could not place the page where the original quote is coming from.

Abstract-Summary

"The paper examines the connection between online security and the protection of civil rights from a legal viewpoint, that is, considering the different types of rights and interests that are at stake in national and international law and whether, and to what extent, they concern matters of balancing."

"At least in Western legal systems, it should be clear that either civil rights prevail over security (no balancing), or such balance has to satisfactorily protect individual rights (proportionality)."

Introduction

"By drawing the attention to the hierarchical structure of the law and how we should interpret the nature and logic of the principles of the system, think of multiple types of rights, such as (a) Natural rights, in accordance with the philosophical tradition of contractualism and liberalism, e.g. John Locke's Two Treatises of Government (b) Human rights in the field of international law, e.g. the European Convention on Human Rights ("ECHR") from 1950 (c) Constitutional rights of national legal systems, such as the civil rights of the US tradition or, conversely, the fundamental rights of the EU law and its Member States, and (d) Rights established by statutory laws This differentiation is fruitful so as to start appreciating the different levels of protection concerning the rights of the individuals."

National Security and Civil Rights

"Pace Hobbes, the limits of political power are legally defined by the law of nature (de iure), rather than factual arguments (de facto); they concern what today's scholars define as human rights in the international law field, civil rights in the US constitutional law field, fundamental rights in the EU law field, etc. In a nutshell, according to Locke, "freedom of Men under Government is, to have a standing Rule to live by, common to every one of that Society, and made by the Legislative Power erected in it; a Liberty to follow my own Will in all things, where the Rule prescribes not; and not to be subject to the inconstant, uncertain, unknown, Arbitrary Will of another Man: as Freedom of Nature is, to be under no other restraint but the Law of Nature" (Second Treatise, Ch."

Private Security and Other Civil Rights

"After the "Security first" and "Liberty first" theses on the political foundations of legal interaction, e.g. the Hobbesian and Lockeian versions of the social covenant, the second polarization of the debate concerns the legal interaction of private actors, that is, how the protection of private (rather than public, or national) security is related to the protection of further individual rights."

"At one end of the spectrum, online security and civil rights pit against each other in a zero-sum game: consider the aim of private companies to protect their rights through the use of self-enforcing technologies, e.g. DRM devices."

"Many cases suggest that the aim to guarantee individual security, and safety, goes hand in hand with the protection of further civil rights of the individual."

"There is no room for balancing: in addition to rights that are absolute in the public law field, security and safety are typical rights of the individual that, although "relative", do not necessarily entail a zero-sum game between private actors."

Security by Design and the Protection of Civil Rights on the Internet

"At one end of the spectrum, security trumps civil rights through the use of such filtering systems, because no balancing would be possible between the aim to ensure online security via this technique and the protection of some basic rights, such as data protection, freedom of speech and of information, freedom to conduct a business, and so forth."

"In the case of Netlog, the national court would have had to issue an injunction against the social network requiring the latter to install a system that, in the wording of the EU Court of Justice, should filter: (a) "Information which is stored on its servers by its service users; (b) Which applies indiscriminately to all of those users; (c) As a preventive measure; (d) Exclusively at its expense; and (e) For an unlimited period; Which is capable of identifying electronic files containing musical, cinematographic or audio-visual work in respect of which the applicant for the injunction claims to hold intellectual property rights" (C-360/10)."

Zero-Sum Games and Win–Win Approaches
"By embedding privacy safeguards into the design of ICTs, the full functionality of the principle would allow a positive-sum or win–win game, making trade-offs unnecessary between security and civil rights (Cavoukian [217])."

"The aim of guaranteeing security can fit like hand to glove with the protection of civil rights through the approach of the principle of privacy by design, e.g. making personal data anonymous in public transportation systems through video surveillance that must be designed in such a way that faces of individuals cannot be recognizable."

"Striking a balance between security and civil rights seems necessary in a number of legal cases that fall in between the ends of the spectrum, that is, between zero-sum games and win–win scenarios."

New Scenarios
"Whilst the "neutrality" of the service, on which the responsibility of ISPs depends, is legally recast by the evolution of automated systems for the processing and filtering of huge amounts of data through search engines, data mining, or cloud computing, it is thus an open question whether today's clauses of immunity should protect gigantic information distributors that either control the architecture of the system with its apps or determine control over content of the communication."

"Going back to the proposed new Police and Criminal Justice Data Protection Directive, the EU Commission significantly refers to "the level of protection of the rights and freedoms of individuals with regard to the processing of personal data by competent authorities for the purposes of prevention, investigation, detection or prosecution of criminal offences or the execution of criminal penalties" which "must be equivalent in all Member States" (Seventh Considerandum of the proposed directive)."

Conclusions
"This paper has dwelt on three possible outcomes of today's debate on online security and civil rights, that is, the opposite views that deem as unnecessary any trade-off between such rights and interests in the name of the "security first" and "liberty first" theses on the political foundations of legal interaction and the idea endorsed by scholars and legislative drafts, or statutes, that aim to strike a proper balance in such fields as constitutional law, international and transnational law, etc. This bifurcation has been traced back to the alternative ways in which the connection between online security and civil rights is understood, namely, grasping that "and" in either a disjunctive or a conjunctive way."

"The use of self-enforcing technologies and filtering systems on the internet should be considered as the exception, or last resort option, to guarantee online security and, even in this case, such a use should be strictly balanced against the protection of basic civil rights, e.g. freedom of information."

7.2.12 The German Facebook Case: The Law and Economics of the Relationship Between Competition and Data Protection Law [305]

This is a machine-generated summary of:

Kerber, Wolfgang; Zolna, Karsten K.: The German Facebook case: the law and economics of the relationship between competition and data protection law [305]

Published in: European Journal of Law and Economics (2022)

Link to original: https://doi.org/10.1007/s10657-022-09727-8

If you want to cite the papers, please refer to the original.

For technical reasons we could not place the page where the original quote is coming from.

Abstract-Summary

"Can competition law consider effects on privacy, or should privacy concerns of data-collecting behaviour only be dealt with by data protection law?"

"Since the traditional approach of a strict separation of both policies is no longer feasible, a more integrative and collaborative policy approach for competition law and data protection law might be necessary."

"With respect to the substantive issue in the Facebook case, i.e. protecting a minimum standard of choice for consumers regarding their personal data vis-a-vis dominant digital platform firms, the recent decision by the German Federal Court of Justice in this case and the proposed Digital Markets Act have led to new perspectives for dealing with privacy concerns in competition law and new forms of ex-ante regulation."

Introduction

"The objective of this article is to analyse the question of whether the negative effects of the data-collecting behaviour and privacy terms of large digital platform firms on competition and privacy can be considered in competition law, or whether—as the critics of the Facebook decision by the FCO claim—privacy concerns should be exclusively dealt with through data protection law (or consumer law)."

"Due to manifold interaction effects between the two market failures and between the two policies on competition and privacy, competition law and data protection law are becoming deeply intertwined, leading to a complex relationship between the two policies."

"Chapter 2 offers a brief overview of the discussion on the Facebook case and its implications for the relationship between competition and data protection law Chapter 3 presents, from an economic perspective, a theoretical framework about the two market failures, two policies, and their interaction effects, applied to the effects of data-collection behaviour and privacy terms on competition and privacy."

The controversial debate about the German Facebook case

"After arguing why Facebook is dominant in the market for social media in Germany, the FCO argued that these privacy terms lead to exploitative abuse because they violate the GDPR, i.e. the FCO used the infringement of EU data protection law as a benchmark to determine the abusive character of these terms of service."

"The court then used this as an argument for qualifying this requirement as "abusive" in competition law Additionally, a number of contributions can be found that welcome the Facebook case and its discussion, because it opens a much broader discussion about the need to develop a wider policy perspective on the manifold, complex, interlinked problems of the data economy, in which the market power problems of large digital platform firms (with gatekeeper positions), data concentration, and unsolved privacy problems of consumers might be directly intertwined with each other."

Competition policy and privacy protection: two market failures, two policies, and their interaction

"We want to demonstrate, from an economic perspective, why in situations of the simultaneous existence of two market failures and of two policies (competition law and data protection law), in digital markets, manifold interaction effects can arise between these policy areas."

"Although competition policy uses, in its application, the consumer welfare standard and therefore also protects consumers, it is the task of consumer policy (and hence consumer law) to protect consumers against the negative effects of the market failures of information and behavioural problems (e.g. asymmetric and misleading information, exploiting behavioural bias)."

"We claim that such a comprehensive analysis, which focuses simultaneously on the effects of competition problems and information and behavioural problems, as well as the effects of competition law and data protection (and consumer) law might also be crucial in digital markets."

The negative effects of competition problems on privacy: a task for competition law or data protection law?

"This implies that any negative effects of competition problems on privacy (e.g. with respect to larger privacy risks due to the excessive collection and use of personal data through firms with market power or through a cartel or merger) cannot be dealt with by the GDPR as long as such firms still comply with these minimum standards."

"Since it is an important principle in EU data protection law that all firms are treated equally in terms of their data-collection and privacy terms, data protection law cannot distinguish between firms with and without market power or whether firms are under effective competition or not."

"If the data-collecting behaviour and privacy terms of dominant firms lead to greater privacy risks than those of non-dominant firms, then a different treatment of the privacy policies of firms with market (or gatekeeper) power might also be defendible in EU data protection law Nevertheless, such an approach is far away from the GDPR's current jurisprudence."

Assessing the abusive behaviour of dominant firms with regard to data-collecting practices and privacy terms

"If we take into account, in competition law, the negative effects of competition problems on the privacy of consumers, as was also intended by the FCO in the Facebook case, the infringement of data protection law is neither a sufficient nor necessary condition for the abusive character of the behaviour of a dominant firm."

"These German decisions in the Facebook case and the DMA proposal have put a new remedy on the agenda of the competition policy discussion, namely, obliging firms with market (or gatekeeper) power to grant consumers a minimum standard of choice (and therefore control) over the use of their personal data."

"We think that with the German Facebook case, the decision of the German Federal Court of Justice and the DMA proposal, this concept of a minimum standard of choice for consumers with respect to their decisions about the collection and use of their personal data vis-a-vis firms with market or gatekeeper power can be a new powerful tool for protecting the privacy and the informational self-determination of consumers."

Towards a broader and more integrative policy approach

"From the perspective of our theoretical framework in Chapter 3, with manifold interaction effects between these policies, a more integrative, collaborative approach between competition law and data protection law (which could include the interaction effects) may help to deal with the difficult issues that arise through the complex intertwining of both policies in digital markets."

"A close collaboration among authorities might lead to a much better analysis and understanding of the problems, as well as through an exchange of information and mutual discussions from different perspectives on competition law and data protection law This collaboration might allow for a better assessment of what role each of these policies could play in specific cases and what combination of remedies might be most effective for solving these problems."

7.2.13 Digital Technology in Family Justice Systems: Global Benefits, Opportunities and Concerns [306]

This is a machine-generated summary of:

Hodson, David: Digital technology in family justice systems: global benefits, opportunities and concerns [306]

Published in: ERA Forum (2021)

Link to original: https://doi.org/10.1007/s12027-021-00660-y

If you want to cite the papers, please refer to the original.

For technical reasons we could not place the page where the original quote is coming from.

Abstract-Summary

"The author surveys the global position on digital technology in family matters and reflects on the challenges, expectations and opportunities presently available."

"The paper examines the existing work by the Hague Conference to use digital technology for cross-border family maintenance and praises the December 2020 EU paper on digital technology in civil justice."

"He concludes by examining three very different centres of population and different adaption of technology for users of family justice."

Three challenges for our discussion

"The ERA conference on digital technology in family matters was organized with the purpose to share, learn and make progress using digital technology for the better conduct of family justice nationally and worldwide, however we must take full account of those for whom digital progress is uncomfortable, disconcerting and threatening."

"Even in England in 2020 where we have made huge progress in remote hearings in the family courts, the voice of Ned Ludd is still heard."

"The reports were accompanied by strong protests and desperate pressing of the brake pedal in the use of this key element of family justice digital technology."

"As we move into far greater use of digital systems, we now have another category: those without access to digital justice."

"Our digital developments are creating a new category of disadvantaged: without access to digital technology to use the systems being put in place in our family justice."

"Because the digital technology we are discussing is undoubtedly itself an answer to access to justice."

Three potential obstacles in family law

"Too often in too many countries digital technology has been a late arrival in family law work."

"One reason is the size of the law firms conducting family law work."

"Progress in digital technology in family justice has a significant problem worldwide in who are the practitioners and their consequential lack of use of IT."

"Another reason for lack of progress in family law itself is the argument that there is such a huge multiplicity of situations presenting families that any digital solutions are impossible."

"One reason for being troublesome in the EU civil law environment was that we are a strongly common law jurisdiction, hugely committed to a discretionary form of justice."

Three instances of transformation not automation

"Our role at the vanguard of digital technology is to work out how the public are living their digital lives and make sure our family justice adapts."

"One consequence of underfunding by many governments in family justice over too many years is the complete lack of any research data into outcomes from court applications, judicial decision-making and similar."

"One glowing example of digital technology in family law is the iSupport scheme ancillary to the 2007 Hague Maintenance Convention."

"For us wealthier countries, there is a call to help fund poorer countries who want to play a full part in joined up digital family justice."

"I suggest we should recognise these Hague and EU initiatives for the opportunities for macro global co-working and collaboration on micro-individual family cases."

A tale of three cities

"The bittersweet experience in England has been the most incredible innovation and acceleration of digital technology by family lawyers and courts."

"I explained that I was a family court judge from London and he showed me around."

"It had more screens, cameras and technology than the most advanced family court in London."

"Either the judges go on the road and we have pop-up courts, in community centres a couple of days a week, or we have local court access points, where safety, security and speed of Internet access is assured, with access to justice."

"I arranged to meet one of the family court judges and had a tremendous tour of their family justice centre."

"The family courts and the family lawyers have adapted, by which I mean transformed, delivery of family justice for a digitally joined up population."

Conclusion

"This has been an overview of some of the topics around digital technology in family matters."

"We must discuss today how we can harness the digital technology which is transforming our world and is in the hands or on the wrists of the public."

"The transformation itself must be international and global."

"There is much in Europe we need to learn from the global experience."

7.2.14 On the Imperative of Thinking Through the Ethical, Health Equity, and Social Justice Possibilities and Limits of Digital Technologies in Public Health [307]

This is a machine-generated summary of:

Gómez-Ramírez, Oralia; Iyamu, Ihoghosa; Ablona, Aidan; Watt, Sarah; Xu, Alice X. T.; Chang, Hsiu-Ju; Gilbert, Mark: On the imperative of thinking through the ethical, health equity, and social justice possibilities and limits of digital technologies in public health [307]

Published in: Canadian Journal of Public Health (2021)

Link to original: https://doi.org/10.17269/s41997-021-00487-7

If you want to cite the papers, please refer to the original.

For technical reasons we could not place the page where the original quote is coming from.

Abstract-Summary

"The COVID-19 pandemic has demonstrated both the positive and negative use, usefulness, and impact of digital technologies in public health."

"Digital technologies are thus—in some areas of public discourse—presented as being both necessary and inevitable requirements to address routine and emergency public health issues."

"Neither a sense of digital technological optimism and determinism nor the demands of addressing pressing public health issues should override critical assessment before development and implementation."

"The urgency of addressing public health emergencies such as the ongoing COVID-19 pandemic requires prompt and effective action, including action facilitated by digital technologies."

Introduction

"Responses to the pandemic have demonstrated the utility and effectiveness of digital technologies (DTs) in core public health functions of epidemiological surveillance and emergency management and response (e.g., Global Public Health Intelligence Network)."

"Using COVID-19 as an example, but referring to public health broadly, we argue that DTs' usefulness must be assessed in relation to their short- and long-term impacts on ethical, health equity, and social justice considerations."

"The critical assessment of DTs should not be overridden by a sense of digital technological optimism (the belief that DTs can and will solve all issues), determinism (the notion that DTs are the one and only way forward), nor the demands of addressing pressing public health issues."

"Despite pressure on public health practitioners to expedite DT development and implementation in response to public health emergencies like COVID-19, we posit that these assessments remain crucial and feasible before introducing DTs and as an ongoing practice."

Ethics

"Discussions before COVID-19 focused on acknowledging the mounting ethical implications arising with DTs, such as preserving data privacy and security, misusing information, and the consequent withdrawal of target end-users from participating in potentially beneficial public health interventions (Gilbert and others [219]; Marckmann [220])."

"The primary attention to the ethics of digital interventions as public-facing tools for end-users, however, somewhat eclipsed scrutiny of DTs as means within public health shaping the interests of various interest stakeholders (Brall and others [221])."

"Martinez-Martin and others ([222]) question traditional ethics models focused on individuals' privacy protection and software output features, in a time when private tech companies are leading the production of digital tools in the absence of a robust public health framework."

"Transparent public discussions about whether, when, for whom, and under which circumstances to use internal and public-facing DTs in public health must involve all stakeholders, including those providing the data behind many DTs."

Health equity

"Discussions about health equity in digital public health considered digitalization as a tool to address existing inequities and emphasized literacy as critical for access, uptake, and use of DTs (Azzopardi-Muscat and Sørensen [223])."

"Considerations about the ways and extent to which other DTs (e.g., big data or artificial intelligence) address health equity issues are evolving (Chauvin and others [224])."

"The use, usefulness, and impact of DTs in public health should involve, at least, asking whether the short- and long-term benefits and burdens are equally distributed across socio-demographic groups and countries."

"Developers and implementers must ask how and to what extent DTs address the root causes of existing health inequities and enable upstream action for health equity."

Social justice

"An extensive and detailed discussion of social justice in the context of DTs in public health—unlike ethics and health equity—is still materializing."

"With COVID-19, debates about the fair distribution of financial, social, and environmental burdens and advantages of new DTs are at an all-time high."

"Given this background, assessing DTs in public health using an anti-oppression lens must consider context, profit distribution, misuse, and public goods."

"Fair distribution of financial profits among contributors to the development and implementation of DTs, including those providing health data, must be considered."

"One must also reflect on secondary uses and potential misuses of the data collected through DTs, with explicit elucidation on how it could be erroneously redirected for corporations' profit, political gain, government surveillance, privacy erosion, or social control rather than public health purposes."

Conclusion

"Public health experts must critically consider the use, usefulness, and impact of digital technologies in public health."

"Digital tools in public health should only be the medium to the pursuit of public health goals and functions, rather than a technological end-goal on their own."

"While DTs can help advance and sustain some core public health functions, their intensified and normalized use, as well as their potential benefits in times of public health emergency, cannot and should not be a substitute for thoughtful consideration of their broad-ranging repercussions."

7.3 Digital Equality

Machine generated keywords: gender, inequality, read, sexual, equality, game, man, literacy, software, computer, male, accessible, document, female, woman

7.3.1 Systematic Literature Review of Digital Resources to Educate on Gender Equality [308]

This is a machine-generated summary of:

Yañez, Alma Gloria Barrera; Alonso-Fernández, Cristina; Fernández-Manjón, Baltasar: Systematic literature review of digital resources to educate on gender equality [308]

Published in: Education and Information Technologies (2023)

Link to original: https://doi.org/10.1007/s10639-022-11574-8

If you want to cite the papers, please refer to the original.

For technical reasons we could not place the page where the original quote is coming from.

Abstract-Summary

"Videogames and digital resources have proven their effectiveness as tools to educate, prevent and raise awareness about social problems."

"This article presents a systematic literature review of digital resources such as videogames, apps and simulations that address gender issues including violence and stereotypes."

"The main goal of this review is to present the status of gender-focused digital resources, their evaluation studies, including the metrics used and samples, as well as the acceptance and impact of their application."

"Most of the studies reviewed aimed to raise awareness about gender-based violence using serious games targeted at teenagers."

"Many of the projects reviewed did not have evaluation studies or the resources were not openly available, thus limiting their massive application and their potential impact on society."

"We consider that our results provide a starting point to better understand the role of digital resources in raising awareness about gender issues, highlighting their current limitations, and providing recommendations for future research in gender-based digital resources."

Introduction

"Gender-related discrimination and violence additionally have a serious impact on children's development, mainly on those of primary school age, whose formative years define and ingrain the individual's future values, beliefs, and behaviours in society."

"Digital resources such as serious games, virtual reality simulations or mobile apps can contribute to address these gender issues, mainly for prevention and awareness."

"This article presents a systematic literature review of studies presenting serious games and other digital resources addressing gender issues (including stereotypes, violence, etc.)."

"The main goals of this review are: To present a compilation of existing works on games and interactive digital resources that address gender issues, describing their main characteristics and, if applicable, the evidence on their effectiveness."

"To present the current state of research to set an academic and informative precedent for those who are developing or researching games and digital resources to address gender issues."

Related work

"Some serious games and other digital resources have started to be used to address gender-based violence and gender-related issues."

"Games to address gender issues may incorporate all features, techniques, and mechanics of serious games in general, to increase empathy, or raise awareness about gender differences like other social issues covered in serious games."

"An important issue regarding gender-related serious games and digital resources is that in most cases these resources have not been scientifically validated, and there are no related scientific articles presenting these resources or providing evidence on their effectiveness."

"To cope with this problem and find such resources, in a previous study a search of general gaming websites was conducted to locate serious games that aimed to educate on a variety of gender topics (Barrera and others, [226])."

"We decided to carry out this systematic literature review, to find such validated and tested games and digital resources addressing gender issues."

Methodology

"We classified the data from the selected studies according to the following measures: Focus and approach to the gender issue (addressing RQ1) Target audience and platform (addressing RQ2) Type of digital resources and features/mechanics (addressing RQ3) Instruments used to study the impact of the resources, such as pre-post questionnaires (addressing RQ4) Sampling used in the studies, including sample size, ages, geographic location, social context (addressing RQ5) Results obtained from the studies (addressing RQ6) Limitations of the studies (addressing RQ7)"

Results

"Serious games and digital resources addressing gender issues are primarily aimed at raising awareness (13 studies, including (Fonseca and others, [227])), changing player behavior (13, including (Jouriles and others, [228])), and teaching or training skills (10, including (Potter and others, [229]))."

"Another large group of games and tools (11 studies) uses perspective-taking to address gender issues."

"Of the 27 studies analyzed that contained an evaluation of the games, 18 reported positive results, while 5 obtained mixed results and 4 stated that were still in the analysis process at the publication time."

"Other studies reported work in progress: a study with preliminary evaluation results yet to be analyzed of a game to increase awareness of gender-based violence (Pabón-Guerrero and others, [230]); a discussion between game designers and participants about the characteristics (characters, environments, story, etc.) of a game to increase awareness of intimate partner violence (Pearson and others, [231]); a preliminary evaluation with a focus group to provide feedback (e.g., should clarify signs of abusive relationships, provide resources for immediate help) on an app for safety planning and teen dating violence (Debnam & Kumodzi, [232]); and a theoretical framework that uses empathy and creativity to design serious games, illustrated with an example of a serious game to raise awareness of domestic violence (Marda and others, [233])."

"The limitations found in the studies included small sample size (9 studies), other limitations in the sample (6 studies), technical problems (8) and other limitations in the game (8), the lack of follow-up studies (6), and the lack of results (5)."

Discussion

"This may be since, compared to post-questionnaire interviews, or the applicator's annotations of each of the digital resources, the conjunction of pre- and post-questionnaires makes it possible to quantify the impact of the tools."

"Unlike these pre-post questionnaires, the tools that only use post-game questionnaires have nothing to contrast the results with and could lose valuable information on the change in participants' perception of the topics addressed."

"The significant gap found between the high relevance of the social problem and the scarcity of resources for its treatment is also present in other fields: such is the case of cyberbullying, a relatively recent problem but of great impact on both the physical and emotional health of the victims (Kim & Leventhal, [234]) for which there are also very few digital resources that address the issue."

Limitations

"We tried to minimize the limitations, regarding the databases, by including a wide range of alternative terms (for both gender issues and digital resources) and modifying the query according to the needs of each database."

"Some of the resources found were not scientifically validated."

"A possible reason to explain the lack of associated scientific studies is the need for more economic support, since this situation occurs more frequently in resources developed in countries that are not considered to be of high or very high human development, according to the United Nations Development Program."

"As a final limitation, the diversity and variety of the tools and results of the studies included in our review limits the possibility to issue definitive general conclusions."

Conclusions
"The results obtained in the review are heterogeneous but can open a relevant analysis, as gender issues are increasingly visible and new tools are being developed to address them, particularly, digital resources like serious games."

"We found no other systematic literature reviews regarding games and digital resources to address gender issues."

"The evaluation studies carried out with the digital resources were varied, including the use of pre- and post-game questionnaires to provide a means to measure the effect of each intervention."

"Further research is needed regarding digital resources for gender education, validating the resources developed and establishing clear means to assess the impact they may have on their target players."

"These requirements would ensure that the benefits of games and digital resources for learning and awareness are adequately translated into their application in the field of gender equality education."

7.3.2 Digital Games, African Development, and Gender Equality: A Comparative Analysis of Family Values and Moraba [309]

This is a machine-generated summary of:

Fisher, Jolene: Digital Games, African Development, and Gender Equality: a Comparative Analysis of Family Values and Moraba [309]

Published in: Studies in Comparative International Development (2017)

Link to original: https://doi.org/10.1007/s12116-017-9244-z

If you want to cite the papers, please refer to the original.

For technical reasons we could not place the page where the original quote is coming from.

Abstract-Summary
"Digital Development Games (DDGs) have been designed to tackle a variety of international development issues, including gender inequality."

"This study uses critical textual analysis and a political economic approach to highlight how the development goals embedded in two games, Family Values and Moraba, both aimed at gender equality in Africa, were shaped by the large donor organizations that funded them."

"As smaller nongovernmental organizations (NGOs) and independent game developers look to large development institutions to support the production of

DDGs, research on the ways in which the ideologies of funding organizations get mapped onto games becomes increasingly important."

Resource Transfers and Power Dynamics: Money and Ideas Flow Downhill
"These big donor/small NGO relationships are not new: Since the New Policy Agenda, there has been a sustained emphasis on partnerships between bilateral and multilateral donor agencies and NGOs and grassroots organizations (GROs) in pursuing the work of development, specifically in terms of poverty alleviation, social welfare, and the development of civil society (Edwards and Hulme [235])."

"The relationships between institutional donors or international nongovernmental organizations (INGOs) and NGOs often result in asymmetric relationships among agents, with NGOs "placed in a subordinated position," to the larger funding institutions (Contu and Girei [236]: 222)."

"Even when compliance with a donor's specific ideology is not explicitly demanded, unequal power dynamics within the relationship makes it difficult for NGOs to separate out material and ideological resources (Contu and Girei [236]; Dar and Cooke [237]; Lewis [238]; Gwin [239])."

"Both donor/NGO relationships, as well as the individual organizations themselves, must be understood within the context of an international development industry that operates within a global political and economic system."

Family Values and USAID: Why Girls Should Help Themselves
"If the player chooses for Mercy to look at her father's business accounts (against his wishes), she finds an error in his bookkeeping and helps the family get more money, to which the father says, "I always said it...Keeping that girl in school would benefit us all!" "

"It is also evident that securing access to an education is the responsibility of the individual: The game encourages the player to make Mercy question the logic of her father, work around her family's poverty, and dedicate herself to studying extra hours in school."

"While the game's presentation of family dynamics that both lead to or preclude Mercy's access to education is at times quite nuanced, it is apparent that the only solution to gender equality offered by Family Values is for an individual woman to build her individual capacity by securing an education."

Moraba and the UN: When It Comes to Equality, Rights Are Key
"The games designed by Afroes address a range of development issues including gender-based violence, child abuse, and environmental and civil rights."

"To Family Values in which protagonists are young women only, players in Moraba get to choose and name a male or female character at the start of each game, creating a more inclusive environment for considering gender inequality and gender rights."

"Is a sampling of the questions, which cover issues of gender violence, emotional abuse, physical abuse, and legal rights: Q1.) Which question is not a sign of control from a partner?"

Conclusion
"This study works to illuminate the continuation of this process in the production of DDGs, tempering the narrative of games as "revolutionary" tools and instead examining how these games work to reinforce mainstream, and at times problematic, development thinking."

"Further, while this game builds on the UN's perspective that gender inequality is embedded within historical, political, and cultural structures, it asks individual players to navigate a complicated legal system fraught with obstacles related to gender, class, and race—thus, while assuming structural backing, it nevertheless predicates its solutions on individual action."

"To paint DDGs in broad strokes as "revolutionary" or "world changing" is to ignore the importance of the narrative the individual game is disseminating, as well as the historical use of mass communication tools within the international development industry."

"While the continued growth of mobile phones in the developing world presents opportunities for game use, mobile-based DDGs also present a new set of gendered challenges."

[Section 5]
"The possibilities have not been lost on the international development industry: at least 30 games designed specifically as development tools were created from 2005 to 2015, many of them supported by institutions such as the World Bank (WB), the United Nations (UN), and the United States Agency for International Development (USAID) (Fisher [240])."

"As more and more smaller nongovernmental organizations (NGOs) and independent game developers look to these institutions for financial support for the production of DDGs, research on the ways in which the broader development ideologies of funding organizations get mapped onto these games becomes increasingly important."

"The analysis presented here makes visible the ways in which the specific ideologies of each game's donor organization influence the issues and solutions to gender inequality that they present, and their implications for women in the developing world."

7.3.3 "It's Getting Difficult to Be a Straight White Man": Bundled Masculinity Grievances on Reddit [310]

This is a machine-generated summary of:

Hanson, Kenneth R.; Pascoe, C. J.; Light, Ryan: "It's Getting Difficult to Be a Straight White Man": Bundled Masculinity Grievances on Reddit [310]

Published in: Sex Roles (2022)

Link to original: https://doi.org/10.1007/s11199-022-01344-8

If you want to cite the papers, please refer to the original.

For technical reasons we could not place the page where the original quote is coming from.

Abstract-Summary

"This article examines a case of internet posts discussing social issues affecting men and masculinity."

"Analysis of 500 posts containing masculine coded language on the subreddit r/unpopularopinion suggests that masculinity, especially when intersected with straightness and whiteness, is discursively constructed in an imagined social hierarchy where the plight of straight white men is invisible."

"Three key findings emerged from this analysis: First, regardless of the particular social issue discussed, efforts to reduce social inequality were negatively evaluated on average."

"These findings suggest that criticisms of social equality are embedded within a discourse of threatened masculinity, straightness, and whiteness."

Changes in Masculinity

"For certain groups of American men, global economic shifts have been particularly painful."

"Given the association of American masculinity with occupational success, whiteness, heterosexuality, and economic self-sufficiency, this decline has been particularly sharp for working-class white men."

"Carlson suggests that contemporary American white men are trying to regain "masculine dignity" ([241], p. 24) in a culture where they no longer have access to what felt like a simpler time associated with a mythical town that reflects a certain set of racial, classed, gendered and sexual norms."

"For these men, this era has felt like one of increasing gun control, government regulation, and too many laws protecting gender, sexual, and racial/ethnic minorities (Hochschild ([242])."

"Some groups of white men frame white masculinity as a stigmatized identity in a way that clouds the gendered and racial advantages that often accompany such an identity (Bridges, [243])."

Masculinity Threat

"The concept of masculinity threat suggests that when a masculine identity is threatened men respond with compensatory behaviors, or behaviors that restore a claim to normative masculinity (Cassino & Besen-Cassino, [244]; Konopka and others, [245]; Vandello and others, [246])."

"In examining narratives about threatened masculinity among young men who espouse egalitarian ideas about gender, Munsch and Gruys ([247]) find that these men still support the subordination of women and other men in response to these threats."

"We suggest a similar model for understanding masculinity threat — under conditions of perceived threat at a large scale (e.g., cultural shifts), men engage a system of meaning that is organized and recognizable to in-group and out-group members alike, although perhaps in different, contested ways, that taps into the concerns of multiple identities and yet overall, its effect is to bolster claims to a shared idea about the status of men in society."

Studying Reddit

"Social scientists have been drawn to Reddit as a research site (Amaya and others, [248]; Davis & Graham, [249]; Duguay, [250])."

"Users, even without a subscription, can explore a range of general interest and topical subreddits, and research suggests that Reddit's user base is growing (Duguay, [250])."

"To studying Reddit's users, scholars have examined discursive contestations on Reddit over myriad social issues."

"Users on subreddits such as r/The_Red_Pill and other Manosphere sites argue that men are disadvantaged in modern society because of liberalism and feminism."

"In finding community with other like-minded users, these subreddits become hubs for anti-feminist sentiments."

"While not altogether untrue, as more Reddit users are men than women, this social dynamic often leads to misogynistic online behavior (Amaya and others, [248]; Massanari, [251]; Phillips, [252])."

"We study the subreddit r/unpopularopinion as a way to examine how Reddit users express unpopular, or canceled, views on relevant social topics."

Method

"We limited the sample to posts including masculine words."

"These data do not include posts where masculine words were absent from the title or main text but were included in the comments."

"To post data, we also wanted to quantitatively analyze user engagement."

"Given the scrape collected posts based on text specificity and not contextual usage, many posts did not relate to masculinity per se, or words were used in ways that matched their gendered meaning but were not germane to the study."

"Each post was also coded according to its valuation of the social issue(s) it discussed: positively, negatively, mixed, and random."

"Analysis began with assessing the distribution of codes within the sample of posts."

"We attend to the content of the posts in terms of their general themes and analyze how themes intersect with valuation codes and engagement."

Results

"The first is that the modal post expressed negative opinions about increasing social equality in ways that suggested white, straight cisgender men are victims of increasing social equality."

"The second type of negative post argues that social inequalities do exist but inverts typical claims about inequality such that privileged groups who benefit from social inequality (e.g., men, white people, straight people) are framed as victims."

"More broadly, these posts discussed sexuality, race, and relationships in ways suggesting that white, straight, cisgender men are the true victims of social equality."

"A post entitled, "Just because I'm a male, doesn't mean that my problems are invalid or any less important," suggests that men need support like other minoritized groups, Guys get too much shit for no reason."

Discussion

"Qualitative analysis of these posts reveals that masculinity, whiteness, and straightness are often framed as social disadvantages."

"Posts often invert claims of inequality to suggest that white, straight, cisgender men are the ones who suffer from inequality the most because of aggressive social shifts towards equality."

"Negative evaluations of social equality are amplified by the hundreds, and sometimes thousands, of comments on posts that argue straight white men are victims of cancel culture."

"Post after post suggests that whiteness is central to understanding how men understand their experiences of increasing equality as a gendered threat."

"These posts show how in non-experimental conditions, pressing social issues like gender, racial and sexual equality are yoked together with institutions and practices in a series of bundled grievances."

"The findings also reveal a persistent masculinity threat, which when bundled with whiteness and straightness, is articulated as straight white men feeling vilified by cultural narratives."

Conclusion

"This paper shows how one of Reddit's most visited subreddits, r/unpopularopinion, is a digital space where users voice concerns about current events and related social equality efforts."

"Although users discuss a wide array of topics, systematic analysis reveals that negative perceptions of increasing social equality are patterned in identifiable ways."

"Examination of these patterns suggests that social grievances are bundled together such that specific issues are used as evidence of the cultural devaluation of straight white men writ large."

"An essential step for understanding why people react negatively to perceived and actual social equality is realizing how social progress is weighed against the perceived threat it poses to distinct combinations of privileged social identities."

[Section 8]
"From Men's Rights Activists and the rise of misogynist internet spaces, what Ging ([253]) calls the Manosphere, to mass shootings fueled by male anger (Silva and others, [254]), to gendered critiques of what some refer to as the rise of "cancel culture," it seems that these shifting gendered meanings are not equally embraced by all men."

"To better understand the gendered dynamics of cancel culture, we examine what, exactly, some of these opinions might be by analyzing posts related to men and masculinity on an internet message board, or subreddit, entitled "Unpopular Opinions.""

"The discursive practices in this space may shed light on shifting definitions of masculinity that might be harder to capture in interview and observational data where men may present themselves as more politically liberal than they are, perhaps due to fears of social condemnation (Pettigrew and others, [255])."

7.3.4 Gender Differences in Teens' Digital Propensity and Perceptions and Preferences With Regard to Digital and Printed Text [311]

This is a machine-generated summary of:

Seok, Soonhwa; DaCosta, Boaventura: Gender Differences in Teens' Digital Propensity and Perceptions and Preferences With Regard to Digital and Printed Text [311]

Published in: TechTrends (2016)
Link to original: https://doi.org/10.1007/s11528-016-0134-4

If you want to cite the papers, please refer to the original.

For technical reasons we could not place the page where the original quote is coming from.

Abstract-Summary
"Predictors of digital propensity were investigated along with gender differences in the context of digital propensity and perceptions and preferences toward the reading of digital and printed text."

"Results revealed significant regression equations, showing that gender was a significant predictor of digital propensity, with males showing a stronger propensity for information and communication technology; the mean of males' digital propensity was significantly higher than that of females."

"The mean of females' perceptions toward digital reading was overall higher than that of males, and in the context of digital reading preferences, females read online more for entertainment and learning purposes, and read more selectively, whereas males read digital text more for comprehension purposes."

Introduction

"Studies have compared the digital reading experiences of the two formats at the postsecondary level (Eden and Eshet-Alkalai [256]; Kolikant [257]; Schmidt [258]; Stonier [259]), and although findings have been mixed, a number of studies have pointed to results favoring digital text in terms of perceptions, preferences, and learning outcomes."

"Kolikant [257], for example, found that students' reading preferences favored digital texts."

"Stonier [259], in turn, noted that teachers' perceptions were increasingly more positive toward digital literacy, integrative texts, and the use of technology tools."

"More research is needed that examines differences regarding perceptions, preferences, and learning outcomes of reading digital vs. printed text."

"Toward this goal, this study explored differences between the reading of digital and printed text in the context of gender."

"We then explored gender differences in the context of digital propensity and perceptions and preferences toward the reading of digital and printed text."

Method

"The study was conducted in 2014 at seven secondary schools in and near Seoul, South Korea."

"Among other reasons, South Korea was chosen as the site of the study because of its forward-thinking attitude towards technological advancement and promotion."

"South Korea offered an excellent setting from which to draw technically adept participants who also spoke English."

Materials

"Instruments comprised the Test of Silent Contextual Reading Fluency (2nd ed.) (Hammill, Wiederholt, and Allen [260]); the Digital Propensity Index (Nasah and others [261]); the Reading Observation Scale (Wiederholt, Hammill, and Brown [262]); and a survey designed to collect information on students' preferences for digital or printed text as well as students' demographics."

"The TOSCRF measures silent general reading comprehension, including vocabulary, of elementary- to secondary-level students (Hammill and others [260])."

"It allows teachers to rate their perceptions of their students' reading of printed text (Wiederholt and others [262])."

"An adapted version of the ROS was used in this investigation, letting students rate their own digital and printed text reading perceptions."

"The survey was specifically developed for this investigation to measure students' preferences for digital or printed text."

"Items comprising the survey were generated from a comprehensive synthesis of research measuring digital and printed material."

"English teachers participated because the students read both English and Korean, and to aid in the proper administration of the TOSCRF, which is in English."

Results

"The independent variables were age, gender, family annual income, education level, grade point average (GPA), native language reading grade point, and TOSCRF scores; the dependent variable was propensity toward the use of digital technology."

"For phase I, the independent variables were age, gender, income, education level, GPA, and reading grade point."

"Multiple statistical approaches were used to investigate this area of the study, to include an independent-sample t-test to explore gender differences in the context of digital propensity and digital reading perceptions."

"The digital reading perceptions of females had significantly higher means than those of males (with the exception of 4 out of 14 significant items)."

"The items for which gender showed a stronger preference for digital over printed texts were those related to reading for entertainment purposes, being selective with content, and skimming for information (nonlinear) when reading."

Discussion

"This study explored differences between the reading of digital and printed text in the context of gender."

"We next explored gender differences in the context of digital propensity and perceptions and preferences toward the reading of digital and printed text."

"With regard to the findings revealing that even though differences in gender may continue to favor male students regarding the use of ICT in general, in terms of reading digital material, females may show higher perceptions toward reading in this medium."

"Much more research is needed to examine the role of digital propensity in the context of perceptions and preferences of digital vs. printed text as a high propensity towards technology may not necessarily equate with a high reading level."

"Further research, however, is needed to generalize these findings in the context of digital propensity and the reading of digital vs. printed text to other people in different parts of the world."

Conclusions
"The findings presented here help strengthen results reported in existing research, while also contributing new information."

"The results revealed that gender was a significant predictor of digital propensity, with males showing a stronger propensity for technology and females holding higher perceptions toward digital reading while the two groups expressed different preferences in the reading of digital text."

"Most important, the findings contribute to a larger body of research supporting the suggestions that the gender gap may be narrowing, at least with regard to the use of technology in the context of reading."

7.3.5 Accessible Software Development: A Conceptual Model Proposal [312]

This is a machine-generated summary of:

Silva, João Sousa e; Gonçalves, Ramiro; Branco, Frederico; Pereira, António; Au-Yong-Oliveira, Manuel; Martins, José: Accessible software development: a conceptual model proposal [312]

Published in: Universal Access in the Information Society (2019)
Link to original: https://doi.org/10.1007/s10209-019-00688-5

If you want to cite the papers, please refer to the original.

For technical reasons we could not place the page where the original quote is coming from.

Abstract-Summary
"Despite the existing poor results on e-accessibility compliance, the mentioned citizens insist on using digital devices in their daily lives."

"Even though, in the last decade, multiple standards and regulations have been published towards indicating how to develop accessible digital user interfaces, there are still two major issues surrounding its implementation: the complexity and disparity of the documents containing the abovementioned norms, and also the lack of e-accessibility know-how by software experts."

"A proposal for an accessible software development model that encompasses e-accessibility incorporation as one of the development process activities has been presented."

"This model might represent a very interesting support tool for software development organizations and a novel resource for learning and training institutions to be able to improve their computer science and informatics students' skills on e-accessibility."

Introduction

"According to Bohman [263] and Klironomos and others [264], the majority of software experts have been demonstrating a serious knowledge gap when it comes to understanding the techniques and regulations associated with e-accessibility, thus creating barriers for those with some sort of disability or impairment, as they are unable to use the inaccessible software products and content, hence limiting their ability to become equal members of society."

"With this context in mind, the present research has been conducted to, drawing on a set of recommendations to be incorporated into software engineering classrooms and IT organizations, propose an accessible software development model."

"To achieve this goal, an analysis of both the existing scientific literature and the current international accessibility regulations and best practices has been done and merged with a focused analysis on existing software development models."

Related studies

"Despite the fact that the majority of existing research and practical work has been focusing on web-based software and computer software, the reality is that currently ICT users are giving particular attention to mobile-based applications that also need to be accessibility compliant [265]."

"As argued by Draffan and others [266] and Gonçalves and others [267], digital accessibility is also not suitably taught, nor applied, in academic environments, and hence, the need to change mentalities and develop not only the existing theory on the topic, but also to present improvements and developments to existing practical approaches when developing software."

"Continuing with the same example of visually impaired people, it is now possible to read a newspaper, accessing news at the same time as a sighted person."

"Several groups of people are now able to communicate, using social media, either because they are accessible by screen readers, or because it is possible to use written communication, or it is simply convenient if, for some reason, the person cannot leave the house."

Accessible software development: existing norms, regulations and standards

"Considering not only the abovementioned but also the existing literature arguments on the topic [268], we pose that the availability of documents regarding e-accessibility and the awareness on its relation with the type of user interface they "regulate" is extremely important not only to better understand the rules of engagement but also to improve the developers' perception on how to implement and validate e-accessibility."

"It is relevant to mention that when a developer keeps the same graphic style from the host operating system on his/her application, he/she is already increasing the level of usability (and by inherence the accessibility), since the interaction will be like the rest of the system [269]."

"When choosing not to use standard graphical components, the developer must be aware that he/she is wasting a lot of work that someone else has already done, during the platform's user interface (UI) development."

Developing accessible software

"As argued by Sánchez-Gordón, Moreno [270], despite the known relevance of the software accessibility topic, it is extremely important to acknowledge the necessity of not only incorporating it in software quality assurance processes and procedures, but also to understand how developers can simply merge their knowledge on how to use the accessibility tools and features of their software development tools, and the accessibility requirements that have been imposed early on in the project."

"In order to complement developers' compliance with existing accessibility and usability standards, software development teams need to have experts on testing not only the main software features but most importantly its conformity with existing regulations, standards and guidelines [271]."

"As one can easily understand from the abovementioned, software developers must not only understand the usability and accessibility guidelines, but also possess the necessary knowledge on how to test their developments against existing usability and accessibility standards and guidelines, thus ensuring their end-product is of the upmost quality."

An accessible software development model proposal

"Combining both agile and user-centred approaches makes sense, given the relevance of allowing for software analysts, software developers, software testers, user interface designers and software "owners", to actively and continuously collaborate in the pursuit of including accessibility concerns and patterns in software development projects [272]."

"The referred teams should also consider the need for the developed software to have a high level of maintainability, thus making it easy to ensure a continuous fulfilment of all requirements (including those related with accessibility and usability) [273]."

"Those who are responsible for performing the maintenance tasks should also help those who analyse, design and develop the software by giving them feedback on their efforts for compliance with existing e-accessibility and usability norms, and on the possibilities of evolution that might be considered for future developments."

Conclusions

"And highlighted, despite the relevance associated with the e-accessibility topic and the existence of multiple research projects focused on the topic, the presentation of the motives behind organizations' and governments' willingness to adopt e-accessibility and the combination of these motives with experts' perspectives on the topic might serve as the basis for future research aimed at not only further developing the techniques, methods and tools to develop accessible software, but also to further develop existing norms."

"This conceptual model could be used not only for supporting software development projects aimed at developing accessible software, but also as a basis for software engineering and development teaching and training activities, thus helping to reduce the gap between existing accessibility regulations, guidelines, tools and evaluation procedures and computer science and informatics students and IT specialists."

Bibliography

1. Smith, A. (2009). The Internet's Role in Campaign 2008 | Pew Research Center. Pew Internet. Retrieved February 10, 2019 (https://www.pewinternet.org/2009/04/15/the-internets-role-in-campaign-2008/).
2. Choi, J., Lee, J. K., & Metzgar, E. T. (2017). Investigating the effects of social media news sharing on the relationship between network heterogeneity and political participation. Computers in Human Behavior, 75, 25–31.
3. Lindberg, S. I., & Morrison, M. K. C. (2005). Exploring voter alignments in Africa: Core and swing voters in Ghana. The Journal of Modern African Studies, 43(4), 565–586.
4. Ichino, N., & Nathan, N. L. (2013). Crossing the line: Local ethnic geography and voting in Ghana. American Political Science Review, 107(2), 344–361.
5. Carlson, E. (2015). Ethnic voting and accountability in Africa: A choice experiment in Uganda. World Politics, 67(2), 353–385.
6. Lindberg, S. I., & Morrison, M. K. C. (2008). Are African voters really ethnic or clientelistic? Survey evidence from Ghana. Political Science Quarterly, 123(1), 95–122.
7. Pinkleton, B. E., Austin, E. W., & Fortman, K. K. J. (1998). Relationships of media use and political disaffection to political efficacy and voting behavior. Journal of Broadcasting & Electronic Media, 42(1), 34–49.
8. Alexander, M i G, & Wood, W. (2000). "Women, men, and positive emotions: A Social Role Interpretation. In A. H. Fischer (Ed.), gender and emotion (pp. 189–210). Cambridge: Cambridge University Press
9. Hamlin, A. (2005). Political dispositions and dispositional politics. ISSN 1749–9747. Manchester.
10. Valentino, N. A., Hutchings, V. L., Banks, A. J., & Davis, A. K. (2008). Is a Worried Citizen a Good Citizen? Emotions, Political Information Seeking, and Learning via the Internet. Political Psychology, 29(2), 247–273.
11. Protevi, J. (2009). Political affect: Connecting the social and the somatic. Minneapolis: University of Minnesota Press.
12. Parker, M. T., & Isbell, L. M. (2010). how i vote depends on how i feel. Psychological Science, 21(4), 548–550.
13. Iyengar, S., Sood, G., & Lelkes, Y. (2012). Affect, not ideologya social identity perspective on polarization. Public Opinion Quarterly, 76(3), 405–431.
14. Gan, C., Lee, F. L. F., & Li, Y. (2017). Social media use, political affect, and participation among university students in Urban China. Telematics and Informatics, 34(7), 936–947.
15. Bosnjak, M., Galesic, M., & Klicek, B. (2007). Determinants of online political participation in Croatia. Drustvena Istrazivanja, 17(4–5), 747–769.
16. Moreri, K., Fairburn, D. & James P. (2015). Technological solutions for citizens' participation into cadastral mapping. Paper presented at 27th International Cartographic Conference, Rio de Janeiro.
17. Lemmen, C., Bennet, R., McLaren, R, & Enemark, S. (2015b). A New Era in Land Administration Emerges. GIM International, January 2015.

18. Datta, M. (2015) All for a win-win relationship. GeoSpatial World. Accessed on March 17, 2016 at http://geospatialworld.net/uploads/magazine/March-2015-Geospatial-World-Magazine/files/24.html.
19. Makau, J., Rojas-Williams, S., Dobson, S., Ouma, S., Selebalo, C., Nyamweru, H., Antonio, D., Gitau, J., & Njogu, S. (2015) Empowering slum communities through geospatial technologies: experiences from Colombia, Kenya and Uganda. Paper prepared for presentation at the "2015 World Bank conference on land and poverty".
20. McLaren, R. (2013). Technology to promote transparency around land acquisitions. Evidence on demand helpdesk report. London: DFID.
21. Hagen, E. (2011). Mapping change: Community information empowerment in Kibera. Innovations, 6(1), 69–94.
22. Byamugisha, F. (2013). Securing Africa's land for shared prosperity: A program to scale up reforms and investments. Washington, DC: World Bank.
23. World Bank. (2011). ICT in agriculture e-sourcebook, accessed online at http://www.ictinagriculture.org/sourcebook/module-14-ict-land-administration-and-management.
24. Sæbø, O. (2012). E-government in Tanzania: Current Status and Future Challenges in: Scholl, H. J. et al (eds) Electronic Government: Proceedings of the 1th IFIP WG 8.5 International Conference, EGOV 2012, Kristiansand, Norway, September 3–6, 2012. Berlin Heidelberg: Springer.
25. Schuppan, T. (2008). E-Government in developing countries: Experiences from sub-Saharan Africa. Government Information Quarterly, 26(1), 118–127.
26. Otiso, K. M. (2005). Kenya's secondary cities growth strategy at a crossroads: Which Way Forward? GeoJournal, 62(1), 117–128.
27. Makoro, D. (2015). Use of land management information system (NLIMS). A case study of ministry of land, housing and urban development, Nairobi. Master's thesis, Technical University of Kenya.
28. Mbaria, C. N. (2015). Automation of Kenya's Land records. Powerpoint presentation, accessed online at http://slideplayer.com/slide/6005827/.
29. WTV. (2015). Groots Kenya Tackles Land Grabbing Solutions. Accessed online on November 4, 2015, at https://www.youtube.com/watch?v=C0f0D71obaA.
30. Auyero, J. (2000). Poor people's politics: Peronist survival networks and the legacy of Evita. Duke University Press.
31. Holland, A. C., & Palmer-Rubin, B. (2015). Beyond the machine: Clientelist brokers and interest organizations in Latin America. Comparative Political Studies, 48(9), 1186–1223.
32. Stokes, S. C., Dunning, T., Nazareno, M., & Brusco, V. (2013). Brokers, voters, and clientelism: The puzzle of distributive politics. Cambridge University Press.
33. Bourdieu, P. (1977). Outline of a theory of practice (R. Nice, Trans.). Cambridge University Press. https://doi.org/10.1017/CBO9780511812507
34. Martin, J. L. (2003). What is field theory? American Journal of Sociology, 109(1), 1–49.
35. Sewell, W. H. (2005). Logics of history: Social theory and social transformation. University of Chicago Press.
36. Gutiérrez, F. (2019). Clientelistic warfare: Paramilitaries and the state in Colombia (1982–2007). Peter Lang.
37. Castells, M. (2015). Networks of outrage and hope: Social movements in the Internet age. John Wiley & Sons.
38. Tufekci, Z. (2017). Twitter and tear gas: The power and fragility of networked protest. Yale University Press.
39. Epstein, B. (2018). The only constant is change: Technology, political communication, and innovation over time. Oxford University Press.
40. Howard, P. N. (2006). New media campaigns and the managed citizen. Cambridge University Press.
41. Karpf, D. (2012). The MoveOn effect: The unexpected transformation of American political advocacy. Oxford University Press.

42. Stromer-Galley, J. (2014). Presidential campaigning in the Internet age. Oxford University Press.
43. Kreiss, D. (2012). Taking our country back: The crafting of networked politics from Howard Dean to Barack Obama. Oxford University Press.
44. Chadwick, A. (2017). The hybrid media system: Politics and power. Oxford University Press.
45. Mahrenbach, Laura (2018) 'The Promises and Challenges of Digitalization and Development in Major Emerging Markets', Global Policy November: 1–11.
46. Ngetich, Dalmas (2018) 'Forget Bitcoin, This Is How China Plans to Dominate the $102 Trillion Blockchain Space', Ethereum World News, available at https://ethereumworldnews.com/china-dominate-blockchain/ (last accessed on 29 July, 2020).
47. Vnesheconombank (2018) 'Vnesheconombank and BRICS Development Banks Will Research the Implementation of Blockchain Technology', PR News Wire, available at https://www.prnewswire.com/news-releases/vnesheconombank-and-brics-development-banks-will-research-the-implementation-of-blockchain-technology-811880894.html (last accessed on 29 July, 2020).
48. Dahl, Robert A., and Edward R. Tufte. 1973. Size and democracy. Stanford: Stanford Univ. Press.
49. Peixoto, Tiago, and Jonathan Fox. 2016. Background paper—when does ICT-enabled citizen voice lead to government responsiveness? World development report.
50. Macintosh, Ann. 2004. Characterizing e-participation in policy-making. In Proceedings of the 37th Annual Hawaii International Conference on System Sciences, ed. Ralph H. Sprague. Los Alamitos: IEEE Computer Soc.
51. Mossberger, Karen, Caroline J. Tolbert, Daniel Bowen, and Benedict Jimenez. 2012. Unraveling different barriers to Internet use. Urban Affairs Review 48:771–810.
52. The World Bank. (2006). Kazakhstan e-government program and the road ahead. Washington, DC: World Bank (a joint economic research program with the Government of Kazakhstan).
53. Brown, D. (2005). Electronic government and public administration. International Review of Administrative Sciences, 71(2), 241–254.
54. Tandon, H. (2005). e-governance: an Indian perspective. Policy & Society, 24(3), 1–28.
55. Agarwal, P. (2008). Economic growth and poverty reduction: evidence from Kazakhstan. Asian Development Review, 24(2), 90–115.
56. UNDP (United Nations Development Program). (2007). Human development report 2007/2008 [Kazakhstan]. Available at: http://hdrstat.undp.org/countries/country_fact_sheets/cty_fs_KAZ.html, last accessed November 13, 2008.
57. Haque, M. S. (2002). E-Governance in India: its impact on relations among citizens, politicians and public servants. International Review of Administrative Sciences, 68, 231–250.
58. Hasse, D. (2002). E-Government: The future of federal and municipal institution. An unpublished Master's thesis. Magdeburg: University of Magdeburg.
59. Heeks, R. (1998). Information technology and public sector corruption. Information System and Public Sector Management Working Paper Series. No. 4. Manchester: University of Manchester.
60. Hesmondhalgh, D., and L. Meier. 2015. Popular Music, Independence, and the Concept of the Alternative in Contemporary Capitalism. In Media Independence: Working with Freedom or Working for Free?, ed. J. Bennett and N. Strange, 94–70. New York: Routledge.
61. Carter, D., and I. Rogers. 2014. Fifteen Years of 'utopia': Napster and Pitchfork as Technologies of Democratization. First Monday. https://doi.org/10.5210/fm.v19i10.5543 .
62. Hatch, C. 1962. Music for America: A Critical Controversy of the 1850s. American Quarterly 14 (4): 578–586.
63. Gennari, J. 2006. Blowin' Hot and Cool: Jazz and Its Critics. Chicago: The University of Chicago Press.
64. Hersch, C. 1995. "Let freedom ring!" : Free Jazz and African-American Politics. Cultural Critique 32: 97–123.

65. Bonotti, Matteo. 2017. Partisanship and political liberalism in diverse societies. Oxford: Oxford University Press.
66. Rawls, John. 2001. Justice as fairness. A restatement. Cambridge: Belknap Press. Edited by Erin Kelly.
67. Hoffmann, Anna Lauren. 2020. Rawls, information technology, and the sociotechnical bases of self-respect. In The oxford handbook of philosophy of technology, ed. Shannon Vallor. Oxford: Oxford University Press. https://doi.org/10.1093/oxfordhb/9780190851187.013.15.
68. Tufekci, Zeynep. 2014. Engineering the public: big data, surveillance and computational politics. First Monday 19 (7). https://doi.org/10.5210/fm.v19i7.4901.
69. Bay, Morten. 2018. The ethics of psychometrics in social media: a Rawlsian approach. ACM Transactions on Social Computing. https://doi.org/10.1145/3281450 .
70. Luhmann, Niklas. 2013. Theory of Society. Vol. 2. Stanford: Stanford University Press.
71. Münte, Peter. 2017. Improving modern society: governing science and technology by engineered participation. In Society, regulation and governance: new modes of shaping social change?, ed. Regine Paul, Marc Mölders, Alfons Bora, Michael Huber, and Peter Münte, 166–180. Cheltenham: Edward Elgar.
72. Bröckling, Ulrich. 2016. Zukunftsmanagement zwischen Planung, Selbstorganisation und Prävention. In Die neue Wirklichkeit. Semantische Neuvermessungen und Politik seit den 1970er-Jahren, ed. Ariane Leendertz, Wencke Meteling, 269–280. Frankfurt a. M.: Campus.
73. Mölders, Marc. 2014. Irritation expertise. Recipient design as instrument for strategic reasoning. European Journal of Futures Research 2(1):32.
74. Hinnant, C., & Moon, J. (2001) The pursuit of e-Government: What determines perceptions of e-Government effectiveness. 6th National Public Management Research Conference, Indiana University, Bloomington, IN.
75. Dunleavy, P., & Margetts, H. 2000. The Advent of Digital Government: Public Bureaucracies and the State in the Internet Age. Paper for the Annual Conference of the American Political Science Association, Washington, 4 September, 2000.
76. Huse, M. 2003. Renewing management and governance: New paradigms of governance? Journal of Management and Governance, 7: 211–221.
77. Mayntz, R. 1998. New Challenges to governance theory. Jean Monet Chairs. Papers No. 50. European University Institute.
78. Mayntz. 2001. El Estado y la sociedad civil en la gobernanza moderna. Reforma y Democracia, 21: 7–22.
79. Rhodes, M. 1997. Understanding governance. Policy networks, governance, reflexivity and accountability. Buckingham/Philadelphia: Open University Press.
80. Grandori, A. 2000. Conjectures for a new research agenda. Journal of Management and Governance, 4: 1–9.
81. Bimber, B. 1999. The internet and citizen communication with government: Does the medium matter? Political Communication, 16: 409–428.
82. Barber, B. 1998. Three scenarios for the future of technology and strong democracy. Political Science Quarterly, 113(4): 573–589.
83. Fountain, J. 2001. Building the virtual state information technology and institutional change. Washington, D.C: Brookings Institution Press.
84. Mayntz, R. 2005. "Nuevos desafíos para la teoría de la gobernanza", in Cerrillo, Agusti (Ed) La gobernanza hoy: 10 textos de referencia. Madrid: INAP.
85. Jain, A. (2004). Using the Lens of Max Weber's Theory of Bureaucracy to Examine E-Government Research. Proceedings of the 37th Hawaii International Conference on System Sciences, January 4–8, Big Island, Hawaii.
86. Fountain, J. 2005. Central Issues in the Political Development of the Virtual State. Paper presented on The Network Society and the Knowledge Economy: Portugal in the Global Context, Lisbon, March 4–5.
87. Lee, T. (2005). The impact of perceptions of interactivity on customer trust and transaction intentions in mobile commerce. Journal of Electronic Commerce Research, 6(3), 165.

88. Mossberger, K., Wu, Y., & Crawford, J. (2013). Connecting citizens and local governments? Social media and interactivity in major US cities. Gov Inf Q, 30(4), 351–358.
89. Sundar, S. S., Bellur, S., Oh, J., Jia, H., & Kim, H. S. (2014). Theoretical importance of contingency in human-computer interaction effects of message interactivity on user engagement. Communication Research, 0093650214534962.
90. Teo, H. H., Oh, L. B., Liu, C., & Wei, K. K. (2003). An empirical study of the effects of interactivity on web user attitude. International journal of human-computer studies, 58(3), 281–305.
91. Burns, R. (2015). Rethinking big data in digital humanitarianism: practices, epistemologies, and social relations. GeoJournal, 80(4), 477–490.
92. Crawford, K., & Finn, M. (2015). The limits of crisis data: analytical and ethical challenges of using social and mobile data to understand disasters. GeoJournal, 80(4), 491–502.
93. Read, R., Taithe, B., & Mac Ginty, R. (2016). Data hubris? Humanitarian information systems and the mirage of technology. Third World Q, 37(8), 1314–1331.
94. Taylor, L., & Broeders, D. (2015). In the name of development: power, profit and the datafication of the global south. Geoforum, 62, 229–237.
95. Waters, R., & Feneley, K. L. (2013). Virtual stewardship in the age of new media: have nonprofit organizations' moved beyond Web 1.0 strategies? International Journal of Nonprofit Voluntary Sector Margeting, 18(3), 216–230.
96. Edwards, H. R., & Hoefer, R. (2010). Are social advocacy groups using web 2.0 effectively? Journal of Policy Practice, 9, 220–239.
97. Heberle, R. (1951). Social movements: Introduction to political sociology. New York: Appleton-Century-Crofts, Inc.
98. Horsely, J. P. (2007). Chapter 2: Toward a more open China? In A. Florini (Ed.), The right to know: Transparency for an open world (pp. 54–91). New York: Columbia University Press.
99. Zhou, H. (2007). Chapter 3: Open government in China. In A. Florini (Ed.), The right to know: Transparency for an open world (pp. 92–115). New York: Columbia University Press.
100. Zhou, H., Sheng, K., Ouyang, W., & Qin, H. (2007). Reviews on the disclosure of government information regulation. Journal of Legal and Economic Studies, 8, 202–228.
101. Fung, A., Graham, M., & Weil, D. (2008). Full disclosure: The perils and promise of transparency. Journal of Policy Analysis and Management, 27(1), 218–221.
102. Carter, M. P., & Yanbin, L. (2008). Comparative research on the system of access to government information in the EU and China. Beijing: China Law Press.
103. Tan, Y. (2014). Transparency without democracy: The unexpected effects of China's environmental disclosure policy. Governance, 27(1), 37–62.
104. Wang, X. (2011). Practice of information disclosure system and its external environment. Nankai Journal (Philosophy, Literature and Social Science Edition), 2, 65–71.
105. Zhou, H. (2009). Some problems and policy suggestions concerning the implementation of "Regulation of Government Information Disclosure" . Chinese Public Administration, 7, 11–14.
106. Sun, Y. (2009). Government information disclosure, public policy agenda and governance based on participation. Chinese Public Administration, 2, 64–67.
107. Zhu, H., & Yu, K. (2012). Citizen satisfaction assessment of government information disclosure. Library and Information Service, 56(3), 130–134.
108. The World Bank: Sustainable Development in a Dynamic World: Transforming Institution, Growth and Quality of Life. The World Bank, Washington, DC (2003)
109. Jaeger, P.T.: Deliberative democracy and the conceptual foundation of electronic government. Gov. Inf. Q. 22, 702–719 (2005)
110. Kim, S., Kim, H.J., Lee, H.: An institutional analysis of an e-government system for anti-corruption: the case of OPEN. Gov. Inf. Q. 26(2009), 42–50 (2009)
111. Gunter, B.: Advances in e-democracy: overview. In: Gunter, B. (ed.) Special issue of Aslib Proceedings: New Information Perspectives, vol. 58, no. 5, pp. 361–370. Emerald Group Publishing Limited, Great Britain (2006)

112. Gauld, R., Gray, A., McComb, S.: How responsive is E-Government? Evidence from Australia and New Zealand. Gov. Inf. Q. 2009(26), 69–74 (2009)
113. Bello, I. E., & Rilwani, L. M. (2016). Quantitative assessment of remotely sensed data for landcover change and environmental management. Indonesian Journal of Geography, 48(2), 135–144. https://jurnal.ugm.ac.id/ijg/article/view/17629. Accessed Dec. 1, 2016.
114. Dekolo, S. O., & Oduwaye, A. O. (2011). Managing the Lagos Megacity and Its Geospatial Imperative. In International Archives of the Photogrammetry, Remote Sensing and Spatial Information Sciences, 38-4/C21, 121–128, 28th Urban Data Management Symposium (UDMS 2011), September 28–30, 2011, Delft, The Netherlands.
115. Guo, H., Huang, Q., Li, X., Sun, Z. & Zhang, Y. (2013). Spatiotemporal analysis of urban environment based on the vegetation–impervious surface–soil model. Journal of Applied Remote Sensing, 8(1), 084597. https://doi.org/10.1117/1.JRS.8.084597 .
116. Obanuso, A., & Salihu, R. M. (2015). The role of civil society in enhancing democracy in Nigeria. International Journal of governance and development, 4(2), 171–186.
117. Groot, R. (1998). The role of national mapping agencies in the development of national spatial data infrastructure.In Paper presented to the International Conference of Spatial Information Science and Technology, Wuhan.
118. Bello, I. E., Nwagwu, C. J. & Edobor, W. W. (2013b). Application of Geo-Information Technology to National Security, Conflict Resolution and Transborder Cooperation: A Case Study of Nigeria-Sao Tome and Principe. International Affairs and Global Strategy, 17, 25–37. Accessed December 2, 2016 from http://www.iiste.org/Journals/index.php/IAGS/article/view/9614/9726.
119. Lagos State Government. (2011). Lagos State Press Breifing. Retrieved March 10, 2011, from Lagos State Government: http://www.lagosstate.gov.ng/. Accessed December 16, 2016.
120. Omofonmwan, S. I. (2016). Spatio-social indicators model: A panacea for realistic measurement of socio-economic development. In 56th Inaugural lecture delivered on the 27th of October, 2016 at Ambrose Alli University, Ekpoma, Nigeria.
121. Campbell C (2017) Anita Sarkeesian's Astounding "Garbage Human" Moment. Polygon. 27 June 2017. https://www.polygon.com/features/2017/6/27/15880582/anita-sarkeesian-garbage-human-vidcon-interview. Accessed 23 April 2020
122. Dewey C (2016) 'This horrifying and newly trendy online-harassment tactic is ruining careers'. The Washington Post. https://www.washingtonpost.com/news/the-intersect/wp/2016/04/11/this-horrifying-and-newly-trendy-online-harassment-tactic-is-ruining-careers/. Accessed 23 April 2020
123. McKibben B (2016) 'embarrassing photos of me, thanks to my right-wing stalkers'. The New York Times. https://www.nytimes.com/2016/08/07/opinion/sunday/embarrassing-photos-of-me-thanks-to-my-right-wing-stalkers.html?smid=tw-share&_r=1. Accessed 23 April 2020
124. IANS (2019) Facebook now 'fifth estate', take it seriously: Mark Zuckerberg. Gulf News. https://gulfnews.com/technology/facebook-now-fifth-estate-take-it-seriously-mark-zuckerberg-1.1571384459114. Accessed 23 April 2020
125. Jacquot, Sophie, and Cornelia Woll. 2003. Usage of European integration-Europeanisation from a sociological perspective. European Integration Online Papers (EIoP) 7 (12).
126. Calista, D. J., & Melitski, J. (2007). E-government and e-governance: Converging constructs of public sector information and communications technologies. Public Administration Quarterly, 87–120.
127. Moon, M. J. (2018). Evolution of co-production in the information age: Crowdsourcing as a model of web-based co-production in Korea. Policy and Society, 37(3), 294–309.
128. Manoharan, A. P., & Mossey, S. (2019). CITIZEN PARTICIPATION. E-Government and Information Technology Management: Concepts and Best Practices, 98.
129. Holzer, M., & Manoharan, A. (2016). Digital governance in municipalities worldwide. N.p.: National Center for Public Performance.
130. Manoharan, A. (2013). A study of the determinants of county e-government in the United States. The American Review of Public Administration, 43(2), 159–178.

131. Moon, M. J. (2002). The evolution of e-government among municipalities: rhetoric or reality? Public Administration Review, 62(4), 424–433.
132. Zheng, Y., & Manoharan, A. P. (2016). The influence of government capacity on e-services diffusion at municipal level in New Jersey. International Journal of Public Administration in the Digital Age (IJPADA), 3(4), 1–9.
133. Melitski, J., & Calista, D. (2016). E-government and E-governance best practices in cities and countries compared between 2003 and 2012: Fad or diffused innovation?. Public Administration Quarterly, 913-948.
134. Youngs, Gillian (2007) Global Political Economy in the Information Age: Power and Inequality, London: Routledge.
135. De Hert, P. J. A., & Sprokkereef, A. C. J. (2012). Second generation biometrics: The ethical, legal and social context. In E. Mordini & D. Tzovaras (Eds.), Biometrics, Privacy and Agency (pp. 81–101). Berlin: Springer.
136. Spahn, A. (2013). Ideas in motion. Moralizing mobility? Persuasive technologies and the ethics of mobility. Transfer, 3(2), 108–115.
137. Willems SM et al (2019) The potential use of big data in oncology. Oral Oncol 98:8–12. https://doi.org/10.1016/j.oraloncology.2019.09.003
138. Cato KD, Bockting W, Larson E (2016) Did i tell you that? Ethical issues related to using computational methods to discover non-disclosed patient characteristics. J Empir Res Hum Res Ethics 11(3):214–219. https://doi.org/10.1177/1556264616661611
139. Nicoletta M et al (2018) Intelligenza Artificiale e Big Data in ambito medico: prospettive, opportunità, criticità. JAMD 21(3). Available at: https://www.jamd.it/wp-content/uploads/2018/11/2018_03_03.pdf
140. Crenshaw K (1989) Demarginalizing the intersection of race and sex: a black feminist critique of antidiscrimination doctrine, feminist theory and antiracist politics. Univ Chicago Legal Forum 140:139–167. Available at: https://chicagounbound.uchicago.edu/uclf/vol1989/iss1/8
141. Gilligan C (1993) In a different voice: psychological theory and women's development. Harvard University Press, Cambridge, MA
142. Cirillo D et al (2020) Sex and gender differences and biases in artificial intelligence for biomedicine and healthcare. npj Digit Med 3(81). https://doi.org/10.1038/s41746-020-0288-5
143. Ferryman K, Pitcan M (2018) Fairness in precision medicine. Data & Society. Available at: https://datasociety.net/output/fairness-in-precision-medicine/
144. Casper MJ, Moore LJ (2009) Missing bodies: the politics of visibility. NYU Press, New York
145. Simmler M, Canova G, Schedler K (2021) Smart criminal justice: phenomena and normative requirements. Int Rev Admin Sci. https://doi.org/10.1177/00208523211039740
146. Završnik A (2019) Algorithmic justice: algorithms and big data in criminal justice settings. Eur J Criminol. https://doi.org/10.1177/1477370819876762
147. Carvalho H (2016) The preventive turn in criminal law. University Press, Oxford
148. Ienca, M., & Vayena, E. (2020). On the responsible use of digital data to tackle the COVID-19 pandemic. Nature Medicine, 26, 463–464. https://doi.org/10.1038/s41591-020-0832-5 .
149. McGee, P., Murphy, H. and Bradshaw, T. (2020). Coronavirus apps: the risk of slipping in to a surveillance state. Financial Times, April 28 https://www.ft.com/content/d2609e26-8875-11ea-a01c-a28a3e3fbd33. Accessed 25 June 2020.
150. Verily. (2020). New Baseline COVID-19 Research Project launches to advance scientific understanding of virus, with initial focus on antibody testing. May 18 https://blog.verily.com/2020/05/new-baseline-covid-19-research-project.html?utm_source=STAT+Newsletters&utm_campaign=3fdf688da3-health_tech_COPY_01&utm_medium=email&utm_term=0_8cab1d7961-3fdf688da3-151653869. Accessed 25 June 2020.
151. Frohlich, H., et al. (2018). From hype to reality: Data science enabling persnalized medicine. BMC Medicine, 16, 150. https://doi.org/10.1186/s12916-018-1122-7 .
152. Kelion, L. (2020). NHS rejects Apple-Google coronavirus app plan. BBC, April 27. https://www.bbc.com/news/technology-52441428. Accessed 25 June 2020.

153. Leprince-Ringuet, D. (2020). The world's first contact-tracing app using Google and Apple's API goes live. ZDNet, May 28. https://www.zdnet.com/article/the-worlds-first-contact-tracing-app-using-google-and-apples-api-goes-live/. Accessed 25 June 2020.
154. Parker, M. J., Fraser, C., Abeler-Dörner, L., & Bonsall, D. (2020). Ethics of instantaneous contact tracing using mobile phone apps in the control of the COVID-19 pandemic. Journal of Medical Ethics, 46(7), 427–431.
155. Machado, G.M. 1989. O Discurso Realista em Flaubert e em Machado de Assis. Revista de Letras 29: 55–70.
156. Miles, Th.J., and C.R. Sunstein. 2008. The New Legal Realism. University of Chicago Law Review 75 (2): 831–851.
157. Leiter, B. 2011. Naturalized Jurisprudence and American Legal Realism Revisited. Law and Philosophy 30: 499–516.
158. Leiter, B. 2013. Legal Realisms, Old and New. Valparaiso University Law Review 47 (4): 67–81.
159. Martin, M. 1997. Legal realism: American and Scandinavian. New York: Peter Lang.
160. Pound, R. 1931. The Call for a Realist Jurisprudence. Harvard Law Review 44 (5): 697–711. https://doi.org/10.2307/1331791 .
161. Gromova, E.A., and S.A. Petrenko. 2023. Quantum Law: The beginning. Journal of Digital Technologies and Law 1: 1.
162. Aletras, N., D. Tsarapatsanis, D. Preoţiuc-Pietro, and V. Lampos. 2016. Predicting Judicial Decisions of the European Court of Human Rights: a Natural Language Processing perspective. PeerJ Computer Science 2: e93. https://doi.org/10.7717/peerj-cs.93 .
163. Liu, M., Ch. Li, Sh. Wang, and Q. Li. 2023. Digital Transformation, Risk-taking, and Innovation: Evidence from Data on Listed Enterprises in China. Journal of Innovation & Knowledge 8: 100332. https://doi.org/10.1016/j.jik.2023.100332 .
164. Ferreira, Daniel, Cristiane Giovannini, Elizaveta Gromova, and Gustavo Schmidt. 2022. Arbitration Chambers and Trust in Technology Provider: Impacts of Trust in Technology Intermediated Dispute Resolution Proceedings. Technology in Society 2 (68): 101872. https://doi.org/10.1016/j.techsoc.2022.101872 .
165. Nascimento, S.N., and D.G.D. Martins. 2022. Smart Contracts: Security Issues and Further Development in Brazil. International Journal of Law in Changing World 1 (2): 26–45. https://doi.org/10.54934/ijlcw.v1i2.22.
166. Ambrogi, R. 2018. This Tech Can Turn the Tables in Litigation. Above The Law. https://bit.ly/358X4qn. Accessed 28 Jan 2023.
167. Brayne, S. (2017). Big Data Surveillance: The Case of Policing. American Sociological Review, 82(5), 977–1008. https://doi.org/10.1177/0003122417725865
168. Eubanks, V. (2017). Automating Inequality. How High-Tech Tools Profile, Police, and Punish the Poor. St. Martin's Press.
169. Ferguson, A. G. (2017). The Rise of Big Data Policing: Surveillance, Rance, and the Future of Law Enforcement. New York University Press.
170. Osoba, O., & Welser IV, W. (2017). An intelligence in our image : the risks of bias and errors in artificial intelligence. https://search.ebscohost.com/login.aspx?direct=true&scope=site&db=nlebk&db=nlabk&AN=1510300
171. Garfinkel, H. (1967). Studies in Ethnomethodology (1st ed.). Prentice Hall.
172. Mair, M., Greiffenhagen, C., & Sharrock, W. W. (2015). Statistical Practice: Putting Society on Display. Theory, Culture & Society, 33(3), 51–77. https://doi.org/10.1177/0263276414559058
173. Bittner, E. (1967). The Police on Skid-Row: A Study of Peace Keeping. American Sociological Review, 32(5), 699–715. https://doi.org/10.2307/2092019
174. Bratton, W. J. (1999). Great Expectations: How Higher Expectations for Police Departments Can Lead to a Decrease in CrimeNo Title. In R. H. Langworthy (Ed.), Measuring What Matters. Proceedings From the Policing Research Institute Meetings (p. 224). U.S. Department of Justice. Office of Justice Programs.

175. Manning, P. (2008). The technology of policing: Crime mapping, information technology, and the rationality of crime control. The Technology of Policing: Crime Mapping, Information Technology, and the Rationality of Crime Control, 1–322.
176. Ferguson, A. G. (2014). Big Data and Predictive Reasonable Suspicion. Criminal Law EJournal.
177. Benson, D. (1993). The police and information technology. In G. Button (Ed.), Technology in working order: Studies of work, interaction and technology (pp. 81–97). Routledge.
178. UN General Assembly. 2015. Transforming our world: The 2030 Agenda for sustainable development. https://sustainabledevelopment.un.org/post2015/transformingourworld. Accessed 18 April 2018.
179. Broadband Commission. 2016. The state of broadband: Broadband catalyzing sustainable development. http://www.broadbandcommission.org/Documents/reports/bb-annualreport2016.pdf. Accessed 18 April 2018.
180. ITU. 2017. ICT facts and figures 2017. https://www.itu.int/en/ITU-D/Statistics/Documents/facts/ICTFactsFigures2017.pdf. Accessed 18 April 2018.
181. Goldstein, Joshua. 2014. The Internet "Access Trap" in Developing Countries. https://freedom-to-tinker.com/2014/01/04/the-internet-access-trap-in-developing-countries/. Accessed 18 April 2018.
182. Iansiti, Marco and Karim R. Lakhani. 2017. Managing our hub economy. https://hbr.org/2017/09/managing-our-hub-economy. Accessed 18 April 2018.
183. Jain, Mayank. 2017. Fintech tracker: Your phone and FB account can help track your credit score. https://www.bloombergquint.com/business/2017/01/28/fintech-tracker-your-phone-and-facebook-account-can-help-improve-your-credit-score. Accessed 18 April 2018.
184. Scholz, Trebor. 2016. Platform cooperativism: Challenging the corporate sharing economy. New York: Rosa Luxemburg Stiftung, http://www.rosalux-nyc.org/wp-content/files_mf/scholz_platformcoop_5.9.2016.pdf. Accessed 18 April 2018.
185. Turner, Adair. 2017. Capitalism in an age of robots. http://azimpremjiuniversity.edu.in/SitePages/pdf/ADAIRS-FINAL.pdf. Accessed 18 April 2018.
186. Saxer, Mark. 2017a. The future of work in Asia: How can India create livelihoods in the digital age? New Delhi: Friedrich-Ebert-Stiftung India Office. http://library.fes.de/pdf-files/bueros/indien/13226.pdf. Accessed 18 April 2018.
187. Vaughan, Adam. 2015. How viral cat videos are warming the planet. https://www.theguardian.com/environment/2015/sep/25/server-data-centre-emissions-air-travel-web-google-facebook-greenhouse-gas. Accessed 18 April 2018.
188. Hern, Alex. 2017. Bitcoin mining consumes more electricity than the whole of Ireland. https://www.theguardian.com/technology/2017/nov/27/bitcoin-mining-consumes-electricity-ireland. Accessed 18 April 2018.
189. APWLD. 2013. Bangkok civil society declaration: From inclusive to just development. http://apwld.org/bangkok-civil-society-declaration-from-inclusive-to-just-development/. Accessed 18 April 2018.
190. Saxer, Marc. 2017b. The Human Economy: The social democratic path to decent livelihoods and development in digital capitalism. New Delhi: Friedrich-Ebert-Stiftung India. http://library.fes.de/pdf-files/bueros/indien/13561.pdf. Accessed 18 April 2018.
191. Singh, Parminder Jeet. 2017. Digital industrialisation in developing countries—A review of the business and policy landscape. London: Commonwealth Secretariat, http://www.itforchange.net/sites/default/files/1468/digital_industrialisation_in_developing_countries.pdf. Accessed 18 April 2018.
192. Web Foundation. 2017. Algorithmic accountability: Applying the concept to different country contexts. https://webfoundation.org/docs/2017/07/Algorithms_Report_WF.pdf. Accessed 18 April 2018.
193. Cohen, Julie. 2017. Law for the platform economy. (22 June 22). UC Davis Law Review, Forthcoming. https://ssrn.com/abstract=2991261.

194. James, Deborah. 2017b. MSME associations reject MSMEs work program and E-commerce rules in WTO. http://ourworldisnotforsale.net/2017/R_MSMEs_reject.pdf. Accessed 18 April 2018.
195. Maclean, Don, Ben Akoh and Bjonar Egede-Nissen. 2010. ICTs, sustainability and the green economy. Global Information Society Watch 2010—Focus on ICTs and Environmental Sustainability, https://www.giswatch.org/sites/default/files/gisw2010thematicictssustainability_en.pdf. Accessed 18 April 2018.
196. Mobbs, Paul. 2010. Introduction., Global information society watch 2010—Focus on ICTs and environmental sustainability. https://giswatch.org/report-introduction/2010-icts-and-environmental-sustainability/introduction. Accessed 18 April 2018.
197. Gurumurthy, Anita and Nandini Chami. 2017. A feminist action framework on development and digital technologies, APC-201705-WRP-I-EN-DIGITAL-267. https://www.apc.org/sites/default/files/FeministActionFrameworkOnDevelopmentAndDigitalTechnologies.pdf. Accessed 18 April 2018.
198. Christofides E, Muise A, Desmarais S (2009) Information disclosure and control on Facebook: are they two sides of the same coin or two different processes? CyberPsychol Behav 12(3):341–345. https://doi.org/10.1089/cpb.2008.0226
199. Luchenski, S., Balasanthiran, S., Marston, C., Sasaki, K., Majeed, A., Bell, D., et al. (2012). Survey of patient and public perceptions of electronic health records for healthcare, policy and research: Study protocol. BMC Medical Informatics and Decision Making,. https://doi.org/10.1186/1472-6947-12-40 .
200. Mill, J. S. (1860). On liberty (Vol. 25). Harvard: Harvard Classics.
201. European Court of Justice (2010). Judgment of 9 November 2010, Volker und Markus Schecke and Eifert, Joined Cases C-92/09 and C-93/09. European Court Reports 2010 Page I-11063.
202. European Court of Justice, Judgment of 24 November 2011, Asociación Nacional de Establecimientos Financieros de Crédito (ASNEF) and Federación de Comercio Electrónico y Marketing Directo (FECEMD) v Administración del Estado, Joined cases C-468/10 and C-469/10. Not yet published in the European Court Reports.
203. Mak, C. (2008). Fundamental rights in European contract law. A comparison of the impact of fundamental rights on contractual relationships in Germany, the Netherlands, Italy and England. Alphen aan den Rijn: Kluwer Law International.
204. Commission Proposal (2008), Proposal for a directive of the European Parliament and of the Council on consumer rights, COM (2008) 614/3.
205. Commission (2006). Green paper on the review of the consumer acquis, 8 February 2007, COM (2006) 744 final.
206. Lonn, S., Aguilar, S. J., & Teasley, S. D. (2013). Issues, challenges, and lessons learned when scaling up a learning analytics intervention. Presented at the LAK '13: Proceedings of the Third International Conference on Learning Analytics and Knowledge.
207. Daniels, G. (1952). The "Average Man" ? (pp. 1–12). Air Force Aerospace Medical Research Lab Wright-Patterson Afb OH.
208. Worsley, M., & Blikstein, P. (2015). Leveraging multimodal learning analytics to differentiate student learning strategies (pp. 360–367). Presented at the Proceedings of the Fifth International Conference on Learning Analytics And Knowledge, New York: ACM Press. https://doi.org/10.1145/2723576.2723624.
209. Diana, N., Eagle, M., Stamper, J., Grover, S., Bienkowski, M., & Basu, S. (2017). An instructor dashboard for real-time analytics in interactive programming assignments (pp. 272–279). Presented at the Seventh International Learning Analytics & Knowledge Conference, New York: ACM Press. https://doi.org/10.1145/3027385.3027441.
210. Verbert, K., Duval, E., Klerkx, J., Govaerts, S., & Santos, J. L. (2013). Learning analytics dashboard applications. American Behavioral Scientist, 57(10), 1500–1509. https://doi.org/10.1177/0002764213479363 .
211. West, D. M. (2012). Big data for education: data mining, data analytics, and web dashboards. Brookings (pp. 1–11).

212. Cutumisu, M., Blair, K. P., Chin, D. B., & Schwartz, D. L. (2015). Posterlet: a game-based assessment of children's choices to seek feedback and to revise. Journal of Learning Analytics, 2(1), 49–71.
213. Roll, I., & Winne, P. H. (2015). Understanding, evaluating, and supporting self-regulated learning using learning analytics. Journal of Learning Analytics, 2(1), 7–12.
214. Gusky, T. R. (2007). Closing achievement gaps: revisiting Benjamin S. Bloom's 'Learning for Mastery'. Journal of Advanced Academics, 19(1), 8–31.
215. Hoel, T., & Chen, W. (2016). Privacy-driven design of learning analytics applications: exploring the design space of solutions for data sharing and interoperability. Journal of Learning Analytics, 3, 139–158. https://doi.org/10.18608/jla.2016.31.9.
216. Keller, F. S. (1968). Good-Bye Teacher. Journal of Applied Behavior Analysis, 1, 79–89.
217. Cavoukian, A. (2010). Privacy by design: the definitive workshop. Identity in the Information Society, 3(2), 247–251.
218. Kofler, N., & Baylis, F. E. (2020). Ten reasons why immunity passports are a bad idea. Nature, 581, 379–381. https://doi.org/10.1038/d41586-020-01451-0 .
219. Gilbert, M., Swenson, L., Unger, D., Scheim, A., & Grace, D. (2016). Need for robust and inclusive public health ethics review of the monitoring of HIV phylogenetic clusters for HIV prevention. Lancet HIV, 3, e461. https://doi.org/10.1016/s2352-3018(16)30156-4 .
220. Marckmann, G. (2020). Ethische Fragen von Digital Public Health [Ethical implications of digital public health]. Bundesgesundheitsblatt, Gesundheitsforschung, Gesundheitsschutz, 63, 199–205. https://doi.org/10.1007/s00103-019-03091-w .
221. Brall, C., Schroder-Back, P., & Maeckelberghe, E. (2019). Ethical aspects of digital health from a justice point of view. European Journal of Public Health, 29, 18–22. https://doi.org/10.1093/eurpub/ckz167 .
222. Martinez-Martin, N., Wieten, S., Magnus, D., & Cho, M. K. (2020). Digital contact tracing, privacy, and public health. Hastings Center Report, 50, 43–46. https://doi.org/10.1002/hast.1131 .
223. Azzopardi-Muscat, N., & Sørensen, K. (2019). Towards an equitable digital public health era: Promoting equity through a health literacy perspective. European Journal of Public Health, 29, 13–17. https://doi.org/10.1093/eurpub/ckz166 .
224. Chauvin, J., Perera, Y., & Clarke, M. (2016). Digital technologies for population health and health equity gains: The perspective of public health associations. Journal of Public Health Policy, 37, 232–248. https://doi.org/10.1057/s41271-016-0013-4 .
225. *Cabrera-González, M. Á. C., & Florido, B. P. (2020). "Diana frente al Espejo": Una Propuesta de Videojuego para la Prevención de la Violencia de Pareja entre jóvenes. In CEUR Workshop Proceedings.
226. Barrera Yañez, A. G., Alonso-Fernandez, C., & Fernandez Manjon, B. (2020). Review of serious games to educate on gender equality. Eighth International Conference on Technological Ecosystems for Enhancing Multiculturality. https://doi.org/10.1145/3434780.3436592
227. *da Fonseca, R. M. G. S., dos Santos, D. L. A., Gessner, R., Fornari, L. F., de Oliveira, R. N. G., & Schoenmaker, M. C. (2018). Gender, sexuality and violence: Perception of mobilized adolescents in an online game. Revista Brasileira De Enfermagem, 71(suppl 1), 607–614. https://doi.org/10.1590/0034-7167-2017-0561
228. *Jouriles, E. N., Kleinsasser, A., Rosenfield, D., & McDonald, R. (2016a). Measuring bystander behavior to prevent sexual violence: Moving beyond self reports. Psychology of Violence, 6(1), 73–81. https://doi.org/10.1037/a0038230
229. *Potter, S. J., Demers, J. M., Flanagan, M., Seidman, M., & Moschella, E. A. (2021). Can videogames help prevent violence? An evaluation of games promoting bystander intervention to combat sexual violence on college campuses. Psychology of Violence, 11(2), 199–208. https://doi.org/10.1037/vio0000365
230. *Pabón-Guerrero, A., Tobar-Munozt, H., & Gonzalez, C. (2019). Tsiunas: A videogame for increasing gender-based violence awareness. In 2019 International Conference on Virtual Reality and Visualization (ICVRV) (pp. 189–192). IEEE. https://doi.org/10.1109/ICVRV47840.2019.00044

231. *Pearson, J., Wu, S., Royston, H., Smailes, H., Robinson, N., Cowell, A., & Jones, A. (2020). Designing a serious game to raise awareness of intimate partner violence among adolescents in the UK: The use of 'Good Games' principles for effective behavioural change (pp. 458–476). https://doi.org/10.1007/978-3-030-53294-9_33
232. *Debnam, K. J., & Kumodzi, T. (2021). Adolescent perceptions of an interactive mobile application to respond to teen dating violence. Journal of Interpersonal Violence, 36(13–14), 6821–6837. https://doi.org/10.1177/0886260518821455
233. *Marda, M., Economou, D., & Bouki, V. (2018). Enhancing deeper learning using empathy and creativity in serious games role-play simulations. In Proceedings of the European Conference on Games-based Learning.
234. Kim, Y. S., & Leventhal, B. (2008). Bullying and suicide. A review. International Journal of Adolescent Medicine and Health, 20(2), 133–154. https://doi.org/10.1515/IJAMH.2008.20.2.133
235. Edwards M, Hulme D. Too close for comfort? The impact of official aid on nongovernmental organizations. Wrld Dev. 1996;24(6):961–73.
236. Contu A, Girei E. NGOs management and the value of 'partnerships' for equality in international development: what's in a name? Hmn Rltns 2014. 2014;67(2):205–32.
237. Dar S, Cooke B, editors. The new development management. London: Zed Books; 2008.
238. Lewis D. The management of non-governmental development organizations. 2nd ed. Oxon: Routledge; 2007.
239. Gwin C. IDA's partnership for poverty reduction. Washington, DC: The World Bank; 2002.
240. Fisher J. Digital development and international aid: are games changing the world? In: Gangadharbatla H, Davis DZ, editors. Emerging research and trends in gamification. Hershey, PA: IGI Global; 2015.
241. Carlson, J. (2015). Citizen-protectors: The everyday politics of guns in an age of decline. Oxford University Press.
242. Hochschild, A. R. (2016). Strangers in their own land: Anger and Mourning on the American right. The New Press.
243. Bridges, T. (2021). Antifeminism, profeminism, and the myth of white men's disadvantage. Signs: Journal of Women in Culture and Society, 46(4), 663–688. https://doi.org/10.1086/712076
244. Cassino, D., & Besen-Cassino, Y. (2021). Gender threat: American masculinity in the face of change. Stanford University Press.
245. Konopka, K., Rajchert, J., Dominiak-Kochanek, M., & Roszak, J. (2021). The role of masculinity threat in homonegativity and transphobia. Journal of Homosexuality, 68(5), 802–829. https://doi.org/10.1080/00918369.2019.1661728
246. Vandello, J. A., Bosson, J. K., Cohen, D., Burnaford, R. M., & Weaver, J. R. (2008). Precarious manhood. Journal of Personality and Social Psychology, 95(6), 1325–1339.
247. Munsch, C. L., & Gruys, K. (2018). What threatens, defines: Tracing the symbolic boundaries of contemporary masculinity. Sex Roles, 79(7–8), 375–392. https://doi.org/10.1007/s11199-017-0878-0
248. Amaya, A., Bach, R., Keusch, F., & Kreuter, F. (2021). New data sources in social science research: Things to know before working with Reddit data. Social Science Computer Review, 39(5), 943–960. https://doi.org/10.1177/0894439319893305
249. Davis, J. L., & Graham, T. (2021). Emotional consequences and attention rewards: The social effects of ratings on Reddit. Information, Communication & Society, 24(5), 649–666. https://doi.org/10.1080/1369118X.2021.1874476
250. Duguay, P. A. (2022). Read it on Reddit: Homogeneity and ideological segregation in the age of social news. Social Science Computer Review, 40(5), 1186–1202. https://doi.org/10.1177/08944393211001053
251. Massanari, A. (2017). #GamerGate and the fappening: How Reddit's algorithm, governance, and culture support toxic technocultures. New Media & Society, 19(3), 329–346. https://doi.org/10.1177/1461444815608807

252. Phillips, W. (2015). This is why we can't have nice things: Mapping the relationship between online trolling and mainstream culture. MIT Press.
253. Ging, D. (2019). Alphas, betas, and incels: Theorizing the masculinities of the manosphere. Men and Masculinities, 22(4), 638–657. https://doi.org/10.1177/1097184X17706401
254. Silva, J. R., Capellan, J. A., Schmuhl, M. A., & Mills, C. E. (2021). Gender-based mass shootings: An examination of attacks motivated by grievances against women. Violence against Women, 27(12–13), 2163–2186. https://doi.org/10.1177/1077801220981154
255. Pettigrew, S., Pescud, M., Jarvis, W., & Webb, D. (2013). Teens' blog accounts of the role of adults in youth alcohol consumption. Journal of Social Marketing., 3(1), 28–40. https://doi.org/10.1108/20426761311297216
256. Eden, S., & Eshet-Alkalai, Y. (2013). The effect of format on performance: editing text in print versus digital formats. British Journal of Educational Technology, 44(5), 846–856.
257. Kolikant, Y. B.-D. (2012). Using ICT for school purposes: is there a student-school disconnect? Computers & Education, 59(3), 907–914.
258. Schmidt, H. C. (2013). Media literacy education from kindergarten to college: a comparison of how media literacy is addressed across the educational system. Journal of Media Literacy Education, 5(1), 295–309.
259. Stonier, W. F. (2012). The impact of intensive experience on prospective teachers' perception of the uses of digital, interactive text among K-12 students. Unpublished doctoral dissertation, Old Dominion University, Norfolk, VA.
260. Hammill, D. D., Wiederholt, J. S., & Allen, E. (2006). Test of Silent Contextual Reading Fluency (TOSCRF). Austin, TX: Pro-Ed.
261. Nasah, A., DaCosta, B., Kinsell, C., & Seok, S. (2010). The digital literacy debate: an investigation of digital propensity and information and communication technology. Educational Technology Research and Development, 58(5), 531–555.
262. Wiederholt, J. L., Hammill, D. D., & Brown, V. L. (2009). Reading Observation Scale. Austin, TX: Pro-Ed.
263. Bohman, P.: Teaching Accessibility and Design-for-All in the Information and Communication Technology Curriculum: Three Case Studies of Universities in the United States, England, and Austria. Utah State University, Utah (2012)
264. Klironomos, I., Antona, M., Basdekis, I., Stephanidis, C.: White paper: promoting design for all and e-accessibility in Europe. Univ. Access Inf. Soc. 5(1), 105–119 (2006). https://doi.org/10.1007/s10209-006-0021-4
265. Serra, L., Carvalho, L., Ferreira, L., Vaz, J., Freire, A.: Accessibility evaluation of e-government mobile applications in Brazil. Proc. Comput. Sci. 67, 348–357 (2015)
266. Draffan, E., Wald, M., Dickens, K., Zimmermann, G., Kelle, S., Miesenberger, K., Petz, A.: Stepwise approach to accessible MOOC development. Stud. Health Technol. Inf. 217, 227 (2015)
267. Gonçalves, R., Martins, J., Pereira, J., Santos, V., Cota, M.: Can i access my school website? Auditing accessibility of the Portuguese teaching institutions websites. J. Univ. Comput. Sci. 19(18), 2639–2655 (2013)
268. Putnam, C., Dahman, M., Rose, E., Cheng, J., Bradford, G.: Best practices for teaching accessibility in university classrooms: cultivating awareness, understanding, and appreciation for diverse users. ACM Trans. Access. Comput. (TACCESS) 8(4), 13 (2016)
269. Corbett, E., Weber, A.: What can I say?: addressing user experience challenges of a mobile voice user interface for accessibility. In: Proceedings of the 18th International Conference on Human–Computer Interaction with Mobile Devices and Services, pp. 72–82. ACM (2016)
270. Sánchez-Gordón, M., Moreno, L.: Toward an integration of web accessibility into testing processes. Proc. Comput. Sci. 27, 281–291 (2014). https://doi.org/10.1016/j.procs.2014.02.031
271. Saldaña-Ramos, J., Sanz-Esteban, A., García-Guzmán, J., Amescua, A.: Design of a competence model for testing teams. IET Softw. 6(5), 405–415 (2012)
272. Brhel, M., Meth, H., Maedche, A., Werder, K.: Exploring principles of user-centered agile software development: a literature review. Inf. Softw. Technol. 61, 163–181 (2015). https://doi.org/10.1016/j.infsof.2015.01.004

273. Hegedűs, P., Kádár, I., Ferenc, R., Gyimóthy, T.: Empirical evaluation of software maintainability based on a manually validated refactoring dataset. Inf. Softw. Technol. 95, 313–327 (2018)
274. Kofi Frimpong, Adasa Nkrumah; Li, Ping; Nyame, Gabriel; Hossin, Md Altab The Impact of Social Media Political Activists on Voting Patterns. Political Behavior (2020). https://doi.org/10.1007/s11109-020-09632-3
275. Huggins, Christopher; Frosina, Natasha ICT-driven projects for land governance in Kenya: disruption and e-government frameworks. GeoJournal (2016). https://doi.org/10.1007/s10708-016-9710-6
276. Torres-Echeverry, Nicolás Social media, meet old politics: preservation and innovation in Colombian presidential elections, 2010–2018. Theory and Society (2023). https://doi.org/10.1007/s11186-022-09509-w
277. Campbell-Verduyn, Malcolm; Rodima-Taylor, Daivi; Hütten, Moritz Technology, small states and the legitimacy of digital development: combatting de-risking through blockchain-based re-risking?. Journal of International Relations and Development (2020). https://doi.org/10.1057/s41268-020-00198-5
278. Kneuer, Marianne; Datts, Mario E-democracy and the Matter of Scale. Revisiting the Democratic Promises of the Internet in Terms of the Spatial Dimension. Politische Vierteljahresschrift (2020). https://doi.org/10.1007/s11615-020-00250-6
279. Bhuiyan, Shahjahan H. E-Government in Kazakhstan: Challenges and Its Role to Development. Public Organization Review (2009). https://doi.org/10.1007/s11115-009-0087-6
280. Hanrahan, Nancy Weiss Music and democracy in America: historical perspectives on 'democratization' in the digital age. American Journal of Cultural Sociology (2021). https://doi.org/10.1057/s41290-021-00138-x
281. Odzuck, Eva; Günther, Sophie Digital campaigning as a policy of democracy promotion: Applying deliberative theories of democracy to political parties. Zeitschrift für Politikwissenschaft (2021). https://doi.org/10.1007/s41358-021-00308-w
282. Mölders, Marc; Schrape, Jan-Felix Digital Deceleration. Protest and Societal Irritation in the Internet Age. Österreichische Zeitschrift für Soziologie (2019). https://doi.org/10.1007/s11614-019-00354-3
283. Welp, Yanina; Urgell, Ferran; Aibar, Eduard From Bureaucratic Administration to Network Administration? An Empirical Study on E-Government Focus on Catalonia. Public Organization Review (2007). https://doi.org/10.1007/s11115-007-0038-z
284. Armstrong, Crystal; Butcher, Charity Digital Civil Society: How Nigerian NGOs Utilize Social Media Platforms. International Journal of Politics, Culture, and Society (2017). https://doi.org/10.1007/s10767-017-9268-4
285. Caliskan, Ozgur Digital Pride on the Streets of the Internet: Facebook and Twitter Practices of the LGBTI Movement in Turkey. Sexuality & Culture (2021). https://doi.org/10.1007/s12119-021-09818-9
286. Yu, Sun; Shengcui, Dang; Bin, Fang; Qiuyan, Zhao Public Access to Government Information in the Digital Age: Findings from China. Korean Social Science Journal (2014). https://doi.org/10.1007/s40483-014-0010-9
287. Saghafi, Fatemeh; Zarei, Behrouz; Fadaei, Maryam The conceptual model for the effect of technology on the political and social aspects of Iran's e-government. Quality & Quantity (2015). https://doi.org/10.1007/s11135-015-0234-9
288. Agbaje, Ganiy I.; Bello, Innocent E.; Ojo, Adebayo Gbenga Empirical assessment of the role of geospatial technology in delivering governance and strengthening democracy in Nigeria. GeoJournal (2017). https://doi.org/10.1007/s10708-017-9797-4
289. Hoverd, William James; Salter, Leon; Veale, Kevin The Christchurch Call: insecurity, democracy and digital media - can it really counter online hate and extremism?. SN Social Sciences (2020). https://doi.org/10.1007/s43545-020-00008-2
290. De Blasio, Emiliana; Selva, Donatella Implementing open government: a qualitative comparative analysis of digital platforms in France, Italy and United Kingdom. Quality & Quantity (2018). https://doi.org/10.1007/s11135-018-0793-7

291. Defacqz, Samuel; Dupuy, Claire Usages of an E-participation platform by legislators: lessons from the French parliament. French Politics (2021). https://doi.org/10.1057/s41253-021-00155-9
292. Manoharan, Aroon P.; Melitski, James; Holzer, Marc Digital Governance: An Assessment of Performance and Best Practices. Public Organization Review (2022). https://doi.org/10.1007/s11115-021-00584-8
293. Youngs, Gillian Digital Globalization and Democracy: The territorial walls come tumbling down?. Development (2011). https://doi.org/10.1057/dev.2011.31
294. Royakkers, Lambèr; Timmer, Jelte; Kool, Linda; van Est, Rinie Societal and ethical issues of digitization. Ethics and Information Technology (2018). https://doi.org/10.1007/s10676-018-9452-x
295. Carnevale, Antonio; Tangari, Emanuela A.; Iannone, Andrea; Sartini, Elena Will Big Data and personalized medicine do the gender dimension justice?. AI & SOCIETY (2021). https://doi.org/10.1007/s00146-021-01234-9
296. Simmler, Monika; Brunner, Simone; Canova, Giulia; Schedler, Kuno Smart criminal justice: exploring the use of algorithms in the Swiss criminal justice system. Artificial Intelligence and Law (2022). https://doi.org/10.1007/s10506-022-09310-1
297. Sharon, Tamar Blind-sided by privacy? Digital contact tracing, the Apple/Google API and big tech's newfound role as global health policy makers. Ethics and Information Technology (2020). https://doi.org/10.1007/s10676-020-09547-x
298. Ferreira, Daniel Brantes; Gromova, Elizaveta A. Hyperrealistic Jurisprudence: The Digital Age and the (Un)Certainty of Judge Analytics. International Journal for the Semiotics of Law - Revue internationale de Sémiotique juridique (2023). https://doi.org/10.1007/s11196-023-10015-0
299. Buscariolli, André Making Crime Visible in the Digital Age: The Ethnomethods of Data Policing. Critical Criminology (2023). https://doi.org/10.1007/s10612-023-09683-y
300. Gurumurthy, Anita; Chami, Nandini Development Justice in the Digital Paradigm: Agenda 2030 and Beyond. Development (2019). https://doi.org/10.1057/s41301-019-00203-6
301. Büschel, Isabell; Mehdi, Rostane; Cammilleri, Anne; Marzouki, Yousri; Elger, Bernice Protecting Human Health and Security in Digital Europe: How to Deal with the "Privacy Paradox" ?. Science and Engineering Ethics (2014). https://doi.org/10.1007/s11948-013-9511-y
302. Mak, Chantal Fundamental Rights and the European Regulation of iConsumer Contracts. Journal of Consumer Policy (2008). https://doi.org/10.1007/s10603-008-9084-3
303. Aguilar, Stephen J. Learning Analytics: at the Nexus of Big Data, Digital Innovation, and Social Justice in Education. TechTrends (2017). https://doi.org/10.1007/s11528-017-0226-9
304. Pagallo, Ugo Online Security and the Protection of Civil Rights: A Legal Overview. Philosophy & Technology (2013). https://doi.org/10.1007/s13347-013-0119-6
305. Kerber, Wolfgang; Zolna, Karsten K. The German Facebook case: the law and economics of the relationship between competition and data protection law. European Journal of Law and Economics (2022). https://doi.org/10.1007/s10657-022-09727-8
306. Hodson, David Digital technology in family justice systems: global benefits, opportunities and concerns. ERA Forum (2021). https://doi.org/10.1007/s12027-021-00660-y
307. Gómez-Ramírez, Oralia; Iyamu, Ihoghosa; Ablona, Aidan; Watt, Sarah; Xu, Alice X. T.; Chang, Hsiu-Ju; Gilbert, Mark On the imperative of thinking through the ethical, health equity, and social justice possibilities and limits of digital technologies in public health. Canadian Journal of Public Health (2021). https://doi.org/10.17269/s41997-021-00487-7
308. Yañez, Alma Gloria Barrera; Alonso-Fernández, Cristina; Fernández-Manjón, Baltasar Systematic literature review of digital resources to educate on gender equality. Education and Information Technologies (2023). https://doi.org/10.1007/s10639-022-11574-8
309. Fisher, Jolene Digital Games, African Development, and Gender Equality: a Comparative Analysis of Family Values and Moraba. Studies in Comparative International Development (2017). https://doi.org/10.1007/s12116-017-9244-z

310. Hanson, Kenneth R.; Pascoe, C. J.; Light, Ryan "It's Getting Difficult to Be a Straight White Man": Bundled Masculinity Grievances on Reddit. Sex Roles (2022). https://doi.org/10.1007/s11199-022-01344-8
311. Seok, Soonhwa; DaCosta, Boaventura Gender Differences in Teens' Digital Propensity and Perceptions and Preferences With Regard to Digital and Printed Text. TechTrends (2016). https://doi.org/10.1007/s11528-016-0134-4
312. Silva, João Sousa e; Gonçalves, Ramiro; Branco, Frederico; Pereira, António; Au-Yong-Oliveira, Manuel; Martins, José Accessible software development: a conceptual model proposal. Universal Access in the Information Society (2019). https://doi.org/10.1007/s10209-019-00688-5

Digitalization and Spaces of Knowledge and Power

8

Introduction

Knowledge and power are intricately but closely related to each other. Knowledge creates power and power in turn creates its own knowledge, and thus, they mutually reinforce each other. Further, power and knowledge are also closely associated with wealth. Knowledge leads to wealth and power and wealth often leads to power.

Knowledge is attained through the processing of information. The distribution of power and wealth in a society often shapes access to information and gives rise to information asymmetry. Before the digital revolution, information dissemination and diffusion were limited and often there was an enormous transaction cost to information as they were geographically and socially unevenly distributed. However, the digital revolution (in terms of the internet, mobile technology, laptops and desktops, and the ability to use the same in a large section of the population) has made to a great extent information ubiquitous in virtual spaces where geography does not matter but only the access to digital technologies and internet. In other words, improved internet and the rise of social media to a large extent have democratized information availability leading to a significant impact on the distribution of power and knowledge in any society. In democratic societies, citizens are now able to access crucial information and able to question their elected leaders or state, while in authoritarian regimes often they can, as was seen during Arab Spring, launch their demand for the democratization of the society, power and information. Though digital technologies like any other technologies have their class impacts, they have also played significant role in massify and eroding the hills and mountains of power and knowledge and smoothen their uneven topographies.

Digitalization has also played a significant role in reducing the epistemic divide and helped in the globalization of local knowledge or from margins, and also the

A. Shaban (ed.), *Digital Geographies—Theory, Space, and Communities*,
https://doi.org/10.1007/978-981-97-4734-4_8

co-production of knowledge and cultures. There has been a platformalization of information distribution and production. YouTube and other social media sites have helped significantly in knowledge dissemination and collaborations between different groups across geographies and socio-economic groups. Zoom, Google Meet, Microsoft Teams, Webx, etc., today, play an important role in shaping knowledge dissemination, and in fact because of digital technologies the educational system kept on working during the difficult time of COVID-19. These technologies now also giving way to Smart schools and universities. Digitalization is also leading to the technology transfer in terms of dissemination of know-how helping massively marginal and local industries.

Among the social scientists, French Sociologist Michell Foucault has written well-articulated the high and positive relationship between knowledge and power. Today's democratic society which needs to empower the citizens equally must attempt to democratize the information and knowledge as well. The knowledge often emerging from privileged sections of societies has been preferred as better information or knowledge, while the knowledge from the marginalized sections has often been ignored. As the mainstreaming of knowledge in the economy and governance systems is important for the socio-economic mobilities, the marginalized sections whose knowledge gets excluded also remain excluded from the development. Many have termed this asymmetry and dominance of knowledge from certain sections as epistemic violence, and advocate for the co-creation of knowledge. Digitalization in many ways has made it possible to also put in the public sphere the knowledge from the less privileged groups on social media. This has helped people from different sections to glean them, though the inclusion of such knowledge in everyday governance and educational systems remains still in question.

The existing social and economic inequalities and regional development are furthering the digital development inequalities. One often finds that economically better-developed social groups and regions also have better digital infrastructure and digital literacy rates. They also often have better access and affordability to digital devices. There has been a significant presence of digital divides in some of the societies that are churning out new inequalities or new core and peripheries. Additionally, as digital technologies are largely emerging from developed countries, they are becoming new centers of power as their servers store information from across the world benefiting these countries in several spheres from understanding the consumer market to manipulating international relations. With the monopoly on digital technologies, the empire has again struck back on the global south. However, a great power contest is also underway in terms of control of digital technologies and data. This contest is well visible between the West (especially the USA) and China, and India's effort to regulate the digital platform giants like Google, Twitter, Facebook, Amazon, etc., and restrict the data flow out of Indian physical borders. India and so many other countries have not only brought several legislations to regulate the data and platforms but also have created their platforms to secure the misuse of data. However, some of countries have been over-protective and in the name of regulations, they are compromising the right to civil liberties and freedom to access the information through internet.

We have reviewed several interesting papers in this chapter. The chapter is divided into three sections. Section 8.1 discusses the studies on digitalization and knowledge production, while Sect. 8.2 deals with the studies on digitalization and knowledge dissemination. Section 8.3 discusses papers related to digitalization and power.

In Sect. 8.1, a paper on 'The Role of Support Units in Digital Transformation: How Institutional Entrepreneurs Build Capacity for Online Learning in Higher Education' argues that for higher education institutions to successfully implement digital transformation, there must be supporting leadership. The paper on 'Indigenous Wisdom, Capital, Technology and Education', contrasts certain elements of indigenous worldviews and knowledge with particular features of capitalism, colonization, corporate interests, technology, and education, arguing that there is a basic conflict of values between the two. It concludes that a more critical and nuanced view of capitalism and technology, as well as the insights of indigenous knowledge, would benefit education. Another paper on 'The Presence of the Body in Digital Education: A Phenomenological Approach to Embodied Experience' claims that the virtual experience is essentially embodied and completely real, building on Merleau-Ponty's phenomenological findings.

A paper in Sect. 8.2 on 'Promises and pitfalls of digital knowledge exchange resulting from the COVID-19 pandemic' argues that the ability of people to form and retain meaningful social connections in the absence of regular face-to-face contact, as well as efforts to address the continuation of the digital gap, are two interrelated elements that will determine the long-term viability and effectiveness of digital knowledge exchange (KE), while another paper in Sect. 8.3 on 'Great Power Contest in a New (Digitalized) Guise' examines the so-called 'Tech Cold War', a competition between China and the United States for global technological supremacy.

I am sure that the students, teachers, and researchers will find the summaries of the papers interesting and helpful them in understanding the ongoing impacts of digital technologies on knowledge, power, wealth geographies and emerging geopolitics and data war.

Machine generated summaries

Disclaimer: The summaries in this chapter were generated from Springer Nature publications using extractive AI auto-summarization: An extraction-based summarizer aims to identify the most important sentences of a text using an algorithm and uses those original sentences to create the auto-summary (unlike generative AI). As the constituted sentences are machine selected, they may not fully reflect the body of the work, so we strongly advise that the original content is read and cited. The auto generated summaries were curated by the editor to meet Springer Nature publication standards. To cite this content, please refer to the original papers.

Machine generated keywords: education, learn, school, learning, educational, student, teacher, digitalization, covid, teaching, learner, pandemic, high education, classroom, production.

8.1 Digitalisation and Knowledge Production

Machine generated keywords: education, school, learn, learning, student, educational, digitalization, classroom, teacher, learner, teaching, pandemic, production, covid, material

8.1.1 The Role of Support Units in Digital Transformation: How Institutional Entrepreneurs Build Capacity for Online Learning in Higher Education [265]

This is a machine-generated summary of:

Langseth, Inger; Jacobsen, Dan Yngve; Haugsbakken, Halvdan: The Role of Support Units in Digital Transformation: How Institutional Entrepreneurs Build Capacity for Online Learning in Higher Education [265]

Published in: Technology, Knowledge and Learning (2022)

Link to original: https://doi.org/10.1007/s10758-022-09620-y

If you want to cite the papers, please refer to the original.

For technical reasons we could not place the page where the original quote is coming from.

Abstract-Summary

"This study used New Institutional Theory to explore how entrepreneurial activities in support units contribute to digital transformation in higher education in Norway."

"We describe how entrepreneurs initiated and operationalized support for Massive Open Online Courses (MOOCs), micro-credentials and fully online courses in pockets of innovation within existing institutional arrangements."

"The informants were chosen through strategic sampling from support units and faculties, where they filled different roles in the production of such online courses."

"Located in pockets of innovation, the entrepreneurs provided open digital platforms, a pedagogy for online course design and support for faculties who engaged in online course production."

"The findings also confirm previous research pointing out how the lack of supportive leadership may impede successful digital transformation in higher education institutions."

"The model can also support entrepreneurial processes in online environments."

Introduction

"While faculties lean on support staff to enhance project results, such as online courses on open platforms, Norwegian support units seem, so far and to a large extent, to be top-down, experience-based and policy driven (Lyby and others, [2]; Tømte and others, [3])."

"From our position in a research project that is driven by a strong belief in open access to online education for all and entails disruption in terms of MOOCs and online course offerings on open platforms with comprehensive support for faculties, our goal in this study is to contribute to a deeper understanding of entrepreneurial bottom-up activities in and in relation to support units."

"New Institutional Theory and current research on Massive Open Online Courses (MOOCs) will then guide us as to where support units are situated and how their activities are framed within institutional arrangements in HEIs in Norway."

A Theoretical Perspective on MOOC Entrepreneurs

"We have outlined the current research literature to frame how MOOC entrepreneurs in the Norwegian HEI sector engage with online education and capacity building in spaces where maneuverability is limited."

"The second stream relates to an emerging horizon where researchers explore MOOCs mainly in a Norwegian context, and in particular studies focusing on support units engaged in MOOC production."

"The second body of research, which is not limited to the Norwegian context, goes deeper into frameworks and learning designs in MOOC production processes."

"From an inside perspective in a previous study, we found major differences in the institutional approach to encourage MOOC entrepreneurship and facilitate and organize support for fully online course offerings (Langseth and others, [4])."

"Operationalizing a systematic framework and a learning design in support for faculties who embark on developing fully online courses, such as MOOCs and micro-credentials, seems to be a question of both entrepreneurship, research and strategic and institutional leadership (Selznick, [5])."

Method

"Out of their roles in the support units, they had also developed an interest in supporting a digital infrastructure for MOOC initiatives, develop research informed competence in technology and pedagogy and give support for online course production in their organization."

"We needed informants with first-hand knowledge of the issue and purposefully selected informants involved in digital support and online course production."

"The first guide was mainly based on our experiences from developing and supporting online offerings, as the second also considered issues that came up during the primary analysis of the first sample."

"From the initial raw codes, they encourage researchers to go on to identify important patterns in the informants' contributions and to develop and refine these into an overall story describing the data."

"This first analysis also informed the development of the second interview guide."

Findings

"It (online course) easily gets very heavy text wise, … And I understand very well the change for many faculties when they have to stand in front of the camera and they have to say things briefly, to break things down and … Most informants (n = 11) reported that the support team was working very hard to contribute to the instructional design; the way the academics were thinking about teaching and learning, content and new target groups, on online platforms."

"One informant (n = 11) described how a loosely coupled network of administrative staff at his department and the immediate organizational levels supported both online course production and digital competence development in an ad-hoc manner: We have built a small network of sorts with a support function, so that each subject area has a person who can help them, but who is not formally employed to do this work."

Discussion

"Adopting the lense of Institutional Entrepreneurship, the study also focused on how stakeholders can benefit from entrepreneurial activities to build strong support units and capacity for digital transformation across institutional barriers in Norwegian HEIs."

"The organizations in our study seemed to be loosely coupled, in the sense that MOOC entrepreneurs and other important stakeholders tended to adhere to different logics that represent different belief systems, values and educational practices related to digitalization (Irgens, [6])."

"As MOOC entrepreneurs operationalized opportunities for change inspired by scalability and open access to education, top management and leadership often seemed to exercise a powerful influence in support of a campus-based educational policy and digital competence linked to the lecture halls and closed platforms (e.g., LMS)."

"In line with Osmundsen and others, [7], we found that MOOC entrepreneurial activities related to digital transformation affect several organizational dimensions, which also entail a redefinition of strategies and the creation of new organizational processes."

Conclusion

"We initially asked: How do institutional entrepreneurs at Norwegian universities build capacity for developing fully online courses?"

"and What characterizes the collaborative processes between support staff, faculties and other colleagues during fully online course production?"

"We found that most MOOC entrepreneurs operate outside formal institutional arrangements, where they collaborate across institutional barriers to explore new digital technologies."

"A main finding is that MOOC entrepreneurial activities are still located in pockets of innovation and have not yet been legitimized in formal institutional arrangements in Norwegian HEIs."

"Our findings suggest that the value of MOOC entrepreneurial activities in and in relation to support units has not been fully acknowledged in HEIs."

"Their entrepreneurial activities were linked to support units, where they collaborated on MOOCs and online course production."

"Out of these data, we were able to describe entrepreneurial processes in support units in HEIs."

Recommendations

"Entrepreneurs play an important role in the development of digital maturity in HEI."

"Systematic support for entrepreneurial ideas will help faculties and staff develop new educational practices that can move the HEI forward."

"Support units should collaborate with faculties on bottom-up entrepreneurial activities and keep management informed to facilitate the transition of entrepreneurial activities into strategy and policy documents."

"Support units play an important role."

"We also propose a new model to measure digital maturity, with possible implications for the redefinition of support units."

Limitations

"Hopefully, stakeholders and future entrepreneurs can benefit from using the insights from this study in decisions to move digitalization in their organization forward and build strong support units with entrepreneurial power."

"The dominant logics pertaining to existing arrangements in the digitalization of education have been outside the scope of this article, as our focus has been on MOOC entrepreneurial activities."

"An obvious limitation in the study is that we did not collect data to explore the phenomena from the management and leadership perspectives."

"Even if we collected data from a strategic sample of MOOC entrepreneurs, the data are limited to a Norwegian higher education context."

[Section 9]
"A landscape is a series of named locales, a set of relational places linked by paths, movements and narratives."

"It is a 'natural' topography perspectivally linked to the existential Being of the body in societal space."

"It is a cultural code for living, and anonymous 'text' to be read and interpreted, a writing pad for inscription, a scape for human praxis, a mode of dwelling and a mode of experiencing. (Tilley, p. 34 in Wittgenstein A Phenomenology of Landscape)."

8.1.2 Programming the Postdigital: Curation of Appropriation Processes in (Collaborative) Creative Coding Spaces [266]

This is a machine-generated summary of:

Ackermann, Judith; Egger, Benjamin; Scharlach, Rebecca: Programming the Postdigital: Curation of Appropriation Processes in (Collaborative) Creative Coding Spaces [266]

Published in: Postdigital Science and Education (2019)

Link to original: https://doi.org/10.1007/s42438-019-00088-1

If you want to cite the papers, please refer to the original.

For technical reasons we could not place the page where the original quote is coming from.

Abstract-Summary
"The learning ecologies GitHub, Stack Overflow, and Pouët present digital infrastructures that curate and support such appropriation-related activities."

"The paper presents a qualitative analysis of the structure of these spaces, and the appropriation-related communication taking place."

"It shows how crucial the varying designs of the learning ecologies and the implemented interaction possibilities are for the appropriation activities taking place and draws a line to the desire for stronger participation in postdigital curation."

Introduction

"To live in postdigital times means that the ways we live, think, and interact are shaped by digital configurations."

"In the field of the arts, the term postdigital is applied analytically and does not describe a concrete genre, but refers to the altered status of media and arts after their digitalization, and the digitalization of the channels they are communicated through (Cramer [8]: 19)."

"They propose new forms of critical examination of digital, networked media and their influence on the arts (Berry and Dieter [9]; Kholeif [10])."

"Postdigital artworks are generated via digital methods or at least informed by digital technology, even if they do not necessarily have to create a digital output (Ackermann and others [11]: 183)."

"One of the earliest examples of digital art practices in the history of technology is the demoscene that can be described as the beginning of creative coding."

Appropriation and Postdigital Art

"They apply code to new fields and employ it for artistic expression, creating specific, rather uncommon usage patterns that rely on expert knowledge and creative workarounds to be seen, for example, in the p5.js community."

"With regard to the innovative ways of using coding skills to solve (and raise) esthetical questions to be found in creative coding, it becomes clear that relying on the basic affordances is not sufficient, but a lot of appropriation activities are needed."

"This leads us to the field of curation, which is a very important concept in the art sphere but also bears usefulness for appropriation activities and resources of knowledge: how disseminated information can be gathered and enriched with contexts to be perceived as knowledge is one of the most virulent questions in the postdigital age of networked human and artificial entities."

Learning Ecologies Surrounding Postdigital Art

"We decided to take a closer look at the digital learning ecologies connected to postdigital art and creative coding, to understand how and where appropriation activities take place and in which ways information is curated by the actors involved."

"Another general learning ecology focusing on coding is Stack Overflow, which describes itself as 'a question and answer site for professional and enthusiast programmers' and follows the plan to 'build a library of detailed answers to every question about programming' together with the community."

"To identify creative coding-related utterances in the two general coding-related spaces (GitHub and Stack Overflow), we narrowed down our research to communication and information dealing with the programming language 'processing' that was especially designed for creative coding."

"Our analysis focuses on the following three questions: How do the varying designs of the learning ecologies differ in each creative coding community space?"

Overall Structure of the Learning Ecologies

"Just like GitHub, Stack Overflow includes a business model offering the categories basic (5$/user/month), business (11$/user/month), and enterprise (17$/user/month) reaching from allowing users to have private questions as a team to including various support activities depending on the chosen category to hosting one's own cloud server for business users."

"Other than in the case of GitHub, Stack Overflow does not allow you to follow other users in order to get in contact, but only offers the possibility of observing tags in the sense of topics."

"The page gives an easy overview on what it offers but it is quite complicated to familiarize oneself with the scene's own terms that are necessary for the communication and usage of the platform: 'prod' for example is short for Production and means Demo, 'glöps' are bonus points users receive for being active and 'BBS' is short for bulletin board system describing the forum area."

Detailed Analysis of Appropriation Processes

"To analyze strategies of solving problems or setting topics in the respective user interactions, we zoomed into the communication practices of the three forums."

"Since GitHub, Stack Overflow, and Pouët allow and organize communication flows differently, it was not possible to proceed methodologically the same way in all three cases."

"The spaces were examined for comparable categories, especially in order to strengthen the peculiarities and differences of the three learning ecologies."

The GitHub Awesome List of Creative Coding

"The GitHub awesome lists present an opportunity for users to collect, organize, and share learning resources and information concerning various topics."

"Within the general list of awesome lists, it is the only one that deals with this topic and includes all languages and tools that can be used for creative coding."

"The participation in the list is not synonymous with activity on GitHub in general: 23 of the users contributing to the list present themselves in their profiles as experts that work professionally as developers or coders."

"The overall additions demonstrate the curatorial power of the creator over the list and highlight the different roles or levels of participation to be found among the users: other contributors stick to the creator's structures and rather suggest small changes to improve the list."

"The given structures of the awesome list have general validity for the curating process of every awesome list covering a specific topic like creative coding."

Stack Overflow

"Eight questions solely used the tag 'Processing.'"

"In the corpus of analyzed questions, only twelve were generated as a true (complete) question, like the following: 'Why is there an error while importing Processing-core to Java?' (Q-ID 8) or 'What does _serial_.bufferUntil(byte) do, and how does it synergies with serialEvent?' (Q-ID 17)."

"Only two of the 54 analyzed questions do not refer to the person asking it as 'I.' These are Q-ID 51 using phrases like 'when you drag' and 'how to make' which are at least still referring to a person and Q-ID 38 which reads 'Objective: When the red candles are clicked they should turn green."

"Looking at the people answering the questions, one finds a totally different picture: concerning the 50 answers given by 27 unique users, the reputation points are 7000 on average."

Pouët

"As the analysis is focused on appropriation, the research of the BBS on Pouët only considers the code topic area, to which users independently assign their threads."

"It is noticeable that two especially long threads with more than 1800 replies found in the code section do not deal with problem-oriented topics, but have to be classified in the categories entertainment ('Random line of code thread') and presentation of demos ('Raymarching Beginners' Thread')."

"Considering the relation between the categories, one would find about 19% threads on non-appropriation-related topics (e.g., 'Demos using Vulkan,' 'The demoscene sourcecode thread,' 'A.I. related stuff'), while about 81% of initial posts are devoted to questions of appropriation (e.g., 'How to render SDF grid volumes?,'"

"The analysis of appropriation-related threads constituting more than three-quarters of the corpus reveals a differentiation into four types: Open question on a specific need (34), Learning resource to be discussed (23), Specific problem to be solved (18), and Technology based recommendation (14)."

Discussion

"In analyzing the concrete appropriation activities in the field of creative coding, we looked at the collaborative interactions in the different learning ecologies to understand how users communicate with each other and what resources they use/produce."

"The analysis revealed that the three spaces address different appropriation needs: while GitHub serves as an archive or a collection of knowledge—a wikipedia for creative coding, where users contribute to collect high-ranked learning resources - the communication activities found there are more connected to improving the material than to discussing actual problems in handling a technology or similar."

"All three learning ecologies correspond to their overall goals that are gathering lots of (and various) user-generated content for GitHub, fixing appropriation needs in a very fast way for Stack Overflow and creating a sense of community, where appropriation happens among lots of other activities for Pouët."

Conclusion

"The combination of looking at the learning ecologies' structures, in terms of design, technical organization, and the way the users are addressed on the one hand, and at the specific appropriation-related communication taking place on the other hand, reveals that the visibility and presentation of interaction modalities, the introduction to a certain appropriation space and the implicit hierarchies trigger certain user behavior."

"The mentioned conditions, in which the analyzed processes of (creative) participation through appropriation take place, show how each desire of shaping the analyzed spaces is constrained by affordances given a priori and the technical, esthetic, and organizing structures bearing them."

"This meets the aims and strategies of curators in the field of postdigital art, who also mention that they want to find ways of allowing visitors to participate in the different curatorial processes—but formulate those ideas very vague and frightful (PKKB [12], 3: 92), imagining user curation with a very limited and predefined scope."

8.1.3 Creating Learning Bridges Through Participatory Design and Technology to Achieve Sustainability in Local Crafts: A Participatory Model to Enable the Transfer of Tacit Knowledge and Experience Between the Traditional Craftsmanship and Academic Education [267]

This is a machine-generated summary of:

Aytekin, Behiç Alp; Rızvanoğlu, Kerem: Creating learning bridges through participatory design and technology to achieve sustainability in local crafts: a participatory model to enable the transfer of tacit knowledge and experience between the traditional craftsmanship and academic education [267]

Published in: International Journal of Technology and Design Education (2018)
Link to original: https://doi.org/10.1007/s10798-018-9454-3

If you want to cite the papers, please refer to the original.

For technical reasons we could not place the page where the original quote is coming from.

Abstract-Summary
"While technology-based developments have produced a wealth of information and experience, the rate of interaction between the academic institutions and craftsmanship establishments is observed to decrease gradually."

"The purpose of this study is to enable the transfer of tacit knowledge and experience between these two foundations by creating "learning bridges" through the use of technology in order to achieve sustainability in local crafts in long term."

"The study indicated that "learning bridges" could be formed and thus facilitate the transfer of tacit knowledge and experience between the academy and workshops through the use of technology."

Introduction

"The field study was completed as a participatory design project bringing together master Telkari craftsmen, instructors at Mardin Artuklu University's Faculty of Fine Arts, and the first-year students of the Faculty's Painting Department."

"As one of the units in the university, Faculty of Fine Arts is also aimed to support local cultural progress and sustainability with the use of artistic and scientific knowledge of the academy."

"The aim was to enable applied academic education, which has moved away from traditional culture and production knowledge."

"Academic education in applied fields such as design, architecture and fine arts has grown into a system that moves away from traditional production methods and does not allow opportunities for students to draw from their cultural resources."

"It is also crucial for applied academic education, which has moved away from traditional culture and production knowledge, to benefit once again from the knowledge and experience of traditional crafts."

Related research and literature review

"The greatest advantage of the traditional workshop education system is that deeply internalized knowledge, and the experience that is unable to be verbalized can be transferred from the craftsman to the apprentice."

"The most significant advantage of the traditional workshop education system is that deeply internalized knowledge, and the experience that is unable to be verbalized can be transferred from the craftsman to the apprentice."

"The knowledge and experience gap between the modern academic and traditional workshop spheres was bridged with the proper management of tacit knowledge transfer processes, and a new interaction was achieved."

"The process of transferring tacit knowledge and experience among communities enabled the observation of the formation of learning bridges (Wood [13, 14]; Wood and Rust [15]; Wood and others [16])."

"Learning bridges can be defined as connections in the level of consciousness formed and developed as continuous interaction is achieved during the transfer of knowledge and experience between two reciprocal groups such as craftsman-apprentice and teacher-student."

Methods

"Since this interaction makes it possible to acquire first-hand data, the research process was split into three major phases: the preliminary field study as the preparation stage, main field study as the design and production stage, and post-field study as the analysis phase."

"During the research, each process was carried out by participatory design method and naturally structured according to the approval of the participants."

"This collaborative research approach involved the use of surveys and in-depth interviews in obtaining a qualitative analysis of the processes before and after the field study."

"All of these methods are known as the participation and design-based multi-method approach developed during the study by the researcher."

"The design of the main field study, which embraces the participatory design method, followed the example of similar studies that were previously conducted in different formats such as industrial product design-based research and application processes (Kaya [17]; Şişman and others [18]; Ünlühisarcıklı [19]; Yağız and Kaya [20])."

Results and discussion

"The findings obtained from the interviews indicated that it was possible to bring together the academy and workshop in a participatory design project in the Mardin field with the help of technological enablers, as well as to have them interact with one another in the tacit knowledge and experience transfer process (Polanyi [21])."

"The simultaneous evaluation of the students' designs by master Telkari craftsmen created a great sphere of interaction and experience given the conditions of the field and geography."

"This finding was interpreted as an indication of the emergence of new opportunities for craftsmen, instructors and students to gain knowledge and experience where the workshop and Academy spheres were able to converge and interact with one another, as intended by the study (Lave and Wenger [22])."

"To the experience of the initial meeting with the craftsman, the researcher observed that the craftsman had an easier time transferring his tacit knowledge due to his increased interaction and shared experience with the students and course instructor."

Conclusions

"The findings of the descriptive and qualitative analysis, which were collected through a multi-method approach at the main fieldwork, namely participation-and design-based multi-method approach, indicated that knowledge could be transferred via the learning bridges (Wood [13]) formed during tacit knowledge transfer (Polanyi [21, 23]) processes."

"As the project progressed, the groups contributing actively to the participatory design-based field study were observed to turn into communities of practice (Lave and Wenger [22]) with the major effect of a technological enabler, namely Facebook."

"The model includes cone-shaped sets that represent the study's three CoP. These sets are not self-enclosed and fixed; on the contrary, they are presented as structures where there is continued flow of knowledge and interaction, or to refer to Lave and Wenger's [22] CoP where the situated learning process is ongoing in each set."

"The goal is to ensure that university students receiving application-based education possess the advantage of knowledge and experience upon graduation, thanks to interactive opportunities provided by studies similar to this one (Hennessy and Murphy [24]; Kvan [25]; Lahti and Seitamaa-Hakkarainen [26]; Yantaç [27]; Kaya [28]; Yağız and Kaya [20])."

8.1.4 Why Flipping the Classroom is not Enough: Digital Curriculum Making After the Pandemic [268]

This is a machine-generated summary of:

Backes, Susanne; Baumann, Isabell; Harion, Dominic; Sattler, Sabrina; Lenz, Thomas: Why flipping the classroom is not enough: Digital curriculum making after the pandemic [268]

Published in: PROSPECTS (2021)

Link to original: https://doi.org/10.1007/s11125-021-09555-9

If you want to cite the papers, please refer to the original.

For technical reasons we could not place the page where the original quote is coming from.

Abstract-Summary

"These measures offer an opportunity to reexamine conventional teaching and learning arrangements, test new digital and analogue concepts, and provide essential inspiration for curriculum making in the twenty-first century."

"The question of how school closures and digitally supported teaching settings may affect an increase in educational inequalities is investigated using an international comparison."

"The pedagogical and didactical implications of distance learning and a digital teaching culture, which constitute the foundation for digital curriculum making, are examined."

The state and the schools

"The educational researchers Ramirez and Ventresca [29] summed up the relationship between the individual, the state, and the school as an institution as follows: "Mass schooling becomes the central set of activities through which the reciprocal links between individuals and nation-states are forged" (p. 24)."

"While home-based instruction is not allowed in Germany, large church-run schools exist alongside the public school system, Luxembourg has pursued a joint state-church school policy that allows homeschooling in principle, even if parents rarely decide to take up this option."

"In the United States, private schools and homeschooling play a much larger role."

"In the United States, homeschooling is made possible by the historically explicable notion of the teacher and a specific kind of instruction (instructional design), which is produced in the form of curricula and can be distinguished from the continental European tradition of Didaktik."

Educational inequality and barriers

"It is therefore not surprising, especially during this pandemic, that scientists and the media have examined the connection between homeschooling or digital schooling and educational inequalities."

"It is worth taking a closer look at homeschooling, digital learning, and distance learning with regard to their impact on educational inequalities."

"Since primary school children in times of homeschooling are more dependent on parental support, and not only with regard to purely technical access to learning materials and tasks, we can assume that the younger the school population, the greater the inequality gap during the pandemic."

"Given the dimensions of inequality outlined above, that can result from homeschooling in times of a pandemic, it is best if students return to education within their learning community as soon as possible."

The plurality of teaching and learning cultures, and thinking outside the box: Digital curricula and school crisis intervention in international comparison

"Finland encourages independent and self-confident use of digital media at an early age, and this is firmly anchored in the curriculum: learning to type on the keyboard is part of the standard repertoire of a primary school student, and research with the help of digital tools is taught in the core curriculum for individual subjects."

"Although equipping schools nationwide with digital resources was an important basic step, both Estonia and Finland have harmonized their curricula with regard to digital education, among other things."

"Compared with Luxembourg students, Finnish and Estonian students were generally well equipped for remote teaching even before the coronavirus crisis, which is obviously related to the mandatory inclusion of digital media literacy in the curriculum."

"The development of such a digitally supported curriculum requires a definition of what is meant by digital cultures in school education systems and how their development transforms the pedagogical framework."

Digital cultures and new models of teaching and learning
"Based on the necessity for an educational response within the framework of the Covid-19 intervention strategies and the worldwide switch to forced distance learning and teaching, the framework of digital curricula could be discussed using the example of the internationally different concepts of curriculum making on the one hand and the digital literacy implemented in educational systems on the other."

"Rather, it would be a foundation reflected in the respective cultures of the teaching subjects, which (a) forms a matrix for the subject-related transfer of teaching contents, methods, and also social forms into the change of leading media, and in doing so; (b) takes into account the characteristics of digital culture in terms of referentiality, communality, and algorithmicity; and (c) offers an orientation for teachers and learners."

[Section 5]
"The small country of Luxembourg, where 60% of the student population is composed of non-Luxembourgish students, offers a nearly ideal laboratory setting for examining the processes relevant to virtually and digitally enhanced classroom instruction in multilingual and intercultural contexts during the coronavirus crisis."

"To better understand the processes relevant to digitally enhanced classroom instruction in multilingual and intercultural contexts during the coronavirus crisis, we will first sketch the historical lines of development of the nationalization of schools, as differentiated from "homeschooling" (section 2)."

"We will look at how school closures and digitally supported teaching settings may increase educational inequalities (section 3), as reflected in an international comparison (section 4)."

"We will also consider the pedagogical and didactical implications of distance learning and a digital teaching culture, which constitute the foundation for digital (in contrast with digitally supported) curriculum making (section 5)."

8.1.5 Post-Covid-19 Education and Education Technology 'Solutionism': A Seller's Market [269]

This is a machine-generated summary of:

Teräs, Marko; Suoranta, Juha; Teräs, Hanna; Curcher, Mark: Post-Covid-19 Education and Education Technology 'Solutionism': a Seller's Market [269]

Published in: Postdigital Science and Education (2020)

Link to original: https://doi.org/10.1007/s42438-020-00164-x

If you want to cite the papers, please refer to the original.

For technical reasons we could not place the page where the original quote is coming from.

Abstract-Summary

"The Covid-19 pandemic and the social distancing that followed have affected all walks of society, also education."

"This paper employs a critical lens to reflect on the possible problems arising from hasty adoption of commercial digital learning solutions whose design might not always be driven by best pedagogical practices but their business model that leverages user data for profit-making."

"Already before Covid-19, there has been increasing critique of how ed-tech is redefining and reducing concepts of teaching and learning."

"The paper also challenges the narrative that claims, 'education is broken, and it should and can be fixed with technology'."

"This is a critical moment to reflect how the current choices educational institutions are making might affect with Covid-19 education and online learning: Will they reinforce capitalist instrumental view of education or promote holistic human growth?"

"This paper urges educational leaders to think carefully about the decisions they are currently making and if they indeed pave the way to a desirable future of education."

The Covid-19 Pandemic and the Push to Online Learning

"This quick transition has also revealed gaps and shortcomings in how online learning has or has not been adopted in educational institutions."

"Quickly jumping on board with learning platforms and online learning has also raised concerns about privacy and surveillance and the impact on students' lives and human dignity (Harwell [30])."

"In the moment of crisis, educational organizations should think carefully about their choices regarding online learning and education technology."

"In order to mitigate possible negative impacts, educational organizations should leverage past knowledge of online learning as something that can be more varied than just a way to deliver information."

"Online learning can take many different forms, including those pedagogically more innovative and engaging than commonly used processes of knowledge delivery and assessment."

"Engaging teachers and students in the development, implementation, and use of education technology can affect how successfully technology can support meaningful teaching and learning (Bates and Sangrà [31]; Howland and others [32])."

Education Technology and Datafication: Seller's Market

"Businesspeople, computer enthusiasts, and politicians considered then, as they do now, that information and communications technology (ICT) is a magic bullet that improves teaching and learning, disciplines students as a skilful future workforce, and enhances democracy both in educational institutions and in larger society (Cuban [33]; Cuban and Jandrić [34]: 432)."

"With the rise of ed-tech companies, digital learning platforms have been criticized for redefining, simplifying, and reducing the concept of learning to better fit the education technology revolution narrative (Knox and others [35]; Manolev and others [36])."

"Many have raised important and unanswered questions about educational data ownership, access, and ethics, in addition to the role of student-centred development (Slade and Prinsloo [37]; Drachsler and Greller [38]; Corrin and others [39]; Prinsloo [40]; Tsai and others [41]; Teräs and Teräs [42])."

"Others have made attempts to reframe the use of data and analytics to promote student agency and educational equity (Tsai and others [41])."

"Students and parents often need to consent to data collection policies in order to use applications and platforms presented by the educational institutions."

Critique of 'Disrupting Broken Education'

"The argument of broken education is based on the popular notion that education is somehow 'broken' so it should and can be fixed with technology (Williamson [43])."

"Whereas recent Illich- and Freire-spirited research offer relevant criticisms of broken capitalist education (see Fernback [44]; Fuchs [45]; Jandrić [46]; McLaren and Jandrić [47]), there is also a strand of emerging criticism of 'broken education' that comes from the opposite ideological direction, that is, from entrepreneurial and pro-capitalist actors."

"In the Covid-19 pandemic, the hypothesis of 'broken education' offers an opportunity to ed-tech businesses to sell untested solutions which sometimes have little to do with proper teaching and learning philosophies."

"If the idea of 'broken' is to be defined empirically, it could then be observed in the reactions and activities that have taken place in the educational sector during the Covid-19 pandemic."

What Should Be Done?

"Already before the pandemic, leading scholars have clearly articulated an urgent need for development of critical education technology research (see Suoranta and Vadén [48]; Bulfin and others [49]; Jandrić [46])."

"Will post-Covid-19 education be shaped by visions of public good and designs of critically reflective and holistic learning, or will it be influenced by company interests for new markets (Selwyn and others [50, 51]; Williamson [43, 52])?"

"It is built through the 'collaborative turn', or what is called participatory culture, which includes relatively low barriers to civic engagement and activism, artistic and other sorts of expression, easy access for creating and sharing one's outputs with others, peer-to-peer relations and informal mentorship, as well as new forms of socialization, social connections, collectivism and solidarity. (Suoranta and Vadén [48]: 2.; see also Suoranta [53]) In the world with Covid-19, we must be ready to oppose the looming 'market-based language of profits, privatization, and commercial exchange' (Giroux [54]) in the area of educational technologies as well as taken for granted behaviourism and surveillance mechanisms in various digital learning solutions in the current seller's market dominated by few digital barons."

Discussion and Conclusion

"The motivation may be to develop and optimize ed-tech platforms that leverage datafication as a business model."

"If education is seen as something that should promote holistic human growth for a more democratic and just society, leveraging digital technologies might look somewhat different: connecting people to discuss, learn, and tackle common problems together."

"Critical examination of technology makes us think what online learning should look like in the with Covid-19 world."

"In order to develop a true praxis as critical intervention of our social surroundings and conditions, aiming at a human-centred world (Freire [55]; Fuchs [56]: 338), we need to begin to imagine that a different world is possible and to act towards it."

"We should avoid the temptation to overcome the Covid-19 pandemic educational crisis with ready-made, top down solutions and ed-tech industry's pedagogical colonialism."

[Section 6]

"Pandemics have forced humans to break with the past and imagine their world anew."

"It is a portal, a gateway between one world and the next."

"We can choose to walk through it, dragging the carcasses of our prejudice and hatred, our avarice, our data banks and dead ideas, our dead rivers and smoky skies behind us."

"We can walk through lightly, with little luggage, ready to imagine another world."

8.1.6 Automation Is Coming! Exploring Future(s)-Oriented Methods in Education [270]

This is a machine-generated summary of:

Cerratto Pargman, Teresa; Lindberg, Ylva; Buch, Anders: Automation Is Coming! Exploring Future(s)-Oriented Methods in Education [270]

Published in: Postdigital Science and Education (2022)

Link to original: https://doi.org/10.1007/s42438-022-00349-6

If you want to cite the papers, please refer to the original.

For technical reasons we could not place the page where the original quote is coming from.

Abstract-Summary

"Emerging automated-decision making (ADM) technologies invite scholars to engage with future points in time and contexts that have not yet arisen."

"We ask the following: what are appropriate research methods for studying emerging ADM technologies in education?"

"This constitutes a case in point for reflecting on the research methods applied to address the future of assessment in education."

"Studying the literature on speculative methods in digital education, we illustrate four categories of future(s)-oriented methods and reflect on their characteristics through a backcasting workshop conducted with teachers."

"We conclude by discussing the need to reconsider the methodological choices made for studying emerging technologies in critical assessment practices and generate new knowledge on methods able to contribute to alternative imaginaries of automation in education."

Introduction

"This paper examines the relevance of research methods oriented toward futures in automating teachers' assessment practices."

"We argue that the emerging status of digital automation in education calls for applying research methods and sensitivities able to engage with the 'not-yetness of digital education' and the inherent 'complexity, uncertainty, and risk … of technologies and practices which are unknown and in flux' (Ross [57]: 214)."

"Examining how ADM systems are currently studied and exploring the value of future(s)-oriented social science methods constitutes a research strategy to unpack 'dominant narratives about automation (Pink and others [58]: 15)."

"Against this backdrop, this article is organized as follows: By taking the case of ADM technologies in teachers' assessment practices, we explore the relevance of research methods oriented toward the future to identify narratives of automation that are less discussed in current debate (Selwyn and others [59])."

"Our first-hand experience with future(s)-oriented research methods is recounted to reflect on their relevance to studies on automation in education."

Knowledge Gaps in the Study of Algorithmic-Decision Making in Assessment Practices

"We motivate the interest in the issue of exploring the relevance of research methods for the study of automation, in the specific case of assessment practices with algorithmic-decision making technologies, and more specifically automated grading systems, for several reasons."

"Our inquiry about the relevance of methods applied for the study of automation in assessment practices is motivated by observing: (1) dimensions of assessment that are currently overlooked, although they are key to focus on, given the increasing deployment of automation in secondary and tertiary education, (2) lack of knowledge about the influence of novel machine learning techniques on everyday teachers' assessment practices, (3) dominant grand narratives about automation that need to be confronted with the messy and mundane realities of assessing students with sophisticated tools that are often hard to appropriate."

Which Research Methods for the Study of Algorithmic-Decision Making in Education?

"Social science methods oriented toward the future issued from anticipation (Adams and others [60]; Miller [61]), speculation (Auger [62]; Ross [57, 63]), and critical future studies (Miller and Sandford [64]; Oomen and others [65]) are in this respect promising for the qualitative research of emerging ADM practices in education."

"Future(s)-oriented methods aim to generate speculative accounts of users' educational experiences with technologies where the past is blended with the future points in time."

"By epistemology, Ross [63] refers to the circular, embodied reasoning mobilized with future(s)-oriented methods in education research."

"The epistemological quality of future(s)-oriented methods, explained by Ross, is central to our inquiry about how to study emerging automated assessment practices as it allows us to reflect on how a specific research problem or its space is defined."

"The insights from this illustrative example contribute to discussing how future(s)-oriented methods complement representational ones in the study of emerging automated assessment practices in education."

Putting Future(s)-Oriented Methods into Practice

"We described the activity consisting of group discussions to stimulate speculative thinking and individually to write postcards from the future year 2032 (i.e., backcasting futuring techniques)."

"In small groups (i.e., two or three teachers), the participants discussed the questions we provided and wrote notes regarding their school and profession in 2032."

"The teachers' speculative activity regarding student examination was a real space for the participating teachers to reflect and discuss assessment and digital practices."

"From the teachers' presentations and readings of their postcards, we agreed with the participants on mentioning 17 topics that shared both preferable and possible futures for the teachers."

"The teachers shared visions of possible implications of automated grading in school; such implications went beyond discussions centered on the efficiency and effectivity of ADM in teachers' assessment practices and student performance."

Discussion

"Although both representational and future(s)-oriented methods are applied for the study of sociotechnical practices in education and particularly ask about how actors interact and engage with the socio-material world they inhabit, they differ significantly in how such a socio-material world is understood and analyzed."

"We connect the performativity quality of speculative methods with how educational researchers support others (e.g., teachers) to enact visions of a common future and actions that matter to them in view of living a good life."

"The choice of postcards as the discursive space to shape futures and pasts, the guidelines selected to guide the participants to write the fictive narratives, and the language used in the questions formulated by the researchers to scaffold participants to discuss futures and automation, the images we pre-selected for the postcards, and the general atmosphere and safe space created during the speculative workshop are, among others, designed materials, research tools that configured the teachers' speculative fabulation experience."

Conclusion

"Current studies on emerging ADM technologies in assessment practices need to be attentive to the methodological choices made in selecting research methods so we can contribute knowledge and change in education that matters for those doing education every day and from the ground."

"This is crucial as automated technologies are emerging, and their role and function in educational practices are unsettled and partially dependent on the anticipations and valorizations of educational practitioners' engagements."

"This means that practitioners — students, teachers, administrators, and technology providers — envisioning the role of emerging technologies play a significant role in how they approach, enact future educational practices, and gain agency through their narratives."

"Neglecting to address the non-linearity, abductive, and emergent qualities of practitioners' experiences and actions constitutes a significant flaw in studying the role played by emerging automation technologies in education."

8.1.7 Indigenous Wisdom, Capital, Technology and Education [271]

This is a machine-generated summary of:

Ruth, Damian: Indigenous Wisdom, Capital, Technology and Education [271]
Published in: New Zealand Journal of Educational Studies (2020)
Link to original: https://doi.org/10.1007/s40841-020-00181-1

If you want to cite the papers, please refer to the original.

For technical reasons we could not place the page where the original quote is coming from.

Abstract-Summary
"This article contrasts specific aspects of indigenous world views and wisdom on the one hand with specific themes in capitalism, colonisation, corporate interests, technology and education, on the other and argues that there is a fundamental clash of values between them."

"There is no assumption of a homogenous indigenous wisdom and no claim is made about novel insight into capitalism, colonisation, corporate interests and technology and education."

"The conclusion is that education would be better served by the insights of indigenous wisdom and a more critical and tempered view of capitalism and technology."

Introduction

"This paper explores the clash between a sensibility shaped by indigenous world views on the one hand and capitalist/marketized world views and a technological mindset on the other."

"The implication of Ruth and Marshall's positions is that the world of capitaliszed, marketized and technologized education is incommensurate with an indigenous view of education as a process of developing a holistic awareness of being in and part of the world, a sensibility, or worldview, eloquently describe by Mason Durie, Charles Royal, Carl Mika, Georgina Stewart and others."

"The intent is to develop the case that the summative impact of the world view informed by capital + market + corporation + technology is much greater than its parts, and in its effect, and in contrast to an indigenous perspective, is inimical to the well-being of the earth and all in it."

The Challenge of Contrasting 'Indigenous' and 'Western' Wisdom

"Contrasting indigenous and Western wisdom can be misleading."

"Some general distinctions may sometimes be made between Western science and indigenous knowledge, but they are always problematic."

"One may challenge beliefs surrounding indigenous knowledges but although beliefs such as " (a) knowledge can be neatly divided into indigenous versus Western worldviews, and these epistemologies are opposite and irreconcilable, (b) indigenous knowledge always supports a view of nature and of social life that is inherently benign and based on an ethic of care, and (c) indigenous knowledge is static and immutable, and when indigenous youth adopt ideas or practice exogenous to the culture they lose their sense of identity" (Barca and Arenas [66], p. 5) may be challenged, it is the case that indigenous communities can find common ground on the basis of a human-ecological union with a distinctive environmental ethic (Durie [67], p. 302)."

"'a "formal indigenous culture" is one that is conscious in its relationship with natural world environments' (Royal [68], p. 1, emphasis in the original)."

Capitalism and Markets

"The thrust of his argument is that the neo-liberal ideology of free markets is not capitalism but destroys it."

"Debates about free and fair trade are "essentially about moral values and political decisions … it is not something economists with their technical tool kits are particularly well equipped to rule on" (Chang [69], p. 7)."

"It was the expansion of markets and of market values, into spheres of life where they don't belong… we need to ask whether there are some things money should not buy" (Sandel [70], p. 7)."

"The environment and education can be capitalized, marketized, financialized, but this does not simply clash with an indigenous world view, it destroys it, along with a concept of education connected to human flourishing."

"The dominant current world view marked by concepts of capital and market is predicated on fragmentation, alienation and extraction, and given the state of the earth, is manifestly bankrupt."

Some Distinguishing Features of Indigenous Wisdom

"According to Viergever (1999, in Durie, p. 3) the three distinguishing features of indigenous knowledge are "that it is a product of a dynamic system, it is an integral part of the physical and social environment of communities and it is a collective good The emphasis on the ground, the earth, the environment is repeatedly emphasized in indigenous knowledge relating to "a specific local environment over the long term … settling in and sinking one's roots" (Jolly and others [71])."

"These themes are noted by indigenous intellectuals from different parts of the world."

"I continue to explore indigenous, specifically Māori, values below, but this is an appropriate point to note the importance of "a specific local environment over the long term … settling in and sinking one's roots" and the metaphor of rauru."

Technology

"The controversy over Francis Bacon's torturing of nature or putting her on the rack, is complex (Pesic [72]) and beyond this paper, although there is a point to be made here about the conflation of Other with reference to the white male, in which nature and indigenous/brown/black/female are treated as subordinate with man (sic) as scientist is engaged in heroic struggle with nature in which both are tested and purified."

"Pesic suggests that Bacon's vocation as a lawyer and judge relates to his use of "torture" (or its Latin cognates) to denote excessive and wrongful force; he never speaks of experiment expressly as the "torture of nature.""

"In identifying technique, or technical efficiency, as the key ideological fetish of modern industrialized authoritarian states, Marcuse ([73], p. 138) defined technology as "a mode of production, as the totality of instruments, devices and contrivances which characterise the machine age [which] is thus at the same time a mode of organizing and perpetuating (or changing) social relationships, a manifestation of prevalent thought and behaviour patterns, an instrument for control and domination"."

Indigenous Wisdoms

"The core Māori value of being-in the-world as relational wisdom rests on kaitiakitanga or stewardship through which humans as stewards are endowed with a mandate to use the agency of their mana (spiritual power, authority) to create mauri ora (conscious well-being in a web of reciprocal relationships."

"In the Māori view, a human is born into a relational world as kaitiaki, empowered to be a steward and endowed with obligations."

"In the Māori worldview, the child arrives with mana, and agency to create mauri ora, a conscious well-being which is the well-spring of sustainability."

"Being a steward endowed with mana requires that a child lives, and grows into an adult, in respectful recognition of their own mana and mana in the world around them."

"In explaining relational well-being, wealth and an ethic of care in terms of business, Spiller and others [74, 75] contrast Māori values with independent, rational, self-interested, utility maximizing organizations and individuals."

Capital, the Corporation and Technology—Mergers Made in Hell

"The conceptual process proceeds: it is possible to monetize/technologize something; it is possible to monetize/technologize almost anything; the monetized/technologized form is what is valued; therefore, if we want to value something, we should monetize/technologize it; there are advantages to monetization/technologization; therefore we should monetize/technologize all possible things so that they can be valued; eventually, only those things which can be monetized/technologized can and will be valued."

"Using a machine to efficiently process an investment is not the same as educating a person; the corporatized technologized university (if it still makes sense to call the machine that) does the former."

"What must be emphasized here is the role of the corporation as "an autonomous technical structure that behaves by a system of logic uniquely well suited to its primary function: to give birth and impetus to profitable new technological forms, and to spread techno-logic around the globe" (Mander [76], p. 120)."

"Corporations are major producers and consumers of technology and the quintessential means of this process is the computer."

Technology

"Embedded in every tool is an ideological bias, a predisposition to construct the world as one thing rather than another, to value one thing over another, to amplify one sense or skill or attitude more loudly than another" Postman points out (p. 13) and notes that "New technologies alter the structure of our interests; the things we think about."

"Odell [77] writes of resisting the attention economy, and refuses "the frame of reference in which value is determined by productivity, the strength of one's career, and individual entrepreneurship [and wishes to celebrate] a form of the self that changes over time, exceeds algorithmic description, and whose identity doesn't always stop at the boundary of the individual" (p. xvi)."

"An important feature of production models that Franklin points out is that it is easy to frame the process to allow for externalities to be disregarded."

"What is disregarded in the production model of education is … "an individual's own conscience and discernment" (p. 37)."

Technology in Education

"A general point made by many negative critiques of educational technologies is that they replicate, reify and support existing political, economic and social relations and in this respect their innovativeness is limited."

"As Jeremy Knox ([78], in Selwyn, p. 80) has noted, the limitations of production and consumption are evident in the emphasis of open educational resources on individual self-driven-direction and autonomy."

"The promotion of cheap, just in time learning does not attend to the traditional purposes associated with higher education, specifically "exposure to a broad range of cultural traditions that enable active and reflective participation in civic life" (p. 19)."

"Technology reinforces the common but misguided assumption that the point of education is to promote freedom and individual autonomy as though individuals can exist independently of the cultural natural ecology from which they derive their language and basic sources of energy."

Conclusion

"Benade [79] reflects on "the mental shifts and sometimes painful transitions teachers and leaders are making and experiencing, as they move through uncharted waters, from traditional classroom practices to ones emphasising collaboration, teamwork and the radical de-centring of their personal roles" (p. 2)."

"The sources of pain in educators have been precisely identified by Palmer [80]: "an obsession with educational externals—including relentless and mindless standardized testing (p. xvii) ... a mindset that cares more about weights and measures more than meaning (p. xxi)"; he offers the simple but profound premise that "good teaching cannot be reduced to technique; good teaching comes from the identity and integrity of the teacher" (pp."

"This is the pain that educators feel—not being trusted to realise their vocation."

"The pain radiates between individuals, cultures and societies via educational systems."

8.1.8 Smart Schools on the Way: How School Principals from Catalonia Approach the Future of Education Within the Fourth Industrial Revolution [272]

This is a machine-generated summary of:

Mogas, Jordi; Palau, Ramon; Fuentes, Marta; Cebrián, Gisela: Smart schools on the way: How school principals from Catalonia approach the future of education within the fourth industrial revolution [272]

Published in: Learning Environments Research (2021)
Link to original: https://doi.org/10.1007/s10984-021-09398-3

If you want to cite the papers, please refer to the original.

For technical reasons we could not place the page where the original quote is coming from.

Abstract-Summary

"Learning becomes ubiquitous, and schools assume new roles with systemic changes in communication, administration and management, becoming learning organisations."

"The use of technologies aligned with pedagogical strategies and new methodologies must lead to more-personalised systems."

"A comprehensive definition of smart schools is proposed."

"Smart schools must be endowed with integral management systems, inclusive, sustainable, and adopt new learning methodologies and advances from Industry 4.0 in an efficient way."

"Conception and because research, government policies and business projects are not always in line with research, there is a need for deeper knowledge of how schools are approaching their upcoming transformation."

"Thematic analysis focusing on technological and pedagogical innovations, management systems, inclusion, and sustainability identified some analogies with related research, pointing out that schools are far from implementing advanced technologies."

"Conclusions are drawn to show that, although schools are not yet prepared to cope with the Fourth Industrial Revolution, its impact relies on the technology's level of maturity and ease of use, as well as stakeholders as policymakers."

Introduction

"The approach of advanced technology for smart classrooms presented by Lorenzo and others [82] is broadened and deepened to focus on the convenience of a critical implementation of technologies in schools in terms of inclusion, personalisation as a complement to collective learning, and ethics when introducing advanced technologies."

"The OECD [83] established principles for innovative learning environments and argued for the need to consider the concept of 'learning environments' rather than 'schools' or 'classrooms', reinforcing a focus on the learning organisation in its whole conception, not only on the institution and physical places where learning formally happens."

"Smart schools involve these technological innovations in schools being complemented with pedagogical innovation, improved communication and management processes, and inclusive designs and decisions towards sustainability."

"Advances in the conceptualisation of smart classroom and smart school concepts take into account the need for these spaces to be inclusive, thereby boosting the paradigm change and avoiding repeating the capitalist paradigm through the implementation of technology in education (Means, [84])."

Background

"While offering automated processes is the ultimate goal, the current state of Industry 4.0 allows initial improvements such as the Smart Schools Management System (SSMS), which was developed in Malaysia for school administrators to manage all aspects (resources and processes) required to support teaching and learning, including administrative functions as lesson schedules, programming activities, and computerised student academic records (Salimi & Ghonoodi, [85])."

"Accessible education focuses on aspects such as the physical environment of schools, communication and language, learning materials, assessment of students, support for individual needs, mobility, and assistive devices or technologies (Booth & Ainscow, [86])."

"Smart education allows smart schools to have more-personalised learning within the learning environment."

"Trending methodologies related to smart education in smart classrooms are cooperative inquiry, project or problem-based learning, case study and simulation (Cebrián and others, [87]), which must allow individual and collective personalisation of learning and the active involvement of students in the learning process to promote personal and community well-being and fulfilment."

Method

"A semi-structured interview approach was chosen to allow deep understanding of the problem being researched through documenting participants' experiences (Kvale & Brinkmann, [88]; Minichiello and others, [89])."

"The interviews included main themes for discussion, using open-ended questions, and allowed school principals to talk in-depth about the use of technology within their schools."

"Interview questions were related to: technological and pedagogical innovations carried out; use of technological systems to improve communication with families and the management of and coordination between teachers; inclusion initiatives that affect school facilities and resources; innovations in school facilities towards sustainability; and the needs, challenges and resistances identified in the adoption of innovative methods supported by the use of ICT."

"This approach was chosen because it allowed a deep analysis and explanation of the data and the creation of key themes that participants considered relevant to how sustainable technological and pedagogical innovations can be implemented within their schools."

Findings

"Some interviewees commented on frequent changes and updating: We have promoted the implementation of technologies, with a long journey: first projectors in all classrooms, which were replaced by interactive whiteboards and now by interactive panels; computers in all classrooms, mini laptops, two computer rooms, tablets, etc and all that its use entails, which is a change in the way it works. (Participant 10) It is usual to report advances in basic technology advances in schools (Vincent-Lancrin and others, [90])."

"Some school principals even reflected awareness about the need to follow the Universal Design for Learning (UDL): This course is already the second one that, according to the strategies of the management project specified in the Annual General Program, the Universal Design for Learning is being worked on with the help of the EAP [...] The teaching staff work to design classroom activities to reduce the barriers to participation that each class have."

Conclusions

"Educational research is very focused on specific advances and so no holistic proposals of smart schools are available yet, considering the economic costs and lack of professional training."

"Key implications for further research and action to embed smart education and technological advances within schools are drawn from this study: The investment in school infrastructure and equipment should be increased and should be considered from a holistic point of view as a smart school."

"Because school principals have limited understanding and vision of smart schools' potentialities and benefits, training and professional development programmes in smart education are needed to help them in the process of implementation."

"Industry 4.0 must develop educational technologies and make them available to smart schools: improvements in artificial intelligence, cloud computing and Internet of Things are needed to produce a clear impact in smart education, with solutions being accessible and easy to use."

8.1.9 Two Comparative Studies of Computer Simulations and Experiments as Learning Tools in School and Out-of-School Education [273]

This is a machine-generated summary of:

Krüger, Johanna T.; Höffler, Tim N.; Wahl, Martin; Knickmeier, Katrin; Parchmann, Ilka: Two comparative studies of computer simulations and experiments as learning tools in school and out-of-school education [273]

Published in: Instructional Science (2022)

Link to original: https://doi.org/10.1007/s11251-021-09566-1

If you want to cite the papers, please refer to the original.

For technical reasons we could not place the page where the original quote is coming from.

Abstract-Summary

"Interactive computer simulations and hands-on experiments are important teaching methods in modern science education."

"For the communication of complex current topics with social relevance (socio-scientific issues), suitable methods in science education are of great importance."

"This paper presents two studies of direct comparisons of hands-on experiments and interactive computer simulations as learning tools in science education for secondary school students in two different learning locations (Study I: school; Study II: student laboratory)."

"Using a simple experimental research design with type of learning location as between-subjects factor ($N_{\text{Study I}} = 443$, $N_{\text{Study II}} = 367$), these studies compare working on computer simulations versus experiments in terms of knowledge achievement, development of situational interest and cognitive load."

"Independent of the learning location, the results showed higher learning success for students working on computer simulations than while working on experiments, despite higher cognitive load."

"Working on experiments promoted situational interest more than computer simulations (especially the epistemic and value-related component)."

Introduction

"Shouldn't educators, therefore, use simulations instead of traditional hands-on experiments especially when communicating complex and current topics like socioscientific issues?"

"This paper presents two studies of direct comparisons of hands-on experiments and interactive computer simulations for learning complex marine ecology issues."

"Findings regarding the influence of hands-on experiments or simulations on students' learning outcomes are inconclusive so far."

"Empirical studies report several cognitive and metacognitive difficulties for students learning with computer simulations (Jong, [95]; Jong & van Joolingen, [96]; Köck, [97]; Mayer, [98])."

"With regard to performing hands-on activities, previous studies (e.g., Scharfenberg and others, [99]; Winberg & Berg, [100]) indicated that students might also experience increased extraneous load during."

"Zendler and Greiner [101] showed no general difference, in a direct comparison of experiments and simulations in chemistry education, in students' learning outcomes on reactions of metals."

The present study

"This paper presents findings of two studies offering direct comparisons of hands-on experiments and interactive computer simulations as learning tools for students from Grade 10 to Grade 13."

"The two studies investigated the effects of working on these learning methods focusing on content knowledge achievement, cognitive load and changes in situational interest."

"These studies often compare at the level of the entire learning environment (e.g., different laboratories) rather than at the level of the individual modality (experiments or simulation)."

"Some studies exist with a direct comparison of the two methods primarily in the fields of physics or chemistry education, focusing on the comparison of virtual vs. analogous experiments of low complexity (e.g., Evangelou & Kotsis, [102]; Madden and others, [103])."

"Our study presents a direct comparison of differently performed experiments with the same content and high complexity."

Research Questions and Hypotheses

"We expected that both methods would improve students' content knowledge, since the topic was new to most of them and both experiments (Ekwueme and others, [104]; Tunnicliffe, [105]) and simulations (Rutten and others, [93]; Smetana & Bell, [94]) have been shown to promote learning."

"We expected a medium students' cognitive load for both methods."

"We expected a low level of prior knowledge of the students, since the topic addressed was not a mandatory topic in the German curriculum, which could have a negative impact on the intrinsic cognitive load (Jong, [95]; Jong & van Joolingen, [96])."

"We expected higher cognitive load when working with the computer simulations compared to working with the experiments. (c) To what extent do interactive computer simulations / hands-on experiments increase situational interest of the students?"

"Through the work on computer simulations we expected a higher increased situational interest because of the motivation character of a digital medium."

Methods and design

"The supervisor assisted the students both in performing the experiments or working through the simulation and in answering the questions."

"The structure of the questions was based on the content areas of the methods and thus reflect the topics covered in the experiments and simulation: pH-value (3 questions), reasons for and extent of the increase in global atmospheric carbon dioxide (CO_2) concentration (2 questions), equilibrium reaction of CO_2 in seawater (3 questions) and the effect of acidification on calcifying organisms (1 question)."

"For questions 2 and 3, students were able to score two points each for causes, process, and effects of the change."

"For the last three open questions, students had to (a) describe the effects of all four predicted changes in combination for the year 2100 at the organism level, (b) the resulting impacts on the ecosystem as well as (c) on the societal level."

Results

"The analysis of the prior knowledge questionnaire showed that students achieved an average of 1.3 points ($SD = 1.05$) out of possible 4 points in the first study."

"In the second study students achieved an average of 3.38 points ($SD = 2.46$) out of 18 possible points for the three open questions in the pre-test."

"The t-tests of the dependent variables of both studies confirmed that on the one hand, learning with the simulation compared to learning with experiments resulted in higher students' content knowledge achievement ($d_{\text{Study I}} = -0.19$; $d_{\text{Study II}} = -1.05$) and a higher level of cognitive load ($d_{\text{Study I}} = 0.19$; $d_{\text{Study II}} = 0.21$)."

"Students who worked with the experiments showed a higher level of the epistemic ($d_{\text{Study I}} = -0.33$; $d_{\text{Study II}} = -0.54$) and value-related ($d_{\text{Study I}} = -0.35$; $d_{\text{Study II}} = -0.37$) component of the situational interest by than by working with the simulation."

Discussion

"The goal of this research was to investigate the differences between learning with computer simulations and experiments concerning (a) the contribution of content knowledge, (b) the extent of affected cognitive load, and (c) the extent of situational interest to find out which method is better suited to communicate current and complex topics."

"We conducted two studies using a simple experimental research design with type of learning location as between-subjects factor."

"The following sections show detailed explanations of the assumptions for each research question and resulting implications."

Content knowledge

"Besides the different modality itself (simulation vs. experiments), there are several indications on why the simulation in this case facilitated learning the interaction of multiple variables: We suggest, that the focus of the students' tasks in the two methods could have an impact on the content learned (Puntambekar and others, [106])."

"Since the students did not have to deal with physical activities during working on the simulation, but had more time to investigate processes and contexts, it stands to reason that they were also able to learn this content better."

"The division of labor in the experiments might have been more enjoyable for the students, as the interdependence in the different steps of the work led to a higher interactivity within the group and supported social learning more than in the simulations."

Limitations

"Of all, the results and implications presented here are of course only valid for the interventions described in this paper."

"The specific context of the activities (complex, current scientific topics) and characteristics of the methods (instructional support, collaboration forms, etc) influenced the results."

"The two studies differed not only in the learning location but also in the duration of the interventions, which is a theoretical limitation."

"Since the results show a high degree of similarity, we conclude that this difference did not have a significant impact on the outcomes."

"We have used this measurement because it has been shown that this subjective evaluation scale is the most sensitive measure available to differentiate the cognitive load of different teaching methods (Sweller, [107])."

"This is especially noticeable for the epistemic and value-related situational interest scales, which show a significant difference between the methods."

Conclusion

"This paper provides insights into some students' motivational and cognitive effects while working on experiments and computer simulations."

"We could show repeatedly the advantages of interactive computer simulations with regard to the communication of complex contents as well as the importance of hands-on learning activities to promote students' situational interest (especially with regard to the value-related and epistemic components)."

"Subsequent to our findings, Puntambekar and others [106] indicated that learning with one method can complement and complete the weaknesses of the other (different focus of the content learned), and also point out to combine both strategically."

"Smetana and Bell [94] as well as Lichti and Roth [108] also pointed out the combination of laboratory experiments and model-based simulations as a potentially powerful learning tool."

"De Jong and others [109] emphasize that a combination of virtual and non-virtual learning elements seems to enable a deeper and more nuanced understanding, especially when teaching complex issues."

8.1.10 Online Distance Mathematics Education in Brazil: Research, Practice and Policy [274]

This is a machine-generated summary of:

Maltempi, Marcus Vinicius; Malheiros, Ana Paula dos Santos: Online distance mathematics education in Brazil: research, practice and policy [274]

Published in: ZDM – Mathematics Education (2010)

Link to original: https://doi.org/10.1007/s11858-009-0231-3

If you want to cite the papers, please refer to the original.

For technical reasons we could not place the page where the original quote is coming from.

Abstract-Summary

"We address online distance mathematics education research and practice in Brazil, which are relative newcomers to the educational scene."

"The national context of education in Brazil, highlighting the organization of the educational system, and also a summary of national legislation on distance education and an overview of digital inclusion in the country."

"With respect to research and practice in online mathematics education, we present support for research, examples of studies and highlight different aspects being addressed, including its essential components."

Introduction

"We address distance education in Brazil, conducted over the Internet in particular, with a focus on mathematics education, portraying various initiatives and research trends."

"Public access to the Internet began in 1995; however, online distance education is recent, and the number of studies that focus on online mathematics education is still small and restricted to a few groups."

"This increase has provided incentives for research in online distance education, including in the field of mathematics."

"The discussion regarding the quality of distance education courses is unquestionably relevant; however, we have not progressed further on this topic due to the lack of data and research in this field, especially in mathematics."

"Our focus is to show the progress in research about online mathematics distance education in Brazil, portraying its various initiatives and trends."

"We then discuss research in online mathematics education, which is strongly associated with the practice of distance education."

The national context in education

"Data from INEP also show that students enrolled in college-level distance courses differ from students enrolled in traditional face-to-face programs in that they are more likely to be non-white, older (average 34 years old), poorer, married, already employed, and their parents have lower educational levels."

"It can thus be said that Brazil is currently experiencing a rapid expansion of college-level distance education programs, serving mainly older students, probably due to disadvantages accumulated in elementary and high school, to which access was made more difficult by their economic condition."

"In college-level programs in pedagogy, specifically, designed to prepare teachers to teach at the pre-school level and the first 5 years of elementary school, there has been a 4% decrease in enrollment in traditional face-to-face programs, while enrollment in distance courses increased 183%."

Research and practice

"Research on different aspects of mathematics distance education has been conducted in association with practice, i.e., based on online courses proffered by researchers interested in the theme or undergraduate distance courses offered by institutions of higher learning."

"To the symbolic notation, Borba and others [110] point to difficulties with the discussion of geometry in distance classes and present important demands for software developers to strengthen online mathematics education, among them the need to facilitate the use of mathematical symbols in all the text-based communication tools."

"Bairral [111] discusses issues related to the evaluation process during continuing education in distance courses proffered to mathematics teachers."

"Much of the research on online distance education cited in this article encourages and values interaction, searching, collaboration, and investigation, the same characteristics emphasized in the mathematical modeling approach."

"Some research has been carried out on the use of mathematical modeling in online distance education in Brazil."

Conclusions and trends

"Online distance education can address the urgent need to prepare more and better teachers in diverse fields of knowledge, mathematics among them, and the data we have presented show this, as long as it is done well."

"This technology, the VLE in the case of online distance education, conditions the process through the tools it makes available to the teachers and students."

"These researchers are studying possibilities for synergy between mathematics education and theories originating from the arts found in the virtual world and analyze how these issues can also be used in the context of online distance education."

"We call attention to the fact that, in the process of reviewing research on online distance mathematics education in Brazil, we noted a lack of research in Brazil (Litto and Formiga [112]) as well as internationally."

8.1.11 The Sublation of Digital Education [275]

This is a machine-generated summary of:

Malott, Curry: The Sublation of Digital Education [275]
Published in: Postdigital Science and Education (2019)
Link to original: https://doi.org/10.1007/s42438-019-00083-6

If you want to cite the papers, please refer to the original.

For technical reasons we could not place the page where the original quote is coming from.

Abstract-Summary

"I argue that teachers' efforts to defend education against reactionary forms of digitalization and from complete automation should be supported by progressives."

"Because sublation occurs both within entities as they develop and out of entities as they are negated, I contend that sublating digitized education both within capitalism and beyond capitalism should be pursued simultaneously."

[Section 1]

"Logical and compelling case against machines, Marx demonstrates that those same machines used to destroy the working class under capitalism can liberate workpeople from the need to labor under socialism."

"While they correctly identify sublation as the process through which the coercive and exploitative consequences of increasing productivity under capitalism are overcome and transformed into benefits for the producers, they do not specifically consider the way the shift from capitalist to socialist purposes and values might sublate the content and structure of digitalization itself."

"They address how the labor-saving properties of technologies will be preserved from capitalism to socialism with the benefits transferred from the capitalist class to the working class."

"Fuchs and Sandoval offer a very general overview of the types of labor dominant in capitalism that might be unnecessary without the very specific process of producing value capitalistically from managers, executives, the military, marketers, entertainers, professional athletes, and so on."

Algorithmic Culture in Education

"The use of algorithms in capitalism will likely become increasingly significant since 'more and more aspects of education...are being conducted through software programs that have been written in code, and that rely on algorithms for their functioning' (Williamson [113]: xv)."

"Because algorithms rely on an immense amount of 'machine-readable digital data' about students and their schooling, Williamson ([113]: xv) describes the outcome of this process as the datafication of education."

"Test scores, attendance records, inspection reports, and 'clickstream data from online courses' are easily converted into digital data that can be inserted into algorithmic programs, contributing to the digitization of education (Williamson [113]: 5)."

"The idea is that the algorithm, functioning as a labor-saving 'teacher bot,' is not only pre-loaded with student data, but processes additional data from test scores and other inputs providing individualized curriculum and instruction in real time."

"Like videogames and teacher bots, behavior modification algorithms such as ClassDojo are both object and process."

Posthuman Sublation

"Bayne (Bayne and Jandrić [114]: 211) discusses massive open online classrooms (MOOCs) that, on one hand, pose a still unrealized threat to universities by dramatically decreasing the number of professors and teachers needed to educate a given quantity of students."

"In making her case against 'the assumption that presence is the privileged mode' and the most 'authentic' mode of higher education, Bayne (Bayne and Jandrić [114]: 201) maintains that 'we should think about what it means to have a global campus' extending 'beyond the material into the digital.'"

"Contributing to the challenge against privileging so-called in-person education, Bayne (Bayne and Jandrić [114]: 205) mobilizes a number of useful concepts: 'bounded space, networked space, fluid space, and space of fire.'"

"Without dismissing the usefulness of the concept of the network, Bayne (Bayne and Jandrić [114]: 205) stresses that 'the network is not enough, so we need to think how it relates to other kinds of connections which may be more fluid, flickering, and volatile.'"

Creativity as Resistance

"The tendency within the education community to view digital learning with skepticism and caution could be an issue that drives teachers to push their unions not in a more Luddite direction, but in a more radical, creative direction."

"Jeremy Knox ([115]: 358) contributes to the calls for creatively transforming digital education by challenging the long-held anthropocentric assumption that 'authentic education has tended to be understood in terms of "pure" human relationships between teachers and their students.'"

"Knox [115] points to three broad areas of the digitalization of education in need of creative transformation: the move to enclose an open web to a platform model using closed apps more conducive for capturing surplus value; the impact of the datafication of education leading administrators, for example, to focus on the narrow data inputted in algorithms to boost their rankings; and the environmental and labor costs of creating the technological infrastructure thereby challenging the myth that immaterial production is paving the way for an environmentally sustainable, exploitation-free capitalism."

Sublating Policy Projections

"Considering capitals' educational policy projections as responses to the way they frame looming disruptions in employment stemming from labor-saving digitalization offers further insight into possible future contours of a sublated digital education."

"Alexander Means [116], for example, examines three major policy and research organizations' framing of future risks or disruptions in employment as a direct result of digitalization."

"Policy institutes are changing their discourse from framing education as the path to growth and development to framing it as mitigating or countering future disruption."

"Means [116] notes how corporate policy projections convey both a sense of fear regarding disruptions to social stability and profitability and a sense of optimism for the economic promises attached to algorithmic culture specifically and labor-displacing digitalization more generally."

"Means [116], for example, notes that some institutes argue that algorithms designed to respond to rapidly changing conditions from environmental catastrophe to social disruptions are crucial to most effectively mitigate against such crises."

Hacking: a Method to Sublate

"At its core then, the progressive hacking of algorithmic education is about developing education in a way that would make it better suited to meet the people's learning needs."

"Galloway's ([117]: 107–108) categorization offers a useful framework for considering the sublation of algorithmic approaches to education: …at the level of its visual design, substituting new level maps, new artwork, new character models, and so on; (2) at the level of the rules of the game, changing how gameplay unfolds—who wins, who loses, and what the repercussions of various gamic acts are; or (3) at the level of its software technology, changing character behavior, game physics, lighting techniques, and so on."

"Applied to algorithmic education, gamic modification, endowed with a progressive political orientation, presents itself as a useful framework for reflecting on the possibilities of how sublation might be enacted."

Conclusion

"To take a non-deterministic form of educational sublation as a serious consideration is to not only acknowledge the need for more social justice within capitalism, but it is also to acknowledge the need to transcend capitalism more completely."

"The various forms of digital education explored in this essay and considering the possible ways they may be sublated offer suggestions for moving forward."

"The critical digital education scholars' many insights regarding the sublation of digitalized education demonstrate what can be achieved within capitalism as a place of departure to think beyond capitalism."

"To facilitate this thinking, an ongoing engagement with digital education policy projections and how they may be sublated seems to be a potentially fruitful endeavor."

[Section 8]

"This emphasis on educators thinking creatively about a social justice approach to digitalization within capitalism offers many important insights for an approach to digitalization connected to the revolutionary process against capitalism."

"There is a trend among critical educators pushing back against new forms of digital social control and student manipulation and the prospects of losing their jobs to machines as a contemporary manifestation of the Luddite tradition."

"Critical educators who overcome disruptive, labor-saving (i.e. efficient) aspects of digital online forms of education preserve them in an altered socially just form within capitalism."

"After laying out this theoretical framework, I delve into various forms of digital education considering the possible ways they may be sublated."

"I also engage critical digital education scholars for insights regarding the sublation of digitalized education."

8.1.12 Digitalization of Education: Commodification Hidden in Terms of Empowerment? [276]

This is a machine-generated summary of:

Armila, Päivi; Sivenius, Ari; Stanković, Biljana; Juutilainen, Lauri: Digitalization of Education: Commodification Hidden in Terms of Empowerment? [276]

Published in: Postdigital Science and Education (2022)

Link to original: https://doi.org/10.1007/s42438-022-00347-8

If you want to cite the papers, please refer to the original.

For technical reasons we could not place the page where the original quote is coming from.

Abstract-Summary

"With our empirical case, we bring into this discussion the contemporary reforms of formal education that lean on an imaginary idea of digitalization as an equalizing power without recognizing the criticism directed towards it."

"Within this case, the article covers postdigitalism in education where digitalization is not enough to make a difference."

"In our data collected from authorities, imaginaries of harmonization and equalization can be recognized where educational equality and empowerment are believed to be achievable via digital structures and patterns of schooling."

"Selected examples from the concrete educational stratification of the Serbian population show that digitalization per se cannot be a solution: for example, poverty hierarchies within the population are tough and should be solved first."

Introduction

"Digitalization, under this scrutiny, means changes in many fields of life, e.g., in education where teaching and learning practices are moving from personal encounters to more distant and algorithm-based ways to operate."

"We look at digitalization of education as one manifestation of sociotechnological imaginaries."

"The article takes part in the contemporary discussions of postdigitalism where digitalization is neither ignored nor taken as a self-evident and questioned reality per se, but reflected with connections to economic rationality, commodification, or 'metrification' of educational aims (e.g., Knox [115]: 357)."

"To prove that their educational systems and practices are 'modern', the societies need to appear convincingly digitized."

"All this means that the digitalization of education gets paid a lot of state-level attention (Senić Ružić [118])."

"We approach the digitalization of education with references to the rationality and reason of technology and technocapitalism."

Educational Policy as a Means for Societal Change: Serbia and Imaginaries of Digitalization

"Besides the above perspective, the digitalization of education offers lenses through which borders, boundaries, and gaps between formal educational policies and concrete educational realities can be analyzed."

"EU support for Serbia aims to harmonize its educational system with the standards, policies and practices of EU countries and is intended for (1) capacity building of educational institutions and employees in education, (2) harmonization of education systems with labor market needs through the development of qualifications systems, (3) introduction of innovative and inclusive teaching practices, and (4) equipping schools with modern teaching aids (e.g., Senić Ružić [118])."

"All these intentions are connected by optimistic policies of digitalization that are believed to change educational imaginaries into lived educational realities."

"One contemporary contradiction is a combination of educational policies (educational equality), educational philosophy (pedagogical ethics), and a tendency called techno-optimism (technologization of education) (Piattoeva and Saari [119])."

Case Scrutiny as an Approach: Confronting Imaginaries and Realities

"We follow the logic of case analysis in exploring what kinds of meanings are given to digitalization as a formal educational reform—and how these relate to the political and pedagogical 'strings' of education, as well as to an unequal pedagogical reality."

"The main data set for this analysis consists of policy makers' answers to a qualitative questionnaire where information was collected concerning the following themes: the phases of the national reform of education and schooling, governmental goals for the process, and professional perceptions of digital education based on the informants' working positions."

"To be capable of contextualizing our data, we have familiarized ourselves with the Serbian discussion around educational policies, digitalization, educational legislation, and state level curricula (e.g., UNESCO [120])."

"We present a thematic analysis of meanings given to digitalization as part of the educational reform by Serbian professionals."

Imagining the Digitalization of Education

"The digitalization of education in terms of fundamental change, and imaginaries dealing with it, become realized in our informants' expressions which we can interpret as reflections of a new pedagogical Zeitgeist that believes in education in a wide sense: in terms of a shared learning society."

"The issue is not only about the formal level of national regulation but is also a question of a new national, digital identity (cf. Senić Ružić [118]) via modernization of education: our children have a chance to transform their image."

"Digitalization is given a transformative role within a process where education is seen as one key for the society to become European, to be a member of an international community (both formally and informally), and to be internationally competitive with competitive and technologically capable citizens."

The Other Side of the Mirror: Educational Realities in a Digitalizing Society

"We look at some concrete examples of Serbian education today and at its stratified nature: the Roma people's educational exclusion on one hand, and an elite group of digital students on the other."

"The general poverty of the Roma people in Serbia, which in 2017 was estimated as five times higher than the rest of the population, can keep them completely away from digital educational opportunities (World Bank [121])."

"The Committee of Roma people, for example, has gathered Roma NGOs to strengthen their missions and goals in gaining equal living conditions—before and much more than new technologies and digitalization of education."

"Another kind of educational reality becomes visible when we look at another side of Serbian educational reality: at elite private high schools that concentrate on economics and trade and organize their teaching in a totally digitized manner."

Conclusion: A Need for Postdigital Policy and Practice

"In Serbia, like in any other western society today, educational policy, as well as teaching aids, teaching capabilities, learning, and studying relate to the megatrend of digitalization and its imaginaries."

"The state-level applying to 'Europeanism' (e.g., McCormick [122]), as well as the calming of internal tensions when national minorities are taken 'in', are intertwined imaginaries connected to digitizing education."

"Digitalization as a huge reform hardly solves these problems, and minorities' positions in education hardly change with digitalization."

"If we accept the principle that education is a human right, and that equal access to education with an ethical quality is something that should be advanced, there is a lot of empirical evidence for a need to look critically at digitalization as the main (and almost only) part of educational reforms."

8.1.13 Digitalization and Uncertainty in the University: Coherence and Collegiality Through a Metacurriculum [277]

This is a machine-generated summary of:

Johnson, Mark William; Suvorova, Elena A.; Karelina, Alina A.: Digitalization and Uncertainty in the University: Coherence and Collegiality Through a Metacurriculum [277]

Published in: Postdigital Science and Education (2022)

Link to original: https://doi.org/10.1007/s42438-022-00324-1

If you want to cite the papers, please refer to the original.

For technical reasons we could not place the page where the original quote is coming from.

Abstract-Summary

"While conceived as a movement from the present of education to an imagined future where envisaged curricula embrace data-oriented and computational practices across subjects, we ask: whose present?"

"We revisit the concept of a metacurriculum (first conceived in the 1990s) as a way of addressing this question, while not avoiding the challenges inherent in the adaptation of education to an increasingly complex postdigital environment."

"Using the uncertainty concept, we present a practical co-designed and dialogical approach supporting the student and teacher journeys towards the transdisciplinary opportunities opened out by technology, based on a cybernetic model of intersubjectivity."

"In analyzing data from one of the central activities of this course, we focus on the teacher collegiality and the students' mechanisms of selection in navigating the transdisciplinary space, and how these mechanisms may provide deeper insight into the dialogical underpinnings of education."

Introduction: The Digitalization Metacurriculum

"While a metacurriculum is an approach to address the transdisciplinary challenges, practical questions arise as to how to create the conditions where teachers can support it, the assessment system can recognize it, and students and their parents and teachers can be given the confidence that transdisciplinary approaches are worthwhile when specialization is increasingly fetishized."

"Our approach to a metacurriculum partly echoes Lauer's [123] ideas about asking questions across the curriculum: teaching students to think in a transdisciplinary way, reflecting on their learning, while also introducing them to concepts around technology which present transdisciplinary connections."

"Since human development is intrinsically connected to the environment, we suggest that a metacurriculum is an acknowledgement that technology (digital or otherwise) is intrinsically connected to thinking and learning, much in the way that some philosophers of technology relate processes of 'individuation' to technology (Simondon [124]; Hui [125])."

Uncertainty and Postdigital Digitalization

"In the face of the explosion of technology in the environment produced by the Internet, AI, and social media, the challenge for individuals is to ask: which options do I choose to organize myself effectively?"

"Our uncertainty perspective looks beyond this, and to the challenges that face any teacher or learner in the face of increasing technological options and environmental complexity."

"As a middle-out process, digital phenomena exist in the environment, and result in increasing uncertainty (uncertainty 2) to which individuals and institutions have to adapt through dialogue."

"The different environmental contexts for dialogue (for example, topics, activities, technologies, face-to-face or online) must be considered, since each context carries different levels of uncertainty: understanding the institutional dimensions of uncertainty sheds further light on the ways learners and teachers navigate and construct a dialogic space."

The Institutional Dimensions of Uncertainty

"While these functions serve to manage different areas of uncertainty, they also serve to reinforce the 'reality' of the distinction between the different practices and artifacts of education: classrooms, timetables, curricula, textbooks, computers, teachers, budgets, buildings, catering, etc This process of reinforcement is particularly noticeable within curriculum subjects, and it presents one of the challenges for any viable metacurriculum."

"This shows the varying specifics and generalities that relate institutionally organized disciplines to technologies and transdisciplinarity."

"Within specific discourses (at the bottom of the diagram), uncertainties can be kept at a minimum, reinforced by traditional methods of teaching, assessment, timetable, and the need to concentrate on disciplinary specifics."

"The movement beyond the uncertainty management of particular groups requires the establishment of new kinds of intersubjective engagement, free from the constraints which operate at the lowest level of curriculum-specific organization."

Collegiality, Uncertainty, and Linearity: Whose Present? Whose Future?
"Asked how they felt at the beginning of this process, they said that there was some confusion as to how the course connected with other disciplines, subject areas, and activities: 'it was hard switch the students to a different rhythm [in their education]', one teacher commented."

"Another noted that it was the teachers who had more learning to do: 'the teachers are moving to a future which is already present for the students'."

"The online-only experience provided a new dimension in understanding the relationship to technology, and teachers spoke of how some activities appeared to be deeper in the online medium (particularly programming, and in the extensive use of video in learning activities) where, while the rich dynamics of the classroom could not be reproduced in the same way, other forms of collaboration took their place including collaborative video-making, alongside opportunities for more rapid formative feedback to students."

Measurement and Metacurriculum
"While classroom discussion is difficult to examine in detail, we devised an activity and technique which would: (a) Create a lot of data about student reflection in relation to a range of topics; and (b) Provide a way of integrating the different themes and experiences of the course."

"The exercise of comparison and reflection served the dual pedagogic purpose of encouraging the students to read around the subjects they were engaging in classroom, while providing some cohesion to the range of topics and activities which they encountered."

"The comparison exercise elicited thoughtful responses from students, producing data not only in the text of their responses, but in the specific documents chosen (and those not chosen), the broader disciplinary areas of those topics, and in the overall ranking of the popularity of particular topics over others."

Measuring the 'Fit' in Student Reflection with Cosine Similarity
"The high cosine similarity for Education and Science is based on a small sample, and the contributions of one particular student who chose to defend their document choice by summarising the contents of both documents before defending their judgement: Brief retelling: 1 Article 1: Technology."

"The extent to which the cosine similarity score reflects the level of depth of reflection raises a deeper question: is the preferred document chosen because its pattern of communication already maps onto the pattern of communication within the student prior to them writing their justification?"

"If the cosine similarity score between student reflections and the original texts is an indicator of similarity between selection mechanisms for utterances (i.e., between an author and the student), then the weighting of the number of students with similar choices is a further indicator of the likelihood that within a body of students, certain conversations about certain documents will stimulate an ongoing dialogue."

Conclusion: Education, Meaning, and Digitalization

"Asking about whose constructions matter, and how diverse constructions of 'present' and 'future' can be coordinated, can help institutions align themselves better to a fast-changing technological environment, taking learners and teachers on a journey that embraces new practices."

"We have argued that a metacurriculum, as a means of navigating the space between disciplinary specifics and technological transdisciplinarity, entails such a process of coordinating expectations as constructions between and among teachers and learners."

"The Global Scientific Dialogue approach, as an example of a metacurriculum, looks to create the conditions where possible constructions—both of the present and the future—can be explored by foregrounding the essential uncertainties that are presented by technology."

"In its focus on dialogue, Global Scientific Dialogue highlights the importance of facilitating the student and teacher journeys from the specifics of their disciplinary studies, towards the transdisciplinary opportunities presented by technology."

8.1.14 Learning by Collaborative Designing: Technology-Enhanced Knowledge Practices [278]

This is a machine-generated summary of:

Seitamaa-Hakkarainen, Pirita; Viilo, Marjut; Hakkarainen, Kai: Learning by collaborative designing: technology-enhanced knowledge practices [278]

Published in: International Journal of Technology and Design Education (2008)

Link to original: https://doi.org/10.1007/s10798-008-9066-4

If you want to cite the papers, please refer to the original.

For technical reasons we could not place the page where the original quote is coming from.

Abstract-Summary

"Knowledge Forum (KF), a virtual platform for the interchange of information, provided tools and practices that facilitated creative working within participating students and teachers."

"The analysis of the teacher's project diary and the contributions to the KF database provides an overview of the knowledge practices enacted during the project."

"The distribution of knowledge-creation activities during the project reveals that the teacher assumed a role of organizer concerning shared knowledge practices instead of controlling all aspects of students' learning."

"However, students were not left working without guidance; the teacher structured their collaborative efforts with the help of KF and by directing classroom activities."

Introduction

"Present-day students are expected to engage in designing, creating, sharing, and manufacturing complex cultural artefacts by capitalizing on versatile collaborative technologies."

"Cognitive competencies required by the emerging advanced knowledge society develop only through taking part in corresponding collaborative practices of working with knowledge ([126]; Paavola and Hakkarainen [127]); cultivation of corresponding personal and collective capabilities should start from the very beginning of education."

"We will show how the ideas in the model of learning by collaborative designing (LCD) were implemented in practice at the 4th and 5th grade of a Finnish elementary school."

"The aim of the longitudinal study project, "The Artefact Project—the Past, the Present, and the Future", was to engage pupils in creating both new conceptual and material artefacts in collaboration with one another."

"According to the evidence we will present, the technology-enhanced learning environments appear to provide tools and practices that elicit participation in meaningful, socially and culturally rich, and creative working with knowledge."

From knowledge building to knowledge practices

"When knowledge-creating learning is understood to be dependent of materially embodied rather that mere conceptual experiences, it follows that genuine facilitation of learning requires (a) engaging students in solving complex problems coming from outside of the educational institution (rather than address mere pre-structured problems, fragmented according to subject-domains); (b) focusing inquiries on parallel work with material and conceptual artefacts (being, simultaneously, both "hands on" and "minds on"); (c) leading students, in a concrete way to break boundaries between communities (so as to make different roles accessible to them and make them reflectively aware of their own knowledge practices); (d) initiate encounters and productive collaboration between students and experts (so as to provide students an access to expert's knowledge practices); (e) deliberately to break boundaries of educational institutions and the surrounding society and address tensions and challenges related to transitions between school, home, friends, and other fields of cultural practices (Nespor [128]; Zittoum [129])."

Learning by collaborative designing

"Designing has conceptual and material aspects: It is not only focused on developing the participants' ideas through taking part in knowledge-creating inquiries, but also has a practical component, creating design prototypes and material products."

"The model emphasizes that the collaboration should occur in all stages of the design process by creating shared design contexts, analysing design constraints, collecting and sharing new knowledge, prototyping, and providing feedback for the artefacts being designed."

"An essential aspect of the process is materialization of the ideas through making and exploring design prototypes."

"The efforts of the participants are organized toward developing shared design ideas (conceptual artefacts), embodying and explicating those ideas in visual sketches (graphic artefacts or inscriptions), and giving the ideas a material form as prototypes or end results (e.g., produced products)."

"Pursuit of research problems, generating working theories, developing design ideas, prototyping, and defining design constraints are such practices regarding creative working with knowledge."

Objectives of the study

"We will report a longitudinal case study of a teacher's and researchers' effort to create classroom activities and social practices that support genuine participation in knowledge-creating inquiry."

"In pursuing the detailed analysis of learning activities within their social context, our main interest was in specifying how the teacher, who is familiar with knowledge building theories and has a long experience in implementing those theories in her class, organizes the collaborative designing and inquiry process."

"How were the classroom practices organized socially, and how were they supported with the technology-enhanced learning environment used?"

"How were the learning activities organized around the KF and the material and conceptual artefacts that the students created during the project?"

Research method

"The project integrated many school subjects, i.e. history, mother tongue, science, design and technology education, and thus, the students worked with a wide variety of knowledge objects, including hybrid artefacts, and experienced various phenomena related to artefacts."

"The teacher created a view (shared working space) for each team where the students collected and processed all the information of the artefact being investigated during this phase of the project."

"During the Artefact Project, students generated problems and research questions, provided diverse theories, ideas and explanations through face-to-face knowledge building discussions, conducted self-generated science experiments, made observations related to phenomena, and shared new information extracted from reading materials, Internet, and various resources."

"In the first phase of the project, the students worked in the "home teams" (about 4 students per group), which investigated the chosen artefacts and produced knowledge to the team views of KF."

Data collection and data analysis

"The teacher filled one project diary template several times in the weeks during which the class took part in project activities."

"It consisted of four sub-categories which may be explained as follows: (a) The Creating category describes how the participants produced their own questions, ideas and explanations of knowledge or created their own product designs. (b) Searching and sharing are the activities where the new information was searched for or shared or the process and its achievements was shared with the rest of the community. (c) Organising represents the activities of process organization or categorization of the information; and d) in Evaluating are the activities in which students assessed solutions and the process."

"In order to examine how the learning activities were organized around artefact creation and how the use of KF mediated these activities, the present investigators used multiple data sources: KF's database and the teacher's project diaries."

Results

"The teacher-driven activities (5 lessons) and the collective class work (18 lessons) were most strongly represented in the first phase of the project."

"Of the project, the amount of the teacher-driven activities (3 lessons) and collective class work (8 lessons) decreased."

"In order to do that we analysed the nature of the learning activities and the artefacts worked on (notes, pictures, photos etc), the numbers of views and the notes created in different phases of the project."

"The aim of this stage of the project was to introduce the working ideas and main principles of KF and to guide students to create their own collective classification framework of the material artefacts around them, with the help of the technology-enhanced learning environment."

Discussion

"The present longitudinal study was an effort to foster young students' knowledge-creating inquiry; they were to learn through collaborative designing, which involved working simultaneously with conceptual and material artefacts."

"The study indicated how knowledge practices involving pursuit of a long and sustained research project (compare Gruber [130]) can integrate learning at several levels: this was a project that cut across school subjects; we created a classroom inquiry culture with field visits to external expert cultures; there were interactions between students and expert-consultants outside and inside the classroom."

"Creative externalization involved in the project provided the participants a clear sense of pride concerning the collective cultural knowledge of artefacts created; a separate investigation of student learning is in progress."

"The students' engagement in in-depth cycles of inquiry throughout the project would not have been possible, in our view, without teacher-controlled and classroom-level activities that organized, encouraged, and provided required conceptual guidance in the process."

Conclusions

"A central aspect of the present approach was to explore collaborative design with special focus on the participants' parallel pursuit of material and conceptual artefacts."

"Material activities carried out during the project guided the participants' transition between levels of knowing rather than move within a closed circle of conceptual knowledge."

"Toward that end, the participants engaged in various artefact-related activities; they organized an exhibition typical of a historian's knowledge practices and designed their own artefacts under guidance of the professional interior designer."

"Visual organization of ideas within KF's shared space allowed the participants to represent material artefacts examined or created by them."

"KF not only brought the students' own ideas into a central role in the process, but also allowed transforming material entities (exhibition and prototypes) as digital artefacts to be collectively reflected on and discussed."

8.1.15 What Has Been the Impact of COVID-19 on Driving Digitalization, Innovation and Crisis Management of Higher Education and Quality Assurance?—A Taiwan Case Study in Alignment with the INQAAHE Virtual Review [279]

This is a machine-generated summary of:

Hou, Angela Yung Chi; Lu, I-Jung Grace; Hill, Christopher: What Has Been the Impact of COVID-19 on Driving Digitalization, Innovation and Crisis Management of Higher Education and Quality Assurance?—A Taiwan Case Study in Alignment with the INQAAHE Virtual Review [279]

Published in: Higher Education Policy (2022)
Link to original: https://doi.org/10.1057/s41307-022-00267-z

If you want to cite the papers, please refer to the original.

For technical reasons we could not place the page where the original quote is coming from.

Abstract-Summary
"In order to respond to the 2020 pandemic, The International Network for Quality Assurance Agencies in Higher Education (INQAAHE) swiftly adopted a virtual mode of the GGP review exercise for the GGP alignment applicants."

"The Higher Education Evaluation and Accreditation Council of Taiwan (HEEACT) was the first case that underwent a thorough virtual review process of GGP alignment during the 2020 pandemic."

"This study aims to outline the impact of the pandemic in Taiwan higher education as well as provide the meta-analysis of the virtual review process of the INQAAHE GGP alignment by using HEEACT as a case study."

Introduction
"Virtual technologies have long been used by HEIs and quality assurance agencies (QAA) to facilitate teaching, learning and knowledge production on campuses and to ensure quality of education."

"Founded in 1999, the International Network for Quality Assurance Agencies in Higher Education (INQAAHE) aims to be a platform for information-sharing on good practices for quality improvement in HE between QAA (International Network for Quality assurance Agencies in Higher Education (INQAAHE), [132])."

"11 national QAA have been recognized as the GGP aligned agencies (International Network for Quality assurance Agencies in Higher Education (INQAAHE), [133])."

"This study aims to outline the impact of the pandemic on Taiwanese higher education, as well as provide the meta-analysis of the virtual review process of the INQAAHE GGP alignment by using HEEACT as a case study."

"What was the impact of the pandemic on Taiwanese higher education, according to the perspectives of university administrators and faculty members participating in the HEEACT INQAAHE GGP Review?"

Literature Review
"In support of QAA and HEIs dealing with the emergency situation, INQAAHE published the statement "INQAAHE's affirmation of support during COVID-19 pandemic" on April 23, 2020,with eight principles for crisis management in QA,

including adherence to integrity in assuring quality in HEIs; maintaining student learning experiences and engagement; ensuring equity and access as the top priorities; remaining close to higher education institutions and sharing quality guidelines with them; launching clear communication plans for confidence building over QA agencies; continuing to enhancing the role of enhanced partnership and sharing good practices; and adapting to unforeseen circumstances in the unprecedented crises (International Network for Quality Assurance in Higher Education (INQAAHE), [134])."

"Over past two decades, HEEACT has learned and developed its accreditation processes to meet changing global trends and national requirements (International Network for Quality Assurance in Higher Education (INQAAHE), [135])."

Methodology

"Via multiple sources of data, the researchers have genuine interests to gain a thorough understanding of the Covid-19 pandemic on Taiwan higher education, as well as to evaluate the implementation of the INQAAHE virtual review."

"As the first INQAAHE GGP virtual review case, first, this case study employed both quantitative and qualitative approaches to collect the feedback of the participants of the HEEACT GGP review, including INQAAHE review panel and coordinators, HEEACT working group, HEEACT senior administrators, HEEACT Board of Trustees, and selected university administrators, reviewers, and student representatives during the virtual visit."

"Based on the three QA experts' consultations, after the INQAAHE review decisions was announced publicly on Dec. 30, 2020 on the website, the study firstly conducted an online survey with a five-point Likert scale from January 1 to February 5, 2021 to perceive the participant opinions over COVID-impacts on Taiwan higher education, their attitude about actual implementation of INQAAHE virtual review, and the future development of the new EQA mode after COVID-19."

Major Findings

"The survey showed that HEEACT respondents highly agreed over the virtual onsite arrangement by the INQAAHE GGP coordinator and the imminent feedback over SAR report from the review panel."

"Seven interviewees all thought that the governments, QA agencies, and the HEIs should quickly respond to the impact of COVID in terms of keeping on assuring the quality in HE as well as operation of virtual QA exercises."

"If done properly, it is highly possible for site visit to be conducted virtually instead of the traditional practices (Q1)."

"One of the interviewees was worried that the failure of IT system while the review is conducting, weak staff IT capacity as well as different time zones could pose a potential threat to the implementation of a virtual QA mode effectively (Interviewee Q2)."

Discussions

"It was also found that governments, institutions and QA immediately made a swift adjustment over teaching and learning in higher education on a basis of crisis management model, while quality assurance was not considered as the first priority."

"The concerns about how virtual QA can guarantee quality of teaching and learning will remain strong if the core standards of assessment are not refined."

"Some expert interviewees are quite concerned that the abrupt shift to online education and virtual quality assurance had resulted in the widening gap of ICT capacity among varying higher education stakeholders accordingly."

"It can be challenging to maintain the integrity of virtual EQA while quality assurance agencies are striving harder to maintain QA exercises in order to survive."

Conclusion

"Under the disruptive era, digitalization not only provides an alternative for HE and QA practice but overwhelmingly drives innovation."

"The HEEACT case study, not only provides a new insightful practice for QAA and accrediting bodies who were used to carry out physical onsite visit but also pressure them to rethink immediately of a new set of suitable standards and QA procedures for universities in order to assess quality of their online programs, digital instruction and student learning outcomes in an appropriate manner."

"Government, institutions and QAA should acknowledge that the appropriate use of digitalization and innovation of QA would benefit HE quality and maintain QA exercises under such a crisis."

"Awareness of the relevant issues that digitalization may cause is considered as the key element to the good practices of online education and virtual QA in future."

8.1.16 From Learner-Centered Education (LCE) to Emancipatory Learner-Centered Education (ELCE): A Comparative Case Study of Language Education for Adult Migrants in Four European Countries [280]

This is a machine-generated summary of:

Gravani, Maria N.; Slade, Bonnie; Brown, Maria; Jõgi, Larissa; Borg, Carmel: From Learner-Centered Education (LCE) to Emancipatory Learner-Centered Education (ELCE): A comparative case study of language education for adult migrants in four European countries [280]

Published in: PROSPECTS (2023)

Link to original: https://doi.org/10.1007/s11125-023-09633-0

If you want to cite the papers, please refer to the original.

For technical reasons we could not place the page where the original quote is coming from.

Abstract-Summary

"This article discusses the findings of qualitative case-study research that looks at pedagogical contexts of adult education programs for migrants in Cyprus, Scotland, Malta, and Estonia."

"The goal of this research is to understand how Learner-Centered Education (LCE) is promoted within a human-rights framework and implemented as an approach to emancipatory social change."

"Informed by these findings, this article builds a case for Emancipatory Learner-Centered Education (ELCE)—a pedagogical approach that goes beyond course adaptations, to engage migrants in consciousness-raising and confidence-building, while fostering communal action between migrants and adult educators."

Learner-Centered Education

"Learner-Centered Education (LCE) can be considered a concept, an approach, a narrative, and a traveling policy (Schweisfurth, [136, 137])."

"Schweisfurth [137] sees educational practice as existing along a continuum, from less learner-centered to more learner-centered, with LCE at one end of the continuum."

"At the LCE end, learners have more control, not just over the content of learning, but also over how they learn, what is learned, and how the teaching-learning process is shaped by the learners' needs, capacities, and interests."

"Schweisfurth identifies several elements on this continuum that comprise LCE practice—techniques, relationships, motivation, and epistemology [137]."

"These elements were employed heuristically in the present study as a framework to unveil the extent to which LCE is used as an approach for social change in the context of adult education language-learning programs for migrants in four European countries."

"The second element, relationships, refers to the range of possible relationships between educators and learners."

Adult education for social change

"With roots in popular education and movements advocating for social transformation, such as the Plebs League and the Highlander Folk School in Tennessee, adult education for social change foregrounds critical reflection and collective action (Borg, [138])."

"Guo and Sork [139], who researched the role of adult education programs in bringing about social change in migrant communities, argue that adult education is also an important forum for building inclusive citizenship."

"Manninen and others, ([140], p. 5) introduced an alternative conception of the term change-oriented adult education, arguing that it encompasses an approach, philosophy and set of teaching and learning methods that seek to create individual and/or social change."

"Various concepts of adult education for social change converge around active participation and an educational space that generates hope and possibility."

Theoretical framework

"LCE is understood here as a pedagogical approach that gives learners, and demands from them, a relatively high level of active control over the contents and processes of learning as well as an educational practice and approach to social change."

"The cognitive narrative, suggesting that the learner has control over the content and process of learning."

"The emancipation narrative, according to which learners not only have more control over what they learn and the process of learning but are encouraged to question critically canons of received knowledge and the unequal structures of society that they support."

"This echoes Freire's [141] work on adult education, which argued that learners' questioning of their own realities is a central aspect of learning."

Aims and research questions

"The research informing this article contributes to adult (migrant) education studies by revealing how LCE is enacted, implemented, or neglected in specific educational contexts."

"To what extent are the adult education programs learner-centered, particularly in terms of techniques, relationships, motivation, and epistemology (Schweisfurth, [137])?"

"Is the capacity for emancipatory social change a learning outcome of the adult education programs under study?"

Methodology

"At a cultural level, the study featured multiple official languages: Scotland (English, Scottish, and Gaelic), Malta (Maltese and English), Cyprus (Greek, Turkish, and English), and Estonia (Estonian and Russian)."

"Malta's National Lifelong Learning Strategy 2020 (Ministry of Education and Employment [MEDE], [142]) considered increased participation of migrants and provision of special programs targeting migrants among its aims."

"At the time of the study, Cyprus and Estonia lacked targeted policy to address the issues of adult-migrant learners."

"An evaluation of cross-case comparability, issues of access to the adult education programs and their participants, and time limitations informed the selection of a specific language program for adult-migrant learners in each of the four countries."

"Greek Language for Foreigners', a program that was open to all non-native Greek speakers, including Europeans who work in Cyprus, and mainly attended by migrants, featured in the Cypriot case study."

Findings

"It's very important for communicate [sic]. (EFL1 Learner, Malta) Curricular relevance and the extent to which courses were built on migrants' knowledge and skills (Schweisfurth, [137]) varied with the type of adult education provider."

"If the subject wasn't new, she usually asked the students to start the dialogue. (Learner 2, Estonia) The Glasgow case study yielded evidence of language-learning through music and song, role play, and group work that integrated the migrants' backgrounds; in addition, it showed how these activities positively impacted the achievement of learning outcomes as well as broader social processes of social capital, integration, and cohesion: I've learnt songs for my Romanians, I know some phrases, it's a great icebreaker."

"Everybody felt free, you could say anything. (Learner 1, Estonia) In the case of Glasgow and Malta, participants identified the relationships nurtured by the staff as a key strength, responsible for helping to build a culture of mutual respect and equality among the migrant learners and between them and the educators."

Discussion

"ELCE seeks to transform education spaces into possibilitarian sites of liberatory practice, where individual needs become communal needs, marginalized voices become centered voices in all aspects of the curricular experience, and education for survival is transformed into education for active engagement with the world (Borg, [138])."

"While focusing on the immediate practical needs of learners, ELCE-driven programs challenge perceptions of migrants and migration and the view that the expert is the only source of knowing; they problematize assumptions about learning in the context of migration and generate empathy with all migrants, but especially those who are struggling as a result of social injustices, by building confidence in the possibility that human injustices can be reversed through collective engagement with the world (Borg, [138]).”

“ELCE programs call for long-term, collaborative, programmatic commitments—from inception to delivery—that serve the ethical and pedagogical need for educators to familiarize themselves with the challenges and realities that define migration and understand how the receiving context intersects with the past and present lived realities of migrants.”

Conclusions

“The comparative cross-case research discussed in this article queried the extent to which LCE is used in adult-migrant education as a tool for social change across four country/city contexts.”

“Relationships emerged as sources of enrichment in the adult-migrant learners' educational and life journeys—though they may be taxing for educators when coupled with other psycho-social issues, such as poverty, social exclusion, trauma, or racism.”

“The cross-national collaboration informed the researchers' self-reflection concerning their assumptions about LCE as a pedagogical approach and field of research, migration as a social process, and migrant adult students as co-creators of educational programs.”

“The findings of comparative cross-case research enabled the researchers to elaborate Schweisfurth's [137] “emancipation narrative” into an empirically founded advocacy for inductive ELCE programs in adult (migrant) education.”

[Section 9]

“Far fewer studies treat pedagogy as a central theme for investigation, with an interest in the intersection of migration and adult education.”

“Their editorial “Migration, adult education and learning” in the special issue of Studies in the Education of Adults (Morrice, and others, [143], p. 133) also indicates the need for further research into practices that embrace the learning of both migrants and longer-term settled communities and enable migrants to determine their own learning needs and set their own goals (p. 133).”

“Using a comparative, qualitative investigation of adult education language learning programs for migrants in four European countries (Cyprus, Scotland, Malta, and Estonia), our research explored the extent to which Learner-Centered Education (LCE) was being enacted and implemented in these contexts as a driver of social change.”

8.1.17 Using Digital Board Games for Genuine Communication in EFL Classrooms [281]

This is a machine-generated summary of:

Wu, Chia-Jung; Chen, Gwo-Dong; Huang, Chi-Wen: Using digital board games for genuine communication in EFL classrooms [281]

Published in: Educational technology research and development (2014)
Link to original: https://doi.org/10.1007/s11423-013-9329-y

If you want to cite the papers, please refer to the original.

For technical reasons we could not place the page where the original quote is coming from.

Abstract-Summary

"This study examines the use of digital board game language learning set in a task-collaborative platform, digital learning playground, designed for classroom learning."

"Ninety-six high school Taiwanese students (n = 96), who were randomly assigned to an ordinary instruction group, a board game language-learning group, or a digital board game language-learning group, engaged in a 50-min English class."

"The quantitative data presents a higher learning performance for the instruction set at digital learning playground."

"The qualitative data confirms computerized simulation adapted from board game entails context-relevant immersion, encouraging communication and efficient game instruction management for EFL language learning."

Introduction

"Knowledge passed on to students in ordinary learning environments, schools, could be dull and unsuccessful when it is not relevant and reflective to learners."

"Besides, schools are sometimes unlikely to accommodate students with dynamic learning materials due to several reasons, such as resources and policy."

"Since the impact of technology and globalization has accelerated English into a powerful language and tool to communicate from field to field, many learners have found that their traditional English learning at school didn't prepare them for real world use of the language."

"The aim of this research is to develop an optimal English learning environment for EFL learners that not only reinforces the function of classroom learning, such as guidance, face to face interaction and low frustration, but also incorporates the learning materials that are real-life connected and hands-on oriented into the classroom instructional design."

Literature review

"Simulation/gaming is a game genre that highlights the strong connection between a real-life reference game environment and knowledge-in-use playing mechanic."

"The challenge for simulation instruction is transferring authentic game resources to comprehensible pieces for learners to play without too much distraction from language barriers."

"Another challenge is to map simulations and games into curriculum that applies to a teacher's role, classroom management and learners' interaction (Jones [144]; Moseley and Jones [145]; Waston and Sharrock [146])."

"Mayer and Harris [147] addressed some examples of simulation board games with relation to English and language arts, like Once Upon a Time and Nonfictionary."

"The effectiveness of board game use in language classrooms has been left in doubt because the overall strategy-oriented game play lets students easily neglect language itself."

"Our research attempted to extend the intriguing context of a board game to a knowledge-utilizing platform that is suitable for classroom learning."

Development of the digital board game for language learning

"The instruction design guideline is centered on communicative language teaching (CLT) and task-based language teaching (TBLT) in order to equip students to carry language performance through game tasks in classrooms."

"The following demonstrates how the instructional design corresponded to the critical implemented guidelines. (a) The setting is originated from a commercial board game, Fresco, designed by Marco Ruskowski and Marcel Süßelbeck in 2010, that illustrates a day for a master painter during the Renaissance. (b) The learning objectives are responding to people's request, making polite requests, expressing purchasing willingness, and requesting a refund. (a) The theme plot is the daily routine of a painter buying paint at the market, making money at their own studio, restoring the frescos at the cathedral, and watching a show at the theater for entertainment. (b) The roles include a buyer, a painter, and a studio owner. (c) The conversations include making purchase, requesting a refund, and dealing with customers."

Interfaces of the game

"The hardware in digital Learning Playground applied mixed reality technology."

"A display and a single-touch whiteboard as a tabletop formed a theatre-like setting."

"The environment setting required a computer and a microphone headset."

"We also applied My SQL database to collect user's personal data and testing data."

Research design

"We hypothesized that English instruction adapted from simulative board game content within the Digital Learning Playground setting provides students with a knowledge-in-use playing experience."

"Does learners' motivation in language learning under computerized simulative board games differ from those in simulation instructions without a digital device?"

"The experiment primarily focused on comparing motivation and learning performance between ordinary learning, simulative board-game blended language learning and computerized language learning simulation."

"The experiment design used one experimental group, DLP group, and two comparison groups, board game group and ordinary teaching group."

"The DLP group operated our core design- digital board game-based language learning."

"To minimize the experimental variations, the instructor in each group was the same one who had gone through the learning materials, the game-based learning instructions with the original board game and DLP before this experiment."

Results

"The first research question was to examine whether there were any differences in the learning performances between the three methods of instruction—ordinary teaching group, board game group and DLP group."

"($F = 3.27$, $p = 0.000$), there was a statistical difference in the learning performances between the three groups; that is, the students who learned under digital board game learning showed significantly better learning performance than those who learned with the traditional teaching and the original board game blended teaching."

"No significant differences between the ordinary teaching group and the board game group shows that the learning performance was not much different under the ordinary instruction or board game supplement."

"From the questionnaire, we unexpectedly discovered that the board game group had the highest tension during the learning activity."

Conclusion and future research

"Of the experiment, the students who learned language with a digital task-collaborative board game platform, digital learning playground, achieved significantly better communication ability in comparison with those who learned in an ordinary teaching and a non-digital board game adaptive language teaching."

"The positive findings in this experiment may have been contributed to two factors: classroom group learning and, digitalization and visualization presented through board games."

"Group learning in classrooms not only easily executes a board game's group-playing mechanic with similar level peers but also has assistance from a teacher."

"The primary concern is how to jam a game mechanic into an existing class mechanic where teachers and students already have a set pace and fixed perception of learning."

"Besides, having assessments such as in-class performance or group performance evaluation assures these games blending into learning being implemented in classrooms."

8.1.18 Active Learning and Education 4.0 for Complex Thinking Training: Analysis of Two Case Studies in Open Education [282]

This is a machine-generated summary of:

Patiño, Azeneth; Ramírez-Montoya, María Soledad; Buenestado-Fernández, Mariana: Active learning and education 4.0 for complex thinking training: analysis of two case studies in open education [282]

Published in: Smart Learning Environments (2023)

Link to original: https://doi.org/10.1186/s40561-023-00229-x

If you want to cite the papers, please refer to the original.

For technical reasons we could not place the page where the original quote is coming from.

Abstract-Summary

"This article focuses on empirically analyzing the final products designed by 147 academics from 11 countries who participated in an international open education movement workshop by answering the research questions "What are the techno-pedagogical components of the products designed by the participants to encourage the open educational movement?"

"The findings of this study reveal that (A) the use of emerging and 4.0 technologies in initiatives of the open education movement continue to increase; (B) most of the open education initiatives designed by academics participating in the workshops were focused on the production of OER; and (C) inclusive access to education and continuing professional development of teachers is a constant concern addressed in open education initiatives."

"The results of this research suggest that training and development interventions implying the creation or design of open education initiatives should focus on encouraging all kinds of open education practices (i.e. use, production, dissemination and mobilization)."

Introduction
"Lebrun [148] provides an inventory of the characteristics of active pedagogy that includes: the personal nature of learning; the catalytic role of prior knowledge; the motivational factors; the importance of the available resources; the role of context, environment and concrete experience; the high-level skills to be exercised; the research approach and questioning in learning; the conceptual change (awareness, imbalance, reformulation); the need for feedback on one's activities; the interactive and cooperative nature of learning; the link between personal, professional, study and life projects; the importance of construction, and production; and the role of reflection on the learning process."

"Smart education is considered as a growing emerging field of research that represents the integration of (a) smart systems and environments; (b) smart technologies; (c) state-of-the-art educational programs and tools; (d) innovative pedagogies, teaching strategies and learning methodologies based on advanced technology (Uskov, [149])."

Materials and method
"This work adopts a case study research approach for data collection and analysis."

"Case study research has been widely used in educational research to enhance the understanding of contexts, communities, and individuals (Hamilton & Corbett-Whittier, [150])."

"This article focuses on empirically analyzing the cases and the final products designed by academics participating in an international open education movement workshop to answer the following research questions: What are the techno-pedagogical components of the products designed by the participants to encourage the open educational movement?"

"To provide answers to the previously mentioned research questions, we analyzed the collected data and classified it into categories that describe (1) the technological components of the products designed by the participants that emerged from the data and (2) the practice of open education (Ramirez-Montoya, [151]) as described below: Use refers to using or implementing existing OER for research, teaching, and learning purposes."

Case studies
"During the workshop, speakers and specialists who shared research, case studies and practical applications of open education carried out various activities such as attending conferences, preparing oral presentations, and participating in a collaborative project."

"The presented initiatives designed by the participants proposed activities categorized as production (n = 8), dissemination (n = 3) and mobilization (n = 3) practices of the open educational movement."

"Objectives The main objective of the workshop was to analyze components and practices of the open education movement, with particular attention to UNESCO's recommendations for open education, through a series of webinars and an international workshop with speakers and specialists to share research, case studies and practical applications, to project educational ventures that support innovation and the mobilization of open educational practices."

"The presented initiatives designed by the academics participating in the workshop proposed activities categorized as production (n = 11), dissemination (n = 4) and mobilization (n = 6) practices of the open educational movement."

Discussion

"The use of emerging and 4.0 technologies in initiatives of the open education movement continues to increase."

"As pointed out by Zhou and others [152], Lane [153], Ahel and Lingenau [154] practitioners and researchers continuously seek strategies based on technological innovation to ensure their integration into academic and professional development and in OER to increase access and boost open education."

"Special attention must be paid to training initiatives to include objectives related to the four identified open education practices (i.e. use, production, dissemination, and mobilization)."

"Inclusive access to education and continuing professional development is a constant concern in open education initiatives."

"Plus, open education has always been associated with practices, resources, and open-access technologies that involve the production, use, dissemination, and mobilization of training (Ramirez-Montoya, [151])."

[Section 5]

"Play2Train4C is a new game-based intervention proposal based on the results of the previous workshops that aims to support academics in designing teaching and learning activities for complex thinking skill training in university students."

"This intervention proposal relies on active learning and digital game-based learning as a theoretical framework underpinning the proposed instructional activities."

"This step-by-step methodology aims to facilitate decision-making concerning game-based activities and modalities that better fit the learners' needs in their specific contexts."

"The HEXA-GBL methodology will also be implemented to accompany academics in designing game-based solutions with the potential to train students in complex thinking skills."

"Activities A real-world immersive experience in collaborative game design for complex thinking skills training will be proposed to academics participating in the 2023 edition of the workshop."

"Theoretical Foundations Active learning pedagogies such as digital game-based learning and collaborative-, project- and inquiry-based learning underpin the envisaged instructional strategies."

Conclusion
"The open educational movement practices identified in the final products designed by the participants are production, dissemination, and mobilization of OER."

"The findings of this study reveal that (A) the use of emerging and 4.0 technologies in initiatives of the open education movement continue to increase; (B) most of the open education initiatives designed by academics participating in the workshops were focused on the production of OER; and (C) inclusive access to education and continuing professional development of teachers is a constant concern addressed in open education initiatives."

"Training and development interventions involving the creation or design of open education initiatives should focus on promoting all types of open education practices (i.e., use, production, dissemination, and mobilization)."

"Future studies might focus on analyzing the practice of the open educational movement executed in specific smart learning environments and examining the impact of open education initiatives depending of the smart learning environments."

8.1.19 The Presence of the Body in Digital Education: A Phenomenological Approach to Embodied Experience [283]

This is a machine-generated summary of:

Willatt, Carlos; Flores, Luis Manuel: The Presence of the Body in Digital Education: A Phenomenological Approach to Embodied Experience [283]

Published in: Studies in Philosophy and Education (2021)

Link to original: https://doi.org/10.1007/s11217-021-09813-5

If you want to cite the papers, please refer to the original.

For technical reasons we could not place the page where the original quote is coming from.

Abstract-Summary
"We aim to critically challenge this assumption by addressing the phenomenon of virtuality in a more original sense: as a fundamental dimension of corporeality itself."

"Drawing on Merleau-Ponty's phenomenological insights, we argue that the experience of the virtual is inherently embodied and fully real."

"We understand corporeality and virtuality as entangled and multidimensional phenomena."

"This leads us to rethink the digital as only one possible medium in which virtuality can appear as inherent to embodied experience."

"From our critical perspective, assumptions about a disembodied digitality and a devirtualized corporeality represent extremes within which educational discourses currently move."

"We outline the heuristic concept of the virtual body (Merleau-Ponty) to describe corporeality and virtuality as entangled and multidimensional phenomena that encompass extension, intercorporeality and intermediality."

"We make some remarks on the relationships of corporeality and virtuality considering the challenges of digital education."

Introduction

"Within these processes, little room is left for critical reflections on the relationships between corporeality, virtuality and digitality."

"This means that the virtual is thought of from a device-centered perspective, namely as a specific technological medium: the digital."

"In the educational field, there are many discourses that overlook the inherent virtuality of embodied experience due to the predominance of a technological, device-centered understanding of the virtual and its identification with the digital."

"The assumption of a disembodied digitality in which the body does not play any significant role and it appears as one technological device among others and, on the other hand, the aforementioned assumption that corporeality is a mere physical phenomenon opposed to virtuality."

"We approach the phenomenon of corporeality by addressing its inherent virtuality."

"We make some remarks on the relationship of corporeality and virtuality considering the challenges of digitally-mediated education today."

Disembodied Digitality

"Since pre-pandemic times, the drive for digitalization in the educational field has been closely linked to at least two assumptions."

"The integration of digital technologies into educational processes is a necessary response to the ubiquity of such technologies in the life-world (Lebenswelt) of schools and higher education students."

"Since students live in a "cyberculture" (Lévy [155]) and use digital technologies in multiple ways, a rapid transformation of traditional educational institutions (school, university) should be implemented."

"At the higher education level, the adoption of digital technologies also emerges as a 'solution' for the heterogeneous and changing realities of students, who may be facing complex work and family obligations (Berry and Hughes [156])."

"Prensky's notion, instead, neglects the distance between school and everyday life to the extent that everyday experiences of the digital are no longer seen as such in their specificity."

Devirtualized Corporeality
"This would be the case of pedagogical experiences in virtual classrooms where the physical absence of both the body and the classroom appears as the most evident deficit."

"As long as the face-to-face classroom situation remains the standard of pedagogical authenticity a typical situation of digitally-mediated education such as those in fully synchronous virtual classrooms via videoconferencing seems to be deficient in several ways: a) Body fragmentation and disproportion: online meetings are usually face-centered situations that make the rest of the body disappear."

"Certain experiences and practices in the virtual classroom can develop a specific sense of place that is not so much based on sharing a physical space, but on sharing a feeling of 'being there' and 'being together' involved in both individual and collective activities (Lehman and Conceiçao [157]; Osler [158])."

Virtualizing Embodied Experience
"Merleau-Ponty claims that what counts in the embodied experience is not the body: as it in fact is, as a thing in objective space, but as a system of possible actions, a virtual body with its phenomenal 'place' defined by its task and situation."

"My body is wherever there is something to be done. (Merleau-Ponty [159], p. 291) In other words, the virtual body transgresses the domain of the physical body that we have (Körper) as a given thing in a measurable space and time, but without overcoming it."

"Considering the experience of transposition, Merleau-Ponty claims that the virtual body "ousts the real one to such an extent that the subject no longer has the feeling of being in the world where he actually is" (Merleau-Ponty [159], p. 291)."

"In the Phenomenology of perception, Merleau-Ponty implicitly relates the presence of the virtual body as a projection to an idea of technology as an extension of the human body."

Final Remarks
"By considering virtuality as constitutive of embodied experience, the question of the body and its presence in education shifts to the question of specific media and modes in which virtuality appears."

"It is important to think of ways to pedagogically address the complexity of the human sensorium by making other intersensory, individual and social experiences of non-digital virtuality possible, for example, by means of olfactory and gustatory experiences both inside and outside the classroom."

"In the experience of baking bread at home, virtuality can appear in different ways: in relation to corporeal movements while kneading the dough, when varying the ingredients, in the memories and emotions evoked by a family recipe, etc This experience allows us to observe and address a wide range of physical, biological and chemical processes of varying complexity."

8.1.20 A Smart Learning Ecosystem Design for Delivering Data-Driven Thinking in STEM Education [284]

This is a machine-generated summary of:

Benita, Francisco; Virupaksha, Darshan; Wilhelm, Erik; Tunçer, Bige: A smart learning ecosystem design for delivering Data-driven Thinking in STEM education [284]

Published in: Smart Learning Environments (2021)

Link to original: https://doi.org/10.1186/s40561-021-00153-y

If you want to cite the papers, please refer to the original.

For technical reasons we could not place the page where the original quote is coming from.

Abstract-Summary

"This paper proposes an Internet of Things device (IoT)-based ecosystem that can be leveraged to provide children and adolescent students with STEM educational activities."

"Our framework is general and scalable, covering multi-stakeholder partnerships, learning outcomes, educational program design and technical architecture."

Introduction

"Various stakeholders within the educational context have proposed digital technologies such as IoT devices in the in- and out-of-school learning settings for children and adolescent students' education (Ito and others, [160])."

"In a nutshell, our smart learning ecosystem i) promotes STEM education and Data-driven Thinking in a student-friendly manner with emphasis on collaborative and experiential learning; ii) integrates various stakeholders (such as pedagogical institutes, educators, funding bodies or research agencies) for a large-scale deployment, and; 3) is based on a wide range of (flexible) services and components, ranging from cloud computing to IoT devices, design of experiments and to analytic platforms."

"The NSE initiative adopted our smart learning ecosystem with the aim of delivering Data-driven Thinking and educating children and adolescent students to be globally aware of STEM subjects."

Background

"The concept of Smart Education is based on smart learning through, but not limited to, IoT devices and other Information and Communication Technologies (ICT), and it is closely related to the literature on Smart Cities (Lee, Zo, & Lee, [161])."

"Project-based learning and collaborative learning have been shown to be effective strategies to engage young students in STEM education (Kelley & Knowles, [162])."

"Although there are many student-centred teaching and learning approaches, project-oriented problem-based learning is more useful in the context of delivering Data-driven Thinking in STEM education (Boss & Krauss, [163])."

"These experiential activities: (i) motivate and increase commitment among students; (ii) are problem-oriented and not subject-oriented; (iii) are based on learning process and methodologies designed to find solutions rather than recall knowledge, and; (iv) promote team work, social and communication skills."

"Some (Meluso, Zheng, Spires, & Lester, [164]) argue that game-based learning provides intrinsically motivating environments enhancing STEM education."

A smart learning ecosystem for enabling Data-driven Thinking in STEM education

"By engaging stakeholders in the various stages of the educational initiative, the proposed framework is tasked to establish, organize, operate and maintain a smart learning ecosystem that promotes Data-driven Thinking in STEM."

"Researchers and developers, pedagogical institutes, and service providers represent main operators of the smart learning ecosystem."

"After educational outcomes are clearly set out, funding provided by different entities, including government agencies, professional organizations, industries, and education institutions would help ensure meeting STEM program's goals and objectives."

"The specific responsibilities of pedagogical institutes include the following: designing, supervising and conducting learning activities, and; developing Data-driven Thinking-related curriculum pedagogical content knowledge and materials (e.g., blogs, websites, teaching materials, etc)."

"Schools serve as the physical and institutional backbone of the smart learning initiative."

"They support students in their Data-driven Thinking endeavors by developing digital functionalities of the smart learning environment."

Data-driven Thinking in STEM education

"Our ecosystem is specially designed for learning through STEM-based Data-driven Thinking."

"To develop cognitive skills (cognitive logical and analytical thinking, see Wing [165] and Grover and Pea [166]) and get comprehensive insight into the usefulness of data to draw effective problem solutions. (ii) Data collection from internal (smart learning ecosystem) and/or external sources (public databases, repositories, social media, etc). (iii) Data analysis and processing."

"To transform text-based data into visually stimulating 2D or 3D charts, maps, graphs, or networks (Benita and others, [167])."

"Students must explain and show how data served to test and validate their hypotheses."

The National Science Experiment as case study

"NSE offered two main types of experiential learning, namely: Data Collection and Big Data Challenge."

"The exposure of students to Data-driven Thinking was higher but the total number of participants was lower than that envisioned in Data Collection."

"During Big Data Challenge, teachers and other mentoring figures actively engaged students in learning through group and project work."

"NSE was conceived and shaped accordingly with the third Master Plan (MOE, [168]) which aims to enrich and transform the learning environment to enable students to develop a critical digital expertise."

"Using different sampling rates in delivering Data-driven Thinking in STEM education is important because higher sampling rates add computational and cognitive complexity (He, Ji, & Bobbie, [169]), thus, allowing elaborated designs of the learning environment."

"The set up ensured out-of-school and off-line functionalities, e.g., students collecting data at any time in any place."

Results

"Data Collection 2 was carried out during 2016 and promoted active learning by including the happy button which students were required to press whenever they felt happy."

"Big Data Challenge 1 connected students with scientists from researcher and developer institutions to come up with innovative STEM applications by using the data collected during Data Collection 2."

"The connection between Data Collection periods and Big Data Challenges is that the former exposed students to get to track their carbon footprint, travel mobility patterns or amount of time they spend indoors/outdoors."

"Through Data Collection, students learned about IoT and Big Data while teachers were able to leverage the data to develop interesting physics lessons and teach concepts such as humidity, linear kinematics and pendulum motion through hypotheses testing and hands-on experiments."

Discussion
"We have presented a general and scalable framework for designing, maintaining, and operating a smart learning ecosystem in STEM education."

"Our work describes the experience of Singapore's National Science Experiment, the world's largest Smart Education initiative where thousands of students and hundreds of teachers and staff got involved in an ecosystem that enabled Data-driven Thinking."

"Although the case study is based on Singapore, the proposed learning ecosystem and findings could have broad implications for other large cities with Smart Education initiatives worldwide."

"SENSE IT is probably the closest Smart Education initiative to NSE due to the aim of promoting STEM education in schools by offering an innovative learning experience through sensors."

"In nations lagging behind other countries in the fields of STEM, lessons learned from NSE, particularly the adoption of Data-driven Thinking, could provide a valuable knowledge base for the creation of (scalable) high-quality youth development programs."

8.1.21 Digitalisation in the Curricula of Vocational Schools: Text Mining as an Instrument of Curricula Analysis [285]

This is a machine-generated summary of:

Delcker, Jan: Digitalisation in the Curricula of Vocational Schools: Text Mining as an Instrument of Curricula Analysis [285]

Published in: Technology, Knowledge and Learning (2022)

Link to original: https://doi.org/10.1007/s10758-022-09591-0

If you want to cite the papers, please refer to the original.

For technical reasons we could not place the page where the original quote is coming from.

Abstract-Summary

"The influence of digitalisation on society and the workplace require stakeholders in school development to incorporate digital competencies into school curricula."

"This study examines the occurrence of digital components in 831 curricula of the vocational schools in Baden-Württemberg, Germany with a text mining approach and the analysis of keywords in context."

"Ongoing reforms of school curricula need to widen their view on digital competencies and integrate these competencies more broadly into the different types of vocational schools."

Introduction

"Teacher education is influenced by the content of school curricula to a large extent and in-service teachers develop their classroom teaching in accordance to the requirements of the school curricula (Chan, [172])."

"The influence of digitalisation on the modern society requires students and teachers to develop digital competencies as a prerequisite for participation processes in the society and the workplace (Fraillon and others, [170]; Roll & Ifenthaler, [171])."

"School curricula are a strong external influence on school development and must contain digital competencies, especially in the curricula of vocational schools (Balsmeier & Woerter, [173]; Seeber & Seifried, [174]; Wild & Schulze Heuling, [175]; Wuttke and others, [176])."

"The different types of vocational schools, the complexity of digital competencies and their characteristics are being prioritized in the research process."

Theoretical Framework

"In Germany, the education system and therefore the curricula of vocational schools are based on the cultural sovereignty of the federated states (Arends, [177])."

"The combination of theoretical and practical knowledge as well as the possibility to advance through the educational system differs between the types of vocational schools, which requires the assignment of a level range, rather than a fixed level."

"The strong interlocking between practical workplace competencies and theoretical knowledge in vocational schools demands a constant alignment with the requirements of the work environment."

"Regarding the slow speed and the often-criticized misalignment of curricula development with changes in the society (Goller and others, [178]), the following research questions concentrate on the role of digitalisation in the curricula of vocational schools in Baden-Württemberg, Germany."

"RQ2: To what degree do current curricula at vocational schools already align to the goals of the strategy concept Education in the Digital World?"

Methods

"The corpus for this research project consists of the curricula of all types of vocational schools in the federate state of Baden-Württemberg, which have been acquired through the process of web scraping (Ignatow, [179])."

"Under those divisions, single subjects (e.g., math, biology, chemistry) can be found, which are themselves segmented into the curricula of different grades, depending on the school type."

"This dictionary consists of 62 different terms (e.g., the German words for digitalisation, digital print, digital photography, analog–digital-transformation, see appendix) which have been identified within the corpus."

"It is unnecessary to include features which exists outside of the corpus (e.g., the term digital currency, in German Digitalwährung), because the curricula are compared between each other and not with other texts or corpora."

"The dictionary has been used to analyze the frequency of digital terms within the corpus as well as the distribution within and between different types of schools."

Results

"The column Rate shows how many percent of curricula of a school contain a digital term."

"The column Terms includes the number of times a digital term appears in the curricula and the percentage value in relation to the total number of digital terms in all documents."

"From the 200 curricula that contain a digital term, 44.5% belong to the Berufsschule."

"With a rate of 29.5%, close to a third of the curricula of the Berufsschule contain a digital term."

"Only a small amount of the curricula of the Berufsfachschule (12.2%), the Berufsoberschule (15%) and the Berufskolleg (17.3%) contain digital terms."

"In terms of occurrences, the curriculum is followed by the curricula of Commercial IT (Kaufmännische IT Berufe 2019, 90 occurrences), Media Agent (Medienkaufmann/frau 2014, 84 occurrences) and Digital Print (Mediengestaltung DigiPrint 2004, 55 occurrences), which all belong to the Berufliche Schule."

Conclusion

"The permeability of the dual system and students' capability to choose their own educational paths requires an equal representation of digitalisation in the curricula of the different types of vocational schools (Virdia & Schindler, [180])."

"The difference between the intended digital competency goals (KMK, [181, 182]) and the current implementation in the curricula of vocational schools can also be found in the findings of RQ2."

"Dissolving insecurities about the importance of digital contents in curricula supports the efforts of teachers to cooperate across subjects, classes, years, vocational programs and the different types of vocational schools."

"Students at vocational school will profit the most from the further integration of the digital competency fields into the curricula."

"The combined effects of the integration of competency fields into the curricula of vocational schools will enable all stakeholders to facilitate the development of student's digital competencies over all vocational schools (Delcker & Ifenthaler, [183])."

8.1.22 Design, Content Validity, and Inter-Observer Reliability of the 'Digitization of Cultural Heritage, Identities, and Education' (DICHIE) Instrument [286]

This is a machine-generated summary of:

Ortega-Sánchez, Delfín; López-Sanvicente, Amparo Bernal: Design, content validity, and inter-observer reliability of the 'Digitization of Cultural Heritage, Identities, and Education' (DICHIE) instrument [286]

Published in: Humanities and Social Sciences Communications (2023)

Link to original: https://doi.org/10.1057/s41599-023-01550-z

If you want to cite the papers, please refer to the original.

For technical reasons we could not place the page where the original quote is coming from.

Abstract-Summary

"The present study, on the one hand, seeks to offer the validated design of the Digitization of Cultural Heritage, Identities and Education (DICHIE) instrument, applicable in focus-group contexts, to extract qualitative data on both the needs and the challenges relating to the digitization of the cultural heritage of Castile and Leon (Spain)."

"The results of the instrument were optimal, both in terms of internal reliability and complete global and intra-group content validity, for its application to professional areas of heritage conservation and education and digital cultural management."

Introduction

"The implementation of interactive multimodal technologies for understanding cultural heritage and for its conservation and educational promotion has burdened the user with a leading role in access to and the creative production of digital content and narratives aimed at improving their cultural experiences (Roussou and Katifori, [184])."

"The design, implementation and evaluation of virtual environments and resources, and digital and technological media for access, cultural understanding, interaction and heritage learning, and its connections with education for citizenship (Castrillo and others, [185]; Delgado-Algarra and Cuenca-López, [186]; Guillate and others, [187, 188]; Hernàndez-Cardona and others, [189]), has also constituted one of the most recognized educational research and innovation objectives in the Spanish context, led by recognized research groups such as DIDPATRI (University of Barcelona), DESYM (University of Huelva), DiCSO (University of Murcia), GIPyPAC (University of the Basque Country) and ARGOS (University of Zaragoza), among others."

Methods

"The experts rated the clarity (clarity of writing), internal coherence, bias (induced responses), language (linguistic adequacy), and precision (measures what is to be measured) of the wording of the questions."

"In a first investigative phase, each professional group conducted an independent evaluation of the consistency and the coherence of the set of guiding questions."

"The results were evaluated, in a second phase, in the rewording of the questions that had not achieved a satisfactory degree of compliance with the evaluation criteria (clarity, coherence, bias, language, and precision)."

"Considering the ordinal level of measurement of the variables and the number of expert judges who had been selected ($n = 20$), we calculated Kendall's concordance coefficient, W, (Kendall, [190]), Fleiss's Kappa (Fleiss and others, [191]), and Krippendorff's α alpha (Krippendorff, [192]), to examine the degree of concordance reached between the experts and the inter-observer reliability."

"The independent inter-judge valuation was in that way composed of five specific Evaluation Criteria (EC_e), applied to the whole instrument and, in an individualized way, to each question: EC_{e1}."

Results

"These values were based on satisfactory general scores ($M_e \geq 3$) in all the specific criteria, except in the application of criteria 1 (clarity) and 2 (coherence) to question 5."

"In the two rounds of evaluation, statistically significant inter-group differences were recorded in the application of EC_{e2} (coherence) to question 7, in EC_{e3} (bias) to question 9, in EC_{e3} (bias) to question 12, and in EC_{e5} (precision) to question 13."

"It may be confirmed that the cultural educators and managers assigned general scores that were lower than the scores given by those interested in the conservation of cultural heritage under the four specific criteria."

"The significative group differences were found for the scores that the conservationists and the cultural managers assigned to the seventh and to the ninth key questions, and for the scores that the conservationists and the educators assigned to the twelfth and to the thirteenth key questions."

Discussion and conclusions

"This evidence has covered the necessary dimensions for an integral evaluation of the needs and the professional challenges relating to the conservation, management, and educational promotion of the digital cultural heritage of Castile and Leon."

"DICHIE aims to contribute to the advancement of methodological rigor for obtaining quality results on the qualitative evaluation of processes, resources and means of conservation, management and education of digital cultural heritage."

"From the perspective of participative culture, not only will the inter-professional collaboration of the agents involved in digital cultural heritage be necessary, but public participation will also be needed in the creation of "heritage communities" and complementary digital narratives or alternatives to institutional narratives (Morgan and Pallascio, [193]; McCandlish and McPherson, [194]), as clearly stated both in recent studies (Bonacini, [195]) and in the inter-rater evaluation of this study."

"As is evident from the results relating to the identification of models with technological designs and contextualized heritage interventions, the professional intersection constitutes, in effect, a recognizable key for action in the field of digital cultural heritage."

8.1.23 Medical Studies in Times of a Pandemic—Concepts of Digital Teaching for Orthopaedics and Trauma at German Universities [287]

This is a machine-generated summary of:

Mielke, Anna-Maria; Ghanem, Mohamed; Back, David Alexander; Fröhlich, Susanne; Herbstreit, Stephanie; Seemann, Ricarda Johanna: Medical studies in times of a pandemic – concepts of digital teaching for Orthopaedics and Trauma at german universities [287]

Published in: BMC Medical Education (2023)
Link to original: https://doi.org/10.1186/s12909-023-04213-4

If you want to cite the papers, please refer to the original.

For technical reasons we could not place the page where the original quote is coming from.

Abstract-Summary
"Due to the Covid-19 pandemic, on-site classroom teaching became limited at most German medical universities."

"How the transfer from classroom to digital teaching or digitally assisted teaching was conducted was decided by each university and/or department individually."

"Specific challenges in designing digital teaching concepts were expected to arise."

"A questionnaire with 17 items was designed and sent to the professors in charge of organising the teaching in Orthopaedics and Trauma at each medical university."

"Each university reported a substantial reduction of their classroom teaching and efforts to transfer their teaching to digital formats."

"Three sites were able to switch to digital teaching concepts completely, whereas others tried to enable classroom and bedside teaching at least for students of higher edcuational levels."

"One year into the pandemic significant differences concerning the proportions of classroom and digital teaching for Orthopaedics and Trauma can be observed."

"Huge differences in concepts used to create digital teaching are present."

"Since a complete suspense of classroom teaching was never mandatory, several universities developed hygiene concepts to enable hands-on and bedside teaching."

Introduction
"The definition of digital or digitally assisted teaching includes all forms of teaching using electronic or digital media for presentation and distribution of study material or communication between people [196]."

"This means digital media may be used for teaching as well as exams at medical universities."

"Starting April 1st 2020, amendments to the German regulations of undergraduate medical education were made to enable further teaching and exams under the new conditions of the Covid-19 pandemic."

"How the transfer from classroom to digital teaching or digitally assisted teaching was conducted was subject to each university and/or department individually."

"Several publications over the past years were able to show successful approaches to integrate digital teaching at different universities [197–199]."

"With this study we tried to create an overview concerning the extent and forms of digital teaching for Orthopaedics and Trauma at German universities under current regulations for the pandemic."

Methods
"The authors developed a questionnaire to collect information on digital teaching currently used under Covid-19 regulations in Orthopaedics and Trauma at German universities."

"It included questions on potentials and pitfalls in (digital) teaching arising from the pandemic in retro- and prospective."

"The questionnaire was then distributed via the German Association for Orthopaedics & Trauma (DGOU) and sent to the professors in charge of organising the teaching in this discipline at each medical university."

"Additional interviews and open questions were used to gather opinions on potentials and pitfalls concerning the current teaching situation in this discipline."

Results
"While some teaching formats were almost completely switched to digital concepts, e.g. lectures and seminars, others like bed side classes were only transformed into hybrid concepts, allowing hands-on-training and patient contact of some extent."

"Other teaching concepts, such as additional voluntary classes were continued at 9 universities and were said to be an essential part in the education and commitment of students to the field of orthopaedics and trauma."

"Hybrid concepts were able to provide theoretical knowledge through digital teaching and increase the time for interactive practice during classroom teaching sessions."

"Equally, basic conditions were criticised, such as lack of time to develop digital concepts and to transfer existing teaching concepts into digital classes, lack of infrastructure and its expansion for proper digital teaching, shortage of staff for classroom teaching due to the splitting of groups into several smaller groups and the omission of internships and rotations for medical students."

Discussion

"Since this process was subject to each university and/or department individually, a wide variety of digital teaching concepts for lectures, seminars and practical training, as well as the used platforms were found."

"To train students on being digitally competent, it will also be essential to the field of Orthopaedics and Trauma to partake in the organization and teaching of classes."

"Further studies may consider involving students for post-pandemic evaluation of digital teaching and training in Orthopaedics and Trauma surgery, as well as their personal experiences and academic performance during the pandemic."

"This provides attractive teaching of high quality and will additionally improve the engagement of students in the field of Orthopaedics and Tauma [200]."

"The exchange between universities and departments on experiences with old and new digital teaching concepts should be encouraged to provide overall improvement of teaching in this field."

Conclusion

"One year into the pandemic, we can observe significant differences concerning the proportions of classroom and digital teaching for Orthopaedics and Trauma."

"Since a complete suspense of classroom teaching was never mandatory, several universities developed hygiene concepts to enable some extent of practical experience for medical students."

"Most universities saw potential in the use of digital formats, since it could offer consistency in teaching quality and independence in time and location for students as well as lecturers."

"The use of digital teaching formats is, without doubt, essential to the future of medical studies and has gained importance not just due to the pandemic, but due to the general digitalization of our world."

"Realizing the potentials of digital teaching, it can and should be used complementary to the practical training of medical students."

8.1.24 Student Access to Technology at Home and Learning Hours During COVID-19 in the U.S. [288]

This is a machine-generated summary of:

Ogundari, Kolawole: Student access to technology at home and learning hours during COVID-19 in the U.S [288]

Published in: Educational Research for Policy and Practice (2023)

Link to original: https://doi.org/10.1007/s10671-023-09342-7

Copyright of the summarized publication:

If you want to cite the papers, please refer to the original.

For technical reasons we could not place the page where the original quote is coming from.

Abstract-Summary

"In light of this, the study investigates the effect of technology access at home on student learning hours during the COVID-19 pandemic and across racial and ethnic groups in the U.S. The Household Pulse Surveys (HPS), conducted by the United States Census Bureau and administered from August 19, 2020, to March 29, 2021, were used for the analysis."

"This means the intensity at which households in the sample have access to technology is higher among the Asian and White students, followed by Hispanic, Black, and other races in that order."

"The estimated effect of technology access on the student learning hours during COVID-19 based on the Tobit regression model shows about a 3.1 unit points increase over the whole sample."

"Further analysis reveals variation at which access to the technology impacts learning hours across race and ethnicity groups."

"We find that access to technology significantly increased learning hours by about 3.5, 1.6, 2.2, and 3.4 unit points among White, Black, Hispanic, and Asian students, respectively."

"The observed differing effect of access to technology on learning hours further highlights the racial disparities in American society's digital divide, which reveal how access to technology disproportionately impacts student learning hours during the COVID-19 pandemic across race and ethnicity."

Introduction

"Understanding how access to computers and the internet impacts student learning hours during the COVID-19 pandemic and across races and ethnic groups is crucial to the ongoing discussion on the digital divide in the American educational system."

"What effect does access to technology have on student learning hours during the COVID-19 pandemic?"

"Are there differences in the effect of technology access on student learning hours across races/ethnicity during the COVID-19 pandemic?"

"Arising from the research question, we believe this study is critical because it provides insights into how access to the technology impacts student learning hours during the COVID-19 pandemic."

"Given this, we believe the finding highlights the racial disparities in the digital divide in American society, which reveal how access to technology disproportionately impacts student learning hours during the COVID-19 pandemic across race and ethnicity."

"The present study contributes to the sparse literature on the effect of technology access on student learning hours, focusing on the COVID-19 pandemic."

The data sources and description

"The study employed the Household Pulse Surveys (HPS) conducted by the United States Census Bureau and administered from August 19, 2020, to March 29, 2021."

"The HPS covers all 50 U.S. states, the District of Columbia, and U.S. territories."

"The present study's focus is educational disruption, which covers computer and internet availability at home, student learning hours per week, and virtual learning experiences."

"The HPS also covers household income levels, marital status, gender, educational levels, and age of household head (constructed using birth year provided in HPS and survey year), the number of children and adults, ethnicity/race, and regions where they reside."

"The HPS offers supplement weights for the individual and household to represent the sample nationally."

Estimation strategy

"We employed principal component analysis (PCA) to generate a composite index of technology access using the 5-point Likert scale."

"The PCA calculates an uncorrelated set of variables called principal components, ordered so that the first few principally composed factors preserve most of the variations present in the original variables (Hardle & Simar, [201])."

"The transformed component employed to generate a composite index of technology (internet and computer) access at home can be defined as, The transformation means the computed indices of technology access at home range from 0 to 1, with 0 indicating lack of technology access and 1 suggesting access to technology at home."

"The output in the present study is the students' learning hours, while the input is the index of technology access and other potential confounding factors considered in the study."

Results and discussion

"The results show that access to technology significantly increased student learning hours by about 3.5, 1.6, 2.2, and 3.4 unit points at a 5% significance level among White, Black, Hispanic, and Asian students, respectively."

"We believe policies that expand internet connectivity and provide computers for students to use at home for learning purposes, especially students of color, should be considered important in promoting education equity in the U.S. This, however, includes policies that prioritize investment in broadband connectivity to address affordability and broadband speed and student access to a computer at home for different subpopulations to ensure that every student in America can learn from a distance."

"Students' hours of learning vary significantly across races and ethnic groups, with students in white households reporting higher learning hours than black, Asian, and other races."

Concluding remarks

"The digital divide among students during COVID-19 has been a concern to the parents, members of the community, and policymakers on how children will receive the same level of education on their own at home when not all students have the same access to computers and the internet resources in the U.S. In light of this, the study investigates the effect of technology access at home on students' learning hours and across race and ethnic groups during the COVID-19 pandemic in the United States."

"This study provides insights into how access to technology impacts students' learning hours differently across races and ethnicity during the COVID-19 pandemic in the U.S. In light of this, we believe policymakers need to prioritize policies that address the digital divide in education as part of the ongoing efforts to strengthen education equity in the country."

8.2 Digitalisation and Knowledge Dissemination

Machine generated keywords: teacher, high education, learn, education, educational, covid, adoption, teaching, course, learner, pandemic, open, ict, platform, learning

8.2.1 Institutional Entrepreneurship in Loosely Coupled Systems: The Subject Position of MOOC Entrepreneurs and Their Interpretive Struggles in a Norwegian Context [289]

This is a machine-generated summary of:

Langseth, Inger Dagrun; Jacobsen, Dan Yngve; Haugsbakken, Halvdan: Institutional Entrepreneurship in Loosely Coupled Systems: The Subject Position of MOOC Entrepreneurs and Their Interpretive Struggles in a Norwegian Context [289]

Published in: Technology, Knowledge and Learning (2023)

Link to original: https://doi.org/10.1007/s10758-023-09647-9

If you want to cite the papers, please refer to the original.

For technical reasons we could not place the page where the original quote is coming from.

Abstract-Summary
"While technological change in organizations is fast and eminent to most people, the adoption of Massive Open Online Courses, micro-credentials, and flexible and scalable online courses, appear to be comparatively slow in Higher Education in the Nordic countries."

"The informants were strategically selected among employees who had been involved in open platform technology, MOOC production and support for faculties."

"Rather than seeing technological innovations as "more of the same", the entrepreneurs embraced the possibilities emerging in new educational practices."

"Inspired by New Institutionalism, we focused on the organizational conditions for MOOC production."

"The entrepreneurs often entered interpretive struggles at higher organizational levels in competition with other stakeholders."

"Our study points to discrepancies in understanding the disruptive and transformative change that new technology can bring to study programs and lifelong learning."

Introduction

"In Higher Education Institutions (HEI), access to Massive Open Online courses (MOOC), study programs and micro-credentials (short courses with badges or ECTS), have grown considerably over the last decade (Class Central, [202])."

"We explore the adoption of such online learning opportunities in HEIs, while focusing on MOOC entrepreneurial activities among staff and faculties in Norwegian HEIs."

"The modulization of online course content (OER) makes it possible to integrate MOOCs and micro-credentials in campus programs, or they may function as supplemental learning for students, who want an alternative perspective on their learning material (Rivard, [203])."

"The impact that the explorative use of new technology in digital teaching and learning had on faculties and organizational contexts during the pandemic in the Nordic countries, supported the already ongoing transition to online education (Laterza and others, [204])."

Theoretical Perspectives

"These research reviews are prone to emphasizing different explanations as to why learners drop out of MOOCs, which can be attributed to ineffective online course design, lack of belonging, time factors, and hidden costs, etc Even so, we observe that educational researchers are deeply engaged in exploring the pedagogical effectiveness of MOOCs from different perspectives."

"There is not an established research stream that examines the organizational conditions for MOOCs at universities, moreover, which exists to theorize and connect MOOC research to digitalization and organizational theory and to frame MOOC research from a national perspective."

"To formulate our theoretical perspective, we use concepts and perspectives from New Institutionalism in organizational research (DiMaggio & Powell, [205]; Meyer & Rowan, [206]), and connect them to emerging research on digitalization of the Norwegian HEI and MOOC in Norway."

Research Focus and Research Questions

"We use New Institutional theory to explore the subject position of Norwegian MOOC entrepreneurs and account for their interpretive struggles, which is marginally represented in the research literature (Langseth and others, [207]; Tungesvik, [208])."

"Our overall goal is to show how entrepreneurs experience and engage in complex organizational processes when attempting to promote MOOCs in Norwegian HEIs."

"In fall 2020, we interviewed employees from different Norwegian HEIs who had been actively involved in the development of MOOCs since the early 2010s."

"We focused on the adoption of MOOCs to explore how the informants occupied the subject position of MOOC entrepreneurs through actions and practices that supported that role."

"We also intended to describe how they negotiated their belonging to the subject position as MOOC entrepreneurs by mobilizing resources."

"What experiences do MOOC entrepreneurs draw from their engagement in educational change?"

Method

"To approach our research topic, we decided to interview informants who were involved in MOOC production and support."

"This means that the findings were established based on the transcribed interviews and what the informants told us about their activities, in this case MOOC entrepreneurship."

"To avoid researchers being overwhelmed by the data, qualitative methods limit the number of informants."

"We found that ten informants would be a manageable sample and selected these from ten institutions based on their visibility in the Norwegian MOOC landscape."

"Purposeful sampling is a widely used technique in qualitative research to identify and select information-rich samples "whose study will illuminate the questions under study" (Patton, [209]: 273)."

"The researchers and informants in this study all have Norwegian as their first language and all the interviews were conducted in Norwegian."

"Thematic Analysis (Attride-Stirling, [210]; Braun & Clark, [211]) explains the main purpose of coding as a process where the researcher identifies significant themes from the interviews."

Findings

"The informants occupied the subject position of MOOC entrepreneurs while framing their educational activities and practices on (1) the choice of target groups, (2) the language of instruction (Norwegian or English), and (3) the kind of platforms that were available to faculties (open platform and/or LMS) in the HEIs."

"Positive feedback on project proposals that resulted in external funding was common, as feedback on possible long-term outcome scenarios for these MOOC projects seemed to be less informed, organized and systematized at higher levels in the HEIs."

"Only a few informants had managed to occupy more powerful subject positions (Törrönen, [212]), where they had established trust and could confidentially argue the value of open platforms and instructional design in MOOCs and micro-credentials to stakeholders in the field."

Discussion

"We theorize project management and the concept of pocket of innovation, while we outline how MOOC entrepreneurs contribute to digital transformation and a paradigm shift in HEIs."

"That said, we will further theorize pockets of innovation and describe a model that can potentially explain the organizational dynamics and why MOOC entrepreneurs struggle to change the educational practices."

"As opposed to scientific research, where institutions recognize strategic support for start-ups and funding for long-term outcomes, educational development related to MOOCs tended to capture innovations in a series of projects and dissemination strategies (e.g., project evaluations, reports, and presentations) in the organizations."

"To entrepreneurial activities in other fields in HE (Feld & Hathaway, [213]), MOOC entrepreneurs developed the innovative ideas, but they lacked the necessary resources to disseminate research-based initiatives outside their pockets of innovation."

Summary and Conclusion

"We have explored how MOOC-entrepreneurs in HEIs can initiate and support transformative change in online teaching and learning."

"Three research questions served as steppingstones for our qualitative study of ten HEIs in Norway: (1) What characterizes MOOC entrepreneurial activities in the current Norwegian sample? (2) What experiences do MOOC entrepreneurs draw from their involvement in educational change strategies? (3) How can transformative models describe digitalization processes in higher education?"

"MOOC entrepreneurial activities were found in initiatives characterized by change agency, hands-on experiences, and timely initiatives."

"MOOC entrepreneurial experiences were characterized by challenges related to both feedback and funding and to institutional roles and positions in in their organizations."

"MOOC-entrepreneurial activities were often based on what we have described as single-loop entrepreneurship, a series of projects without a long-term plan and with limited feedback from management and stakeholders, which narrowed the potential to contribute to organizational change."

Recommendations

"Another finding points to the problem that innovations are organized as one project at the time."

"Organizations must find new ways to evaluate project results to consider outcomes in the organization and avoid project overlap and subsequent burnout."

"Organizations should consider new roles to make sure that successful innovations are followed up and human capital taken care of in new reorganization processes."

"We also see the need for more research in organizational theory."

Limitations

"The study focuses on patterns and possibilities, where informants tell their story."

"One perspective that is not explored is the leadership's take on these processes."

"These perspectives were, to a certain degree, intercepted from informants' descriptions of their subject position and how they interacted with other stakeholders."

8.2.2 Educational Quality and Technological Complexity on Recognition of Enhanced Learning Platform in Developing Countries Using PLS-SEM in a Post COVID Era [290]

This is a machine-generated summary of:

Idoga, Patience E.; Oluwajana, Dokun Iwalewa; Adeshola, Ibrahim: Educational quality and technological complexity on recognition of enhanced learning platform in developing countries using PLS-SEM in a post COVID era [290]

Published in: Educational technology research and development (2022)

Link to original: https://doi.org/10.1007/s11423-022-10150-8

If you want to cite the papers, please refer to the original.

For technical reasons we could not place the page where the original quote is coming from.

Abstract-Summary

"The purpose of this study is to examine instructors' acceptance of e-learning in Nigerian universities."

"The study results indicate the factors affecting instructors' adoption of e-learning platforms and the subsequent impact on instructors' behavioral intention, and consequenly, an impact on the acceptance of the e-learning platform in Nigeria."

"Based on the study, self-efficacy, educational quality, and ease of use, perceive usefulness, and behavioral intention were found to be predictors for instructors' acceptance of the e-learning platform in Nigeria."

"The results drawn from the study revealed that there are individual challenges that come from the use of the e-learning platform."

Introduction

"The reduced level of interaction techniques, mastering challenging tasks, trust, and absence of a reward system hinder the use of educational technology in Nigerian universities (Adewale & Ghavifekr, [214]; Yakubu & Dasuki, [215]) stated that Western African countries such as Nigeria, are still under-researched in terms of e-learning adoption."

"To ensure proper implementation of e-learning technology, it is necessary to identify the factors that contribute to instructors' acceptance of e-learning at Nigerian universities, as existing studies have indicated little research on the factors responsible for instructors' adoption of e-learning in Nigerian academic environments."

"The principal purpose of this study is to identify the factors that facilitate instructors' motivation to adopt e-learning based on a new conceptual framework."

"The conceptual study model adopts constructs drawn from the TAM and the motivational factors of e-learning adoption that contribute to the present knowledge by examining the motivational factors that influence instructors' adoption of e-learning in Nigerian universities."

Review of literature

"Few studies have adopted the TAM model, and other models have based their influence on students' behavioral intention to use e-learning in Nigeria (Yakubu & Dasuki, [215, 216]; Yakubu and others, [217])."

"No study on the adoption of e-learning studies in Nigeria universities has examined the motivational factor of the instructor's efficacy, technological complexity, and educational quality on the technology acceptance of e-learning in Nigeria."

"Inadequate professional skills, access, and difficulties associated with the understanding of recent systems and the various interrelating technologies have been stated to represent a significant limitation to instructors' acceptance of e-learning platforms (Wang and others, [218]; Chin & Lin, [219])."

"This study defined perceived ease of use as the extent to which an instructor's use of an e-learning platform is perceived as easy or effortless."

"It is suggested that ease of use represents a strong construct that motivates users' intention to accept and use e-learning (Cheng, [220])."

Research methodology

"In the context of e-learning, this study use factors such as "self-efficacy," "educational quality," and "technological complexity" with the concept of TAM to contribute to the adoption of e-learning among instructors."

"This study examines the factors affecting instructors' adoption of e-learning platform in Nigeria during Covid-19 pandemic."

"The predictive power of perceived ease of use (32%) and perceived usefulness (34%) both proved to be a strong predictor of behavioral intention (47%) of instructors' adoption of the e-learning platform."

"The study measures the data analysis by testing the reliability and validity consist of factor loadings, Average Variance Extracted, Composite Reliability, and Cronbach's alpha (Construct Reliability) values."

"The use of Partial Least Squares (PLS) to measure the estimated value of measured variables and several latent constructs (Oluwajana and others, [221]): With the use of composite PLS, model's goodness of fit: need to be tested with the unweighted least squares discrepancy (d_ULS), the geodesic discrepancy (d_G), and the Standardized Root Mean Squared Residual (SRMR)."

Discussions

"This study used structural equation modeling (SEM) to examine the factors affecting instructors' acceptance and adoption of e-learning platforms as new educational technology tools in the Nigerian education sector."

"The study outcome shows that instructors' perception of the application's use shows that self-efficacy has a significantly positive effect on instructors' perceived ease of use in adopting and using the e-learning platform, which is in line with Chen and Tseng [222] and Scherer and others [223]."

"The study also showed that self-efficacy had an insignificant effect on instructors' perceived usefulness of adopting and using e-learning platforms."

"The results indicate that educational quality has a significant positive effect on instructors' perceived ease of use and usefulness of adopting and utilizing e-learning platforms."

"Instructors' perception of technology complexity exerts a positive influence on perceived ease of use but an insignificant effect on the perceived usefulness of adopting the e-learning platform."

Conclusion

"Technology acceptance of e-learning platforms can be added to existing traditional methods to improve the learning process and enhance the instructors' acceptance of e-learning platforms."

"This study used the TAM as a theoretical framework to examine the factors affecting instructors' acceptance of educational technology, especially e-learning platforms during Covid-19 pandemic."

"There is a need to determine instructors' self-efficacy and technological complexity, increase their awareness of the use of e-learning platforms, and provide instructors with adequate training on various services offered by e-learning platforms."

"The results indicate a need to develop strategic plans and provide guidelines for instructors' technology complexity and self-efficacy acceptance to improve modern driving factors to meet the educational goal of using e-learning platforms."

8.2.3 Faculty Perceptions, Awareness and Use of Open Educational Resources for Teaching and Learning in Higher Education: A Cross-Comparative Analysis [291]

This is a machine-generated summary of:

Marín, Victoria I.; Zawacki-Richter, Olaf; Aydin, Cengiz H.; Bedenlier, Svenja; Bond, Melissa; Bozkurt, Aras; Conrad, Dianne; Jung, Insung; Kondakci, Yasar; Prinsloo, Paul; Roberts, Jennifer; Veletsianos, George; Xiao, Junhong; Zhang, Jingjing: Faculty perceptions, awareness and use of open educational resources for teaching and learning in higher education: a cross-comparative analysis [291]

Published in: Research and Practice in Technology Enhanced Learning (2022)

Link to original: https://doi.org/10.1186/s41039-022-00185-z

If you want to cite the papers, please refer to the original.

For technical reasons we could not place the page where the original quote is coming from.

Abstract-Summary

"This paper explores faculty's perspectives and use of open educational resources (OER) and their repositories across different countries by conducting a multiple case study to find similarities and differences between academics' awareness, perceptions and use of OER, as well as examining related aspects of institutional policy and quality that may influence individual views."

"Findings show the impact on individual OER adoption with regard to the individual control of diverse factors by faculty members; of institutional policies and quality measures on the externally determined factors (by the institution); and of institutional professional development and provision of incentives in more internally determined factors (by the faculty members themselves)."

Introduction
"Prior research focusing on factors influencing OER adoption by individual faculty members has found, for example, appropriate institutional support to be an enabler, while inadequate support hinders OER practice (Baas and others, [224]; Bates and others, [225]; Belikov & Bodily, [226]; Bossu and others, [227]; Schuwer & Janssen, [228])."

"As part of the research project "Digital educational architectures: Open learning resources in distributed learning infrastructures - EduArc" (https://uol.de/coer/research-projects/projects/eduarc), and as a follow-up to the macro- and meso-level studies (i.e., national systems and institutional infrastructures and organization (see Marín and others, [229]; under review), this study aimed to analyze aspects related to the use, creation, remix and sharing of OER in HE at the micro-level of teaching and learning (that is, faculty members' experiences) across countries."

"The framework of content analysis for the macro- and meso-level studies was adopted, with the unique features of the micro-level also taken into account, namely infrastructure (local environment), quality (quality of OER), policy (local policies) and change (incentives and faculty support)."

Methodology
"While much research around OER relies on one or two data sources, this study draws from multiple and complementary data sources contained in each case report, in order to enable a more nuanced analysis of the topic under investigation."

"In the first iteration, the data were categorized into main codes based on the four elements of the reports described in the research questions (Infrastructure, Policy, Quality and Change)."

"This process resulted in an enhanced model that combines the OER Adoption Pyramid and the four elements of the research project at the micro-level based on data from the international reports."

"As each country expert (or group of experts) was able to organize their case as they wanted, according to the project's main research questions, we could not report on homogeneous data collection methods as the basis for the reports."

Findings and discussion
"A high percentage of the Spanish and German faculty participants in the survey studies did not know about the existence of OER repositories in their institutions (27.4% and 36.8%, respectively)."

"A survey study in China with 172 participants from eight universities identified a lack of supervision to ensure that faculty members implemented online teaching and OER of a high quality (Xu, [230])."

"Previous studies also echo these findings in relation to the lack of institutional and departmental policies for the development and use of OER repositories as a barrier for the uptake of OER initiatives (Bates and others, [225]; Mtebe & Raisamo, [231]; Yuan and others, [232])."

"The survey studies in Spain and Germany showed that many faculty members did not use OER from repositories (68% and 54%), search for OER in them (48% and 55.3%), and neither published in them (48% and 43.4%) nor in non-institutional repositories (68.8% and 86.8%)."

Conclusion

"In RQ2, we explored the awareness of faculty members of institutional procedures related to OER QA and QA agents across countries."

"Similar findings were reported in RQ3 regarding faculty awareness of institutional OER policies and involvement in policymaking: a lack of institutional policies along with a low awareness of them was common."

"In RQ4, we identified the need for encouragement measures to be implemented by HE institutions, in order to motivate faculty to use OER."

"Concrete implications of the study for HE institutions to foster faculty use of OER could point toward the improvement in the availability of OER in institutional repositories and the development of measures for dissemination and faculty involvement in OER policy and QA."

"We believe that these institutional measures may address most of the factors identified for individual volition and encourage faculty to use, remix, share and publish OER at the micro-level."

8.2.4 China's Emerging Informatization Strategy [292]

This is a machine-generated summary of:

Hanna, Nagy K.; Qiang, Christine Zhen-Wei: China's Emerging Informatization Strategy [292]

Published in: Journal of the Knowledge Economy (2010)

Link to original: https://doi.org/10.1007/s13132-009-0001-z

If you want to cite the papers, please refer to the original.

For technical reasons we could not place the page where the original quote is coming from.

Abstract-Summary

"The authors adopt a holistic framework of China's e-development (informatization), covering the enablers of the knowledge economy and the application of information and communication technology (ICT) to government, business, and society."

A holistic framework of e-development

"In analyzing China's informatization experience, we adopt the framework of e-development as an integrated approach to leveraging ICT for development or

ICT4D—pursuing mutually reinforcing ICT-enabled development initiatives at the national or regional level [233]."

"This framework is about creating an information society or knowledge economy "ecosystem"—a holistic approach that engenders integrated leadership and shared vision, facilitates partnerships among stakeholders, and maps the connections and shapes the relationships among diverse players."

"E-development is defined most holistically so as to facilitate a way of systematically thinking about ICT as enabler of development, of strategically managing informatization programs, of tapping synergies among interdependent elements of ICT, and of communicating to a broad community of practice."

"The broad areas of applications are e-government, e-business, e-community, and e-rural."

Assessment of past achievements

"An overall assessment of China's informatization efforts and achievements would cover the institutions leading these efforts, the legal and regulatory reforms necessary to create an ICT-enabled economy, the massive development of the telecommunications infrastructure, the promotion of the information technology (IT) industry as a key engine for export, and the diverse application of ICT in government, business, and rural development."

"These projects were designed to facilitate China's transition from strictly administrative use of ICT applications to more public service-oriented uses; in 1997 the Outline of the National Informatization Development Plan established a framework for promoting informatization, supporting the development of related policies and regulations, information resources (content), network infrastructure, applications, the IT industry, and human resources."

"Three levels of ICT human resources are critical to informatization: A general public able to use ICT applications at work and home Informatization managers who lead ICT development in government and business ICT professionals experienced in network design, software development, and R&D China's compulsory basic education has dramatically increased literacy—reaching more than 90% in 2007."

Future directions for informatization

"China's future economic growth will likely come from three main sources: Encouraging the development of services and domestic demand for them Increasing firm competitiveness, particularly through ICT and foreign direct investment (FDI) Facilitating the movement of workers out of agriculture and to cities To sustain the momentum of economic growth, China needs to closely examine three interwoven factors during the 11th Five-Year Plan (2006–2010) and well beyond."

"To pursue ICT-enabled transformation of government, China needs to develop an e-leadership institutional model capable of promoting enterprise architecture, interoperability, public–private partnerships (PPPs), effective outsourcing and contract management, exchange of experience, information sharing, and prioritization across e-development and e-services [234]."

"China's legal and regulatory framework needs to provide certainty and predictability to help the country realize its informatization strategy by fostering a business-friendly climate where competition can flourish, foreign and domestic sponsors can invest with confidence, and ICT development can take root."

Conclusion: tapping synergies and managing cross-cutting issues
"To address these challenges, China needs to take key cross-cutting and synergistic actions: develop e-leadership capabilities and institutions, develop human resources to support national ICT diffusion priorities, promote affordable and shared access to ICT, and orient the national innovation system and ICT industry development to support local adaptation and local users."

"In the case of rural informatization, efforts to build capacity and raise awareness must span many levels: policymakers and administrators of various concerned departments (such as the Ministry of Agriculture, Ministry of Culture and Ministry of Information Industry), provincial decision-makers and administrators, managers, owners, and entrepreneurs of information centers, central and local content and service providers, rural communities and farmer groups, rural commercial and agricultural enterprises, farmer and professional cooperatives, and ICT technical staff and information intermediaries."

"The Chinese government needs to provide incentives and support for R&D in application development, in support of local ICT adaptation for both business and government users."

[Section 5]
"Since the mid-1990s, China has devoted considerable resources to developing its informatization strategy and conducted extensive research to determine the appropriate framework and goals for building an information society."

"Strong leadership at the national, provincial, and municipal levels, combined with a sustained long-term focus from the highest levels of the Chinese government on developing information and communication technology (ICT), have given ICT a prominent role in sustainable development strategies and ensured its application to core economic and social goals."

"The first part takes a holistic view of China's informatization (e-development) strategy."

"It describes the status of China's informatization strategy in terms of key ICT enablers and application areas."

"This e-development framework is used to identify key gaps and weaknesses in each enabler as well as synergies and cross-cutting coordination issues that need to be managed to support ICT application priorities and achieve national development goals."

8.2.5 Promises and Pitfalls of Digital Knowledge Exchange Resulting from the COVID-19 Pandemic [293]

This is a machine-generated summary of:

Nguyen, Vivian M.; Bell, Christina; Berseth, Valerie; Cvitanovic, Christopher; Darwent, Ray; Falconer, Matthew; Hutchen, Jenna; Kapoor, Tyreen; Klenk, Nicole; Young, Nathan: Promises and pitfalls of digital knowledge exchange resulting from the COVID-19 pandemic [293]

Published in: Socio-Ecological Practice Research (2021)

Link to original: https://doi.org/10.1007/s42532-021-00097-0

If you want to cite the papers, please refer to the original.

For technical reasons we could not place the page where the original quote is coming from.

Abstract-Summary

"We aim to inform how best to maintain effective KE practices and processes in a digital-first world."

"The acceleration in digital innovation and culture has thus resulted in new tools and diversified approaches for the KE toolbox to inform decisions and practices."

"The long-term sustainability and effectiveness of digital KE depend on two interconnected factors: addressing the persistence of the digital divide and people's abilities to make and maintain meaningful social connections in the absence of regular face-to-face contact."

"We thus offer three considerations to guide KE efforts and initiative in a digital-first world: (1) consider both digital divide and equity; (2) revisit user needs and preferences for KE to address the diversity of users, and (3) leverage the diversification of KE approaches and innovations."

Introduction: knowledge exchange—an interpersonal activity

"A recent investigation of one of the worlds oldest boundary organizations focused on improving KE at the interface of science and policy found that efforts to build trust between individuals and organizations is best achieved through a combination of formal and informal in-person, face-to-face interactions, particularly among stakeholders that have differing worldviews (Cvitanovic and others [235], p. 184)."

"Knowledge exchange research and practices must be re-examined, especially when digital KE is more common now than before."

"Our Perspective Essay examines how knowledge exchange principles may need to be re-examined through COVID lenses by discussing potential short-

and long-term implications of this digital transformation of KE spurred by the COVID-19 pandemic."

"By combining our perspectives with the case study, we are able to identify and discuss bright spots and opportunities for the digital transformation of KE."

"Is there an optimal way to leverage opportunities of digital interaction for improved KE access and participation?"

The case: the Canadian Forest Service KE practitioners

"CFS KE practitioners work in six regional research centres across Canada in addition to their head office located in Canada's Capital, Ottawa."

"Prior to the pandemic, CFS KE practitioners conducted their work in both digital and in-person formats, where digital formats could include webinars, videos, and newsletters, and in-person work could involve field visits to demonstrate technology, conferences, and collaborating with knowledge users to co-develop project objectives."

"CFS KE practitioners represent an ideal case study to begin to explore the risks, promises, and pitfalls of digital knowledge exchange in a post-COVID-19 pandemic world."

"We use responses from interviews with nine CFS KE practitioners to provide examples."

"In the interviews, the CFS KE practitioners discussed their work on KE and their perspective on the lasting impacts of COVID-19."

What we heard: perspectives about digital KE from the CFS KE practitioners

"The importance of in-person interactions for building relationships, strengthening communication with users, and the integral role it plays in KE was emphasized by many CFS KE practitioners, along with how KE is, and will be affected by COVID-19."

"One CFS KE practitioner summarized this issue: I think it's important to listen carefully to people."

"The analysis of interviews also led to the identification of personal disconnection and loss of social learning as other challenges that CFS KE practitioners have faced in relation to the digital transformation of KE."

"Another CFS KE practitioner described their thoughts on an alternative plan to exchange knowledge for an event that had traditionally been in-person: The strategy on that [work with industry on transfer or building capacity with climate models] is—normally a 2–3-day workshop here, bring people here."

Bright spots and opportunities for digital knowledge exchange

"One of the interviewed CFS KE practitioners offers their perspective on how the CFS has adapted and adopted digital platforms, which can be seen as an opportunity for other organizations to change: I think we'll have to have new tools or new ways to produce or to do e-lecture or virtual days."

"The pandemic and the forced digital acceleration may thus be an opportunity for individuals, groups and organizations to adapt and innovate new ways to exchange knowledge and meet desired outcomes."

"A commonly identified challenge to implement KE in pre-COVID times at the CFS was the lack of travel budgets to engage participatory research as well as building and maintaining relationships."

"Many of these tools addressed KE problems identified by CFS respondents—the need for personal connection, maintaining knowledge and communication flows, and stimulating engaging discussions."

"With increased access, skill, and culture of meeting and exchanging knowledge digitally, there are opportunities for more frequent interactions without travelling."

Looking into the future: sustainability and effectiveness of digital knowledge exchange

"We cannot consider the potential benefits in isolation of ongoing challenges, which makes the long-term sustainability of digital KE highly uncertain."

"In our view, the long-term efficacy of digital KE likely has less to do with software features than with two interconnected factors: (1) the stubborn persistence of the digital divide, and (2) people's abilities to make and maintain meaningful social connections and trusting relationships in the absence of regular in-person, face-to-face contact."

"On the second point, a major challenge to digital KE over the long term is the fact that knowledge sharing and exchange as processes rely so strongly on social connections and trust (Reed and others [236])."

"If digital KE is to persist long-term in a digital-first world, people will need to find a way to replicate the social nuances of trust building and contextual understanding that we are accustomed to doing in the physical presence of others."

Conclusion—considerations for post-pandemic knowledge exchange

"Leverage the diversification of KE approaches and innovations The acceleration in both digital innovation and of user access, skills and acceptance of digital work offers new tools and approaches for the KE toolbox."

"KE practitioners may leverage frequency, rapidity, and innovative approaches of gathering people virtually to better respond to user needs and support practice, policy, and decisions."

"We conclude our essay with a quote from a KE practitioner on the implications for KE in a digital-first world: Knowledge exchange is going to be important for us to get out there [on the ground/in-person]."

"The pandemic has introduced challenges to KE work but has pushed boundaries for diversification of KE approaches, which will enhance knowledge exchange among diverse audiences and user groups."

8.2.6 Teachers' Experiences of Teaching Online During COVID-19: Implications for Postpandemic Professional Development [294]

This is a machine-generated summary of:

González, Carlos; Ponce, Daniel; Fernández, Violeta: Teachers' experiences of teaching online during COVID-19: implications for postpandemic professional development [294]

Published in: Educational technology research and development (2023)

Link to original: https://doi.org/10.1007/s11423-023-10200-9

If you want to cite the papers, please refer to the original.

For technical reasons we could not place the page where the original quote is coming from.

Abstract-Summary

"Of the Great Online Transition (GOT) that occurred during COVID-19, it is increasingly necessary to understand the digital competencies that are required for online and blended learning in the postpandemic era."

"Postquarantine, higher education institutions must return to on-campus face-to-face learning, a situation which raises questions concerning how to retain the lessons learned from this period of the forced acquisition and innovation of online teaching competencies."

"We present the results of an interview-based study of teachers' experiences of online teaching during the pandemic."

"Our results show that despite the problems faced due to the disruption caused by the pandemic, several lessons were learned: teachers employed an array of digital tools to maintain content delivery and promote interaction, deepened their understanding of course design and assessment, and developed an empathic disposition to understand students' situations."

Introduction

"Prepandemic research on teachers' digital competencies aspired to generate frameworks to advance the pedagogical use of digital technologies."

"A helpful summary of the competencies dimensions included in several frameworks is that provided by Esteve-Mon and others [237], who affirm that the core elements include (1) basic digital skills—the use of digital tools for interacting with content and with others, which entails the capacity to create content such as documents, multimedia, or other formats; (2) the pedagogical application of digital technologies—application to the learning processes, incorporating them into the design, enactment, and evaluation of learning experiences; (3) the use of technology to support continuous professional development—which implies the capacity to participate in formal professional development activities and networks in pursuit of this purpose; and (4) the ability to further the digital competencies of higher education students—facilitating technical, communicative and multimedia skills in students, both for learning purposes and for students' professional development."

Methodology

"Pragmatic research suits our aim of projecting disruptive experiences for post-GOT teachers' professional development, thus allowing us to develop recommendations for action."

"This approach is useful in such a study given that at least part of the teachers' experiences, due to the exceptional circumstances, were unlikely to be captured by an a priori codebook."

"Semistructured interviews were conducted by experienced research assistants via videoconferencing to collect data."

"This approach allowed us to conduct interviews that would otherwise not have been possible considering the circumstances."

"Team members coded approximately one-third of the interviews using this codebook."

"One team member who did not directly participate in the coding process reviewed the analysis to determine whether the themes fit the data."

"These meetings allowed the team to develop a shared understanding of the codes, calibrate the research assistants' coding processes, analyze complex transcripts, and discuss coding decisions to ensure consistency."

Results

"I think students know digital technologies better [] they are inserted in new technologies, they are inserted in social networks, all the time, during classes, before, after [] If you use these tools for learning I think for them it will be closer (to their everyday lives) (I27) Regardless of students' high level of social network use, teachers also generally perceived that they were less confident and struggled with the use of digital technologies for learning and professional purposes."

"Teachers noted that it is not apparent that students know how to employ the digital technologies officially provided by their institutions (such as learning management systems, digital libraries, or videoconferencing) or other technologies that are not provided by the institutions but are used to promote interaction (for example, Padlet or Kahoot)."

"Concerning the capacity to establishment a safe online learning environment, teachers claimed that it is essential to provide students with the confidence to interact academically."

Discussion

"We investigated higher education teachers' experiences of teaching online during COVID-19 with the aim of contributing to current debates on teachers' digital competency development needs during this disruptive era."

"Teachers also valued professional development during the lockdown, which was aimed at the development of digital technology competencies and an improved understanding of student learning processes in this scenario."

"This situation was also a period of tremendous learning, particularly with regard to advancing the incorporation of digital technologies into teaching and learning, prioritizing content, introducing new forms of assessment, and developing empathic relationships with students."

"The ability to further the digital competencies of higher education students Given that teachers realized that students need support to use digital technologies, teachers are closely involved in the task of furthering students' digital competencies, particularly in learning and professional settings."

Conclusion

"Our objective was to contribute to current debates on the digital competencies that teachers will need in the post-GOT era."

"Our results showed that teachers faced a very difficult situation."

"They simultaneously obtained a tremendous amount of learning: they expanded the range of digital technologies in use to maintain content delivery and promote interaction, developed innovations in terms of course design and assessment, and developed an empathic disposition to understand students' situations."

8.2.7 Teachers' Digital Literacy and Digital Activity as Digital Divide Components Among Basic Schools in Ghana [295]

This is a machine-generated summary of:

Quaicoe, James Sunney; Pata, Kai: Teachers' digital literacy and digital activity as digital divide components among basic schools in Ghana [295]

Published in: Education and Information Technologies (2020)

Link to original: https://doi.org/10.1007/s10639-020-10158-8

If you want to cite the papers, please refer to the original.

For technical reasons we could not place the page where the original quote is coming from.

Abstract-Summary

"This survey-based study explores the nature of Digital Divide among Ghana's basic (primary and junior high) schools from the perspective of Teachers' Digital Literacy and their usage of digital technologies in school."

"The ICT-enhanced Teacher Standards for Africa (ICTeTSA) Framework comprising digital attitude, knowledge, skills and application components was used for measuring the perceived Teacher Digital Literacy, while, the European Union's rubrics for measuring ICT frequency in schools was adopted for estimating the extent of teachers' digital technology usage/activities."

"Based perceptions of the Teacher Digital Literacy two significantly different Digital Divide clusters were found among schools that diverged based on the teachers' digital application component of Teacher Digital Literacy."

"The study reveals several critical issues of teachers' digital empowerment for technology in Ghana's basic schools, of which school-based management (SBM) governance be adapted to address them."

Introduction

"This study looks at the DD in public basic (primary and junior high) schools in Ghana focusing on the teachers' perceived digital literacy and the corresponding digital activities associated with teacher professional roles."

"There are many unanswered questions about ICT in schools in the context of teachers' digital literacy and the digital activities (which is the actual ICT usage), and how those factors account for the digital divide in the basic schools in Ghana."

"This paper looks into the subject of DD in the public basic schools in Ghana as informed by the Teachers Digital Literacy (TDL) components in the form of Teacher Digital Knowledge (TDkn), Teacher Digital Skills (TDsk), Teacher Digital Attitude (TDAt), and Teacher Digital Application (TDAp)."

"4) RQ4: Which of the TDL components (teachers' digital knowledge, skills, attitude, and application) determine the disparities in Ghana's basic schools?"

Conceptual background of the study

"That as school autonomy misses out in the managerial practices in the educational sector, it would be replaced with shirked responsibilities on the part of the school actors; and as school leadership and staff experiences some degree of autonomy, school innovative strategies are jointly designed, visions are shared, ownership of policies or strategies enhanced and the sense of accountability heightened."

"In the same vein the authors of this paper hold the view school-based designed digital agenda could be an effective way for schools to take ownership of the implementation of the ICT in Education policy as advanced by various governments."

"That role of the Regional, District and Ministry of Education are to gather these school-based ideas as a bottom-up approach to policy implementation and provided the needed resources to the schools."

"The latter is not be designed to kill school-based innovations, rather to enhance it through the framework of the national ICT in education."

Methodology

"Two sets of paper-based questionnaire which were distributed using the face-to-face mode, measured the following variables: 1) more general indicators of TDL components (Digital knowledge, skills, attitude, and application) as estimated by the teachers themselves; and 2) TDA, which explored teacher claims in the use of digital applications in accomplishing activities such as preparing, delivering and assessing student learning, communicating student learning outcomes, professional development, building learning repositories among and other others."

"Participants for the study were teachers from 45 schools selected from six districts in the Western Region of Ghana."

"Various analytical methods were adopted in processing the collected data – using the statistical tool SPSS; Descriptive statistics (Pallant [238]) was used in finding the ICT statuses of the schools in the context of TDL and TDA."

"Fisher's test was conducted to find out if schools location (district) influenced their DD status."

Results

"Teachers digital activities (TDA) in schools were found to be performed with rather low frequency in the districts studied."

"To identify the respective factors influencing the DD the classified schools was used as group variable, and together with the TDL components {Teacher Digital Knowledge (TDkn), Teacher Digital Skills (TDsk), Teacher Digital Attitude (TDAt), and Teacher Digital Application (TDAp)} as the predictor variables in running a Discriminant Function Analysis."

"The corresponding equation of the function: Deducing from the outcome of the canonical discriminant function results, Teacher Digital Application (TDAp) is the principal component informing the DD across the school cluster, more than the other TDL components – Teacher Digital Knowledge (TDkn), Teacher Digital Skills (TDsk), and Teacher Digital Attitude (TDAt)."

"For individual TDL components, the Teacher Digital Application (TDAp) was found to be the principal component informing the DD (disadvantaged group ($M = 2.95$, $SD = 0.326$, and advantaged group ($M = 3.83$, $SD = 0.320$), t (43) = , $p = 0.001$, Eta squared = 0.659)."

Discussion

"The findings in this paper indicated that schools in Western Ghana Region are digitally divided based on Teacher Digital Literacy (TDL)."

"The study by Edumadze and Owusu [239] found a correlation between the ICT related courses and development of students' ICT competencies, suggesting that the basic school teachers without TDL might fail to function as digital change agents for today's learners."

"The case of high perceptions of TDL did not commensurate the professional digital engagements of the teachers, because the results of the study indicated low frequency of TDA in schools."

"Agyei [240] confirms most of the findings of this current paper regarding TDA and DD in Ghana's schools when it comes integrating ICT into pedagogy within non-ICT related subjects, lack of teacher educational applications and lack of professional ICT integration training for teachers."

Conclusion

"A low level of teachers' application of digital technologies was found in half of the schools we explored."

"The study found out that there are discrepancies between the above average estimation of their TDL among teachers and the low frequency of digital activities teachers actually conduct in the schools."

"Combining this results with our fieldwork in the overlapping pilot sample and the previously complimenting findings, there is the need for the government (District Assemblies and Educational Directorate) level empowerment and school level empowerment in the form of coordinated school-based management (Caldwell [241]; David [242]; Mihai and Nieuwenhuis [243]; Patrinos and others [244]) to intervene in the obstacles that hinder teachers conducting lessons with digital technologies."

"There is the need for lifelong digital literacy trainings for teachers that will help schools to develop ICT agendas for maintaining digital learning despite the lack of resources."

8.3 Digitalisation and Power

Machine generated keywords: china, chinese, power, politic, security, international, consumer, political, global, justice, bind, vision, digitalization, company, service.

8.3.1 The Empire Strikes Back: Digital Control of Unfair Terms of Online Services [296]

This is a machine-generated summary of:

Micklitz, Hans-W.; Pałka, Przemysław; Panagis, Yannis: The Empire Strikes Back: Digital Control of Unfair Terms of Online Services [296]

Published in: Journal of Consumer Policy (2017)

Link to original: https://doi.org/10.1007/s10603-017-9353-0

If you want to cite the papers, please refer to the original.

For technical reasons we could not place the page where the original quote is coming from.

Abstract-Summary

"The authors argue that it is possible to partly automate the process of abstract control of fairness of clauses in online consumer contracts."

"This type of automation would not replace human lawyers but would assist them and make their work more effective and efficient."

The Problem in Context

"Unfair contractual term legislation was adopted in numerous European states in the second part of the twentieth century (Micklitz and others [245]), and finally harmonized and introduced throughout the European Union with Directive 93/13/EEC on Unfair Terms in Consumer Contracts, which is still in effect."

"The UCTD requires that, through their legal systems, Member States, ensure that: "adequate and effective means exist to prevent continued use of unfair terms in contracts concluded with consumers by sellers or suppliers" (art."

"7.1.) and: "the means referred to in paragraph 1 shall include provisions whereby persons or organizations, having a legitimate interest under national law in protecting consumers, may take action according to the national law concerned before the courts or before competent administrative bodies for a decision as to whether contractual terms drawn up for general use are unfair, so that they can apply appropriate and effective means to prevent continued use of such terms" (art."

Information Technology in the Service of Consumer Organizations

"The authors provide the argument for the positive claim of the paper, namely that it is possible to automate certain tasks which currently need to be performed by humans in the process of abstract control of fairness of clauses in consumer contracts."

"To appreciate the significance of the automation of this step, imagine the difference between receiving a set of documents with certain clauses highlighted and the command "please read the highlighted parts and assess if we should claim they are unfair" and simply receiving the set of documents with the command "please read all these contracts and look for clauses that you think we should claim to be unfair.""

"To bring this theory to the research task considered here, successfully implementing the machine learning approach in the annotation of potentially unfair terms, would require several hundred, if not several thousand actual contracts with all the potentially unfair clauses annotated and categorized."

A Peek into the (Possibly Near) Future: What Else Does the Technology Have to Offer?

"Imagine that whenever a consumer body or organization, governmental or non-governmental, or a court, or researcher, or a legally conscious user found a clause that is potentially unfair in some contract, they allowed the machine to learn about it."

"As of now, we have automated the reading of terms of service and the detection of potentially unfair clauses, but a consumer-protection authority/organization would still choose what terms of what service should be scanned."

"It is possible for consumer-protection authorities/organizations to independently construct a database containing all the terms of use of online services out there, or in a particular market (this can be designed by the institutions themselves), with an indication of the potentially unfair clauses in the documents."

"The e-mail would inform the business that potentially unfair clauses were detected in their terms, inform them that the consumer-protection authority/organization had also been notified and sent the report, and propose a way to proceed."

Conclusion

"The authors have argued that several tasks which currently need to be performed by humans in the process of abstract control of fairness of clauses can be automated."

"What is relevant from a policy perspective is to convey the message that consumers can be better protected, and the resources of consumer authorities better used, if research continues on this path."

"Closer collaboration between legal scholars, engineers, and consumer-protection people is needed."

"We can literally revolutionize the enforcement of the law regarding unfair contractual terms, and we hope the research presented here is one of the sparks that will start the revolution."

[Section 5]

"To prove the positive claim, the authors propose a theoretical model of tasks which need to be conducted by the consumer-protection organizations before actual legal proceedings commence and present the range of technological options available in the automation of these tasks."

"Further, the authors provide empirical evidence that one of the tasks, i.e., reading contracts in search of potentially unfair clauses, can largely be automated."

"The authors claim that there are several tasks that could be undertaken by machines, and demonstrate that at least one, e.g., reading of contracts and preparing legal material for the creative work of lawyers can already be automated."

"The theoretical explanation of how to automate one of these tasks, i.e., reading the terms of service and annotating the potentially unfair clauses, is offered."

8.3.2 Power in the Age of Datafication: Exploring China's Global Data Power [297]

This is a machine-generated summary of:

Huang, Ying; Mayer, Maximilian: Power in the Age of Datafication: Exploring China's Global Data Power [297]

Published in: Journal of Chinese Political Science (2022)

Link to original: https://doi.org/10.1007/s11366-022-09816-0

If you want to cite the papers, please refer to the original.

For technical reasons we could not place the page where the original quote is coming from.

Abstract-Summary

"Anxieties about China's growing data power have begun to drive geopolitical and technological competition."

"Using the concepts of compulsory, institutional, infrastructural, and productive data power, we assess how China exercises different forms of data power and what shapes and limits the increase of Chinese data power."

"The study concludes that China's data power is less grandiose than often assumed."

"By contrast, infrastructural and productive dimensions remain limited, largely due to China's high compulsory and institutional data power."

Introduction

"The White House's consideration of blocking the app entirely from US markets put geopolitical anxieties about China's growing data power on full display."

"There is no handy conceptual approach for measuring China's data power."

"If data is conceptualized through the resource lens, the unspecified emphasis on volume makes it difficult to assess, for instance, how much China's data power has increased."

"Our argument about China's global data power develops a "state-centric" view, although the power that stems from collecting, storing, and analyzing data is not limited to nation-states."

"The control over access to data has become a key battleground between the dominant Internet power, the US, and a rising China [246]."

"Four different manifestations of China's data power are explored by drawing on qualitative analysis and various quantitative indicators."

"The limitations of China's overall data power are discussed."

Conceptualizing Data Power

"The three concepts – data, information, and knowledge – are closely related and need to be distinguished in definitions of power [247]."

"As data power refers to the capacity to access, collect and process global data and convert large amounts of data into information, theoretical frameworks that highlight relational and structural forms of power are most relevant to make sense of China's global data power [248–251]."

"The key to a better understanding of data power lies in the dense web of global information technologies [252]."

"Institutional data power refers to the regulatory control a cyberspace actor possesses over data resources through formal and informal institutions."

"Fourth, productive data power refers to the constitution of social identities and capabilities of actors in cyberspace through discursive processes and practices related to information."

"This conceptual framework of data power, even though it engages different actors, assumes that technology firms ultimately remain subject to national jurisdiction and state authority."

Assessing China's Global Data Power

"The Chinese government exercises a mix of compulsory and institutional data power to cope with the influence of structural data power by US companies that dominate the Internet information services and data infrastructures."

"Chinese companies are affected by the globalized institutional data power of other countries."

"Although many countries adopt an approach similar to China's stricter controls over the internet, data, and the flow of information, they do it because of the rising threats to their national security, digital sovereignty, and privacy rather than because of the normative influence of Chinese diplomacy [253]."

"It is hard to find examples of Chinese cyber norms shaping social identities and narratives elsewhere, even though China's state-centric vision of data sovereignty seems attractive as governments worldwide attempt to control the flow of data and limit the power of big technology companies [253]."

Conclusion: Limitations of Chinese Data Power

"While institutional and compulsory data power is growing – mainly domestically – infrastructural and productive data power, most relevant for global hegemonic competition, are smaller than often assumed."

"There is an intriguing tension at the heart of China's data power."

"Intensifying geopolitics and economic decoupling trends may even shrink China's future global infrastructural and productive data power."

"Infrastructural and institutional data power is relatively stable since a country can neither enhance its control or access to global data infrastructures nor create new international institutions and shape commonly shared norms in a short period."

"While there are diverging timescales at play, the different types of data power still can influence each out."

"Our findings suggest, however, that government policies are critical intermediate factors that can change all four types of data power."

8.3.3 Maintaining the Ideals of Co-production During Rapid Digitalisation: A Comparative Case Study of Digital Restorative Services in Estonia, Finland, Ireland and Portugal [298]

This is a machine-generated summary of:

Surva, Laidi: Maintaining the Ideals of Co-production During Rapid Digitalisation: A Comparative Case Study of Digital Restorative Services in Estonia, Finland, Ireland and Portugal [298]

Published in: VOLUNTAS: International Journal of Voluntary and Nonprofit Organizations (2022)

Link to original: https://doi.org/10.1007/s11266-022-00502-6

If you want to cite the papers, please refer to the original.

For technical reasons we could not place the page where the original quote is coming from.

Abstract-Summary

"With the exceptional COVID-19 circumstances in early 2020, public service co-production went through a push towards digitalisation."

"Using normalisation process theory as the basis for analysis, the article looks at the immediate effects of digitalisation on restorative practices, which is a co-produced approach to delivering justice."

"The launch of digital restorative services depended more on service providers' readiness to try digital solutions and less on service experience before digitalisation."

Introduction

"Carroll and Conboy [254] argue that technology-driven practices will form a part of 'the new normal' and organisations need to normalise the use of technology to accomplish service goals, possibly resulting in long-lasting effects on public service co-production."

"Restorative practices, which by their nature require physical interaction and active participation, offer a unique opportunity to understand how digitalisation influences the ability of service professionals, individuals and the community to contribute to highly interactive and sensitive public services such as restorative justice."

"Showing what happens to restorative practices when they are co-produced digitally can shed light on the consequences of abrupt digitalisation of human connection-dependent service processes, which the digital co-production literature so far has not focused on."

"Considering that online restorative practices are a recent phenomenon and there are not much data available, the article cannot provide far-reaching conclusions or in-depth analysis on long-term impacts; however, it can shed light on the emerging processes taking place when digitalising a very sensitive co-produced service."

Method

"Longer restorative justice experience comes into play is with regard to the approach to the co-production process, i.e. when Finland and Ireland wanted to introduce standardised training, a digital mediation process and solve security issues before going online, then Estonia opted for learning by doing to ensure service continuity."

"Although the mediation process did not seemingly change, there was an intuitive change in the co-production process, which meant a more directive role for the mediators and more responsibility for the citizens in organising the service context, i.e. ensuring a safe and quiet place to talk and the necessary equipment to attend the online meetings."

"A short guideline was developed for the mediators, outlining the service process, including how to make sure citizens are comfortable with having an online meeting and that they participate voluntarily (FIN2)."

Discussion

"The answer to the first research question is that there were changes in the roles that mediators and citizens play in the co-production process, as the mediators had to be more directive than is recommended in restorative practices and citizens on their part were much more responsible for creating the needed environment for restorative justice to take place."

"With the new practices introduced with digitalisation, the facilitators attributed an identity to the digital version of mediation as something that is not as good as face-to-face interaction, but something that is needed to be able to sustain restorative practices during the pandemic."

"The launch of digital restorative justice in Estonia, Finland and Ireland meant supported enrolment for both mediators and citizens; they all had to engage in the new practice."

Conclusion

"The article looked at the immediate effects of digitalisation on service co-production, the mechanisms used to safeguard the underlying principles of restorative practices and how new practices were embedded in the existing co-production process."

"The main mechanisms used in launching digital mediation were learning-by-doing, active exchange of best practice, and integrating digitalisation with existing service processes, which helped to embed the new format alongside the old one."

"One thing the case study showed was that the launch of digital restorative practices depends most of all on the readiness to try digital solutions and somewhat less on prior service experience."

"What the case taught us is that setting up work practices that aim to keep alive the inherent nature of the co-production process of highly sensitive services, even though there is depletion in interaction quality, is worthwhile if the other alternative is no services at all."

8.3.4 Governing Through Cybersecurity: National Policy Strategies, Globalized (in-)Security and Sociotechnical Visions of the Digital Society [299]

This is a machine-generated summary of:

Haddad, Christian; Binder, Clemens: Governing through cybersecurity: national policy strategies, globalized (in-)security and sociotechnical visions of the digital society [299]

Published in: Österreichische Zeitschrift für Soziologie (2019)

Link to original: https://doi.org/10.1007/s11614-019-00350-7

If you want to cite the papers, please refer to the original.

For technical reasons we could not place the page where the original quote is coming from.

Abstract-Summary

"Cybersecurity policies present sites of political articulation and intervention where the very contours of an emerging digital society and the socio-technical relationships of power and control deemed necessary to govern its emergence are (re-)assembled."

"Our research prompts a rethinking of the relationships between cybersecurity and the digital society to the extent that cybersecurity represents a new globalizing form and rationality of security that encodes and enables new forms of control and intervention, but also new responsibilities at the interface between the state, society, and individuals."

Introduction
"Policy discourses thus articulate the need for novel regimes of (cyber-)security that protect citizens, consumers, and entrepreneurs from the multiple and inherently novel kinds of risks and threats that occur in a society increasingly interlaced with digital technology."

"More often than not, in these visions, the possibility for a desirable digital future has become increasingly tied to concerns for cybersecurity that articulate the need to envision and bring into existence a veritable "culture of cybersecurity" not only composed of safe technologies and robust institutions but also of digitally prudent and skilled subjects—citizens and professionals who act competently and responsibly in cyberspace."

"Sensitized by research perspectives from critical studies of security as well as from Science & Technology Studies, we argue that these various interrelated efforts to build a novel regime of cybersecurity and, correspondingly, a "culture of cybersecurity" (ENISA [256]) constitute a techno-political experimental field where the contours of the emerging digital society are collectively imagined, articulated, negotiated, and acted into existence."

Research framework
"Doing so enabled us to complete four processes: to discern the multiple meanings of cybersecurity for and in (visions of) an emerging digital society; to analyze problematizations of cybersecurity in line with the various solutions suggested to confront these problems; to examine the policy strategies articulated to bring about a "culture of cybersecurity"; and to identify the different forms of knowledge, the institutional arrangements and policy instruments, and the forms of subjectification invoked and mobilized in the making of cybersecurity in Austria."

"We analyze how visions of the digital society are performed through strategies of cybersecurity and how particular notions of power, governance, and control are articulated in and through the policy concept of a "culture of cybersecurity" that needs to be established to safeguard a desirable digital future."

Problematizing cyberspace through the lenses of security
"Under the auspices of the Board of Trustees for a Secure Austria (Kuratorium Sicheres Österreich, KSÖ), a stakeholder consultation of various actors from the public and private sector preoccupied with digital safety and security, found various shortcomings and lamented the lack of an explicit cybersecurity policy as a governmental responsibility."

"These structural deficits were addressed in the ensuing Austrian Cyber Security Strategy (BKA [257]), which has moved cybersecurity into the center of Austria's general security policy."

"The general security strategy of 2013 explicitly turns to cybersecurity, stating that cyber-crime, cyber-attacks, the misuse of the internet for extremist purposes and network security are serious new challenges for all stakeholders and require wide-reaching cooperation as part of a comprehensive policy (BKA [258], p. 11)."

"Through the lenses of critical security studies, this emergence of cybersecurity in Austria can be captured as a process of securitization of cyberspace (cf. Hansen and Nissenbaum [259])."

Articulating a strategy for a safe and secure digital future

"Computer Emergency Response Teams (CERTs)—those units that populate the operational level of the cybersecurity architecture—are particularly crucial in this regard as they constitute branch-specific hubs that monitor, process, and report cyber incidents."

"Increasing digitalization has strengthened the role of CERTs as integral parts of public and private organizations that specialize in the study of IT network security risks."

"With the European Network and Information Security Directive (EU [260]) in force, it has become legally mandatory to implement and run effective information security management systems (ISMS) in all areas deemed critical infrastructures."

"Elements of an operational ISMS should be fine-tuned with the specificities and requirements of any particular system and include management of risks and auditing programs at the systemic, process, and product-level, security analyses, and the management of regulatory guidelines."

"If gradually established in all critical infrastructure sectors and in all organizations deemed security-relevant, ISMS spans a truly global network of digitally safe and secure organizations."

Promoting digital safety and security through knowledge, skills, and innovation

"To a considerable extent, the vision of a safe and secure digital society is grounded in the aspiration to advance research and development (R&D) to produce safe digital technologies and secure digital infrastructures."

"In Austria, this emphasis on security R&D and technologies and devices that are deemed conductive to the emergence of safe and secure digital practices at all levels of society was recently reaffirmed in the Digital Roadmap (BKA [261]) as well as in the strategy of the newly elected government that aspires to render Austria as one of the main hubs of security technology development (Republic of Austria [262], p. 32)."

"Creating safer and more secure technologies is part of an effort to promote the responsible use of digital technologies and, for the entire digital society, to reduce risks that might be caused through harmful actions of a few users."

"Knowledge on risk mitigation, therefore, is not only produced through technology R&D strictu senso, but through educational and training practices that should assist in creating knowledge among all practitioners within the digital society."

Discussion

"The Austrian experience with the (re-)imagination and conceptualization of digital safety and security regimes exemplifies how cybersecurity does not merely present a discrete agenda within security policy but rather amounts to a "globalizing" form of security animating, and animated by, an emerging sociotechnical regime of control."

"Cybersecurity presents a testing ground where the relationships between an ever-emerging digitalizing society and their corresponding conceptions of security are problematized and reordered."

"Given this predicament, we conclude by suggesting that the problematiziations that guide visions of an incipient digital future can be grasped by exploring the nexus between digitalization and society in terms of an emerging globalizing regime of digital safety and security."

"As rationalities of cybersecurity proliferate in all fields of society, the visions of an open, inclusive, and truly innovative digital future risk becoming haunted by the demons of a fully-fledged security society."

8.3.5 Great Power Contest in a New (Digitalized) Guise [300]

This is a machine-generated summary of:

Sahin, Kaan: Great Power Contest in a New (Digitalized) Guise [300]
Published in: Zeitschrift für Außen- und Sicherheitspolitik (2020)
Link to original: https://doi.org/10.1007/s12399-020-00797-7

If you want to cite the papers, please refer to the original.

For technical reasons we could not place the page where the original quote is coming from.

Abstract-Summary

"The technology-driven great power contest between the U.S. and China is often titled as a Tech Cold War."

Introduction

"The increasing significance of technologies for international politics can be notably discerned in the context of the apparent great power battle over global influence between the United States of America and the People's Republic of China."

"The ongoing debate whether Western countries should allow Huawei equipment – a state-connected Chinese tech giant – into their 5G infrastructure, shows vividly how the technological confrontation between the U.S. and China has affected all regions."

"This great power rivalry between Beijing and Washington – especially while emphasizing the growing significance of technologies – is increasingly dubbed with Cold War analogies."

"After outlining these new parameters, selected examples of the consequences resulting from the new circumstances of the technology-driven great power contest, for both the U.S., China and third countries on a global scale, will be presented."

[Section 2]

"The danger of resorting to analogies, describing current events in international politics, has been discussed in the literature and, for instance, "ready-to-use analogies could make people believe everything is doomed, and therefore not make strong efforts to uphold peace and to create new opportunities for reconciliation" (Wang [263])."

"The influence of political framing, for instance by applying analogies or metaphors, on how the public views certain issues and how policymakers come up with solutions should not be underestimated and can have severe effects on decision-making."

"Heralding a Tech Cold War in the strict sense is dangerous since the circumstances surrounding the technological confrontation between Washington and Beijing are different from the division of two blocks during the Cold War."

"One reading is that the U.S. and the West were able to win the Cold War over its Soviet counterpart – among other aspects – due to its technological edge."

New parameters

"Washington and Beijing are intertwined concerning technologies as well, for instance in terms of the research and development (R&D) of AI systems, China's dependence on the U.S. semiconductor industry and software development (e.g. for mobile applications) or the reliance of U.S. companies on Chinese manufacturing technologies."

"A first difference to the old Cold War days is that technological advances are now primarily produced by the private sector: Most of the emerging technologies are no longer innovated in self-contained national systems, which existed in the form of the military-industrial complex – in a working relationship with national research institutions and state funded labs – in the U.S. Since the Eisenhower administration from the late 1950s to the 1990s, technological innovations such as the internet or microchips resulted mainly from these state-connected collaborations (Lewis [264], pp)."

Preliminary implications

"Considering the more complex environment – be it driven by the onset of emerging technologies and their related conditions or shifts in the political and economic order – since the beginning of the 1990s, the Tech Cold War analogy is not helpful to understand the current technological confrontation between the U.S. and China."

"Certain conclusions of the further technological confrontation between the U.S. and China, the related impact on third countries and regions, as well as for the interplay of digitalization and international politics, can be drawn."

"Given that Beijing has established close economic ties with countries all over the world and aspires to innovate cutting-edge technologies, the technological clash between the U.S. and China means a continual competition and negotiation mode towards third countries for both actors."

Bibliography

1. Røvik, K. A. (2016). Knowledge transfer as translation: Review and elements of an instrumental theory. International Journal of Management Reviews, 18(3), 290–310. https://doi.org/10.1111/ijmr.12097. https://onlinelibrary.wiley.com
2. Lyby, L., Fevolden, A. M., & Tømte, C. (2018). Organisering av etter-og videreutdanningstilbudet ved NTNU. NIFU 2018:30 https://nifu.brage.unit.no/nifu-xmlui/bitstream/handle/11250/2570764/NIFUrapport2018-30.pdf
3. Tømte, C. E., Laterza, V., Pinheiro, R. M., & Avramovic, A. (2020). Is there a Scandinavian model for MOOCs? Nordic Journal of Digital Literacy, 15(04), 234–245. https://doi.org/10.18261/issn.1891-943x-2020-04-02
4. Langseth, I., Jacobsen, D. Y., & Haugsbakken, H. (2021). MOOCs for Flexible and Lifelong Learning in Higher Education. EMOOCs 2021 proceedings, p. 63–78
5. Selznick, P. (2011). Leadership in administration: A sociological interpretation. Quid Pro Books
6. Irgens, E. J. (2016). Skolen. Organisasjon og ledelse, kunnskap og læring. Fagbokforlaget
7. Osmundsen, K., Iden, J., & Bygstad, B. (2018). Digital Transformation: Drivers, Success Factors, and Implications. In MCIS (p. 37)
8. Cramer, F. (2015). What is 'post-digital'? In D. M. Berry & M. Dieter (Eds.), Postdigital aesthetics. London: Palgrave Macmillan.
9. Berry, D. M., & Dieter, M. (Eds.). (2015). Postdigital Aesthetics. London: Palgrave Macmillan.
10. Kholeif, O. (2017). You are here. Art After the Internet. London: Cornerhouse Publications.
11. Ackermann, J., Doerk, M., & Seitz, H. (2019). Postdigitale Kunstpraktiken: Ästhetische Begegnungen zwischen Aneignung, Produktion und Vermittlung [Postdigital Art Practices: Aesthetical Encounters between Appropriation, Production and Education]. In B. Jörissen, S. Kröner, & L. Unterberg (Eds.), Forschung zur Digitalisierung in der Kulturellen Bildung [Research on Digitalisation in Cultural Education] (pp. 183–194). München: Kopaed.
12. PKKB (2018) Postdigitale Kunst aus der Perspektive von Künstler*innen und Kurator*innen - eine qualitative Interviewstudie im Rahmen des BMBF-Projekts Postdigitale Kunstpraktiken in der Kulturellen Bildung (PKKB). Transkribiertes Interviewkorpus [Postdigital Art from the Perspective of Artists and Curators - A Qualitative Interview Study as Part of the BMBF-Research Project Postdigital Art Practices in Cultural Education (PKKB). Transcribed Corpus of Interviews](unpublished).
13. Wood, N. (2006). Transmitting craft knowledge: Designing interactive media to support tacit skills learning. Unpublished Ph.D. thesis, Sheffield Hallam University, UK.

14. Wood, N. (2007). Discourse through making: Eliciting knowledge to support craft skills learning. In European academy of design conference. İzmir, Turkey.
15. Wood, N., & Rust, C. (2003). Design for tacit learning: An investigation of design strategies for multimedia supported learning in the crafts. In Proceedings of European academy of design conference, University of Barcelona.
16. Wood, N., Rust, C., & Horne, G. (2009). A tacit understanding: The designer's role in capturing and passing on the skilled knowledge of master craftsmen. International Journal of Design, 3(3), 65–78.
17. Kaya, Ç. (2009). Tasarım Olarak Ev Eksenli Üretim. 4. Ulusal Tasarım Kongresi: Kriz Veya Tasarım, İstanbul Teknik Üniversitesi, 8–9 Ekim 2009, İstanbul.
18. Şişman, O., Ünlü, C. E., & Kaygan, H. (2009). Fate of Turkish traditional crafts: A case of economic, legal and political marginalization. Making Futures, 1, 96–104.
19. Ünlühisarcıklı, Ö. (2001). Training on the job in İstanbul: A study of skills acquisition in carpentry and car-repair workshops. International Review of Education. Internationale Zeitschrift Fur Erziehungswissenschaft. Revue Internationale de Pedagogie, 47(5), 443–458.
20. Yağız, B. Y., & Kaya, Ç. (2013). Tasarım Eğitiminde Zanaatın Yeri Üzerine Bir Önerme. Unpublished manuscript, İstanbul Technical University, İstanbul, Turkey.
21. Polanyi, M. (1966). The tacit dimension. Gloucester, MA: Peter Smith.
22. Lave, J., & Wenger, E. (1991). Situated learning. Legitimate peripheral participation. Cambridge: Cambridge University Press.
23. Polanyi, M. (1958). Personal knowledge: Towards a post-critical philosophy. London: Routledge & Kegan Paul.
24. Hennessy, S., & Murphy, P. (1999). The potential for collaborative problem solving in design and technology. International Journal of Technology and Design Education, 9(1), 1–36.
25. Kvan, T. (2001). The pedagogy of virtual design studios. Automation in Construction, 10, 345–353.
26. Lahti, H., & Seitamaa-Hakkarainen, P. (2005). Towards participatory design in craft and design education. CoDesign, 1(2), 103–117.
27. Yantaç, A. E. (2009). Engelli Bir Ortamda Kullanılan Mobil Bilgi-Destek Araçlarının İdeal Arayüz Tasarımı için Özel Grafik Arayüz Tasarım Kurallarının Belirlenmesi. Unpublished Ph.D. thesis, Yıldız Teknik Üniversitesi, İstanbul, Turkey.
28. Kaya, Ç. (2011). Designer as enabler: A methodology of intervention for designers. İstanbul: İstanbul Teknik Üniversitesi.
29. Ramirez, F. O., & Ventresca, M. (1992). Building the institution of mass schooling. In B. Fuller & R. Runinson (Eds.), The political construction of education (pp. 20–59). New York, NY: Praeger.
30. Harwell, D. (2020). Mass school closures in the wake of the coronavirus are driving a new wave of student surveillance. Washington Post, 1 April. https://www.washingtonpost.com/technology/2020/04/01/online-proctoring-college-exams-coronavirus/ . Accessed 22 April 2020.
31. Bates, T., & Sangrà, A. (2011). Managing technology in higher education: strategies for transforming teaching and learning. San Francisco: Jossey-Bass.
32. Howland, J. L., Jonassen, D. H., & Marra, R. M. (2011). Meaningful learning with technology (4th ed.). Boston: Pearson Education.
33. Cuban, L. (2004). The blackboard and the bottomline. Why schools can't be businesses. Cambridge, MA: Harvard University press.
34. Cuban, L., & Jandrić, P. (2015). The dubious promise of educational technologies: Historical patterns and future challenges. E-Learning and Digital Media, 12(3–4), 425–439. https://doi.org/10.1177/2042753015579978 .
35. Knox, J., Williamson, B., & Bayne, S. (2020). Machine behaviourism: Future visions of 'learnification' and 'datafication' across humans and digital technologies. Learning, Media and Technology, 45(1), 31–45. https://doi.org/10.1080/17439884.2019.1623251 .

36. Manolev, J., Sullivan, A., & Slee, R. (2019). The datafication of discipline: ClassDojo, surveillance and a performative classroom culture. Learning, Media and Technology, 44(1), 36–51. https://doi.org/10.1080/17439884.2018.1558237 .
37. Slade, S., & Prinsloo, P. (2013). Learning analytics: Ethical issues and dilemmas. American Behavioral Scientist, 57(10), 1510–1529. https://doi.org/10.1177/0002764213479366 .
38. Drachsler, H., & Greller, W. (2016). Privacy and learning analytics - It's a DELICATE issue. In Proceedings of the Sixth International Conference on Learning Analytics & Knowledge - LAK '16 (pp. 89–98). https://doi.org/10.1145/2883851.2883893.
39. Corrin, L., Kennedy, G., French, S., Shum, B., Simon, K., Pardo, A., West, D., Mirriahi, N., & Colvin, C. (2019). The ethics of learning analytics in Australian higher education. A Discussion Paper. https://melbourne-cshe.unimelb.edu.au/research/research-projects/edutech/the-ethical-use-of-learning-analytics . Accessed 15 June 2020.
40. Prinsloo, P. (2019). Learning analytics: Mapping a critique and agenda. Journal of Learning Analytics, 6(3), 20–24.
41. Tsai, Y., Perrotta, C., & Gašević, D. (2019). Empowering learners with personalised learning approaches? Agency, equity and transparency in the context of learning analytics. Assessment and Evaluation in Higher Education, 45, 554–567. https://doi.org/10.1080/02602938.2019.1676396 .
42. Teräs, H. & Teräs, M. (2019). Student-centered learning analytics development in higher education: Initial observations from needs analysis. In J. Theo Bastiaens (Ed.), Proceedings of EdMedia + Innovate Learning (pp. 488–492). Amsterdam, Netherlands: Association for the Advancement of computing in education (AACE). https://www.learntechlib.org/primary/p/210419/ . Accessed 15 June 2020.
43. Williamson, B. (2020b). New pandemic edtech power networks. Code acts in education, 1 April. https://codeactsineducation.wordpress.com/2020/04/01/new-pandemic-edtech-power-networks/ . Accessed 22 April 2020.
44. Fernback, J. (2018). Academic/digital work: ICTs, knowledge capital, and the question of educational quality. tripleC, 16(1), 143–158. https://doi.org/10.31269/triplec.v16i1.878 .
45. Fuchs, C. (2020a). Communicative socialism/digital socialism. tripleC, 18(1), 1–31. https://doi.org/10.31269/triplec.v18i1.1144 .
46. Jandrić, P. (2017). Learning in the Age of Digital Reason. Rotterdam: Sense.
47. McLaren, P., & Jandrić, P. (2020). Postdigital Dialogues. New York: Bloomsbury.
48. Suoranta, J., & Vadén, T. (2010). Wikiworld. London: Pluto.
49. Bulfin, S., Johnson, N., & Bigum, C. (Eds.). (2015). Critical perspectives on technology and education. New York: Palgrave.
50. Selwyn, N., Hillman, T., Eynon, R., Ferreira, G., Knox, J., Macgilchrist, F., & Sancho-Gil, J. M. (2020a). What's next for Ed-Tech? Critical hopes and concerns for the 2020s. Learning, Media and Technology, 45(1), 1-6. https://doi.org/10.1080/17439884.2020.1694945 .
51. Selwyn, N., Pangrazio, L., Nemorin, S., & Perrotta, C. (2020b). What might the school of 2030 be like? An exercise in social science fiction. Learning, Media and Technology, 45(1), 90–106. https://doi.org/10.1080/17439884.2020.1694944 .
52. Williamson, B. (2020a). Making markets through digital platforms: Pearson, edu-business, and the (e)valuation of higher education. Critical Studies in Education. https://doi.org/10.1080/17508487.2020.1737556 .
53. Suoranta, J. (2020). Critical pedagogy and Wikilearning. In S. Steinberg & B. Down (Eds.), The SAGE handbook of critical pedagogies (pp. 1126–1138). London: Sage.
54. Giroux, H. A. (2020b). The plague of neoliberalism and the politics of pandemics. In B. Evans (Ed.), The quarantine files. Thinkers in Self-Isolation. Los Angeles: Los Angeles Review of Books. https://lareviewofbooks.org/article/quarantine-files-thinkers-self-isolation/ . Accessed 15 May 2020.
55. Freire, P. (2018). Pedagogy of the oppressed (50th Anniversary ed.). New York: Bloomsbury.
56. Fuchs, C. (2020b). Communication and capitalism: A critical theory. London: University of Westminster Press. https://doi.org/10.16997/book45 .

57. Ross, J. (2017). Speculative method in digital education research, Learning, Media and Technology, 42(2), 214-229. https://doi.org/10.1080/17439884.2016.1160927 .
58. Pink, S., Berg, M., Lupton, D., & Ruckenstein, M. (2022). Everyday Automation: Experiencing and Anticipating Emerging Technologies. London: Routledge. https://doi.org/10.4324/9781003170884 .
59. Selwyn, N., Hillman, T., Bergviken Rensfeldt, A., & Perrotta, C. (2021). Digital Technologies and the Automation of Education—Key Questions and Concerns. Postdigital Science and Education. https://doi.org/10.1007/s42438-021-00263-3 .
60. Adams, V., Murphy, M., & Clarke, A. E. (2009). Anticipation: Technoscience, life, affect, temporality. Subjectivity, 28(1), 246–265. https://doi.org/10.1057/sub.2009.18 .
61. Miller, R. (2012). Anticipation: the discipline of uncertainty. In A. Curry (Ed.), The Future of Futures (pp. 39–43). Houston, TX: Association of Professional Futurists. https://www.academia.edu/en/8329146/Anticipation_The_Discipline_of_Uncertainty Accessed 4 October 2022.
62. Auger, J. (2013). Speculative design: Crafting the speculation. Digital Creativity, 24(1), 11–35. https://doi.org/10.1080/14626268.2013.767276 .
63. Ross, J. (2022). Digital Futures for Learning. Speculative Methods and Pedagogies. New York: Routledge.
64. Miller, R., & Sandford, R. (2018). Futures Literacy: The Capacity to Diversify Conscious Human Anticipation. In R. Poli (Ed.), Handbook of Anticipation (pp.1–20). Cham: Springer. https://doi.org/10.1007/978-3-319-31737-3_77-1 .
65. Oomen, J., Hoffman, J., & Hajer, M. A. (2022). Techniques of futuring: On how imagined futures become socially performative. European Journal of Social Theory, 25(2), 252–270. https://doi.org/10.1177/1368431020988826 .
66. Barca, D., & Arenas, A. (2012). Words of caution on indigenous knowledge and education. Educational Practice and Theory, 32(1), 5–28.
67. Durie, M. (2005). indigenous knowledge within a global knowledge system. Higher Education Policy, 18, 301–312.
68. Royal, C. (2004). Exploring "Indigenous'. Retrieved September 10, 2020, from https://www.charles-royal.nz/papers-reports .
69. Chang, H.-J. (2010). 23 things they don't tell you about capitalism. London: Penguin.
70. Sandel, M. J. (2012). What money can't buy: The moral limits of markets. London: Penguin.
71. Jolly, F., Whiteman, G., Atkinson, M., & Radu, I. (2011). Managing and educating outside: A Cree Hunter's perspective on management education. Journal of Management Education, 35(1), 27–50. https://doi.org/10.1177/1052562910386112
72. Pesic, P. (1999). Wrestling with Proteus: Francis Bacon and the "Torture" of Nature. Isis, 90(1), 81–94.
73. Marcuse, H. (1941). Some social implication of modern technology. In A. Arato & E. Gebhardt (Eds.), The essential Frankfurt school reader. Oxford: Basil Blackwell.
74. Spiller, C., Erakovic, L., Henare, M., & Pio, E. (2011). Relational well-being and wealth: Māori businesses and an ethic of care. Journal of Business Ethics, 98, 153–169.
75. Spiller, C., Pio, E., Erakovic, L., Henare, M. (2011) Wise Up: Creating organizational wisdom through an ethic of Kaitiakitanga. Journal of Business Ethics, 104(2), 223–235. https://www.jstor.org/stable/41476082 .
76. Mander, J. (1991). In the absence of the sacred: The failure of technology and the survival of the Indian Nations. San Francisco: Sierra Book Club.
77. Odell, J. (2019). How to Do Nothing: Resisting the Attention Economy. Brooklyn; London: Melville House.
78. Knox, J. (2012). Five critiques of the open educational resources movement. Retrieved from http://jeremyknox.net/2012/03/28/five-critiques-of-theopen-educational-resources-movement .
79. Benade. . (2017). Being a teacher in the 21st century: A critical New Zealand research study. Singapore: Springer Nature.

80. Palmer, P. J. (1998). Foreword, in Mary Rose O'Reilly The Garden At Night: Burnout and Breakdown in the Teaching Life. Portsmouth, NH: Heinemann.
81. Mogas, J., Palau, R., Lorenzo, N., & Gallon, R. (2020). Developments for smart classrooms: Schools' perspective and needs. International Journal of Mobile and Blended Learning, 12(4), 34–50. https://doi.org/10.4018/IJMBL.2020100103
82. Lorenzo, N., Gallon, R., Palau, R., & Mogas, J. (2021). New objectives for smart classrooms from industry 4.0. Technology, Knowledge and Learning. https://doi.org/10.1007/s10758-021-09527-0
83. OECD. (2017b). The OECD handbook for innovative learning environments. OECD Publishing. https://doi.org/10.1787/9789264277274-en
84. Means, A. J. (2018). Learning to save the future: Rethinking education and work in an era of digital capitalism. Routledge. https://doi.org/10.4324/9781315450209
85. Salimi, L., & Ghonoodi, A. (2012). WCLTA 2011 The study of functional elements of management system in smart schools. Procedia Social and Behavioral Sciences, 31, 140–144. https://doi.org/10.1016/j.sbspro.2011.12.031
86. Booth, T., & Ainscow, M. (2002). Index for inclusion: Developing learning and participation in schools. Centre for Studies on Inclusive Education (CSIE). https://www.eenet.org.uk/resources/docs/Index%20English.pdf .
87. Cebrián, G., Palau, R., & Mogas, J. (2020). The Smart Classroom as a means to the development of ESD methodologies. Sustainability, 12(7), 3010. https://doi.org/10.3390/su12073010
88. Kvale, S., & Brinkmann, S. (2009). Interviews: Learning the craft of qualitative research interviewing (2nd ed.). SAGE Publications, Inc.
89. Minichiello, V., Aroni, R., Timewell, E., & Alexander, L. (1990). In-Depth Interviewing: Researching people. Longman Cheshire.
90. Vincent-Lancrin, S., Urgel, J., Kar, S., & Jacotin, G. (2019). Measuring innovation in education 2019: What has changed in the classroom? OECD Publishing. https://doi.org/10.1787/9789264311671-en
91. Bennett, R. E., Persky, H., Weiss, A., & Jenkins, F. (2010). Measuring problem solving with technology: A demonstration study for NAEP. The Journal of Technology, Learning and Assessment, 8(8), 4–89.
92. Quellmalz, E. S., DeBarger, A. H., Haertel, G., Schank, P., Buckley, B. C., Gobert, J., et al. (2008). Exploring the role of Technology-Based Simulations in Science Assessment: The Calipers Project: Assessing Science Learning: Perspectives From Research and Practice.
93. Rutten, N., van Joolingen, W. R., & van der Veen, J. T. (2012). The learning effects of computer simulations in science education. Computers & Education, 58(1), 136–153.
94. Smetana, L. K., & Bell, R. L. (2012). Computer simulations to support science instruction and learning: A critical review of the literature. International Journal of Science Education, 34(9), 1337–1370.
95. de Jong, T. (2010). Cognitive load theory, educational research, and instructional design: Some food for thought. Instructional Science, 38(2), 105–134.
96. de Jong, T., & van Joolingen, W. R. (1998). Scientific discovery learning with computer simulations of conceptual domains. Review of Educational Research, 68(2), 179–201.
97. Köck, H. (2018). Fachliche Konsistenz und Spezifik in allen Dimensionen des Geographieunterrichts. In A. Rempfler (Ed.), Unterrichtsqualität Wirksamer Geographieunterricht 5 (pp. 110–121). Schneider Verlag Hohengehren GmbH.
98. Mayer, R. E. (2004). Should there be a three-strikes rule against pure discovery learning? The case for guided methods of instruction. The American Psychologist, 59(1), 14–19.
99. Scharfenberg, F. J., Bogner, F. X., & Klautke, S. (2007). Learning in a gene technology laboratory with educational focus: Results of a teaching unit with authentic experiments. Biochemistry and Molecular Biology Education, 35(1), 28–39.
100. Winberg, T. M., & Berg, C. A. R. (2007). Students' cognitive focus during a chemistry laboratory exercise: Effects of a computer-simulated prelab. Journal of Research in Science Teaching, 44(8), 1108–1133.

101. Zendler, A., & Greiner, H. (2020). The effect of two instructional methods on learning outcome in chemistry education: The experiment method and computer simulation. Education for Chemical Engineers, 30, 9–19.
102. Evangelou, F., & Kotsis, K. (2019). Real vs virtual physics experiments: Comparison of learning outcomes among fifth grade primary school students: A case on the concept of frictional force. International Journal of Science Education, 41(3), 330–348.
103. Madden, J., Pandita, S., Schuldt, J. P., Kim, B., Won, S., & Holmes, N. G. (2020). Ready student one: Exploring the predictors of student learning in virtual reality. PLoS ONE, 15(3), 0229788.
104. Ekwueme, C. O., Ekon, E. E., & Ezenwa-Nebife, D. C. (2015). The impact of hands-on-approach on student academic performance in basic science and mathematics. Higher Education Studies, 5(6), 47.
105. Tunnicliffe, S. D. (2017). Overcoming 'Earth Science Blindness' earth science in action in natural history dioramas. In M. F. P. D. C. M. Costa & J. B. Vázquez Dorrío (Eds.), Hands-on science: Growing with science. Universidade de Vigo.
106. Puntambekar, S., Gnesdilow, D., Dornfeld Tissenbaum, C., Narayanan, N. H., & Rebello, N. S. (2020). Supporting middle school students' science talk: A comparison of physical and virtual labs. Journal of Research in Science Teaching, 58(3), 392–419.
107. Sweller, J. (2011). Cognitive Load Theory. In Psychology of learning and motivation (55th edn., pp. 37–76). Academic Press.
108. Lichti, M., & Roth, J. (2018). How to foster functional thinking in learning environments using computer-based simulations or real materials. Journal for STEM Education Research, 1(1–2), 148–172.
109. Jong, T., de Linn, M. C., & Zacharia, Z. C. (2013). Physical and virtual laboratories in science and engineering education. Science (new York, NY), 340(6130), 305–308.
110. Borba, M. C., Maltempi, M. V., & Malheiros, A. P. S. (2005). Internet Avançada e Educação Matemática: novos desafios para o ensino e aprendizagem on-line. Revista Novas Tecnologias na Educação-RENOTE, 3(1). http://www.cinted.ufrgs.br/renote/maio2005/artigos/a28_edumatematica.pdf . Accessed 25 February 2009.
111. Bairral, M. A. (2007b). Avaliação na Formação Continuada de Professores a Distância: construindo estratégias didático-metodológicas e emergindo funções do formador. In IX Encontro Nacional de Educação Matemática, 9, 2007, Belo Horizonte. Anais… Belo Horizonte: Centro Universitário de Belo Horizonte.
112. Litto, F. M., & Formiga, M. (2009). Educação a Distância: o estado da arte. São Paulo: Pearson Education do Brasil.
113. Williamson, B. (2017). Big data in education: the digital future of learning, policy and practice. New York: Sage.
114. Bayne, S., & Jandrić, P. (2017). From anthropocentric humanism to critical posthumanism in digital education. Knowledge Cultures, 5(2), 197–216. https://doi.org/10.22381/KC52201712 .
115. Knox, J. (2019). What Does the 'Postdigital' Mean for Education? Three Critical Perspectives on the Digital, With Implications for Educational Research and Practice. Postdigital Science and Education, 1(2), 357–370. https://doi.org/10.1007/s42438-019-00045-y .
116. Means, A. (2019). Hypermodernity, automated uncertainty, and education policy trajectories. Critical Studies in Education, 1–16. https://doi.org/10.1080/17508487.2019.1632912.
117. Galloway, A. (2006). Gaming: essays on algorithmic culture. London: University Minnesota Press.
118. Senić Ružić, M. (2021). Digital transformation of education in Serbia / Issue of digital literacy or digital competence. Belgrade: University of Belgrade, Faculty of Philosophy.
119. Piattoeva, N., & Saari, A. (2020). Rubbing against data infrastructure(s): methodological explorations on working with(in) the impossibility of exteriority. Journal of Education Policy, 37(2), 165-185. https://doi.org/10.1080/02680939.2020.1753814 .

120. UNESCO. (2021). The Law of the Fundamentals of the Education System 2021. Paris: UNESCO. https://planipolis.iiep.unesco.org/en/2009/law-fundamentals-education-system-5205 . Accessed 3 October 2022.
121. World Bank (2017). Annual Report. https://documents.worldbank.org/en/publication/documents-reports/documentdetail/143021506909711004/world-bank-annual-report-2017 . Accessed 3 October 2022.
122. McCormick, J. (2020). Europeanism. Oxford: Oxford University Press. https://doi.org/10.1093/acprof:oso/9780199556212.001.0001 .
123. Lauer, R. M. (1996). A meta curriculum based upon critical thinking. ETC: A Review of General Semantics, 53(4), 374–387.
124. Simondon, G. (2016). On the mode of existence of technical objects. Trans. C. Malaspina. Minneapolis, MN: Univocal Publishing.
125. Hui, Y. (2019). Recursivity and Contingency. Lanham, MD: Rowman & Littlefield.
126. Hakkarainen, K., Palonen, T., Paavola, S., & Lehtinen, E. (2004). Communities of networked expertise. Professional and educational perspectives. Oxford: Elsevier.
127. Paavola, S., & Hakkarainen, K. (2005a). The knowledge creation metaphor—an emergent epistemological approach to learning. Science & Education, 14, 537–557.
128. Nespor, J. (1997). Tangled up in school: Politics, bodies, and signs in the educational process. Mahwah, NJ: LEA.
129. Zittoum, T. (2006). Transitions: Development through symbolic resources. Greenwich, Advances in Cultural Psychology: Constructing human development. Connecticut: IAP.
130. Gruber, H. (1989). Networks of enterprise in creative scientific work. In B. Gholson, W. R. Shadish, R. A. Neimayer, & A. C. Houts (Eds.), Psychology of science: Contributions to metascience (pp. 246–274). Cambridge, Ma: Cambridge University Press.
131. Higher Education Evaluation & Accreditation Council of Taiwan (HEEACT) (2020b) HEEACT self-Assessment report. Taipei: HEEACT.
132. International Network for Quality assurance Agencies in Higher Education (INQAAHE) (2018) Guidelines for Good Practices (GGP). INQAAHE, https://www.inqaahe.org/sites/default/files/GGP-Procedural-Manual-2018.pdf , accessed 1 June 2018.
133. International Network for Quality assurance Agencies in Higher Education (INQAAHE) (2021) GGP aligned agencies. INQAAHE, https://www.inqaahe.org/ggp-aligned-agencies , accessed 1 June 2021.
134. International Network for Quality Assurance in Higher Education (INQAAHE) (2020c) INQAAHE principles for crisis management. INQAAHE, https://www.inqaahe.org/sites/default/files/INQAAHE-Principles-for-crisis-management.pdf , accessed July 2021.
135. International Network for Quality Assurance in Higher Education (INQAAHE) (2020d) Guidelines of Good Practice (GGP) External Review Report. INQAAHE, https://www.inqaahe.org/sites/default/files/INQAAHE-GGP-Review-Report.pdf , accessed July 2021.
136. Schweisfurth, M. (2013a). Learner-centred education in international perspective. Journal of International and Comparative Education, 2(1), 1–8. https://doi.org/10.14425/00.45.70 .
137. Schweisfurth, M. (2013b). Learner-centred education in international perspective: Whose pedagogy for whose development? Routledge. https://doi.org/10.4324/9780203817438 .
138. Borg, C. (2021). Reading the migrants' world through emancipatory learner-centred education. In M. N. Gravani & B. Slade (Eds.), Learner-centred education for adult migrants in Europe: A critical comparative analysis (pp. 166–181). Brill.
139. Guo, S., & Sork, T. J. (2005). Adult education for social change: Deconstructing programs and services for adult immigrants. Adult Education Research Conference. https://newprairiepress.org/aerc/2005/papers/37
140. Manninen, J., Jetsu, A., & Sgier, I. (2019). Change-oriented adult education in the fields of democracy and digitalization. FutureLabAE. http://flf.dk/wp-content/uploads/2019/11/FutureLabAE_analysis_report_final.pdf
141. Freire, P. (1972). Pedagogy of the oppressed. Penguin.
142. MEDE [Ministry for Education and Employment] (2016). Malta national lifelong learning strategy 2020. https://lifelonglearning.gov.mt/dbfile.aspx?id=37

143. Morrice, L., Shan, H., & Sprung, A. (2017). Migration, adult education and learning. Studies in the Education of Adults, 49(2), 129–135. https://doi.org/10.1080/02660830.2018.1470280 .
144. Jones, K. (1982). Simulations in language learning. New York: Cambridge University Press.
145. Moseley, A., & Jones, R. (2012). Mapping games to curricula. In N. Whitton & A. Moseley (Eds.), Using games to enhance learning and teaching: A beginner's guide (pp. 109–123). New York: Routledge.
146. Waston, D. R., & Sharrock, W. W. (1990). Realities in simulation gaming. In D. Crookall & R. L. Oxford (Eds.), Simulation, gaming, and language learning (pp. 231–245). New York: Newbury House Publishers.
147. Mayer, B., & Harris, C. (2010). Libraries got game. Chicago: American Library Association.
148. Lebrun, M. (2007). Théories et méthodes pédagogiques pour enseigner et apprendre: quelle place pour les TIC dans l'éducation? De Boeck Université.
149. Uskov, V., Howlett, R. J., & Jain, L. C. (Eds.). (2015). Smart education and smart e-learning (1st ed. 2015). Springer.
150. Hamilton, L., & Corbett-Whittier, C. (2013). Using case study in education research. SAGE Publications Ltd. https://doi.org/10.4135/9781473913851
151. Ramirez-Montoya, M. S. (2020). Challenges for open education with educational innovation: A systematic literature review. Sustainability, 12(17), 7053. https://doi.org/10.3390/su12177053
152. Zhou, L., Rudhumbu, N., Shumba, J., & Olumide, A. (2020). Role of higher education institutions in the implementation of sustainable development goals. In G. Nhamo & V. Mjimba (Eds.), Sustainable development goals and institutions of higher education (pp. 87–96). Springer International Publishing.
153. Lane, A. (2017). Open education and the sustainable development goals making change happen. Journal of Learning for Development. https://doi.org/10.56059/jl4d.v4i3.266
154. Ahel, O., & Lingenau, K. (2020). Opportunities and challenges of digitalization to improve access to education for sustainable development in higher education. In W. Leal Filho, A. L. Salvia, R. W. Pretorius, L. L. Brandli, E. Manolas, F. Alves, U. Azeiteiro, J. Rogers, C. Shiel, & A. De Paco (Eds.), Universities as living labs for sustainable development (pp. 341–356). Springer International Publishing. https://doi.org/10.1007/978-3-030-15604-6_21
155. Lévy, Pierre. 2001. Cyberculture. Minneapolis: University of Minnesota Press.
156. Berry, Gregory, and Heidi Hughes. 2019. Integrating work—life balance with 24/7 information and communication technologies: the experience of adult students with online learning. American Journal of Distance Education 34 (2): 1–15.
157. Lehman, Rosemary M., and O. C. Simone, and Conceiçao. 2010. Creating a sense of presence in online teaching. How to "be there" for distance learners. San Francisco, CA: Jossey-Bass.
158. Osler, Lucy. 2020. Feeling togetherness online: a phenomenological sketch of online communal experiences. Phenomenology and the Cognitive Sciences 19 (3): 569–588.
159. Merleau-Ponty, Maurice. 2005. Phenomenology of perception. London and New York: Routledge.
160. Ito, M., Soep, E., Kligler-Vilenchik, N., Shresthova, S., Gamber-Thompson, L., & Zimmerman, A. (2015). Learning connected civics: Narratives, practices, infrastructures. Curriculum Inquiry, 45(1), 10–29. https://doi.org/10.1080/03626784.2014.995063 .
161. Lee, J., Zo, H., & Lee, H. (2014). Smart learning adoption in employees and HRD managers. British Journal of Educational Technology, 45(6), 1082–1096. https://doi.org/10.1111/bjet.12210 .
162. Kelley, T. R., & Knowles, J. G. (2016). A conceptual framework for integrated STEM education. International Journal of STEM Education, 3(1), 1–11.
163. Boss, S., & Krauss, J. (2014). Reinventing project-based learning: Your field guide to real-world projects in the digital age, (2nd ed.,). International Society for Technology in Education.

164. Meluso, A., Zheng, M., Spires, H. A., & Lester, J. (2012). Enhancing 5th graders' science content knowledge and self-efficacy through game-based learning. Computers & Education, 59(2), 497–504. https://doi.org/10.1016/j.compedu.2011.12.019 .
165. Wing, J. M. (2006). Computational thinking. Communications of the ACM, 49(3), 33–35. https://doi.org/10.1145/1118178.1118215 .
166. Grover, S., & Pea, R. (2013). Computational thinking in k–12: A review of the state of the field. Educational Researcher, 42(1), 38–43. https://doi.org/10.3102/0013189X12463051 .
167. Benita, F., Perhac, J., Tunçer, B., Burkhard, R., & Schubiger, S. (2020). 3D-4D visualisation of IoT data from Singapore's National Science Experiment. Journal of Spatial Science, 1–19. https://doi.org/10.1080/14498596.2020.1726219.
168. MOE (2008). Masterplan for ICT in education (2009–2014). Ministry of Education (MOE).
169. He, J. S., Ji, S., & Bobbie, P. O. (2017). Internet of things (IoT)-based learning framework to facilitate stem undergraduate education. In Proceedings of the SouthEast Conference, (pp. 88–94). ACM.
170. Fraillon, J., Ainley, J., Schulz, W., Friedman, T., & Duckworth, D. (2020). Preparing for life in a digital world. In Preparing for life in a digital world. Springer International Publishing.
171. Roll, M., & Ifenthaler, D. (2020). Lernortübergreifende Kompetenzentwicklung in der Industrie 4.0: Die Entwicklung digitaler Handlungskompetenz in der dualen Berufsausbildung aus der Ausbilderperspektive. Zeitschrift Für Berufs- Und Wirtschaftspädagogik, Beiheft, 29, 185–209.
172. Chan, J.K.-S. (2010). Teachers' responses to curriculum policy implementation: Colonial constraints for curriculum reform. Educational Research for Policy and Practice, 9(2), 93–106. https://doi.org/10.1007/s10671-010-9082-5
173. Balsmeier, B., & Woerter, M. (2019). Is this time different? How digitalization influences job creation and destruction. Research Policy, 48(8), 103765. https://doi.org/10.1016/j.respol.2019.03.010
174. Seeber, S., & Seifried, J. (2019). Herausforderungen und Entwicklungsperspektiven der beruflichen Bildung unter veränderten Rahmenbedingungen. Zeitschrift Für Erziehungswissenschaft, 22(3), 485–508. https://doi.org/10.1007/s11618-019-00876-2
175. Wild, S., & Schulze Heuling, L. (2020). How do the digital competences of students in vocational schools differ from those of students in cooperative higher education institutions in Germany? Empirical Research in Vocational Education and Training, 12(1), 5. https://doi.org/10.1186/s40461-020-00091-y
176. Wuttke, E., Seifried, J., & Niegemann, H. M. (2020). Editorial. Vocational education and training in the age of digitization (pp. 9–14). Verlag Barbara Budrich.
177. Arends, H. (2017). Equal living conditions vs. Cultural sovereignty? Federalism reform, educational poverty and spatial inequalities in Germany. Publius: the Journal of Federalism, 47(4), 673–706. https://doi.org/10.1093/publius/pjw042
178. Goller, M., Caruso, C., Berisha-Gawlowski, A., & Harteis, C. (2020). Digitalisierung in der Landwirtschaft: Gründe, Optionen und Bewertungen aus Perspektive von Milchviehlandwirtinnen und -landwirten. In D. Heisler & J. Meier (Eds.), Digitalisierung am Übergang Schule Beruf Ansätze und Perspektiven in Arbeitsdomänen und beruflicher Förderung (Vol. 56, pp. 53–80). wbv Media GmbH & Co. KG.
179. Ignatow, G. (2019). Text mining. SAGE Publications Ltd.
180. Virdia, S., & Schindler, S. (2019). Educational upgrading, career advancement, and social inequality development from a life-course perspective in Germany. Research in Social Stratification and Mobility, 60(2), 29–38. https://doi.org/10.1016/j.rssm.2019.02.002
181. KMK. (2017a). Bildung in der digitalen Welt - Strategie der Kultusministerkonferenz. Sekretariat der Kultusministerkonferenz. Retrieved from https://www.kmk.org/fileadmin/Dateien/pdf/PresseUndAktuelles/2018/Digitalstrategie_2017a_mit_Weiterbildung.pdf
182. KMK. (2017b). The Standing Conference's "Education in the Digital World" strategy - Summary. Secretariat of the Standing Conference of the Ministers of Education and Cultural Affairs of the Länder in the Federal Republic of Germany. Retrieved from https://www.kmk.org/fileadmin/Dateien/pdf/PresseUndAktuelles/2017b/KMK-Strategie_Bildung_in_der_digitalen_Welt_Zusammenfassung_en.pdf

183. Delcker, J., & Ifenthaler, D. (2020). Teachers' perspective on school development at German vocational schools during the Covid-19 pandemic. Technology, Pedagogy and Education. https://doi.org/10.1080/1475939X.2020.1857826
184. Roussou M, Katifori A (2018) Flow, staging, wayfinding, personalization: evaluating user experience with Mobile Museum narratives. Multimodal Technol Interact 2:32. https://doi.org/10.3390/mti2020032
185. Castrillo J, Gillate I, Luna Ú, Ibáñez-Etxeberria A (2021) Developing social and civic competence via apps: The role of historical memory in the initial teacher training. In: Gómez-Carrasco C, Miralles P, López-Facal R (eds) Handbook of research on teacher education in history and geography. Peter Lang, Switzerland, pp. 367–390
186. Delgado-Algarra E, Cuenca-López JM (eds) (2020) Handbook of research on citizenship and heritage education. IGI Global, Pennsylvania
187. Gillate I, Ibáñez-Etxeberria A, Cuenca-López JM (2018) Competencia digital y patrimonio: posibilidades de adquisición a través de videojuegos y apps. In: Gómez-Carrasco C, Miralles P (eds) La educación histórica ante el reto de las competencias. Octaedro, Barcelona, pp. 27–38
188. Gillate I, Castrillo J, Luna U, Ibáñez-Etxeberria A (2022) Temas controvertidos y apps para el desarrollo de la Competencia Social y Cívica. Análisis de la efectividad del proyecto 1936 en la Formación Inicial Docente. Revista de Psicodidáctica. https://doi.org/10.1016/j.psicod.2022.10.002
189. Hernàndez-Cardona FX, Sospedra-Roca R, Íñiguez-Gracia D (2021) Educational illustration of the historical city, education citizenship, and sustainable heritage. Sustainability 13(10):5706. https://doi.org/10.3390/su13105706
190. Kendall MG (1948) Rank correlation methods. Griffin, London
191. Fleiss JL, Levin B, Paik MC (2003) Statistical methods for rates and proportions, 3rd edn. Wiley, New York
192. Krippendorff K (1980) Content analysis, an introduction to its methodology. SAGE, Beverly Hills
193. Morgan C, Pallascio PM (2015) Digital media, participatory culture, and difficult heritage: online remediation and the trans-atlantic slave trade. J Afr Diaspora Archaeol Heritage 4(3):260–278. https://doi.org/10.1080/21619441.2015.1124594
194. McCandlish A, McPherson G (2021) Promoting tangible and Intangible Hidden Cultural Heritage: local communities influencing civic decision-making and International Cultural Policy. Int J Cult Policy 27(5):683–698. https://doi.org/10.1080/10286632.2020.1822348
195. Bonacini E (2019) Engaging participative communities in cultural heritage: using digital storytelling in Sicily (Italy). Int Inform Libr Rev 51(1):42–50. https://doi.org/10.1080/10572317.2019.1568786
196. Sangrà A, Vlachopoulos D, Cabrera N. Building an inclusive definition of e-learning: an approach to the conceptual framework. Int Rev Res Open Distance Learn. 2012;13(2):145–59. https://doi.org/10.19173/irrodl.v13i2.1161 .
197. Schöbel T, Zajonz D, Melcher P et al. Podcasts as a teaching tool in orthopaedic surgery: Is it beneficial or more an exemption card from attending lectures? Orthopade. Published online 2020. https://doi.org/10.1007/s00132-020-03956-y
198. Back DA, von Malotky J, Sostmann K, Hube R, Peters H, Hoff E. Superior Gain in Knowledge by Podcasts Versus text-based learning in teaching orthopedics: a Randomized Controlled Trial. J Surg Educ. 2017;74(1):154–60. https://doi.org/10.1016/j.jsurg.2016.07.008 .
199. Back DA, Haberstroh N, Sostmann K, et al. High efficacy and students' satisfaction after voluntary vs mandatory use of an e-learning program in traumatology and orthopedics - A follow-up study. J Surg Educ. 2014;71(3):353–9. https://doi.org/10.1016/j.jsurg.2013.11.007 .
200. Alkatout I, Günther V, Brügge S et al. Involvement of medical students in a surgery congress: impact on learning motivation, decision-making for a career in surgery, and educational

curriculum.Wiener Medizinische Wochenschrift. Published online January14, 2021:1–12. https://doi.org/10.1007/s10354-020-00802-w
201. Hardle, W., & Simar, L. (2015). Applied statistical analysis (4th ed.). Springer.
202. Class Central (2021). The Second Year of the MOOC: A Review of MOOC Stats and Trends in 2020. https://www.classcentral.com/report/the-second-year-of-the-mooc/
203. Rivard, R. (2013). Measuring the MOOC dropout rate Inside Higher Education. 8, 2013. https://www.insidehighered.com/news/2013/03/08/researchers-explore-who-taking-moocs-and-w hy-so-many-drop-out .
204. Laterza, V., Tømte, C. E., & Pinheiro, R. M. (2020). Guest Editorial: Digital transformations with "Nordic characteristics"? Latest trends in the digitalisation of teaching and learning in Nordic higher education. Universitetsforlaget. https://doi.org/10.18261/issn.1891-943x-2020-04-01
205. DiMaggio, P., & Powell, W. W. (1983). The Iron Cage Revisited - Institutional Isomorphism and Collective Rationality in Organizational Fields. American Sociological Review, 48(2), 147–160
206. Meyer, J. W., & Rowan, B. (1977). Institutionalized Organizations - Formal-Structures as Myth and Ceremony. American Journal of Sociology, 83(2), 340–363
207. Langseth, I., Jacobsen, D. Y., & Haugsbakken, H. (2022). The Role of Support Units in Digital Transformation: How Institutional Entrepreneurs Build Capacity for Online Learning in Higher Education. Technology, Knowledge and Learning. https://doi.org/10.1007/s10758-022-09620-y
208. Tungesvik, R. (2021). Digital tilstand 2021. Støttemiljøers arbeid med digitalisering og utdanning. Digital tilstand. Issue. https://diku.no/rapporter/digital-tilstand-2021-stoettemiljoeers-arbeid-med-digitalisering-og-utdanningskvalitet-ved-universiteter-og-hoeyskoler .
209. Patton, M. Q. (2002). Two decades of developments in qualitative inquiry: A personal, experiential perspective. Qualitative social work, 1(3), 261–283. https://doi.org/10.1177/1473325002001003636 .
210. Attride-Stirling, J. (2001). Thematic networks: an analytic tool for qualitative research. Qualitative research, 1(3), 385–405.
211. Braun V, Clarke V (2006) Using thematic analysis in psychology. Qual Res Psychol 3(2):77–101
212. Törrönen, J. (2001). The concept of subject position in empirical social research. Journal for the theory of social behaviour, 31(3), 313–329. https://doi.org/10.1111/1468-5914.00161 .
213. Feld, B., & Hathaway, I. (2020). The Startup Community Way: Evolving an entrepreneurial ecosystem. John Wiley & Sons.
214. Adewale, A. S., & Ghavifekr, S. (2019). Leadership self-efficacy and staff organizational citizenship behavior in higher education institutions: Experience from Nigeria. International Journal of Leadership in Education, 22(1), 116–133
215. Yakubu, M. N., & Dasuki, S. I. (2019). Factors affecting the adoption of e-learning technologies among higher education students in Nigeria: A structural equation modelling approach. Information Development, 35(3), 492–502. https://doi.org/10.1177/0266666918765907
216. Yakubu, N., & Dasuki, S. (2018). Measuring e-learning success in developing countries: Applying the updated DeLone and McLean model. Journal of Information Technology Education: Research
217. Yakubu, M. N., Dasuki, S. I., Abubakar, A. M., & Kah, M. M. (2020). Determinants of learning management systems adoption in Nigeria: A hybrid SEM and artificial neural network approach. Education and Information Technologies, 1–25
218. Wang, S. C., Cowie, B., & Jones, A. (2008). Challenges of e-learning for university instructors in Taiwan. In The 16th international conference on computers in education (ICCE 2008), Taipei, Taiwan (pp. 229–236)
219. Chin, J., & Lin, S. C. (2016). A behavioral model of managerial perspectives regarding technology acceptance in building energy management systems. Sustainability, 8(7), 641
220. Cheng, Y. (2012). Effects of quality antecedents on e-learning acceptance. Internet Research, 22(3), 361e390

221. Oluwajana, D., Idowu, A., Nat, M., Vanduhe, V., & Fadiya, S. (2019). The adoption of students' Hedonic motivation system model to gamified learning environment. Journal of Theoretical and Applied Electronic Commerce Research, 14(3), 156–167
222. Chen, H., & Tseng, H. (2012). Factors that influence acceptance of web-based e-learning systems for the in-service education of junior high school teachers in Taiwan. Evaluation and Program Planning, 35, 398–406
223. Scherer, R., Siddiq, F., & Tondeur, J. (2019). The technology acceptance model (TAM): A meta-analytic structural equation modeling approach to explaining teachers' adoption of digital technology in education. Computers & Education, 128, 13–35
224. Baas, M., Admiraal, W., & van den Berg, E. (2019). Teachers' adoption of open educational resources in higher education. Journal of Interactive Media in Education, 1(9), 1–11. https://doi.org/10.5334/jime.510
225. Bates, M., Loddington, S., Manuel, S., & Oppenheim, C. (2007). Attitudes to the rights and rewards for author contributions to repositories for teaching and learning. ALT-J, 15(1), 67–82. https://doi.org/10.1080/09687760600837066
226. Belikov, O. M., & Bodily, R. (2016). Incentives and barriers to OER adoption: A qualitative analysis of faculty perceptions. Open Praxis, 8(3), 235–246. https://doi.org/10.5944/openpraxis.8.3.308
227. Bossu, C., Brown, M., & Bull, D. (2014). Adoption, use and management of open educational resources to enhance teaching and learning in Australia. Sydney. https://bit.ly/36xqK3d .
228. Schuwer, R., & Janssen, B. (2018). Adoption of sharing and reuse of open resources by educators in higher education institutions in the Netherlands: A qualitative research of practices, motives, and conditions. The International Review of Research in Open and Distributed Learning. https://doi.org/10.19173/irrodl.v19i3.3390
229. Marín, V. I., Bond, M., Zawacki-Richter, O., Aydin, C., Bedenlier, S., Bozkurt, A., et al. (2020). A Comparative Study of National Infrastructures for Digital (Open) Educational Resources in Higher Education. Open Praxis, 12(2), 241–256. https://doi.org/10.5944/openpraxis.12.2.1071 .
230. Xu, Y. (2018). A study on the effectiveness of online open course construction policy in colleges and universities. Master's dissertation, Southwest University.
231. Mtebe, J., & Raisamo, R. (2014). Challenges and instructors' intention to adopt and use open educational resources in higher education in Tanzania. International Review of Research in Open and Distributed Learning, 15(1), 249–271. https://doi.org/10.19173/irrodl.v15i1.1687
232. Yuan, L., MacNeill, S., & Kraan, W. (2008). Open educational resources—Opportunities and challenges for higher education. Educational Cybernetics: Reports. Institute for Educational Cybernetics, University of Bolton.
233. Hanna, Nagy K (2008) Why a holistic e-development framework? In: Information technology and international development journal, vol. 4, no. 4. MIT Press, Cambridge, pp. 1–7
234. Hanna NK, Qiang C (2009) National e-government institutions in information and communications for development 2009: extending reach and increasing impact. World Bank, Washington, D.C
235. Cvitanovic C, Shellock RJ, Mackay M, van Putten EI, Karcher DB, Dickey-Collas M, Ballesteros M (2021) Strategies for building and managing 'trust' to enable knowledge exchange at the interface of environmental science and policy. Environ Sci Policy 123:179–189. https://doi.org/10.1016/j.envsci.2021.05.020
236. Reed MS, Stringer LC, Fazey I, Evely AC, Kruijsen JH (2014) Five principles for the practice of knowledge exchange in environmental management. J Environ Manag 146:337–345. https://doi.org/10.1016/j.jenvman.2014.07.021
237. Esteve-Mon, F. M., Llopis-Nebot, M. Á., & Adell-Segura, J. (2020). Digital teaching competence of university teachers: A systematic review of the literature. IEEE Revista Iberoamericana de Tecnologias del Aprendizaje, 15(4), 399–406. https://doi.org/10.1109/RITA.2020.3033225

238. Pallant, J. (2001). SPSS survival manual: A step by step guide to data analysis for windows (version 10 and 11). Maidenhead: Open University Press Retrieved from https://www.openup.co.uk/spss .
239. Edumadze, J. K. E., & Owusu, A. A. (2013). Use of information and communication Technology for Teaching and Learning in Ghanaian universities : Case of University of Cape Coast. International Journal of Computing Academic Research, 2(6), 266–277.
240. Agyei, D. D. (2019). Needs analysis to large-scale implementation: Using collaborative design to support ICT integration. In J. Pieters, J. Voogt, & N. P. Roblin (Eds.), Collaborative Curriculum Design for Sustainable Innovation and Teacher Learning (springer O, pp. 105–114). Cham: Springer Nature Switzerland AG. https://doi.org/10.1007/978-3-030-20062-6_6 .
241. Caldwell, B. J. (2005). School-based management. Education Policy Series. Brussels: International academy of Education (IAE) and the International Institute for Educational Planning (IIEP) Retrieved from http://www.unesco.org/iiep .
242. David, J. L. (1989). Synthesis of research on school-based management. Education Leadership, 45–53.
243. Mihai, M., & Nieuwenhuis, J. (2015). Management challenges in an information communication technology (ICT) network in rural schools. South African Journal of Education, 35(4), 1–10. https://doi.org/10.15700/saje.v35n4a1203 .
244. Patrinos, H. A., Barrera-Osorio, F., Fasih, T., & Santibáñez, L. (2009). Decentralized decision-making in schools: The theory of evidence on school-based management. Decentralized decision-making in schools. Washington, DC: The International Bank for Reconstruction and Development / The World Bank. https://doi.org/10.1596/978-0-8213-7969-1 .
245. Micklitz, H.-W., Reich, N., Rott, P., & Tonner, K. (2014). European Consumer Law (2nd ed.). Cambridge: Intersentia.
246. Schiller, Dan. 2011. Geopolitical-economic conflict and network infrastructures. Chinese Journal of Communication 4 (1): 90–107.
247. Park, S.: Digital Capital. Palgrave Macmillan, London (2017). https://doi.org/10.1057/978-1-137-59332-0
248. Barnett, Michael and Raymond Duvall. 2005. Power in international politics. International Organization 59 (1): 39–75.
249. Dahl, Robert. 1957. The concept of power. Behavioral Science 2 (3): 201–215.
250. Hills, Jill. 1994. Dependency theory and its relevance today: International institutions in telecommunications and structural power. Review of International Studies 20 (2): 169–186.
251. Strange, Susan. 1987. The persistent myth of lost hegemony. International Organization 41 (4): 551–574.
252. Harvey, Penelope. 2012. The topological quality of infrastructural relation: An ethnographic approach. Theory, Culture & Society 29 (4–5): 76–92.
253. Segal, Adam. 2020. China's vision for cyber sovereignty and the global governance of cyberspace. In An Emerging China-Centric Order. China's Vision for a New World Order in Practice, ed. Nadège Rolland, 86–100. NBR Special Report no. 87. Washington, DC: The National Bureau of Asian Research.
254. Carroll, N., & Conboy, K. (2020). Normalising the "new normal": Changing tech-driven work practices under pandemic time pressure. International Journal of Information Management, 55, 1–6.
255. Deleuze, Gilles. 1990. Post-script on the societies of control. October 59:3–7.
256. ENISA. 2018c. Cyber Security Culture in organisations. https://www.enisa.europa.eu/publications/cyber-security-culture-in-organisations . Accessed 23 July 2018.
257. BKA. 2013a. Austrian cyber security strategy. http://archiv.bundeskanzleramt.at/DocView.axd?CobId=52251 . Accessed 3 Mar 2018.
258. BKA. 2013b. Austrian security strategy. Security in a new decade—shaping security. http://archiv.bundeskanzleramt.at/DocView.axd?CobId=52251 . Accessed 3 Mar 2018.

259. Hansen, Lene, and Helen Nissenbaum. 2009. Digital disaster, cyber security, and the copenhagen school. International Studies Quarterly 53(4):1155–1175. https://doi.org/10.1111/j.1468-2478.2009.00572.x .
260. EU. 2016. Directive (EU) 2016/1148 of the European Parliament and of the Council of 6 July 2016 concerning measures for a high common level of security of network and information systemsacross the Union. https://eur-lex.europa.eu/legal-content/EN/TXT/PDF/?uri=CELEX:32016L1148&from=EN . Accessed 21 Mar 2018.
261. BKA. 2017. Digital Roadmap 2017. https://www.digitalroadmap.gv.at/fileadmin/downloads/digital_road_map_broschuere.pdf . Accessed 10 May 2018.
262. Republic of Austria. 2017. Zusammen. Für unser Österreich. Regierungsprogramm 2017–2022. https://www.bundeskanzleramt.gv.at/documents/131008/569203/Regierungsprogramm_2017%E2%80%932022.pdf/b2fe3f65-5a04-47b6-913d-2fe512ff4ce6 . Accessed 11 May 2018.
263. Wang, Z. (2014, 25 July). The dangers of history analogies. The Diplomat. https://thediplomat.com/2014/07/the-dangers-of-history-analogies/ . Accessed 28 Jan. 2020.
264. Lewis, J. A. (2019, 04 Sep.). Emerging technologies and managing the risk of tech transfer to China. Center for Strategic & International Studies. https://www.csis.org/analysis/emerging-technologies-and-managing-risk-tech-transfer-china . Accessed 28 Jan. 2020.
265. Langseth, Inger; Jacobsen, Dan Yngve; Haugsbakken, Halvdan The Role of Support Units in Digital Transformation: How Institutional Entrepreneurs Build Capacity for Online Learning in Higher Education. Technology, Knowledge and Learning (2022). https://doi.org/10.1007/s10758-022-09620-y
266. Ackermann, Judith; Egger, Benjamin; Scharlach, Rebecca Programming the Postdigital: Curation of Appropriation Processes in (Collaborative) Creative Coding Spaces. Postdigital Science and Education (2019). https://doi.org/10.1007/s42438-019-00088-1
267. Aytekin, Behiç Alp; Rızvanoğlu, Kerem Creating learning bridges through participatory design and technology to achieve sustainability in local crafts: a participatory model to enable the transfer of tacit knowledge and experience between the traditional craftsmanship and academic education. International Journal of Technology and Design Education (2018). https://doi.org/10.1007/s10798-018-9454-3
268. Backes, Susanne; Baumann, Isabell; Harion, Dominic; Sattler, Sabrina; Lenz, Thomas Why flipping the classroom is not enough: Digital curriculum making after the pandemic. PROSPECTS (2021). https://doi.org/10.1007/s11125-021-09555-9
269. Teräs, Marko; Suoranta, Juha; Teräs, Hanna; Curcher, Mark Post-Covid-19 Education and Education Technology 'Solutionism': a Seller's Market. Postdigital Science and Education (2020). https://doi.org/10.1007/s42438-020-00164-x
270. Cerratto Pargman, Teresa; Lindberg, Ylva; Buch, Anders Automation Is Coming! Exploring Future(s)-Oriented Methods in Education. Postdigital Science and Education (2022). https://doi.org/10.1007/s42438-022-00349-6
271. Ruth, Damian Indigenous Wisdom, Capital, Technology and Education. New Zealand Journal of Educational Studies (2020). https://doi.org/10.1007/s40841-020-00181-1
272. Mogas, Jordi; Palau, Ramon; Fuentes, Marta; Cebrián, Gisela Smart schools on the way: How school principals from Catalonia approach the future of education within the fourth industrial revolution. Learning Environments Research (2021). https://doi.org/10.1007/s10984-021-09398-3
273. Krüger, Johanna T.; Höffler, Tim N.; Wahl, Martin; Knickmeier, Katrin; Parchmann, Ilka Two comparative studies of computer simulations and experiments as learning tools in school and out-of-school education. Instructional Science (2022). https://doi.org/10.1007/s11251-021-09566-1
274. Maltempi, Marcus Vinicius; Malheiros, Ana Paula dos Santos Online distance mathematics education in Brazil: research, practice and policy. ZDM – Mathematics Education (2010). https://doi.org/10.1007/s11858-009-0231-3
275. Malott, Curry The Sublation of Digital Education. Postdigital Science and Education (2019). https://doi.org/10.1007/s42438-019-00083-6

276. Armila, Päivi; Sivenius, Ari; Stanković, Biljana; Juutilainen, Lauri Digitalization of Education: Commodification Hidden in Terms of Empowerment?. Postdigital Science and Education (2022). https://doi.org/10.1007/s42438-022-00347-8
277. Johnson, Mark William; Suvorova, Elena A.; Karelina, Alina A. Digitalization and Uncertainty in the University: Coherence and Collegiality Through a Metacurriculum. Postdigital Science and Education (2022). https://doi.org/10.1007/s42438-022-00324-1
278. Seitamaa-Hakkarainen, Pirita; Viilo, Marjut; Hakkarainen, Kai Learning by collaborative designing: technology-enhanced knowledge practices. International Journal of Technology and Design Education (2008). https://doi.org/10.1007/s10798-008-9066-4
279. Hou, Angela Yung Chi; Lu, I-Jung Grace; Hill, Christopher What Has Been the Impact of COVID-19 on Driving Digitalization, Innovation and Crisis Management of Higher Education and Quality Assurance?—A Taiwan Case Study in Alignment with the INQAAHE Virtual Review. Higher Education Policy (2022). https://doi.org/10.1057/s41307-022-00267-z
280. Gravani, Maria N.; Slade, Bonnie; Brown, Maria; Jõgi, Larissa; Borg, Carmel From Learner-Centered Education (LCE) to Emancipatory Learner-Centered Education (ELCE): A comparative case study of language education for adult migrants in four European countries. PROSPECTS (2023). https://doi.org/10.1007/s11125-023-09633-0
281. Wu, Chia-Jung; Chen, Gwo-Dong; Huang, Chi-Wen Using digital board games for genuine communication in EFL classrooms. Educational technology research and development (2014). https://doi.org/10.1007/s11423-013-9329-y
282. Patiño, Azeneth; Ramírez-Montoya, María Soledad; Buenestado-Fernández, Mariana Active learning and education 4.0 for complex thinking training: analysis of two case studies in open education. Smart Learning Environments (2023). https://doi.org/10.1186/s40561-023-00229-x
283. Willatt, Carlos; Flores, Luis Manuel The Presence of the Body in Digital Education: A Phenomenological Approach to Embodied Experience. Studies in Philosophy and Education (2021). https://doi.org/10.1007/s11217-021-09813-5
284. Benita, Francisco; Virupaksha, Darshan; Wilhelm, Erik; Tunçer, Bige A smart learning ecosystem design for delivering Data-driven Thinking in STEM education. Smart Learning Environments (2021). https://doi.org/10.1186/s40561-021-00153-y
285. Delcker, Jan Digitalisation in the Curricula of Vocational Schools: Text Mining as an Instrument of Curricula Analysis. Technology, Knowledge and Learning (2022). https://doi.org/10.1007/s10758-022-09591-0
286. Ortega-Sánchez, Delfín; López-Sanvicente, Amparo Bernal Design, content validity, and inter-observer reliability of the 'Digitization of Cultural Heritage, Identities, and Education' (DICHIE) instrument. Humanities and Social Sciences Communications (2023). https://doi.org/10.1057/s41599-023-01550-z
287. Mielke, Anna-Maria; Ghanem, Mohamed; Back, David Alexander; Fröhlich, Susanne; Herbstreit, Stephanie; Seemann, Ricarda Johanna Medical studies in times of a pandemic – concepts of digital teaching for Orthopaedics and Trauma at german universities. BMC Medical Education (2023). https://doi.org/10.1186/s12909-023-04213-4
288. Ogundari, Kolawole Student access to technology at home and learning hours during COVID-19 in the U.S. Educational Research for Policy and Practice (2023). https://doi.org/10.1007/s10671-023-09342-7
289. Langseth, Inger Dagrun; Jacobsen, Dan Yngve; Haugsbakken, Halvdan Institutional Entrepreneurship in Loosely Coupled Systems: The Subject Position of MOOC Entrepreneurs and Their Interpretive Struggles in a Norwegian Context. Technology, Knowledge and Learning (2023). https://doi.org/10.1007/s10758-023-09647-9
290. Idoga, Patience E.; Oluwajana, Dokun Iwalewa; Adeshola, Ibrahim Educational quality and technological complexity on recognition of enhanced learning platform in developing countries using PLS-SEM in a post COVID era. Educational technology research and development (2022). https://doi.org/10.1007/s11423-022-10150-8
291. Marín, Victoria I.; Zawacki-Richter, Olaf; Aydin, Cengiz H.; Bedenlier, Svenja; Bond, Melissa; Bozkurt, Aras; Conrad, Dianne; Jung, Insung; Kondakci, Yasar; Prinsloo, Paul;

Roberts, Jennifer; Veletsianos, George; Xiao, Junhong; Zhang, Jingjing Faculty perceptions, awareness and use of open educational resources for teaching and learning in higher education: a cross-comparative analysis. Research and Practice in Technology Enhanced Learning (2022). https://doi.org/10.1186/s41039-022-00185-z
292. Hanna, Nagy K.; Qiang, Christine Zhen-Wei China's Emerging Informatization Strategy. Journal of the Knowledge Economy (2010). https://doi.org/10.1007/s13132-009-0001-z
293. Nguyen, Vivian M.; Bell, Christina; Berseth, Valerie; Cvitanovic, Christopher; Darwent, Ray; Falconer, Matthew; Hutchen, Jenna; Kapoor, Tyreen; Klenk, Nicole; Young, Nathan Promises and pitfalls of digital knowledge exchange resulting from the COVID-19 pandemic. Socio-Ecological Practice Research (2021). https://doi.org/10.1007/s42532-021-00097-0
294. González, Carlos; Ponce, Daniel; Fernández, Violeta Teachers' experiences of teaching online during COVID-19: implications for postpandemic professional development. Educational technology research and development (2023). https://doi.org/10.1007/s11423-023-10200-9
295. Quaicoe, James Sunney; Pata, Kai Teachers' digital literacy and digital activity as digital divide components among basic schools in Ghana. Education and Information Technologies (2020). https://doi.org/10.1007/s10639-020-10158-8
296. Micklitz, Hans-W.; Pałka, Przemysław; Panagis, Yannis The Empire Strikes Back: Digital Control of Unfair Terms of Online Services. Journal of Consumer Policy (2017). https://doi.org/10.1007/s10603-017-9353-0
297. Huang, Ying; Mayer, Maximilian Power in the Age of Datafication: Exploring China's Global Data Power. Journal of Chinese Political Science (2022). https://doi.org/10.1007/s11366-022-09816-0
298. Surva, Laidi Maintaining the Ideals of Co-production During Rapid Digitalisation: A Comparative Case Study of Digital Restorative Services in Estonia, Finland, Ireland and Portugal. VOLUNTAS: International Journal of Voluntary and Nonprofit Organizations (2022). https://doi.org/10.1007/s11266-022-00502-6
299. Haddad, Christian; Binder, Clemens Governing through cybersecurity: national policy strategies, globalized (in-)security and sociotechnical visions of the digital society. Österreichische Zeitschrift für Soziologie (2019). https://doi.org/10.1007/s11614-019-00350-7
300. Sahin, Kaan Great Power Contest in a New (Digitalized) Guise. Zeitschrift für Außen- und Sicherheitspolitik (2020). https://doi.org/10.1007/s12399-020-00797-7

The manufacturer's authorised representative in the EU is Springer Nature Customer Service Centre GmbH, Europaplatz 3, 69115 Heidelberg, Germany. If you have any concerns regarding our products, please contact ProductSafety@springernature.com

Printed and bound by CPI Group (UK) Ltd, Croydon, CR0 4YY
01/10/2026
02225926-0020